AF334041

Supramolecular Systems in Biomedical Fields

Monographs in Supramolecular Chemistry

Series Editors:
Philip Gale, *University of Southampton, UK*
Jonathan Steed, *Durham University, UK*

Titles in this Series:
 1: Cyclophanes
 2: Calixarenes
 3: Crown Ethers and Cryptands
 4: Container Molecules and Their Guests
 5: Membranes and Molecular Assemblies: The Synkinetic Approach
 6: Calixarenes Revisited
 7: Self-assembly in Supramolecular Systems
 8: Anion Receptor Chemistry
 9: Boronic Acids in Saccharide Recognition
10: Calixarenes: An Introduction, 2nd Edition
11: Polymeric and Self Assembled Hydrogels: From Fundamental Understanding
    to Applications
12: Molecular Logic-based Computation
13: Supramolecular Systems in Biomedical Fields

How to obtain future titles on publication:
A standing order plan is available for this series. A standing order will bring
delivery of each new volume immediately on publication.

For further information please contact:
Book Sales Department, Royal Society of Chemistry, Thomas Graham House,
Science Park, Milton Road, Cambridge, CB4 0WF, UK
Telephone: + 44 (0)1223 420066, Fax: + 44 (0)1223 420247,
Email: booksales@rsc.org
Visit our website at http://www.rsc.org/Shop/Books/

Supramolecular Systems in Biomedical Fields

Edited by

Hans-Jörg Schneider
Saarland University, Saarbrücken, Germany
Email: h-j.schneider@mx.uni-saarland.de

RSCPublishing

Monographs in Supramolecular Chemistry No. 13

ISBN: 978-1-84973-658-9
ISSN: 1368-8642

A catalogue record for this book is available from the British Library

Published by The Royal Society of Chemistry,
Thomas Graham House, Science Park, Milton Road,
Cambridge CB4 0WF, UK

Registered Charity Number 207890

Visit our website at www.rsc.org/books

Preface

The basis of all living systems is, to an essential degree, nothing but supramolecular chemistry. Inspiration from nature has indeed always been a driving force for the development of host-guest chemistry. Previously, Emil Fischer's lock-and-key proposal addressed the function of proteins and Friedrich Cramer's early studies in the middle of the last century with cyclodextrin complexes were stimulated by the idea of biocatalytic models. Many functions in biological systems rely on non-covalent interactions, including, for instance, the actions of hormones, neurotransmitters, metabolites, nucleotides and, last but not least, drugs. Clearly, a book on the role of supramolecular chemistry in the life sciences cannot possibly encompass all these aspects, but must concentrate on the achievements reached with synthetic ligands interacting with biological systems. The scope of the book has not allowed the illustration of all aspects of non-covalent assemblies and, for instance, only mentions micelles, vesicles, *etc.* and drug design, for which there are also special books available. Nevertheless, the unlimited possibilities of using intermolecular interactions in living systems should become visible. After the foundations of supramolecular chemistry have been laid, practical applications become increasingly the focus of many researchers, particularly in fields which are in nature close to non-covalent interactions. At the same time, scientific contributions to societal needs are in demand, such as for health care, and this is also at the forefront of research focus.

The editor is most grateful to the authors that contributed to this monograph; he also appreciates their willingness to make compromises in view of space limitations and to take the editor's suggestions into account. For any mistakes, the editor takes responsibility, whereas the hopeful success of the book goes to the credit of the chapter authors.

Hans-Jörg Schneider
Saarland University, Saarbrücken

Monographs in Supramolecular Chemistry No. 13
Supramolecular Systems in Biomedical Fields
Edited by Hans-Jörg Schneider
© The Royal Society of Chemistry 2013
Published by the Royal Society of Chemistry, www.rsc.org

Contents

Monographs in Supramolecular Chemistry No. 13
Supramolecular Systems in Biomedical Fields
Edited by Hans-Jörg Schneider
© The Royal Society of Chemistry 2013
Published by the Royal Society of Chemistry, www.rsc.org

Chapter 6   Interactions of Calix[*n*]arenes and Other Organic Supramolecular Systems with Proteins   140
Florent Perret and Anthony W. Coleman

Chapter 7   Cucurbiturils in Drug Delivery And For Biomedical Applications   164
Na'il Saleh, Indrajit Ghosh and Werner M. Nau

Chapter 8 Nucleic Acids as Supramolecular Targets   **213**
Enrique García-España, Ivo Piantanida and Hans-Jörg Schneider

Chapter 9   Biomolecular Interactions of Platinum Complexes   260

Benjamin W. Harper, Feng Li, Rhys Beard,
K. Benjamin Garbutcheon-Singh, Neville S. Ng and
Janice R. Aldrich-Wright

Chapter 13 Constitutional Dynamic Chemistry for Bioactive Compounds   **397**

Yan Zhang, Lei Hu and Olof Ramström

Chapter 14 Molecular Imprinted Polymers for Biomedical Applications   **419**

Adnan Mujahid and Franz L. Dickert

CHAPTER 1

Introduction

HANS-JÖRG SCHNEIDER

Universität des Saarlandes, FR Org. Chemie, D 66041 Saarbrücken
Email: ch12hs@rz.uni-sb.de

Non-covalent interactions dominate central parts of living systems, and provide a major role for chemistry in both healthcare and biotechnology. Modern synthetic methods have made it possible to prepare host compounds for virtually every target molecule, including those which function in the natural medium, water.[1] In Chapter 13 Ramström *et al.* discuss how dynamic combinatorial chemistry can lead to an unlimited number of optimal ligands, inhibitors and potential drugs for biological targets. The action and development of drugs is the realm of medicinal chemistry, for which a large number of monographs and reviews is available. Nevertheless, the understanding of many biological functions, and particularly the rational design of new drugs or bioorganic self-assemblies, can take advantage of insights from the study of synthetic host–guest complexes.[2] Thus, cation–π[3] or anion–π[4] interactions and their role in living systems first became apparent in artificial complexes; the same holds for many other interactions with aromatic moieties.[5] Weak hydrogen bonds such as with C–H bonds[6] or with organic halogens[7] first became accessible to detailed elucidation with synthetic complexes; this applies also to, for instance, the interaction between halogen atoms and Lewis bases,[8] and to dispersive forces. A remarkable limitation of Emil Fischer's lock-and-key principle was observed first with synthetic complexes, in which usually only 50% of the available space in a binding cavity is used;[9] this rule was shown later to apply also to enzyme complexes.[10]

Perhaps the most important aspect of using synthetic complexes is the possibility not only to elucidate the mechanism of all contributions to molecular recognition, but also to clarify geometric constraints—in particular

Monographs in Supramolecular Chemistry No. 13
Supramolecular Systems in Biomedical Fields
Edited by Hans-Jörg Schneider

Published by the Royal Society of Chemistry, www.rsc.org

to assign discrete energy values to them.[11] This can help to develop energy scoring functions for drug design.[12] It would have been tempting to devote some chapters to these mechanistic contributions for the understanding of non-covalent binding with biological systems. Instead the reader is pointed to the above-mentioned leading references; the present monograph tries to summarize practical applications of host–guest complexes in life sciences by means of chapters written by well-known specialists in applications, which can be grouped as follows.

1.1 Sensing/Diagnostics/Imaging

From the beginning of supramolecular chemistry, analytical applications have played a major role. Supramolecular complexes can eventually lead to direct sensing of targets in the biological matrix without sample pretreatment, which often is still necessary[13] in order to extract, isolate and concentrate the analytes of interest. In Chapter 2 Prins *et al.* illustrate how very high sensitivity can be achieved with amplification pathways combining catalysis and multivalency. Based on the use of synthetic catalysts containing recognition sequences one can, for example, detect DNA targets with a sensitivity down to 5 nM; with cationic polythiophenes forming triple helices with DNA targets, an affinity corresponding to 3×10^{-21} M detection limit can be reached. The use of host molecules which in sensors exhibit sensitive signals upon recognition of biomedically important analytes is discussed in Chapter 3 by Magri and Mallia with respect to metal ions, and in Chapter 4 by Schneider for organic and biological compounds. Magri describes the implementation of several inter-action sites within hosts functioning also as logical gates; this leads to 'lab-on-a-molecule' devices. Confocal microscopy with fluorescent ligands allows, for example, imaging of Cd^{2+} ions in living cells. Sensing of organic analytes comprises—from metabolites, through alkaloids to drugs and toxins— a large variety of structures as illustrated with the typical examples presented in Chapter 4. For a highly sensitive and selective detection a myriad of synthetic host compounds has been designed, which bear suitable units for mostly optical signalling. In Chapter 14 Dickert and Mujahid show that molecular imprinting (MIP) techniques not only allow economical separations, but in particular highly selective recognition, which now extends also to antibodies, cells, viruses (including HIV) and even bacteria. They demonstrate that MIP receptors possess the potential to substitute natural antibodies, and allow detection of, for example, drug metabolites from complex matrices including blood and urine. Several chapters are centred not on particular targets or methods, but on host molecules which are used most often in biomedical applications, such as cyclodextrins (CDs; Chapter 5 by Ortiz Mellet *et al.*), calixarenes (Chapter 6 by Coleman and Perret), and cucurbiturils (CBs; Chapter 7 by Saleh *et al.*). Others host molecules such as crown ethers, cyclophanes, molecular tweezers, and porphyrins also play an increasingly important role in biomedical fields; they are mentioned in several chapters, in particular Chapter 4. Cyclodextrin-based assemblies can serve for the detection of pathogens or allergens, and are more cost-effective than the immunosorbent ELISA assay. CD–mannopyranosyl conjugates on a Ru(II) fluorescent core allowed identification of mannose-specific receptors presenting cells from *Escherichia coli*, for example. Cucurbituril derivatives have been used for the detection of amino acids, peptides, biogenic amines, alkaloids, or of cancer-associated nitrosamines, after immobilizaton, also on biochip sensors (chapters 7 and 12). As outlined in Chapter 12 by Hennig, such cucurbituril complexations can be used to monitor enzymatic reactions by supramolecular tandem assays. Hennig describes how for almost all enzyme classes fluorescence-based, label- and antibody-free supramolecular systems can be applied, based on chemosensors for products, membrane transport systems and tandem assays. Imaging with the help of supramolecular complexes, mostly involving fluorescence tomography, is discussed in Magri's contribution (Chapter 3) and also by Gupta and Pandey (Chapter 15); it is an emerging technique for the non-invasive, real-time visualization of biochemical events at the molecular level within living cells, tissues and/or intact organs. Gupta and Pandey show how nanoparticles allow the loading of multiple agents such as near-infrared fluorophores or radio-tracers and photosensitizers for tumour detection. Chapter 10 by Schatz and Schüle highlights the recent progress in medical MRI diagnostics with supramolecular metal complexes, in particular Gd^{3+} complexes, which has enabled the control of their relaxivity, possible toxicity and biodistribution, with tumour cells as target, for example. They also discuss optical imaging methods, including the use of supramolecular reporter units for selective imaging with Quantum dots and radioimaging with, for example, technetium complexes.

1.2   Interaction with Proteins and Nucleic Acids

As mentioned above, the action of drugs on proteins can be considered, medicinally, to be the most important part of supramolecular chemistry, but is treated elsewhere in many books. Closer to traditional supramolecular approaches is the use of macrocycles such as calixarenes modulating protein functions, as highlighted in Chapter 6 by Coleman and Perret, for cucurbiturils in Chapter 7 by Saleh *et al.*, and for cyclodextrins, in Chapter 5 by Ortiz Mellet *et al.*, and with some other hosts in Chapter 4 by Schneider. The supramolecular action even of small molecules can effectively compete with protein–protein interactions, which holds great promise for development of new drugs.[14] Coleman and Perret illustrate the use of calixarenes for enzyme protection or activation and inhibition, as anticoagulants such as antithrombotic agents, specific binding to lectins, detection of, for instance, the pathogenic prion protein, and other 'theragnostic' applications. Nucleic acids lend themselves particularly well to studies of supramolecular complexation and drug interference in view of their regular structures and information content. In Chapter 8 Garcia-Espagna *et al.* show how synthetic polyamines offers new ways to differentiate groove binding and effect gene delivery; macrocyclic derivatives can lead to base flipping, to unfolding helices and to quadruplex stabilization in telomers. Metal complexes with allosteric control lead to new bioactive DNA ligands, and allow intriguing sequence selective cleavage, now partially surpassing the performance of natural restriction enzymes. The most established application of metal complexes concerns tumour therapy with platinum derivatives, for which recent developments are outlined in Chapter 9 by Aldrich-Wright and co-workers. They illustrate how addition of functional groups in such complexes can greatly enhance their efficiency—for example, by adding intercalating units—and hold promise with respect to biological targeting. New Pt(IV) instead of Pt(II) derivatives can help to solve toxicity problems; selectivity may be gained, for example, with ligands that bind to receptors that are overexpressed in tumours.

1.3 Drug Protection, Release and Targeting, Gene Delivery

Encapsulation of drugs in suitable host compounds was recognized early on as an efficient way to increase their water solubility and bioavailability. In particular cyclodextrins (Chapter 5 by Ortiz Mellet *et al.*), with already 30 different products on the market and more in clinical phases, can enhance drug absorption efficiency, alter pharmacokinetics, and enhance stability against oxidation or enzymatic degradation. Nanocapsules with amphiphilic CDs have been shown to decrease toxicity of, for instance, cytostatic drugs. Polycationic cyclodextrins can act as on-viral gene vectors, multivalent CD conjugates can control carbohydrate–protein and cell surface interactions, and restore correct folding of peptides involved in diseases like Alzheimer's. Antitumour agents such as taxol derivatives have been targeted, for example, towards the mannose receptor of macrophages. As shown in Chapter 7 by Saleh *et al.*, cucurbiturils recently emerged as carriers serving the same purposes as cyclodextrins, but often with different characteristics. Photocontrolled release of antibiotics is possible with a photo-base as auxiliary material. CB-based nanoparticles decorated non-covalently with, for example, a folate-spermidine conjugate can target human ovarian carcinoma cells. Experiments with various cancer cell lines have demonstrated that CBs are not toxic and retain the pharmacological activity of drug loads. Gupta and Pandey describe in Chapter 15 how selectivity in photodynamic therapy can be enhanced by binding the photosensitizer to molecular delivery systems or by conjugating sensitizers with targeting agents such as monoclonal antibodies, integrin antagonists, carbohydrates and other moieties with high affinity to target tissues. They show that biodegradable polyacrylamide-based nanoparticles hold much promise for both tumour detection and therapy. Gels formed by supramolecular aggregation of small molecules allow interesting biomedical applications. In Chapter 11 Escuder and Miravet show how implementation of, for example, antibiotics in such gels can enhance considerably their antibiotic activity. Peptide amphiphile-derived gels have already been used to treat a mouse model of spinal cord injury, demonstrating the amplification of properties associated to fibre formation—this is an example of tissue engineering. Hydrogel-based drug delivery can be used for oral, rectal, ocular, epidermal and subcutaneous application, which includes, for example, synthetic octapeptides that mimic natural hormones such as somatostatin.

Targeting in biological systems with the help of supramolecular complexation in macromolecules is also highlighted in Chapter 16 by Leblond *et al.* Polymers engineered with suitable binding functions can, by non-covalent interactions, mediate inflammation, block cellular receptors, immune responses or viral epitopes; they can damage bacterial membranes, inhibit adsorption and serve as toxin scavengers. For example, polymeric scavengers targeting bacterial toxins—virulence factors responsible for the symptoms associated with bacterial infections—are already being tested clinically. Polymeric binders can also help the body to fight against infections by targeting the pathogen: Virus epitopes can be treated with shielding polymers such as a polylysine dendrimer, impairing in this way the colonization capacity of viruses.

It is hoped that the selection of topics mentioned above provides a foretaste of the different chapters, which span a wide range of possible supramolecular applications in the life sciences.

References

1. G. V. Oshovsky, D. N. Reinhoudt and W. Verboom, *Angew. Chem. Int. Ed.*, 2007, **46**, 2366; see also R. N. Dsouza, U. Pischel and W. M. Nau, *Chem. Rev.*, 2011, **111**, 7941.
2. (a) M. Zürcher and F. Diederich, *J. Org. Chem.*, 2008, **73**, 4345; (b) D. K. Smith, *J. Chem. Educ.*, 2005, **82**, 393; (c) D. A. Uhlenheuer, K. Petkau and L. Brunsveld, *Chem. Soc. Rev.*, 2010, **39**, 2817.
3. (a) J. C. Ma and D. A. Dougherty, *Chem. Rev.*, 1997, **97**, 1303; (b) N. Zacharias and D. A. Dougherty, *Trends Pharmacol. Sci.*, 2002, **23**, 281.
4. (a) A Frontera, P. Gamez, M. Mascal, T. J. Mooibroek and J. Reedijk, *Angew Chem. Int. Ed.*, 2011, **50**, 9564; (b) B. L. Schottel, H. T. Chifotides and K. R. Dunbar, *Chem. Soc. Rev.*, 2008, **37**, 68; (c) B. P. Hay and V. S. Bryantsev, *Chem. Commun.*, 2008, 2417.
5. (a) E. A. Meyer, R. K. Castellano and F. Diederich, *Angew. Chem. Int. Ed. Engl.*, 2003, **42**, 1210; (b) L.M. Salonen, M. Ellermann and F. Diederich, *Angew. Chem. Int. Ed.*, 2011, **50**, 4808.
6. (a) G. R. Desiraju and T. Steiner, *The Weak Hydrogen Bond*; Oxford University Press, Oxford, 1999; (b) T. Steiner, *Angew. Chem. Int. Ed.*, 2002, **41**, 48; (c) R. K. Castellano, *Curr. Org. Chem*, 2004, **8**, 845.
7. See, for example: (a) C. Ouvrard, M. Berthelot and C. Laurence, *J. Phys. Org. Chem*, 2001, **14**, 804; (b) E. Carosati, S. Sciabola and G. Cruciani, *J. Med. Chem.*, 2004, **47**, 5114.
8. (a) A. C. Legon, *Phys. Chem. Chem. Phys.*, 2010, **12**, 7736; (b) M. R. Scholfield, C. M. Vander Zanden, M. Carter and P. S. Ho, *Prot. Sci.*, 2013, **22**, 139; (c) M. Erdelyi, *Chem. Soc. Rev.*, 2012, **41**, 3547;

(d) E. Parisini, P. Metrangolo, T. Pilati, G. Resnati and G. Terraneo, *Chem. Soc. Rev.*, 2010, **40**, 2267.

9. S. Mecozzi and J. Rebek, Jr., *Chem. Eur. J*, 1998, **4**, 1016.
10. M. Zürcher, T. Gottschalk, S. Meyer, D. Bur and F. Diederich, *Chem. Med. Chem.*, 2008, **3**, 237.
11. H.-J. Schneider, *Angew. Chem. Int. Ed. Engl.*, 2009, **48**, 3924.
12. H.-J. Böhm and G. Schneider ed. *Protein–Ligand Interactions,* Wiley-VCH, Weinheim, 2003; ch. 1 by H.-J. Böhm, pp. 3–20; ch. 2 by H.-J. Schneider, pp. 21–50.
13. H. Kataoka, *TrAC: Trends Anal. Chem.*, 2003, **22**, 232.
14. J. A. Wells and C. L. McClendon, *Nature*, 2007, **450**, 1001.

Signalling Techniques in Supramolecular Systems

JEALEMY GALINDO MILLÁN AND LEONARD J. PRINS*

Department of Chemical Sciences, University of Padova, Via Marzolo 1, 35131 Italy
*Email: leonard.prins@unipd.it

2.1 Introduction

2.1.1 Background

A growing need exists for sensing techniques and strategies that permit accurate detection and quantification of analytes at ultralow concentrations.[1,2] In the biomedical field, analyte detection with high sensitivity is vital since it permits the diagnosis of diseases at (very) early stages, leading to a higher chance of successful treatment.[3] In addition, sensing techniques also play a key role in the detection of explosives in transportation systems (airports, railways systems, *etc.*), cargo and landmines,[4,5] heavy metals in soil, water, environment and food[6–8] and DNA in forensic sciences.[9] Analyte detection limits are continuously being pushed down by the improvement of existing analytical technologies. In fact, the single molecule detection level can now be reached using various techniques, which include transmission electron microscopy (TEM),[10] scanning electron microscopy (SEM),[11] atomic force microscopy (AFM),[12,13] surface plasmon resonance (SPR),[14] surface enhanced Raman spectroscopy (SERS)[15] and advanced fluorescence spectroscopy,[16] the latter also permitting an application for sensing purposes. Nonetheless, the ultimate challenge in sensing is the development of a methodology that allows naked-eye

Monographs in Supramolecular Chemistry No. 13
Supramolecular Systems in Biomedical Fields
Edited by Hans-Jörg Schneider

Published by the Royal Society of Chemistry, www.rsc.org

detection of a single molecule in a complex mixture, that is, without the need for sophisticated instrumentation. In fact, the availability of simple and inexpensive techniques would favour an implementation in social contexts where advanced instrumentation or care centres are less available. A bottom-up approach to sensing is a perfect task for chemists, as it requires the ability to master molecular properties both on the molecular and collective levels. Such an analytical approach must evidently rely on signal amplification, implying that a weak signal originating from the analyte molecule is reinforced to a detectable intensity.

2.1.2 Signal Amplification in Nature

Nature does not have access to advanced instrumentation, but is nevertheless able to act very efficiently and selectively on weak impulses using very effective signal amplification pathways. This will be illustrated by two examples that cover two conceptually different approaches to signal amplification, which are also at the basis of the synthetic amplification schemes discussed in this chapter. Signal amplification in Nature occurs either through the analyte-triggered activation of a catalytic cascade leading towards the formation of multiple reporters (catalytic signal amplification) or the analyte-triggered alteration release of many reporters from a multivalent structure (multivalent amplification).

The mitogen-activated protein (MAP) kinase cascade is an excellent example of the power of catalytic signal amplification in Nature.[17] A multi-tiered pathway is responsible for the activation of a catalytically multifunctional MAP kinase (MAPK), where each step consists of the phosphorylation of threonine/serine and/or tyrosine residues performed by an upstream kinase. In a typical three-tiered MAPK pathway, phosphorylation of threonine and tyrosine residues is carried out by MAP2 kinases which, in turn, are activated by phosphorylation of serine/threonine residues by MAP3 kinases. Signal amplification in such complex pathways occurs when the number of successive proteins exceeds that of its regulators, as has been observed for extracellular signal-regulated kinase (ERK 1/2) pathways. In this way, a single cell-surface binding event induced by an extracellular entity leads to a cascade of phosphorylations that harmonically regulate cell survival, apoptosis and mitosis, among other processes.

On the other hand, cell-signalling processes nicely illustrate a natural example of amplification due to multivalency.[18] In neurons, neurotransmitters such as glutamate, serotonin and epinephrine are confined in synaptic vesicles, which basically consist of a number of self-assembled proteins forming a sphere that acts as a container for these small organic molecules. These vesicles, situated underneath the axon terminal of neurons, are responsible for the release of neurotransmitters, which occurs by means of diffusion across the synaptic cleft, and is usually triggered by an action potential (*e.g.*, from voltage-gated calcium channels). Each synaptic vesicle releases a large number of neurotransmitters, which then bind (extracellularly) to specific receptors in cell membranes, which then transmit information to internal signalling pathways. In this way, neurotransmitters act as messengers to modulate and amplify signals between neurons and cells.

2.1.3 Biological Assays

The advance of signal amplification in a chemical context is nicely illustrated by the development of biological assays. More precisely, this example gives a good insight into the importance of the correlation between the analyte concentration and the strength of the output signal which eventually dictates the sensitivity of any analytical protocol. The first assays developed for biosensing relied on the direct detection of analytes using radioactive labels in order to generate a measurable output signal.[19] In cases where this correlation was not strong enough to obtain a detectable output, signal amplification offered a novel alternative. Excellent examples are enzyme-linked immunosorbent assays (ELISAs),[20–22] where catalytic moieties replaced radioactive reporters. In short, in an ELISA an enzyme is co-immobilized on a surface following a molecular recognition event between the analyte and a surface bound receptor.[23] Interestingly, the fact that one enzyme, recruited by a single analyte, is able to generate multiple reporter molecules is the key to the amplified signal. Typically, this signal is detected using convenient techniques such as fluorescence or ultraviolet–visible (UV/Vis) spectroscopy. The magnitude of the amplification depends on the number of reporter molecules generated per enzyme, which is dictated by the turnover number (or, for practical reasons the turnover frequency) of the enzyme. Despite the evident benefits of signal amplification, the lower detection limit for proteins using ELISAs is usually in the picomole range, still far above the single molecule detection level. However, the replacement of the enzyme tag by a DNA tag, followed by amplification using the polymerase chain reaction (PCR),[24] has made detection at such low concentration levels possible.[25] The strong amplification in these immuno-PCR assays results from the exponential growth curve of DNA through the PCR reaction and can lead up to a 10^5-fold increase in sensitivity compared with classical ELISAs. Using this approach, Cantor *et al.*[25] reported the reliable and reproducible detection of as few as 580 antigen molecules in a 45 µL volume.

2.1.4 Signalling Strategies in Synthetic Systems

Inspired by Nature, many artificial signalling systems have been developed along two conceptual lines, namely catalysis and multivalency (Figure 2.1). In this chapter, we intend to discuss key examples that best illustrate how these concepts have been applied in synthetic systems. In particular, this literature survey focuses on sensing systems aimed at the detection of biologically relevant species, which are DNA, bioactive small molecules and proteins. For each class, examples are provided that rely either on catalysis or multivalency. This excludes a discussion on highly sensitive analytical techniques (SERS, SPR, *etc.*) or methods relying on changes in physical properties

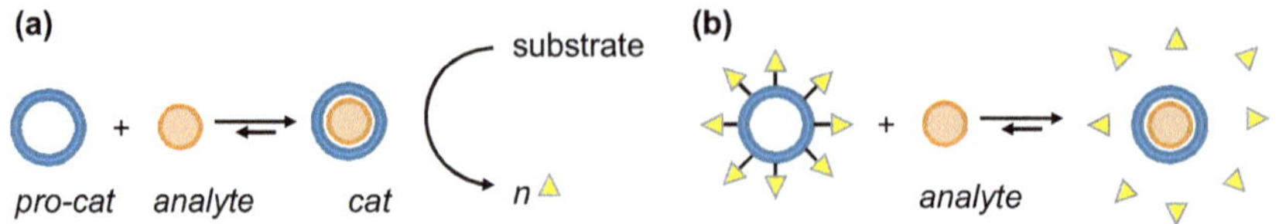

Figure 2.1 Schematic representations of signalling strategies in synthetic systems. Approaches can be conceptually divided in two classes: (a) catalysis and (b) multivalency. In the first approach, an analyte activates a catalyst, which results in the production of reporter molecules. In the second approach, an analyte triggers a change in the property of multiple reporter molecules.

(due to nanoparticle aggregation, for example). For obvious reasons, the treatment of this argument cannot be comprehensive and many beautiful examples could not be included.

2.2 DNA Detection

The sensing of DNA is facilitated by the extremely reliable *and* unique molecular recognition between complementary oligonucleotide strands.[26] In addition, DNA can be used as a structural element to form receptors (aptamers) or catalysts (DNAzymes). Furthermore, DNA is, synthetically, easily accessible and can be conjugated to other (bio)molecules. Finally, the biological machinery is available for processing and transforming DNA. In this section, supramolecular signalling systems will be discussed, in which the presence of the DNA target sets off an amplification mechanism (or a cascade) leading to a detectable signal.[27,28] Similarly as for the following sections, the examples are grouped according to the way signal amplification is achieved: through catalysis or multivalency.

2.2.1 Catalysis

2.2.1.1 Enzymes

An elegant example was provided by Ghadiri and co-workers,[29] who reported on the use of a protease for the amplified sequence specific detection of single-stranded (ss) DNA. A construct was engineered with a 24-mer ssDNA probe separating a zinc-metalloprotease (Cereus neutral protease-CNP) and a phosphoramidite inhibitor (Figure 2.2). The choice of the enzyme was motivated by its compatibility with the DNA tether and, importantly, its high intrinsic catalytic activity ($k_{cat} = 165\,s^{-1}$) for a rapid signal generation. The working principle of this system lies in the fact that in the absence of a complementary DNA strand, the flexible ssDNA linker permits the inhibitor to block the active site of the enzyme. The presence of a complementary DNA strand results in the formation of a rigid duplex which causes the removal of the inhibitor from the enzyme. The enzyme is thus activated and generates a fluorescent product by hydrolysing an appropriate substrate. The approach

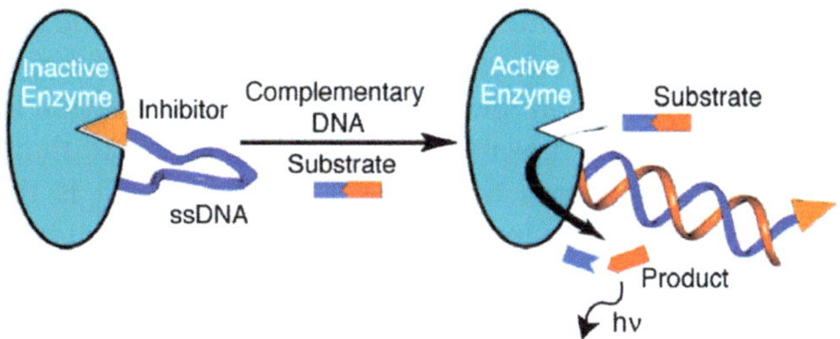

Figure 2.2 Schematic representation of an inactivated inhibitor-DNA-enzyme, which is activated upon the addition of the complementary DNA target. (Reproduced with permission from Saghatelian *et al.*[29] Copyright 2003 American Chemical Society.)

was validated by the observation that the addition of a 100 pM solution of the complementary DNA strand gave a detectable signal after just 3 minutes. A 10 pM solution required an elongated period of 80 minutes to be detected. Sequence selectivity was demonstrated by the absence of fluorescent signal in the case of non-complementary DNA being added, even at 10 μM concentrations.

A related approach, but employing a synthetic catalyst, was used by Krämer's group[30] for DNA sensing. A single-stranded 20-mer DNA oligonucleotide was functionalized at both the 3′ and 5′ termini with terpyridine ligands. In the presence of Cu(II), both terpyridine moieties fold back and form an intrastrand complex with DNA. Addition of the complementary ssDNA target results in the formation of double-stranded (ds) DNA and disrupts the intramolecular $Cu(II) \cdot (tpy)_2$ complex. Under these conditions, 1,10-phenantroline (phen) is able to scavenge the Cu(II) ion to constitute the active catalyst $Cu(II) \cdot phen$, which then catalyses the oxidation of 2′,7′-dichlorodihydrofluorescein into the fluorescent product, 2′,7′-dichlorofluorescein. The DNA target could be detected down to 5 nM concentrations and a turnover number of 5 with respect to the DNA target was reported. The low turnover number was overcome in a second generation system by displacing the Zn(II) from the intrastrand complex using an activated apo-carbonic anhydrase.[31] It was calculated that a single DNA target strand was amplified into 10 000 CO_2 molecules in just 30 seconds.

2.2.1.2 DNAzymes

DNAzymes are non-natural oligonucleotide-based catalysts, which, compared to enzymes, typically exhibit a higher thermal stability. They can be easily conjugated to other molecules, and may be prepared in large quantities using PCR. The underlying idea of using DNA as a source library for the formation of catalysts relies on the large structural variation in the type of folded oligonucleotides available and the enormous combinatorial libraries that can be assessed (typically up to 10^{14} or 10^{16} members). DNAzymes are obtained through an exponential enrichment process (SELEX),[32,33] which relies on the isolation of the fraction of library members able to perform a given reaction (typically using affinity chromatography).[34] The enriched fraction is multiplied using PCR and screening is repeated. Repetitive cycles make that the library composition converges towards the active sequence, which is then isolated and characterized. A DNAzyme which has been amply used in sensing applications is a sequence that mimics horse-radish peroxidase (HRP).[35] Complexation of hemin by the guanine-rich sequence causes folding of the oligonucleotide in a G-quadruplex structure that catalyzes the oxidation of 2,2′-azino-bis(3-ethylbenzothiazoline-6-sulfonic acid) (ABTS) by H_2O_2 to give the coloured product $ABTS^{\bullet+}$ [36] or, alternatively, the oxidation of luminol by H_2O_2 to give chemiluminescence.[37]

The use of DNAzymes for DNA detection is attractive, since molecular recognition and signal generation can be combined in the same oligonucleotide. In a straightforward example, a hairpin structure was designed containing two regions A and B, that in an open configuration form the G-quadruplex with haemin to give a catalytic complex (Figure 2.3).[38] In the resting state, formation of the DNAzyme is prohibited because region B is hybridized to give the hairpin structure. The DNA target sequence hybridizes with the stem region of the hairpin, liberating the B region. Upon opening up, the DNAzyme is formed and oxidation of ABTS to give $ABTS^{\bullet+}$ occurs. Hybridization and hairpin opening can be detected spectroscopically following the accumulation of the coloured product. DNA target concentrations in the low micromolar region could be detected.

DNA sensing systems of much higher complexity have been obtained by taking advantage of natural DNA processing mechanisms.[39] An illustrative example is the report of Mao and colleagues[40] in which a cascade of amplification events was induced by the target DNA sequence (Figure 2.4). The first amplification step is the analyte-induced activation of polymerase-mediated rolling circle amplification (RCA). RCA is an effective method for the preparation of repetitive, ssDNA molecules up to tens of thousands of bases

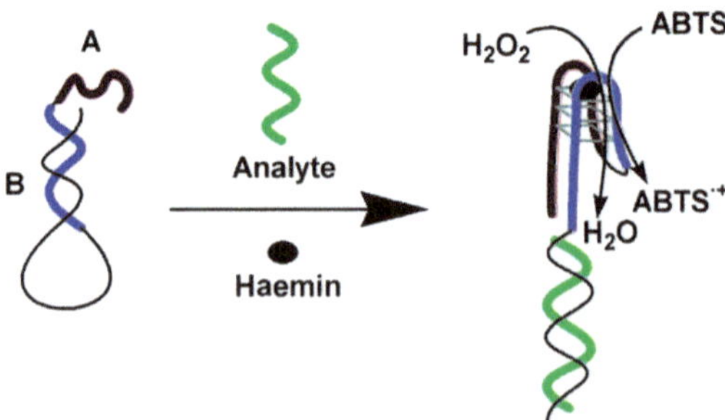

Figure 2.3 Detection of DNA by opening of a hairpin oligonucleotide and the activation of the DNAzyme.
(Reproduced with permission from Xiao *et al.*[38] Copyright 2004 American Chemical Society.)

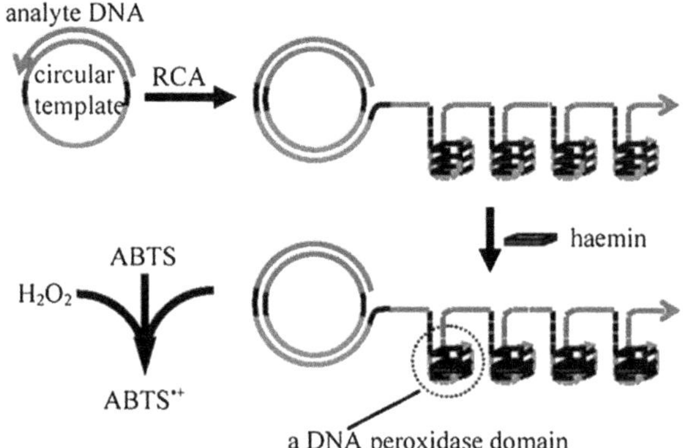

Figure 2.4 Target-induced activation of rolling cycle amplification leading towards the production of multiple DNAzymes. RCA = rolling circle amplification. (Reproduced with permission from Tian *et al.*[40] Copyright 2006 John Wiley and Sons.)

long. Mao *et al.* used a single-stranded circular RCA template containing an analyte DNA-recognition sequence and a sequence complementary to the HRP-like DNAzyme. Hybridization of the target DNA with the recognition site initiates DNA polymerization producing a long, single DNA strand containing multiple sequences encoding for the DNAzyme. The second amplification step involves the activation of the DNAzyme in the presence of haemin, ABTS and H_2O_2. As illustrated before, this is a reaction with multiple turnovers, implying that each enzyme generates multiple copies of the coloured product. It was found that analyte DNA concentrations down to 1 pM were able to generate a distinguishable signal compared to a control sample treated in the same way but in the absence of analyte. An advantage of this design is that it does not require thermal cycles as in PCR.

2.2.2 Multivalency

Highly sensitive non-catalytic systems for DNA detection have been developed in which the analyte affects the properties of multiple reporter molecules. In particular, solid surface-based (both 2D and 3D) techniques have been developed.

2.2.2.1 Macromolecules

Boudreau, Leclerc and co-workers[41,42] have provided an example in which polymer fluorescence was turned on by the addition of a DNA target sequence. The principal constituent of their system is a water-soluble, cationic poly-thiophene, of which the chromogenic and fluorogenic properties strongly change upon the addition of single-stranded or double-stranded oligo-nucleotides (Figure 2.5). For example, a yellow-to-red colour change and a

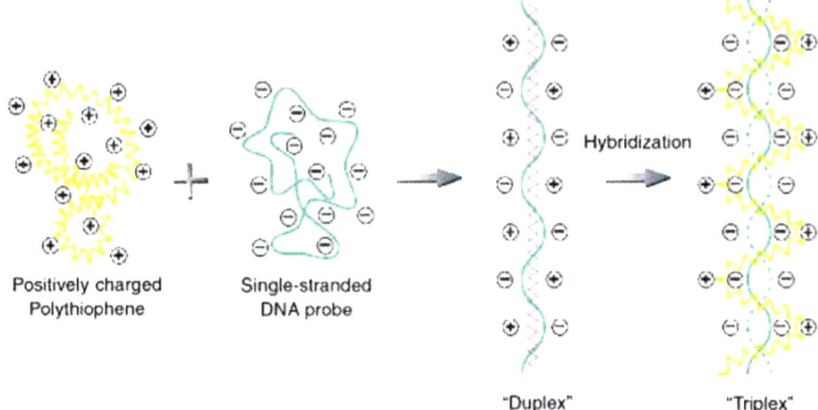

Figure 2.5 Addition of a capture DNA strand induces a planar, highly conjugated conformation of a cationic polythiophene. The change in conformation is accompanied with a yellow-to-red colour change and a complete fluorescence quenching. Upon addition of the complementary target DNA strand a triplex structure is formed in which the polythiophene is wrapped in a helical fashion around the DNA duplex, resulting in a blue shift and an increase in fluorescence.
(Reproduced with permission from Ho *et al.*[41] Copyright 2004 John Wiley and Sons.)

complete fluorescence quenching is observed upon the stoichiometric addition of the capture probe 5′-CATGATTGAACCATCCACCA-3′. This is ascribed to the formation of a planar, highly conjugated form of the polythiophene backbone. The addition of the complementary DNA target strand results in a blue shift and an increase in fluorescence. Spectroscopic data suggested the formation of a superstructure in which the oligothiophene was wrapped in a helical fashion around the DNA duplex (less conjugated, less planar). The sensing system gave excellent results in terms of sensitivity and selectivity. A two base-pair (bp)-mismatch 20-mer DNA sequence was unable to induce triple helix formation, whereas a single bp-mismatch could be discriminated under kinetic monitoring. For 20-mer oligonucleotide targets, a lower detection limit of around 310 molecules per $150\,\mu$L sample ($3.6\,$aM; $1\,$aM$\,=1\times10^{-18}\,$M) could be achieved. Importantly, the multivalency of the system permitted the application of a new readout mechanism based on FRET (Förster resonance energy transfer).[43] Labelling of the DNA capture probe with a fluorophore (Alexa Fluor 456) that absorbs at the emission wavelength of the polymer ($530\,$nm) results in the instalment of FRET upon formation of the triple helix. Interestingly, starting from a large number of probes ($\sim10^{10}$ copies), the addition of just 30 copies of the 20-mer target in $3\,$mL could be easily detected. A lower detection limit of 5 molecules in $3\,$mL (corresponding to $3\,$zM; 1 zM$\,=1\times10^{-21}\,$M) was estimated for this system.

As pointed out by the authors, signal amplification not only arises from the high optical density of the polymer, but presumably also from a fast and efficient energy transfer from the helical and well-structured polythiophenes to many neighbouring chromophores. This process is referred to as 'superlighting' or 'fluorescence chain reaction' (FCR). Evidence for the formation of nanoaggregates of the duplexes (also after hybridization) was obtained from dynamic light-scattering measurements, which indicated that energy transfer between different oligothiophenes could indeed play a role in the signal amplification process.

2.2.2.2 Surfaces

Highly sensitive analytical techniques such as SERS, SPR, and electrochemical detection all rely on interactions occurring between the analyte and a metal surface. In particular SERS can give a signal amplification as high as 10^{10}, which permits detection even at the single molecule level.[15] However, as will be illustrated in this section, also these sensitive surface-based detection methodologies benefit from the use of signal amplification strategies to improve their sensitivity. In particular, we will illustrate different approaches aimed at improving the detection of oligonucleotides using sandwich assays (Figure 2.6a).[44] The classical sandwich assay relies on the capture of the target oligonucleotide by a *capture* oligonucleotide probe immobilized on a surface. Subsequently, a second probe labelled with a signalling unit is added to hybridize with the target probe. Depending on the method of analytical detection, the signalling unit can either be a catalyst, a fluorophore or an electrochemically active unit. The main advantages of the sandwich assay are, first, that it does not require labelling of the target probe and, second, no signal is generated in the absence of the target. Clearly, the sandwich assays bears much resemblance to an ELISA, which relies on the surface-immobilization of an enzyme following a recognition event between a surface-bound receptor and an analyte, often in the form of sandwich type complexes. Similar to ELISAs, a limitation of sandwich assays in terms of sensitivity is the fact that each recognition event results in the capture of only one signalling unit. A methodology that allows multiple signalling units to be captured through a single binding event allows sensitivity limits to be pushed further down.

The electrochemical detection of analytes is highly appealing because signal generation is straightforward and does not require sophisticated instrumentation.[45] An electrochemical readout requires the presence of a redox-active moiety in the signal probe that induces a flow of electrons from the electrode upon capture by the target. In a recent study, Zuo, Plaxco, Heeger *et al.*[46] illustrated a simple, but ingenious approach to improve the sensitivity of the sandwich assay (Figure 2.6b). The key novelty of this so-called super sandwich assay is the use of signal probes (DNA sequences functionalized with the redox-active compound methylene blue) able to hybridize to two regions on the target DNA. However, the probes are engineered such that hybridization occurs more readily between complementary regions on two distinct target sequences rather than two regions on the same target molecule (inspired by the work by Matile *et al.*,[47] see Section 2.3.2.2). Consequently, an extended

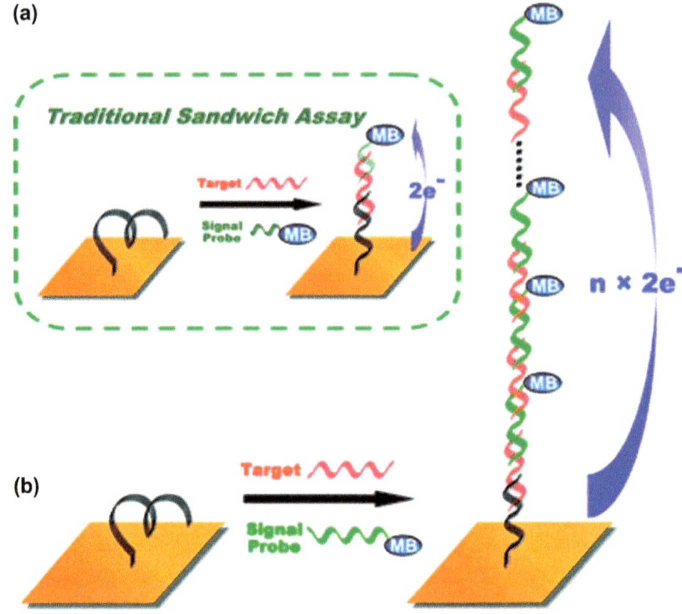

Figure 2.6   (a) In the classical sandwich assay, the target DNA oligonucleotide strand is captured by the complementary capture strand immobilized on a surface. Subsequent hybridization with a DNA strand containing a signalling unit completes the assay. (b) The super sandwich assay reported by Zuo, Plaxco, Heeger and co-workers[46] relies on the use of signalling probes which are able to hybridize to complementary regions on two distinct target sequences, resulting in the formation of DNA polymers composed of alternating target and signalling DNA sequences. (Reproduced with permission from Xia *et al.*[46] Copyright 2010 American Chemical Society.)

superstructure composed of alternating target molecules and signalling units is formed on the surface of the electrode. The capture of a large number of methylene blue moieties results in a strongly enhanced Faraday current compared to a classical sandwich assay. The use of the super sandwich assay permitted target DNA detection down to 100 fM, which is 3 orders of magnitude lower compared to the classical assay. This approach was recently further developed by Chen, Yang and co-workers,[48] reaching sensitivity levels down to 100 aM.

An entirely different surface-based approach towards an ultrasensitive DNA detection system was reported by Mirkin *et al.* based on the use of nanoparticles (NPs) and bio-bar-codes (Figure 2.7).[49] The assay developed for the detection of DNA relies on the use of two probes: (1) 30 nm-sized gold (Au) NPs coated with numerous copies of hybridized oligonucleotides (the bio-barcodes) and a small amount of ssDNA complementary to part of the

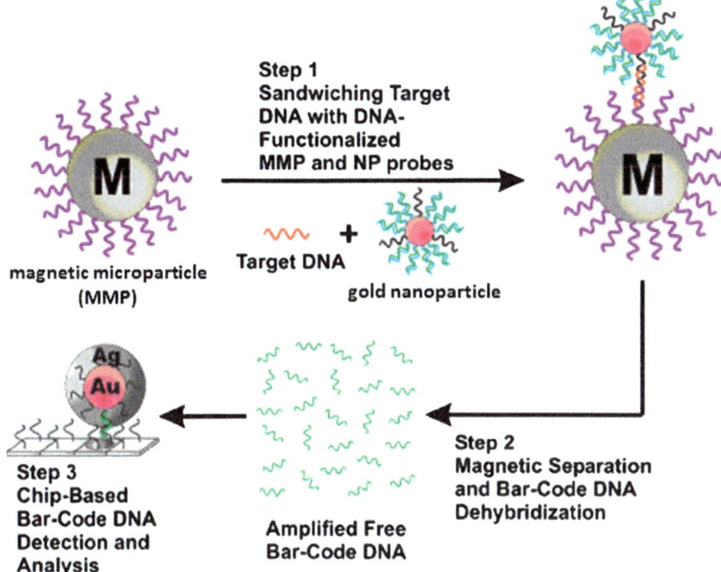

Figure 2.7 Schematic representation of bio-barcode-based signal amplification. The target DNA is sandwiched between magnetic microparticles and gold nanoparticles, both functionalized with complementary DNA strands. The aggregated systems are isolated, the DNA barcodes released and quantified using DNA-chip methodology.
(Reproduced with permission from Nam *et al.*[49] Copyright 2003 American Chemical Society.)

target sequence (in total around 360 oligonucleotides with a $\sim 70:1$ ratio of bio-barcodes to target binding DNA); (2) 1 µm-sized magnetic microparticles (MMPs) containing a magnetic iron oxide core functionalized with a DNA sequence complementary to another part of the target sequence. The protocol relies on the sandwiching of target DNA between the MMPs and the NPs. The resulting superstructures embed the key element of this approach concerning signal amplification, since they combine a limited number of DNA targets with a multitude of bio-barcodes, *i.e.* reporter molecules. These superstructures are then isolated through the application of a magnetic field and the bio-barcodes are released through dehybridization and then quantified. Among various DNA detection methods (gel electrophoresis and fluorescence labelling), the scanometric DNA detection method combined easy implementation with high sensitivity. In short, this detection relies on the capture of the bio-barcodes on DNA-chips to which subsequently oligonucleotide containing Au NP probes are attached, which are then enlarged through the reduction of Ag^+ on the Au colloids.[50] Final detection occurs through measurement of light scattering from the developed spots using specialized instrumentation. A lower DNA detection limit of 500 zM was reported, corresponding to around 10 copies of target DNA per 30 µL sample.

2.2.3 Catalysis and Multivalency

Impressive amplifications can be obtained in the event that the presence of the analyte sets off a cascade of amplification events, just as in Nature. The sensitivity of a catalytic assay can be increased if a single recognition event results in the immobilization of multiple catalysts rather than one. For that purpose, NPs are a perfect platform, as demonstrated by Willner *et al.*[52] They engineered Au NPs functionalized with multiple (96) copies of a ssDNA strand containing an internal DNAzyme sequence able to generate chemiluminescence in the presence of haemin, O_2 and luminol, and a terminal DNA sequence complementary to the ssDNA strand to be detected (Figure 2.8). In the first instance, the analyte is hybridized to a single strand immobilized on a Au surface resulting in the exposure of a sticky end able to capture the DNAzyme-functionalized Au NPs. The subsequent addition of the DNAzyme substrates then results in chemiluminescence which can be readily detected. Compared to ELISAs, a second amplification step is present since a single recognition event now can recruit in principle up to 96 catalysts. The assay was applied for the detection of telomerase activity, which is a versatile marker for cancer cells. For this purpose, a Au surface was functionalized with a DNA sequence which was telomerized in the presence of HeLa cell extracts (containing telomerase). Since the extent of telomers

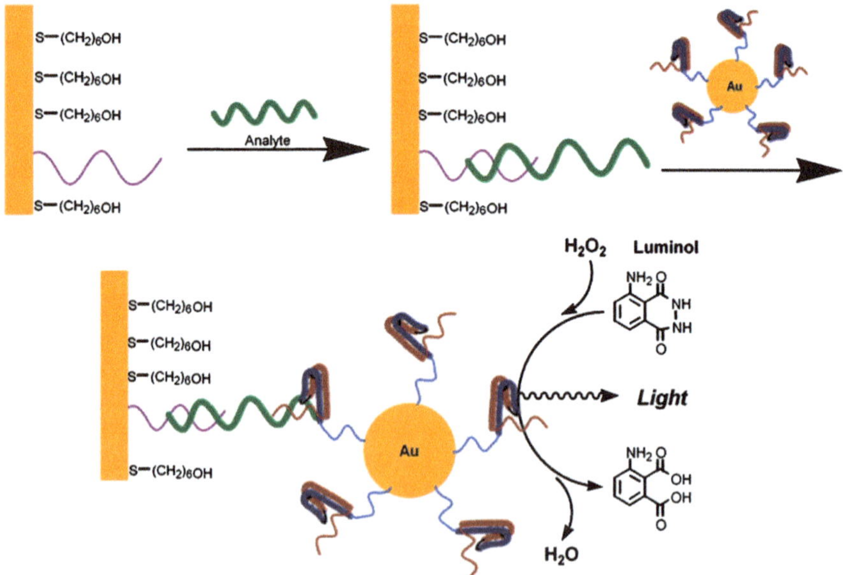

Figure 2.8 Amplified chemiluminescence detection of DNA using DNAzyme functionalized gold (Au) nanoparticles.
(Reproduced with permission from Niazov *et al.*[52] Copyright 2004 American Chemical Society.)

formed on the surface is controlled by the concentration of telomerase in the cell extracts, the amount of DNAzyme labeled Au NPs that are captured on the surface, and thus the intensity of the generated light, is directly related to the concentration of the telomerase. In fact, a linear relation was observed between the intensity of the emitted light and the number of HeLa cells within a range between 1000 and 10 000. The sensitivity of the reported method was 10^2–10^4-fold higher when compared to chemiluminescence assays for DNA analysis or telomerase assays. A conceptually related approach has recently been used by Wang and co-workers[53] for the detection of α-fetoprotein, which is a protein cancer biomarker. Their approach relied on the α-fetoprotein-induced aggregation of DNAzyme-functionalized Au NPs and MMPs functionalized with α-fetoprotein monoclonal antibodies.

2.3 Small (bio)Molecule Detection

2.3.1 Catalysis

2.3.1.1 *Aptamers and DNAzymes*

Aptamers are nucleic acids exhibiting specific recognition properties towards low-molecular-weight substrates and are obtained in a similar manner as DNAzymes (Section 2.2.1.2).[54] For sensing purposes, sophisticated oligonucleotide constructs encoding both an analyte recognition site (aptamers) and a catalytic site (DNAzyme) have been designed. The general working principle of these systems relies on the activation of the catalytic site upon the analyte-induced refolding of the sequence. The application of an aptamer-DNAzyme construct for sensing of small molecules is nicely illustrated by the following example by Willner and co-workers.[55] A sensor for adenosine monophosphate (AMP) was developed using the hairpin-DNAzyme structure given in Figure 2.9. It consists of an AMP aptamer sequence and the HRP-mimicking DNAzyme sequence. The aptamer and DNAzyme sequences are bridged at the $5'$- and $3'$-end, respectively, by an additional sequence. Also, a short nucleic acid strand is added to the $3'$-end of the aptamer. This composition ensures that the DNAzyme sequence cannot fold in the catalytically active G-quadruplex/ haemin complex. In the presence of AMP, the AMP–aptamer complex is stabilized, resulting in the opening of the hairpin structure and, consequently, the activation of the DNAzyme. Measurement of $ABTS^{\bullet +}$ production as a function of time showed increased rates as a function of the concentration of AMP. A calibration curve was used to determine the detection threshold of a 50 µM AMP solution. Interestingly, the relevance of the DNAzyme (and thus signal amplification) was demonstrated using a related DNA hairpin sensing system relying on fluorescence quenching. Here, AMP binding to the aptamer sequence opens the hairpin, which results in the separation of the fluorophore and the quencher and, thus, an increase in fluorescence intensity. Apart from the fact that nonlinear behaviour was observed for high analyte concentrations (because of fluorescence quenching induced by the analyte),

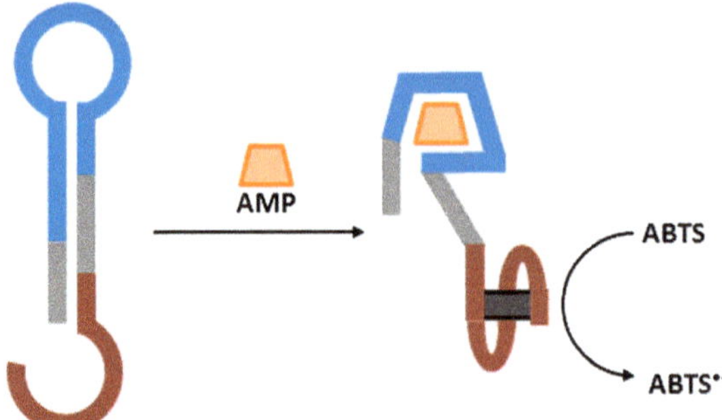

Figure 2.9   A DNAzyme–aptamer construct for the sensing of AMP. In the presence of the analyte the sequence opens up and the DNAzyme is activated.

signal amplification was evidenced by the readout intensities after normalization, which were 3.4 times higher in intensity for the DNAzyme-aptamer construct in a 25 µM analyte solution.

The use of DNA as constructing element also allows the development of more complicated sensing systems that exploit the particular folding and hybridization patterns of DNA *and* the biological available machinery such as polymerases. The result is intricate hybrid bio-synthetic sensing systems. For instance, Willner and co-workers[56] reported the detection of cocaine using a similar system (Figure 2.10). The key component of the system is an oligonucleotide sequence containing an aptamer region (I) for cocaine, a nicking site for the enzyme Nt.BbvC I (II) and a region encoding for the product used for signal transduction (III). In the absence of cocaine, the structure cannot fold into the active form because of the presence of a small complementary strand. The presence of analyte causes the aptamer to fold in the active conformation, separating the blocking double strand. The aptamer·cocaine complex contains a 7-base duplex structure that serves as an initiation point for oligonucleotide replication by a polymerase. Replication yields the duplex strand including the nicking site for Nt.BbvC I. Scission of the replicated strand results in a new replication site for the polymerase. In this way, multiple copies of the complementary DNA strand are produced by a single cocaine-sensor. The key to signal generation is a hairpin DNA-structure containing fluorophores at both the 3′ and 5′, 6-carboxyfluorescein (FAM) and 6-carboxytetramethylrhodamine (TAMRA), respectively. In the closed form, excitation of the FAM-fluorophore results in fluorescence resonance energy transfer to TAMRA ($\lambda_{em} = 580$ nm). Upon hybridization with the formed products, however, the hairpin opens up and prohibits FRET. The result is an increase in the fluorescence intensity of FAM ($\lambda_{em} = 520$ nm). Using a time interval of 60 minutes for operating the machine, a detection limit of 5 µM was determined for cocaine, which is comparable to the sensitivity of enzyme-linked immunoassays.

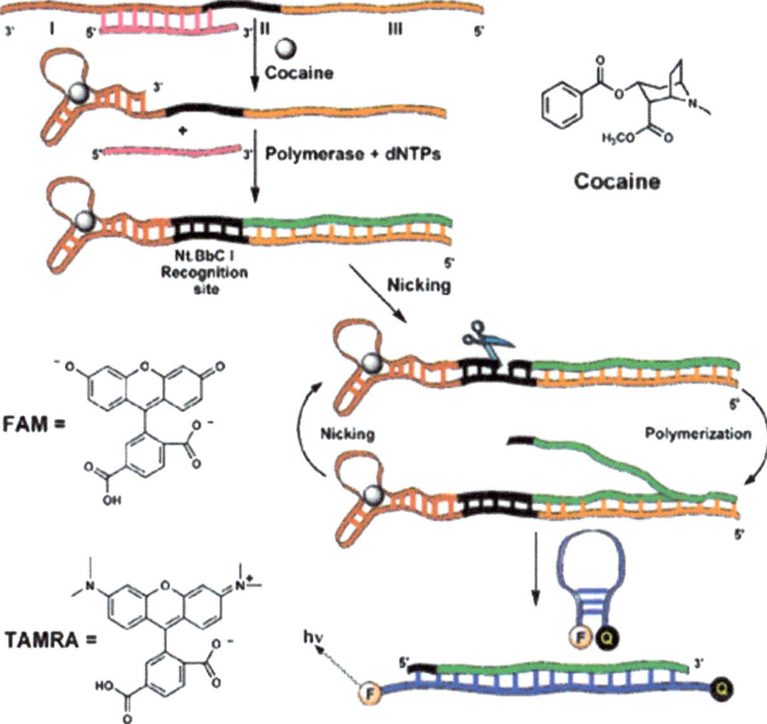

Figure 2.10 A DNA-aptamer machine for the amplified detection of cocaine. (Reproduced with permission from Shlyahovsky *et al.*[56] Copyright 2007 American Chemical Society.)

2.3.1.2 Synthetic Catalysts

The development of entirely synthetic catalysts is attractive as it permits the screening of small molecules also under non-physiological conditions (*e.g.* in organic solvents or at highly acidic/basic pH). Here, two examples that illustrate two different ways in which catalysis is used for small molecule detection will be discussed, namely: either through allosteric regulation of a catalyst or by triggering an autocatalytic pathway. Allosteric regulation refers to systems in which an analyte changes the catalyst from an active to an inactive state.[57] Related allosteric and autocatalytic systems have also been reported for the detection of metal ions, but these analytes fall outside the scope of this review.[58]

Mirkin *et al.*[59] prepared catalysts according to the so-called 'Weak-Link' approach. Catalysts of this type are characterized by two structural domains containing Rh(I) metal centres and a catalytic domain in which two Zn(II) ions act cooperatively (Figure 2.11). The 'Weak-Link' refers to the labile thioether-Rh(I) bond, which breaks in the presence of CO and Cl$^-$. Although the phosphine-Rh(I) bonds remains intact, their position changes from *syn* to *anti*, which induces an increased distance between the catalytic Zn(II) ions. It was observed that in the 'open' conformation of the catalyst, the acyl transfer from acetic anhydride to pyridyl carbinol is accelerated up to 25 times through a catalytic pathway that involves both Zn(II) ions. This enabled the possibility to detect Cl^- ions through a catalytic readout, which was confirmed by measuring the rate of formation of 4-acetoxymethylpyridine as a function of the concentration of Cl^- ions (under CO saturation). To facilitate readout, the catalytic amplification step was coupled to a pH-sensitive fluorophore which permitted a visual and spectrophotometrical monitoring of the amplification process. Chloride concentrations down to 800 nM could be detected in a straightforward manner. Successively, it was shown that the system was also amenable to the detection of small molecules such as di-imines and isocyanides.[60] In particular, the observation that also the acetate ion, *i.e.* the product of the amplification reaction, had an allosteric effect created the possibility to chemically couple the allosteric and amplification processes.[61] In a

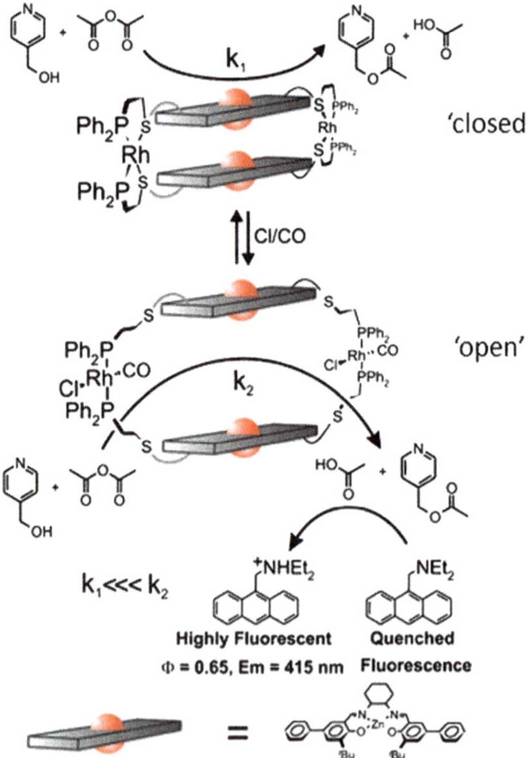

Figure 2.11 Supramolecular allosteric catalytic signal amplifier. In the presence of Cl/CO the catalyst opens up and accelerates the transacylation reaction. (Reproduced with permission from Gianneschi *et al.*[59] Copyright 2005 American Chemical Society.)

process that can be regarded as a truly artificial version of the polymerase PCR, an initially small amount of acetate ion activates the catalyst, resulting in the production of more acetate, which then activates more catalyst, and so on. Plots of the substrate conversion in time as a function of the initial analyte concentration gave sigmoidal profiles, typical of PCR-like and self-replication processes (see below). Slow induction followed by rapid exponential amplification, linear growth, and then eventual saturation (after $\sim 80\%$ of the substrate had been consumed) is observed, regardless of initial analyte concentration. As with PCR, the time at which the exponential step turns-on correlates with the analyte concentration. As a result, acetate concentrations in the 20–80 µM concentration range were detected in a straightforward manner. Other autocatalytic sensing systems for metal ions have been reported by Phillips *et al.*[62]

2.3.2 Multivalency

In this section sensing systems for small molecule detection are discussed in which a single analyte induces the response of multiple reporters clustered in a macromolecule or a supramolecular aggregate.

2.3.2.1 *Macromolecules*

In the past decades, conjugated polymers have been widely exploited as chemosensor components, because of their unique electronic and photophysical properties.[63,64] In particular, the electrical conductivity is highly sensitive to very small perturbations induced by the interaction of an analyte with recognition units in the polymer. In addition, the fluorescence emission of conjugated polymers is often dominated by energy migration to local minima in their band structures. Swager and co-workers have applied these properties for the development of fluorescent chemosensors in a so-called 'molecular wire approach' (Figure 2.12).[65,66] The molecular wire is a strongly fluorescent cyclophane-containing conjugated polymer with a penyleneethylene-backbone. Fluorescence quenching occurs upon the addition of the electron transfer quenching agent paraquat **PQ^{2+}**. A dramatic quenching enhancement was observed when compared to the same concentration of monomeric cyclophane units. For example, the addition of 3.45×10^{-4} M of **PQ^{2+}** to a 3.69×10^{-6} M concentration of an isolated cyclophane unit resulted in a 30% quenching of the fluorescence of the monomer. However, a near-quantitative quenching was observed when the cyclophane was incorporated in the polymer. The improved quenching enhancement in the molecular wire originates from the migration of the excited state through the polymer backbone, thus permitting a sampling of a multitude of cyclophane receptors, rather than just one. If only one of the receptors has formed a complex with **PQ^{2+}**, quenching of the excited state occurs. Quantitative fluorescence studies confirmed that quenching occurs through a static process in accordance with a host–guest complexation. In this setup, quenching enhancement is favoured both by the long lifetime of the excited state and a high rate of energy migration in the polymer backbone,

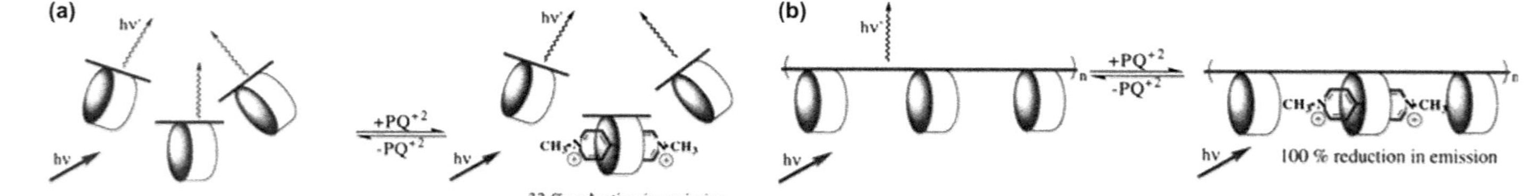

Figure 2.12 Illustration of the enhanced sensitivity by the 'molecular wire' sensing system (b) as opposed to a conventional molecular sensing system (a).
(Reproduced with permission from Zhou and Swager.[65] Copyright 1995 American Chemical Society.)

which permits a sampling of more receptor units. In an ideal case, where the association constants for complex formation are additive and the diffusion length of the excitation state exceeds the length of the polymer, the maximum signal enhancement is determined by the degree of polymerization. A further increase in sensitivity was observed in cases where the polymers were deposited as films, because in the solid state communication between polymer backbones is also possible.[67]

A different application of polymers for sensing purposes originates from the concentration of receptor molecules in the small volume provided by a polymeric matrix.[68] This is referred to as compartmentalization of the host–guest system.[69] Schneider *et al.*[70] employed various chemomechanical polymers that change in size upon the addition of analyte. Interestingly, the sensitivity of these systems increases as the particle size decreases, since in that case a smaller number of analyte molecules is required to induce the same relative volume change in the material. Consequently, miniaturization permits the detection of analytes at higher dilutions provided that two conditions are met: (1) the number of receptor sites in the polymer must be large in order to attract a large number of analytes from the surrounding medium, and (2) the affinity between analyte and receptor must be high enough to ensure complex formation also at low concentrations.

As an example, a sensing system able to discriminate between the enantiomers of *O,O'*-dibenzoyltartaric acids (DBTA) will be discussed.[71] Schneider and co-workers used chitosan as hydrogel because it contains aminoglucose units as chiral components. The gel contains 47% of water after swelling which increased to 96% when chitosan was pretreated with acetic acid. Exposure of the gel to D-DBTA at low millimolar concentrations triggered a large volume contraction of 95%. Much higher concentrations of L-DBTA were needed to induce a similar volume change. Importantly, it was shown that the concentration of D-DBTA required to induce an 80% change in volume required about 1.2 mM D-DBTA when the particle volume was 7.5 mm^3, but only about 0.2 mM with a particle volume of 0.2 mm^3, illustrating the advantage of miniaturization. The importance of compartmentalization is expected to increase with the advance of analytical chemistry on the femtolitre scale.[72]

Self-immolative dendrimers are designed to fully disassemble through a domino-like cascade of cleavage reactions induced by a single reaction that occurs at the dendritic core.[73] Such materials are interesting as prodrugs, because multiple peripheral drug molecules can be locally released upon a single enzymatic cleavage. In an elegant approach, Shabat and co-workers have exploited self-immolative dendrimers for the exponential signal amplification of an input signal.[74] Their approach, referred to as dendritic chain reaction, relies on the use of the AB$_3$ dendron, designed to detect the analyte hydrogen peroxide (Figure 2.13). The dendron contains a phenyl boronic acid as analyte-sensitive trigger, two choline units and one *p*-nitroaniline reporter unit. In the presence of hydrogen peroxide the boronic acid is oxidized, which then initiates a series of intramolecular cascade reactions, ultimately leading to the release of the two cholines and yellow *p*-nitroaniline. The two choline units are oxidized

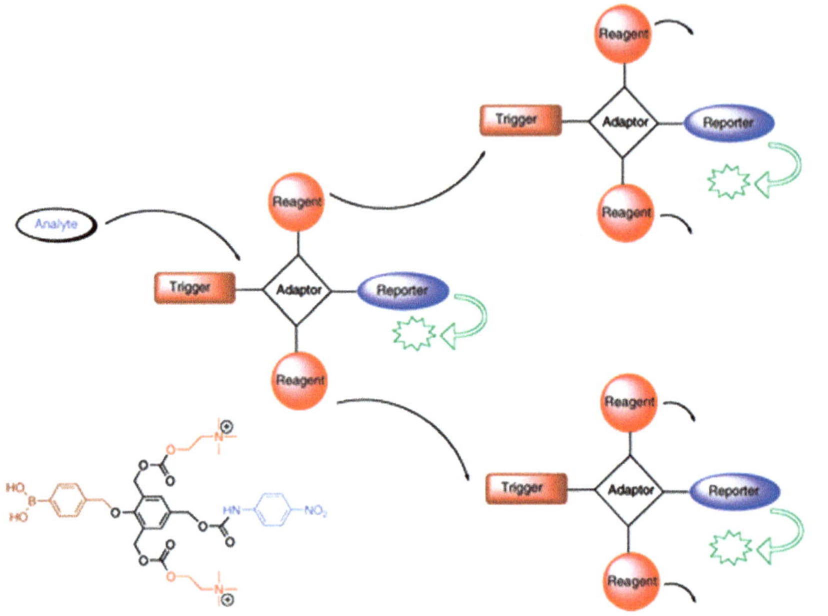

Figure 2.13   Schematic representation of the dendritic chain reaction. The presence of an initial analyte triggers the cleavage of the dendron. This sets off a cascade of reactions leading towards an exponential increase of *p*-nitroaniline reporter molecules (indicated as green stars). (Reproduced with permission from Sella and Shabat.[74] Copyright 2009 American Chemical Society.)

by choline oxidase (COX), which gives four new hydroperoxide molecules as side product. In turn, these trigger the destruction of four more dendrons, which is the source of the exponential amplification. In fact, sigmoidal curves characteristic of exponential growth were observed when the release of *p*-nitroaniline was measured in the presence of low initial concentration of hydrogen peroxide. For an initial concentration of $5\,\mu M$ of hydrogen peroxide an output signal could be generated that was 53 times stronger than a conventional probe. As illustrated by the authors, the system can be adapted for other analytes in a relatively straightforward manner, especially by separating the trigger and reporter release on two different dendrons.[75] Nonetheless, considering that spontaneous hydrolysis triggers the same cascade events, the clear challenge in this approach is the development of dendrimers that self-destruct only in the presence of analyte.

For chiral signal transduction in polymers, the 'sergeants-and-soldiers' principle has emerged as a very potent technique.[76] This principle relies on the observation that the chirality of a (helical) macromolecule is induced nearly quantitatively by a small fraction of chiral monomers (sergeants). This is by definition an amplification mechanism as the achiral monomers (soldiers) behave as if they were chiral. For sensing purposes, it is important that the macromolecule has a physical property that is affected by the cooperative interaction between chiral and achiral monomers. Frequently, this physical property is the induced circular dichroism (ICD) originating from the polymer backbone. The use of the sergeants-and-soldiers principle for sensing purposes requires a non-covalent interaction between the (chiral) analyte and the racemic polymer.[77] Such a system developed for detecting very small enantiomeric excesses in α-amino acids was reported by Yashima and co-workers.[78] A stereoregular *cis-transoidal* poly(phenylacetylene) ($M_n = 19.7 \times 10^4$) containing peripheral crown ethers was used, in which the helical polyacetylene backbone serves as the chromogenic reporter unit and the bulky crown ethers serve both as amino acid receptors and as rigidifying elements in order improve cooperative behaviour (Figure 2.14). The addition of either one of the 19 natural amino acids (with the exception of L-Pro) in dimethyl sulfoxide resulted in intense ICDs, reaching a constant value upon the addition of 1 equivalent of amino acid (with respect to the crown ether) at $-10\,°C$. For L-Ala it was observed that the addition of 0.1 equivalent was sufficient for the nearly complete induction of single handedness in the polymer, and even for 0.01 equivalents an appreciable ICD was obtained. This is an illustration of the 'sergeants-and-soldiers' principle in which the empty crown ethers follow the handedness induced by the occasional crown ether/ L-Ala complex. A lower detection limit of 70 ng of L-Ala per mL was determined at $25\,°C$, corresponding to 0.005

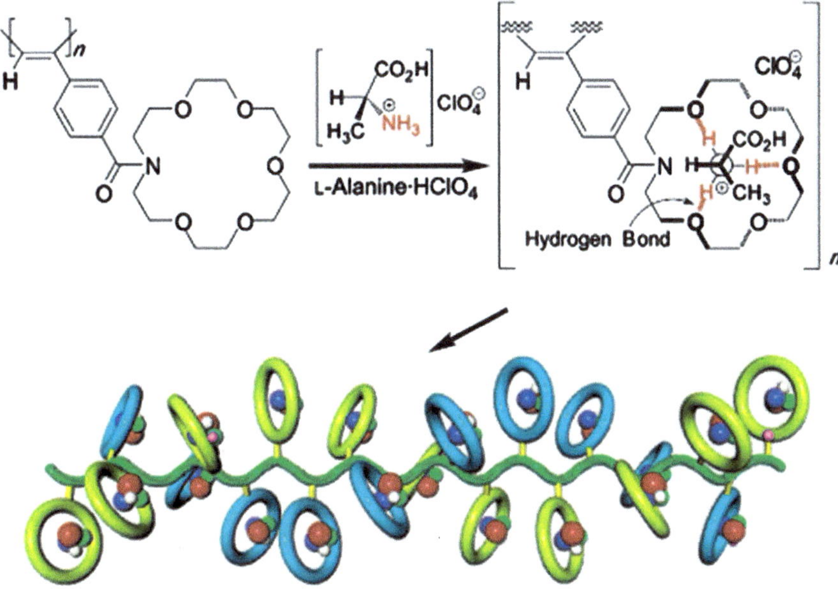

Figure 2.14 Induction of a single-handedness in a polymer induced by non-covalent interactions between an amino acid and pending crown ethers. Counterions have been omitted for clarity.
(Reproduced with permission from Nonokawa and E. Yashima.[78] Copyright 2003 American Chemical Society.)

equivalents per crown-ether, which makes it one of the most sensible synthetic amino acid receptors reported until this day. Subsequently, the ability of the system to detect the enantiomeric excess (*ee*) of amino acids according to the 'majority rule' was studied. An *ee* of L-Ala as low as 5% was sufficient to induce a CD of the same intensity as the enantiopure sample. The impressive ability of this system to reliably amplify chiral information was evidenced by the observed linear relationship between the *ee* of Ala mixtures (in the range from 0.1% to 0.005%!) and the measured CD intensity.

2.3.2.2   *Supramolecular Aggregates*

Matile and co-workers[79] have been particularly involved in the development of conceptually new approaches towards sensing and signal amplification. In a first example, they used a vesicle-based system for the sensing of ATP (Figure 2.15).[47] Here, the working principle is the ability of DNA to act as a carrier for hydrophilic counterions across lipid bilayers, on the prerequisite that DNA is initially coated with amphiphilic counterions in order to 'dissolve' DNA in the apolar bilayer. The activity of DNA polyion transporters increases with the number of charges in an overadditive manner because of multivalency effects. For instance, dsDNA is a better cation transporter than ssDNA. These characteristics were

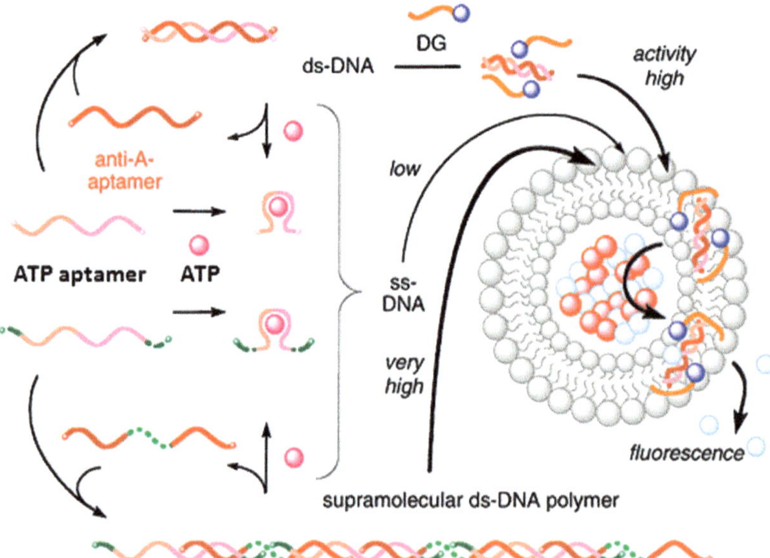

Figure 2.15   Activated by countercations (here dodecylguanidinium [DG]), DNA can act as cation transporter in fluorogenic vesicles. For sensing with DNA aptamers, the differences in activity of ss-DNA (low), ds-DNA (high), and supramolecular ds-DNA polymers (very high) are used. The disassembly of aptamer/antiaptamer duplexes in response to the binding of analyte ATP thus results in a decrease of the fluorescence emission. (Reproduced with permission from Takeuchi and Matile.[47] Copyright 2009 American Chemical Society.)

used to set up an assay for the detection of ATP by preparing vesicles loaded with the anionic fluorophore 8-hydroxy-1,3,6-pyrenetrisulfonate and the cationic quencher *p*-xylene-bis-pyridinium bromide. DNA mediated removal of the cationic quencher from the interior of the vesicles results in an increase in fluorescence intensity. An ATP-aptamer was used in order to render the system responsive to the analyte. In particular, the high affinity of ATP for the aptamer results in dissociation of the double-stranded helix formed between the aptamer and the complementary anti-aptamer strand. Upon dissociation, the ability of the DNA strand to transport the quencher across the bilayer diminishes, thus causing a decrease of the fluorescence intensity ('on-off' sensor). Nonetheless, deactivation of the aptamer·anti-aptamer strand by ATP occurred with an IC_{50} value of 2.1 mM and did not reach completion ($\Delta Y = 50\%$). Therefore, the DNA-sequences were redesigned in such a way that, in the resting state, the aptamer sequence was present as a supramolecular dsDNA polymer with a much higher ion-transport capability. This permitted a three-fold reduction of the IC_{50}-value to 0.9 mM and a response, ΔY, of 87%. The characteristics of aptamers (rather poor sensitivity, excellent selectivity) are reflected by the obtained results for this system (ATP detection in the low millimolar range, but with excellent selectivity over GTP).

Likewise, the same group makes use of analyte-induced transport across lipid bilayers based on the concept of counterion-exchange, but in a much different way.[80] Here, the authors were challenged by developing a responsive system towards steroids, which are important targets in medicinal diagnostics and environmental monitoring. However, their poor solubility in aqueous media and their tendency to partition in the membrane bilayer of pore sensors renders the development of sensing systems challenging. In this work, anionic fragments were conjugated to a series of steroids through hydrazone formation. Subsequently, these amphiphilic conjugates were used as amphiphilic counteranions of pR (poly-arginine), a cationic cell-penetrating peptide. Similar to the first example, this system is able to activate ion transport (in this case anions) across the lipid bilayer through a two-step counterion exchange (hydrophilic to amphiphilic and back). In this case, an outflux of 5(6)-carboxyfluorescein (CF) from the vesicle with a concomitant increase in fluorescence intensity ('off-on' sensor) was observed. EC_{50} values were in the low micromolar concentration regime and could be further reduced using triton-X 100 as an additive.

2.4 Protein Detection

2.4.1 Catalysis

Evidently, enzymes are intrinsic signal amplifiers and for that reason form the signal generating component of enzyme-linked immunosorbent assays (ELISAs). Nonetheless, the efficiency in signal generation depends on the availability of pro-fluorogenic or pro-chromogenic substrates that are cleaved with a high turnover frequency (TOF). Preferably, the substrates should also have a low K_M value (high affinity) in order to ensure that the enzyme operates at the highest rate

(V_{max}). Often, not all these prerequisites are met or are completely absent (no enzyme activity). In these cases, the development of innovative assays is vital.

Recently, Rotello and co-workers presented a highly innovative sensing system for proteins referred to as enzyme-amplified array sensing.[81] Key components of the assay are the enzyme β-galactosidase (β-Gal), responsible for signal generation by hydrolysing the fluorogenic substrate 4-methyl-umbelliferyl-β-D-galactopyranoside (MUG), and monolayer-protected Au colloids (Au MPCs), which inhibit the activity of β-Gal through complex formation with the enzyme (Figure 2.16). The Au MPCs are displaced from β-Gal upon the addition of analyte proteins, which then restores the catalytic activity of the β-Gal enzyme, thereby generating a fluorescent output signal. Complexation between the positively charged Au MPCs and β-Gal is driven by electrostatics and the effectiveness of the enzyme inhibition was evidenced by

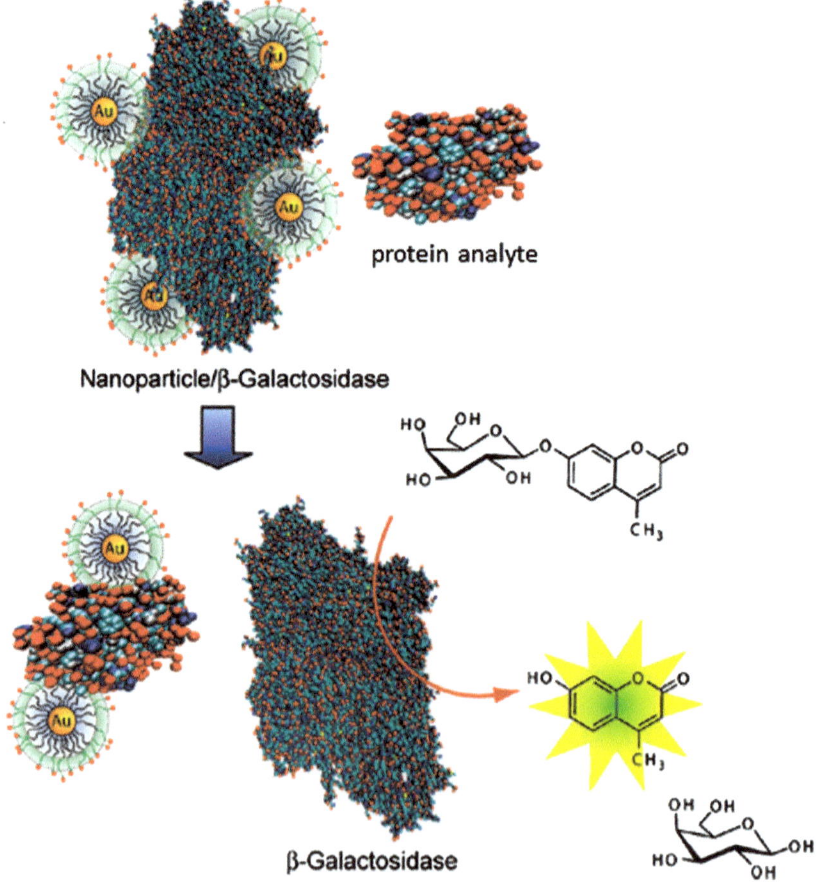

Figure 2.16 The activity of the the β-Gal enzyme is restored upon the displacement of Au nanoparticles by the protein analyte.
(Reproduced with permission from Schneider and Kato.[71] Copyright 2010 American Chemical Society.)

the progressive decrease in enzymatic activity in the presence of increasing concentrations of Au MPCs. Given the aspecific nature of the interaction between Au MPCs and β-Gal, a clear challenge in this setup is the issue of analyte selectivity. This was resolved by using a displacement assay for sensing the analyte proteins. It was observed that the inhibitory power of a series of six Au MPCs containing differently charged headgroups was dependent on the nature of the headgroup (linear–branched, aliphatic–aromatic). The addition of a set of nine analyte proteins, with varying isoelectric points, to the individual β-Gal/Au MPC complexes gave different responses, resulting in specific response patterns for each analyte protein. On this basis, a fluorescent displacement assay was set up in which the overall response was correlated to the analyte input through linear discriminant analysis (LDA). A canonical score plot of the three major factors revealed nine distinct clusters for each of the analytes. Importantly, enzymatic amplification of the input signal permitted differentiation of the analytes in the 1 nM range, which is significantly lower compared with other methods (1–350 μM). An identification accuracy of 92% was obtained from a screening of 60 samples of unknown analyte composition at 1 nM concentrations. The ability to perform these analyses also in desalted human urine marks a big step towards diagnostic applications.

The versatility of the sensing system has been further demonstrated in a recent contribution in which the β-Gal/Au MPC was used for the colorimetric sensing of bacteria such as *Escherichia coli*, *Streptomyces griseus* and *Bacillus subtilis*.[82] In a similar manner as before, in the presence of bacteria the cationic Au MPCs are displaced from the surface of β-Gal enzyme resulting in its activation. It was shown that bacterial levels down to 100 cells mL^{-1} gave a distinguishable, reproducible and colorimetric response when the assay was performed in solution. Moreover, the assay could also be performed on a paper strip, which is important for practical applications. On these strips, a quantitative assessment was possible by analyzing the RGB profiles of the images. On paper, bacteria concentrations down to 1×10^4 per mL could be distinguished using this method.

Although enzymes can often be detected at low concentration levels because of their intrinsic catalytic activity, those thresholds can be pushed to lower limits by coupling the enzyme activity to a second amplification event. Such an example was recently reported by Scrimin, Prins and colleagues employing catalytic Au NPs for the detection of enzyme activity.[83] The catalytic Au NP catalyses very efficiently the transphoshorylation of 2-hydroxypropyl-4-nitrophenyl phosphate (HPNPP), exhibiting a rate acceleration over 30 000 (k_{cat}/k_{uncat}) under saturation conditions in buffer (pH 7.5) (Figure 2.17).[84] Catalysis originates from the cooperative action of two TACN·Zn(II) complexes on the periphery of the Au NP. It was observed that oligoanionic species, such as the tripeptide Ac-DDD-OH and ATP, act as inhibitors for catalysis because of complex formation with the positively charged surface. Binding studies revealed that the free energy of binding between inhibitor and the Au NP increases linearly as a function of the number of negative charges. This implies that the system could act as a sensor for enzymes that are able to

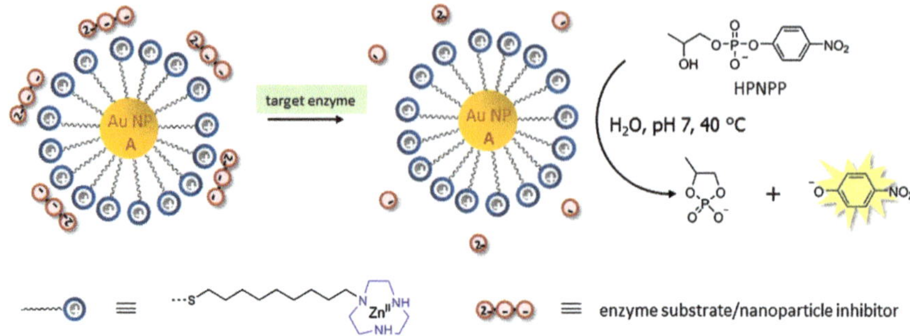

Figure 2.17 The catalytic activity of the gold nanoparticle (Au NP) is inhibited in the presence of oligoanions (such as tri-Asp or ATP).[83] The addition of an enzyme, able to hydrolyse the substrate in small fragments, restores the catalytic activity of the Au NP. HPNPP = 2-hydroxypropyl-4-nitro-phenyl phosphate.

switch-off the inhibitory ability of the inhibitors by converting these into smaller fragments. This was illustrated by incubating a 5 µM solution of Ac-DDD-OH with the non-specific protease subtilisin A in concentrations down to 66 nM for 1 hour. Measurement of the absorbance at 405 nm after the addition of the Au NP and HPNPP gave indeed a correlation between the amount of enzyme present and the rate of hydrolysis of HPNPP. By relying on a cascade of two catalytic events, the final strength of the output signal depends on two time frames: first, the exposure time of the substrate to the enzyme (or, alternatively and more important from an analytical point of view, the concentration of enzyme present) and, second, the time permitted for production of the *p*-nitrophenol reporter molecule. The system has a general applicability because enzyme selectivity can be introduced by varying the peptide sequence of the inhibitor. A signal amplification factor of 18 (compared to a conventional assay) was obtained when targeting caspase 1.

2.4.2 Multivalency

As shown in Section 2.3.2.2., synthetic pores in vesicles can be used for the sensing of small molecules if these analytes alter the flux of reporter molecules through such pores. These sensing systems can be applied also for the sensing of enzymes if the enzymatic activity changes the ability of the small molecules to (de)activate the synthetic pores. Matile and co-workers first illustrated this concept by engineering synthetic pores based on rigid-rod β-barrels in the lipid bilayers of large unilamellar vesicles (LUVs) composed of egg yolk phos-phatidylcholine loaded with 5(6)-carboxyfluorescein (Figure 2.18).[85] The β-barrels spontaneously form through the self-assembly of *p*-octiphenyl units containing small pending peptides.[86] The pores were designed such to contain arginine–histidine dyads lining up in the interior of the pore. Fluorescence titration experiments revealed that ATP was complexed inside the pores with a dissociation constant of around 2 µM, compared to 66 µM for the lesser

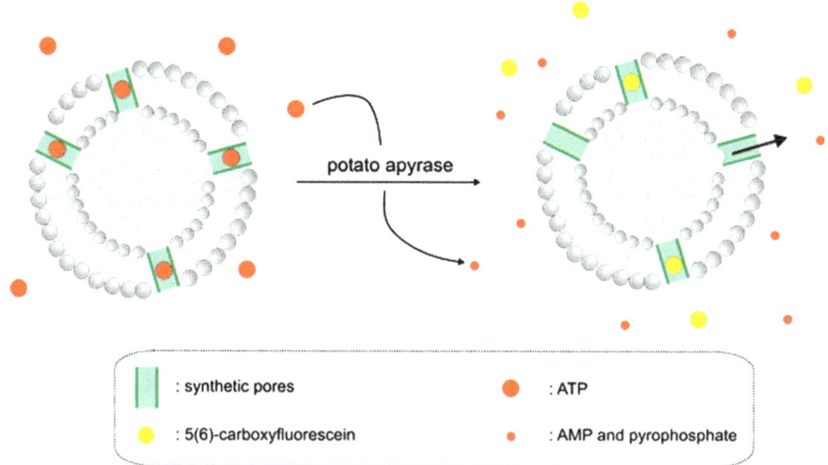

Figure 2.18 ATP blocks the synthetic pores in the vesicle bilayer by preventing the export of the internal 5(6)-carboxyfluorescein. Potato apyrase converts ATP into the poor pore blockers AMP and pyrophosphate. Consequently, the fluorophore exits the vesicle, resulting in an increase in fluorescence intensity.[85]
(Reproduced with permission from Scrimin and Prins.[2] Copyright 2011 Royal Chemical Society.)

charged AMP. Consequently, detection of potato apyrase, a non-specific ATPase that converts ATP into AMP and pyrophosphate, was possible. Advantages of the assay include the use of straightforward detection methodology as well as unlabelled enzyme substrates. The system can be easily adapted to various enzymes simply by changing the substrate. An obvious limitation of this first setup is posed by the fact that the enzyme substrate and product need to have different affinities for the pores. This limitation was overcome by using so-called reactive signal amplifiers, which are compounds that, upon selective reaction with the products of enzymatic reactions, turn into pore blockers thus reducing the fluorescence intensity.

Mirkin *et al.* implemented the bio-bar-code approach discussed in Section 2.2.2.2 for the detection of prostate-specific antigens (PSA).[51] Unlike the system used for DNA detection, in this case the Au NPs were loaded with polyclonal antibodies for PSA recognition, together with numerous copies of the bio-bar-codes. On the other hand, the magnetic microparticles (MMPs) were functionalized with PSA monoclonal antibodies. The protocol relied on the initial binding of PSA to the MMPs, followed by the subsequent capture of multiple NPs on the surface of each MMP. As previously discussed, the superstructures are then isolated through the application of a magnetic field and the bio-barcodes are released through dehybridization and quantified. Using this protocol, target PSA concentrations down to 30 aM could be detected. The lower detection limit could be further pushed down to 3 aM by performing PCR on the released bio-barcodes prior to their quantification.

These results are highly impressive considering that conventional assays for the detection of PSA had sensitivity limits in the order of 3 picomolar, which is 6 orders of magnitude higher.

2.5   Conclusions

Driven by the need to develop sensing systems with ever-increasing sensitivity and selectivity, chemists have approached this challenge with enormous creativity. In cases where the analyte concentration to be detected is far below the sensitivity limit of the analytical instruments used, signal amplification mechanisms have proven to be essential. This is clearly illustrated by the variety of systems discussed in this chapter which, in particular, has focused on supramolecular sensing systems that break with the traditional 1 : 1 relationship between analyte and reporter. Signal amplification takes place when the presence of an analyte sets of a chemical cascade resulting in the production of a multitude of reporter molecules. Signal amplification mechanisms become essential in case the analyte concentrations drop below the sensitivity limit of the analytical instrument. Analogous to Nature, there are two conceptually different approaches to signal amplification. An analyte can activate a catalyst and trigger the production of numerous reporters or an analyte can directly activate multiple reporter molecules connected in a multivalent architecture. In the examples shown here, some general trends with respect to the future developments in this area are highlighted. With regard to catalysis, it appears evident that the analyte-induced activation of autocatalytic cascades can indeed lead towards PCR-like sensitivities. The challenge here will be to develop systems that can be easily adjusted to various analytes and that do not suffer from spontaneous activation. Another trend is the use of orthogonal cascades of amplification events, for instance, by combining catalytic and multivalent approaches. It has been shown that such approach enables a near single molecule detection limit. Challenges in the future also include the development of selective receptors for small biomolecules that are able to trigger such signal amplification events.

Acknowledgement

Financial support from the European Research Council under the European Community's Seventh Framework Programme (FP7/2007-2013)/ERC Starting Grant agreement n° 23 9898 is acknowledged.

References

1. D. A. Giljohann and C. A. Mirkin, *Nature*, 2009, **462**, 461–464.
2. P. Scrimin and L. J. Prins, *Chem. Soc. Rev.*, 2011, **40**, 4488–4505.
3. N. L. Rosi and C. A. Mirkin, *Chem. Rev.*, 2005, **105**, 1547–1562.
4. M. E. Germain and M. J. Knapp, *Chem. Soc. Rev.*, 2009, **38**, 2543–2555.
5. V. K. K. Upadhyayula, *Anal. Chim. Acta*, 2012, **715**, 1–18.

6. P. C. Ray, H. Yu and P. P. Fu, *J. Environ. Sci. Heal., C*, 2011, **29**, 52–89.

7. M. T. Albelda, J. C. Frías, E. García-España and H.-J. Schneider, *Chem. Soc. Rev.*, 2012, **41**, 3859–3877.

8. Y.-W. Lin, C.-C. Huang and H.-T. Chang, *Analyst*, 2011, **136**, 863–871.

9. H. Peng, L. Zhang, C. Soeller and J. Travas-Sejdic, *Biomaterials*, 2009, **30**, 2132–2148.

10. L. Reimer and H. Kohl, *Transmission Electron Microscopy; Physics of Image Formation*, Springer, New York, 5th edn, 2008.

11. L. Reimer, *Scanning Electron Microscopy: Physics of Image Formation and Microanalysis*, 2nd edn., Springer, New York, 1998.

12. F. J. Giessibl, *Reviews of Modern Physics*, 2003, **75**, 949–983.

13. D. J. Müller, J. Helenius, D. Alsteens and Y. F. Dufrene, *Nat. Chem. Biol.*, 2009, **5**, 383–390.

14. H. Raether, *Surface Plasmons on Smooth and Rough Surfaces and on Gratings*, Springer-Verlag, Berlin and Heidelberg, 1988.

15. S. Nie and S. R. Emory, *Science*, 1997, **275**, 1102–1106.

16. S. Weiss, *Science*, 1999, **283**, 1676–1683.

17. G. Pearson, F. Robinson, T. B. Gibson, B. E. Xu, M. Karandikar, K. Berman and M. H. Cobb, *Endocrine Rev.*, 2001, **22**, 153–183.

18. G. Kraus, *Biochemistry of Signal Transduction and Regulation*, Wiley-VCH, Weinheim, 2nd edn, 2001.

19. R. S. Yalow and S. A. Berson, *J. Clin. Invest.*, 1960, **39**, 1157–1175.

20. E. Engvall and P. Perlmann, *Immunochemistry*, 1971, **8**, 871–874.

21. B. K. van Weemen and A. H. W. M. Schuurs, *FEBS Lett.*, 1971, **15**, 232–236.

22. S. Avrameas and B. Guilbert, *C. R. Acad. Sci. Hebd. Seances Acad. Sci. D*, 1971, **273**, 2705–2707.

23. B. B. Haab, *Curr. Opin. Biotechnol.*, 2006, **17**, 415–421.

24. C. M. Niemeyer, M. Adler and R. Wacker, *Trends Biotechnol.*, 2005, **23**, 208–216.

25. T. Sano, C. L. Smith and C. R. Cantor, *Science*, 1992, **258**, 120–122.

26. J. W. Liu, Z. H. Cao and Y. Lu, *Chem. Rev.*, 2009, **109**, 1948–1998.

27. I. Willner, B. Shlyahovsky, M. Zayats and B. Willner, *Chem. Soc. Rev.*, 2008, **37**, 1153–1165.

28. J. Liu, Z. Cao and Y. Lu, *Chem. Rev.*, 2009, **109**, 1948–1998.

29. A. Saghatelian, K. M. Guckian, D. A. Thayer and M. R. Ghadiri, *J. Am. Chem. Soc.*, 2003, **125**, 344–345.

30. N. Graf, M. Goritz and R. Krämer, *Angew. Chem. Int. Ed.*, 2006, **45**, 4013–4015.

31. N. Graf and R. Krämer, *Chem. Commun.*, 2006, 4375–4376.

32. C. Tuerk and L. Gold, *Science*, 1990, **249**, 505–510.

33. A. D. Ellington and J. W. Szostak, *Nature*, 1990, **346**, 818–822.

34. R. R. Breaker and G. F. Joyce, *Chem. Biol.*, 1995, **2**, 655–660.

35. P. Travascio, P. K. Witting, A. G. Mauk and D. Sen, *J. Am. Chem. Soc.*, 2001, **123**, 1337–1348.

36. P. Travascio, Y. F. Li and D. Sen, *Chem. Biol.*, 1998, **5**, 505–517.
37. Y. Xiao, V. Pavlov, R. Gill, T. Bourenko and I. Willner, *ChemBioChem*, 2004, **5**, 374–379.
38. Y. Xiao, V. Pavlov, T. Niazov, A. Dishon, M. Kotler and I. Willner, *J. Am. Chem. Soc.*, 2004, **126**, 7430–7431.
39. O. I. Wilner and I. Willner, *Chem. Rev.*, 2012, **112**, 2528–2556.
40. Y. Tian, Y. He and C. Mao, *ChemBioChem*, 2006, **7**, 1862–1864.
41. H. A. Ho, M. Boissinot, M. G. Bergeron, G. Corbeil, K. Dore, D. Boudreau and M. Leclerc, *Angew. Chem. Int. Ed.*, 2002, **41**, 1548–1551.
42. K. Dore, S. Dubus, H. A. Ho, I. Levesque, M. Brunette, G. Corbeil, M. Boissinot, G. Boivin, M. G. Bergeron, D. Boudreau and M. Leclerc, *J. Am. Chem. Soc.*, 2004, **126**, 4240–4244.
43. H. A. Ho, K. Dore, M. Boissinot, M. G. Bergeron, R. M. Tanguay, D. Boudreau and M. Leclerc, *J. Am. Chem. Soc.*, 2005, **127**, 12 673–12 676.
44. M. M. C. Cheng, G. Cuda, Y. L. Bunimovich, M. Gaspari, J. R. Heath, H. D. Hill, C. A. Mirkin, A. J. Nijdam, R. Terracciano, T. Thundat and M. Ferrari, *Curr. Opin. Chem. Biol.*, 2006, **10**, 11–19.
45. T. G. Drummond, M. G. Hill and J. K. Barton, *Nat. Biotechnol.*, 2003, **21**, 1192–1199.
46. F. Xia, R. J. White, X. L. Zuo, A. Patterson, Y. Xiao, D. Kang, X. Gong, K. W. Plaxco and A. J. Heeger, *J. Am. Chem. Soc.*, 2010, **132**, 14346–14348.
47. T. Takeuchi and S. Matile, *J. Am. Chem. Soc.*, 2009, **131**, 18 048–18 049.
48. X. Chen, Y.-H. Lin, J. Li, L.-S. Lin, G.-N. Chen and H.-H. Yang, *Chem. Commun.*, 2011, **47**, 12 116–12 118.
49. J. M. Nam, S. I. Stoeva and C. A. Mirkin, *J. Am. Chem. Soc.*, 2004, **126**, 5932–5933.
50. T. A. Taton, C. A. Mirkin and R. L. Letsinger, *Science*, 2000, **289**, 1757–1760.
51. J. M. Nam, C. S. Thaxton and C. A. Mirkin, *Science*, 2003, **301**, 1884–1886.
52. T. Niazov, V. Pavlov, Y. Xiao, R. Gill and I. Willner, *Nano Lett.*, 2004, **4**, 1683–1687.
53. W. H. Zhou, C. L. Zhu, C. H. Lu, X. C. Guo, F. R. Chen, H. H. Yang and X. R. Wang, *Chem. Commun.*, 2009, 6845–6847.
54. T. Hermann and D. J. Patel, *Science*, 2000, **287**, 820–825.
55. C. Teller, S. Shimron and I. Willner, *Anal. Chem.*, 2009, **81**, 9114–9119.
56. B. Shlyahovsky, D. Li, Y. Weizmann, R. Nowarski, M. Kotler and I. Willner, *J. Am. Chem. Soc.*, 2007, **129**, 3814–3815.
57. L. Kovbasyuk and R. Krämer, *Chem. Rev.*, 2004, **104**, 3161–3187.
58. L. Zhu and E. V. Anslyn, *Angew. Chem. Int. Ed.*, 2006, **45**, 1190–1196.
59. N. C. Gianneschi, S. T. Nguyen and C. A. Mirkin, *J. Am. Chem. Soc.*, 2005, **127**, 1644–1645.
60. M. S. Masar, N. C. Gianneschi, C. G. Oliveri, C. L. Stern, S. T. Nguyen and C. A. Mirkin, *J. Am. Chem. Soc.*, 2007, **129**, 10 149–10 158.

61. H. J. Yoon and C. A. Mirkin, *J. Am. Chem. Soc.*, 2008, **130**, 11 590–11 591.
62. M. S. Baker and S. T. Phillips, *J. Am. Chem. Soc.*, 2011, **133**, 5170–5173.
63. D. T. McQuade, A. E. Pullen and T. M. Swager, *Chem. Rev.*, 2000, **100**, 2537–2574.
64. S. W. Thomas, G. D. Joly and T. M. Swager, *Chem. Rev.*, 2007, **107**, 1339–1386.
65. Q. Zhou and T. M. Swager, *J. Am. Chem. Soc.*, 1995, **117**, 12 593–12 602.
66. Q. Zhou and T. M. Swager, *J. Am. Chem. Soc.*, 1995, **117**, 7017–7018.
67. I. A. Levitsky, J. S. Kim and T. M. Swager, *J. Am. Chem. Soc.*, 1999, **121**, 1466–1472.
68. R. Kopelman and S. Dourado, *Proceed. SPIE-Intern. Soc. Optical Engin*, 1996, **2836**, 2–11.
69. H. J. Schneider, *Angew. Chem. Int. Ed.*, 2009, **48**, 3924–3977.
70. H. J. Schneider, L. Tianjun and N. Lomadze, *Chem. Commun.*, 2004, 2436–2437.
71. H. J. Schneider and K. Kato, *Angew. Chem. Int. Ed.*, 2007, **46**, 2694–2696.
72. H. H. Gorris and D. R. Walt, *Angew. Chem. Int. Ed.*, 2010, **49**, 3880–3895.
73. R. J. Amir, E. Danieli and D. Shabat, *Chem. Eur. J*, 2007, **13**, 812–821.
74. E. Sella and D. Shabat, *J. Am. Chem. Soc.*, 2009, **131**, 9934–9936.
75. E. Sella, A. Lubelski, J. Klafter and D. Shabat, *J. Am. Chem. Soc.*, 2010, **132**, 3945–3952.
76. M. M. Green, M. P. Reidy, R. J. Johnson, G. Darling, D. J. O'Leary and G. Willson, *J. Am. Chem. Soc.*, 1989, **111**, 6452–6454.
77. E. Yashima, K. Maeda and T. Nishimura, *Chem. Eur. J*, 2004, **10**, 42–51.
78. R. Nonokawa and E. Yashima, *J. Am. Chem. Soc.*, 2003, **125**, 1278–1283.
79. S. Matile, A. V. Jentzsch, J. Montenegro and A. Fin, *Chem. Soc. Rev.*, 2011, **40**, 2453–2474.
80. S. M. Butterfield, T. Miyatake and S. Matile, *Angew. Chem. Int. Ed.*, 2009, **48**, 325–328.
81. O. R. Miranda, H. T. Chen, C. C. You, D. E. Mortenson, X. C. Yang, U. H. F. Bunz and V. M. Rotello, *J. Am. Chem. Soc.*, 2010, **132**, 5285–5289.
82. O. R. Miranda, X. Li, L. Garcia-Gonzalez, Z.-J. Zhu, B. Yan, U. H. F. Bunz and V. M. Rotello, *J. Am. Chem. Soc.*, 2011, **133**, 9650–9653.
83. R. Bonomi, A. Cazzolaro, A. Sansone, P. Scrimin and L. J. Prins, *Angew. Chem. Int. Ed.*, 2011, **50**, 2307–2312.
84. G. Zaupa, C. Mora, R. Bonomi, L. J. Prins and P. Scrimin, *Chem. Eur. J*, 2011, **17**, 4879–4889.
85. G. Das, P. Talukdar and S. Matile, *Science*, 2002, **298**, 1600–1602.
86. N. Sakai, J. Mareda and S. Matile, *Acc. Chem. Res.*, 2005, **38**, 79–87.

Metal Ion Sensing for Biomedical Uses

DAVID C. MAGRI* AND CARL J. MALLIA

Department of Chemistry, University of Malta, Msida, MSD 2080, Malta
*Email: david.magri@um.edu.mt

3.1 Introduction

Fluorescence sensing is a powerful tool for investigating the concentration of various analytes in organized media (micelles, colloids, polymers, *etc.*) as well as in living cells and biological fluids and tissues.[1] For biomedical uses, fluorescent sensors must meet several strict requirements.[2] Most important, they must be selective for a specific metal ion, even in the presence of other metals ions found at a higher concentration.[3] They also must be sensitive to the pathological concentration range—the metal-responsive probe is involved in equilibria which must be matched with the dissociation constant (K_d). For example, the concentration of sodium ions is significantly different in blood and urine than in intracellular locations. Principles of coordination chemistry are important for the rational design of selective sensors, including preferred donor numbers and ligand geometries and hard–soft acid–base considerations.[4] The sensors must also be compatible with the biological matrix and generally water-soluble.[5] Typically, a 'turn-on' emission response or a wavelength shift is preferred over a 'turn-off' emission quenching response for maximizing spatial resolution, notably with a light microscope. For cell biology applications, higher fluorescence brightness is advantageous as less of the intrusive

Monographs in Supramolecular Chemistry No. 13
Supramolecular Systems in Biomedical Fields
Edited by Hans-Jörg Schneider

Published by the Royal Society of Chemistry, www.rsc.org

indicator is needed, which minimizes the possibility of toxicity and altering the cellular environment.[6] Despite these many criteria, fluorescent sensors are advantageous over other types of sensors in their potential to demonstrate superior selectivity, sensitivity, rapid response and high spatial resolution in imaging applications. The World Health Organization (WHO) has set guidelines for rapid testing of sexually transmitted infections for use in developing countries,[7] but these guidelines also serve a useful purpose in metal ion sensing applications. The WHO insists that the ideal test should be ASSURED: **a**ffordable, **s**ensitive, **s**pecific, **u**ser-friendly, **r**apid, **e**quipment-free and **d**eliverable.

Living things require a regular intake of minerals in order to maintain good health. Some minerals are required in bulk amounts while others only in trace amounts. The metal ions in greatest abundance in living things come from the alkaline and alkaline earth families, Na^+, K^+, Mg^{2+} and Ca^{2+}. These species are readily soluble in aqueous media and are thus found in most fluids throughout organisms. Most of the first row d-block metal ions, excluding scandium and titanium, are also believed to be essential in trace amounts for various catalytic and enzymatic processes.

Within a narrow concentration range for each analyte, an organism maintains a homeostatic state of optimal health. A limited deficiency of a mineral may result in biological impairment, as too will a marginal excess. Far too much of a mineral, though, can be poisonous, if not fatal. The tolerance between survival and toxicity is often no more than a 10-fold range. However, maintaining good health also means staying away from minerals that are poisonous. Heavy metals such as mercury, cadmium and lead have a notorious reputation for persistence in the environment and eventual bioaccumulation. Hence, early detection of metals ions in the water supply and food chain is a standard concern. There has been considerable motivation for developing novel fluorescent probes for trace-level environmental and clinical detection of toxic heavy metals,[8,9] particularly in biological samples. Fortunately, the on-going development of selective and sensitive fluorescent probes for metal ion detection is an active area of research in many research laboratories around the world.[10–12]

This chapter highlights various molecular and supramolecular fluorescence sensors for metal ions. Examples have been selected from those with proven application in clinical chemistry or cell imaging, promising prototypes in aqueous media or probes that demonstrate a novel concept or paradigm. The script presented is restricted mostly to synthetic molecular probes, noting that other complementary alternatives based on fluorescent proteins[13,14] and luminescent semiconductor nanoparticles,[15,16] for example, contribute to the topic. The metal ions in greatest abundance are discussed according to their general categorization in the periodic table. The reader wishing for more background information on fluorescence sensing of metal ions has an assortment of literature sources to choose from.[1,3,17–26]

3.2 Sensing Mechanisms

3.2.1 Photoinduced Electron Transfer (PET)

The basic design of a PET sensor involves a 'fluorophore–spacer–receptor' format as illustrated in Figure 3.1.[27] PET sensor operation is based on the competition between photoinduced electron transfer and fluorescence. Such chemosensors have three components: (i) a luminescent component (fluorophore), (ii) a receptor for binding the analyte and (iii) a spacer connecting (i) and (ii). The spacer keeps the fluorophore and the receptor at a fixed distance of a few angstroms. This allows for a fast intramolecular PET in the 'off' state, which results in quenching of the fluorescence.[28] In the 'off' state, upon irradiation of the fluorophore, electron transfer occurs from the unbound receptor to the fluorophore, resulting in essentially no fluorescence. However, in the 'on' state, the analyte is bound by the receptor preventing electron transfer and enhancing the fluorescence. The ideal PET probe causes an 'off–on' switching of the emission intensity with no change in the wavelength.

A simple molecular diagram illustrating the two states of an 'off–on' PET sensor is shown in Figure 3.2. After the excitation of the fluorophore, an electron from the highest occupied molecular orbital (HOMO) is promoted to the lowest unoccupied molecular orbital (LUMO) of the fluorophore. When the lone electron pair in the HOMO of the unbound receptor has a slightly higher energy than the lowest singly occupied molecular orbital (SOMO) of the fluorophore, a fast intramolecular PET occurs resulting in quenching of the fluorescence.[29] Alternatively, when the receptor binds to the analyte, the oxidation potential of the donor is increased so that the HOMO of the bound receptor becomes lower in energy than the lowest SOMO of the fluorophore. The intramolecular PET process is now not feasible and quenching is suppressed, representing the 'on' stage of the fluorescent sensor. Hence, a competition between electron transfer and fluorescence is the basis of PET sensors.[30]

The key parameter that relates the fluorescence signal to the ion concentration is the binding constant (also known as the dissociation constant, K_d). The law of mass action dictates that the concentration of the indicator needs to be close to the K_d. For intracellular studies this is not always a simple matter as

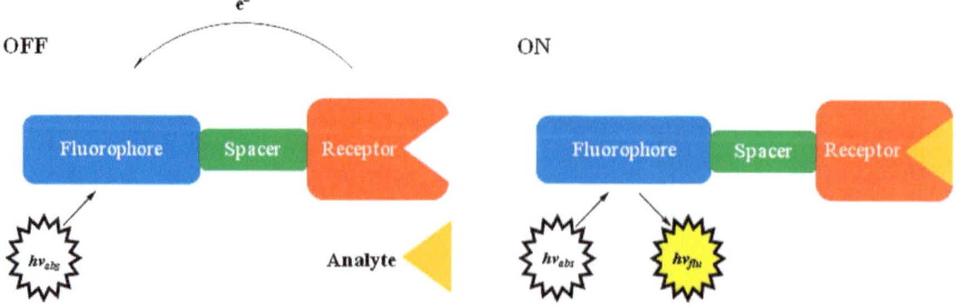

Figure 3.1 A simplified 'fluorophore–spacer–receptor' diagram.

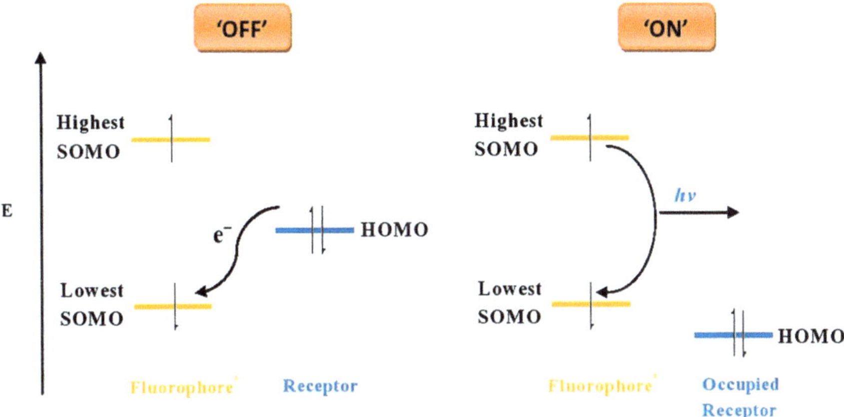

Figure 3.2 Photoinduced electron transfer mechanism in terms of the molecular orbitals of the excited fluorophore and the receptor.

the intracellular concentration can reach 10 times, or even up to 100 times, more than the applied concentration depending on the microenvironment. The concentration range over which an indicator operates is approximately plus or minus 1 log unit. For ratiometric measurements, the response range is also determined by wavelength-dependent parameters. For example, polyamino carboxylic acids are known to be sensitive to many environmental parameters including pH, temperature and ionic strength.

The loading and calibration of intracellular ion indicators requires the localization of the indicator in the region of interest and the need to quantitatively calibrate the concentration of free ion. The most popular technique for non-invasive loading of indicators is the use of acetoxymethyl ester indicator derivatives as they change the hydrophilicity of the indicator by making it more hydrophobic. Typically, phenoxy and carboxy moieties are derivatized as acetoxymethyl or acetate esters rendering the indicator permeable to the cell membranes and insensitive to the ion concentrations. Once inside the cell, ubiquitous intracellular esterases hydrolyse the side chains.

3.2.2 The State-of-the-Art in Point-of-Care Technology

The first commercial *in vitro* application of PET fluorescent sensors resulted in the development of the Osmetech/Roche OPTI critical care analyzer (Figure 3.3).[31] The device can simultaneously measure six critical care analytes on a $120\,\mu L$ sample of whole blood in about two minutes. There are five types of OPTI disposable cassettes available, each tailored to measure up to eight parameters: pH, pCO_2, tHb, pO_2, SO_2, glucose, Na^+, K^+, Ca^{2+} and Cl^- levels.[32] The successful implementation of the sensors required them to satisfy four 'S characters', these being: **selectivity, sensitivity, stability and speed**.[33] The pH and CO_2/tHb sensor exploit the physicochemical properties of pyranine (HPTS, defined below). The measuring of O_2 and Cl^- concentration is based on

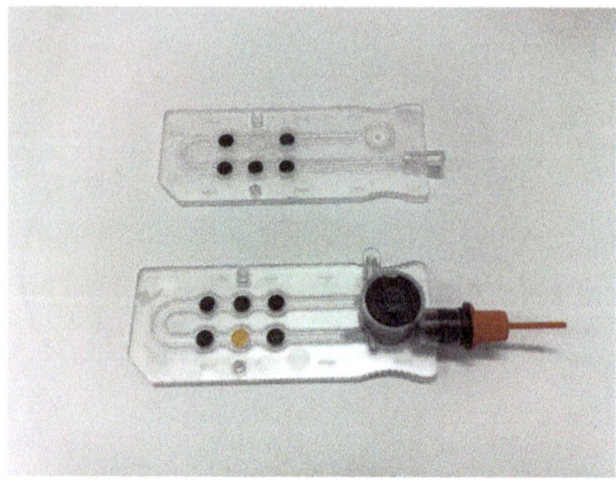

Figure 3.3 The OPTI LION (top) and OPTI R (bottom) cassettes. The black spots are the organic polymer fibre mats covalently attached to the appropriate sensor molecules. Na^+, K^+, Ca^{2+}, pH and CO_2 are measured by the fluorescence of these spots. The orange spot is the sensor for O_2.
(Adapted from de Silva[30] and reproduced by permission of the Royal Society of Chemistry.)

a collisional quenching mechanism. Selective recognition of Na^+, K^+ and Ca^{2+} is achieved by carefully designed fluoroionophores consisting of a naphthalimide fluorophore attached to an aza-crown for Na^+, a cryptand for K^+ and a chelator for Ca^{2+} at one end of the fluorophore and a polymer support at the other.[34] These optic probes for alkali and alkaline earth cations will be discussed in more detail in the forthcoming sections.

3.3 Alkali and Alkaline Earth Ion Chemosensors

As already highlighted, the monitoring of monovalent and divalent metal ions, notably Na^+, K^+, Mg^{2+} and Ca^{2+}, in blood[31] and urine are of major medical diagnostic importance.[2] The normal concentration range of these cations in biological fluids can be significantly different, particularly for potassium, which has a typical concentration of 4 mM in blood and 70 mM in urine (see Table 3.1). There is an even greater difference for calcium, for example, on comparing the concentration in blood serum to that inside a living cell. While the concentration range of Ca^{2+} in blood plasma is within the millimolar range, it is orders of magnitude lower inside living cells, ranging from the micromolar to the nanomolar level. Hence, well-known sensors for calcium such as Indo-1[35] and Fura-2[36] are suitable for cell biology, but not for medical diagnostic applications. The monitoring of sodium and potassium blood serum levels is routinely done for patients with high blood pressure; the monitoring of sodium and potassium in urine is generally a concern for kidney problems. For patients undergoing treatment for manic depression with lithium salts, monitoring of lithium levels is a concern. Table 3.1 lists typical concentration levels for the main alkali and alkaline earth cations in blood and urine. The upper and lower limits in the table are the critical values, also known as the

Table 3.1 Typical concentrations of alkali and alkaline earth cations in blood and urine.[2]

Cation	Blood Serum[a](mM)	Lower Limit(mM)	Upper Limit(mM)	Urine[b](mM)
Na^+	140	120	160	120
K^+	4.0	2.8	6.2	70
Ca^{2+}	1.2	1.0	3.2	4
Mg^{2+}	0.8	0.4	1.9	4

[a]At physiological pH between 7.35 and 7.42.
[b]Typical pH between 6 and 7.

panic values, in blood serum that indicate a life-threatening condition for the patient. A close look at the table indicates that the tolerance limits in the blood serum are within a narrow range. The tolerance in urine is generally greater and highly dependent on the diet of the patient.

3.3.1 Sodium (Na^+)

Sodium is the most abundant cation in the extracellular fluid, while potassium is the most abundant intracellular cation. Methods for the determination of sodium were once commonly done by atomic absorption or flame emission spectroscopy. However, many ion-selective electrode methods[37] based on macrocyclic ionophores of crown ethers[38] or cryptands[39] are commonly used. Perhaps the most state-of-the-art technology, though, is based on fibre optics.[40]

The fluoroionophore **1** on the OPTI cassette was strategically synthesized with a linker for covalent immobilization to a hydrophilic polymer (Figure 3.4).[32] In water or blood serum, the *N*-phenyl aza-15-crown-5 ether reversibly binds Na^+ with a K_d of 119 mM or a $\log \beta_{Na^+}$ of 0.92 at a near-neutral pH of 7.4, physiological temperature of 37 °C and ionic strength of 160 mM. Excitation with a blue light-emitting diode causes a green fluorescence signal from the 4-aminonaphthalimide fluorophore. The proof-of-principle of a PET-based optical dip-sensor with an *N*-phenyl aza-15-crown-5 ether, but lacking the methoxy arm as in **1** has been developed by an Australian team.[41]

Early fluorescent PET sensors with crown ethers were designed with an anthracene fluorophore with synthetic simplicity in mind.[42–44] Molecule **2** was studied in methanol owing to solubility issues in water and found to have a $\log \beta_{Na^+}$ of 2.3, a maximum fluorescence of only 0.06 and a switching on factor of 6.[42] Benzo-15-crown-5-ether systems, being pH-independent, were quite an improvement towards biomedical use with a $\log \beta_{Na^+}$ of 3.0, a fluorescence quantum yield of 0.19 and a switching ratio of 16 in methanol.[43] Gunnlaugsson realized that selectivity in mixed aqueous solution could be achieved by including a methoxy group in the *ortho* position of the aza-crown ether taken from Gokel's work,[44] which assists complexation of the metal ion with a stronger binding constant and better selectivity for sodium over potassium. The anthracene-based sensor **3**, at the time of publication, was a significant achievement displaying a $\log \beta_{Na^+}$ of 2.3, a quantum yield of 0.08 and a

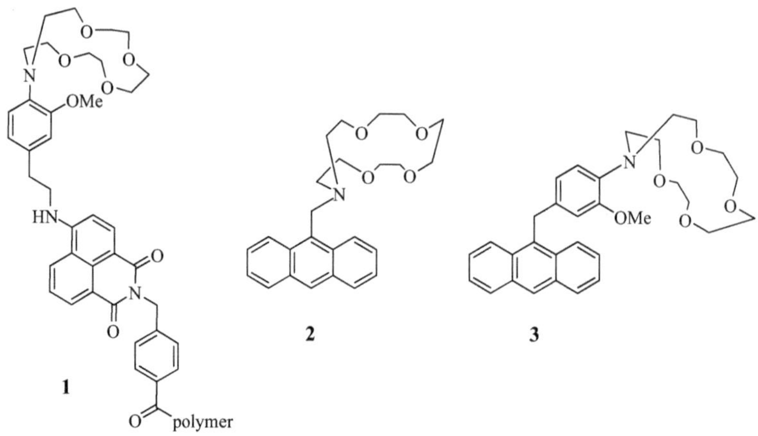

Figure 3.4 Representative chemosensors for Na$^+$ ions.

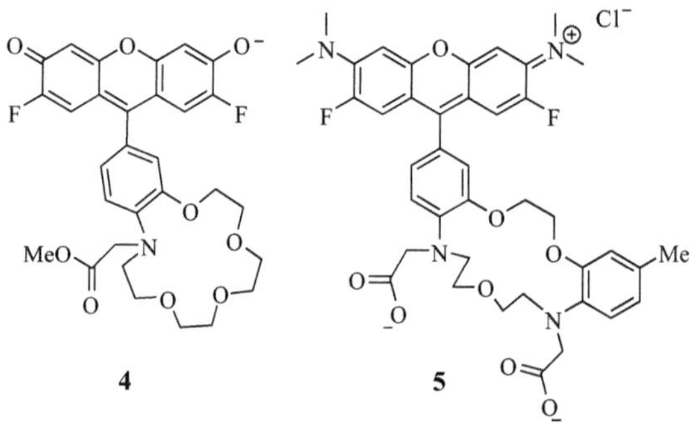

Figure 3.5 Chemosensors for measuring intracellular Na$^+$ ions.

switching factor of 6 in 1:1 methanol–water solution.[45] Advantageously, the 5.3 pK_a of the nitrogen atom is also outside the physiological range. It is no coincidence that the same nitrogen-pivot lariat ether was used in **1**.

CoroNa Green **4**[46] and CoroNa Red Chloride **5**[47] are commercially available molecular probes developed by a team at Molecular Probes Invitrogen for cell imaging (Figure 3.5).[48] CoroNa Green has a fluorinated fluorescein fluorophore attached to a five-atom heteroatom crown ether with an emission maximum at 516 nm and a binding constant of *ca.* 80 mM. It is useful in the presence of other monovalent cations, notably potassium. CoroNa Red is a 1,7-diaza-15-crown-5 cryptand, which is structurally similar to the BAPTA receptor, except that an ether bridge links the two nitrogen atoms to form a three-dimensional cavity.[39] The rhodamine-type fluorophore, with a net positive charge on the fluorophore, helps the indicator to find its way to the mitochondria so that an ester derivative is not required for loading into the cells. With a higher binding constant of 200 mM than most cellular indicators, a fluorescence increase of 15-fold is observed on saturation.

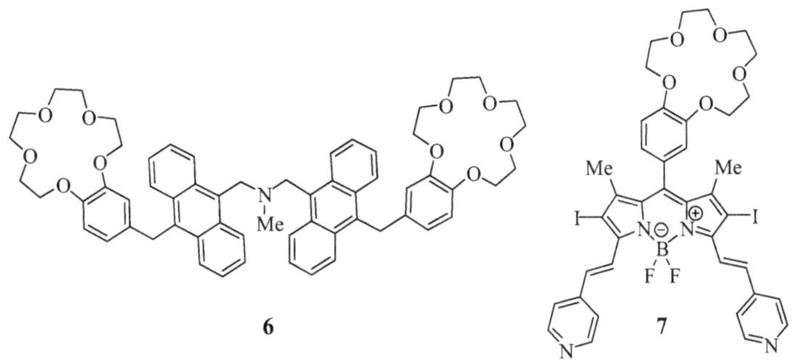

Figure 3.6 Examples of AND logic gates for H^+ and Na^+ ions.

Molecular sensors can also be designed to replicate logic gates. A logic gate is an elementary unit normally associated with digital electronic circuits, usually with two inputs and one output. Logic gates using H^+ and Na^+ as inputs[49,50] and luminescence as the output were the trendsetters in the field of molecular information processing.[51] A related multi-functional logic gate is **6** with two benzo-15-crown-5 ethers, two anthracene fluorophores, and a tertiary amine in the middle (Figure 3.6).[52] On addition of 0.1 M Na^+ and 0.015 M acid in methanol, **6** shows a high fluorescence output with a quantum yield of 0.22 resulting from inhibition of the three PET reactions. An interesting feature of **6** is that it also complexes with Cs^+ when the amine is protonated. Sandwich-type systems resulting from intramolecular complexes of potassium or rubidium ions were previously known.[53] Unexpectedly, no excimer emission was observed on complexation of Cs^+, suggesting that the anthracene units do not align in a stacking orientation.

A potential application of the concept of AND logic could be in selective photodynamic therapy (PDT).[54] A current complication during PDT therapy is that many healthy cells are killed by singlet oxygen along with cancerous cells. The team lead by Akkaya proposed **7** as an AND logic gate with H^+ and Na^+ as the inputs and singlet oxygen as the output.[55] It is known that some types of cancer cells have higher levels of H^+ and Na^+ ion levels in the lysosomes than those of healthy cells. The strategy proposed with **7** is the possibility of targeting cancerous cells only on detection of high levels of H^+ (pH below 4) and Na^+ ions. Acting as an automaton, molecules like **7** could monitor the H^+ and Na^+ ion concentrations in all types of cells and only generate the cytotoxic agent, singlet oxygen, in cancerous cells. This concept could substantial improve the recovery time of patients after PDT treatment.

3.3.2 Potassium (K^+)

Potassium is the second most abundant cation in whole blood after sodium. In tissue cells, the K^+ concentration is typically about 150 mM and while in erythrocytes it is about 105 mM. Intracellular K^+ levels appear to be a diagnostic marker for apoptotic cell death.

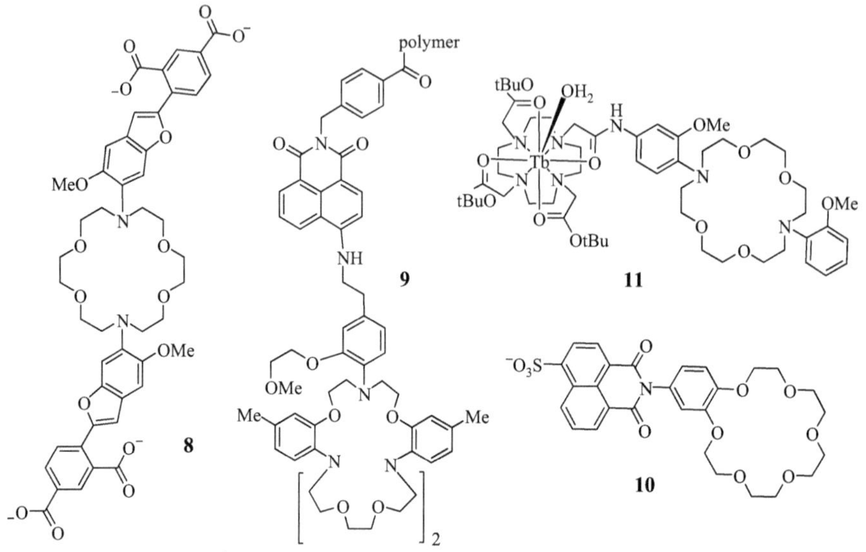

Figure 3.7   Examples of chemosensors for K^+ ions.

The fluorescent indicator **8**, known as potassium-binding benzofuran isophthalate (PBFI), detects for physiological concentrations of K^+ in living cells in the presence of other monovalent cations (Figure 3.7).[56] The structure consists of two benzofuranyl fluorophores linked to a crown ether chelator. On binding K^+, the quantum yield of **8** increases, the excitation peak narrows and shifts to shorter wavelength. Ratiometric measurements are also possible using the intensities at 340 nm and 380 nm. A large Stokes shift in the order of 200 nm is also a characteristic feature of the benzofuranyl fluorophore. The reported binding constant is 5.1 mM. A drawback of **8**, though, is that the molecular probe lacks selectivity in the presence of Na^+.

To alleviate this problem, a team at Arizona State University has developed a highly selective sensor for K^+ over Na^+ in living cells using the identical triazacryptand receptor as in **9** with a concentration range up to 1600 mM.[57] The Tusa team developed **9** for use on the OPTI cassette for reversibly binding K^+ in blood serum.[32] The aminonaphthalimide PET sensor binds K^+ at 17 mM at physiological temperature and pH at 160 mM ionic strength. The naphthalimide fluorophore is stable against hydrolysis and photodegradation. Detection of K^+ results in a green fluorescence output signal, which is detected by the critical care device.

A recent example of another naphthalimide-based K^+ sensor is **10**.[58] Titration with up to 6 mM of potassium chloride in buffered solution results in a five-fold fluorescence enhancement. The molecule is similar to classic examples of PET sensors with an anthracene fluorophore[42] except that this example has a sulfate group making it readily soluble in water. The binding constant of 0.4 mM is ideal, as it is just below the pathological and normal range of K^+ in human blood.

A lanthanide-based sensor **11** has also been demonstrated as an 'off–on' sensor for K^+ in water at simulated physiological conditions.[59] The cyclen diaza-aromatic crown ether conjugate targets K^+ efficiently with up to a 40-fold increase in this system. Operating as a PET system, the delayed Tb(II) emission is switched 'on' occurring as line-like emission bands between 490 and 622 nm. Further discussion on actinide-based sensors is forthcoming in Section 3.5.

3.3.3 Calcium (Ca^{2+})

Most fluorescent probes for Ca^{2+} are polyamino carboxylic acid derivatives of well-known chelators such as BAPTA (1,2-bis(*o*-aminophenoxy)ethanetetraacetic acid), and EGTA (ethyleneglycoltetraacetic acid). They have been extremely valuable for investigating the free Ca^{2+} concentrations in cells by imaging fluorescence spectroscopy.[60,61] A wide selection of fluorescent indicators of various fluorophores is commercially available for detecting intracellular Ca^{2+} changes over a wide concentration range from nanomolar to micromolar.[48] Ratiometric probes exhibiting a spectral shift upon Ca^{2+} binding are known such as the Fura[35] and Indo[36] dyes (Figure 3.8). Indo-1 **12** requires excitation in the UV light region and has dual emission properties at 401 nm and 475 nm, with and without Ca^{2+}, respectively, with a binding constant of 230 nM. Fura-2 emits at 510 nm and the excitation bands at 340 nm and 380 nm can be used for ratiometric measurements.

For applications requiring excitation with visible light, a number of commercially available rhodamine and fluorescein-based probes have been developed such as Calcium Green-1, Calcium Green-2 and Calcium Orange.[48,61] These indicators exhibit an increase in emission intensity on

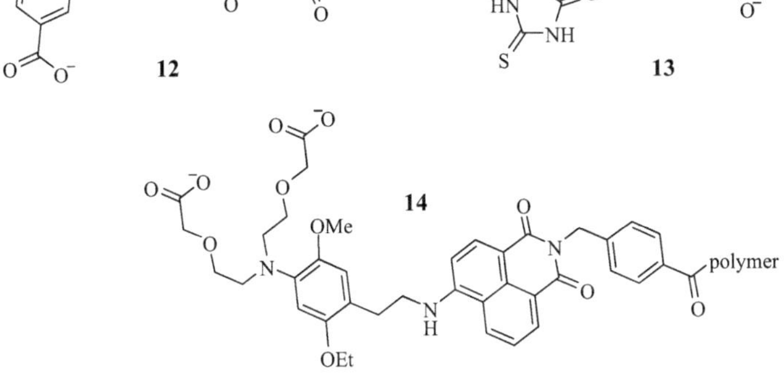

Figure 3.8 Examples of chemosensors for Ca^{2+} ions.

binding Ca^{2+} and are virtually non-fluorescent in the absence of Ca^{2+}, as are the acetoxyester derivatives. The Calcium Green indicators are less toxic to cells, partly due to their substantially greater fluorescence quantum yield, so that lower concentrations of indicator are required to achieve the same luminescent response. Most of these probes are appropriate for measuring cytosolic calcium with a binding constant about 200 nM.

A Ca^{2+} indicator with a different mode of action is Fura Red **13**, which exhibits a *decrease* in fluorescence on binding Ca^{2+}.[48] It is a derivatized fura-2 probe, which exhibits a long wavelength emission maximum of about 660 nm in the absence of Ca^{2+}. The long-wavelength emission minimizes autofluorescence interference from pigments in biological fluids and tissues. The probe is highly sensitive with a binding constant of 140 nM. An enormous stoke shift on the order of 200 nm permits multicolour analysis with fluorescein-based dye using only one excitation wavelength. The absorbance spectrum also has a significant blue-shift of 60 nm on complexation with Ca^{2+} allowing for ratiometric measurements of intracellular Ca^{2+} using the intensities at 420 nm and 480 nm.

Sensing for Ca^{2+} in blood serum is a key role of the OPTI point-of-care instrument.[32] Attached to one side of the naphthalimide sensor **14** is a modified diacetic acid moiety, a phenylcarboxy unit for covalent attachment to an aminocellulose solid support. The design principle exploits the switching between PET and fluorescence for medical diagnostic information. The sensing layer is covered with a black hydrophilic layer for protection and yet allows for the rapid diffusion of ions throughout the sensor. In aqueous solution or blood serum, the chelator reversibly binds Ca^{2+} with a K_d of 1.1 mM at a near-neutral pH and physiological temperature of pH 7.4 and 37 °C, respectively, in a solution with an ionic strength of 160 mM.

3.3.4   Magnesium (Mg^{2+})

Magnesium is the fourth most abundant element in the human body. Concentration ranges in cells typically range from 0.1 to 6 mM. Mg^{2+} is vital to mediating enzymatic reactions, DNA synthesis, and hormonal and muscular functions. The physiological changes of Mg^{2+} tend to be rather small compared to Ca^{2+} so high sensitivity is desirable. Many Mg^{2+} indicators also bind Ca^{2+}; however, the concentration ranges of Ca^{2+} are usually at least 100-fold less, usually peaking at micromolar levels. Compared to Ca^{2+} probes, the choice of readily available Mg^{2+} indicators is rather limited.

Mag-fluo-4 **15** is a visible light PET probe with a virtual spacer (Figure 3.9).[62] It has a binding constant of 4.7 mM for Mg^{2+}, and 22 µM if used for measuring Ca^{2+}. The emission increases dramatically from a flat baseline to an intense fluorescence at 517 nm with a Mg^{2+} binding constant of 1.9 mM. A structural cousin, Mag-fura-2 **16**, is useful as a ratiometric probe in excitation mode with a clear isosbestic point at 350 nm.[63] Mag-Fura-2 has a K_d of 1.9 mM for Mg^{2+} and a K_d of 25 µM for Ca^{2+}.

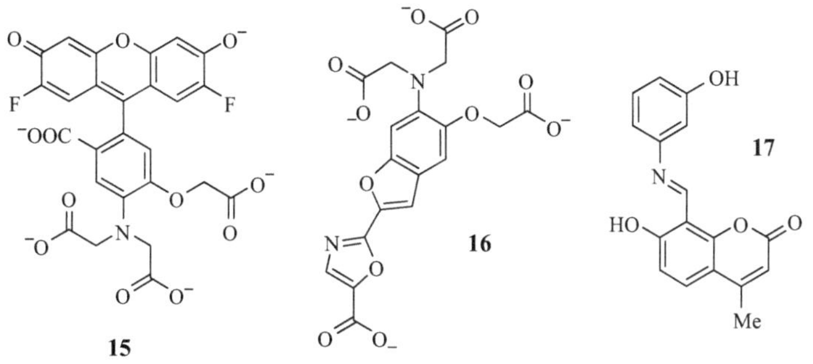

Figure 3.9 Examples of chemosensors for Mg^{2+} ions.

The fluorescent sensor **17** is part of a multi-component sensory system that uses *L*-proline as a promoter agent for Mg^{2+} detection.[64] A 47-fold enhancement is observed on addition of 5 µM Mg^{2+} while other cations exhibit no signal above baseline. Moreover, the emission wavelength of **17** undergoes a blue shift from 539 to 477 nm. The selectivity for Mg^{2+} is outstanding with bright green fluorescence observed on binding of Mg^{2+} allowing for naked-eye detection with 365 nm UV light in acetonitrile. The coumarin-based ligand binds to Mg^{2+} in a 2 : 1 ratio in the presence of *L*-proline. Without the presence of *L*-proline and Na^+ as cofactors, there is no discrimination from other cations. In essence, the synergistic effect operates as a two-input AND logic gate for Na^+ and Mg^{2+}, or as a three-input AND logic gate if *L*-proline is included as an input.

3.3.5 Lithium (Li$^+$)

Lithium is one of the five most biologically important alkali and alkaline earth metal cations and is frequently used in treatment of manic-depressive psychosis.[65] The reliable determination of lithium ion concentration levels in blood samples is important, as patients customarily take the medicine for extended periods lasting several months or often years. As with all drugs, administration of too low a level results in no effect, while an overdose of lithium can lead to life-threatening effects. The lithium ion concentration in blood serum during treatment is a narrow range, typically from 0.6 to 1.2 mM.[2]

Chemosensors for Li^+ that are soluble in water are still rather rare. Most Li^+ sensors rely on nitrogen and/or oxygen-containing crown ethers. A highly sensitive and selective example studied in acetonitrile and methanol–acetonitrile is compound **18** (Figure 3.10).[66] It is designed on a PET sensing mechanism according to an uncommon 'fluorophore–spacer–receptor–spacer–fluorophore' format. The emission at 337 nm from the naphthalene moieties is switched 'on' upon complexation with Li^+ with a fluorescent enhancement factor of 5 from maximum and minimum quantum yields of 0.11 and 0.022, respectively. The small di-aza-crown ether provides a binding constant $\log\beta_{Li^+}$ of 5.4 in acetonitrile. Although of the appropriate cavity size,

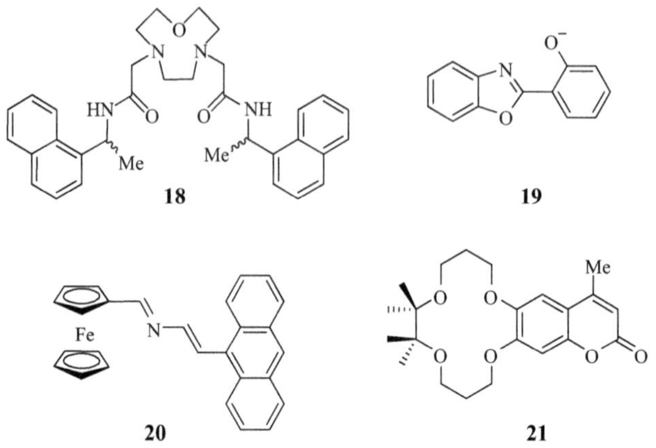

18   **19**

20   **21**

Figure 3.10   Examples of chemosensors for Li$^+$ ions.

the sensor has a pK_a value of 7.2 in water, which complicates complexation at physiological pH 7.4.

The ligand 2-(2-hydroxyphenyl)benzoxazole (HPBO)[67], **19**, exhibits distinct UV–visible and fluorescence characteristics for Li$^+$ at 370 nm and 440 nm, respectively, with high selectivity compared to Na$^+$ and K$^+$ in the alkaline medium of acetonitrile. In ethanol, a distinct emission maximum is observed at 421 nm in the fluorescence spectrum. The addition of base, such as triethylamine, is a prerequisite for complexation. A binding constant $\log \beta_{Li^+}$ of 5.6 was measured.

A simple sensor **20** consisting of an anthracene fluorophore and a ferrocene moiety linked by a 2-aza-1,3-butadiene bridge shows highly selective sensing for Li$^+$ in aqueous acetonitrile/water (7 : 3 ratio) over other monovalent cations.[68] The presence of water is key to the eight-fold fluorescence enhancement observed at pH 5, although the maximum quantum yield is a modest 0.042. The association constant was calculated to be 11.8 M^{-1} when excited at 370 nm. Supported by density functional theory calculations, the authors propose that protonation occurs at the N atom of the bridge, whereas complexation of the Li$^+$ occurs through the ferrocene moiety.

The ligand **21** consists of a tetramethyl-14-crown-4 crown ether as the Li$^+$ binding site and 4-methylcoumarin as the fluorophore.[69] The fluoroionophore was attached to a polymer for application as an optode. The sensing mechanism relies on an intramolecular ICT wavelength shift, which allows for ratiometric measurements. High selectivity for Li$^+$ with a $\log \beta_{Li^+}$ of 2.80 was observed in 99 : 1 water–methanol. A blue shift is observed with Li$^+$ while no response has been observed from other alkaline cations. The sensor is pH-independent over a range of pH 3–10 and allows for the measuring of Li$^+$ concentrations between 0.1 and 10 mM. The sensor was also attached to an optode membrane consisting of three monomers units via an alkyl spacer. The fluorescence properties are generally similar, although the binding strength was found to be poorer with a $\log \beta_{Li^+}$ of 1.95.

Trinuclear [12]metallacrown-3 complexes utilize a known Li^+ receptor combined with a novel signalling mechanism.[70] The ternary complex with the trinuclear [12]metallacrown-3 complex acts as a ditopic receptor for Li^+ with the fluorescent dye 8-hydroxy-1,3,6-pyrenetrisulfonate (HPTS). The ensemble can sense for Li^+ within the pharmacological concentration range of 0.5 to 1.5 mM. A drawback of the assay is its slow response, which requires several hours at room temperature to reach equilibrium.

3.4   Heavy Metal Ion Chemosensors

Iron, zinc and copper are considered essential to health; deficiency or excess results in illness. The monitoring of heavy metal ions in blood, notably Fe^{3+}, Zn^{2+} and Cu^{2+}, are of interest in blood diagnostics. Probe selectivity is of great importance as these metal cations all have similar concentration ranges in blood serum. The normal concentration range of these cations in biological fluids is typically from 10 to 30 µM as listed in Table 3.2.[2] In cells, the concentration levels are typically within the nanomolar or picomolar range.[48]

3.4.1   Iron (Fe^{3+})

Iron is ubiquitous in cells and plays a crucial role in a variety of vital cell functions, including oxygen metabolism and the electron transfer processes associated with DNA and RNA synthesis.[71] Most iron in biological systems, as Fe^{2+} or Fe^{3+}, is tightly bound to enzymes and specialized transport and storage proteins. This is for good reason, as labile or free iron, as Fe^{2+}, is a potential cause of cytotoxicity in cells because it catalyses the formation of reactive oxygen species via the Fenton reaction.[72] Furthermore, the solubility of iron at the physiological pH of 7.4 is rather poor—intracellular concentrations are in the region of 10^{-18} M. Consequently, plants and bacteria require highly effective iron-complexing ligands, known as siderophores, in order to mobilize iron in cells. In the human body, iron overload is a serious complication resulting in β-thalassaemia, which must be treated by iron chelation therapy. A comprehensive review of Fe^{3+} sensors has recently been published;[73] hence we will highlight just a few examples.

A commercially available probe for intracellular Fe^{3+} is Phen Green SK.[48] It responds to both Fe^{2+} and Fe^{3+} with 'turn-off' percentages of 90% and 50%, respectively. As a result, selectivity is not very good. The majority of Fe^{3+}

Table 3.2   Typical concentrations of the three most abundant heavy metals in blood serum.[2]

Cation	Blood Serum[a] (µM)	Lower Limit (µM)	Upper Limit (µM)
Fe^{3+}	19	9.0	31
Zn^{2+}	15	11	19
Cu^{2+}	16	11	22

[a]At physiological pH between 7.35 and 7.42.

'turn-off' probes rely on siderophore receptors. A prototypical example is **22** from the Callan team (Figure 3.11).[74] The naphthalene-based fluorescent probe contains two Schiff base receptors, and displays good selectivity for Fe^{3+} in HEPES buffered solution at pH 7.0 between 5.0 and 80 µM in mixed tetrahydrofuran–water solution. Another interesting 'on–off' probe from the same laboratory uses two aminobisulfonate ligands and a tertiary amine as the receptor connected to a naphthalene fluorophore.[75] The nitrogen pK_a of 6.0 is far enough away from the physiological range and good linearity was observed from 16 to 63 µM. Real-life application will, of course, require the development of probes with spectrum signatures with a longer wavelength in order to eliminate inner filter effects due to the absorbent nature of Fe^{3+}.

One of these, a fluorescent chemosensor with two aza-18-crown-6 moieties linked to a coumarin fluorophore **23**[76] demonstrates remarkable selectivity for Fe^{3+} over 13 other metal cations with a 15-fold fluorescence enhancement. Selectivity is even maintained in the presence of physiologically background metal ion concentrations of 0.135 M Na^+, 0.01 M K^+, 0.001 M Mg^{2+} and 0.001 M Ca^{2+} at pH 7.4. The complex formation is 1:1 (host:Fe^{3+}) with a $log\beta_{Fe^{3+}}$ of 5.2. A red-shift of 8 nm in the absorption spectrum suggests formation of an excimer in the excited state.

Elevated levels of redox-active iron and protons have been linked to various cancers, such as colorectal and liver cancers.[77] Molecule **24** combines concepts from PET pH sensors[78,79] and redox-fluorescent pE switches[25] to exemplify a 'Pourbaix sensor', a new class of sensor that monitors the pE and pH of a solution—or potentially in living cells—according to an AND logic algorithm (Figure 3.12). A three-fold enhanced fluorescence is observed on oxidation of **24** with an electrode at a positively applied potential or on addition of up to one

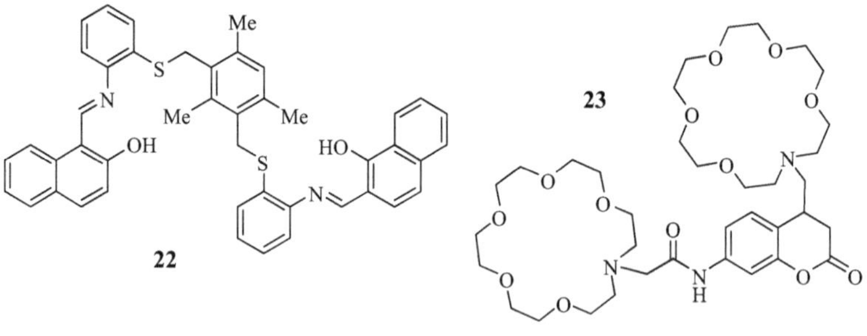

Figure 3.11   Examples of chemosensors for Fe^{3+} ions.

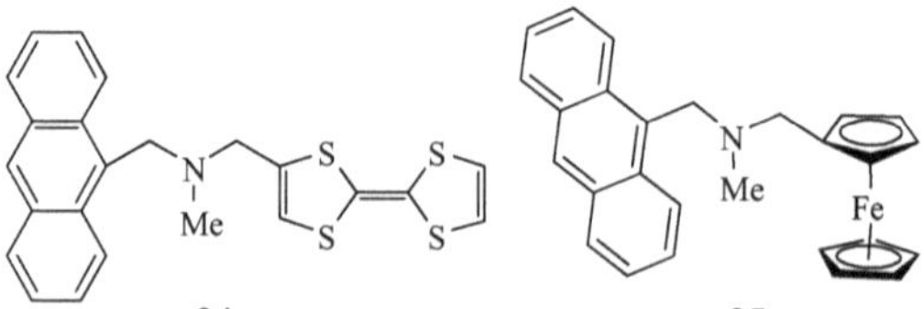

Figure 3.12   Examples of logic gates with H^+ and Fe^{3+} as inputs.

equivalent of Fe^{3+}; however, addition of more than one equivalent of Fe^{3+} results in quenching.[80] Replacement of the tetrathiafulvulene redox donor with ferrocene yields **25**, which exhibits fluorescence centred at 420 nm in the presence of up to 10 equivalents of Fe^{3+}.[81] The applicable concentration range for Fe^{3+} was found to be between 10 and 100 μM. The sensor displays a five-fold 'turn-on' response, which is visually discriminated by a blue fluorescence with the naked eye. The tertiary amino group with a pK_a of 7.8, makes these 'Pourbaix sensors' potentially usable in near-neutral pHs of 7.0–8.5.

3.4.2 Zinc (Zn^{2+})

Zinc (Zn^{2+}) is the second most abundant trace metal ion in living organisms after iron. It is integral to hundreds of various enzymes and plays a key role in the regulation of gene expression and in wound healing for the biosynthesis and integrity of connecting tissue. Many neurological disorders, including epilepsy and Alzheimer's disease, have been linked to this metal ion. The intracellular concentration of free zinc is rather low, typically at nanomolar levels, as most zinc is bound to proteins or nucleic acids. Several reviews on metal ion sensors[5,6,9] and specifically for Zn^{2+} chemosensors[82,83] are available so we will only draw attention to a few representative examples.

FluoZin-3 **26** has a modified BAPTA chelator with one *N*-acetic acid moiety removed (Figure 3.13).[84] The indicator is a PET-type 'off–on' sensor with a low K_d of 15 nM, which is unperturbed by calcium concentrations up to 1 μM. The spectrum in zinc-free solution is negligible; at saturated levels 75-fold fluorescence enhancement results with a peak maximum at 516 nm. The minimum and maximum quantum yields of fluorescence are 0.005 and 0.43, the latter in the presence of 100 nM Zn^{2+}. At saturated Zn^{2+} levels, **26** is stable between pH 6 and 9, but at lower pHs the emission decreases due to protonation of the phenolic hydroxylic group (p$K_a = 4.8$). The probe was used to measure Zn^{2+} in pancreatic *β*-cells. A related probe, RhodZin-3, with its inherent positive charge from a fluorinated rhodamine-type fluorophore, is available for measuring zinc levels in mitochondria.[85]

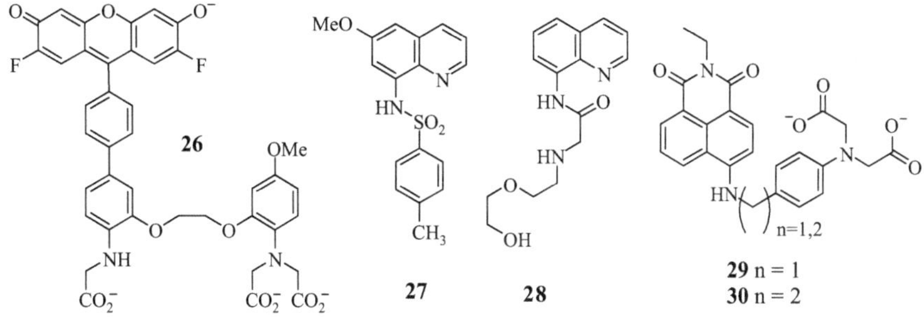

Figure 3.13 Examples of chemosensors for Zn^{2+} ions.

The membrane permeable probe TSQ (*N*-6-methoxy-8-quinolyl)-*p*-toluene-sulfonamide) **27** is based on a quinoline fluorophore.[86] It forms a $1:2$ Zn^{2+} to dye complex, but a $1:1$ complex with metalloproteins. TSQ was the first chemosensor to display selectivity towards physiological Ca^{2+} and Mg^{2+} ions, but is known to be blocked by intracellular dithizone. Other analogues have been developed to improve water solubility and membrane permeability by introducing a carboxylic or ester group.[87] A recent example, **28,** works as a ratiometric fluorescent probe.[88] The 2-(2-hydroxyethoxy)-ethylamino group improves hydrophilicity. Complexation of Zn^{2+} by the quinoline nitrogen results in an eight-fold fluorescence enhancement and a 75 nm red-shift in buffered aqueous solution.

The Gunnlaugsson group developed two 4-amino-1,8-naphthalimide fluor-ophores, **29** and **30**, with benzyl and phenylethyl iminodiacetate ligands.[89,90] These PET sensors show an absorption peak at 450 nm and emit a green fluorescence at 550 nm. The pK_a values of the phenyl iminodiacetate are 3.2 and 2.1 for **29** and **30**, respectively, which are advantageous for the monitoring Zn^{2+} in the physiological pH range. The binding constants ($\log \beta_{Zn2+}$ is 3.9) are identical for a $1:1$ stoichiometry at pH 7.4 in the presence of 135 mM NaCl. The sensor **29** has a fluorescence quantum yield of 0.004 in the absence of Zn^{2+}, which increases to 0.21 in the presence of 5 μM Zn^{2+} with a 56-fold enhancement. As expected, the model with the longer spacer **30** has a lower fluorescence quantum yield of 0.14 on Zn^{2+} binding.

ZnAF-1 **31** from the Nagano laboratory is a fluorescein-based Zn^{2+} probe derivative of FuraZin (Figure 3.14).[91] The first generation of these probes displays a low background fluorescence of 0.02 and a 'turn-on' enhancement in the presence of Zn^{2+} of 17-fold. The K_d is 0.78 nM making this probe useful in the nanomolar range and sensitive enough for use in mammalian cells. Quite possibly, ZnAF-1 is the first zinc detector that can distinguish between Zn^{2+} and Cd^{2+}. Other monovalent and divalent cations (Ca^{2+}, Mg^{2+}, Na^+, and K^+) did not show any fluorescent enhancement at millimolar concentrations. A fluorinated derivative exhibits negligible background fluorescence of 0.004

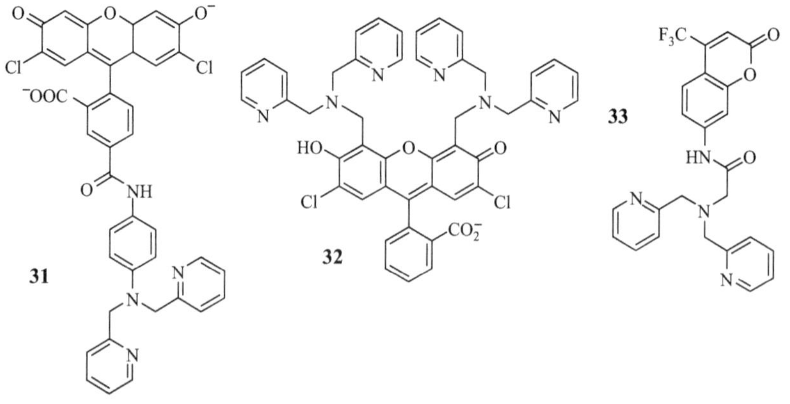

Figure 3.14   More representative examples of Zn^{2+} chemosensors.

and improved 'turn-on' responses at least four times greater.[92] It has been used to image the pH dependence of Zn^{2+} in leukaemia cells.[93]

The Lippard laboratory is also credited with developing a number of fluorescein-based Zn^{2+} chemosensors.[94–97] An example is Zinpyr-1 **32** which consists of two di-2-picolylamine receptors. The probe exhibits a three-fold fluorescence increase upon binding Zn^{2+} at 515 nm with a dissociation constant of 0.7 nM and a bright quantum yield of 0.87. The favourable binding constant, water solubility and ability to be passively loaded into cells makes this genre of probes useful for intracellular studies. However, Zinpyr sensors tend to be sensitive to protons.

A collaborative group demonstrated a naphthalimide fluorescent sensor with an amide-containing di-2-picolylamine receptor selectivity for Zn^{2+}.[98] The sensor showed an enhancement of 21-fold and blue-shifted emission from 483 nm to 446 nm, which was successfully applied to detect Zn^{2+} in living zebrafish embryos. Concerned with the toxicity of the naphthalimide fluorophore, a more biocompatible coumarin-based probe was developed.[99] The newer prototype has binding constants of 18 nM and 0.34 mM for Zn^{2+} and Cd^{2+}, respectively. Ratiometric detection of Zn^{2+} is also possible using the wavelength shift from 425 nm to 505 nm via a Cd^{2+} displacement approach. The sensor is cell-permeable and can be applied to discriminate Zn^{2+} from Cd^{2+} in cells.

3.4.3 Copper (Cu^+ and Cu^{2+})

Copper is the third most abundant trace element in the human body, where it is mainly associated with metalloproteins. Dietary copper is required for the synthesis of haemoglobin and for redox enzyme activity. It is present in two valence states, these being Cu^{2+} or Cu^+. The Cu^{2+} species is a catalytic cofactor for numerous metalloenzymes. It has been linked to many neurological diseases including Menkes syndrome (copper deficiency in infants) and Wilson's disease (copper toxicity in children and young adults) and Alzheimer's disease. The species Cu^+ is the main oxidation state within the reducing environment of the cytosol and disproportionation is quite common for such redox state changes. Cu^+ is also an efficient quencher via electron transfer and energy transfer processes.

There are few examples of fluorescent Cu^+ probes for visualizing labile copper in cellular systems.[5] The first was a pyrazoline-based probe with an azatetrathia-crown receptor.[100] Despite an incorporation of the charged carboxylate moiety on one of the phenyl rings, dynamic light scattering measurements revealed the presence of colloidal aggregates with an average hydrodynamic radius of 100 nm. To address this issue, the Fahrni group synthesized a new series of water-soluble, Cu^+ selective probes with a thia-crown receptor modified with four hydroxymethyl groups and a sulfonated triarylpyrazoline fluorophore (Figure 3.15).[101] The probe **34** quickly dissolves in water up to millimolar concentrations, circumventing the formation of colloidal aggregates. On saturation with Cu^+ at pH 7.2 in MOPS buffer

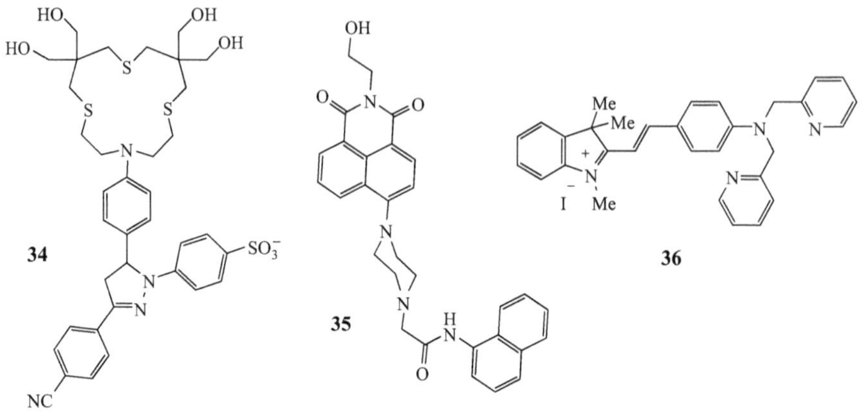

Figure 3.15 Examples of chemosensors for Cu^+ and Cu^{2+} ions

solution, **34** displays a 65-fold fluorescence enhancement at 508 nm with an observed quantum yield of 0.083. Furthermore, the fluorescence response of **34** has proved to be unaffected by other biologically relevant ions.

The ratiometric sensor **35** for Cu^{2+}, which integrates a 1,8-naphthalimide fluorophore with 8-aminoquinoline, has a highly selective ratiometric response in aqueous media.[102] Its practical imaging ability for intracellular Cu^{2+} was tested with human breast adenocarcinoma cells (MCF-7 cells) using a confocal microscope. On excitation at 395 nm, two characteristic fluorescence bands were observed at 435 nm and 526 nm. Titration with Cu^{2+} reveals a decrease in the emission band at 526 nm, and a steady emission band of 435 nm. The binding constant is 29 mM.

Sensor **36** is a merocyanine dye with a di-2-picolylamine.[103] In buffered aqueous media it undergoes a drastic colour change from reddish-purple to yellow in the presence of four equivalents of copper ions. With other cations, **36** displays an absorption maximum at 504 nm, but with Cu^{2+} a 94 nm blue-shift results in a peak maximum at 408 nm at pH 7.4. Addition of Cu^{2+} results in fluorescence emission quenching. This is partly explained by a decrease in the electron-donating abilities of the nitrogen atom of the ligand, thereby inhibiting the internal charge transfer process.

3.4.4 Mercury (Hg^{2+}), Cadmium (Cd^{2+}) and Lead (Pb^{2+})

Fluorescence sensors for the detection of lead, cadmium, and mercury ions have been recently reviewed.[104] However, the chapter would not be complete without highlighting some of the probes for these metal ions. A remarkable example in aqueous solution is **37** (Figure 3.16). It has two aminonaphthalimide fluorophores with a 2,6-bis(aminomethyl)pyridine receptor and two hydroxyethyloxyethyl groups, which aid water solubility. On complexation with Hg^{2+} a red-shift of 8 nm is observed, possibly due to some intramolecular excimer formation. A fluorescence enhancement of 17 and a detection limit

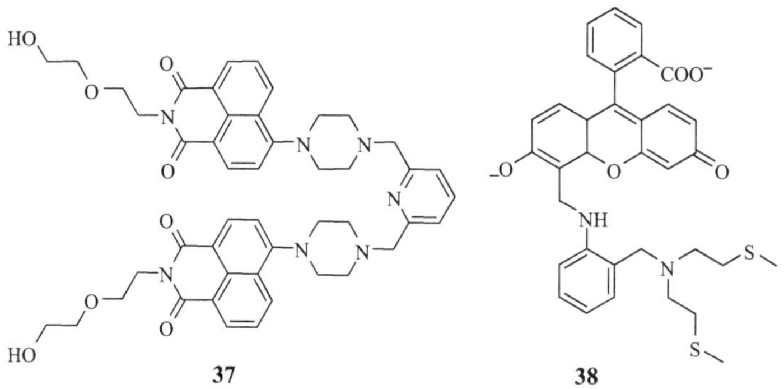

Figure 3.16 Representative chemosensors for Hg^{2+} ions.

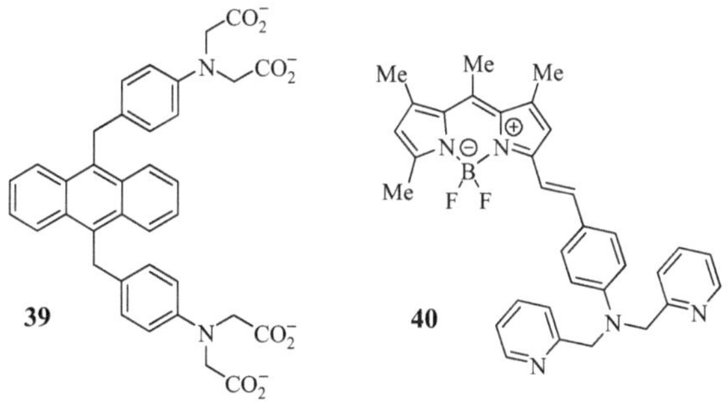

Figure 3.17 Examples of chemosensors for Cd^{2+} ions.

below 0.1 nM are advantageous as is the lower pK_a of 5.2 for the amino groups, which minimizes protonation perturbation when conducting studies around physiological pH.

The team of Nolan and Lippard have developed a series of 'turn-on' and ratiometric fluorescent sensors for Hg^{2+}.[105–107] The fluorescein probe **38** exhibits a five-fold emission in water with high selectivity over other heavy metals.[105] The related seminaphthofluorescein derivative operates over neutral and basic pH ranges with a four-fold emission enhancement of quantum yield 0.032 and dual emission at 524 nm and 612 nm.[106] It has been demonstrated to be useful for detecting mercury levels in contaminated freshwater samples.

Gunnlaugsson reported an anthracene chemosensor **39** with two phenyl-liminodiacetate ligands that responds to both Zn^{2+} and Cd^{2+}, but in different ways (Figure 3.17).[108,109] Both metal ions have similar binding constants, namely $\log \beta_{Cd^{2+}}$ of 3.9 and $\log \beta_{Zn^{2+}}$ of 3.8, in aqueous buffered solution at pH 7.4. However, the fluorescence spectra with Zn^{2+} are rather typical of the anthracene fluorophore centred at 415 nm, while those with Cd^{2+} lack the fine structure with a maximum at 500 nm. A BOPIDY probe **40** with a di-2-picolylamine receptor based on an internal charge transfer mechanism displays a unique emission maximum at 600 nm, which allows for the

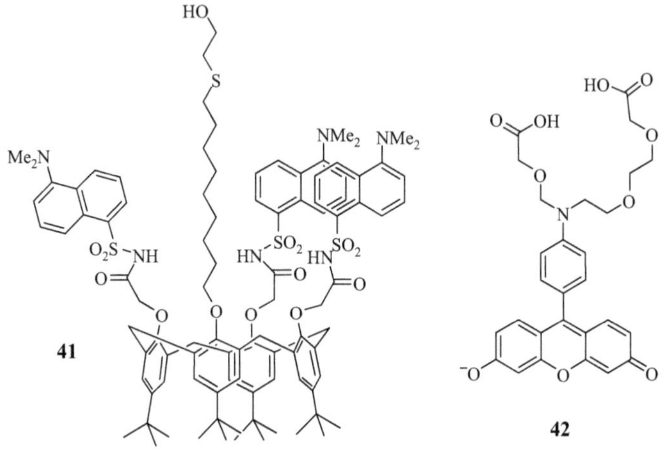

Figure 3.18   Examples of chemosensors for Pb^{2+} ions.

fluorescent imaging of Cd^{2+} in living cells.[110] The last few years have seen the development of a number of novel Cd^{2+} sensors in aqueous solution.[111–113]

A calixarene bearing three dansyl groups and one long alkyl chain, **41,** was reported to selectively detect Pb^{2+}.[114] Demonstration of a portable sensor was attempted by grafting the wall of a polydimethylsiloxane microfluidic device. Absorption and emission maxima were observed at 347 nm and 572 nm. Complexation with Pb^{2+} results in fluorescence quenching with a detection limit down to 0.2 µM (Figure 3.18).

A Pb^{2+} specific chemosensor for living cell imaging is **42.**[115] The xanthenone-based probe with a pseudocrown ether of two different carboxylate ligands shows a weak background quantum yield for fluorescence of 0.001. Under physiological conditions an 18-fold fluorescence enhancement is observed. The probe responds to changes of Pb^{2+} in the cytosol of living mammalian cells.

3.5   Actinide and Lanthanide Ion Chemosensors

Lanthanides are elements finding use in mobile phones, lasers, anticounter-feiting (as markers on European bank notes), nuclear reactors and medical diagnostics.[116] The long-lived nature of the excited state of lanthanide complexes is of interest in biological applications because the longer fluor-escence lifetimes, in the order of milliseconds, allow for time delay studies. In addition, the emission bands are line-like rather than broad: the emission results from f-f transitions. Most of the trivalent ions are luminescent except for La (empty f orbitals) and Lu (full f orbitals). Another notably interesting property is that the emission can extend into the near infra-red region to 1500 nm. A number of reviews highlight new developments with respect to supramolecular luminescent sensors[117,118] and biomedical analysis and imaging.[119]

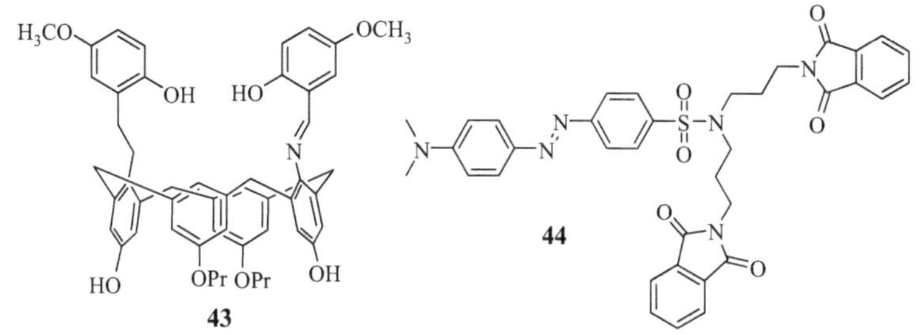

Figure 3.19 Chemosensors for Eu^{3+} and Nd^{3+}.

The separation of actinides (An) from lanthanides (Ln) is an unsolved problem in the management of nuclear waste from nuclear power plants. The spent fuel typically contains moderate levels of long-lived actinides (Np, Am, Cm) along with many fission products from the lanthanides row. Most actinides are radiotoxic—hence direct study of them is not feasible in the standard laboratory. To circumvent this problem, analogous cations are often examined: Eu^{3+} for Pu^{3+}, Am^{3+} and Cm^{3+}; Th^{4+} for U^{4+} and Pu^{4+}; and UO_2^{2+} for NpO_2^{2+} and PuO_2^{2+}. Among actinides, uranium and thorium complexes are the most studied. In the search for extractants of actinides and lanthanides, calixarenes have been studied extensively for separating actinides from lanthanides using liquid–liquid extraction.[120]

Chemosensors for actinides and lanthanides are rather rare. The calixarene **43** is thought to be a PET probe: electron transfer may occur from the phenolate to the imine moiety (Figure 3.19).[121] In the presence of Eu^{3+} a significant chelation-enhancement effect results in a fluorescent enhancement with a peak maximum at 598 nm in dichloromethane. The specificity for erbium over the lanthanides may be a size-fit effect of the cavity.

Selectively is also achieved with ligand **44**, which binds Nd^{3+} even in presence of all other lanthanide ions.[122] In the presence of the ligand, most lanthanide cations turn the solution yellow in acetonitrile; in contrast, Nd^{3+} turns the solution a brilliant red. Furthermore, Nd^{3+} recognition is also communicated by a six-fold increase in the luminescence on addition of 1 mM Nb^{3+}. On subsequent additions of Nd^{3+} the luminescence decreases. Another example worth mentioning is a selective colorimetric receptor for Yb^{3+}.[123]

3.6 Strategies for Multi-Analyte Sensing

3.6.1 'Lab-on-a-Molecule' Systems

Sensing for more than one analyte with a single molecule has been around for two decades, since de Silva and company demonstrated the first two-input molecular AND logic gate.[49] Since then, two-input logic gates[124,125] and many three- and four-input logic gate arrays have been demonstrated at the molecular level.[126]

The first illustration of the potential cross-fertilization between Boolean algebra and biomedical sensing was our report on a 'lab-on-a-molecule'

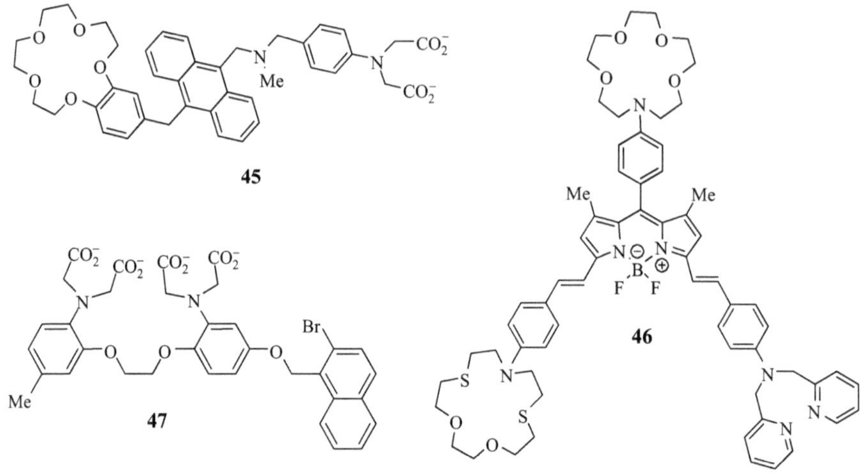

Figure 3.20 Examples of multi-analyte logic gates with three inputs.

prototype—a three-input AND logic gate **45** based on a competition between PET and fluorescence (Figure 3.20).[127] Within a single molecule are three receptors, each selective for a specific biological analyte: a benzo-15-crown-5 ether for Na^+, a tertiary amine for H^+, and a phenyliminodiacetate for Zn^{2+}. In the absence of one, or two, or all three analytes, the fluorescence emission in water is low due to PET from a vacant receptor to the excited state fluorophore. However, when all three analytes are bound in excess of threshold concentrations, a high emission is observed. The quantum yield of fluorescence of **45** in the 'on' state is modest at 0.02. A model compound for proton and zinc exhibited a fluorescence quantum yield of 0.19, providing insight that the low fluorescence output is due to very weak binding by the benzo-15-crown-5-ether in aqueous solution such that PET is not completely deactivated.[128]

The Akkaya group has demonstrated a three-input logic gate **46** for detecting a congregation of metal ions based on the highly fluorescent boron-dipyrromethene dye.[129] Their prototype combines both internal charge transfer and PET processes and operates in acetonitrile. The molecule incorporates three selective receptors that simultaneous detect Ca^{2+}, Hg^{2+} and Zn^{2+}. Advantageously, **46** uses a visible wavelength for excitation at 620 nm, and displays an emission maximum at 656 nm with a quantum yield of 0.266 in the presence of elevated cation concentrations.

It is worth reiterating that for biological systems not only can too much of an analyte be life-threatening, but the same is true of too little. Hence, from a clinical diagnostic viewpoint, three-input **AND** logic coincides with a condition when all three analytes are above threshold levels. However, for diagnostic purposes, we could also consider three-input **INHIBIT** logic: a state when two analytes are above threshold levels and the third is below.

The concept is illustrated by **47**, which consists of a 2-bromonaphthalene phosphor linked to a BAPTA receptor.[130] This supramolecular logic gate complex relies on the bromonaphthalene for phosphorescence output, the BAPTA receptor to bind Ca^{2+}, a β-cyclodextrin to complex with the bromonaphthalene, and the absence of oxygen in solution. When the concentrations of Ca^{2+} and β-cyclodextrin are both high and the solution is free of oxygen, the light output is high. The roles of the Ca^{2+} and β-cyclodextrin inputs are to bind the BAPTA receptor, thus preventing PET from quenching the bromonaphthalene phosphorescence, and maximizing the phosphorescence output, respectively. However, when oxygen finds its way into the solution, it readily quenches the phosphorescence regardless of the threshold concentrations of Ca^{2+} and β-cyclodextrin.

Disease screening is a foreseeable application for the 'lab-on-a-molecule' paradigm. The conventional approach, discussed throughout most of this chapter, is for a practitioner to measure analytes on a one-for-one basis, followed by analysis of the data by a practitioner, who studies the data and comes to a conclusion. In the future, specifically designed molecules could test for key analyte combinations in a single rapid test and perform the diagnosis independently. A vision is that molecular diagnostics and cellular imaging will progress to the digital world, where multi-analyte sensors will monitor for many biologically relevant parameters simultaneously and make an intelligent diagnosis autonomously.[131]

3.6.2 One Receptor for Multiple Analytes

The sensing approach discussed earlier involved designing selective chemosensors following the paradigm of 'one binding site per analyte'. Schmittel and Lin[132] propose a paradigm involving a multi-tasking sensor **48** with only one receptor, which binds non-specifically to many analytes (Figure 3.21).[1] The molecule consists of a ruthenium complex as the optical signalling unit and two aza-crown ethers as the receptor. The crown ethers work cooperatively as a single binding site. Depending on the choice of one of four analytical

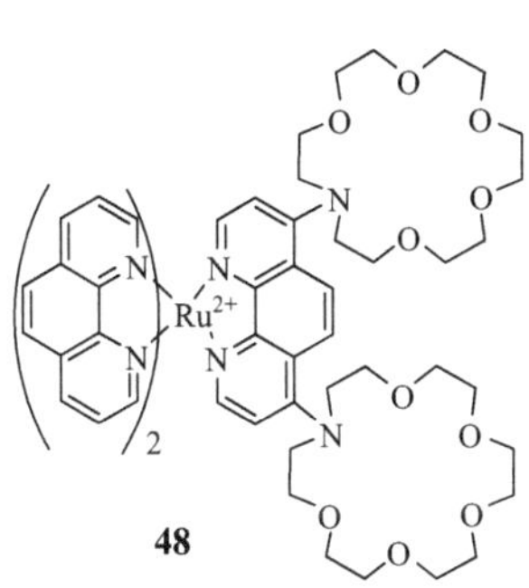

48

Figure 3.21 A chemosensor for three different metal ions.

techniques, quantitative measurements of Pb^{2+}, Hg^{2+} and Cu^{2+} concentrations are recorded. The sensor distinguishes the cations by UV-visible spectroscopy, luminescence spectroscopy, cyclic voltammetry and electrogenerated chemiluminescence. The first technique uses colour to monitor ion binding, the second luminescence, the third redox potential, and the fourth spectroelectrochemistry. The approach is successful, as each channel results from a unique conformational bias of the crown ether receptor site. The same group recently reported a dual-channel chemodosimeter for the anion sensing of acetate and cyanide.[133] A practical engineering challenge, from the viewpoint of point-of-care applications, is the miniaturization of the instrumentation for these analytical techniques in one compact device.

Another multi-analyte chemosensor exploits the solvent conditions to differentiate between Zn^{2+}, Pb^{2+} and Hg^{2+}.[134]

3.7   Summary and Outlook

The current PET sensor technology measures the blood analytes of ions such as Na^+, K^+ and Ca^{2+} using one molecule per specific analyte.[34] In the long term, the development of a suite of 'lab-on-a-molecule' systems could assist practitioners in biomedical diagnostic and point-of-care applications. Rather than designing specific molecules to sense for only one specific analyte, molecules with built-in algorithms will be able to sense for specific disease conditions by monitoring many clinically relevant parameters simultaneously. With the assistance of intelligently designed molecules, a practitioner could get a 'yes' or 'no' digital response for a disease condition in addition to a summary of analog data.

The synthesis and application of fluorescent probes for probing the metal ion concentrations inside cells remains a story still to be fully told.[5] Numerous opportunities are available for discovery with respect to their application to biological assays. Possibilities abound for developing more sensitive and selective probes with improved traits from lower background fluorescence, greater enhancement factors and better targetability to specific subcellular locations and probes with ratiometric properties.

Yet there is still much opportunity for the development of selective metal ion chemosensors, particularly for actinide and lanthanide ions. There are bound to be growing health and environmental concerns, as these lesser understood elements, from a biological viewpoint, become more prominent in our society as more and more applications are found for them. Hence, supramolecular chemistry and chemosensors will play an increasing greater role in monitoring and solving many environmental, industrial and biomedical challenges.

Acknowledgements

We gratefully thank the University of Malta and the Strategic Educational Pathways Scholarship (Malta) program for funding. This Scholarship is part-financed by the European Social Fund (ESF) under Operational Programme II—Cohesion Policy 2007–2013.

References

1. B. Valeur and M. N. Berberan-Santos, *Molecular Fluorescence*, Wiley-VCH, Weinheim, 2nd edn, 2012.
2. *Tietz Fundamentals of Clinical Chemistry*, ed. C. A. Burtis and E. R. Ashwood, WB Saunders Company, Philadelphia, PA, 5th edn, 2001.
3. J. R. Lakowicz, *Principles of Fluorescence Spectroscopy*, 3rd edn, Plenum, New York, NY, 2009.
4. F. A. Cotton, *Advanced Inorganic Chemistry*, Wiley, New York, 6th edn, 1999.
5. D. W. Domaille, E. L. Que and C. J. Chang, *Nat. Chem. Biol.*, 2008, **4**, 168.
6. E. L. Que, D. W. Domaille and C. J. Chang, *Chem. Rev.*, 2008, **108**, 1517.
7. R. W. Peeling, K. K. Holmes, D. Mabey and A. Ronald, *Sex. Transm. Infect.*, 2006, **82**, v1.
8. M. Dutta and D. Das, *Trends Analyt. Chem.*, 2012, **32**, 113.
9. K. Kaur, R. Saini, A. Kumar, V. Luxami, N. Kaur, P. Singh and S. Kumar, *Coord. Chem. Rev.*, 2012, **256**, 1992.
10. M. Formica, V. Fusi, L. Giorgi and M. Micheloni, *Coord. Chem. Rev.*, 2012, **256**, 170.
11. Y. Jeong and J. Yoon, *Inorg. Chem. Acta.*, 2012, **381**, 2.
12. A. P. de Silva, T. S. Moody and G. D. Wright, *Analyst*, 2009, **134**, 2385.
13. B. N. G. Giepmans, S. R. Adams, M. H. Ellisman and R. Y. Tsien, *Science*, 2006, **312**, 217.
14. J. Lippincott-Schwartz and G. H. Patterson, *Science*, 2003, **300**, 87.
15. R. Freeman and I. Willner, *Chem. Soc. Rev.*, 2012, **41**, 4067.
16. X. Michalet, F. F. Pinaud, L. A. Bentolila, J. M. Tsay, S. Doose, J. J. Li, G. Sundaresan, A. M. Wu, S. S. Gambhir and S. Weiss, *Science*, 2005, **307**, 538.
17. A. P. de Silva, H. Q. N. Gunaratne, T. Gunnlaugsson, A. J. M. Huxley, C. P. McCoy, J. T. Rademacher and T. E. Rice, *Chem. Rev.*, 1997, **97**, 1515.
18. J. F. Callan, A. P. de Silva and D. C. Magri, *Tetrahedron*, 2005, **61**, 8551.
19. V. Balzani, M. Venturi and A. Credi, *Molecular Devices and Machines*, Wiley-VCH, Weinheim, 2nd edn, 2008.
20. R. A. Bissell, A. P. de Silva, H. Q. N. Gunaratne, P. L. M Lynch, G. E. M. Maguire, C. P. McCoy and K. R. A. S. Sandanayake, *Top. Curr. Chem.*, 1993, **168**, 223.
21. A. P. de Silva and S. Uchiyama, *Top. Curr. Chem.*, 2011, **300**, 1.
22. A. W. Czarnik, *Acc. Chem. Res.*, 1994, **27**, 302.
23. A. W. Czarnik and J. P. Desvergne, *Chemosensors of Ion and Molecule Recognition*, Kluwer, Dordrecht, 1997.
24. L. Fabbrizzi, M. Licchelli and P. Pallavicini, *Acc. Chem. Res.*, 1999, **32**, 846.
25. L. Fabbrizzi and A. Poggi, *Chem. Soc. Rev.*, 1995, **24**, 197.

26. L. Prodi, F. Bolletta, M. Montalti and N. Zaccheroni, *Coord. Chem. Rev.*, 2000, **205**, 59.
27. R. A. Bissell, A. P. de Silva, H. Q. N. Gunaratne, P. L. M. Lynch, G. E. M Maguire and K. R. A. S Sandanayake, *Chem. Soc. Rev.*, 1992, **21**, 187.
28. A. P. de Silva, T. P. Vance, M. E. S. West and G. D. Wright, *Org. Biomol. Chem.*, 2008, **6**, 2468.
29. G. J. Kavarnos, *Fundamentals of Photoinduced Electron Transfer*, VCH, Weinheim, 1993.
30. A. P. de Silva, *J. Phys. Chem. Lett.*, 2011, **2**, 2865.
31. http://www.cedarinteractive.com/demos/Osmetech/start.html.
32. J. K. Tusa and H. He, *J. Mater. Chem.*, 2005, **15**, 2640.
33. H. He, M.A. Mortellaro, M. J. P. Leiner, S. T. Young, R. J. Fraatz and J. K. Tusa, *Anal. Chem.*, 2003, **75**, 549.
34. H. He, M. A. Mortellaro, M. J. P. Leiner, R. J. Fraatz and J. K. Tusa, *J. Am. Chem. Soc.*, 2003, **125**, 1468.
35. G. Grynkiewicz, M. Poenie and R. Y. Tsien, *J. Biol. Chem.*, 1985, **260**, 3440.
36. A. Malgaroli, D. Milani, J. Meldolesi and T. Pozzan, *J. Cell. Biol.*, 1987, **105**, 2145.
37. K. Suzuki, K. Sato, H. Hisamoto, D. Siswanta, K. Hayashi, N. Kasahara, K. Watanabe, N. Yamamoto and H. Sasakura, *Anal. Chem.*, 1996, **68**, 208.
38. J. W. Steed, *Coord. Chem. Rev.*, 2001, **215**, 171.
39. B. Dietrich, in *Comprehensive Supramolecular Chemistry*; G. W. Gokel, Elsevier, Oxford, 1996, vol. 1, p. 153.
40. O. S. Wolfbeis, *Anal. Chem.*, 2006, **78**, 3859.
41. F. V. Englich, T. C. Foo, A. C. Richardson, H. Ebendorff-Heidepriem, C. J. Sumby and T. M. Monro, *Sensors*, 2011, **11**, 9560.
42. A. P. de Silva and S. A. de Silva, *J. Chem. Soc., Chem. Commun.*, 1986, 1709.
43. A. P. de Silva and K. R. A. S. Sandanayake, *Tetrahedron Lett.*, 1991, **32**, 421.
44. R. A. Schultz, B. D. White, K. A. Dishong, W. Arnold and G. W. Gokel, *J. Am. Chem. Soc.*, 1985, **107**, 6659.
45. T. Gunnlaugsson, M. Nieuwenhuyzen, L. Richard and V. Thoss, *J. Chem. Soc., Perkin Trans.*, 2002, **2**, 141.
46. V. V. Martin, A. Rothe and K. R. Gee, *Bioorg. Med. Chem. Lett.*, 2005, **15**, 1851.
47. V. V. Martin, A. Rothe, Z. Diwu and K. R Gee, *Bioorg. Med. Chem. Lett.*, 2004, **14**, 5313.
48. I. Johnson and M. T. Z. Spence, *The Molecular Probes Handbook: A Guide to Fluorescent Probes and Labeling Technologies*, Life Technologies Corporation, 11th edn, Carlsbad, California, 2010.
49. A. P. de Silva, H. Q. N. Gunaratne and C. P. McCoy, *Nature*, 1993, **364**, 42.

50. A. P. de Silva, H. Q. N. Gunaratne and C. P. McCoy, *J. Am. Chem. Soc.*, 1997, **119**, 7891.
51. E. Katz, *Molecular and Supramolecular Information Processing: From Molecular Switches to Logic Systems*, Wiley-VCH Verlag, Weinheim, 2012.
52. D. C. Magri, G. D. Coen, R. L. Boyd and A. P. de Silva, *Anal. Chim. Acta*, 2006, **568**, 156.
53. M. Takeshita and M. Irie, *J. Org. Chem.*, 1998, **63**, 6643.
54. D. E. J. G. J. Dolmans, D. Fukumura and R. K. Jain, *Nature Rev. Cancer*, 2003, **3**, 380.
55. S. Ozlem and E. U. Akkaya, *J. Am. Chem. Soc.*, 2009, **131**, 48.
56. A. Minta and R. Y. Tsien, *J. Biol. Chem.*, 1989, **264**, 19 449.
57. X. Zhou, F. Su, Y. Tian, C. Youngbull, R. H. Johnson and D. R. Meldrum, *J. Am. Chem. Soc.*, 2011, **133**, 18 530.
58. P. Nandhikonda, M. P. Begaye and M. D. Heagy, *Tetrahedron Lett.*, 2009, **50**, 2459.
59. T. Gunnlaugsson and J. P. Leonard, *Chem. Commun.*, 2003, 2424.
60. A. Minta, J. P. Y. Kao and R. Y. Tsien, *J. Biol. Chem.*, 1989, **264**, 8171.
61. R. M. Paredes, J. C. Etzler, L. T. Watts and J. D. Lechleiter, *Methods*, 2008, **46**, 143.
62. S. Lee, H. G. Lee and S. H. Kang, *Anal. Chem.*, 2009, **81**, 538.
63. R. Martínez-Zaguilán, J. Parnami and G. M. Martinez, *Cell Physiol Biochem.*, 1998, **8**, 158.
64. Y. Dong, X. Mao, X. Jiang, J. Hou, Y. Cheng and C. Zhu, *Chem. Commun.*, 2011, **47**, 9450.
65. C. Baastrup and M. Schou, *Arch. Gen. Psychiatry*, 1967, **16**, 162.
66. T. Gunnlaugsson, B. Bichell and C. Nolan, *Tetrahedron*, 2004, **60**, 5799.
67. S. O. Obare and C. J. Murphy, *New J. Chem.*, 2001, **25**, 1600.
68. A. Caballero, R. Tormos, A. Espinosa, M. D. Velasco, A. Tárraga, M. A. Miranda and P. Molina, *Org. Lett.*, 2004, **6**, 4599.
69. D. Citterio, J. Takeda, M. Kosugi, H. Hisamoto, S.-I. Sasaki, H. Komatsu and K. Suzuki, *Anal. Chem.*, 2007, **79**, 1237.
70. J. Gao, S. Rochat, X. Qian and K. Severin, *Chem. Eur. J.*, 2010, **16**, 5013.
71. S. J. Lippard and J. M. Berg, *Principles of Bioinorganic Chemistry*, University Science Books, Mill Valley, CA, 1994.
72. B. Halliwell and J. M. C. Gutteridge, *Free Radicals in Biology and Medicine*, Oxford University Press, New York, 4th edn, 2007.
73. S. K. Sahoo, D. Sharma, R. K. Bera, G. Crisponi and J. F. Callan, *Chem. Soc. Rev.*, 2012, **41**, 7195.
74. N. Singh, N. Kaur, J. Dunn, M. MacKaya and J. F. Callan, *Tetrahedron Lett.*, 2009, **50**, 953.
75. N. Singh, N. Kaur and J. F. Callan, *J. Fluoresc.*, 2009, **19**, 649.
76. J. Hua and Y. G. Wang, *Chem. Lett.*, 2005, **34**, 98.
77. X. Huang, *Mut. Res.*, 2003, **533**, 153.
78. A. P. de Silva and R. A. D. D. Rupasinghe, *J. Chem. Soc., Chem. Commun.*, 1985, 1669.

79. J. Han and K. Burgess, *Chem. Rev.*, 2010, **110**, 2709.
80. D. C. Magri, *New. J. Chem.*, 2009, **33**, 457.
81. T. J. Farrugia and D. C. Magri, *New. J. Chem.*, 2013, **37**, 148.
82. Z. Xu, J. Yoon and D. R. Spring, *Chem. Soc. Rev.*, 2010, **39**, 1996.
83. P. Jiang and Z. Guo, *Coord. Chem. Rev.*, 2004, **248**, 205.
84. K. R. Gee, Z.-L. Zhou, W.-J. Qian and R. Kennedy, *J. Am. Chem. Soc.*, 2002, **124**, 776.
85. T. Hirano, K. Kikuchi, Y. Urano, T. Higuchi and T. Nagano, *Angew. Chem. Int. Ed.*, 2000, **39**, 1052.
86. C. J. Frederickson, E. J. Kasarskis, D. Ringo and R. E. Frederickson, *J. Neurosci. Methods*, 1987, **20**, 91.
87. P. D. Zalewski, S. H. Millard, I. J. Forbes, O. Kapaniris, A. Slavotinek, W. H. Betts, A. D. Ward, S. F. Lincoln and I. Mahadevan, *J. Histochem. Cytochem.*, 1994, **42**, 877.
88. Y. Zhang, X. Guo, W. Si, L. Jia and X. Qian, *Org. Lett.*, 2008, **10**, 473.
89. T. Gunnlaugsson, T. Clive. Lee and R. Parkesh, *Org. Biomol. Chem.*, 2003, **1**, 3265.
90. R. Parkesh, T. Clive Lee and T. Gunnlaugsson, *Org. Biomol. Chem.*, 2007, **5**, 310.
91. T. Hirano, K. Kikuchi, Y. Urano, T. Higuchi and T. Nagano, *J. Am. Chem. Soc.*, 2000, **122**, 12399.
92. S. Ueno, M. Tsukamoto, T. Hirano, K. Kikuchi, M. K. Yamada, N. Nishiyama, T. Nagano, N. Matsuki and Y. Ikegaya, *J. Cell Biol.*, 2002, **158**, 215.
93. T. Hirano, K. Kikuchi, Y. Urano, T. Higuchi and T. Nagano, *J. Am. Chem. Soc.*, 2002, **124**, 6555.
94. C. J. Chang, E. M. Nolan, J. Jaworski, S. C. Burdette, M. Shang and S. J. Lippard, *Chem. Biol.*, 2004, **11**, 203.
95. C. Burdette, G. K. Walkup, B. Spingler, R. Y. Tsien and S. J. Lippard, *J. Am. Chem. Soc.*, 2001, **123**, 7831.
96. E. M. Nolan, J. W. Ryu, J. Jaworski, R. P. Feazell, M. Sheng and S. J. Lippard, *J. Am. Chem. Soc.*, 2006, **128**, 15517.
97. E. M. Nolan, J. Jaworski, K.-I. Okamoto, Y. Hayashi, M. Sheng and S. J. Lippard, *J. Am. Chem. Soc.*, 2005, **127**, 16812.
98. Z. Xu, K. Baek, H. N. Kim, J. Cui, X. Qian, D. R. Spring, I. Shin and J. Yoon, *J. Am. Chem. Soc.*, 2010, **132**, 601.
99. Z. Xu, X. Liu, J. Pan and D. R. Spring, *Chem. Commun.*, 2012, **48**, 4764.
100. L. C. Yang, R. McRae, M. M. Henary, R. Patel, B. Lai, S. Vogt and C. J. Fahrni, *Proc. Natl. Acad. Sci. USA*, 2005, **102**, 11179.
101. M. T. Morgan, P. Bagchi and C. J. Fahrni, *J. Am. Chem. Soc.*, 2011, **133**, 15906.
102. Z. Liu, C. Zhang, X. Wang, W. He and Z. Guo, *Org. Lett.*, 2012, **14**, 4378.
103. X. Guo, X. Qian and L. Jia, *J. Am. Chem. Soc.*, 2004, **126**, 2272.
104. H. N. Kim, W. X. Ren, J. S. Kim and J. Yoon, *Chem. Soc. Rev.*, 2012, **41**, 3210.
105. E. M. Nolan and S. J. Lippard, *J. Am. Chem. Soc.*, 2003, **125**, 14270.

106. E. M. Nolan, M. E. Racine and S. J. Lippard, *Inorg. Chem.*, 2006, **45**, 2742.

107. E. M. Nolan and S. J. Lippard, *J. Am. Chem. Soc.*, 2007, **129**, 5910.

108. T. Gunnlaugsson, C. T. Lee and R. Parkesh, *Tetrahedron*, 2004, **60**, 11239.

109. T. Gunnlaugsson, C. T. Lee and R. Parkesh, *Org. Lett.*, 2003, **5**, 4065.

110. X. Peng, J. Du, J. Fan, J. Wang, Y. Wu, J. Zhao, S. Sun and T. Xu, *J. Am. Chem. Soc.*, 2007, **129**, 1500.

111. X.-L. Tang, X.-H. Peng, W. Dou, J. Mao, J.-R. Zheng, W.-W Qin, W.-S. Liu, J. Chang and X.-J. Yao., *Org. Lett.*, 2008, **10**, 3653.

112. Y. Wang, X. Hu, L. Wang, Z. Shang, J. Chao and W. Jin, *Sens. Actuator B*, 2011, **156**, 126.

113. Q. Zhao, R.-F. Li, S.-K. Xing, X.-M. Liu, T.-L. Hu and X.-H. Bu, *Inorg. Chem.*, 2011, **50**, 10041.

114. D. Faye, J. P. Lefevre, J. A. Delaire and I. Leray, *J. Photochem. Photobiol. A: Chem.*, 2012, **234**, 115.

115. Q. W. He, E. W. Miller, A. P. Wong and C. J. Chang, *J. Am. Chem. Soc.*, 2006, **128**, 9316.

116. H.-J. Schneider, in *Applications of Supramolecular Chemistry*, ed. H.-J. Schneider, CRC Press, Boca Raton, FL, 2012, p. 159.

117. C. M. G. dos Santos, A. J. Harte, S. J. Quinn and T. Gunnlaugsson, *Coord. Chem. Rev.*, 2008, **252**, 2512.

118. S. Shinoda and H. Tsukube, *Analyst*, 2011, **136**, 431.

119. J.-C. G. Bünzli, *Chem. Rev.*, 2010, **110**, 2729.

120. B. Mokhtari, K. Pourabdollah and N. Dallai, *J. Radioanal. Nucl. Chem.*, 2011, **287**, 921.

121. Z. Liang, Z. Liu and Y. Gao, *Tetrahedron Lett.*, 2007, **48**, 3587.

122. P. Das, A. Ghosh and A. Das, *Inorg. Chem.*, 2010, **49**, 6909.

123. C. Han, L. Zhang and H. Li, *Chem. Commun.*, 2009, 3545.

124. A. P. de Silva and N. D. McClenaghan, *Chem. Eur. J.*, 2004, **10**, 574.

125. A. P. de Silva and S. Uchiyama, *Nat. Nanotechnol.*, 2007, **2**, 399.

126. A. P. de Silva, *Chem. Asian. J.*, 2011, **6**, 750.

127. D. C. Magri, G. J. Brown, G. D. McClean and A. P. de Silva, *J. Am. Chem. Soc.*, 2006, **128**, 4950.

128. D. C. Magri and A. P. de Silva, *New J. Chem.*, 2010, **34**, 476.

129. O. A. Bozdemir, R. Guliyev, O. Buyukcakir, S. Selcuk, S. Koleman, G. Gulseren, T. Nalbantoglu, H. Boyaci and E. U. Akkaya, *J. Am. Chem. Soc.*, 2010, **132**, 8029.

130. A. P. de Silva, I. M. Dixon, H. Q. N. Gunaratne, T. Gunnlaugsson, P. R. S. Maxwell and T.E. Rice, *J. Am. Chem. Soc.*, 1999, **121**, 1393.

131. T. Konry and D. R. Walt, *J. Am. Chem. Soc.*, 2009, **131**, 13232.

132. M. Schmittel and H.-W. Lin., *Angew. Chem. Int. Ed.*, 2007, **46**, 893.

133. M. Schmittel and Q. H. Shu, *Chem. Commun.*, 2012, **48**, 2707.

134. J. Hatai, S. Pal, G. P. Jose, T. Sengupta and S. Bandyopadhyay, *RSC Advances*, 2012, **2**, 7033.

Complexation of Biomedically Important Organic Compounds

HANS-JÖRG SCHNEIDER

FR Organische Chemie, Universität des Saarlandes, D 66041, Saarbrücken
Email: ch12hs@rz.uni-sb.de

4.1 Introduction

The complexation of organic compounds by supramolecular host molecules has been the focus of many investigations. Modern synthetic strategies and insights into the relevant binding mechanisms have made it possible to design host structures that can complex any target which occurs in living systems, or which may be harmful. The most frequently used host compounds are cyclo-dextrins, calixarenes, and cucurbiturils, to which chapters 5, 6 and 7, respectively, are devoted. The present chapter deals in particular with sensing of biomedically interesting organic compounds, mostly with other receptors,[1] as well as with the modulation of biopolymer functions with supramolecular ligands. In this framework we discuss the use of other host compounds, such as crown ethers, cyclophanes, and clefts or molecular tweezers. Emphasis is given to the use of fast and economical optical sensing methods, although techniques such as mass-sensitive detection (Quartz Crystal Microbalance, QCM), electrochemical techniques such as "ion" selective electrodes (ISEs), which are not restricted to ions, photonic crystals (PCs), surface plasmon resonance (SPR), and others are also applied. Immobilization of supramolecular receptor compounds on surfaces can be used for the screening of whole assemblies of target molecules such as substances from combinatorial libraries or from natural sources.[2] Molecular imprinting techniques, discussed in detail in Chapter 14 by

Monographs in Supramolecular Chemistry No. 13
Supramolecular Systems in Biomedical Fields
Edited by Hans-Jörg Schneider

Published by the Royal Society of Chemistry, www.rsc.org

Mujahid and Dickert, offer a particularly versatile approach to any kind of analyte. An intriguing strategy towards flexible, highly specific and luminescent chemosensors for organic analytes is the use of surface imprinting on vesicle membranes.[3] Vesicles from zinc-cyclen complexes, for example, appended with long alkyl chains containing diacetylene units, which are photopolymerized for crosslinking, can be used; amphiphilic dyes such as carboxyfluorescein co-embedded into the membrane secure significant fluorescence intensity changes upon analyte association. In this way, phosphopeptides differing in the number and nature of side groups of amino acids can be differentiated with, for example, either $\log K = 4.1$ or $\log K = 7.4$ as apparent binding constant.

The advantages of synthetic receptor compounds are their high stability and their use in any desired medium, including water.[4] Biosensors derived from biomaterials have been said to have largely failed, thus far, to realize their potential as reagentless, real-time analytical devices, with the possible exception of the glucose monitor.[5] In particular, synthetic sensors can be easily equipped, for instance, with fluorophores, securing in that way simple and sensitive detection.[6] Alternatively, one can use the indicator displacement assay (IDA), in which the signaling dye is displaced *via* competition by the analyte, and thereby changes its emission or absorption.[7] In most cases synthetic receptors possess one binding site securing a high sensitivity, and another one which can be mounted at some distance and is responsible for selectivity. Affinity and selectivity correlate well only in ionophore complexes in which there is only one kind of binding interaction;[8] for organic analytes the affinity contribution necessary for a $1\,\mu M$ detection limit, for example, is with a ΔG value of about $30\,kJ\,mol^{-1}$, which is usually much higher than the one for selectivity, for which a free energy difference $\Delta\Delta G_{A/B}$ between two analytes A and B of, for instance, $10\,kJ\,mol^{-1}$ is enough to discriminate the analytes by a ratio of 100 in binding constants. If the selectivity is insufficient with a single host compound, one can use simultaneously many receptors which may exhibit small selectivity differences; signal processing with principal component analysis (PCA) then allows selectivity with a whole array of receptors as may be likened to a "chemical nose".[9] In what follows, typical applications are illustrated for selected targets of biological or medicinal interest, without aiming at a comprehensive description.

4.2 Amines/Neurotransmitters/Catecholamines

Biogenic amines (Figure 4.1) can be complexed efficiently with anionic host compounds providing several salt bridges, and additional lipophilic recognition sites for the alkyl chains. Instead of, or in addition to, ion pairing with the protonated amino groups or their cation–π interactions, one can use their hydrogen-bonding to crown ethers. Host **1** binds spermine with a dissociation constant of only 22 nM. This host is of interest also as a spermine-targeting anticancer agent; it has been shown that the affinity is high enough to compete with spermine binding to double-stranded DNA.[10] Colorimetric recognition of

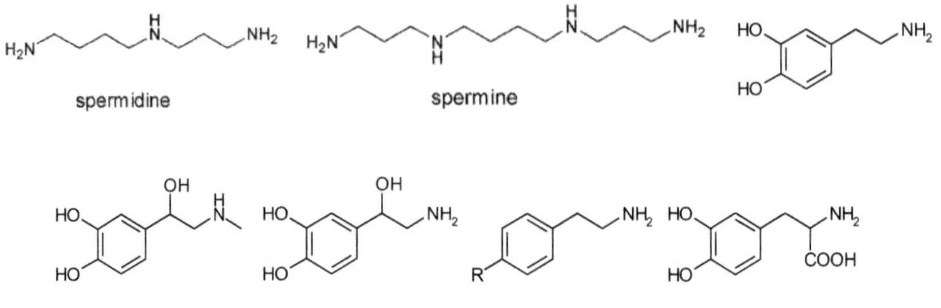

spermidine

spermine

dopamine   epinephrine   norepinephrine   phenethylamine (R = H)   DOPA

(adrenaline)   (noradrenaline)   tyramine (R = OH)

Figure 4.1 Biogenic amines.

both spermine and spermidine is possible with phenolphthalein derivatives of crown ethers such as **2**;[11] the latter's selectivity and sensitivity are high compared with other biogenic amines.

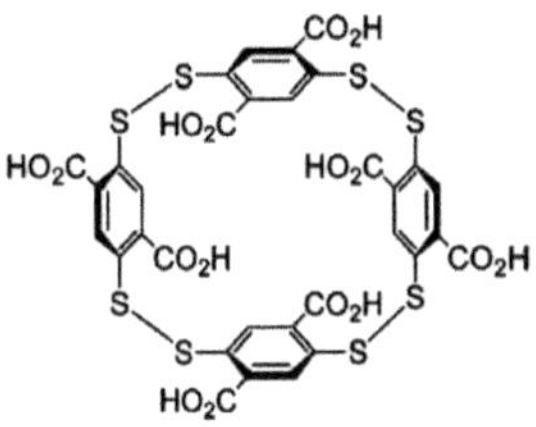

1 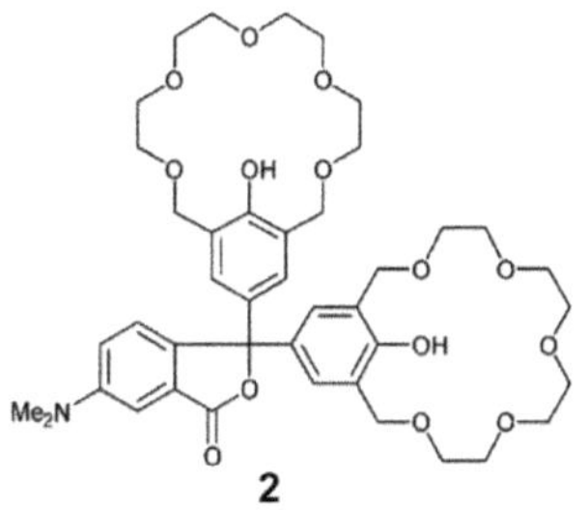   2 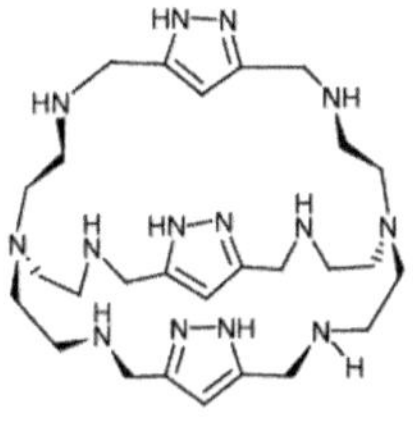   3

Neurotransmitters, such as dopamine, serotonin, and catecholamines, *etc.* (Figure 4.1), play an important role in signal transmission; their deficiency can lead to mental disorders, among other illnesses. Often they are detected with electrochemical methods; however, these suffer from the interference of compounds that are oxidized at similar potentials. The use of cyclodextrins can help to overcome this problem, as shown by the selective detection of dopamine with a gold electrode modified by a β-cyclodextrin/thioctic acid mixed monolayer.[12]

The neurotransmitter serotonin can be measured with an electrochemical sensor with simultaneous quantification of serotonin and dopamine, based on a β-cyclodextrin / poly(*N*-acetylaniline) / carbon nanotube composite modified carbon paste electrode, in which cyclodextrins also serve to preconcentrate the analyte.[13] A fluorometric assay of dopamine relies on 2,7-diazapyrenium dications coated on silica particles, dominated by the interaction of the electron-rich dioxyarene fragment with the electron-deficient fluorophore.[14] Receptor **3** shows with protonated dopamine in water a remarkably high binding constant of about $K = 10^6\ M^{-1}$.[15]

A combination of ion pairing, cation–π, and π–π stacking interactions is responsible for the association of catecholamines to receptors **4** and **5**: the host **4** binds noradrenaline selectively with $K = 1800$ M^{-1} in methanol;[16] receptor **5** also binds noradrenaline with $K = 1250$ M^{-1} in water.[17] By incorporation of host **4** in phospholipid/polydiacetylene vesicles, specific fluorescent detection of noradrenaline becomes possible.[18]

Another way to detect many biologically important target compounds is the simultaneous complexation of the analyte with a fluorescence dye, which has been particularly successful with cucurbiturils such as **6** as host, which is shown below for the detection of amino acids or peptides[19] (Figure 4.2; see also chapters 7 and 12). Fluorescence quenching occurs, for example, with about 1 mM dopamine as analyte and 2,7-dimethyldiaza-pyrenium as electron-poor dye and the cucurbit[8]uril **6**.[20]

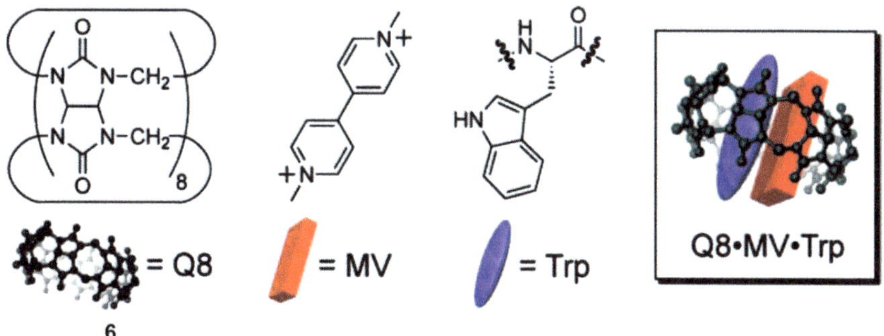

Figure 4.2   Ternary complex of cucurbit[8]uril **6** (= Q8) with methylviologen (MV) and the amino acid Trp.
(Reprinted with permission from *J. Am. Chem. Soc.* 2005, **127**, 14 511. Copyright 2005 American Chemical Society.)

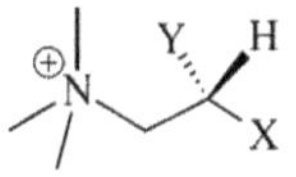

	X	Y
choline	OH	H
acetylcholine	OCOCH$_3$	H
carbamoylcholine	OCONH$_2$	H
L-carnitine	CH$_2$CO$_2^-$	OH
O-acetyl-L-carnitine	CH$_2$CO$_2^-$	OCOCH$_3$

Figure 4.3 Neurotransmitters with trimethylammonium groups.

The most important neurotransmitter, acetylcholine (ACh) (Figure 4.3), can be detected by the interaction of the permanently charged ammonium center with anionic groups and with aryl moieties by cation–π interaction.[21] Tetra-anionic resorcarenes **7**[22] or other macrocycles such as **8**,[23] the calixarene tetrasulfonate **9**,[24] or the speleand **10**[25] bind AcCh in water down to about 0.05 mM concentration. If a pyrene-modified *N*-alkylpyridinium dye is used in an IDA assay in combination with the resorcarene **7**, addition of AcCh at 10^{-4} M concentration leads to fluorescence regeneration, while other amines such as adrenaline give negligible response.[26] The use of 2,3-diazabicyclo[2,2,2]oct-2-ene as indicator instead of the pyrene-modified *N*-alkylpyridinium dye, for example, has the advantage of higher stability and less interference with medium effects; it also allows detection of carnitine, although this is bound 10 times less effectively, due to the presence of the negative carboxylate function.[27]

With [4-(dimethylamino)styryl]-1-methylpyridinium-*p*-toluenesulfonate as indicator, sub-micromolar concentrations of AcCh are detected, also by immobilization of the host on an oxide-containing silicon surface.[28] The deep cavitands **11** (R = Et, X = NH$_2$) bind AcCh with $K = 4000 \, \text{M}^{-1}$, choline with $K = 12\,000 \, \text{M}^{-1}$, and carnitine with $K = 15\,000 \, \text{M}^{-1}$, but only in the aprotic solvent dimethyl sulfoxide (DMSO).[29] The water-soluble cavitand **12** complexes both AcCh and choline with $K > 10^4 \, \text{M}^{-1}$, whereas carnitine with $K = 140 \, \text{M}^{-1}$ again binds more weakly.[30] Dynamic combinatorial chemistry based on dipeptide hydrazones, which reversibly combine through hydrazone

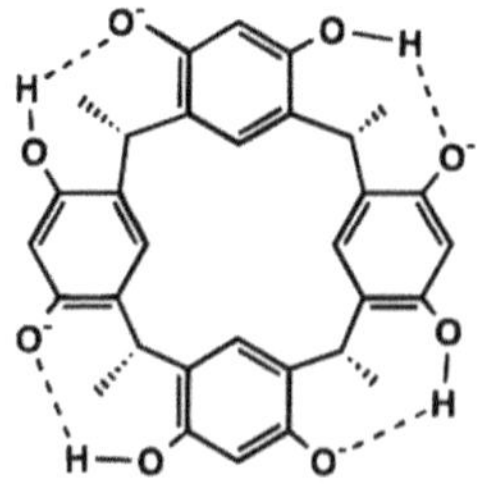

7 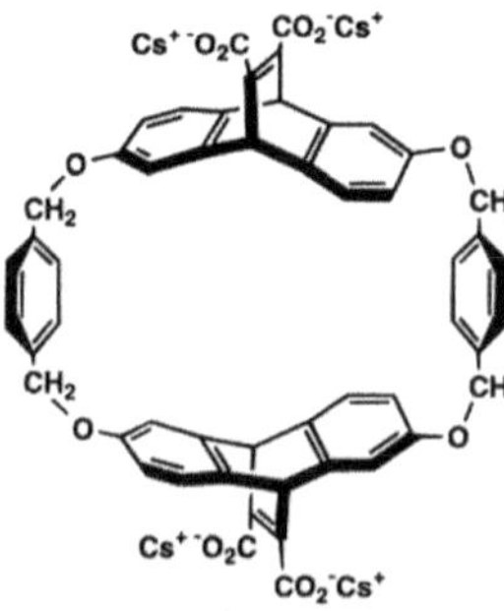   8 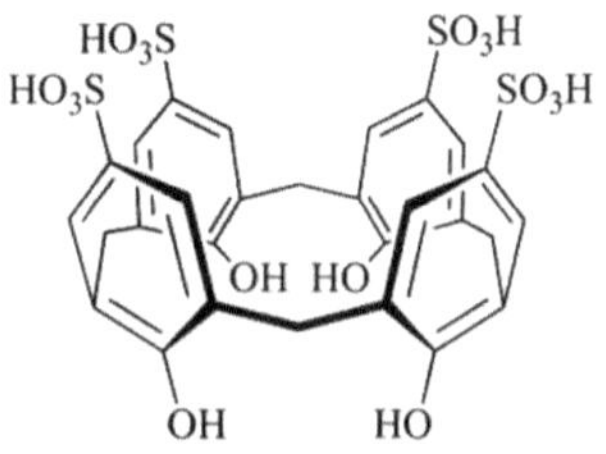   9

linkages, has led to a [2]-catenane consisting of two interlocked macrocyclic trimers, which binds AcCh with a 100 nM affinity.[31] Electrochemical sensing of acetylcholine was achieved with a membrane electrode containing the cucurbituril **6** ($n = 6$), with high selectivity over choline and other ions such as Na^+, K^+, and $^+NH_4$.[32]

4.3   Biologically Important Acids

The obvious way to provide associations for acids is by ion pairing with cationic host compounds, using shape recognition for selectivity as, for instance, with a ternary fluorophore complex with the imidazolium-substituted calix[4]arene **13**. Citrate induces pK_a shifts of an amino group attached to the used fluorophore and thereby protonation of the amino group, switches the intramolecular photoelectron transfer and thus enhances the fluorescence signal.[33] Citrate can with a binding constant $K = 5 \times 10^5$ M^{-1} be selectively detected in presence of malate or tartrate with the triscationic host **14**, with the help of the indicator tris-anioncarboxyfluorescein, allowing naked-eye detection by appearance of the quenched fluorescence.[34]

Gamma-hydroxybutyric acid (GHB), a psychotherapeutic but also dangerous drug of abuse, can be detected in millimolar concentrations with a colorimetric sensor based on supramolecular host–guest complexes of fluorescent dyes with organic capsules fabricated from cucurbiturils (see Chapter 7). The basis is a library combining four fluorescent tricyclic basic dyes as sensing guests and two cucurbiturils (**15**, $n = 7,8$) as hosts, which bind the dyes with constants between $K = 3$ and 15×10^6 M^{-1}.[35]

4.3.1   Amino Acids

The complexation of amino acids and peptides can rely on binding the COO$^-$ and $^+$NH$_3$ termini with anionic and cationic host units, and on side-chain recognition with complementary binding sites in the host. The dicopper host **16** allows glutamate recognition with rhodamine as displacement indicator,[36] the copper derivative **17** with moderate enantioselectivity recognizes Val, Phe, Trp or Leu using pyrocatechol as indicator, with binding constants of up to 10^5 M^{-1} and selectivity ratios K_D/K_L between 1.7 and 2.6.[37] Receptor **18** is a fluorescent probe for ω-aminoacids including gamma-aminobutyric acid (GABA) and L-carnitine ; it behaves as a molecular ruler, changing the yellow fluorescent emission into blue as a function of the distance between the terminal ammonium and the carboxylate termini, down to 10^{-6} M concentration.[38]

Basic amino acids are complexed particularly well with cyclophanes bearing multiple anion sites. The sulfonato-calixarene complex **19** illustrates this, with K values for Arg and Lys of 1520 M^{-1} and 740 M^{-1}, respectively, at pH 8; aliphatic or aromatic amino acids are bound with K values of only 15–65 M^{-1}.[39] The advantage of Arg, with its flat guanidinium residue, over Lys is seen also with the receptor **20**, due to a larger cation–π effect.[40] The cleft **21** binds Lys with $K_d \approx 20$ μM, 10 times less tightly than arginine, and negligibly with most other cationic biomolecules. The selectivity, which also allows sensing of the location of Lys in peptides (see below) , results from threading the alkyl chain of Lys through the cavity of **21**, allowing hydrophobic interactions with host side-walls, and simultaneous ion pairing with the anionic bridgehead groups of **21**.[41] As shown already in Figure 4.2, the combination of cucurbit[8]uril **6** with methylviologen as indicator can be used for the detection of both amino acids and peptides; aromatic amino acids bind with $\log K$ values of 3.3 (Tyr), 3.7 (Phe), or 4.6 (Trp).[19]

4.4 Peptides

The design of artificial receptors for peptides relies on similar principles as for amino acids,[1c,42] but the aim here is to find hosts which are also able to discriminate the amino-acid sequence. Complex **22** illustrates how one can achieve this by placing binding units for the $^+NH_3$ and COO^- termini at the ends of a spacer, which also bears recognition substituents (**L**) at suitable positions for interaction with the side chain. With **22** one observes, for instance, a K value for Gly-Phe-Gly of 1700 M^{-1}, whereas Gly-Gly-Phe or Phe-Gly-Gly as well as Gly-Gly-Ala—or, due to mismatch, all di- or tetrapeptides—bind 10 times less tightly.[43] Sensitive signaling is possible by using the dansyl fluorophore as side chain ligand **L**; binding of the peptide N-terminus to the crown ether oxygens removes the photoelectron transfer and leads to strong emission. Other examples, often with protected peptides, have shown how larger affinities can be reached with, for example, guanidinium groups for the carboxylate binding and additional hydrophobic or hydrogen-bond contributions.[42] The tetrapeptide Ac-Val-Val-Ile-Ala-OH, which represents the C-terminal sequence of the amyloid peptide, can be bound selectively by the guanidiniocarbonyl pyrrole receptor complex **23**, in which hydrophobic interactions secure the selectivity, and ion pairing and hydrogen bonds secure the affinity.[42] Host **20** is very suitable for arginine-based dipeptides, including those with β- and γ-amino acids, which in water bind with free energies of around $\Delta G = 20\,kJ\,mol^{-1}$, depending on the degree of protonation.[40]

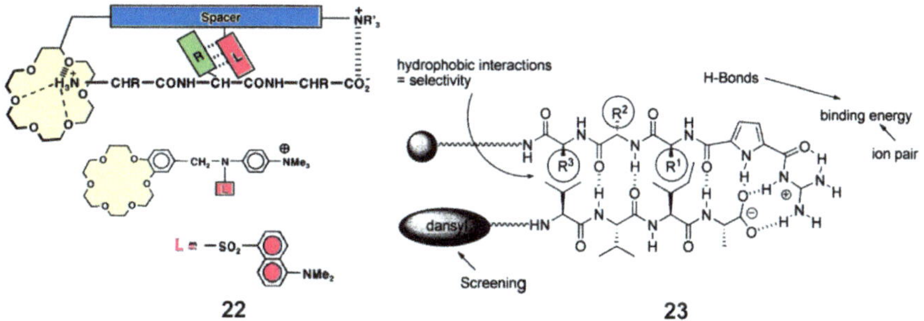

Sequence selective recognition in dipeptides can also be achieved with cucurbit[7]uril (CB[7]);[44] here only dipeptides with aromatic residues at the

N-terminus interact strongly with CB[7] (Figure 4.4a), whereas in the other case (Figure 4.4b) the carboxylate terminus is closer to the negative charges at the CB rim. In consequence one observes for Phe-Gly that $K = 3.0 \times 10^7$ M^{-1}, but for Gly-Phe only $K = 1300$ M^{-1}.

Simultaneous inclusion of an indicator dye and the analyte in cucurbiturils (see Figure 4.2) can also be used for the detection of peptides. Cucurbit[8]uril in combination with methylviologen shows for tripeptides log$K$ values such as 3.5 for Gly-Gly-Trp, 4.3 for Gly-Trp-Gly, and 5.1 for Trp-Gly-Gly; the limitation is that aromatic residues are necessary.[19]

The hemoregulatory peptide Ac-Ser-Asp-Lys-Pro (**24**), which has anti-inflammatory and antifibrotic properties, is bound by a combination of hydrogen bonding and ion pairing with $K = 7 \times 10^3$ M^{-1} in a complex with the bis-guanidinium crown ether (Figure 4.5), exhibiting fluorescence emission;

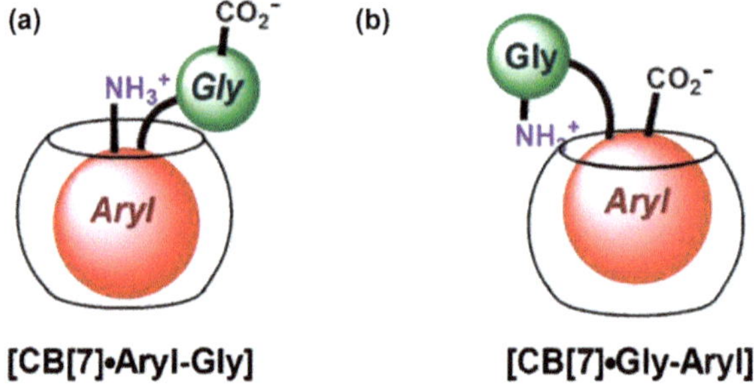

Figure 4.4 Binding of Phe-Gly (a) and Gly-Phe (b) with cucurbit[7]uril, see text for explanation.
(Reprinted from Rekharsky *et al.*[44] with permission, The Royal Society of Chemistry.)

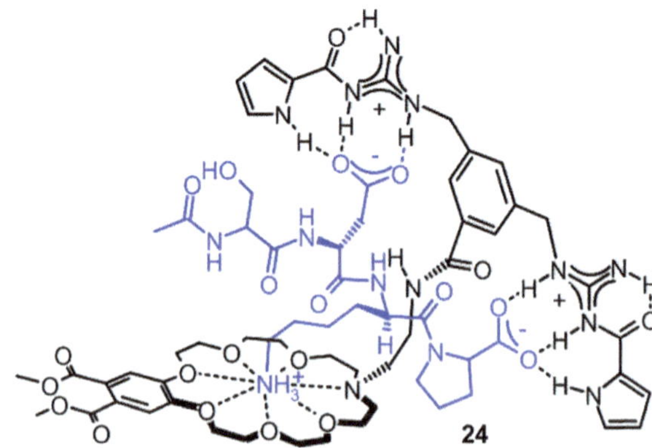

Figure 4.5 Hydrogen bonding and ion pairing in the complex **24** between a bis-guanidinium crown ether and the tetrapeptide Ac-Ser-Asp-Lys-Pro (in blue).
(Reprinted from Späth and B. König[45], with permission from Elsevier, 2010.)

shorter or other tetrapeptides show no or significantly reduced affinity in the complex.[45]

Modulation of protein–protein interactions by small ligands, discussed in Chapter 6 by Perret and Coleman mostly with respect to calixarenes, can rely also on specific recognition of amino-acid sequences. Molecular tweezers such as **21** were found to be capable of inhibiting the aggregation and toxicity of multiple amyloidogenic proteins by binding to Lys residues and disrupting hydrophobic and electrostatic interactions, which are responsible for nucleation, oligomerization and fibril elongation in corresponding proteins. Binding of the tweezer **21** to the Lys residues in the amyloid β-protein occurs at the earliest stages of assembly, and is found to be non-toxic at the level used for interference.[46] Single surface-exposed lysine (Lys 214) of a 14-3-3 protein in the proximity of a central channel inhibits binding of the normal partner protein.[47] In another intriguing application the tweezer **21**, which is known for its strong association with $NADP^+$ and related electron-poor compounds,[48] competes with the Rossmann fold of glucose-6-phosphate dehydrogenase (G6PD) for $NADP^+$ inclusion; thereby the enzyme can only be switched on by addition of large amounts of $NADP^+$.[49] Since the enzyme G6PD is important for cell growth, tweezers such as **21** hold promise to interfere also with proliferation of tumor cells.

4.5 Nucleotides

Nucleotides play a crucial role in many cellular functions, such as transport across membranes, DNA synthesis, cell signaling and energy or electron-transfer processes. Their supramolecular complexation in water relies to a large degree on ion pairing with cationic hosts having suitable receptors, on stacking effects, hydrogen-bonding and metal coordination. Hundreds of artificial receptors for phosphorylated compounds can be found in a recent comprehensive review.[50] Complexations with the cyclophane **25** ($n = 6$) illustrate typical differences:[51] the binding constants increase from 2×10^3 M^{-1} for AMP through 13×10^3 M^{-1} for ADP to 35×10^3 M^{-1} for ATP. The binding free energy, ΔG, differences (in kJ per mol) between nucleotides and the corresponding electro-neutral nucleosides are always around 10: AMP 19 *vs.* 10, GMP 16 *vs.* 6, UMP 17 *vs.* 6, CMP 18 *vs.* 7. NMR spectroscopy and molecular modeling indicates that the nucleobases are always included within the cavity and that the phosphate residue can be in contact with one of the host ^+N centers. The observed constant binding free energy difference of $\Delta\Delta G = 10\,kJ$ agrees with the expected formation of two salt bridges, with an increment of 5 kJ mol^{-1} as found with hundreds of other ion pairs.[52] The interaction between the nucleobase and the host, due to stacking and cation–π interactions, contributes 6–10 kJ mol^{-1} to the total binding. The relatively small selectivity between the bases is typical for most artificial receptors.

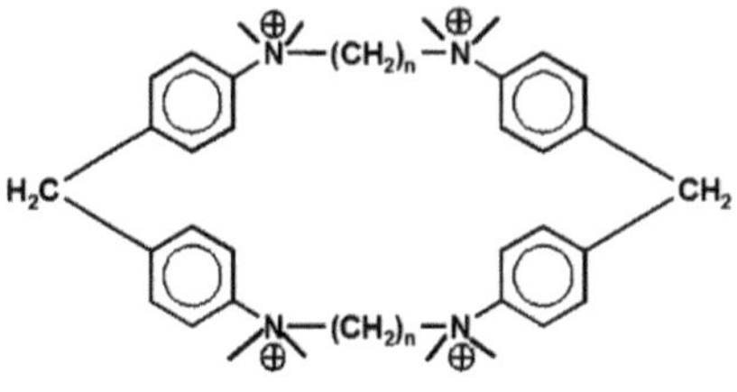

25

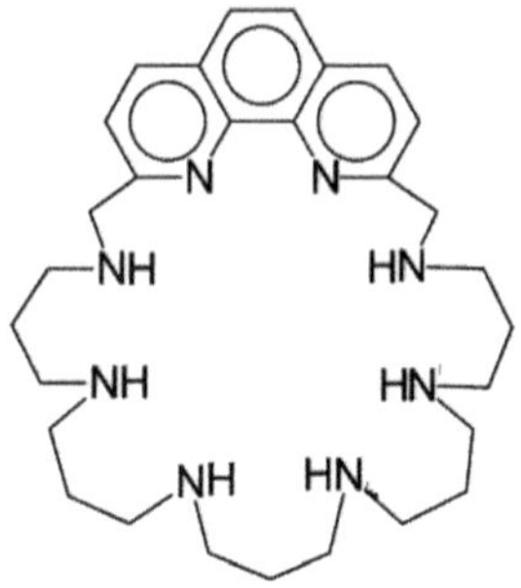

26

Heteroaromatic units such as phenanthroline within a host allows fluorimetric detection, as illustrated with the receptor **26**. Here the binding, for example, at pH around 6.5, amounts to $\log K = 9$ for ATP, $\log K = 8$ for CTP, $\log K = 7.8$ for TTP, and $\log K = 7$ for GTP.[53] Measurements by ^{1}H and ^{31}P-NMR and molecular mechanics calculations indicate that the nucleotide phosphate chains give strong electrostatic and hydrogen-bonding interactions with the ammonium groups of the protonated receptors, while the nucleobases interact either *via* π-stacking with phenanthroline or *via* hydrogen bonding with the ammonium groups, with only moderate selectivity.

The anthracene-based host **27** exhibits GTP chelation-enhanced fluorescence quenching; in contrast, the chelation-enhanced fluorescence effect is seen with ATP, ADP, and AMP. The binding constants for GTP, ATP, ADP, and AMP are $87\,000$, $15\,000$, 610, and $120\,\text{M}^{-1}$, respectively.[54] From a combinatorial benzimidazolium dye library, the receptor **28** was found to be an optimal turn-on fluorescent GTP sensor, with an 80-fold fluorescence increase; in contrast, one observes only a two-fold change with ATP, indicating that the 2-hydroxyl group of GTP is decisive.[55]

The receptor **29** has the advantage of exhibiting fluorescence of the indicator 8-hydroxypyrene-1,3,6-trisulfonic acid, which is quenched as long it stays within the host.[56] Upon binding nucleotides, the fluorescence emission is restored. With GTP a 150-fold emission increase is observed, whereas ATP leads

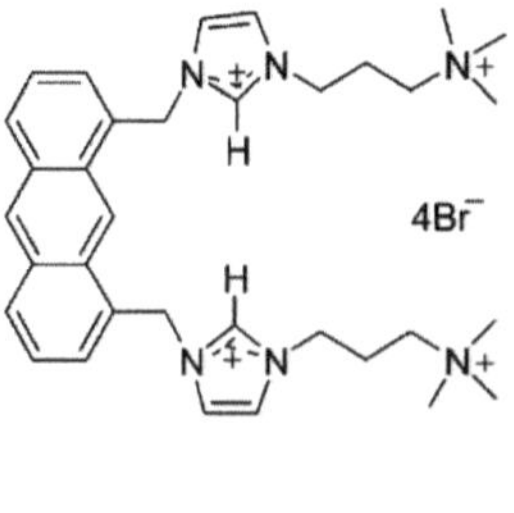

27

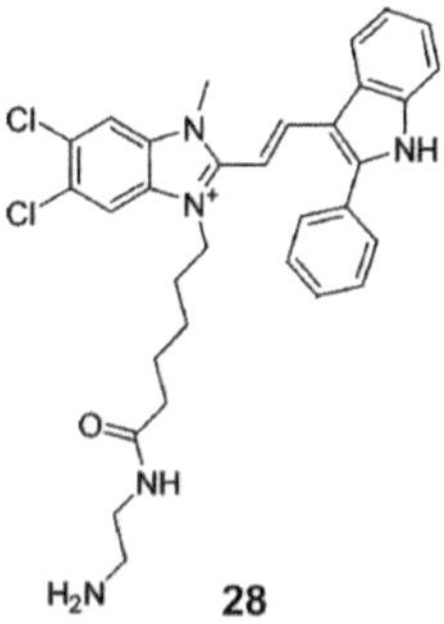

28

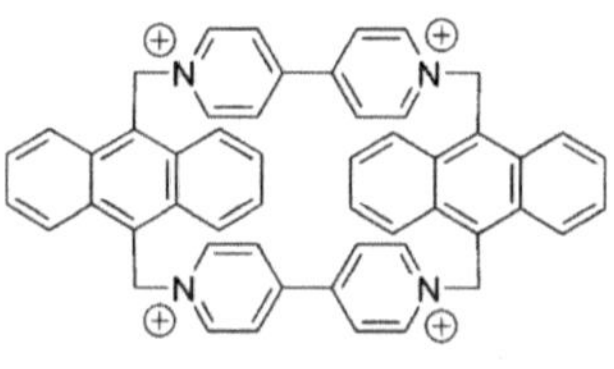

29

to only a 45-fold increase; adenosine, AMP, ADP, CTP and UTP do not cause emission changes.

The host **30** uses xanthene as the fluorescent sensing unit and dizinc-picolylamine as the binding unit for nucleoside polyphosphates; no fluorescence change is induced by monophosphates.[57] This complex allows visualization of ATP in living cells by fluorescence. The combination of the dicopper host **31** with the indicator 6-carboxytetramethylrhodamine allows identifification of ATP among, for example, common neurotransmitters by visual inspection of the sample upon illumination from a UV lamp.[58]

The dizinc-host **32** has higher affinity for thymidine nucleotides than for other nucleobases,[59] likely due to the interaction between the dibasic phosphate and the second Zn^{2+} center, with binding constants of around 10^6 M^{-1}. Zinc complexes such as **32** even hold some promise to interfere with the interaction of viral RNA with proteins; this plays a role in transcription of the HIV-1 genome, which is facilitated by an HIV-1 regulatory protein. Footprinting analysis revealed the UUU bulge in RNA to be strongly protected by **32** in a TAR model sequence, the *trans*-activation responsive RNA element. With a related tris-Zn-cyclen derivative the K_d value for the TAR_{33}-Tat complexation is as low as 15 nM, which may offer a new strategy to fight AIDS.

Ion-sensitive electrodes (ISEs) with host **33** containing a base-pairing cytosine residue give at 10^{-4} M concentration a potentiometric response that is selective for Watson–Crick base-pairing guanosine nucleotides (5'- and 2'-GMP); in contrast, the response to AMP is negligible.[60] Ditopic recognition sites on ISEs with complementary base-pairing and electrostatic interaction hold promise to discriminate among different nucleotides.[61] Receptor **34** binds AMP, but not the cyclic 3',5'-AMP, which can be used for a visual real-time monitoring assay of cyclic nucleotide phosphodiesterase activity.[62]

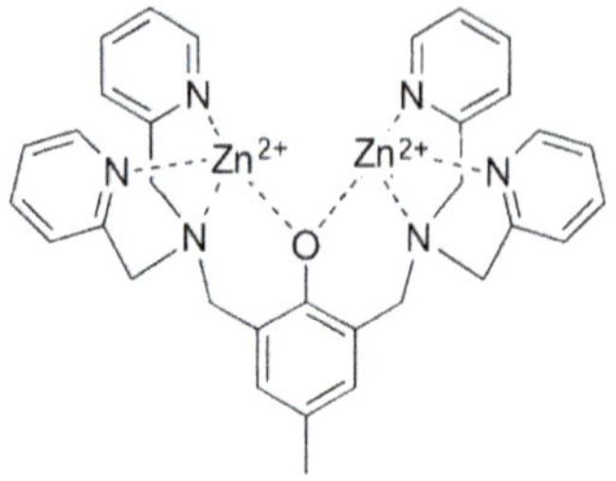

34

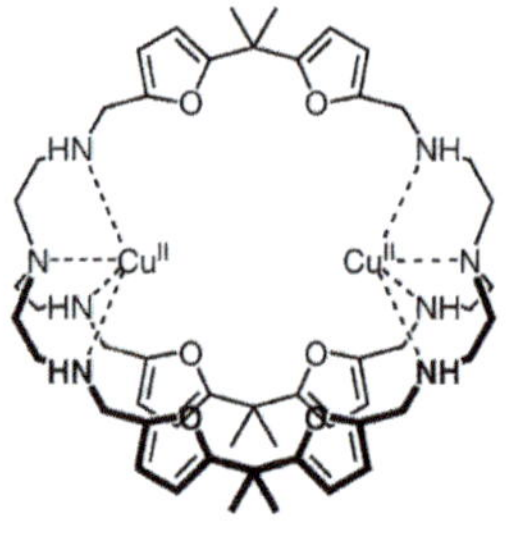

35 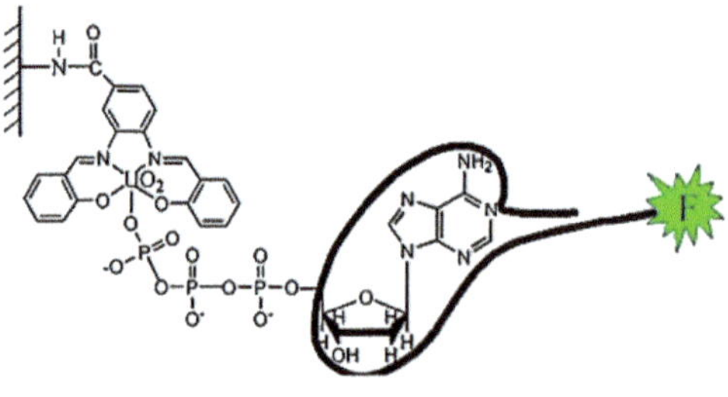   **36**

The dicopper cryptate **35** binds GMP with $\log K = 4.7$, which is visible by displacement of 6-carboxyfluorescein as indicator, whereas AMP, CMP, UMP, and TMP show $\log K$ values between 3.7 and 4.2.[63] The observed differences are ascribed to the nature of nucleobase donor atoms involved in the coordination and their geometric disposition with respect to the Cu–Cu distance. Watson–Crick base-pairing can be used for getting a high base selectivity with an aptamer, which is a hydrogen-bonding single-stranded oligonucleotide and can recognize specifically, for instance, the adenosine group in ATP . The anti-adenosine aptamer is labeled with a fluorescent group and used as a labeled receptor of ATP, which is bound first with a uranyl–salophen complex that is immobilized on the surface of amino-silica gel particles.[64] The detection limit for ATP is 0.037 nmol mL^{-1}, and GTP, CTP, and UTP do not give a fluorescence signal.

The interaction of calixarenes, bearing amino groups, with nucleic acids[65] is discussed in Chapter 8. Calixarenes provide a scaffold for preorganization of polar or lipophilic substituents at their rim, and can form different aggregates with DNA.[66]

Supramolecular metal complexes containing, for instance, several cerium(IV) or cobalt (III) ions can cleave selectively nucleic acids, with rate enhancements up to 10^{11} times; they are discussed in detail in Chapter 8, including their use as artificial restriction enzymes. Chemosensors for pyrophosphates—the product of ATP hydrolysis under cellular conditions and involved in DNA replication—have been reviewed.[67] Nucleotide recognition is also the basis of monitoring nucleotide triphosphate-dependent enzymes with tandem assays, as described in detail in Chapter 12 by Hennig. One can, for instance, use two complementary reporter pairs, such as amino-β-cyclodextrin with 2-anilino-naphtalene-6-sulfonate (ANS) as dye, and a polycationic cyclophane with 8-hydroxy-1,3,6-pyrene with trisulfonate (HPTS) dye to measure the dephosphorylation kinetics of ATP, GTP, TTP, and CTP with different enzymes.[68]

4.6 Ureas, Biotin and Barbiturates

Urea and corresponding derivatives such as barbiturates are not easily accessible to detection by supramolecular complexes, due to their low solubility in common apolar solvents which must be used, as these analytes offer only hydrogen-bonding as possible interactions. Urea itself occurs in urine, blood, bile, milk, and perspiration of mammals; its monitoring plays a diagnostic role in kidney and thyroid diseases, for example, and is usually performed by enzymatic assays. Of the many hosts reported for such analytes,[69] we mention mostly those that allow fluorescence detection.

The pyrene derivative **37** allows binding of urea in chloroform with $K > 10^5$ M^{-1}, leading to fluorescence quenching.[70] The phenanthroline-host **38**

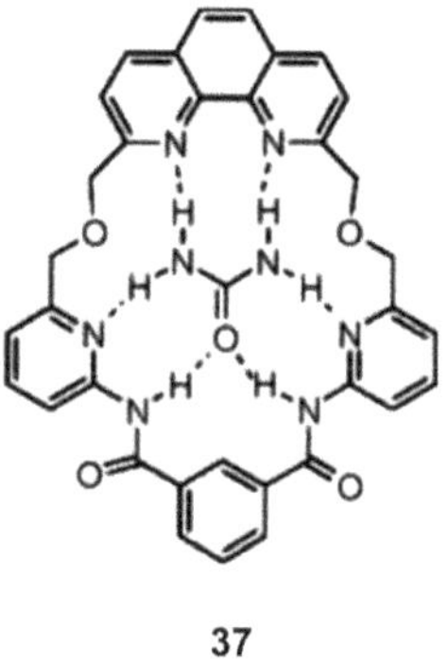

37

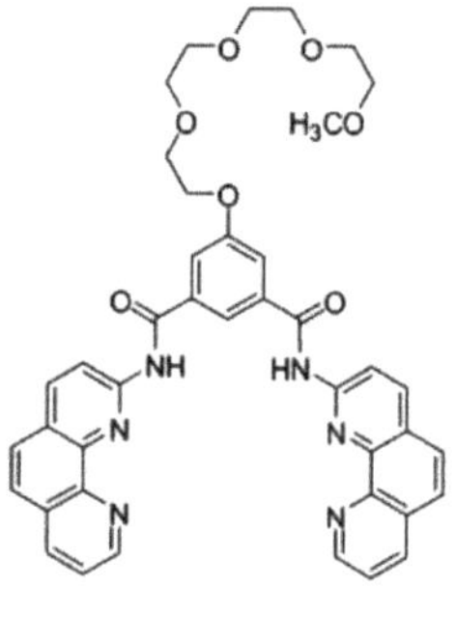

38

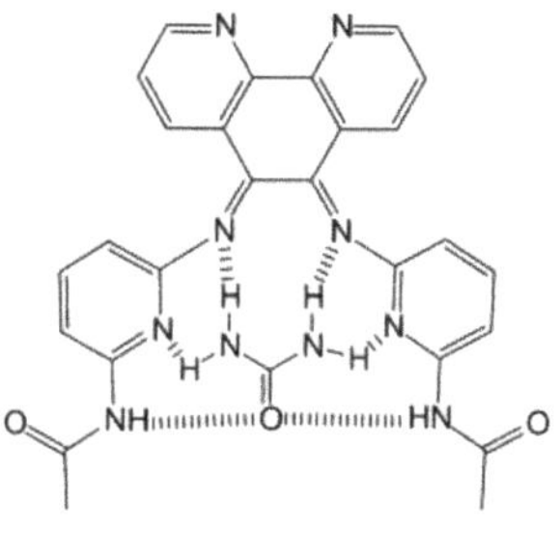

39

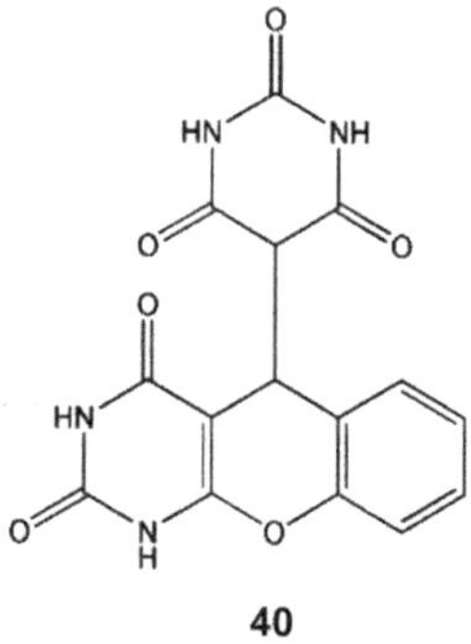

40

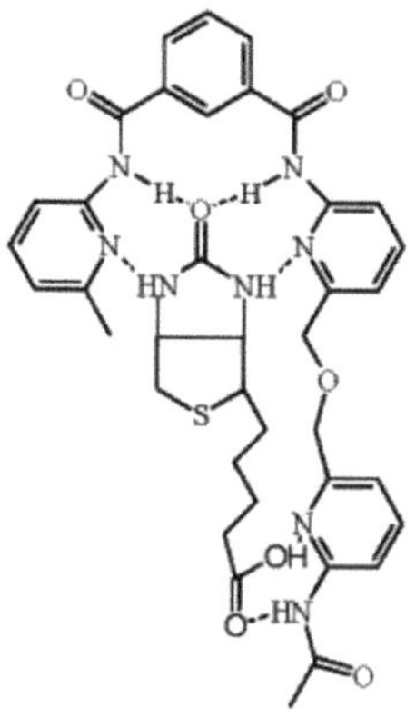

41

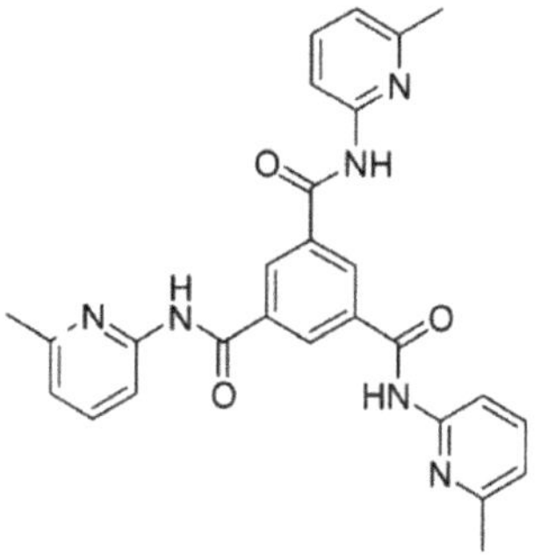

42

can be used for ratiometric sensing of ureas by fluorescence, with $K = 3 \times 10^4$ M^{-1} in CH$_3$CN.[71] Receptor **39** binds urea with $K = 4 \times 10^3$ M^{-1} in CH$_3$CN.[72] The barbiturate derivative **40** is a superior receptor, which binds urea selectively in the micromolar concentration range,[73] measurable by either absorption or emission spectroscopy, with $K = 3 \times 10^6$ M^{-1}.

The vitamin biotin plays a role in gluconeogenesis and fatty acid biosynthetic carboxylation processes; it can be detected with UV/vis spectroscopy with receptor **41**,[74] showing a binding constant of $K = 4.5 \times 10^4$ M^{-1} in (deuterated) chloroform. The hypnotic barbital can be bound with host **42** in CDCl$_3$ with $K = 6 \times 10^3$ M^{-1}; related host compounds bind also urea and biotin derivatives in chloroform with relatively high affinities.[75]

4.7 Carbohydrates

Recognition of carbohydrates is of obvious biomedical interest, but still difficult to achieve with supramolecular complexes in water, due to the weak hydrogen-bonding of OH groups which compete with bulk water.[76] One way to obtain artificial lectins is to provide additional hydrophobic forces, which is the basis of several synthetic receptors for sugars. The resorcarene derivative **43** can extract carbohydrates into chloroform.[77] In chloroform, even open-chain host compounds such as **44** bind, for example *n*-octyl-β-D-glucopyranoside with the high constant of $K = 1.4 \times 10^5$ M^{-1}.[78] Significant progress has been made with the recently reported receptor **45**, which binds glucose efficiently in water, even with remarkable selectivity *versus* other common monosaccharides, with binding constants of $K = 56$ M^{-1} for glucose, 9 for xylose, and 1 or less for galactose, mannose, or fructose.[79] Another great advantage is the built-in signaling system with the anthracene rings, giving rise to a strong and analyte-dependent fluorescence emission.

Alternatively one can use the relatively strong affinity of some metal ions towards saccharides, although usually under rather basic conditions. With the dinuclear copper complex **46** the apparent binding constants with unprotected disaccharides range from 10^3 to 10^4 M^{-1}.[80] The lanthanide complex **47** binds saccharides in the millimolar concentration range and has the advantage of using fluorescence for detection.[81]

46

47

Ionic carbohydrates such as sugar acids can be complexed more easily using ion pairing, as with the cationic receptor **48**. The binding constants are pH-dependent and, for example, amount to 6×10^3 mol^{-1} with galacturonic acid or to 25×10^3 mol^{-1} with glucose-1-phosphate.[82]

48

More sensitive detection of neutral and unprotected carbohydrates in aqueous media is possible with the long-known[83] reaction with boronic acids. Although the ester formation is fast and reversible, these are not really supramolecular systems; we therefore discuss only a few typical applications, and refer to extensive reviews on such so-called boronolectins,[84] also with respect to their diagnostic potential in medicine[85] and their possible miniaturization.[86] Receptor **49** shows how fluorimetric detection of glucose at physiological levels is achieved,[87] as the reaction with glucose places the sugar unit above the anthracene face in the resulting macrocycle **50**.

49

50

With boronic ester residues at the rim of calixarenes one can visually distinguish structurally related saccharides, including glucose phosphates and amino and carboxylic acid sugars, by their different colors;[88] related resorcarene-based boronic acids such as **51** exhibit a high colorimetric fructose selectivity. Fluorimetric glucose monitoring at physiological levels is also possible with such receptors.[89] The host molecule **52** bears on a chiral tetra-peptide scaffold two boronic ester units and can achieve glucose recognition with high affinity and selectivity and an enantioselectivity of $K_D/K_L = 2.1$, with a high apparent binding constant of 2.4×10^4 M^{-1} for D-glucose. Moreover, fluorescence detection is possible by the quenching upon the sugar reaction.[90]

The viologen-based receptor **53** with the aminopyrene **54** can be used to measure glucose-6-phosphate (with $K = 7.5 \times 10^2$ M^{-1}) and fructose (with $K = 5.6 \times 10^2$ M^{-1}); the enzyme phosphoglucomutase, which converts glucose-1-phosphate to glucose-6-phosphate, can be assayed this way. The fluorescence emission is due to the disappearance of the anionic charge upon boronic ester formation;[91] it can be used in multi-well plates suitable for high-throughput screening of enzyme inhibitors, with a detection limit of $50\,\mathrm{ng\,mL^{-1}}$. Fluorogenic affinity gels displaying dose-dependent fluorescence intensity change upon binding fructose and glucose have been reported recently.[92] Another way to use boronic ester formation is to trigger sol–gel transformation in hydrogels containing boronic acids; this has been achieved, for example, for fructose as analyte with modified poly(acrylic acid) and a glucan with a nanotube device.[93]

The reaction of carbohydrates with boronic esters holds promise also for simultaneous monitoring of glucose in blood and automatic delivery of insulin upon change in the glucose level. This can in principle be achieved by implementation of boronic acid-containing polymers in hydrogels, as proposed already in 1992;[94] since then, several related publications and patents have appeared.[95] Figure 4.6 illustrates how the ester formation of glucose, which is known to react in its furanose form, with two boronic acid residues in a polymer gel can lead to crosslinking, with subsequent shrinking of the gel. The efficiency has been demonstrated with tiny tubes, containing a dye for visibility, which has a boronic acid–containing polymer as stopper. In presence of glucose the stopper shrinks, and thus opens the container which releases the interior solution. Such a chemical corkscrew demonstrates how a chemoresponsive material can combine sensor and actuator within a single unit, operating

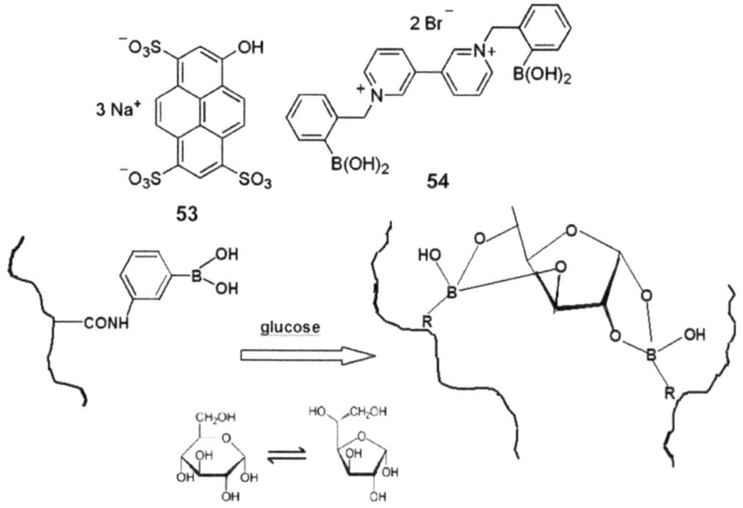

Figure 4.6 Crosslinking and gel contraction by boronic ester formation with glucose.

without external devices such as transducer and power supply. In human blood plasma millimolar concentrations of glucose can lead to contraction within minutes, with negligible effects on other blood components such as fructose;[96] the response time can be varied by the surface-to-volume ratio of such gel particles.[95] The boronic ester reaction can also be applied to biopolymers, such as heparin, which is a biopolymer containing iduronic acid and glucosamine units; it can be measured in serum down to nanomolar concentrations with a fluorescent sensor based on a 1,3,5-triphenylethynylbenzene core with side arms containing boronic acid and ammonium groups.[97]

4.8   Steroids, Bile Acids, Prostaglandins

Terpenoids have been complexed mostly by inclusion in cyclodextrins (CDs), also for the purpose of drug delivery and protection; for these topics we can refer to Chapter 5. If CDs are equipped with fluorophores, they are suitable also for sensing; for instance, a CD with a conjugated helical peptide bearing coumarin and pyrene fluorophores in the oligopeptide side-chains shows intramolecular fluorescence resonance energy transfer (FRET) without quenching of two fluorophores.[98] Addition of hyodeoxy cholic acid, for example, which binds in water with $K = 2 \times 10^5 \, \mathrm{M}^{-1}$ to the CD–peptide, leads to a drastic reduction of the fluorescence emission due to exclusion of the coumarin fluorophore from inside to outside of the CD cavity. L/D-Tryptophan-modified β-cyclodextrins have been employed as fluorescence sensors for bile salts such as deoxycholate.[99]

Terpenoids are notoriously hydrophobic compounds, and their complexation in hydrophilic hosts offers a pharmaceutically interesting way to solubilize them (see Chapter 5). The azoniacyclophane **25** (p. 78) complexes the hormone β-estradiol as anion with a binding constant of 250 M^{-1}; extrapolation to pure water as the medium predicts association constants of up to 10^5 M^{-1}.[100] More effective complexations are achieved with dendritic cyclophanes of the type **55** (R = OH), which bind testosterone and progesterone in 50% aqueous methanol with association constants of about 1000 M^{-1}.[101] The complexation between the double-decker cyclophane **56** and a series of 30 steroids was investigated in CD_3OD by ^{1}H-NMR titrations.[102] The stability constants with different steroids vary in the range 1.3×10^2 to 3×10^3 M^{-1} and seem to increase with the lipophilicity of the analyte. Such lipophilic receptors hold some promise for the extraction of steroids from body liquids into non-polar solvents.

55

56

4.9 Alkaloids

Host–guest complexes with alkaloids as guests are of interest for analytical applications, and for the improvement of their solubility and bioavailability; this has mostly involved native or modified cyclodextrins (see Chapter 5). More recently, cucurbiturils have evolved as superior complexers of alkaloids and drugs, often allowing fluorometric detection, also protection against photo-oxidation, for example; many other examples are presented in Chapter 7 by Nau and Saleh. An extensive compilation of complex formation constants for these and some other hosts with alkaloids can be found in a review of Yatsimirsky.[103] Because of their basic nature, alkaloids are present in neutral aqueous solutions as protonated cationic species, which makes them appropriate guests for cucurbiturils (see Chapter 7 for examples). The alkaloid caffeine can be detected by extraction into non-aqueous media, in which hydrogen-bonding becomes strong enough.[104] The combination of host **57** with an indicator **58**, which competes with the caffeine complexation, leads to a fluorescence signal increase by a factor of up to four.[105] Another way to complex alkaloids is to use stacking and metal coordination, as shown with the complex **59** between a tetracationic Zn(II) porphyrin–peptide conjugate and caffeine, with about 5×10^3 M^{-1} as stability constant.[106] The toxic alkaloid nicotine is bound by π–π and hydrophobic interactions with $K = 3 \times 10^3$ M^{-1} with the host **60**, which was obtained from a dynamic combinatorial library.[107] Toxic pyrrolizidine alkaloids such as retronecine, which occurs in livestock forage plants, can be complexed with $K = 1.5 \times 10^3$ M^{-1} with the calix[6]arene hexasulfonate analogue of **19**.[108]

57

58

60

59

4.10 Antibiotics and Toxins

Cyclodextrins are the host compounds most often used for antibiotics, notably for drug delivery, but they are also quite efficient as antitoxins. Many examples of this application for CDs can be found in Chapter 5; the complexations with cucurbiturils, which are often particularly efficient, are aptly discussed in Chapter 7. We shall mention here only one supramolecular complex **61** which allows the direct detection of shellfish toxins such as saxitoxin in fish extracts by fluorescence emission in the visible range; the system has a detection limit of $40\,\mu M$ and can replace the conventional toxicity tests on mice that were used until now.[109] The test relies on hydrogen-bonding with a crown ether and on boronic ester formation with the BF_2 unit.

Saxitoxin **61**

Acknowledgement

Valuable suggestions from Professor Stefan Kubik, Kaiserslautern are gratefully acknowledged.

References

1. For reviews see E. V. Anslyn, *J. Org. Chem.*, 2007, **72**, 687; L. A. Joyce, S. H. Shabbir, and E. V. Anslyn, *Chem. Soc. Rev.*, 2010, **39**, 3621; A. Späth and B. König, *Beilstein J. Org. Chem.* 2010, **6**, 1.
2. (a) O. E. Beske and S. Goldbard, *Drug Discovery Today*, 2002, **7**, S131; (b) G. Henderson and M. Bradley, *Curr. Opin. Biotechnol.*, 2007, **18**, 326.
3. S. Banerjee and B. König, *J. Am. Chem. Soc.*, 2013, **135**, 2967.
4. G. V. Oshovsky, D. N. Reinhoudt and W. Verboom, *Angew. Chem. Int. Ed. Engl.*, 2007, **46**, 2366.
5. A. A. Lubin and K. W. Plaxco, *Acc. Chem. Res.*, 2010, **43**, 496.
6. (a) R. N. Dsouza, U. Pischel and W. M. Nau, *Chem. Rev.*, 2011, **111**, 7941; (b) L. Basabe-Desmonts, D. N. Reinhoudt and M. Crego-Calama, *Chem. Soc. Rev.*, 2007, **36**, 993.
7. (a) S. L. Wiskur, H. Ait-Haddou, J. J. Lavigne and E. V. Anslyn, *Acc. Chem. Res.*, 2001, **34**, 963; (b) M. Kitamura, S. H. Shabbir and E. V. Anslyn, *Org. Chem.*, 2009, **74**, 4479; (c) L. Fabbrizzi, M. Licchelli and A. Taglietti, *Dalton Trans.*, 2003, 3471; (d) L. Fabbrizzi and A. Poggi, *Chem. Soc. Rev.*, 1995, **24**, 197; for the first use see (e) F. Cramer, W. Saenger and H.-C. Spatz, *J. Am. Chem. Soc.*, 1967, **89**, 14.
8. H.-J. Schneider and A. Yatsimirsky, *Chem. Soc. Rev.*, 2008, 263.
9. (a) A. T. Wright and E. V. Anslyn, *Chem. Soc. Rev.*, 2006, **35**, 14; (b) A. P. Umali and E. V. Anslyn, *Curr. Opin. Chem. Biol.*, 2011, **14**, 685; (c) P. Anzenbacher, P. Lubal, P. Bucek, M. A. Palacios and M. E. Kozelkova, *Chem. Soc. Rev.*, 2010, **39**, 3954; (d) J. J. Lavigne, P. Ciosek and W. Wroblewski, *Analyst*, 2007, **132**, 963.

10. L. Vial, R. F. Ludlow, J. Leclaire, R. Perez-Fernandez and S. Otto, *J. Am. Chem. Soc.*, 2006, **128**, 10 253.
11. D. Tanima, Y. Imamura, T. Kawabata and K. Tsubak, *Org. Biomol. Chem.*, 2009, **7**, 4689.
12. A. Fragoso, E. Almirall, L. Cao, Echegoyen and R. Gonzalez-Jonte, *Chem. Commun.*, 2004, 2230.
13. A. Abbaspour and A. Noori, *Biosensors Bioelectronics*, 2011, **26**, 4674.
14. F. M. Raymo and M. A. Cejas, *Org. Lett.*, 2002, **4**, 3183.
15. L. Lamarque, P. Navarro, C. Miranda, V. J. Aran, C. Ochoa, F. Escarti, E. Garcia-Espana, J. Latorre, S. V. Luis and J. F. Miravet, *J. Am. Chem. Soc.*, 2001, **123**, 10 560.
16. O. Molt, D. Rubeling and T. Schrader, *J. Am. Chem. Soc.*, 2003, **125**, 12 086.
17. O. Molt, D. Rubeling, G. Schofer and T. Schrader, *Chem. Eur. J.*, 2004, **10**, 4225.
18. S. Kolusheva, O. Molt, M. Herm, T. Schrader and R. Jelinek, *J. Am. Chem. Soc.*, 2005, **127**, 10 000.
19. M. E. Bush, N. D. Bouley and A. R. Urbach., *J. Am. Chem. Soc.*, 2005, **127**, 14 511.
20. V. Sindelar, M. A. Cejas, F. M Raymo, W. Z. Chen, S. E. Parker and A. E. Kaifer, *Chem. Eur. J.*, 2005, **11**, 7054.
21. J. C. Ma and D. A. Dougherty, *Chem. Rev.*, 1997, **97**, 1303.
22. H.-J. Schneider, D. Güttes and U. Schneider, *Angew. Chem., Int. Ed. Engl.*, 1986, **25**, 647.
23. D. A. Dougherty and D. A. Stauffer, *Science*, 1990, **250**, 1558.
24. (a) J.-M. Lehn, R. Meri, J.-P. Vigneron, M. Cesario, J. Guilhem, C. Pascard, Z. Asfari and J. Vicens, *Supramol. Chem.*, 1995, **5**, 97; (b) S. Shinkai, K. N. Koh, K. Araki, A. Ikeda and H. Otsuka, *J. Am. Chem. Soc.*, 1996, **118**, 755.
25. M. Dhaenens, D. Lacombe, J.-M. Lehn and J.-P. Vigneron, *J. Chem. Soc., Chem. Commun.*, 1984, 1097.
26. M. Inouye, K. Hashimoto and K. Isagawa, *J. Am. Chem. Soc.*, 1994, **116**, 5517.
27. H. Backirci and W. M. Nau, *Adv. Funct. Mater.*, 2006, **16**, 237.
28. N. Korbakov, P. Timmerman, N. Lidich, B. Urbach, A. Sa'ar and S. Yitzchaik, *Langmuir*, 2008, **24**, 2580.
29. P. Ballester, A. Shivanyuk, A. R. Far and J. Rebek, Jr., *J. Am. Chem. Soc.*, 2002, **124**, 14 014.
30. F. Hof, L. Trembleau, E. C. Ullrich and J. Rebek, Jr., *Angew. Chem. Int. Ed.*, 2003, **42**, 3150.
31. R. T. S. Lam, A. Belenguer, S. L. Roberts, C. Naumann, T. Jarrosson, S. Otto and J. K. M. Sanders, *Science*, 2005, **308**, 667.
32. J. Z. Zhao, H. J. Kim, J. Oh, S. Y. Kim, J. W. Lee, S. Sakamoto, K. Yamaguchi and K. Kim, *Angew. Chem. Int. Ed. Engl.*, 2001, **40**, 4233.
33. A. L. Koner, J. Schatz, W. M. Nau and U. Pischel U., *J. Org. Chem.*, 2007, **72**, 3889.

34. C. Schmuck and M. Schwegmann, *Org. Bio.. Chem.*, 2006, **4**, 836.
35. L. A. Baumes, M. B. Sogo, P. Montes-Navajas, A. Corma and H. Garcia, *Chem. Eur. J.*, 2010, **16**, 4489.
36. M. Bonizzoni, L. Fabbrizzi, G. Piovani and A. Taglietti, *Tetrahedron*, 2004, **60**, 1.
37. J. F. Folmer-Andersen, V. M. Lynch and E. V. Anslyn, *J. Am. Chem. Soc.*, 2005, **127**, 7986.
38. D. Moreno, J. V. Cuevas, G. Garcia-Herbosa and T. Torroba, *Chem. Commun.*, 2011, **47**, 3183.
39. (a) N. Douteau-Guével, F. Perret, A. W. Coleman, J.-P. Morel and N. Morel-Desrosiers, *Perkin Trans.*, 2002, **2**, 524; (b) G. Arena, A. Casnati, A. Contino, A. Magri, F. Sansone, D. Sciotto and R. Ungaro, *Org. Biomol. Chem.*, 2006, **4**, 243.
40. S. M. Ngola, P. C. Kearney, S. Mecozzi, K. Russell and D. A. Dougherty, *J. Am. Chem. Soc.*, 1999, **121**, 1192.
41. M. Fokkens, T. Schrader and F. G. Klärner, *J. Am. Chem. Soc.*, 2005, **127**, 14415.
42. C. Schmuck, *Coord. Chem. Rev.*, 2006, **250**, 3053.
43. A. Md. Hossain and H.-J. Schneider, *J. Am. Chem. Soc.*, 1998, **120**, 11 208.
44. M. V. Rekharsky, H. Yamamura, Y. H. Ko, N. Selvapalam, K. Kim and Y. Inoue, *Chem. Commun.*, 2008, 2236.
45. A. Späth and B. König, *Tetrahedron*, 2010, **66**, 6019.
46. S. Bush, D. H. J. Lopes, Z. M. Du, E. S. Pang, A. Shanmugam, A. Lomakin, P. Talbiersky, A. Tennstaedt, K. McDaniel, R. Bakshi, P. Y. Kuo, M. Ehrmann, G. B. Benedek, J. A. Loo, F.-G. Klärner, T. Schrader, C. Y. Wang and G. Bitan, *J. Am. Chem. Soc.*, 2011, **133**, 16 958.
47. D. Bier, R. Rose, K. Bravo-Rodriguez, M. Bartel, J. M. Ramirez-Anguita, S. Dutt, C Wilch, F.-G. Klärner, E. Sanchez-Garcia, T. Schrader and C. Ottmann, *Nat. Chem.*, 2013, **5**, 234.
48. F.-G. Klärner and B. Kahlert, *Acc. Chem. Res.*, 2003, **36**, 919.
49. M. Kirsch, P. Talbiersky, J. Polkowska, F. Bastkowski, T. Schaller, H. de Groot, F.-G. Klärner and T. Schrader, *Angew. Chem. Int. Ed.*, 2009, **48**, 2886.
50. A. E. Hargrove, S. Nieto, T. Zhang, J. L. Sessler and E. V. Anslyn, *Chem. Rev.*, 2011, **111**, 6603.
51. H.-J. Schneider, T. Blatter, B. Palm, U. Pfingstag, V. Rüdiger and I. Theis, *J. Am. Chem. Soc.*, 1992, **114**, 7704.
52. H.-J. Schneider, *Angew. Chem. Int. Ed. Engl.*, 2009, **48**, 3924.
53. C. Bazzicalupi, A. Bencini, S. Biagini, E. Faggi, S. Meini, C. Giorgi, A. Spepi and B. Valtancoli, *J. Org. Chem.*, 2009, **74**, 7349.
54. J. Y. Kwon, N. J. Singh, H. N. Kim, S. K. Kim, K. S. Kim and J. Y. Yoon, *J. Am. Chem. Soc.*, 2004, **126**, 8892.
55. S. L. Wang and Y. T. Chang, *J. Am. Chem. Soc.*, 2006, **128**, 10 380.
56. P. P. Neelakandan, M. Hariharan and D. Ramaiah, *J. Am. Chem. Soc.*, 2006, **128**, 11 334.

57. A. Ojida, I. Takashima, T. Kohira, H. Nonaka and I. Hamachi, *J. Am. Chem. Soc.*, 2008, **130**, 12 095.
58. N. Marcotte and S. Taglietti, *Supramol. Chem.*, 2003, **15**, 617.
59. S. Aokiand and E. Kimura, *J. Am. Chem. Soc.*, 2000, **122**, 4542.
60. K. Tohda, M. Tange, K. Odashima, Y. Umezawa, H. Furuta and J. L. Sessler, *Anal. Chem.*, 1992, **64**, 960.
61. K. Tohda, R. Naganawa, M. L. Xiao, M. Tange, K. Umezawa, K. Odashima, Y. Umezawa, H. Furuta and J. L. Sessler, *Actuators, B*, 1993, **14**, 669.
62. M. S. Han and D. H. Kim, *Bioorg. Med. Chem. Lett.*, 2003, **13**, 1079.
63. V. Amendola, G. Bergamaschi, A. Buttafava, L. Fabbrizzi and E. Monzani, *J. Am. Chem. Soc.*, 2010, **132**, 147.
64. M. M. Zhao, L. F. Liao, M. L. Wu, Y. W. Lin, X. L. Xiao and C. M. Nie, *Biosensors Bioelectronics*, 2012, **34**, 106.
65. (a) L. Baldini, A. Casnati, F. Sansone and R. Ungaro, *Chem. Soc. Rev.*, 2007, **36**, 254; (b) F. Sansone, L. Baldini, A. Casnati and R. Ungaro, *New J. Chem.*, 2010, **34**, 2715.
66. M. S. Peters, M. Li and T. Schrader, *Nat. Prod. Commun.*, 2012, **7**, 409.
67. S. K. Kim, D. H. Lee, J.-I. Hong and J. Yoon, *Acc. Chem. Res.*, 2009, **42**, 23.
68. M. Florea and W. M. Nau, *Org. Bioorg. Chem.*, 2010, **8**, 1033.
69. For leading references seeD. Santa María, M. Angeles Farran, M. Angeles García, E. Pinilla, M. Rosario Torres, J. Elguero and R. M. Claramunt, *J. Org. Chem.*, 2011, **76**, 6780.
70. S. Goswami, R. Mukherjee and J. Ray, *Org. Lett.*, 2005, **7**, 1283.
71. Y. Engel, A. Dahan, E. Rozenshine-Kemelmakher and M. Gozin, *J. Org. Chem.*, 2007, **72**, 2318.
72. A. K. Mahapatra, P. Sahoo, G. Hazra, S. Goswami and H. K. Fun, *J. Luminesc.*, 2010, **130**, 1475.
73. N. Dixit, P. K. Shukla, P. C. Mishra, L. Mishra and H. W. Roesky, *J. Phys. Chem. A*, 2010, **114**, 97.
74. S. Goswami and S. Dey, *J. Org. Chem.*, 2006, **71**, 7280; for detection of biotin carboxylate see K. Ghosh, A. R. Sarkar and T. Sen, Supramol. Chem., 2010, **22**, 81.
75. F. Herranz, M. D. Santa Maria and R. M. Claramunt, *J. Org. Chem.*, 2006, **71**, 2944.
76. (a) A. P. Davis, *Org. Biomol. Chem.*, 2009, **7**, 3629; (b) S. Kubik, *Nat. Chem.*, 2012, **4**, 697; (c) M. Mazik, *Chem. Soc. Rev.*, 2009, **38**, 935; (d) S. Kubik, *Angew. Chem. Int. Ed.*, 2009, **48**, 1722.
77. R. Yanagihara and Y. Aoyama, *Tet. Lett.*, 1994, **35**, 9725.
78. M. Mazik and M. Kuschel, *Chem. Eur. J.*, 2008, **14**, 2405.
79. C. Ke, H. Destecroix, M. P. Crump and A. P. Davis, *Nat. Chem.*, 2012, **4**, 718.
80. S. Striegler, *Curr. Org. Chem.*, 2007, **11**, 1543.

81. O. Alpturk, O. Rusin, S. O. Fakayode, W. H. Wang, J. O. Escobedo, I. M. Warner, W. E. Crowe, V. Kral, J. M. Pruet and R. M. Strongin, *Proc.Natl. Acad. Sci. USA*, 2006, **103**, 9756.

82. C. Schmuck and M. Schwegmann, *Org. Lett.*, 2005, **7**, 3517.

83. J. Boeseken, *Adv. Carbohydrate Chem.*, 1949, **4**, 189.

84. (a) T. D. James and S. Shinkai, *Top. Curr. Chem.*, 2002, **160**, 218; (b) T. D. James, M. D. Phillips and S Shinkai, *Boronic Acids in Saccharide Recognition*, RSC: Cambridge, 2006.

85. J. Yan, H. Fangand and B. H. Wang, *Med. Res. Rev.*, 2005, **25**, 490.

86. H. S. Mader and O. S. Wolfbeis, *Microchim. Acta*, 2008, **162**, 1.

87. T. D. James, P. Linnane and S. Shinkai, *Chem. Commun.*, 1996, 281.

88. C. J. Davis, P. T. Lewis, M. E. McCarroll, M. W. Read, R. Cuetoand and R. M. Strongin, *Org. Lett.*, 1999, **1**, 331.

89. O. Rusin, O. Alpturk, M. He, J. O. Escobedo, S. Jiang, F. Dawan, K. Lian, M. E. McCarroll, I. M. Warner and R. M. Strongin, *J. Fluorescence*, 2004, **14**, 611.

90. G. Heinrichs, M. Schellenträger,and and S. Kubik, *Eur. J. Org. Chem.*, 2006, 4177.

91. B. Vilozny, A. Schiller, R. A. Wessling and B. Singaram, *Anal. Chim. Acta*, 2009, **649**, 246; (b) B. Vilozny, A. Schiller, R. A. Wessling and B. Singaram, *J. Mater. Chem.*, 2011, **21**, 7589.

92. K. M. A. Uddin and L. Ye, *J. Appl.. Polym. Sci.*, 2013, 1527.

93. S. Tamesue, M. Numata and S. Shinkai, *Chem. Lett.*, 2011, **40**, 1303.

94. Y. Kitano, K. Koyama, O. Kataoka, T. Kazunori, Y. Okano and Y. Sakurai, *J. Control. Release*, 1992, **19**, 161.

95. H.-J. Schneider, K. Kato and R. M. Strongin, *Sensors*, 2007, **7**, 1578.

96. G. K. Samoei, W. H. Wang, J. O. Escobedo, X. Y. Xu, H.-J. Schneider, R. L. Cook and R. M. Strongin, *Angew. Chem. Int. Ed. Engl.*, 2006, **45**, 5319.

97. A. T. Wright, Z. L. Zhong and E. V. Anslyn, *Angew. Chem. Int. Ed. Engl.*, 2005, **44**, 5679.

98. M. A. Hossain, H. Mihara and A. Ueno, *Bioorg. Med. Chem. Lett.*, 2003, **13**, 4305.

99. H. Wang, R. Cao, C.-F. Ke, Y. Liu, T. Wada and Y. Inoue, *J. Org. Chem.*, 2005, **70**, 8703.

100. S. Kumar and H.-J. Schneider, *J. Chem. Soc. Perkin Trans*, 1989, **2**, 245.

101. P. Wallimann, P. Seiler and F. Diederich, *Helv. Chem. Acta*, 1996, **79**, 779.

102. A. Fürer, T. Marti, F. Diederich, H. Künzer and M. Brehm, *Helv. Chim. Acta*, 1999, **82**, 1843.

103. A. K. Yatsimirsky, *Nat. Prod. Commun.*, 2012, **7**, 369.

104. (a) S. Goswami, A. K. Mahapatra and R. Mukherjee, *J. Chem. Soc. Perkin Trans.*, 2001, **1**, 2717; (b) C. Siering, H. Kerschbaumer, Martin Nieger and S. R. Waldvogel, *Org. Lett.*, 2006, **8**, 1471; (c) C. Siering, B. Beermann and S. R. Waldvogel, *Supramol. Chem.*, 2006, **18**, 23.

105. C. Siering, H. Kerschbaumer, M. Nieger and S. R. Waldvogel, *Org. Bioorg. Chem.*, 2006, **8**, 1471.
106. R. Fiammengo, M. Crego-Calama, P. Timmerman and D. N. Reinhoudt, *Chem. Eur. J.*, 2003, **9**, 784.
107. S. Hamieh, R. F. Ludlow, O. Perraud, K. R. West, E. Mattia and S. Otto, *Org. Lett.*, 2012, **14**, 5404.
108. D. Leite da Silva, E. do Couto Tavares, L. de Souza Conegero, Â. de Fátima, R. A. Pilli and S. A. Fernandes, *J. Incl. Phen. Macromol. Chem.*, 2011, **69**, 149.
109. R. E. Gawley, H. Mao, M. M. Haque, J. B. Thorne and J. S. Pharr, *J. Org. Chem.*, 2007, **72**, 2187.

Cyclodextrins for Pharmaceutical and Biomedical Applications

C. ORTIZ MELLET,[*a] J. M. GARCÍA FERNÁNDEZ[b] AND J. M. BENITO[*b]

[a] University of Sevilla, Faculty of Chemistry, Department of Organic Chemistry, C/ Prof. García González, Sevilla, E-41012, Spain; [b] Institute for Chemical Research (IIQ), CSIC – University of Sevilla, Avda. Américo Vespucio 49, Sevilla, E-41092, Spain
*Email: mellet@us.es; juanmab@iiq.csic.es

5.1 Cyclodextrins: The Early Times

Having first been identified from bacterial digests of starch by Villiers[1,2] at the end of 19th century and later isolated by Schardinger during the first decades of the past century,[3–6] cyclodextrins (CDs) largely remained as chemical curiosities for some 50 years. Then, around the middle of the 20th century, Freudenberg and co-workers unveiled their macrocyclic structure composed of $\alpha(1\rightarrow4)$-linked glucopyranose units, featuring a basket-shaped topology in which glucose hydroxyls orient to the outer space flanking the upper and lower rims, while methinic protons (H-5 and H-3) point to the inner cavity (Figure 5.1).[7,8]

By the mid-1950s Cramer had published the first report on the molecular inclusion capabilities of CDs.[9] This seminal observation, followed by a series of papers demonstrating the potential of these capabilities in catalysis[10] or racemic resolution,[11] represented a milestone that certainly altered the course of cyclodextrin research history and steered it towards the second half of the 20th

Monographs in Supramolecular Chemistry No. 13
Supramolecular Systems in Biomedical Fields
Edited by Hans-Jörg Schneider

Published by the Royal Society of Chemistry, www.rsc.org

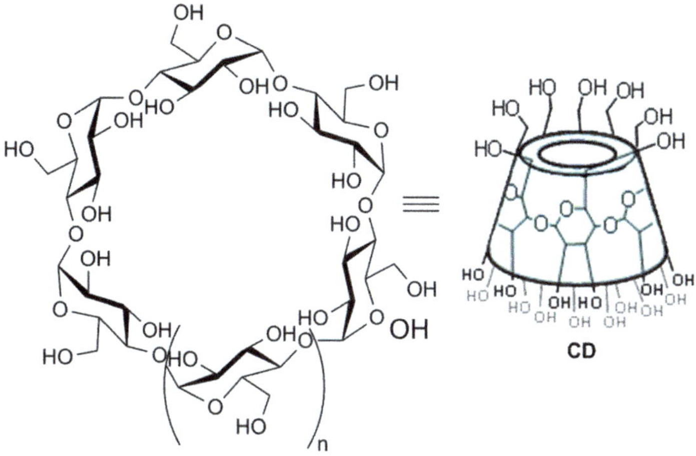

CD	i. d. (Å)	o. d. (Å)	height (Å)
αCD (n = 1)	4.7 - 5.2	14.2 - 15.0	7.9 - 8.0
βCD (n = 2)	6.0 - 6.4	15.0 - 15.8	7.9 - 8.0
γCD (n = 3)	7.5 - 8.3	17.1 - 17.9	7.9 - 8.0

Figure 5.1 General structure and dimensions of native (first generation) cyclodextrins (CDs).

century. Indeed, the academic and (bio)technological interest in cyclodextrins has been historically dominated by the unique ability of these naturally occurring compounds to form inclusion complexes with guest molecules fitting into their hydrophobic cavity.[12–16]

CDs have long occupied a prominent position in most pharmaceutical laboratories as "off-the-shelf" tools to manipulate the pharmacokinetics and dynamics of a broad range of active principles and pharmaceutical ingredients. Nowadays, there is a clear consensus of the *avant garde* role of cyclodextrins in pharmaceutical technology. These more "classical" effects derived from CD–drug supramolecular interactions, and their relevance to the pharmaceutical industry, will be briefly discussed from a historical perspective in the next section. For more comprehensive information, the reader is encouraged to consult some excellent comprehensive accounts that have been published on these topics.[17,18]

This chapter will emphasize the roles that CDs are starting to play in the era of nanotechnology, and recent advances which have made it feasible to address certain shortcomings associated with pharmacological properties (*e.g.* limited solubility or circulating half-life). In this context, CDs might not only be regarded as hosts to modify the properties of an included guest by exploiting their intrinsically organized "inner" space, but also as nanometric platforms, themselves susceptible to selective chemical manipulation forming nano-objects with tailored capabilities towards the outer environment. The amalgamation of chemical, supramolecular and nanotechnological concepts allows an unprecedent and fluent crosstalk between formerly distant research disciplines, from synthetic chemistry to pharmaceutical formulation, or from material science and biophysics to diagnosis, offering alternative solutions to intricate challenges through the cyclodextrin link.[19] A number of selected examples will be described to illustrate how covalent CD modification, self-assembling and supramolecular interactions can be implemented (i) for manipulation of the bioavailability of therapeutic agents, (ii) for the rational design of intelligent systems envisioned for targeted delivery and programmed release of drugs and macromolecules, (iii) for devising new therapeutic alternatives for known maladies, or even (iv) for the engineering of biosensing devices.

5.2 First and Second Generation Cyclodextrins in Pharmaceutical Formulations: Improving Drug Solubility and Bioavailability

The commercial availability of CDs remained very limited for several decades. Then, in the 1970s, key biotechnological advances were made that dramatically improved the production efficiency of CDs.[16] The consequently lowered prices then facilitated investigation of a plethora of technological applications of native (first generation) CDs and a number of chemically modified (second generation) congeners, such as per(di-O-2, O-6)-, per(tri-O-2, O-3, O-6)-, and randomly-methylated CDs (DIMEB, TRIMEB and RAMEB, respectively) or hydroxypropylated derivatives (HPCDs) in food (*e.g.* as thickening and flavor preservative agents),[20,21] cosmetics (*e.g.* as stabilizers or sustained release and anti-irritant agents),[22] environmental applications (*e.g.* adsorbents in waste treatment),[23–26] and the pharmaceutical industry (*e.g.* excipient in drug formulation).[17,18,27,28] Due to more relaxed regulations, Japan witnessed in 1976 the commercialization of the first CD-containing drug: Prostarmon E[TM] (Ono Pharmaceuticals Co.), consisting of a formulation of prostaglandin E2 and βCD.[29] Despite being initially more refractory, European and American agencies also approved CD-containing formulations: Chiesi Farmaceutici's Brexin[®] (piroxicam/βCD tablets) in 1988[30] and Johnson & Johnson's Sporanox[®] (itraconazol/HPβCD complex) in 1997[31] pioneered the field in Europe and USA, respectively. The list of approved CD-containing pharmaceutical formulations has progressively built up, exceeding nowadays 30 different products.[14] Moreover, taking into account the current trends on drug discovery, increasingly selecting poorly soluble candidates (type II and IV compounds, according to Food and Drug Administration's Biopharmaceutical Classification System), this list is expected to enlarge further in the future. The successful irruption of CDs on the market has further stimulated the investigation of other types of macrocyclic hosts that can complement some limitations of CDs, such as the cytotoxicity associated to certain CD family members.[32,33] For example, Chapters 6 and 7 in this volume beautifully illustrate how the supramolecular capabilities of calixarenes and cucurbiturils, which are functional mimics of CDs, have been successfully implemented into exciting biomedical applications.

The roles of CDs in pharmaceutical formulations are several-fold. For this reason, interpretation of the precise effect that CDs exert on drug availability could be controversial at some point.[17,28,34] Cyclodextrins may mitigate the irritant effect of certain drugs or excipients. For instance, HPβCD enables solubilization of tropicamid at physiological pH, avoiding the use of eye-irritating acidic buffers.[35] In another example, the enhanced aqueous solubility of diclofenac in the presence of HPβCD avoids the use of organic co-solvents such as benzyl alcohol and propyleneglycol, known to cause complications such as irritation and thrombophlebitis.[36] Besides, CDs are also reported to enhance drug absorption efficiency, which has benefited a number of eye drop formulations such as Novartis's Voltaren Ophthalmic® (diclofenac/HPβCD solution).[37] Other drugs benefit from enhanced stability, such as amlodipine, a dihydropyridine-type calcium-channel blocker with anti-hypertensive activity, for which photochemical oxidation can be prevented by complexation with CDs.[38]

CDs are also known to alter drug pharmacokinetics, thus exerting a deep impact on drug performance. For instance, βCD complexation of nicotine allows programming for a sustained release and safer doses of the drug, modulation being possible by manipulating the βCD/nicotine ratio.[39] In addition, CDs can also act by perturbing biological membranes through the interaction with third components (bile acid, cholesterol, lipids), which may help to either release the included drug and/or facilitate physiological barrier crossing.[40]

Besides "classical" host–guest interactions with low-molecular-mass species, in the last decades CDs have been shown to participate in various types of supramolecular associations involving biomacromolecular binding.[41] A body of knowledge suggests the utility of native CDs and second-generation derivatives (*e.g.* methylated CDs or HPβCD) to improve the bioavailability of peptides and proteins, thereby preserving their therapeutic potential. Though the molecular basis for such bioavailability improvements are not precisely defined, and probably differ from case to case, CDs are hypothesized to interact with non-polar amino acid side-chains exposed to the solvent, coating peptide and protein surfaces with a hydrophilic shell that shields them from non-specific interactions leading to self-aggregation.[42–44] Other mechanisms, such as non-ionic surfactant-like effects, have also been invoked.[45] CD-stabilized insulin[46] and growth hormone formulations[47–49] were among the first examples evidencing this behavior, but the aggregation–suppression capability of CDs has been further exploited with many other pharmacologically interesting macromolecules, including therapeutically relevant antibodies.[50] The strategy is not only useful in solution formulations, but also in solid state preparations, for example, in lyophilizates and spray-dried solutions.[45]

In addition, CD–protein complex formation has proven useful to recover the correct folding pattern in denaturated proteins—in lysozymes[51,52] or carbonic anhydrases, for example.[53,54] Inclusion of hydrophobic macromolecule residues in CD cavities is at the origin of this molecular chaperone behavior, as demonstrated by the fact that no renaturation was detected in the presence of competing inclusion guests.[55] CD chaperones can be further exploited to assist

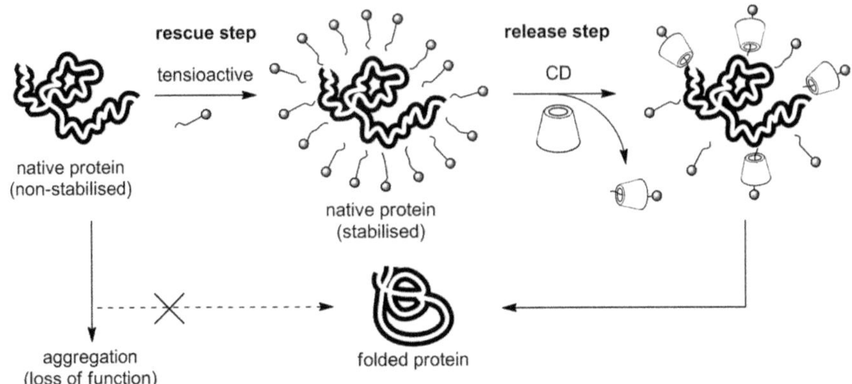

Figure 5.2   Two-step cyclodextrin-assisted protein renaturing methodology described by Gelman.[56]
(Adapted with permission from *Chem. Rev.*, 2007, **107**, 3088. Copyright 2007 American Chemical Society.)

protein (re)folding in biological extracts or biotechnologically produced peptides and proteins, where renaturation is a common limitation to proper pharmacological activity. In a pioneering work, Gelman and co-workers developed a two-step method in which the non-native protein is first rescued from a denaturating environment in the form of a surfactant complex. Then, in the second step, the surfactant is sequestered with a CD derivative to allow proper protein refolding (Figure 5.2).[56,57] Based on this strategy, procedures involving CD-based artificial chaperones for cell-free high-throughput production of proteins have been developed.[58–60]

5.3   Third-generation CDs: Molecular Shuttles for Site-specific Drug Delivery

The potential of CDs to act as molecular carriers was recognized soon after their molecular inclusion properties were discovered.[61] However, implementing molecular designs exhibiting targeting capabilities had to wait a long time. Targeted delivery inherently encompasses milder off-target effects, dose reduction and, therefore, increased efficiency, which is intuitively very attractive from the perspective of aggressive treatments (*e.g.* anticancer chemotherapies). Native CDs and second generation derivatives might be well suited for encapsulation of a variety of drugs, but devoid of recognition capabilities at the level of biological receptors, and the subsequent trafficking of the drug–CD complex would be essentially unspecific. Furnishing CDs with bio-recognizable antennae, then, is a prerequisite for site-directed transport. Obviously, the ligation chemistry bridging the CD carrier and the bio-recognizable antenna may limit the architectural design due the complications associated with macromolecular handling. Implementation of regioselective tools for chemical[62–71] or enzymatic[72] manipulation of the CD topology has largely cleared the way to realizing these preconditions.

In a seminal contribution, a Spanish–French consortium designed a molecularly well-defined multitopic CD host aimed at solubilizing and shuttling the taxane antimitotic drug docetaxel (Taxotere®) specifically to macrophages (Figure 5.3).[73] The drug encapsulating moiety consisted in a βCD dimer assembled by thiourea coupling of 6^I-amino-6^I-deoxy βCD with a diisothiocyanate derivative incorporating a spacer element whose length was adapted to span the distance between the two aromatic rings in the docetaxel molecule, thereby promoting the formation of a chelate-type complex. The linker was further armed with a reactive amino group allowing the incorporation of a hexavalent α-D-mannopyranosyl ligand, exhibiting high affinity towards the

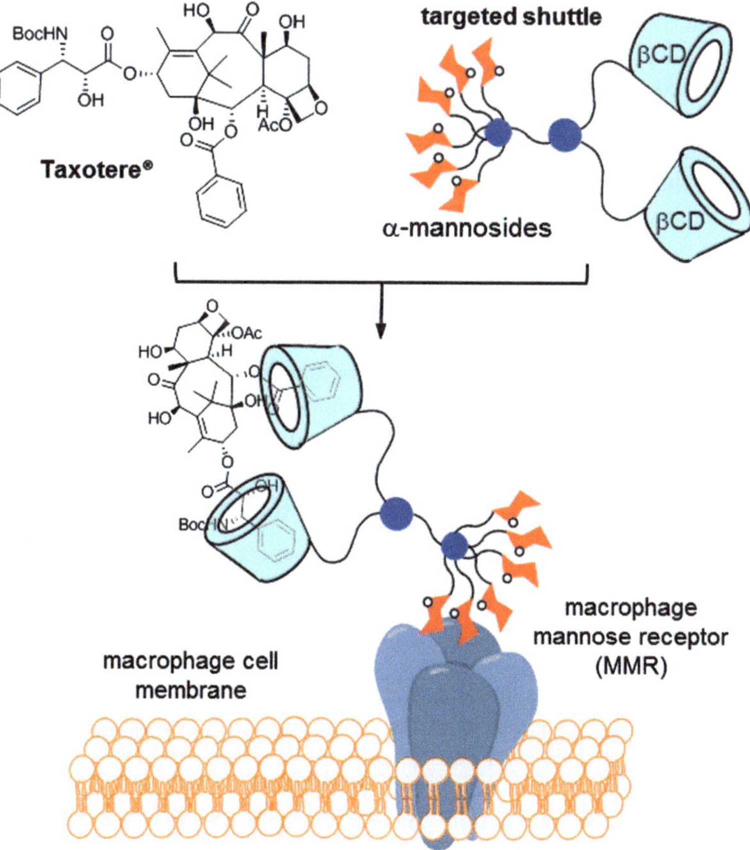

Figure 5.3 Schematic representation of the macrophage mannose receptor (MMR)-targeted Taxotere® drug carrier developed by Ortiz Mellet, Defaye, García Fernández and co-workers.[73]

specific mannose receptor at the surface of macrophages (macrophage mannose receptor, MMR). Cell adhesion experiments confirmed that the CD dimer : drug complex was able to bind to the MMR to form a ternary complex that elicited internalization, supporting the validity of the concept.

Recognition of saccharide antennae by specific cell membrane receptors (lectins) has been further examined by several groups as a general strategy to delivery poorly soluble drugs and probes to different targets.[74] For instance, Hattori and co-workers comparatively assessed a series of hyperbranched galactosylated CDs having different spacer lengths and number of galactose arms as hepatocyte-targeted doxorubicin (DX) shuttles. Long spacers and dense galactoside displays favored specific binding to peanut agglutinin (PNA), used in this work as a model of the asialoglycoprotein receptor (ASGPr) expressed at the hepatocyte cell surface.[75,76] Interestingly, heavy functionalization of the CD primary rim with galactopyranoside glycotopes bearing aromatic linkers, far from disturbing the molecular inclusion capabilities, significantly increased DX encapsulation avidity by enlarging the cavity and providing favorable contacts following an induced-fit process (Figure 5.4A). In a more recent contribution, Seeberger and co-workers demonstrated the suitability of ASGPr-targeted glycodendritic galactosylated CDs (Figure 5.4B) to elicit specific delivery and uptake of DX into liver cells (human hepatocellular carcinoma cells, HepG2).[77] Cell uptake studies demonstrated that galactosylated CDs were preferentially endocytosed as compared to mannose or glucosamine decorated analogues, and efficiently induced apoptosis *in vitro*.

As an alternative to carbohydrate-lectin mediated vectorization, βCD carriers functionalized with folic acid (FA)-βCD have been proposed to specifically deliver antimitotic drugs such as 5-fluorouracil (5-FU) to cancerous tissues, taking advantage of the overexpression of the FA receptor (FAr) in a number of cancer cell lines.[78,79] The authors noted preferential accumulation of 5-FU in human cervix cancer HeLa cells *versus* human lung carcinoma A549 cells, in agreement with the higher levels of FAr expression in the former cell line.

5.4 Cyclodextrin Polymers: Increasing the Drug Loading Capacity

Despite the potential of synthetic chemistry for the design of molecularly well-defined CD derivatives featuring both carrier and targeting capabilities, success is far from evident. Drug loading capacity by monomeric CDs is limited in the best case to one guest molecule for 1 : 1 complexes with very high association constants (K_{as}). This limitation can be overcome by the use of CD-based polymeric platforms.[80,81] As later emphasized in Chapter 16, the presence of a multivalent display of interactions sites (the CD cavities) onto the same polymeric chain generally leads to increase guest avidities for the corresponding $(CD)_n$-polymer:(guest)$_n$ complex. Originally conceived for chromatographic

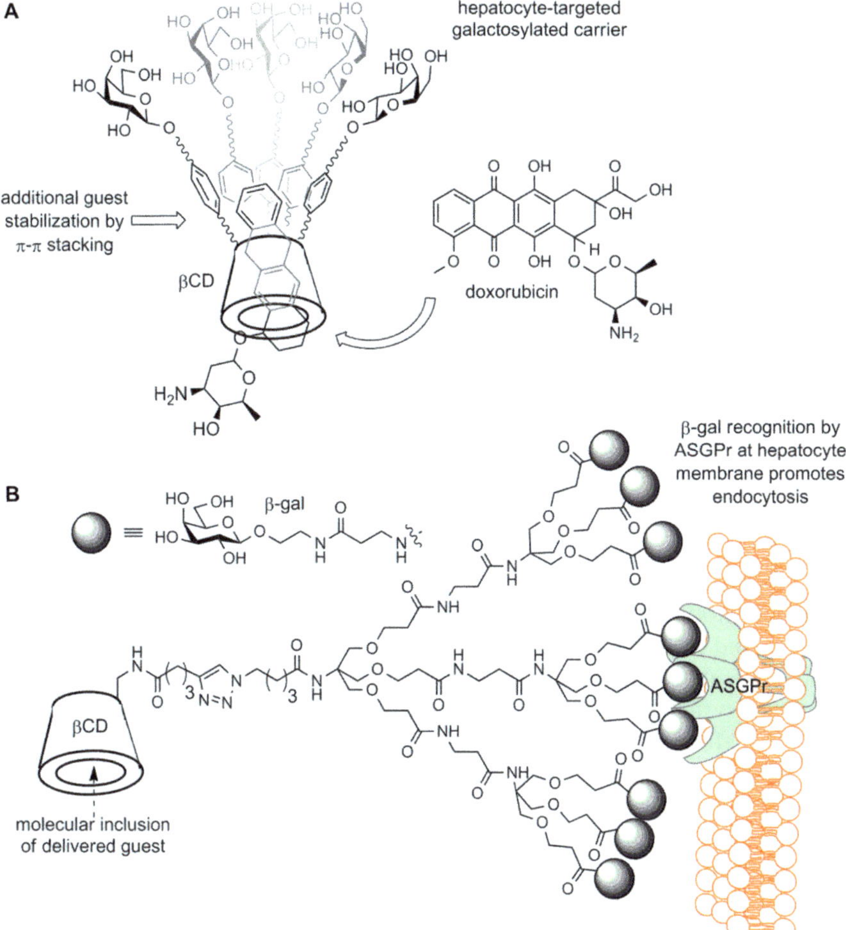

Figure 5.4 Schematic representation of (A) heptagalactosylated β-cyclodextrins (βCDs) reported by Hattori and co-workers,[75,76] illustrating the π–π stacking interactions stabilizing doxorubicin inclusion, and (B) glycodendritic βCD carrier envisioned by Seeberger and co-workers[77] for hepatocyte targeting.

separations,[23–26] CD polymers have been explored in pharmacy for the last 30 years.[82,83] Thus, the properties of pre-existing polymers (*e.g.* chitosan,[84–86] polyglycerol,[87] polyester[88,89] or polyethyleneimine)[90,91] and dendrimers[92,93] have been individually manipulated by covalently grafting CD units onto the parent framework. As an illustrative case, chitosan, a biocompatible naturally-occurring polysaccharide with mucoadhesive properties, has been engineered to design chimeric CD-coated composites envisioned to improve bioavailability of insulin after oral or nasal administration.[94,95] Several groups have demonstrated that the problems associated with the acidic environment or the presence of proteolytic enzymes in the first case and the limited effective absorption in the second route can be largely overcome by entrapping insulin into CD-containing chitosan-based matrices, although the precise mechanisms at work are unclear.[96–98]

For those cases in which the backbone of pre-existing polymers does not fulfill the requirements for a particular application, a great variety of CD-containing materials obtained by polymerization or copolymerization of CD building blocks with one or more co-monomers have also been investigated. The hydrophilic epichlorohydrin (EPH)–crosslinked-βCD polymer described by Mura *et al.*,[99] exhibiting a naproxen loading capacity 30-fold higher than its aqueous solubility, illustrates the potential of these CD-based composites as drug vehicles. In addition to EPH, diisocyanates and more recently acrylates, poly(ethyleneglycols), polyvinyl alcohols or "clickable" derivatives (*e.g.* building blocks suitable for Cu(I)-catalyzed azide-alkyne cycloaddition, CuAAC) have been exploited as crosslinking elements.[80,100,101] Ma and co-workers have recently shown the utility of these CD "click" polymers for encapsulation of both hydrophobic and macromolecular drugs.[102]

CD polymers benefit from an unparalleled ability to furnish nanoscaled systems exhibiting an even larger array of functional characteristics in terms of physical and mechanical properties, stimuli-responsiveness, guest entrapment, tunable circulation times and release features, thereby being very well suited to operate as systemic delivery systems for therapeutic agents. The development of CRLX101 (formerly IT-101), a camptothecin (CPT)-grafted linear βCD–PEG (polyethylene glycol) co-polymer described by Davis and co-workers, is probably one of the cases best illustrating these principles.[103] As many other examples, CPT is a remarkable antimitotic agent whose use has long been restricted in clinics as a consequence of its severe toxicity. A CPT carrier system to ameliorate its off-target effects was, therefore, highly desired. CRLX101 was designed for the purpose of complying with regulatory demands concerning biocompatibility by (i) reacting a bifunctional βCD building block, namely bis-(C-6)-cysteinylated βCD, with an α,ω-carboxylated PEG co-monomer and (ii) covalently attaching CPT *via* amide bonds to the βCD-PEG backbone (Figure 5.5). Optimally performing CPT-loaded polymers are able to self-assemble into virtually non-toxic nanoparticles (average diameter 36 nm) showing long circulation half-lives *in vivo*.[104,105] These features promoted passive tumor accumulation of the nanometric carrier *via* enhanced permeability and retention (EPR)[106] and, at the same time, prevented CPT from degradation (*e.g.* lactone ring opening), thereby achieving drug accumulation at the right spot to enhance its therapeutic activity. Moreover, the authors showed that when CPT is released from the polymer backbone, the nanoparticle disassembles and the polymer is excreted *via* the kidney.[103] Clinical trials focused on the treatment of several types of solid tumors, sponsored by Cerulean Pharma Inc., are ongoing[107,108] and a second candidate from this platform focused at delivering doxorubicin (CRLX301) will enter phase 1 clinical trials by 2013.[109]

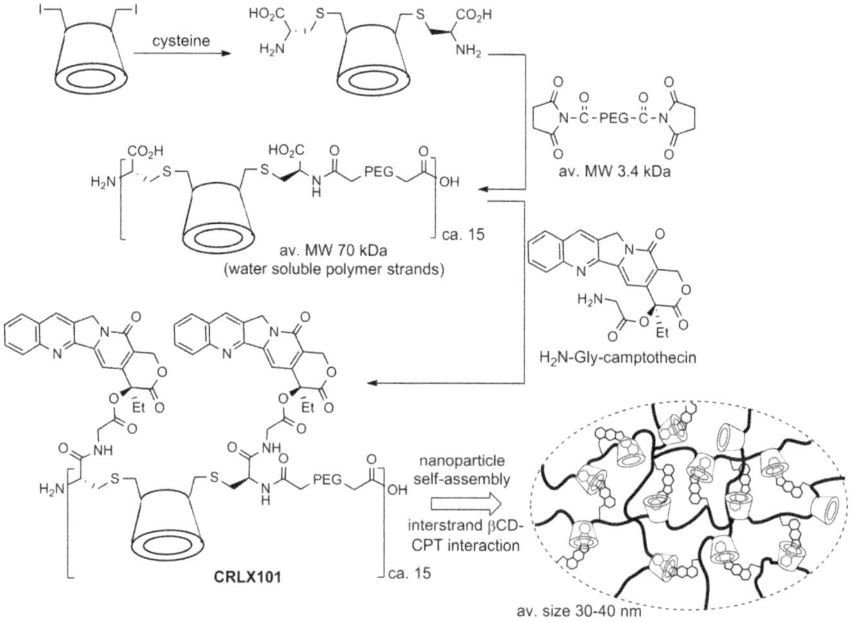

Figure 5.5 Schematic synthesis of the camptothecin (CPT)-loaded polymer CRLX-101 described by Davis.[103]

5.5 Cyclodextrin-based Poly(pseudo)rotaxanes for Sustained Drug Release

Rotaxanation of polymeric chains with CD–drug conjugates represents a versatile strategy to develop sustained drug release systems.[110] In a pioneering work, Yui and co-workers, engineered for that purpose a HPαCD-threaded PEG polyrotaxane capped with phenylalanine onto which theophylline units (a bronchodilator drug) were grafted (Figure 5.6). The authors hypothesized that theophylline would remain inactive while immobilized on the polyrotaxane, either due to steric hindrance or poor membrane permeability. Proteolytic cleavage of the phenylalanine stoppers let the assembly dissociate into the bioactive theophylline–HPαCD conjugates.[111] More recently, Uekama has exploited a similar approach to achieve the sustained release of insulin by taking advantage of the solubility decrease of pegylated insulin upon rotaxanation with αCD.[112] Active PEG–insulin is slowly released into the solution from HPαCD–PEG–insulin polyrotaxane aggregates in biological media, probably by an aggregate surface erosion mechanism.

5.6 Cyclodextrin-Based Hydrogels and Nanogels as Controlled Drug Release Systems

Supramolecular gels are thoroughly discussed by Miravet and Escuder in Chapter 11. Among them, hydrogels and nanoparticulated hydrogels[113–115]

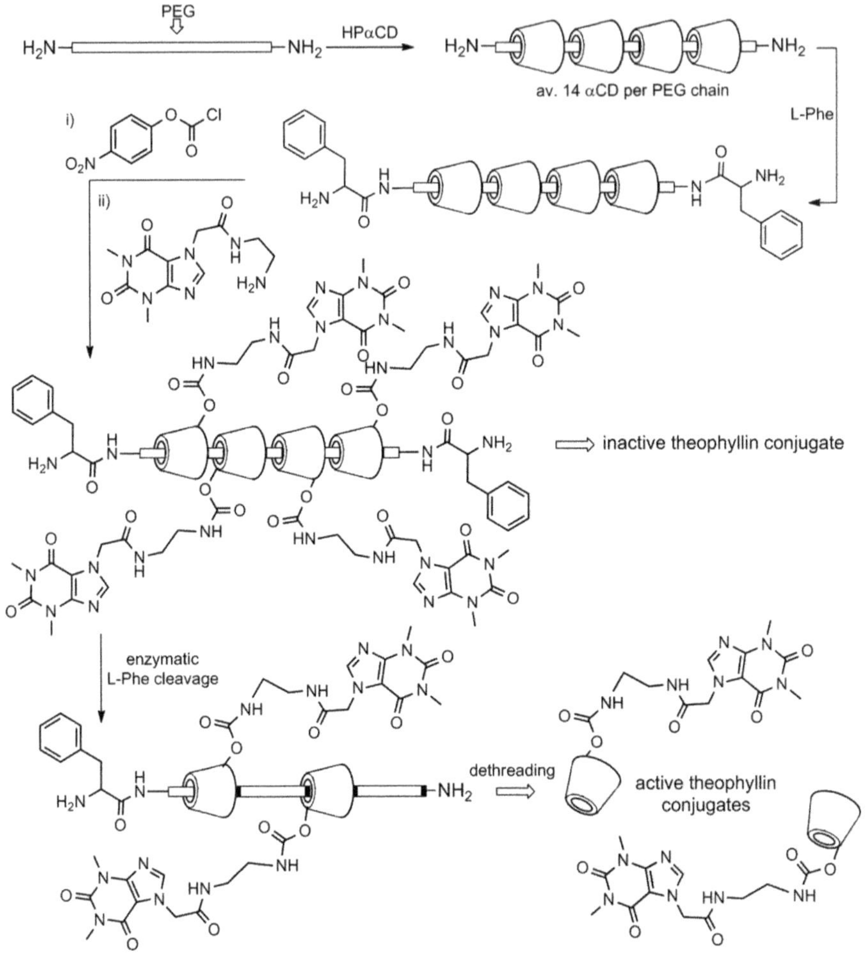

Figure 5.6 Schematic representation of theophyllin release from HPαCD–PEG poly-rotaxane conjugates *via* enzymatic degradation.[111]

(nanogels) based on CDs represent a particularly attractive family of supra-molecularly crosslinked polymeric materials for applications in pharmaceutical technology with outstanding swelling properties.[116] Incorporation of CD moi-eties in the structure of a nanogel furnishes a matrix with affinity-based mech-anism for guest loading, provides hydrophilicity and prevents dilution-induced premature guest release—all desirable features for a drug carrier system (Figure 5.7A).[117] CD-based nanogels may also exhibit reversible assembling/disassembling capabilities. Thus, Guo, Lincoln and co-workers have shown that the hydrogel formed by supramolecular crosslinking between adamantane- and βCD-functionalized polyacrylic acid chains can be broken by addition of com-peting agents such as unbound βCD (Figure 5.7B).[118] The strength and dynamics of the network can be tailored by adjusting the CD content.[119–121]

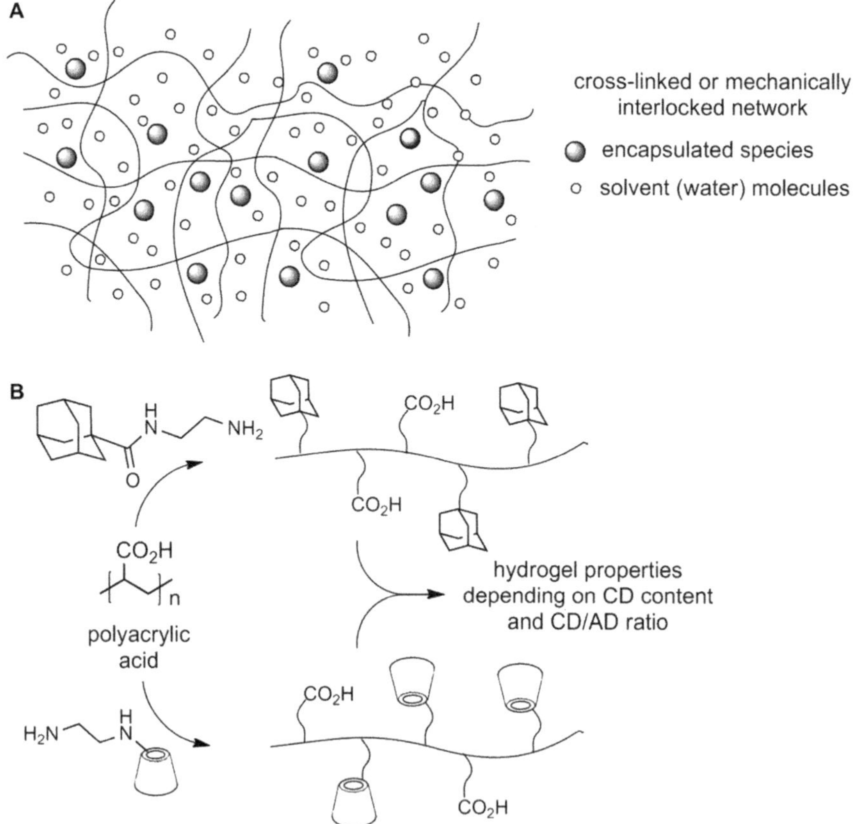

Figure 5.7 (A) Representative structure of the highly swollen network of a hydrogel and (B) schematic structure of the CD/adamantane-interlocked hydrogel described by Guo and Lincoln.[118]

Interestingly, CD-containing nanogel supramolecular networks are usually assembled in mild conditions, thus permitting the presence of sensitive loads during their preparation, which is not always the case for classical polymerization procedures. For instance, tamoxifen (an anticancer drug), benzophenone (a sun screen agent)[122] and Gd(III) complexes (contrast agents for magnetic resonance imaging)[123] have been entrapped efficiently in a biocompatible nanogel with potential practical utility in formulations for local delivery, formed upon mixing hydrophobically modified dextran and a water-soluble epichlorohydrin-crosslinked βCD polymer.

The number of investigational and technological applications of CD-based hydrogels and nanogels largely exceeds those commented on above[81,124] and are not exclusively envisioned at protecting and stabilizing their payload,[125,126] but simultaneously to more ambitious tasks such as controlled release and specific delivery to therapeutic selected targets. The hydrophilic CD-based hydrogels developed by Hennink and co-workers, consisting of mixtures of

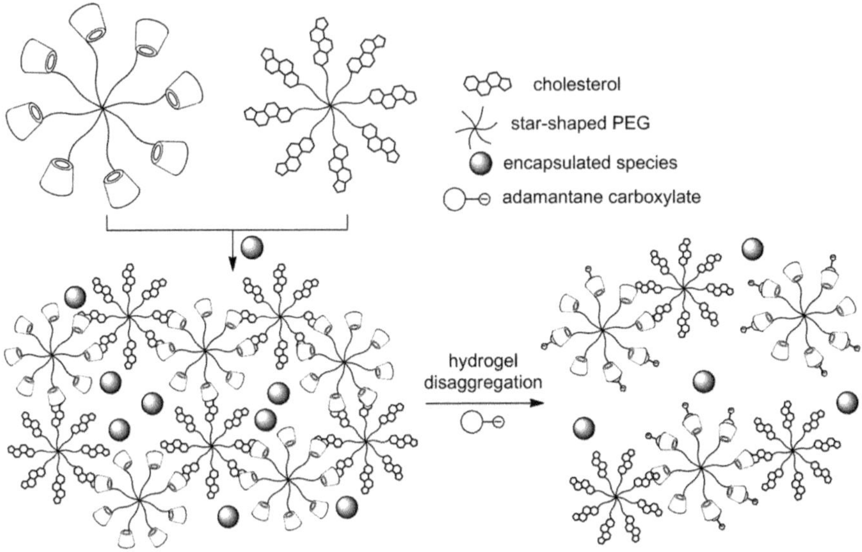

Figure 5.8 Schematic structure of the CD/cholesterol-interlocked network composed by Hennink.[125]

star-shaped PEG polymers coated with either βCD or cholesterol, exemplify this potential.[127] Hydrogelation in these systems is driven by cholesterol–βCD inclusion complex formation and can be tailored by a broad range of parameters—for example, concentration, CD : cholesterol ratio, component architecture, temperature, or addition of competing agents—which allows optimization for encapsulation of a range of proteins (*e.g.* lysozyme, bovine serum albumin and immunoglobulin, of mol. wt 15, 67 and 150 kDa, respectively).[125] The authors demonstrated that hydrogel degradation is mainly due to mechanical erosion, which depends on the network properties and can be tailored to warrant protein release at a linear rate (Figure 5.8).

Photo-responsive CD-based hydrogels have also shown remarkable features for controlled release. In a recent contribution, Kros and co-workers used a mixture of dextrans grafted with either βCD or azobenzene to furnish supramolecularly crosslinked matrices with photo-switchable mesh size.[128] The hydrogels were shown to encapsulate green fluorescent protein (GFP) and to slowly release it while azobenzene moieties were in the ground state (*trans*-configured), well-suited for inclusion in the βCD cavity. GFP release was significantly accelerated upon UV irradiation-induced azobenzene *trans → cis* isomerization, in agreement with the expected disassembly of the hydrogel network due to the known mismatching of the *cis*-azobenzene : βCD pair (Figure 5.9).

Recently, Li and co-workers developed a biodegradable (pseudo)polyrotaxane made by threading αCD on linear polyethyleneglycol-poly(3-hydroxybutyrate)-polyethyleneglycol (PEG–PHB–PEG) triblock polymers (Figure 5.10). Balancing the PEG-to-PHB ratio, the authors obtained a

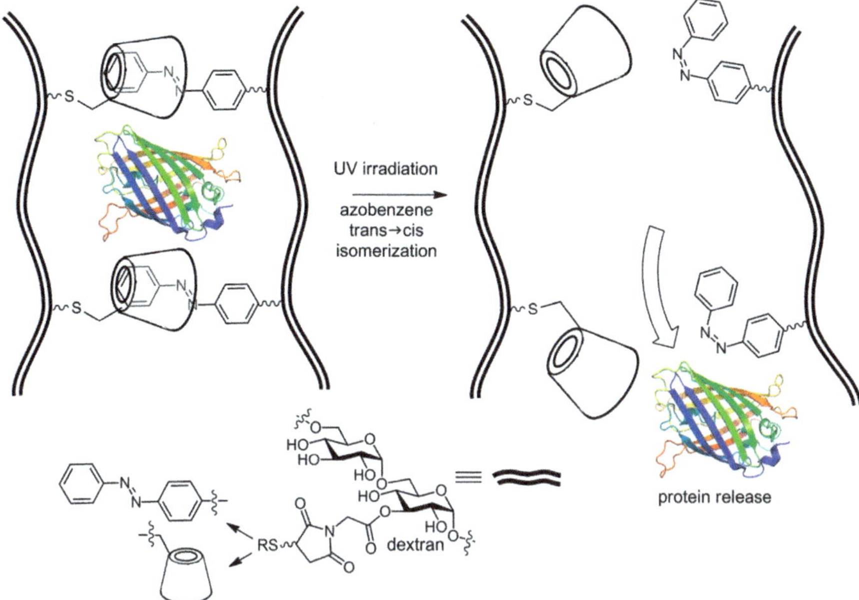

Figure 5.9　Schematic representation of photo-responsive protein release from the βCD/azobenzene-interlocked network. Upon the UV irradiation, azobenzene moieties isomerize from *trans* to *cis* configurations, resulting in the facile release of the entrapped protein.
(Adapted from Peng *et al.*[128] with permission from The Royal Society of Chemistry.)

hydrogel offering a sustained release of fluorescently-labeled dextran, as model macromolecular drug, for as long as 1 month.[129]

5.7　CD-based Coatings and Films for Surgical Applications

In addition to 3D networks, devices controlling drug release from surfaces are urgently demanded in biomedicine. CD-based self-assembled or polymeric films and multilayers are relatively young paradigms making their first steps towards successful applications. Jessel and Ogier pioneered the field when they first reported on carboxylated βCD-piroxicam complexes embedded into a multilayer film composed by alternating poly-L-lysine (PLL) and poly(L-glutamic acid) (PGA) layers that exhibited a sustained anti-inflammatory effect over 12 h (Figure 5.11).[130]

A more efficient mode of control over the release of small hydrophobic drugs was implemented by Hammond's group, who constructed films by alternately depositing carboxylated CD polymers and biodegradable cationic poly(β-aminoesters).[131] Regardless of their nature, guests were released from these films with nearly linear kinetics over much longer periods (up to 1 month),

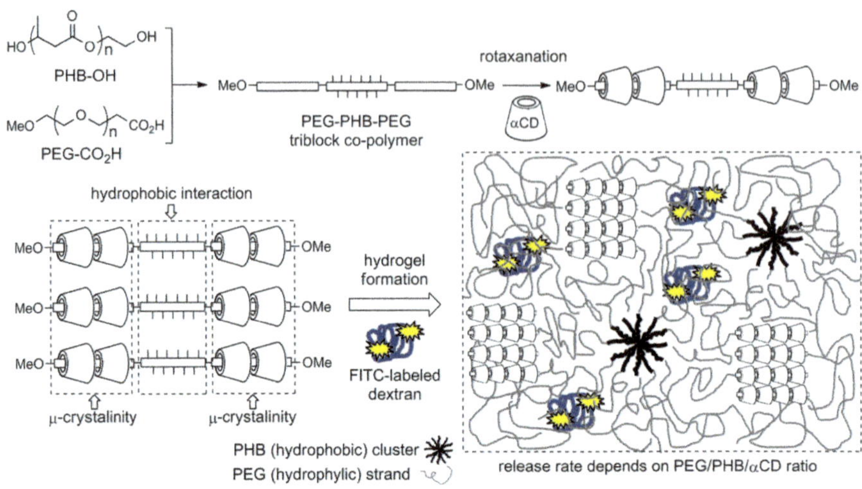

Figure 5.10 Synthesis of the αCD-threaded PEG–PHB–PEG triblock copolymer and schematic representation of its hydrogelation mechanism.
(Adapted from J. Li *et al. Biomaterials*, 2006, **27**, 4132, Self-assembled supramolecular hydrogels formed by biodegradable PEO–PHB–PEO triblock copolymers and α-cyclodextrin for controlled drug delivery, Copyright 2006, with permission from Elsevier.)

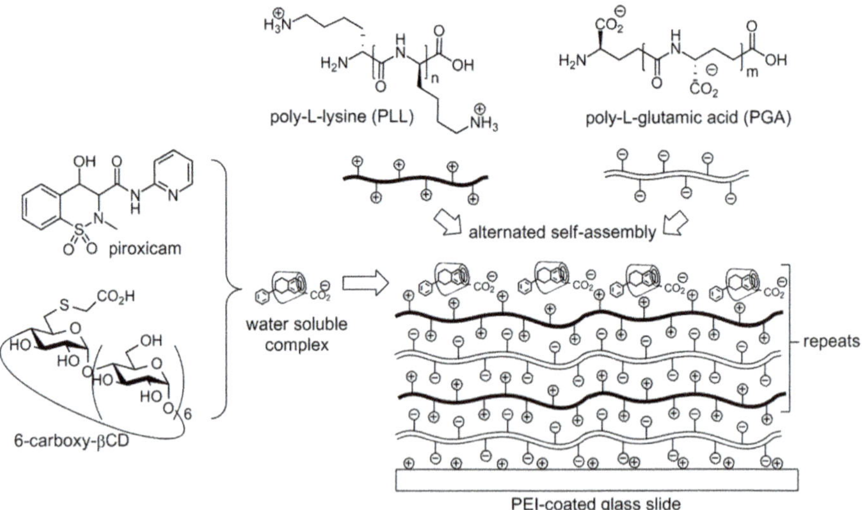

Figure 5.11 Schematic representation of the composition of βCD-piroxicam complex-loaded multilayered films described by Jessel, Ogier and co-workers.[130] PEI = polyethylene imine.

depending on the precise film composition. Taking advantage of this technology, the same team engineered a bifunctional film construct combining a permanent microbicidal film with a biodegradable topping that either provided a bolus release of an antibiotic drug (gentamicin) or a sustained release of an anti-inflammatory drug (diclofenac). Release of gentamicin was demonstrated over hours and that of diclofenac over days, while the microbicidal base film retained its functionality after the biodegradable films had completely degraded, thereby fulfilling the most important requirements for application as a surgical implant coating.[132]

The utility of these self-assembled films has also been demonstrated for sustained release of biomacromolecules.[133,134] Recently Li and co-workers fabricated a multilayer chip by alternating deposition of a cationic CD-centered dendrimer, insulin, and glucose oxidase in order to operate as a switchable insulin release system.[135] When exposed to low glucose levels, the chip remained in the "off" state, featuring slow insulin release kinetics. At higher glucose levels, glucose oxidase catalysed the formation of gluconic acid, thereby decreasing the microenvironmental pH to levels that induced a conformational switch in the CD-centered dendrimer, which in turn enhanced film swelling and accelerated insulin release. The film is reported to operate reversibly even when implanted in animal models, reinforcing its potential as "on-demand" insulin delivery system.

5.8 Cyclodextrin-Based Nanoparticulated Devices from Amphiphilic CDs: Nanospheres and Nanocapsules

Self-assembled nanoparticles are attractive alternatives to the more classical polymeric or liposome constructs as high-capacity tunable drug carriers. The possibility to tailor the discrete components at the molecular level can be put forward to adapt the system to predefined standards (size, loading, stability, responsivenes, *etc.*). CD-based amphiphiles have proven very useful in this regard.[136] Moreover, the presence of the CD platform may promote nanoparticle interaction with biological membranes or enhance encapsulation of hydrophobic drugs. This has been translated into a number of applications for delivery and controlled release of sparingly soluble drugs, such as tamoxifen,[137] paclitaxel[138] or camptothecin,[139] and photosensitizers.[140] In a number of cases, side effects associated to conventional drug formulations were significantly alleviated when the drugs were encapsulated in CD nanoparticles.[138] Self-assembling of CD amphiphiles occurs spontaneously in polar environments, but nanoaggregate manufacturing can usually be rationally tuned depending on the CD structure and formulation conditions. Nanoprecipitation[141] or emulsion/solvent evaporation techniques,[142] eventually in the presence of additional ingredients, can be implemented to produce nanospheres or nanocapsules. In the first case the matrix is formed by the intrinsic amphiphilic CDs in a compact arrangement, whereas in the second case the nanoparticle core has an oily nature, with the CDs at the periphery (Figure 5.12). This is indeed a useful tool to manipulate drug release profiles. Thus, Defaye and co-workers[143] showed that the release kinetics of diazepam can be dramatically altered depending on whether the drug is loaded into nanospheres assembled from a

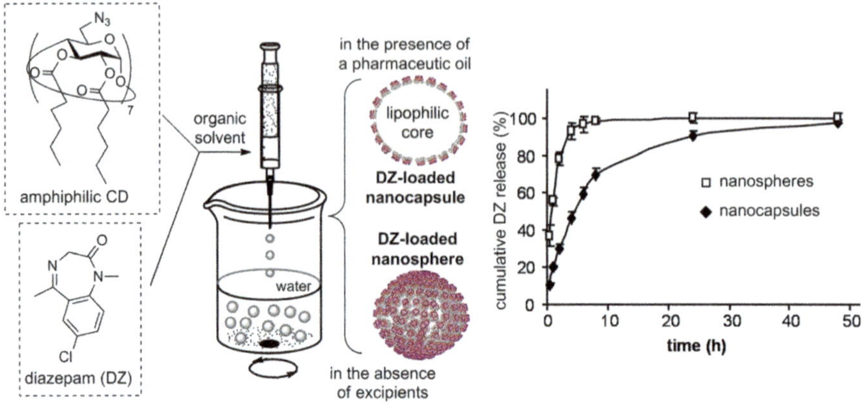

Figure 5.12 Encapsulation of diazepam by nanoprecipitation with an amphiphilic cyclodextrin in the absence and the presence of a pharmaceutically approved oil (nanospheres and nanocapsules, respectively), and their comparative drug release profiles.[143]

per-(C-6)-azido βCD derivative or into nanocapsules formulated from the same CD in the presence of a pharmaceutically approved oil (Figure 5.12).

5.9 Polycationic Cyclodextrins as Non-viral Gene Vectors

The potential of native CDs as transfection enhancers, ascribed at least in part to their ability to host cell membrane components such as cholesterol in their hydrophobic cavity, has been long known and exploited to improve the gene delivery capabilities of polycationic lipidic or polymeric non-viral vectors.[144] New functional materials for gene delivery have been obtained by combining CD motifs and polycationic oligomers, such as the polyamines, surveyed in Chapter 8 by García-España *et al.* The group of Davis pioneered this field, with a focus on the development of targeted carriers for gene therapy.[145,146] The authors conceived a class of cationic polymers based on the controlled condensation of bifunctional CD monomers, such as bis-(C-6)-cysteaminylated βCD, and cationic co-monomers, in order to yield linear chains with alternating CD and cationic segments (CDP in Figure 5.13A).[147,148] Electrostatically driven complexation of the resulting cationic CDPs and negatively charged plasmid DNA (pDNA, ~5 kbp) rendered nanometric complexes (poly-CDplexes; 100–150 nm) featuring *in vitro* cell transfection efficiency comparable to that obtained with the gold standard commercial non-viral vectors polyethylene imine (PEI) and Lipofectamine™, while preserving reduced toxicity.

The above milestone contribution was followed by a series of reports in which the effect of structural modifications of CDPs on the gene delivery capabilities were investigated.[149–153] Most interestingly, supramolecular

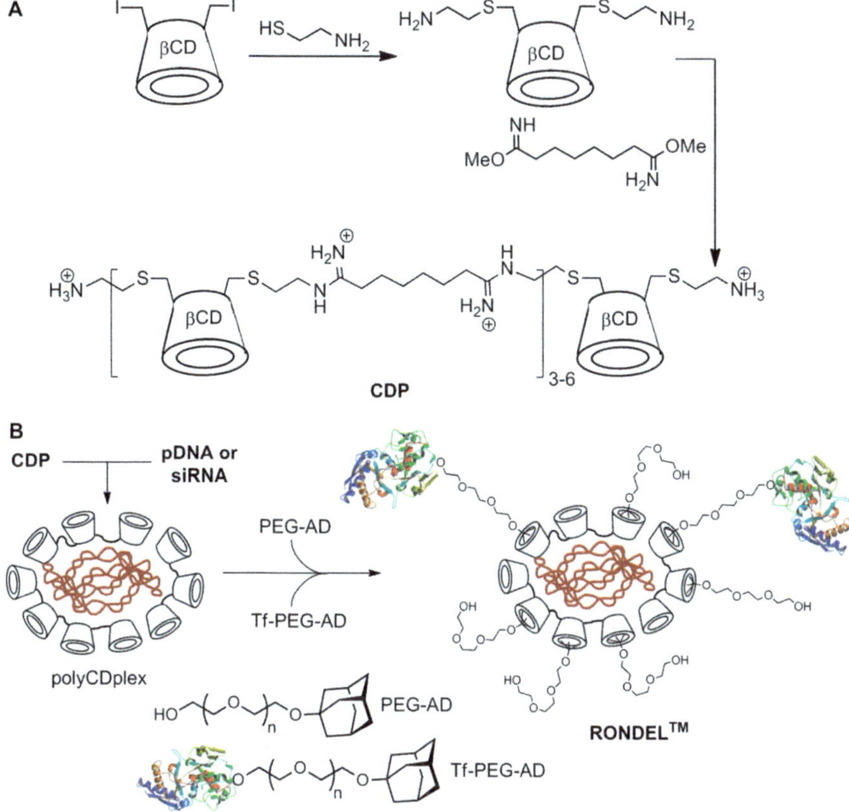

Figure 5.13 Schematic representation of the transferrin-targeted nucleic acid–CDP nanoparticles (RONDEL™) developed by Davis and co-workers.[147] (Adapted from Ortiz Mellet *et al.*[144] with permission from The Royal Society of Chemistry.)

host–guest CD chemistry could be advanced to further decorate the nanoparticle surface. Thus, inclusion of the adamantane (AD) moiety of AD–PEG conjugates into CD cavities of polyCDplexes could prevent non-specific interactions with biological components[154] and also biorecognizable ligands could be installed at the distal end of these PEG chains to aim the polyCDplexes to specific target tissues (Figure 5.13B).[155,156] Using transferrin (Tf) as the peripheral ligand, whose receptor is known to be upregulated in malignant cells, the resulting polyCDplexes were shown to selectively target and efficiently transfect different tumor tissues *in vivo* in mammals (murine[157–159] and primate).[160,161] This tripartite therapeutic concept (CDP as nucleic acid complexing element / AD–PEG as surface shielding element / AD–PEG–Tf as targeting ligand) is named RONDEL®. In June 2008, CALAA-01 (Calando Pharmaceuticals), a version of RONDEL® loaded with a specific oligonucleotide sequence (siRNA) that inhibits tumor growth *via* RNA interference

(RNAi), entered phase I clinical trials to determine its safety and efficacy towards solid tumors refractory to standard-of-cure therapies.[162]

In the light of these stimulating results, research on polycationic polymeric materials incorporating CDs as gene vectors, eventually armed with biorecognizable ligands, has flourished.[163–172] Yet, the above systems still suffer from the inconveniences associated to polydispersity in view of advanced structure–activity relationship (SAR) studies. In the last years, several groups have tried to overcome this limitation by developing monodisperse polycationic clusters after single or dual face selective functionalization of the CD core. In the first case, the cationic centers (amine, guanidine, amino acids) were installed at the primary positions, keeping unmodified the OH-2 and OH-3 hydroxyls, taking advantage of the direct accessibility of the corresponding per-(C-6)-halogenated precursors (Figure 5.14A–D).[173–177] Dual face modification has been brought in to endow the system with facial amphiphilicity properties, a biomimetic design, aiming at improving self-assembling and membrane-crossing capabilities.[178] By equipping the cyclo-oligosaccharide platform with segregated cationic and lipophilic domains, polycationic amphiphilic CDs (paCDs, Figure 5.14E) have ben accessed that efficiently condensed plasmid DNA into stable nanocomplexes (CDplexes) of 40–50 nm diameter with very low polydispersity.[179–181] DNA complexation is assumed to take place though a two-step process involving, first, an electrostatically driven interaction between the anionic polyphosphate chain and the cationic amphiphile and, then, a

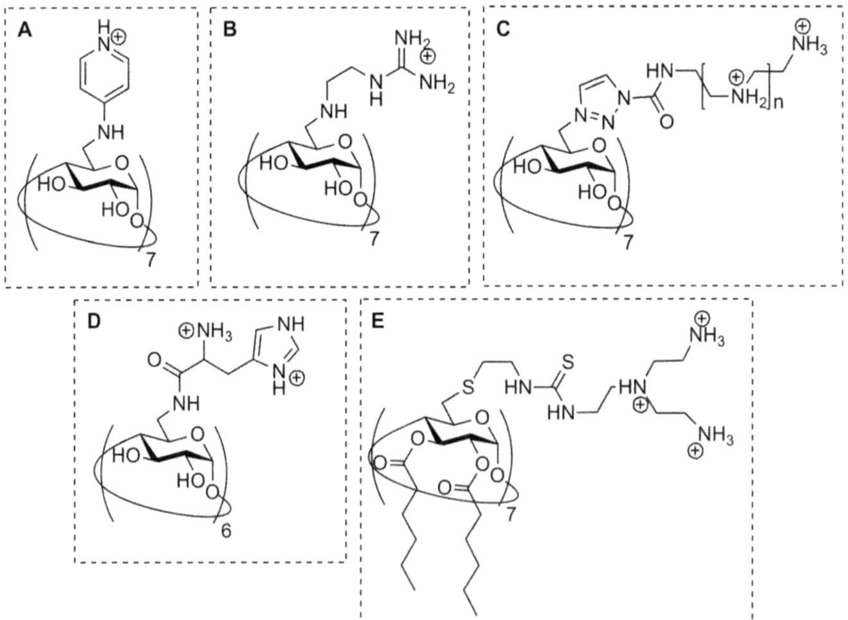

Figure 5.14   Representative examples of homogeneously primary rim-functionalized polycationic cyclodextrins with gene vector capabilities.[173–177]

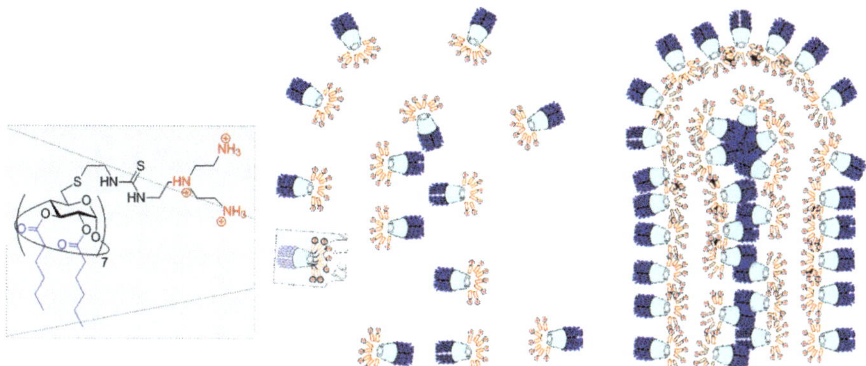

Figure 5.15 Schematic illustration of the proposed two-step mechanism for CDplex formation between paCDs and DNA.

hydrophobically driven compaction (Figure 5.15), a mechanism that is reminiscent of that operating in viral particle assembly.

CDplexes based on paCDs have been shown to be "promiscuous" gene delivery agents, affording high transfection efficiencies in a variety of cell lines, often surpassing that of PEI and Lipofectamine®. Recently, a glycosylated version of paCDs (pGaCDs) has been developed for targeting purposes.[182] The artificial glycocalyx-like surface generated after glycoCDplex assembly can be recognized by complementary cell membrane receptors and/or intracellular trafficking machineries. As a proof of concept, a mannosylated candidate was prepared that succeeded at selectively transfecting MMR-positive cells (RAW264.7 macrophages) *via* the MMR-dependent route (Figure 5.16). The strategy has been further validated for the transfection of hepatocytes with galactosylated pGaCDs *via* specific recognition of the corresponding glyco-CDplexes by the asialoglycoprotein receptor (ASGPr).[183] Using a similar conceptual approach, O'Driscoll and co-workers have designed a βCD derivative featuring facially segregated guanidinium groups and anisamide-terminated PEG chains to exclusively deliver siRNA loads to sigma receptor-positive cells.[184] About 80% knockdown luciferase expression was achieved *in vitro* in prostate cancer cells and significant tumor inactivation levels were observed when the CDplex was intravenously administered into a mouse model.

5.10 Cyclodextrin-Based Hybrid Nanosystems: Programmable Nanocontainers

The design of suitable devices for programmed drug release is probably the field that has most benefited from the synergy between CD chemistry and nanotechnology. The level of refinement achieved is paradigmatically illustrated with CD-gate-kept nanocontainers developed independently in the last few

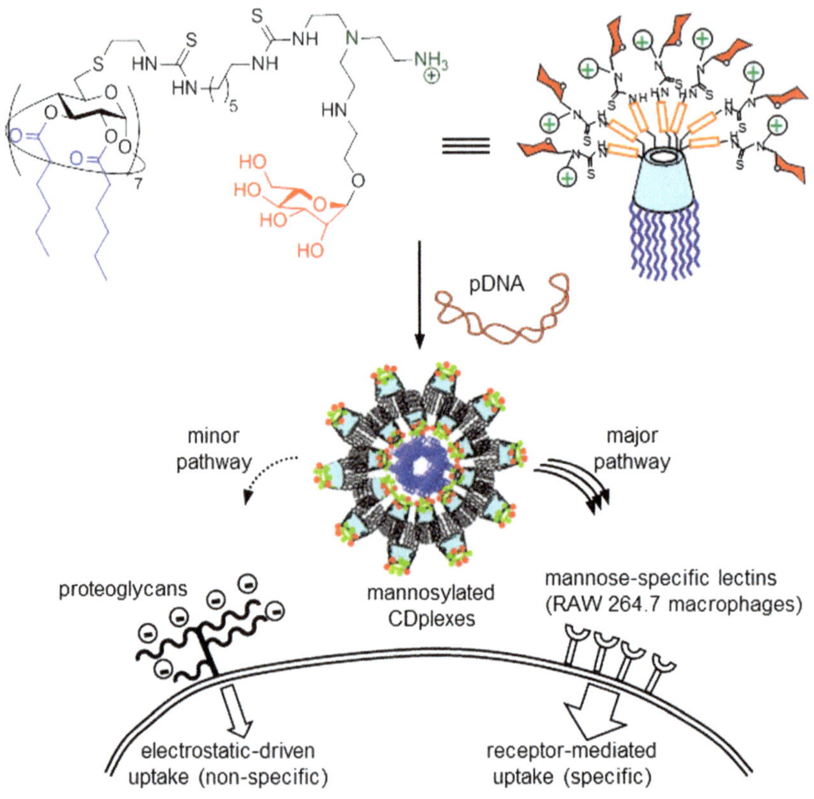

Figure 5.16   Schematic representation of a polycationic glyco-amphiphilic cyclodextrin (pGaCD) and the corresponding glycosylated nanoparticle formed with pDNA (GlycoCDplex) and its specific lectin-mediated cell uptake route.
(Adapted from *Biomaterials*, 2011, **32**, 7263, A. Díaz-Moscoso *et al.*, Mannosyl-coated nanocomplexes from amphiphilic cyclodextrins and pDNA for site-specific gene delivery, Copyright 2011, with permission from Elsevier.)

years by Stoddart and Kim using mesoporous silica nanoparticles (Si-MPs).[185] Inorganic hollow nanoparticles, in particular mesoporous silica nanoparticles (Si-MPs), are relatively new materials which—due to their nontoxicity, ease of surface modification and pore and cavity size control—bear great potential for molecular encapsulation. The key challenge to transform these nanoobjects into useful drug delivery vehicles is engineering switchable pore gatekeepers. The mentioned research groups employed CDs as pivotal functional elements in a series of contributions reporting smart delivery nanocarriers that rank among the most fascinating ever designed. Many of them take advantage of CDs as stimuli- responsive gatekeepers or nanovalves owing to the size-complementarity with the pore internal diameter. Si-MPs chemically furnished with CD-based pH-,[187,188] redox-,[189] photo-[190,191] and enzyme-responsive[192,193] pore gates have been developed. As documented by Saleh *et al.* in Chapter 7, cucurbiturils have also been successfully employed for these tasks. The architectural design of these gate-kept systems is very versatile and is mostly based in two different approaches: (i) CDs units covalently linked to the Si-MPs, close to the pore gates, *via* a photo-,[191] chemo-[189] or enzyme-sensitive tether,[192] which sterically hinders load traffic out of the pore (Figure 5.17A), or (ii) rotaxanized CD units that either move closer (close) or further (open) from the pore entrance or scape from the interlocked

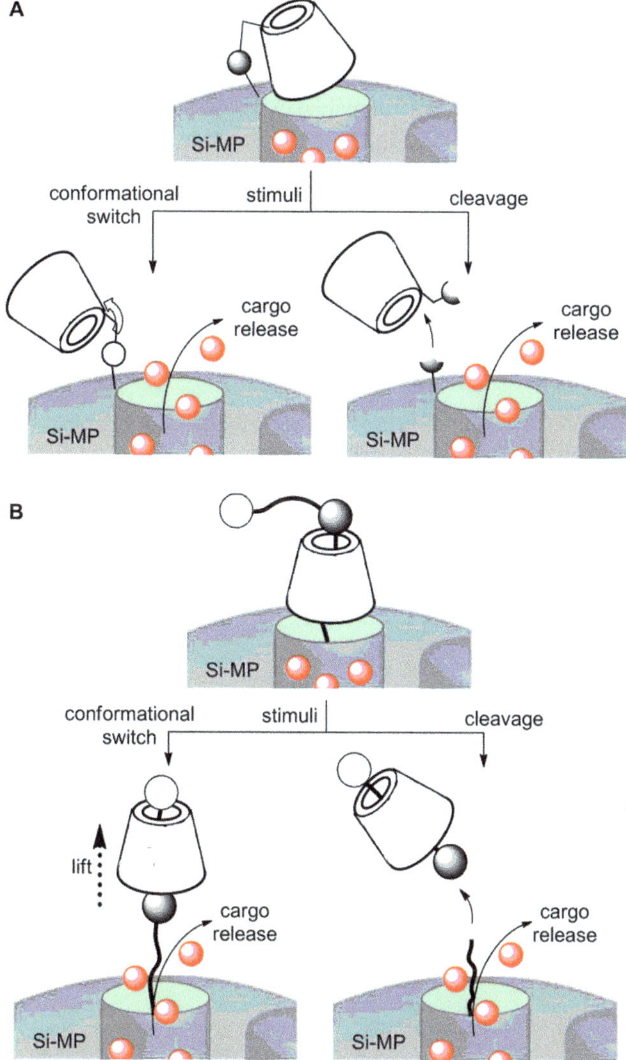

Figure 5.17 Schematic representation of cargo release from SiMPs using (A) covalently-linked and (B) rotaxanized CD-based nanovalves.

structure *via* pH,[186,187] redox,[188] photochemical[190] or enzymatic stimuli[191] (Figure 5.17B).

The sophistication achieved in the degree of control of the functional features of these nanomaterials has allowed, for instance, the preparation of nanovalves that are responsive to endosomal acidification conditions in human cancer cell lines,[187] clearing the way to the implementation of targeted platforms for effective and save delivery of aggressive and poorly bioavailable anticancer drugs. In another version of these carriers, glutathione-induced disulfide reduction has also been shown to efficiently control intracellular release of doxorubicin.[189] More recently, Zink, Stoddart and co-workers have designed a new series of CD-based gatekeepers to fit Si-MPs with enlarged pores (up to 6.5 nm in diameter) that may allow encapsulation and controlled release of macromolecular loads.[193,194] These "megagates" remain closed in neutral conditions, but open below pH 5. Si-MPs furnished with dual gate opening mechanisms have also been designed. Thus, the same teams have reported particles loaded with differently sized cargos that can be released successively: in a first stage a pH-sensitive gate opens up to release sufficiently small molecules to diffuse through a βCD cavity while, in a second stage, a reductive agent cleaves off the disulfide-bonded βCD stopper from the pore gate to let larger molecules be released (Figure 5.18).[195] Regardless of their unquestionable academic interest, the authors also claim that such dual carriers may help to tackle cancer with the often-used combined therapies.

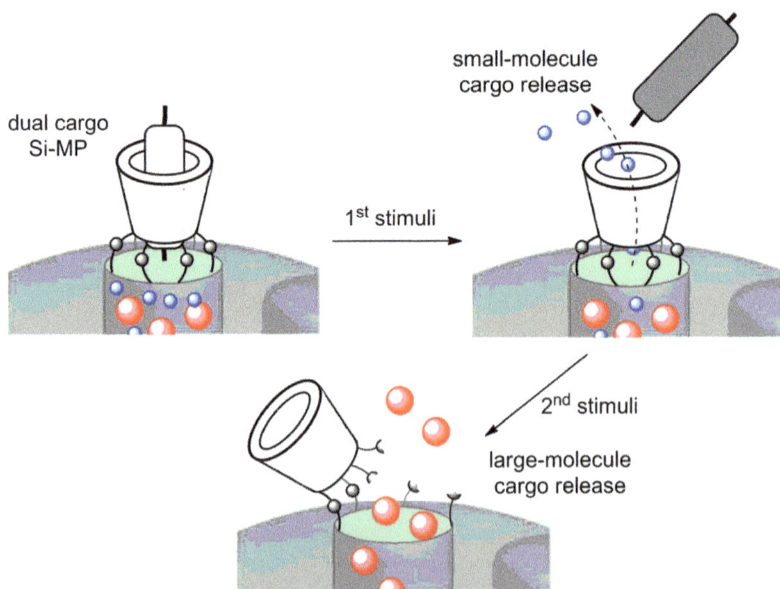

Figure 5.18   Schematic representation of dual-cargo release using the CD gatekeepers described by Stoddart, Zink and collleagues.[195]

5.11 Cyclodextrin-Based Therapeutics

Recent findings have shown that CDs and their derivatives display a range of physiological effects that can be exploited therapeutically. Conceptually, this implies the use of the CD component not as a drug vehicle, but as the drug itself. Both, the molecular inclusion properties and the interactions of peripheral branches with relevant partners can be at the basis of their therapeutic activity. Some examples are herein selected to illustrate the novel approaches under investigation using CD-based drugs.

5.11.1 Cyclodextrins as Regulators of Cholesterol Metabolism

Improper cholesterol management is known to disrupt a number of cellular mechanisms. Given the ability of some CDs to interact with cholesterol, their potential as cholesterol-rebalancing agents is very appealing. More than a decade ago, Furuchi and Anderson[196] demonstrated the correlation between HPβCD-mediated cell-membrane cholesterol depletion and the mitogen-activated protein (MAP) kinase levels that control mitogenesis. This observation has led to a major breakthrough in the development of therapeutic alternatives towards cholesterol metabolism-associated disorders, such as Niemann–Pick (NP) disease. NP disease is a rare inherited neurodegenerative disorder caused by a dysfunction in the proteins involved in cholesterol trafficking from endosomes/lysosomes to cytosolic compartments, which results in an intracellular accumulation of unesterified cholesterol. Very recently, Dietschy and co-workers noticed that injection of a single dose of HPβCD to 7-day-old NP type C mutant mice restored cholesterol traffic and reduced the accumulation of unesterified cholesterol to healthy levels for several weeks, leading to improved liver function, decreased neurodegeneration and prolonged life.[197,198] The mode of action has not been completely elucidated so far, but it seems to be cholesterol-specific, since related lysosomal storage disorders (*e.g.* GM1 gangliosidosis or mucopolysaccharidosis IIIA) were not ameliorated by HPβCD administration.[198] It has been demonstrated that HPβCD internalizes NP endosomes and mobilizes cholesterol from inside rather than withdrawing it from membranes.[199–201] However, Peake and Vance have shown that this role of HPβCD is dose dependent: higher HPβCD doses extract cholesterol from membranes producing exactly the opposite effect.[202] These findings are timely considering that trials in humans have just being initiated[203] after the Food and Drug Administration and European Medicines Agency granted HPβCD the orphan drug status for the treatment of NP type C disease.

5.11.2 Cyclodextrins as Anti-infective Agents

Cholesterol homeostasis has been shown to play also a fundamental role in host–pathogen interactions, which may explain the inhibitory activity of CDs in certain pathogenic infections. Indeed, CD-mediated cholesterol depletion from either cell membranes and/or viral envelopes somehow interferes in the infection processes of a number of viruses. The case of HIV is both the most illustrative and best studied due to its clinical relevance. Two decades ago Moriya and co-workers reported the first evidence pointing to the anti-HIV activity of sulfated CDs, attributing this activity to the inhibition of HIV binding to host membrane.[204,205] Unfortunately, none of the tested compounds featured a clinically relevant absorption rate to be further tested, but this research opened the door for other groups to explore the prophylactic potential of CDs in HIV transmission. Thus, Hildreth and co-workers made a promising breakthrough demonstrating the relevance of cholesterol-rich lipid rafts on viral budding[206] and envisioned the topical application of HPβCD to block vaginal transmission of cell-associated HIV-1 in mouse models.[207,208] In a strategically distinct approach, Boasso, Graham and co-workers have used HPβCD to deplete cholesterol from viral envelop thereby yielding virons with reducing ability to activate dendritic cells. These virons exhibited a powerful memory CD8 T-cell specific response, demonstrating that HIV suppresses antiviral T-cell response *via* dendritic cell activation.[209]

Cell infection *via* cholesterol-rich lipid rafts is also well documented for a number of microorganisms. Aimed at developing an alternative therapeutic option to tackle malaria, Crandall has demonstrated that *Plasmodioum* entry into red blood cells can be prevented with sulfated CDs,[210] though the precise mechanistic interpretation of this effect remains elusive.

5.11.3 Cyclodextrins as Pharmacological Chaperones in the Prevention and Treatment of Folding Diseases

In addition to the their utility as bioavailability enhancers of therapeutically useful peptides and proteins mentioned in Section 5.2, CDs possess in themselves a therapeutic potential as drugs capable of acting as artificial chaperones to prevent misfolding, or restore the correct folding, of endogenous proteins/peptides involved in folding diseases. Alzheimer's disease and amyloid-folding related diseases are probably examples that best exemplify the potential of CDs in this regard. Alzheimer's disease involves a progressive neuronal loss intimately associated to the aggregation of β-amyloid (Aβ), a polypeptide found in extracellular amyloid plaques in the brain, into amyloid fibrils.[211] Simons and co-workers observed over a decade ago that treatment of hippocampal neurons with statistically methylated βCD (RAMEB) completely inhibited formation of Aβ aggregates, which they associated to cholesterol depletion.[212] While they unequivocally demonstrated that cholesterol is required for Aβ aggregation (in fact, it has been shown that individuals carrying certain alleles of the apolipoprotein E gene, associated with higher cholesterol levels, have increased risk of developing Alzheimer's disease[213]), the contribution of the interaction between Aβ aromatic residues and RAMEB was not properly assessed.[214]

Spectroscopic evidences of how the interaction of Aβ with CD derivatives can modulate peptide aggregation have been recently gained.[215,216] In the light of these data, Ando and co-workers embarked in the investigation of the capability of a series of CD derivatives, mostly glycosylated CDs, to avoid transthyretin (TTR) amyloid formation.[217] TTR is a plasma transport protein

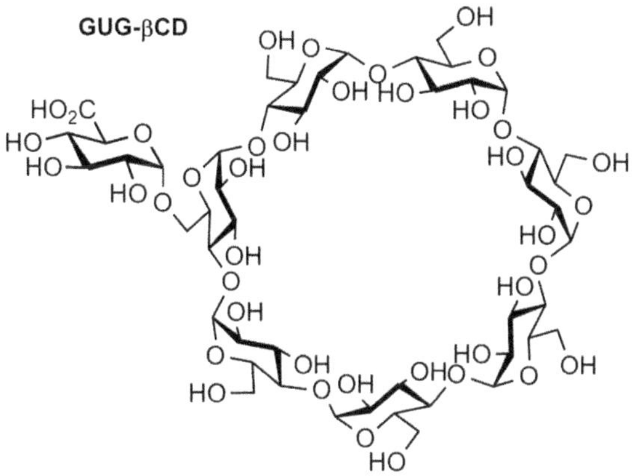

Figure 5.19 Structure of GUG-βCD, the transthyretin amyloid formation inhibitor developed by Ando.[217]

that, when mutated, undergoes aggregation into amyloid fibrils that play a crucial role in the pathogenesis of several amyloidotic neuropathies. One of these compounds, namely 6-*O*-[α-(4-*O*-α-D-glucuronyl)-D-glucosyl]-βCD (GUG-βCD; Figure 5.19) induced a remarkable inhibition in TTR amyloid formation. The mechanism of action involves stabilizing interactions between GUG-βCD and TTR hydrophobic residues, mainly tryptophan, leading to a "conformational freezing" process. Interestingly, the inhibitory effect of GUG-βCD has also been demonstrated *in vivo* in transgenic rats possessing a human variant TTR gene.[217] The biocompatibility and non-toxicity of this CD derivative may enhance the potential of this type of pharmacological chaperones to treat protein folding-related diseases in the near future.

5.11.4 Multivalent Cyclodextrin Conjugates to Control Carbohydrate–Protein Interactions

Grafting multiple copies of a saccharide ligand on CD-based cores has revealed a useful strategy to overcome the inherent weakness of individual protein–carbohydrate interactions in Nature by taking advantage of the so-called *cluster* or *multivalent glycoside effect*.[218] Besides providing instrumental insights at the molecular level on the mechanisms underlying these phenomena,[219–222] CD-scaffolded multivalent glycoclusters bear strong potential to interfere in biological and pathological processes governed by these types of events, which includes bacterial and viral infection, cancer metastasis, immune response and inflammation, among others.[223] The work reported by André and Nishimura on the capability of a set of hepta-galactosylated βCD (Figure 5.20) conjugates to inhibit binding of human lactose/galactose-specific lectins (galectins) to their biological targets illustrates those developments.[224] A nearly 400-fold increase in protein binding avidity per saccharide residue (relative to the free sugar) was observed for some candidates, which could be put forward to prevent undesired

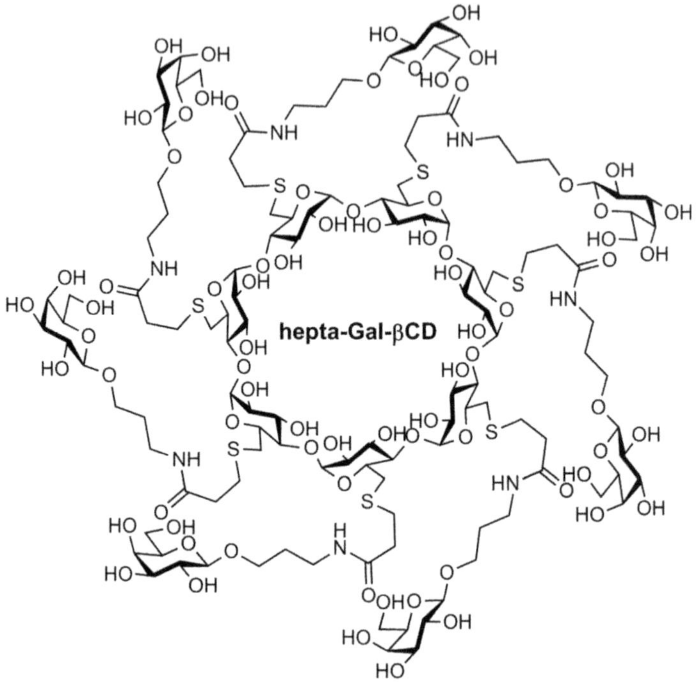

Figure 5.20 Structure of the β-cyclodextrin-centered heptagalactoside inhibitor of galectin-mediated adhesion designed by André and Nishimura.[224]

protein–carbohydrate binding events *in vivo*, for example, antibody-dependent acute rejection of xenotransplants due the immunoreactivity of galactose-bearing xenoantigens. More recently, CD conjugates bearing multiple imino-sugar (*e.g.* 1-deoxynojirimycin, DNJ) copies have been synthesized to investigate the possibility of enhancing the intrinsic glycosidase inhibitory capabilities of the glycomimetic motif (Figure 5.21). An astonishing valency- and architecture-dependent enhancement of the inhibition potency and selectivity towards α-mannosidase was revealed.[225] Although the investigation of multi-valent effects in glycosidase inhibition is in its infancy, the critical role of these enzymes in cellular activities, from glycoprotein biosynthesis to carbohydrate metabolism, warrants a strong development of this new family of CD deriva-tives in the near future, with biomedical applications foreseen in diseases such as cancer or lysosomal storage disorders.[226]

5.11.5 Cyclodextrin-Based Antitoxins

Among the variety of virulence factors secreted by pathogens, there is a sig-nificant group of multicomponent toxins that are capable of forming pores in their host cell membrane and using them to translocate toxicity factors into the cytoplasm. These pores are usually constituted by structurally well-defined protein multimers exhibiting axial symmetry that attach to more or less

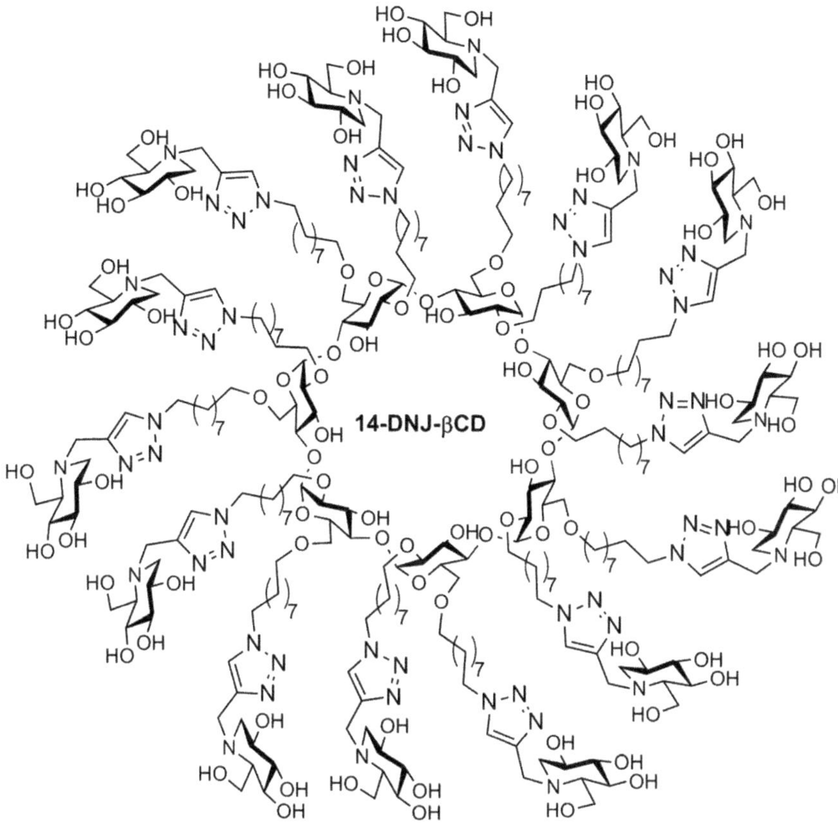

Figure 5.21 Structure of the multivalent mannosidase inhibitor discovered by Compain and Ortiz Mellet.[225] DNJ =1-deoxynojirimycin.

ubiquitous cell membrane receptors.[227] The idea of blocking the pores to inhibit toxin virulence using symmetry and size-complementary compounds represents an attractive goal towards which the use of well-defined CD platforms has greatly contributed.[228] Karginov and co-workers have pioneered this concept for the case of anthrax toxin, probably the most alarming case due to its dreadful potential as biological weapon. The authors showed that βCD derivatives homogeneously substituted with cationic groups can efficiently block the heptameric pores formed by anthrax toxin in the host cell membrane, which exhibit a relatively high density of anionic amino acids (Figure 5.22).[229] Different series of βCD-centered cationic clusters have been synthesized aiming at elucidating the structural basis for efficient anthrax antitoxin activity.[230,231] Some of these compounds have already proven efficient to protect mice from *Bacillus anthracis* infection.[232] Moreover, βCD-scaffolding has also proven useful to achieve specific peptide-mediated anthrax toxin inhibition.[233] Interestingly, this strategy has also being shown to operate for other heptameric pore-forming toxin-mediated infections (*e.g. Clostridium botulinum* and

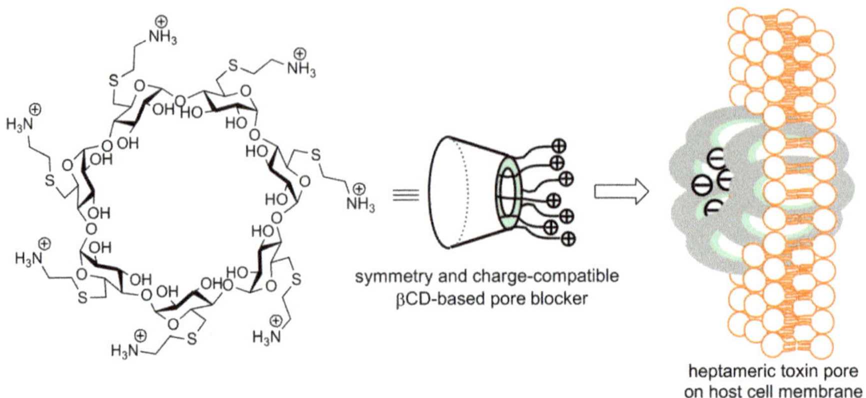

Figure 5.22   Schematic illustration of the symmetry complementarity-guided design of anthrax toxin inhibitors developed by Karginov.[229]

C. perfringens),[234] highlighting the broad applicability of the CD-based "symmetry complementarity" concept in antitoxin design.

5.11.6   Cyclodextrin-Based Antagonists of Neuromuscular Blocking Drugs

A milestone in the therapeutic application of CD-based drugs is the case of sugammadex, an anionic γCD derivative (octakis[6-(3-carboxyethy)thio]γCD; Figure 5.23) commercialized by Merck under the name Bridion® to reverse the action of steroidal neuromuscular blocking drugs (NMBDs), such as rocuronium.[235–237] Nondepolarizing steroidal NMBDs represented a substantial progress in modern anesthesiology, but still not fully devoid of potential unwanted effects during surgical intervention and postanesthesia. To limit the drawbacks, reversal agents with their own potential risks (*e.g.* cholinesterase inhibitors) have long being used. Zhang and co-workers envisioned an innovative strategy relying on the encapsulation of rocuronium with an exogenous CD host antagonist. Interestingly, the *in vitro* potency of the natural CDs to reverse rocuronium-induced neuromuscular block correlated quite well with cavity sizes. γCD was the best fitting size but the binding constant towards rocuronium was far below that required for efficient *in vivo* "sequestration". Guided by molecular modeling and titration experiments, they disclosed the structure of sugammadex, a γCD derivative featuring an expanded hydrophobic cavity and rocuromium-complementary anionic groups.[238,239] Sugammadex exerts its effect by forming a 1 : 1 complex with rocuronium (K_{as} up to 1.8×10^7 M⁻¹, the highest ever published for a γCD complex!) and other steroidal NMBDs. This creates a concentration gradient favoring the movement of the remaining NMBD molecules from the neuromuscular junction back into the plasma, thereby facilitating rapid reversal of neuromuscular blockage.[240] The unparalleled clinical performance of the rocuronium/sugammadex pair has revolutionized anesthesiology and dramatically influenced the preference of

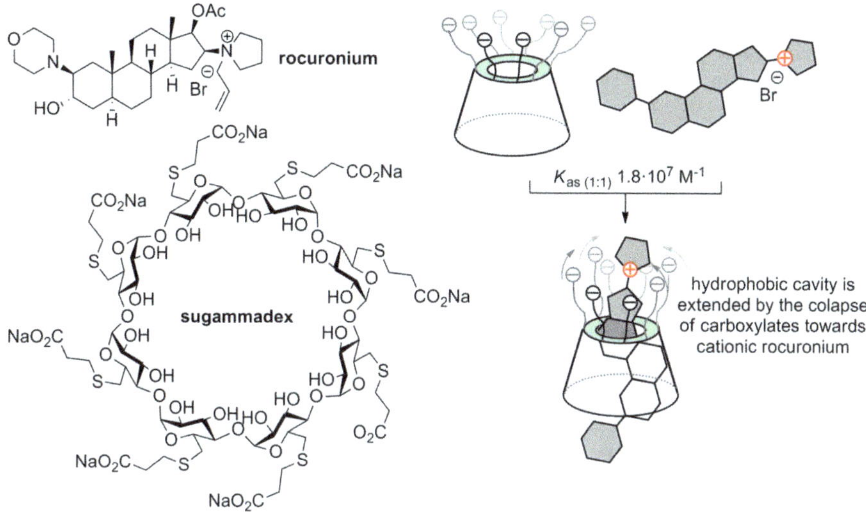

Figure 5.23 Structure of sugammadex, the γCD-based antagonist of rocuronium, and schematic representation mechanism for induced-fit complex formation and sequestration.[238]

anesthesiologists.[241] Despite other drugs in the market featuring a more benign neuromuscular blockage mechanism, the lack of safe and fast reversal antagonists comparable to sugammadex is working against them.

5.11.7 Cyclodextrin-Based Sensitizers in Photodynamic Therapy

An arena that has also benefited from the cross-field interaction with cyclodextrins is photodynamic therapy (PDT). PDT takes advantage of the light-induced excitation of a photosensitizer which, either directly or indirectly, interacts with cells and/or tissues to elicit a preconceived therapeutic effect. In many ways, light represents a invasive stimulus to elicit a therapeutic activity since accurate control of site timing and dose are feasible; yet, *in vivo* applications are limited to areas that can be effectively irradiated. A thorough survey of recent developments in PDT, towards which CDs have substantially contributed by providing suitable vehicles to protect the photosensitizer in biological media,[242,243] is provided in Chapter 15 in this volume. The example commented on in this section relates to a case in which the therapeutically active species is a CD–photosensitizer conjugate, an approach that, in principle, should facilitate a higher degree of sophistication.

Within the framework of a project devoted to the development of intelligent anticancer drugs, Ng and co-workers designed an axially bis-substituted Si(IV)-phthalocyanine derivative bearing two Si-appended permethylated βCD units (Figure 5.24). This conjugate was able to form a stable 1 : 1 complex with a tetrasulfonated porphyrin, the resulting system exhibiting remarkable light-harvesting properties. Irradiation at the absorption wavelength of the

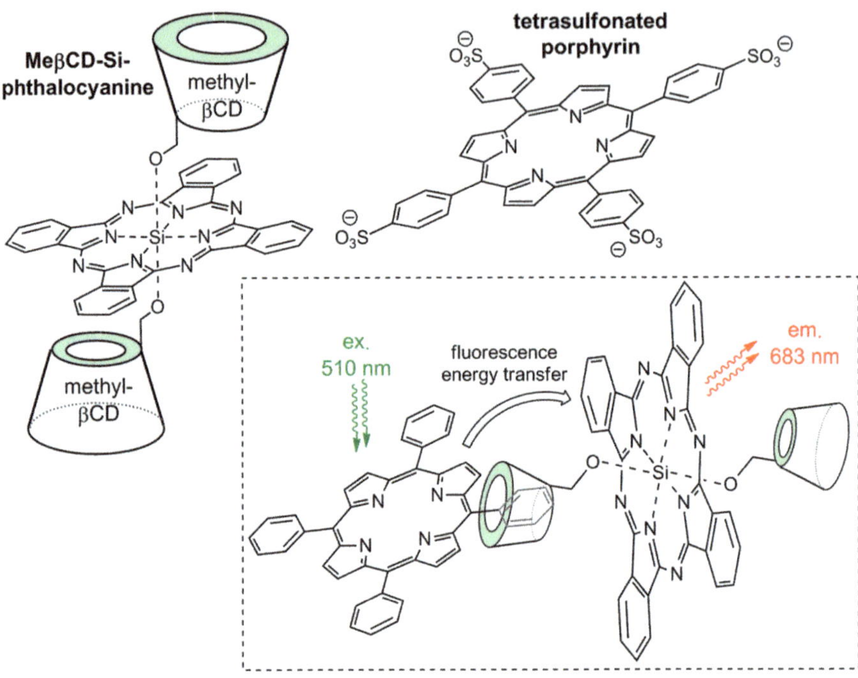

Figure 5.24 Structures of the cyclodextrin-conjugated Si(IV)-phthalocyanine and the tetrasulfonated porphyrin described by Ng[244] and schematic representation of their operational mechanism based on fluorescence resonance energy transfer.

porphyrin component then resulted in fluorescence resonance energy transfer (FRET) to the phthalocyanine moiety, leading to an extremely efficient cytotoxic effect towards HT29 human colon adenocarcinoma cells ($IC_{50} = 90$ nM).[244] The molecular design is amenable to structural modifications and the authors have demonstrated that this can be exploited to further improve efficiency.[245,246]

5.12 Cyclodextrin-Based Sensing Devices

There is considerable interest in the implementation of sensing and diagnostic devices for direct measurement of biological parameters for biomedical, environmental or technological purposes. Indeed, sensing, broadly speaking, has turned out to be one of the leitmotivs of nanosciences and nanotechnologies. The supramolecular and scaffolding abilities of CDs did not go unnoticed in this context. They have been profusely exploited in the design of integrated devices for a wide range of applications. Hereinafter, some recent examples are described, selected to avoid overlap with those discussed in more depth in chapters 2, 4 and 10. Altogether, they illustrate the versatility of CDs to build up useful biomedical sensing devices.

5.12.1 Cyclodextrin-Based Pathogen and Allergen Sensors

Smart implementation of the self-assembling capabilities of CDs has proven very fruitful for the design of integrated systems for the fast and reliable detection of pathogens or allergens. This is remarkably illustrated by the case of the electrochemical biosensor for celiac condition-related antibodies which was recently described by Fragoso and co-workers.[247] These authors constructed a supramolecular platform consisting on a βCD-modified gold surface onto which carboxymethylcellulose that had been simultaneously decorated with adamantane (as high affinity ligands for the βCD cavity) and antigenic fragments of antigliadin antibodies (a diagnostic marker for the gluten-intolerance condition) was supramolecularly attached (Figure 5.25). Specific antibody adsorption on the chip surface could then be amperometrically monitored. The authors demonstrated that the sensor could detect antigliadin antibodies from serum samples from patients with the celiac condition at comparable efficiency to less cost-effective immunosorbent assays (ELISAs).

Taking advantage of specific lectin–carbohydrate interactions and of the complementarity between adamantane moieties and βCD cavities, Seeberger and co-workers have designed a supramolecular fluorescent probe for visualizing strains of *Escherichia coli*.[248] This sensing construct is built by self-assembling primary-face-linked α-D-mannopyranosyl-βCD heptaconjugates onto a Ru(II) fluorescent core grafted with six adamantane units (Figure 5.26). The system succeeded at identifying mannose-specific receptor presenting cells (*e.g. E. coli* ORN178) by confocal microscopy, highlighting the potential of adapting the prototype to sense other bacterial strains.

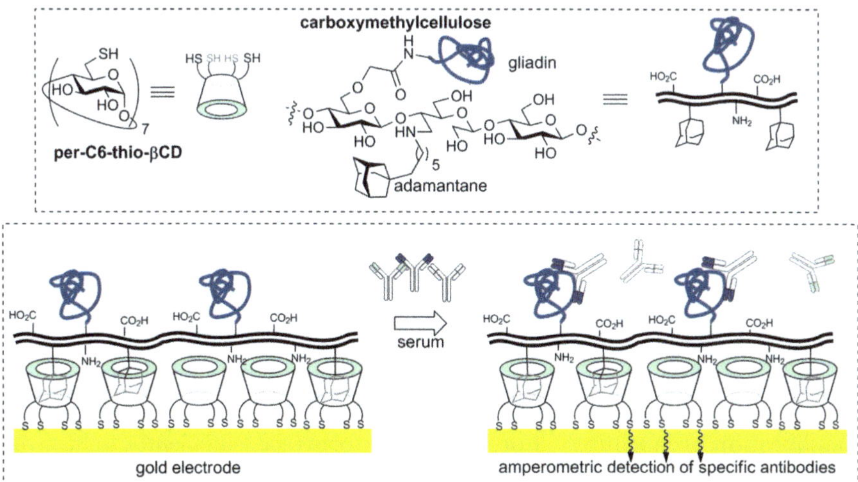

Figure 5.25 Schematic illustration of the celiac-condition biosensor engineered by Fragoso.[247] Structure of the biosensor components (the thiolated cyclodextrin host and the gliadin/adamantane-grafted carboxymethylcellulose polymer) and schematic representation of its construction and operational mechanism.

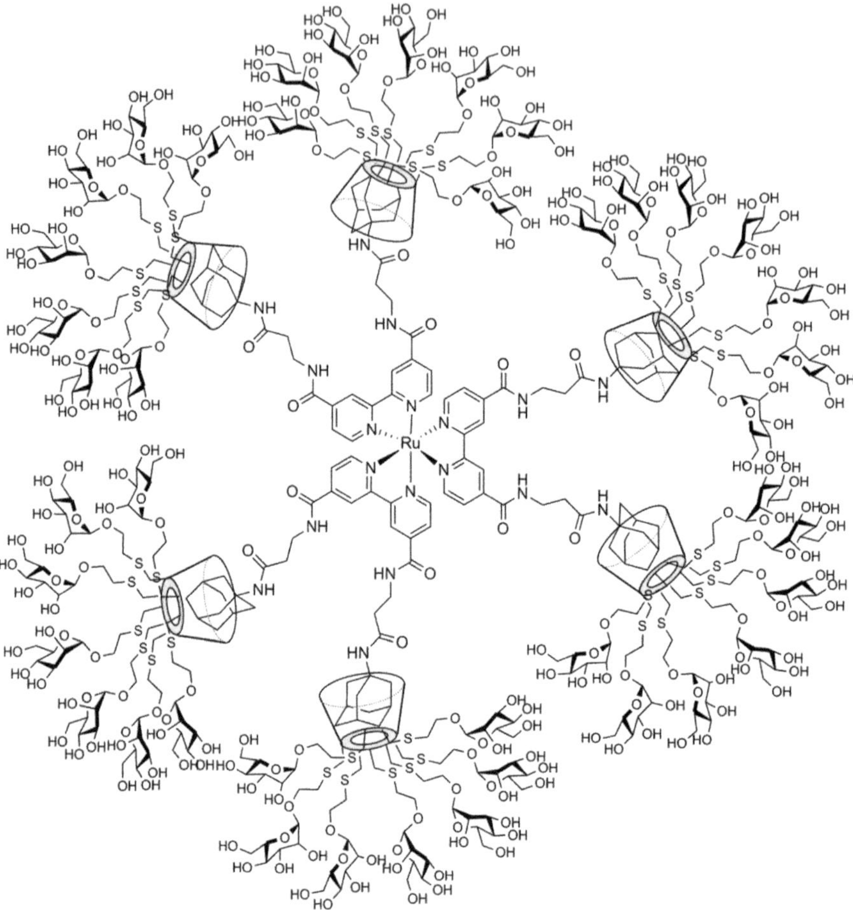

Figure 5.26   Structure of the hypermannosylated supramolacular construct developed by Seeberger's group for bacterial sensing.[248]

5.12.2   Cyclodextrin-Based Nucleotide Sequencers

The last few years have witnessed the emergence of nanopore-based devices for single-molecule detection and analysis. One of the most fascinating applications of this technology is oligonucleotide sequencing by engineering the nanopore topology, so as to provide a highly confined space within which single oligo-nucleotides can be electrophoretically driven and analyzed without the need for amplification or labeling.[249] For optimal performance, sufficiently tight fitting between the nanopore internal surface and the threading oligonucleotide sequence should be achieved. Bayley and co-workers have shown the usefulness of CD adapters for this task. They demonstrated that α-hemolysin (αHL) nanopores with a covalently attached CD moiety at a precise position in its barrel could continuously identify unlabeled nucleoside 5-monophosphate molecules with accuracyover 99% (Figure 5.27).[250] Moreover, the authors have

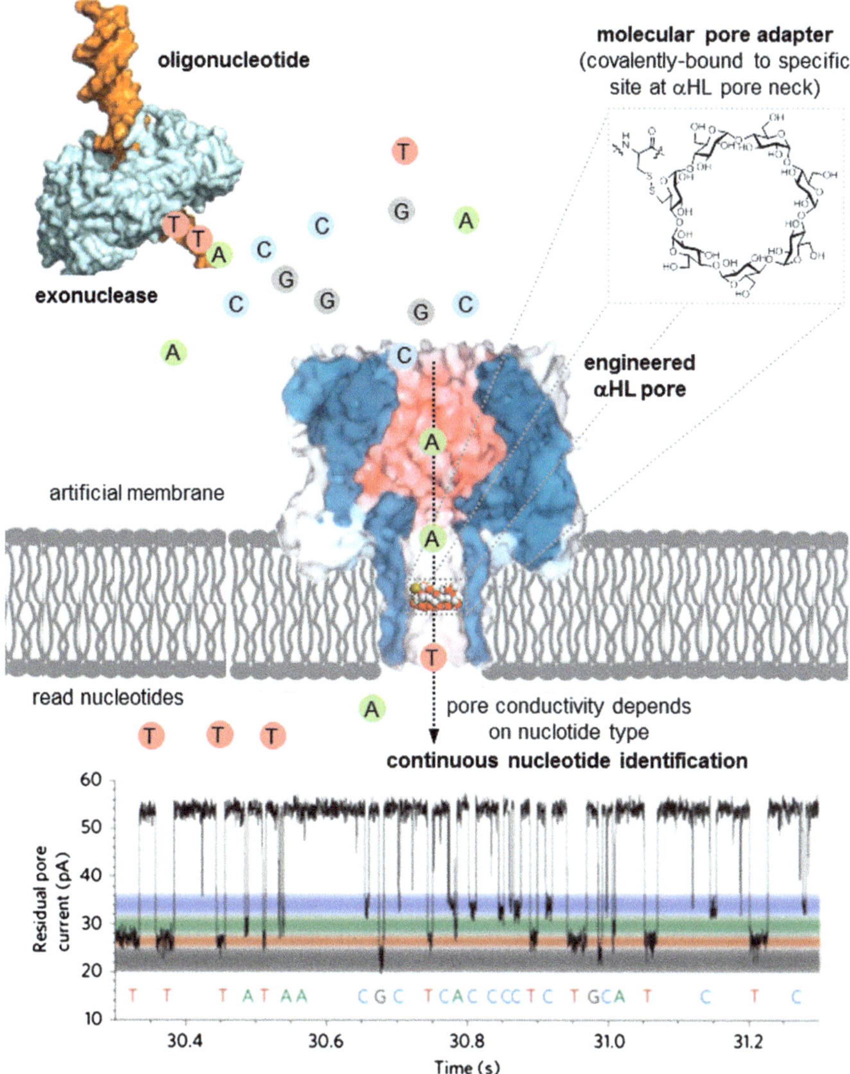

Figure 5.27 Continuous detection of nucleotides, enzymatically cleaved from a single DNA strain by a βCD-engineered αHL nanopore. Residual pore current correlates with nucleotide identity with an accuracy exceeding 99%.[250] (Adapted with permission from Macmillan Publishers Ltd: *Nat. Nanotechnol.*, 2009, **4**, 265, copyright 2009, and *Nat. Biotechnol.*, 2008, **26**, 1146, copyright 2008.)

demonstrated that the discriminating abilities of the CD-modified nanopore can be juggled with the action of exonucleases that would be required to sequentially trim off the terminal nucleotide from an oligonucleotide chain. A nanopore set-up combining both trimming/identifying activities is still to be published, though earlier in 2012, Oxford Nanopore Technologies already announced their success in DNA nanopore sequencing using this approach with a useful initial error rate and kilobase-reading capabilities.[251,252]

5.13 Conclusion and Perspectives

The astonishing array of applications presented in the previous sections illustrates the paradigm of CD research, a case in which sustained efforts on fundamental research over long periods (more than a century!) can transform an investigational oddity into an interconnecting node for many other research fields and an expanding technological arena of yet unforeseen revenues. The accumulated chemical, supramolecular, pharmaceutical and technological knowledge constitutes a valuable arsenal for rational engineering at the molecular level—sophisticated systems programmed to solve unmet problems.

The versatility of the CD platform has also cleared the way to rational design of a number of revolutionary therapeutic candidates. The examples collected in this chapter are just the "tip of the iceberg" of a yet maturing expertise. In the near future, refinement of some of these concepts will probably facilitate the design of CD-based systems for combined therapies, where simultaneous delivery of different therapeutic cargos or multi-functional formulations is required.[253,254] An exciting advance in this regard would be the ability to combine targeted imaging and therapeutic activity within the same system, allowing simultaneous visualizing of the disease target and local delivery. First delineations of such CD-based *theranostic* systems have recently been published.[255,256] Without doubt, enlargement of the available methodologies for selective chemical derivatization will foster the discovery of CD-based drug candidates for known and new therapeutic targets.

Acknowledgements

The Spanish Ministerio de Economía y Competitividad (contract numbers CTQ2010-15848 and SAF2010-15670; co-financed with the Fondo Europeo de Desarrollo Regional, FEDER).

References

1. A. Villiers, *Compt. Rend.*, 1891, **112**, 412.
2. A. Villiers, *Compt. Rend.*, 1891, **112**, 536.
3. F. Schardinger, *Z. Unters. Nahr. Genussm.*, 1903, **6**, 865.
4. F. Schardinger, *Wiener Klin. Wochenschr.*, 1904, **17**, 207.
5. F. Schardinger, *Zentralbl. Bakteriol. Parasitenkd. Abt. II*, 1905, **14**, 772.
6. F. Schardinger, *Zentralbl. Bakteriol. Parasitenkd. Abt. II*, 1911, **29**, 188.
7. K. Freudenberg and M. Meyer-Delius, *Ber. Chem.*, 1938, **71**, 1596.
8. K. Freudenberg, E. Schaaf, G. Dumpert and T. Ploetz, *Naturwissenschaften*, 1939, **27**, 850.
9. F. Cramer, *Angew. Chem.*, 1952, **64**, 136.
10. F. Cramer and W. Kampe, *J. Am. Chem. Soc.*, 1965, **87**, 1115.

11. F. Cramer and W. Dietsche, *Chem. Ber. Rec.*, 1959, **92**, 378.

12. K. H. Frömming and J. Szejtli, *Cyclodextrins in Pharmacy*, Kluwer Academic Publishers, Dordrecht, 1994.

13. H. Doziuk, ed. *Cyclodextrins and Their Complexes*, Wiley-VCH, Weinheim, 2006.

14. T. Loftsson and D. Duchêne, *Int. J. Pharm.*, 2007, **329**, 1.

15. M. E. Davies and M. E. Brewster, *Nat. Rev. Drug Discov.*, 2004, **3**, 1024.

16. J. Szejtli, *Chem. Rev.*, 1998, **98**, 1743.

17. T. Loftsson and M. E. Brewster, *J. Pharm. Pharmacol.*, 2010, **62**, 1607.

18. E. Bilensoy *Cyclodextrins in Pharmaceutics, Cosmetics and Biomedicine*, Wiley-VCH, Weinheim, 2011.

19. F. J. Otero-Espinar, A. Luzardo-Álvarez and J. Blanco-Méndez, *Mini-Rev. Med. Chem.*, 2010, **10**, 715.

20. A. R. Hedges, *Chem. Rev.*, 1998, **98**, 2035.

21. L. Szente and J. Szetjli, *Trends Food Sci. Technol.*, 2004, **14**, 137.

22. D. Duchêne and R. Gref, in *Cyclodextrins in Pharmaceutics, Cosmetics and Biomedicine*, ed. E. Bilensoy, Wiley-VCH, Weinheim, 2011, ch. 20, p. 371.

23. G. Crini, *Prog. Polym. Sci.*, 2005, **30**, 38.

24. G. Crini and M. J. Morcellet, *J. Sep. Sci.*, 2002, **25**, 789.

25. C. Baudin, C. Pean, B. Perly and P. Goselin, *Int. J. Environ. Anal. Chem.*, 2000, **77**, 233.

26. D. Landy, I. Mallard, A. Ponchel, E. Monflier and S. Fourmentin, *Environ. Chem. Lett.*, 2012, **10**, 225.

27. H. Arima, F. Hirayama, C. T. Okamoto and K. Uekama, *Recent Res. Devel. Chem. Pharm. Sci.*, 2002, **2**, 155.

28. K. Cal and K. Centkowska, *Eur. J. Pharm. Biopharm.*, 2008, **68**, 467.

29. K. Uekama and M. Otagiri, *Crit. Rev. Ther. Drug Carrier Syst.*, 1987, **3**, 1.

30. C. R. Lee and J. A. Balfour, *Drugs*, 1994, **48**, 907.

31. D. A. Stephens, *Pharmacotherapy*, 1999, **19**, 603.

32. L. Matilainen, T. Toropainen, H. Vihola, J. Hirvonen, T. Järvinen, P. Jarho and K. Järvinen, *J. Control. Release*, 2008, **126**, 10.

33. Y. Toyama, E. Pais, H. J. Meiselman and T. Alexy, *Clin. Hemorheol. Microcirc.*, 2007, **36**, 173.

34. M. E. Brewster and T. Loftsson, *Adv. Drug Deliv. Rev.*, 2007, **59**, 645.

35. B. Cappello, C. Carmignani, M. Iervolino, M. I. La Rotonda and M. F. Saettone, *Int. J. Pharm.*, 2001, **213**, 75.

36. R. D. Colucci, C. Wright IV, F. H. Mermelstein, D. G. Gawarecki and D. B. Carr, *Acute Pain*, 2009, **11**, 15.

37. O. Reer, T. K. Bock and B. W. Müller, *J. Pharm. Sci.*, 1994, **83**, 1345.

38. G. Ragno, E. Cione, A. Garofalo, G. Genchi, G. Ioele, A. Risoli and A. Spagnoletta, *Int. J. Pharm.*, 2003, **265**, 125.

39. S. Davaran, M. R. Rashidi, R. Khandaghi and M. Hashemi, *Pharmacol. Res.*, 2005, **51**, 233.

40. S. K. Rodal, G. Skretting, O. Garred, F. Vilhardt, B. van Deurs and K. Sandvig, *Mol. Biol. Cell*, 1999, **10**, 961.
41. T. Irie and K. Uekama, *Adv. Drug Deliv. Rev.*, 1999, **36**, 101.
42. M. E. Brewster, J. W. Simpkins, M. S. Hora, W. C. Stern and N. Bodor, *J. Parenter. Sci. Technol.*, 1989, **43**, 231.
43. R. A. Rajewski and V. J. Stella, *J. Pharm. Sci.*, 1996, **85**, 1142.
44. F. L. Aachmann, D. E. Otzen, K. L. Larsen and R. Wimmer, *Protein Eng.*, 2003, **16**, 905.
45. T. Serno, R. Geidobler and G. Winter, *Adv. Drug Deliv. Rev.*, 2011, **63**, 1086.
46. A. K. Banga and R. Mitra, *J. Drug Target*, 1993, **1**, 341.
47. S. Tavornvipas, S. Tajiri, F. Hirayama, H. Arima and K. Uekama, *Pharm. Res.*, 2004, **21**, 2369.
48. M. E. Brewster, M. S. Hora, J. W. Simpkins and N. Bodor, *Pharm. Res.*, 1991, **8**, 792.
49. S. A. Charman, K. L. Mason and W. N. Charman, *Pharm. Res.*, 1993, **10**, 954.
50. T. Serno, J. F. Carpenter, T. W. Randolph and G. Winter, *J. Pharm. Sci.*, 2010, **99**, 1193.
51. A. Desai, C. Lee, L. Sharma and A. Sharma, *Biochimie*, 2006, **88**, 1435.
52. S. Tavornvipas, F. Hirayama, S. Takeda, H. Arima and K. Uekama, *J. Pharm. Sci.*, 2006, **95**, 2722.
53. N. Karuppiah and A. Sharma, *Biochem. Biophys. Res. Commun.*, 1995, **211**, 60.
54. L. Sharma and A. Sharma, *Eur. J. Biochem.*, 2001, **268**, 2456.
55. J. Horský and J. Pitha, *J. Incl. Phenom. Mol. Recogn. Chem.*, 1994, **18**, 291.
56. D. Rozema and S. H. Gellman, *J. Am. Chem. Soc.*, 1995, **117**, 2373.
57. D. Rozema and S. H. Gellman, *J. Biol. Chem.*, 1996, **271**, 3478.
58. D. L. Daugherty, D. Rozema, P. E. Hanson and S. H. Gellman, *J. Biol. Chem.*, 1998, **273**, 33 961.
59. Y. Sasaki, W. Asayama, T. Niwa, S.-i. Sawada, T. Ueda, H. Taguchi and K. Akiyoshi, *Macromol. Biosci.*, 2011, **11**, 814.
60. Y. Nomura, M. Ikeda, N. Yamaguchi, Y. Aoyama and K. Akiyoshi, *FEBS Lett.*, 2003, **553**, 271.
61. K. Uekama, F. Hirayama and T. Irie, *Chem. Rev.*, 1998, **98**, 2045.
62. A. R. Khan, P. Forgo, K. J. Stine and V. T. D'Souza, *Chem. Rev.*, 1998, **98**, 1977.
63. M. J. González-Álvarez, P. Balbuena, C. Ortiz Mellet, J. M. García Fernández and F. Mendicuti, *J. Phys. Chem. B*, 2008, **112**, 13 717.
64. A. J. Pearce and P. Sinay, *Angew. Chem. Int. Ed.*, 2000, **39**, 3610.
65. T. Lecourt, A. Herault, A. J. Pearce, M. Sollogoub and P. Sinay, *Chem. Eur. J.*, 2001, **10**, 2960.
66. P. Balbuena, D. Lesur, M. J. González Álvarez, F. Mendicuti, C. Ortiz Mellet and J. M. García Fernández, *Chem. Commun.*, 2007, 3270.
67. S. Guieu and M. Sollogoub, *J. Org. Chem.*, 2008, **73**, 2819.

68. R. Ghosh, P. Zhang, A. Wang and C.-C. Ling, *Angew. Chem. Int. Ed.*, 2012, **51**, 1548.

69. H. Law, J. M. Benito, J. M. García Fernández, L. Jicsinszky, S. Crouzy and J. Defaye, *J. Phys. Chem. B.*, 2011, **115**, 7524.

70. M. J. González Álvarez, A. Méndez-Ardoy, J. M. Benito, J. M. García Fernández and F. Mendicuti, *J. Photochem. Photobiol. A – Chem.*, 2011, **223**, 25.

71. N. Masurier, O. Lafont, R. Le Provost, D. Lesur, P. Masson, F. Djedaïni-Pilard and F. Estour, *Chem. Commun.*, 2009, 589.

72. K. Matsuda, T. Inazu, K. Haneda, M. Mizuno, T. Yamanoi, K. Hattori, K. Yamamoto and H. Kumagai, *Bioorg. Med. Chem. Lett.*, 1997, **7**, 2353.

73. J. M. Benito, M. Gómez García, C. Ortiz Mellet, I. Baussanne, J. Defaye and J. M. García Fernández, *J. Am. Chem. Soc.*, 2004, **126**, 10 355.

74. C. Bies, C.-M. Lehr and J. F. Woodley, *Adv. Drug Deliv. Rev.*, 2004, **56**, 425.

75. K. Hattori, A. Kenmoku, T. Mizuguchi, D. Ikeda, M. Mizuno and T. Inazu, *J. Incl. Phenom. Macrocycl. Chem.*, 2006, **56**, 9.

76. Y. Oda, H. Yanagisawa, M. Maruyama, K. Hattori and T. Yamanoi, *Bioorg. Med. Chem.*, 2008, **16**, 8830.

77. G. J. L. Bernardes, R. Kikkeri, M. Maglinao, P. Laurino, M. Collot, S. Y. Hong, B. Lepenies and P. H. Seeberger, *Org. Biomol. Chem.*, 2010, **8**, 4987.

78. H. Zhang, Z. Cai, Y. Sun, F. Yu, Y. Chen and B. Sun, *J. Biomed. Mater. Res. Part A*, 2012, **100**, 2441.

79. P. Caliceti, S. Salmaso, A. Semenzato, T. Carofiglio, R. Fornasier, M. Fermeglia, M. Ferrone and S. Pricl, *Bioconj. Chem.*, 2003, **14**, 899.

80. F. van de Manakker, T. Vermonden, C. F. van Nostrum and W. E. Hennink, *Biomacromolecules*, 2009, **10**, 3157.

81. R. Gref and D. Duchêne, *J. Drug Deliv. Sci. Tech.*, 2012, **22**, 223.

82. J. Szeman, E. Fenyvesi, J. Szejtli, H. Ueda, Y. Machida and T. Nagai, *J. Incl. Phenom*, 1987, **5**, 427.

83. E. Fenyvesi, *J. Incl. Phenom*, 1988, **6**, 537.

84. R. Jayakumar, M. Prabaharan, R. L. Reis and J. F. Mano, *Carbohydr. Polym*, 2005, **62**, 142.

85. M. Prabaharan and J. F. Mano, *Carbohydr. Polym.*, 2006, **63**, 153.

86. H. Sashiwa and S.-i. Aiba, *Prog. Polym. Sci.*, 2004, **29**, 887.

87. X. Zhang, X. Zhang, Z. Wu, X. Gao, C. Cheng, Z. Wang and C. Li, *Acta Biomater.*, 2011, **7**, 585.

88. N. Blanchemain, T. Laurent, F. Chai, C. Neut, S. Haulon, V. Krump-Konvalinkova, M. Morcellet, B. Martel, C. J. Kirkpatrick and H. F. Hildebrand, *Acta Biomater.*, 2008, **4**, 1725.

89. M. Weickenmeier and G. Wenz, *Macromol. Rapid Commun.*, 1996, **17**, 731.

90. S. H. Pun, N. C. Bellocq, A. Liu, G. Jensen, T. Machemer, E. Quijano, T. Schluep, S. Wen, H. Engler, J. Heidel and M. E. Davis, *Bioconj. Chem.*, 2004, **15**, 831.

91. J. Suh, S. H. Lee and K. D. Zoh, *J. Am. Chem. Soc.*, 1992, **114**, 7916.
92. H. Wang, N. Shao, S. Qiao and Y. Cheng, *J. Phys. Chem. B*, 2012, **116**, 11217.
93. F. Kihara, H. Arima, T. Tsutsumi, F. Hirayama and K. Uekama, K, *Bioconj. Chem.*, 2002, **13**, 1211.
94. P. Mukhopadhyay, R. Mishra, D. Rana and P. P. Kundu, *Prog. Polym. Sci.*, 2012, **37**, 1457.
95. A. Chaudhury and S. Das, *AAPS PharmSciTech*, 2011, **12**, 10.
96. N. Zhang, J. Li, W. Jiang, C. Ren, J. Li, J. Xin and K. Li, *Int. J. Pharm.*, 2010, **393**, 213.
97. S. Sajeesh and C. P. Sharma, *Int. J. Pharm.*, 2006, **325**, 147.
98. A. F. Soares, R. A. Carvalho and F. Veiga, *Nanomedicine*, 2007, **2**, 183.
99. P. Mura, M. T. Faucci, F. Maestrelli, S. Furlanetto and S. Pinzauti, *J. Pharm. Biomed. Anal.*, 2002, **29**, 1015.
100. J. Zhou and H. Ritter, *Polym. Chem.*, 2010, **1**, 1552.
101. P.-A. Faugeras, B. Boëns, P.-H. Elchinger, F. Brouillette, D. Montplaisir, R. Zerrouki and R. Lucas, *Eur. J. Org. Chem.*, 2012, 4087.
102. J. Zhang, K. Ellsworth and P. X. Ma, *Macromol. Rapid Commun.*, 2012, **33**, 664.
103. M. E. Davis, *Adv. Drug Deliv. Rev.*, 2009, **61**, 1189.
104. J. Cheng, K. T. Khin, G. S. Jensen, A. Liu and M. E. Davis, *Bioconj. Chem*, 2003, **14**, 1007.
105. J. Cheng, K. T. Khin and M. E. Davis, *Mol. Pharmaceut.*, 2004, **1**, 183.
106. H. Maeda, J. Wu, T. Sawa, Y. Matsumura and K. Hori, *J. Control. Release*, 2000, **65**, 271.
107. S. Gaur, L. Chen, T. Yen, Y. Wang, B. Zhou, M. E. Davis and Y. Yen, *Nanomed. Nanotechnol. Biol. Med.*, 2012, **8**, 721.
108. http://clinicaltrials.gov/ct2/show/NCT01380769 (last visited Jan 31st, 2013).
109. http://ceruleanrx.com/ (last visited Jan 31st, 2013).
110. T. Ooya and N. Yui, *J. Control. Release*, 1999, **58**, 251.
111. G. Wenz, B.-H. Han and A. Müller, *Chem. Rev.*, 2006, **106**, 782–817.
112. T. Higashi, F. Hirayama, H. Arima and K. Uekama, *Bioorg. Med. Chem. Lett.*, 2007, **17**, 1871.
113. R. Gref, C. Amiel, K. Molinard, S. Daoud-Mahammed, B. Sébille, B. Gillet, J.-C. Beloeil, C. Ringard, V. Rosilio, J. Poupaert and P. Couvreur, *J. Control. Release*, 2006, **111**, 316.
114. H. S. Choi, T. Ooya, S. Sasaki and N. Yui, *Macromolecules*, 2003, **36**, 5342.
115. H. S. Choi, K. M. Huh, T. Ooya and N. Yui, *J. Am. Chem. Soc.*, 2003, **125**, 6350.
116. M. D. Moya-Ortega, C. Alvarez-Lorenzo, A. Concheiro and T. Loftsson, *Int. J. Pharm.*, 2008, **428**, 152.
117. F. Trotta, R. Cavalli, K. Martina, M. Biasizzo, J. Vitillo, S. Bordiga, P. Vavia and K. Ansari, *J. Incl. Phenom. Macrocycl. Chem.*, 2011, **71**, 189.

118. L. Li, X. Guo, J. Wang, P. Liu, R. K. Prud'homme, B. L. May and S. F. Lincoln, *Macromolecules*, 2008, **41**, 8677.
119. A. Charlot and R. Auzély-Velty, *Macromolecules*, 2007, **40**, 9555.
120. A. Charlot and R. Auzély-Velty, *Macromolecules*, 2007, **40**, 1147.
121. A. Charlot, R. Auzély-Velty and M. Rinaudo, *J. Phys. Chem. B*, 2003, **107**, 8248.
122. S. Daoud-Mahammed, P. Couvreur, K. Bouchemal, M. Chéron, G. Lebas, C. Amiel and R. Gref, *Biomacromolecules*, 2009, **10**, 547.
123. E. Battistini, E. Gianolio, R. Gref, P. Couvreur, S. Fuzerova, M. Othman, S. Aime, B. Badet and P. Durand, *Chem. Eur. J.*, 2008, **14**, 4551.
124. T. Vermonden, R. Censi and W. E. Hennink, *Chem. Rev.*, 2012, **112**, 2853.
125. F. van de Manakker, K. Braeckmans, N. el Morabit, S. C. De Smedt, C. F. van Nostrum and W. E. Hennink, *Adv. Funct. Mater.*, 2009, **19**, 2992.
126. C. Galant, C. Amiel and L. Auvray, *Macromol. Biosci.*, 2005, **5**, 1057.
127. F. van de Manakker, M. van der Pot, T. Vermonden, C. F. van Nostrum and W. E. Hennink, *Macromolecules*, 2008, **41**, 1766.
128. K. Peng, I. Tomatsu and A. Kros, *Chem. Commun.*, 2010, **46**, 4094.
129. J. Li, X. Li, X. Ni, X. Wang, H. Li and K. W. Leong, *Biomaterials*, 2006, **27**, 4132.
130. N. Benkirane-Jessel, P. Schwinté, P. Falvey, R. Darcy, Y. Haïkel, P. Schaaf, J. C. Voegel and J. Ogier, *Adv. Funct. Mater.*, 2004, **14**, 174.
131. R. C. Smith, M. Riollano, A. Leung and P. T. Hammond, *Angew. Chem. Int. Ed.*, 2009, **48**, 8974.
132. S. Y. Wong, J. S. Moskowitz, J. Veselinovic, R. A. Rosario, K. Timachova, M. R. Blaisse, R. C. Fuller, A. M. Klibanov and P. T. Hammond, *J. Am. Chem. Soc.*, 2010, **132**, 17 840.
133. N. Jessel, M. Oulad-Abdelghani, F. Meyer, P. Lavalle, Y. Haïkel, P. Schaaf and J.-C. Voegel, *Proc. Natl. Acad. Sci. USA*, 2006, **103**, 8618.
134. X. Zhang, K. K. Sharma, M. Boeglin, J. Ogier, D. Mainard, J.-C. Voegel, Y. Mély and N. Benkirane-Jessel, *Nano Lett.*, 2008, **8**, 2432.
135. X. Chen, W. Wu, Z. Guo, J. Xin and J. Li, *Biomaterials*, 2011, **32**, 1759.
136. E. Bilensoy and A. A. Hincal, *Expert Opin. Drug Deliv.*, 2009, **6**, 1161.
137. E. Memisoglu-Bilensoy, I. Vural, A. Bochot, J. M. Renoir, D. Duchene and A. A. Hincal, *J. Control. Release*, 2005, **104**, 489.
138. E. Bilensoy, O. Gürkaynak, A. L. Doğan and A. A. Hincal, *Int. J. Pharm.*, 2008, **347**, 163.
139. Y. Cirpanli, E. Bilensoy, A. L. Doğan and S. Çaliş, *Eur. J. Pharm. Biopharm.*, 2009, **73**, 82.
140. S. Sortino, A. Mazzaglia, L. M. Scolaro, F. M. Merlo, V. Valveri and M. T. Sciortino, *Biomaterials*, 2006, **27**, 4256.
141. M. Skiba, D. Wouessidjewe, F. Puisieux, D. Duchêne and A. Gulik, *Int. J. Pharm.*, 1996, **142**, 121.
142. E. Lemos-Senna, D. Wouessidjewe, S. Lesieur and D. Duchêne, *Int. J. Pharm.*, 1998, **170**, 119.

143. A. Méndez-Ardoy, M. Gómez-García, A. Gèze, J. L. Puteaux, D. Wouessidjewe, C. Ortiz Mellet, J. Defaye, J. M. García Fernández and J. M. Benito, *Med. Chem.*, 2012, **8**, 524.

144. C. Ortiz Mellet, J. M. García Fernández and J. M. Benito, *Chem. Soc. Rev.*, 2011, **40**, 1586.

145. D. Putnam, *Nat. Mater.*, 2006, **5**, 439.

146. K. A. Whitehead, R. Langer and D. G. Anderson, *Nat. Rev. Drug Discov.*, 2009, **8**, 129.

147. M. E. Davis, *Mol. Pharmaceut.*, 2009, **6**, 659.

148. H. Gonzalez, S. J. Hwang and M. E. Davis, *Bioconj. Chem.*, 1999, **10**, 1068.

149. S. J. Hwang, N. C. Bellocq and M. E. Davis, *Bioconj. Chem.*, 2001, **12**, 280.

150. T. M. Reineke and M. E. Davis, *Bioconj. Chem.*, 2003, **14**, 247.

151. T. M. Reineke and M. E. Davis, *Bioconj. Chem.*, 2003, **14**, 255.

152. S. R. Popielarski, S. Mishra and M. E. Davis, *Bioconj. Chem.*, 2003, **14**, 672.

153. M. E. Davis, S. H. Pun, N. C. Bellocq, T. M. Reineke, S. R. Popielarski, S. Mishra and J. D. Heidel, *Curr. Med. Chem.*, 2004, **11**, 179.

154. S. Mishra, P. Webster and M. E. Davis, *Eur. J. Cell Biol.*, 2004, **83**, 97.

155. N. C. Bellocq, S. H. Pun, G. S. Jensen and M. E. Davis, *Bioconj. Chem.*, 2003, **14**, 1122.

156. D. W. Bartlett and M. E. Davis, *Bioconj. Chem.*, 2007, **18**, 456.

157. S. H. Pun, F. Tack, N. C. Bellocq, J. Cheng, B. H. Grubbs, G. S. Jensen, M. E. Davis, M. Brewster, M. Janicot, B. Janssens, W. Floren and A. Bakker, *Cancer Biol. Ther.*, 2004, **3**, 641.

158. S. Hu-Lieskovan, J. D. Heidel, D. W. Bartlett, M. E. Davis and T. J. Triche, *Cancer Res.*, 2005, **65**, 8984.

159. D. W. Bartlett and M. E. Davis, *Biotechnol. Bioeng.*, 2008, **99**, 975.

160. J. D. Heidel, Z. Yu, J. Y. Liu, S. M. Rele, Y. Liang, R. K. Zeidan, D. J. Kornbrust and M. E. Davis, *Proc. Natl. Acad. Sci. USA*, 2007, **104**, 5715.

161. M. E. Davis, J. E. Zuckerman, C. H. J. Choi, D. Seligson, A. Tolcher, C. A. Alabi, Y. Yen, J. D. Heidel and A. Ribas, *Nature*, 2010, **464**, 1067.

162. http://clinicaltrials.gov/ct2/show/NCT00689065 (last visited Jan 31st, 2013).

163. K. Wada, H. Arima, T. Tsutsuma, Y. Chihara, K. Hattori, F. Hirayama and K. Uekama, *J. Control. Release*, 2005, **104**, 397.

164. H. Arima, Y. Chihara, M. Arizono, S. Yamashita, K. Wada, F. Hirayama and K. Uekama, *J. Control. Release*, 2006, **116**, 64.

165. H. Arima, S. Yamashita, Y. Mori, Y. Hayashi, K. Motoyama, K. Hattori, T. Takeuchi, H. Jono, Y. Ando, F. Hirayama and K. Uekama, *J. Control. Release*, 2010, **146**, 106.

166. Y. Hayashi, Y. Mori, S. Yamashita, K. Motoyama, T. Higashi, H. Jono, Y. Ando and H. Arima, *Mol. Pharmaceut.*, 2012, **9**, 1645.

167. Q.-Y. Jiang, L.-H. Lai, J. Shen, Q.-Q. Wang, F.-J. Xu and G.-P. Tang, *Biomaterials*, 2011, **32**, 7253.

168. M. Liu, Z. H. Li, F. J. Xu, L. H. Lai, Q.-Q. Wang, G.-P. Tang and W. T. Yang, *Biomaterials*, 2012, **33**, 2240.
169. T. Ooya, H. S. Choi, A. Yamashita, N. Yui, Y. Sugaya, A. Kano, A. Maruyama, H. Akita, R. Ito, K. Kogure and H. Harashima, *J. Am. Chem. Soc.*, 2006, **128**, 3852.
170. J. Li, C. Yang, H. Li, X. Wang, S. H. Goh, J. L. Ding, D. Y. Wang and K. W. Leong, *Adv. Mater.*, 2006, **18**, 2969.
171. J. Li and X. J. Xian, *Adv. Drug Deliv. Rev.*, 2008, **60**, 1000.
172. Y. Zhou, H. Wang, C. Wang, Y. Li, W. Lu, S. Chen, J. Luo, Y. Jiang and J. Chen, *Mol. Pharmaceut.*, 2012, **9**, 1067.
173. S.-A. Cryan, A. Holohan, R. Donohue, R. Darcy and C. M. O'Driscoll, *Eur. J. Pharm. Sci.*, 2004, **21**, 625.
174. S. Srinivasachari, K. M. Fichter and T. M. Reineke, *J. Am. Chem. Soc.*, 2008, **130**, 4618.
175. N. Mourtzis, K. Eliadou, C. Aggelidou, V. Sophianopoulou, I. M. Mavridis and K. Yannakopoulou, *Org. Biomol. Chem.*, 2007, **5**, 125.
176. N. Mourtzis, M. Paravatou, I. M. Mavridis, M. L. Roberts and K. Yannakopoulou, *Chem. Eur. J.*, 2008, **14**, 4188.
177. V. Bennevault-Celton, A. Urbach, O. Martin, C. Pichon, P. Guégan and P Midoux, *Bioconj. Chem.*, 2011, **22**, 2404.
178. C. Ortiz Mellet, J. M. Benito and J. M. García Fernández, *Chem. Eur. J*, 2010, **16**, 6728.
179. A. Díaz-Moscoso, L. Le Gourriérec, M. Gómez-García, J. M. Benito, P. Balbuena, F. Ortega-Caballero, N. Guilloteau, C. Di Giorgio, P. Vierling, J. Defaye, C. Ortiz Mellet and J. M. García Fernández, *Chem. Eur. J.*, 2009, **15**, 12 871.
180. F. Ortega-Caballero, C. Ortiz Mellet, L. Le Gourriérec, N. Guilloteau, C. Di Giorgio, P. Vierling, J. Defaye and J. M. García Fernández, *Org. Lett.*, 2008, **10**, 5143.
181. A. Méndez-Ardoy, M. Gómez-García, C. Ortiz Mellet, N. Sevillano, M. D. Girón, R. Salto, F. Santoyo-González and J. M. García Fernández, *Org. Biomol. Chem.*, 2009, **7**, 2681.
182. A. Díaz-Moscoso, N. Guilloteau, C. Bienvenu, A. Méndez-Ardoy, J. L. Jiménez Blanco, J. M. Benito, L. Le Gourriérec, C. Di Giorgio, P. Vierling, J. Defaye, C. Ortiz Mellet and J. M. García Fernández, *Biomaterials*, 2011, **32**, 7263.
183. N. Symens, A. Méndez-Ardoy, A. Díaz-Moscoso, E. Sánchez-Fernández, K. Remaut, J. Demeester, J. M. García Fernández, S. C. De Smedt and J. Rejman, *Bioconj. Chem.*, 2012, **23**, 1276.
184. J. Guo, J. R. Ogier, S. Desgranges, R. Darcy and C. O'Driscoll, *Biomaterials*, 2012, **33**, 7775.
185. Y.-W. Yang, *Med. Chem. Commun.*, 2011, **2**, 1033.
186. C. Park, K. Oh, S. C. Lee and C. Kim, *Angew. Chem. Int. Ed.*, 2007, **46**, 1455.
187. H. Meng, M. Xue, T. Xia, Y.-L. Zhao, F. Tamanoi, J. F. Stoddart, J. I. Zink and A. E. Nel, *J. Am. Chem. Soc.*, 2010, **132**, 12 690.

188. M. W. Ambrogio, T. A. Pecorelli, K. Patel, N. M. Khashab, A. Trabolsi, H. A. Khatib, Y. Y. Botros, J. I. Zink and J. F. Stoddart, *Org. Lett.*, 2010, **12**, 3304.
189. H. Kim, S. Kim, C. Park, H. Lee, H. J. Park and C. Kim, *Adv. Mater.*, 2010, **22**, 4280.
190. D. P. Ferris, Y. L. Zhao, N. M. Khashab, H. A. Khatib, J. F. Stoddart and J. I. Zink, *J. Am. Chem. Soc.*, 2009, **131**, 1686.
191. C. Park, K. Lee and C. Kim, *Angew. Chem. Int. Ed.*, 2009, **48**, 1275.
192. C. Park, H. Kim, S. Kim and C. Kim, *J. Am. Chem. Soc.*, 2009, **131**, 16 614.
193. K. Patel, S. Angelos, W. R. Dichtel, A. Coskun, Y.-W. Yang, J. I. Zink and J. F. Stoddart, *J. Am. Chem. Soc.*, 2008, **130**, 2382.
194. M. Xue, D. Cao, J. F. Stoddart and J. I. Zink, *Nanoscale*, 2012, **4**, 7569.
195. C. Wang, Z. Li, D. Cao, Y.-L. Zhao, J. W. Gaines, O. A. Bozdemir, M. W. Ambrogio, M. Frasconi, Y. Y. Botros, J. I. Zink and J. F. Stoddart, *Angew. Chem. Int. Ed.*, 2012, **51**, 5460.
196. T. Furuchi and R. G. W. Anderson, *J. Biol. Chem.*, 1998, **273**, 21 099.
197. B. Liu, S. D. Turley, D. K. Burns, A. M. Miller, J. J. Repa and J. M. Dietschy, *Proc. Natl. Acad. Sci. USA*, 2009, **106**, 2377.
198. A. Aqul, B. Liu, C. M. Ramirez, A. A. Pieper, S. J. Estill, D. K. Burns, B. Liu, J. J. Repa, S. D. Turley and J. M. Dietschy, *J. Neurosci.*, 2011, **31**, 9404.
199. C. D. Davidson, N. F. Ali, M. C. Micsenyi, G. Stephney, S. Renault, K. Dobrenis, D. S. Ory, M. T. Vanier and S. U. Walkley, *PLoS ONE*, 2009, **4**, e6951.
200. A. I. Rosenbaum, G. T. Zhang, J. D. Warren and F. R. Maxfield, *Proc. Natl. Acad. Sci. USA*, 2010, **107**, 5477.
201. A. I. Rosenbaum and F. R. Maxfield, *J. Neurochem.*, 2011, **116**, 789.
202. K. B. Peake and J. E. Vance, *J. Biol. Chem.*, 2012, **287**, 9290.
203. http://www.nnpdf.org/Cyclodextrin.html (last visited Jan 31st, 2013).
204. T. Moriya, H. Kurita, K. Matsumoto, T. Otake, H. Mori, M. Morimoto, N. Ueba and N. Kunita, *J. Med. Chem.*, 1991, **34**, 2301.
205. T. Moriya, K. Saito, H. Kurita, K. Matsumoto, T. Otake, H. Mori, M. Morimoto, N. Ueba and N. Kunita, *J. Med. Chem.*, 1993, **36**, 1674.
206. D. H. Nguyen and J. E. K. Hildreth, *J. Virol.*, 2000, **74**, 3264.
207. K. V. Khanna, K. J. Whaley, L. Zeitlin, T. R. Moench, K. Mehrazar, R. A. Cone, Z. H. Liao, J. E. K. Hildreth, T. E. Hoen, L. Shultz and R. B. Markham, *J. Clin. Invest.*, 2002, **109**, 205.
208. D. R. M. Graham, E. Chertova, J. M. Hilburn, L. O. Arthur and J. E. K. Hildreth, *J. Virol.*, 2003, **77**, 8237.
209. A. Boasso, C. M. Royle, S. Doumazos, V. N. Aquino, M. Biasin, L. Piacentini, B. Tavano, D. Fuchs, F. Mazzotta, S. Lo Caputo, G. M. Shearer, M. Clerici and D. R. Graham, *Blood*, 2011, **118**, 5152.
210. I. E. Crandall, W. A. Szarek, J. Z. Vlahakis, Y. Xu, R. Vohra, J. Sui and R. Kisilevsky, *Biochem. Pharmacol.*, 2007, **73**, 632.
211. H. Naiki and Y. Nagai, *J. Biochem.*, 2009, **146**, 751.

212. M. Simons, P. Keller, B. De Strooper, K. Beyreuther, C. G. Dotti and K. Simons, *Proc. Natl. Acad. Sci. USA*, 1998, **95**, 6460.
213. W. J. Strittmatter, K. H. Weisgraber, D. Y. Huang, L.-M. Dong, G. S. Salvesen, M. Pericak-Vance, Donald Schmechel and A. D. Roses, *Proc. Natl. Acad. Sci. USA*, 1993, **90**, 8098.
214. P. Camilleri, N. J. Haskins and D. R. Hewlett, *FEBS Lett.*, 1994, **341**, 256.
215. X.-r. Qin, H. Abe and H. Nakanishi, *Biochem. Biophys. Res. Commun.*, 2002, **297**, 1011.
216. A. Wahlström, R. Cukalevski, J. Danielsson, J. Jarvet, H. Onagi, J. Rebek Jr., S. Linse and A. Gräslund, *Biochemistry*, 2012, **51**, 4280.
217. H. Jono, T. Anno, K. Motoyama, Y. Misumi, M. Tasaki, T. Oshima, Y. Mori, M. Mizuguchi, M. Ueda, M. Shono, K. Obayashi, H. Arima and Y. Ando, *Biochem. J.*, 2011, **437**, 35.
218. J. J. Lundquist and E. J. Toone, *Chem. Rev.*, 2002, **102**, 555.
219. J. L. Jiménez Blanco, C. Ortiz Mellet and J. M. García Fernández, *Chem. Soc. Rev.*, 2013, **42**, 4518.
220. M. Gómez-García, J. M. Benito, D. Rodríguez-Lucena, J.-X. Yu, K. Chmurski, C. Ortiz Mellet, R. Gutiérrez Gallego, A. Maestre, J. Defaye and J. M. García Fernández, *J. Am. Chem. Soc.*, 2005, **127**, 7970.
221. M. Gómez-García, J. M. Benito, R. Gutiérrez Gallego, A. Maestre, C. Ortiz Mellet, J. M. García Fernández and J. L. Jiménez Blanco, *Org. Biomol. Chem.*, 2010, **8**, 1849.
222. M. Gómez-García, J. M. Benito, A. P. Butera, C. Ortiz Mellet, J. M. García Fernández and J. L. Jiménez Blanco, *J. Org. Chem.*, 2012, **76**, 1273.
223. A. Martínez-Ceballos, C. Ortiz Mellet and J. M. García Fernández, *Chem. Soc. Rev.*, 2013, **42**, 4746.
224. S. André, H. Kaltner, T. Furuike, S.-I. Nishimura and H.-J. Gabius, *Bioconj. Chem*, 2004, **15**, 87.
225. C. Decroocq, D. Rodríguez-Lucena, V. Russo, T. Mena Barragán, C. Ortiz Mellet and P. Compain, *Chem. Eur. J.*, 2011, **17**, 13 825.
226. C. Decroocq, D. Rodríguez-Lucena, K. Ikeda, N. Asano and P. Compain, *ChemBioChem*, 2012, **13**, 661.
227. E. M. Nestorovich and S. M. Bezrukov, *Chem. Rev.*, 2012, **112**, 6388.
228. L.-Q. Gu, O. Braha, S. Conlan, S. Cheley and H. Bayley, *Nature*, 1999, **398**, 686.
229. V. A. Karginov, E. M. Nestorovich, M. Moayeri, S. H. Leppla and S. M. Bezrukov, *Proc. Natl. Acad. Sci. USA*, 2005, **102**, 15075.
230. K. Yannakopoulou, L. Jicsinszky, C. Aggelidou, N. Mourtzis, T. M. Robinson, A. Yohannes, E. M. Nestorovich, S. M. Bezrukov and V. A. Karginov, *Antimicrob. Agents Chemother.*, 2011, **55**, 3594.
231. A. Díaz-Moscoso, A. Méndez-Ardoy, F. Ortega-Caballero, J. M. Benito, C. Ortiz Mellet, J. Defaye, T. M. Robinson, A. Yohannes, V. A. Karginov and J. M. García Fernández, *ChemMedChem*, 2011, **6**, 181.

232. M. Moayeri, T. M. Robinson, S. H. Leppla and V. A. Karginov, *Antimicrob. Agents Chemother.*, 2008, **52**, 2239.
233. A. Joshi, S. Kate, V. Poon, D. Mondal, M. B. Boggara, A. Saraph, J. T. Martin, R. McAlpine, R. Day, A. E. Garcia, J. Mogridge and R. S. Kane, *Biomacromolecules*, 2011, **12**, 791.
234. E. M. Nesorovich, V. A. Karginov, M. R. Popoff, S. M. Bezrukov and H. Barth, *PLoS ONE*, 2011, **6**, e23927.
235. M. Naguib, *Anesth. Analg.*, 2007, **104**, 575.
236. M. Shields, M. Giovannelli, R. K. Mirakhur, I. Moppett, J. Adams and Y. Hermens, *Brit. J. Anaesth.*, 2006, **96**, 36.
237. B. A. Baldo, N. J. McDonnell and N. H. Pham, *Mini-Rev. Med. Chem.*, 2012, **12**, 701.
238. A. Bom, M. Bradley, K. Cameron, J. K. Clark, J. van Egmond, H. Feilden, E. J. MacLean, A. W. Muir, R. Palin, D. C. Rees and M.-Q. Zhang, *Angew. Chem. Int. Ed.*, 2002, **41**, 265.
239. J. M. Adam, D. J. Bennett, A. Bom, J. K. Clark, H. Feilden, E. J. Hutchinson, R. Palin, A. Prosser, D. C. Rees, G. M. Rosair, D. Stevenson, G. J. Tarver and M.-Q. Zhang, *J. Med. Chem.*, 2002, **45**, 1806.
240. I. F. Sorgenfrei, K. Norrild, P. B. Larsen, J. Stensballe, D. Ostergaard, M. E. Prins and J. Viby-Mogensen, *Anesthesiology*, 2006, **104**, 667.
241. R. Miller, *Anesth. Analg.*, 2007, **104**, 477.
242. For a recent survey see: A. Mazzaglia, M. T. Sciortino, N. Kandoth and S. Sortino, *J. Drug Deliv. Sci. Technol.*, 2012, **22**, 235
243. N. Kandoth, E. Vittorino, M. T. Sciortino, T. Parisi, I. Colao, A. Mazzaglia and S. Sortino, *Chem. Eur. J.*, 2012, **18**, 1684.
244. X. Leng, C.-F. Choi, P.-C. Lo and D. K. P. Ng, *Org. Lett.*, 2007, **9**, 231.
245. J. T. F. Lau, P.-C. Lo, Y.-M. Tsang, W.-P. Fong and D. K. P. Ng, *Chem. Commun.*, 2011, **47**, 9657.
246. J. T. F. Lau, P.-C. Lo, W.-P. Fong and D. K. P. Ng, *Chem. Eur. J*, 2011, **17**, 7569.
247. M. Ortiz, A. Fragoso and C. K. O'Sullivan, *Anal. Chem.*, 2011, **83**, 2931.
248. D. Grünstein, M. Maglinao, R. Kikkeri, M. Collot, K. Barylyuk, B. Lepenies, F. Kamena, R. Zenobi and P. H. Seeberger, *J. Am. Chem. Soc.*, 2011, **133**, 13957.
249. D. Branton, D. W. Deamer, A. Marziali, H. Bayley, S. A. Benner, T. Butler, M. Di Ventra, S. Garaj, A. Hibbs, X. Huang, S. B. Jovanovich, P. S. Krstic, S. Lindsay, X. S. Ling, C. H. Mastrangelo, A. Meller, J. S Oliver, Y. V Pershin, J. M. Ramsey, R. Riehn, G. V. Soni, V. Tabard-Cossa, M. Wanunu, M. Wiggin and J. A Schloss, *Nat. Biotechnol.*, 2008, **26**, 1146.
250. J. Clarke, H.-C. Wu, L. Jayasinghe, A. Patel, S. Reid and H. Bayley, *Nat. Nanotechnol.*, 2009, **4**, 265.
251. E. C. Hayden, *Nature*, 2012, doi 10.1038/nature.2012.10051.
252. H. Bayley, *Phys. Life Rev.*, 2012, **9**, 161.

253. F. Greco and M. J. Vicent, *Adv. Drug Delivery Rev.*, 2009, **61**, 1203.
254. H. Fan, Q.-D. Hu, F.-J. Xu, W.-Q. Liang, G.-P. Tang and W.-T. Yang, *Biomaterials*, 2012, **33**, 1428.
255. T. Liu, X. J. Li, Y. F. Qian, X. L. Hu and S. Y. Liu, *Biomaterials*, 2012, **33**, 2521.
256. J. Lee, H. Kim, S. Kim, H. Lee, J. Kim, N. Kim, H. J. Park, E. K. Choi, J. S. Lee and C. Kim, *J. Mater. Chem.*, 2012, **22**, 14061.

Interactions of Calix[n]arenes and Other Organic Supramolecular Systems with Proteins

FLORENT PERRET[a] AND ANTHONY W. COLEMAN[*b,c]

[a] Université Lyon 1, ICBMS UMR-CNRS 5246, CSAp, Villeurbanne, F69622, France; [b] I.I.S., University of Tokyo, Tokyo 153-8505, Japan; [c] Université Lyon 1, LMI UMR-CNRS 5615, Villeurbanne, F69622, France
*Email: anthony.coleman@univ-lyon1.fr

6.0   Introduction

Because proteins contain multiple functional groups, recognizing them is challenging. Medical diagnostic, pathogen detection and proteomics require the use of synthetic ligands that can specifically detect a target protein *via* complexation.[1,2] In aqueous media, such ligands can act as protein modulators, as agonists of biological responses, and mediate protein function and dimerization.[3,4] The modification of proteins by non-covalent interactions is favourable because it is a simple and relatively gentle reaction.

In this chapter, we will deal with the interaction of proteins with molecular organic supramolecular systems; for reasons of simplicity the equally relevant subject of protein interactions with supramolecular assemblies, such as membranes[5] or biomineral assemblies,[6] will not be treated here. The interactions of

Monographs in Supramolecular Chemistry No. 13
Supramolecular Systems in Biomedical Fields
Edited by Hans-Jörg Schneider

Published by the Royal Society of Chemistry, www.rsc.org

calix[*n*]arenes with nucleic acids are discussed in Chapter 8. In contrast to the more classical supramolecular chemistry, here the organic host molecules are ligands for various functionalities of proteins, which are much larger; however, the interactions may occur in two distinct ways: complexation of amino-acid side chains by the organic supramolecular molecules or *via* decoration of the organic supramolecular molecules with peptide or carbohydrate functions which are known to be ligands for protein recognition sites. In the first case the organic supramolecular molecules are acting as active biological systems, whereas in the second case they are passive transporters for the ligands, although they may also act to block protein–protein interactions.

In historical terms, non-covalent interactions with supramolecular systems largely antedate supramolecular chemistry, as the cyclodextrins are produced by the action of the enzyme cyclodextrin glycosyltransferease with amylose; indeed the majority of early work on bio-supramolecular chemistry and protein interaction involved the cyclodextrins (*e.g.* see Chapter 5).[7]

It was only in the 1980s that water-soluble calixarenes became available, although Cornforth had shown already in the 1950s the anti-tubercular activity of the calixarenes.[8] The availability of the water-soluble calixarenes stimulated a rapid rise in the study of their biological properties,[9–12] and in 2012 this work was crowned by the determination of the first crystal structure of a protein–calixarene complex, that of cytochrome *c* with para-sulfonato-calix[4]arene by Crowley *et al.*[13] A smaller body of work has been dedicated to the crown ethers, cavitands,[14,15] and very recently the so-called Noria molecules.[16]

In this chapter we have chosen to break the work down by type of protein involved rather than by type of organic supramolecular molecule. This choice reflects the position of the authors, namely that it is the protein under study which dictates the construction of the organic supramolecular system, rather than the other way around. However, in the last section of the chapter we will deal with the multi-protein sensing approach of Schrader.[17]

A number of general reviews on the area of supramolecular systems and proteins have been published. Among these is the work of Martos *et al.*,[18] on the use of supramolecular scaffolds, including calix[4]arenes, porphyrins, rotaxanes, dendrimers, polymers and capped noble metal nanoparticles. The scaffolds described use diverse structures combined with flexibility and multi-valency to impart strong binding to proteins.

Casnati *et al.*[19] have reviewed the use of peptides and glycocalix[*n*]arenes for binding to various proteins; again the calix[*n*]arene skeleton is use as a simple scaffold to present the peptide or glycosyl ligands to the proteins. For example, the authors describe how upper-rim bridged peptidocalix[4]arenes act as vancomycin mimics binding to D-alanyl-D-alanine (D-Ala-D-Ala) residues. As with the work of Martos, the glycocalix[4]arenes show the phenomenon of multivalency in their binding to specific lectins, and those bearing thiourea spacers between the calix[4]arene. The authors note that while hydrogen bonding plays an important role in organic solvents its importance is inexistent in aqueous systems.

With regard to the cyclodextrins (see Chapter 5), emphasis has been on drug delivery and protein stabilization. While not a central theme of the current chapter, the work of Yannakopoulou[20] on the use of cationic cyclodextrins for cell penetration and their use as inhibitors of virulence factors by blocking the pores of pore-forming proteins—and finally acting as cell penetrating cyclodextrins (CPCDs) and DNA delivery agents—is reviewed.

Serno *et al.*[21] have reviewed the research work that has been accomplished highlighting the potential of cyclodextrins as stabilizers of therapeutic proteins. Here the possible mechanisms of the interactions of the cyclodextrins with hydrophobic amino acid residues are described. Of note, the review deals with stabilization in both solution and the solid-state. In a similar vein, Ohtake *et al.*[22] review the use of cyclodextrins to act as protein stabilizers, again in both the solution and solid-state; another review by Otero-Espinar[23] deals with essentially the same topics. Sasaki *et al.*[24] consider the intriguing possibility of using the cyclodextrins as molecular chaperones to assist in the refolding processes of various proteins.

In a more general way, Uhlenheuer *et al.*[25] review the use of supramolecular scaffolds in various biological applications. They describe supramolecular host–guest interactions and recognition motifs, which are functional in water, and their respective association constants. Among systems described are strong host–guest interactions of different synthetically attractive guest molecules with cucurbiturils of different sizes, strong binding *via* optimal hydrophobic recognition of lithocholic acid by β-cyclodextrin, inclusion of a specific peptide motif in a self-assembled supramolecular host, and recognition of negatively charged protein patches by synthetic calix[4]arene ligands (*e.g.* see Chapter 7 by Nau and colleagues).

6.1 Soluble Proteins

6.1.1 Albumins

The serum albumins constitute a class of globular proteins that play a major role in adhesion to surfaces, acting as anchors for subsequent formation of protein films.[26] They are capable of binding both cations and anions, with multiple arginine and lysine residues being the known sites for interaction with anions, including fatty acids, salicylates, and even drugs such as sulfonamides.

6.1.1.1 Calixarenes

ElectroSpray Ionization Mass Spectrometry (ESI-MS), Dynamic Light Scattering (DLS) and Atomic Force Microscopy (AFM) have been used by our group to study the complexation of bovine serum albumin (BSA) with *para*-sulfonato-calix[*n*]arenes (**1–3**; Figure 6.1).[27,28] It has been shown that for **1**, there exist one strong and two weaker binding sites. Studies on the structure of thin films formed by surface deposition of BSA show that the

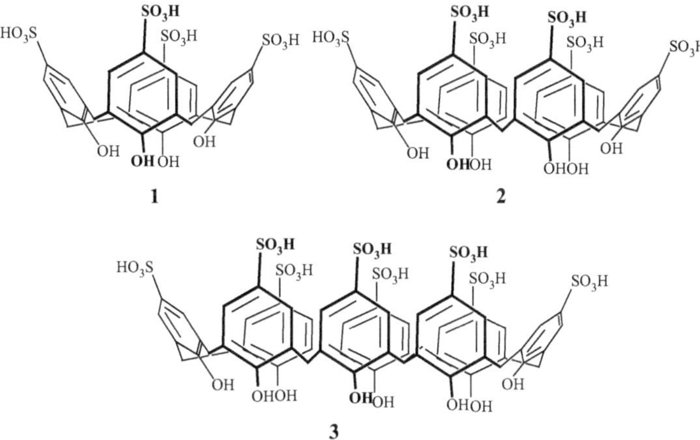

Figure 6.1　Structures of the *para*-sulfonatocalix[*n*]arenes.

para-sulfonato-calix[*n*]arene derivatives act to reticulate the films and produce essentially planar systems.

In a more recent publication,[29] ESI-MS was used in order to determine the association constants and binding stoichiometries for parent *para*-sulfonato-calix[*n*]arenes and their derivatives with BSA: compound **1** forms 1 : 1, 1 : 2 and 1 : 3 (BSA–calix) complexes, whereas **2** forms 1 : 1 and 1 : 2 species and **3** only a 1 : 1 complex. The strength of the interactions between the calixarene and BSA is inversely proportional to the size of the macrocyclic ring: $n=4>n=6>n=8$, and also, due to steric hindrance, the number of available binding sites decreases with the size of the macrocycle. The binding of *para*-sulfonato-calix[4]arenes to a series of different serum albumin proteins (HSA, PSA, SSA, RSA) has also been studied using ESI-MS.[30] Each protein shows different capacities to interact with **1**, including the number of ligands bound (from three to five), the association constants observed (from 1.14 to 0.03×10^6 M^{-1}), and the stoichiometries at which the onset of each binding event was observed to be species dependent.

Recently, Sandor Kunsagi-Mate *et al.*[31] studied the complexation ability of water-soluble *para*-sulfonato- thiacalix[4]arene towards three aromatic amino acids. The high stability of the complexes between individual amino acids and the calixarene derivatives led to the proposition that complexation may also occur when the amino acids are deeply buried in a protein. To test this idea the conversion rate, enthalpy and entropy change associated with a structural transition of BSA was investigated by Differential Scanning Calorimetry (DSC) either in the absence or in the presence of calixarene. The results showed that the presence of calixarene changes significantly both the thermodynamics and the kinetics of the transition of BSA.

Finally, the anionic *para*-octylcalix[8]arenes derivatives (**4–6**) have been synthesized by Jebors *et al.*[32] and their interaction with serum albumins has been studied (Figure 6.2).

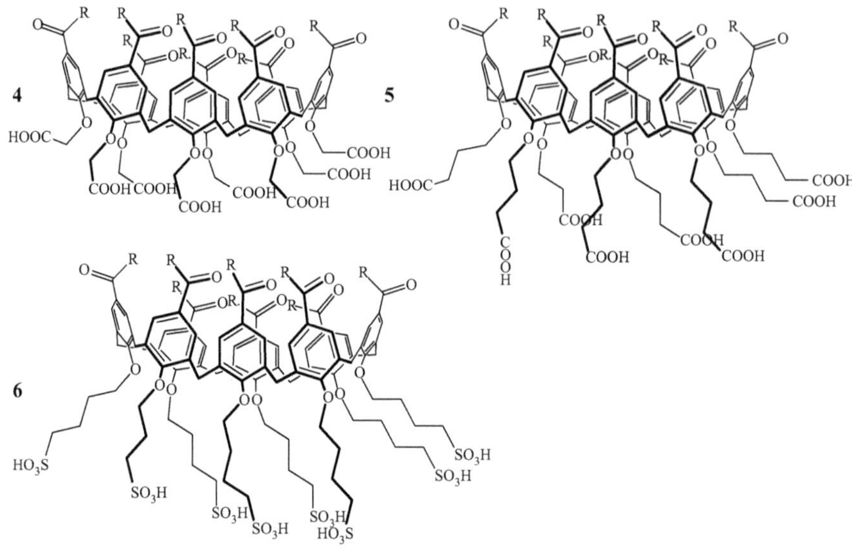

Figure 6.2   Anionic *para*-acylcalix[8]arenes derivatives synthesized by Jebors *et al.*[34]

The *O*-4-butylsulfonate derivative of *para*-octylcalix[8]arene, **6**, which is soluble in water has been demonstrated to be capable of complexing the serum albumin proteins of various animal sources, by forming a monolayer around the various proteins.

Along with *O*-carboxymethoxy-*para*-octanoylcalix[8]arene, the butylsulfonate derivative **6** possesses anticoagulant properties, showing inhibition of coagulation at 20 μM and 100 μM, respectively, but have no haemolytic toxicity at concentrations up to 200 μg/mL.

6.1.2   Enzymes

6.1.2.1   *Phosphatase Inhibitors*

Kalchenko *et al.* have investigated the inhibition of alkaline phosphatase from bovine intestine mucosa and bovine kidney by calix[4]arenes functionalized at the upper rim by one or two methylene bisphosphonic acid fragments (**7,8**; Figure 6.3).[33] The mechanisms of enzyme inhibition have been discussed using a molecular docking approach by computational modelling of inhibitors into active centres of alkaline phosphatase from *Escherichia coli*.

The analogous thiacalix[4]arene derivatives (**9,10**) displayed stronger inhibition properties towards alkaline phosphatases from bovine intestine mucosa, shrimp and human placenta than the calix[4]arene analogues.[34] Concerning this family of calixarenes, inhibition of tyrosine phosphatase from *Yersinia* (Yersinia PTP) by calix[4]arene mono-, bis-, and tetrakis(methylene-bisphosphonic) acids as well as calix[4]arene and thiacalix[4]arene tetrakis-(methylphosphonic) acids has been investigated. The kinetic studies revealed that some compounds in this class are potent competitive inhibitors of Yersinia PTP with inhibition

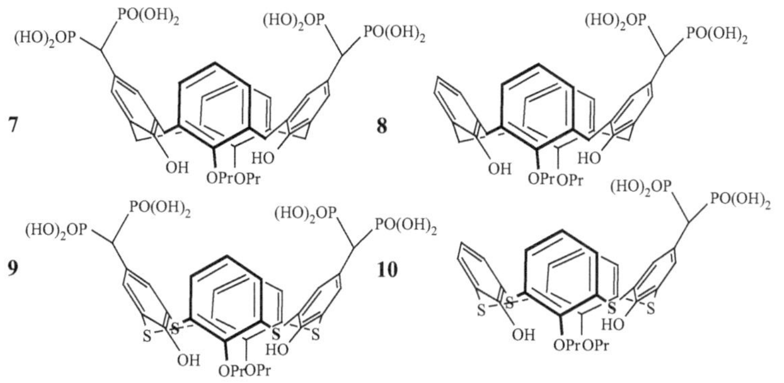

Figure 6.3　Phosphonic acid calixarene derivatives used as phosphatase inhibitors by Kalchenko and co-workers.[35,36]

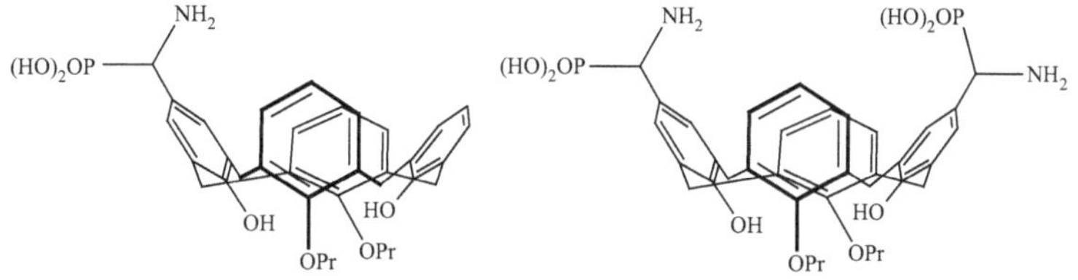

Figure 6.4　Calix[4]arene α-aminophosphonic acid derivatives synthesized by Kalchenko and co-workers.[35]

constants in the low micromolar range. The binding modes of macrocyclic phosphonate derivatives in the enzyme active centre have been explained using a computational docking approach. The results obtained indicate that calix[4]arenes are promising scaffolds for the development of inhibitors of Yersinia PTP.

Calix[4]arenes bearing one or two methylenebisphosphonic acid fragments (**7,8**) displayed stronger inhibition of calf intestine alkaline phosphatase than simple methylenebisphosphonic or 4-hydroxyphenyl methylene-bis-phosphonic acids. The action of these phosphorylated calix[4]arenes is concordant with partial mixed-type inhibition. The replacement of the phosphonic acid moieties on the macrocyle with diethylphosphonate resulted in a sharp decrease of its inhibitory action. Thus pre-organizing phosphonic acid fragments using a calixarene platform provides a promising approach for the design of efficient alkaline phosphatase inhibitors.

Chiral calix[4]arene α-aminophosphonic acid derivatives **11** and **12** have been synthesized and studied by Kalchenko's group. The di-acids obtained showed inhibitory activity towards porcine kidney alkaline phosphatase (PKAP) which depends considerably on the absolute configuration of the α-carbon atoms (Figure 6.4).[35] The *R,R* isomer binds to PKAP about 50 times stronger than the *S,S* isomer, the inhibition constants being 1.17 μM and 86 μM, respectively.

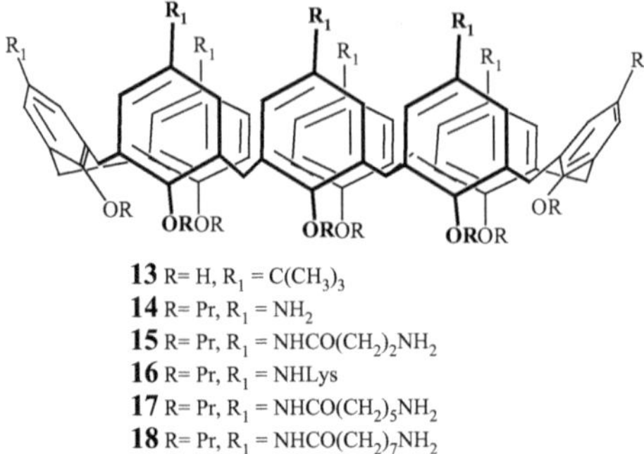

13 R= H, R$_1$ = C(CH$_3$)$_3$
14 R= Pr, R$_1$ = NH$_2$
15 R= Pr, R$_1$ = NHCO(CH$_2$)$_2$NH$_2$
16 R= Pr, R$_1$ = NHLys
17 R= Pr, R$_1$ = NHCO(CH$_2$)$_5$NH$_2$
18 R= Pr, R$_1$ = NHCO(CH$_2$)$_7$NH$_2$

Figure 6.5 Basic amino acid calix[8]arene derivatives used by Consoli's group and shown to display tryptase inhibition.[36]

6.1.2.2 *Tryptase Recognition*

The group of Consoli has developed a series of basic amino acid-derived calix[8]arene derivatives (Figure 6.5) that act as surface receptors with potent and selective inhibition activity for both human and recombinant lung β-tryptase.[36] Kinetic inhibition analysis of recombinant tryptase of lung showed a time-dependent competitive inhibition with both initial and steady state rate constants in the nanomolar range: respectively, 582 nM and 77 nM for **15**, 283 nM and 2 nM for **16**, 99 nM and 6 nM for **17** and 626 nM and 79 nM for **18**.

Human tryptase was inhibited indirectly, owing to the antagonist effect of derivatives **13–18** on the proteoglycan heparin. At the same time, competitive inhibition of recombinant human tryptase was also achieved, supporting the effectiveness of these surface binding receptors. These findings could herald a new approach to the design of artificial enzymatic inhibitors.

6.1.2.3 *Glyceraldehyde-3-phosphate Dehydrogenase (GPDH) Mimetic*

Kalchenko and co-workers have proved that the calix[6]arene sulfonate derivatives (**2** and **19**; Figure 6.6) efficiently accelerate acid-catalysed hydration of 1-benzyl-1,4-dihydronicotinamide (synthetic model of NADH) as much as GPDH, and that the reaction proceeds according to Michaelis–Menten kinetics.[39]

This has been explained as being due to the acidic protons that catalyse the reaction and the anionic sulfonates that stabilize the cationic intermediate. Furthermore, the (CH$_2$COOH) derivative has a greater association constant than that of the OH (1340 M^{-1} *vs.* 564 M^{-1}, respectively), which may be rationalized in terms of the elongated cavity size favouring interaction with the NADH-model molecule.

Comparison of the series of carboxyethylcalix[*n*]arenes (**20–22**) and model monomers in acceleration of acid catalysed hydration of 1-benzyl-1,4-dihydronicotinamide was carried out by Gutsche and Alam (Figure 6.7).[37]

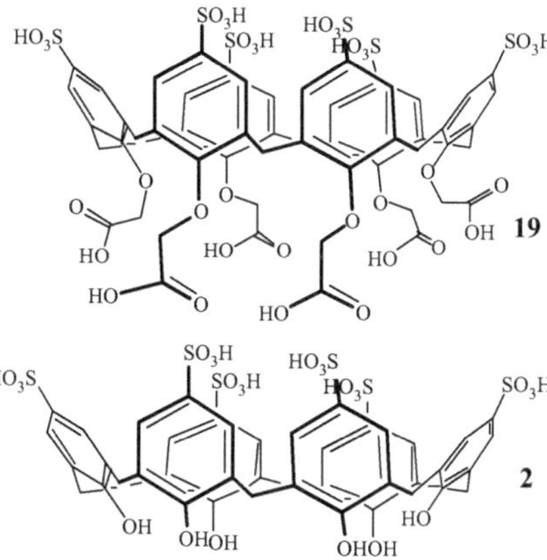

Figure 6.6   Calix[6]arenes sulfonate derivatives used by Kalchenko's group.[39]

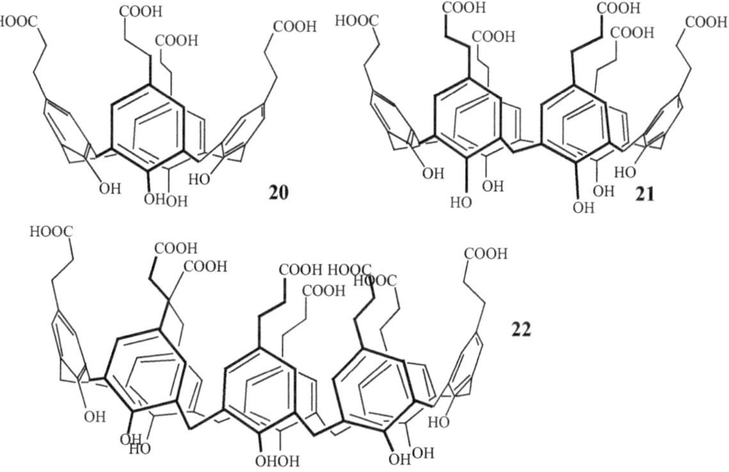

Figure 6.7   Molecules used in the work of Gutsche and Alam.[37]

The results showed that compound **21** was a better catalyst than either the smaller or larger members of the series but was considerably inferior to its sulfonic acid analogue. The calixarene effect is even greater if monomeric phenol is taken as the reference point.

6.1.2.4   *L-Arabinose Isomerase Stabilization by So-called Noria Molecules*

Recently, another type of supramolecular system has been used for studying complexation to proteins, namely the Noria systems. The name Noria is derived from the geometry of these molecules, which resembles the Noria waterwheels (see Figure 6.8). L-Arabinose isomerase is relevant for its industrial

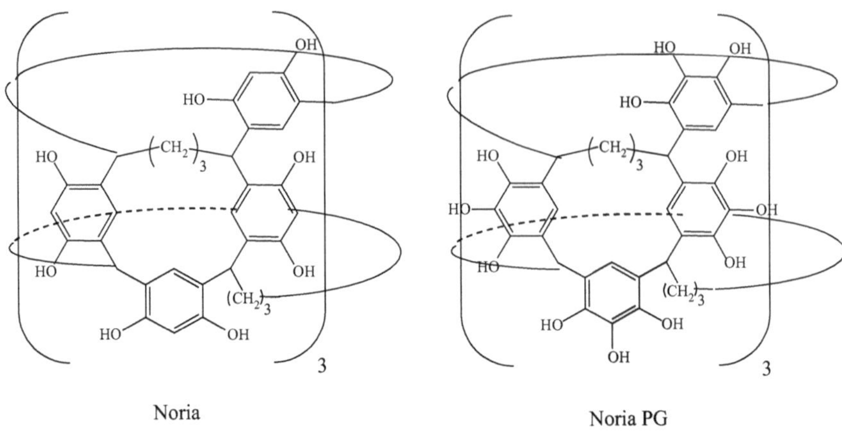

Figure 6.8 Noria derivatives used by Coleman's group.[16]

application as a biocatalyst, a function for which the enzyme needs to be stabilized. In this context, Coleman's group[16] has shown that Noria and NoriaPG (Figure 6.8) are capable of increasing the stability of L-arabinose isomerase not only at high temperatures, but also at low pH.

6.1.2.5 *Cyclodextrin Glycosyltransferase*

Cyclodextrin glycosyltransferase (CGTase) belongs to the α-amylase family and mainly catalyses transglycosylation. It is widely used to produce cyclodextrins and related α-1,4-glucans from starch. The catalytic site of CGTase specifically conserves four aromatic residues, Phe183, Tyr195, Phe259 and Phe283, which are not found in other α-amylases. Production of cyclodextrins based on the use of CGTase has been reviewed by Wang *et al.*[38]

Given the use of this enzyme to produce cyclodextrins, it is not surprising that much effort has been put into the understanding of the structure of CGTase and the nature of its catalytic site. Haga *et al.*[39] showed in the CGTase from the alkalophilic *Bacillus* sp. 1011[40] that the stacking interaction between a tyrosine residue and the sugar ring at the catalytic subsite −1 is strictly conserved in the glycoside hydrolase family of enzymes. Replacing Tyr100 with leucine in CGTase from *Bacillus* sp. 1011in order to prevent stacking significantly decreased all enzymatic activity. The adjacent stacking interaction with both Phe183 and Phe259 on to the sugar ring at subsite +2 is essentially conserved among the CGTases. The F183L/F259L mutant CGTase affects donor substrate binding and/or acceptor binding during transglycosylation.

6.1.2.6 *Xe129 Protein Markers*

Hyperpolarized Xe129 is an excellent NMR probe for the local environment around a supramolecualr system, and it is not surprising that a range of studies have been undertaken on this system.

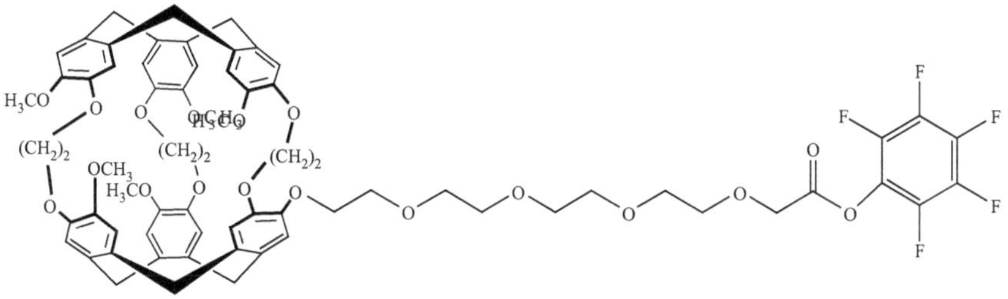

Figure 6.9 Cryptophane carrier for xenon-129 studied by Boutin *et al.*[41]

Various crytophanes have been used as container molecules for studying the protein complexation. For example, Boutin *et al.*[41] used this system to study intracellular interactions with transferrin and showed that these biosensors allowed the in-cell probing of biological events using hyperpolarized xenon (Figure 6.9).

Similarly Seward *et al.*[42] also used a cryptophane biosensor (functionalized with cyclic RGDyK peptide and Alexa Fluor 488 dye) to detect vitronectin binding to $\alpha v\beta 3$ integrin and fibrinogen binding to $\alpha IIb\beta 3$ integrin using hyperpolarized ^{129}Xe NMR spectroscopy.

6.1.3 Blood Coagulation Cascade Proteins

Para-sulfonato-calix[*n*]arenes and six of their *O*-monosubstituted derivatives have also been investigated by the Coleman group for their *in vitro* anti-coagulant activity.[43] The mono-(2-carboxymethoxy)-5,11,17,23,29,35,41,47-octa-sulfonato-calix[8]arene (**25b**) prolongs the activated partial thromboplastin time (APTT) and the thrombin time (TT) significantly more than do the other calixarenes. Thrombin inhibition mediated by antithrombin (AT) and heparin cofactor II (HCII) activation was investigated and compared to the biological activators heparin and dermatan sulfate. The results showed that the mono-(2-carboxymethoxy)-5,11,17,23,29,35,41,47-octa-sulfonato-calix[8]arene (**25b**) and 5,11,17,23,29,35-hexa-sulfonato-calix[6]arene (**24a**) produced activation of HCII at 500 µM, which is comparable to that induced by dermatan sulfate at 100 µM. However, activation of AT by all of the investigated calixarenes was between 10 and 50 times lower than that observed in the presence of heparin. The mechanism of the anticoagulant effect of these calixarenes is as activators of HCII and not as activators of AT (Figure 6.10).

More recently, calix[4]arene derivatives bearing two or four methylene-bis-phosphonic acid groups at the macrocyclic upper rim (Figure 6.11) have been studied by Kalchenko *et al.* with respect to their effects on fibrin polymerization.[44]

Calix[4]arene tetrakis-methylene-bis-phosphonic acid (**26**) has been shown to be the most potent inhibitor: the maximum rate of fibrin polymerization in the fibrinogen + thrombin reaction decreased by 50% at concentrations of 0.52×10^{-6} M (IC_{50}). The authors hypothesized that calixarene **26** blocks fibrin formation by combining with polymerization site 'A' ($A\alpha 17\text{-}19$), which ordinarily initiates protofibril formation in a 'knob–hole' manner. This suggestion was

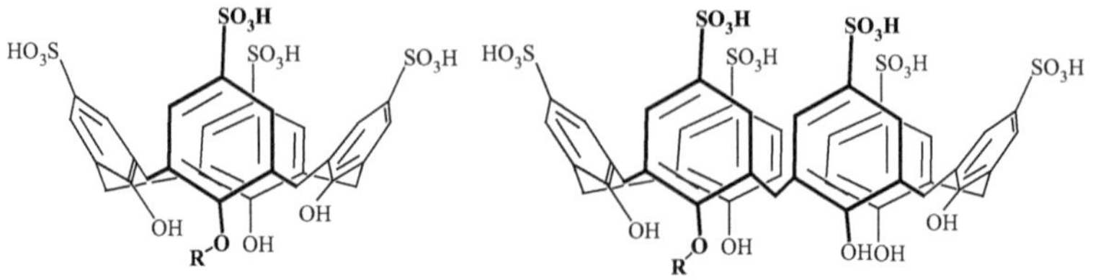

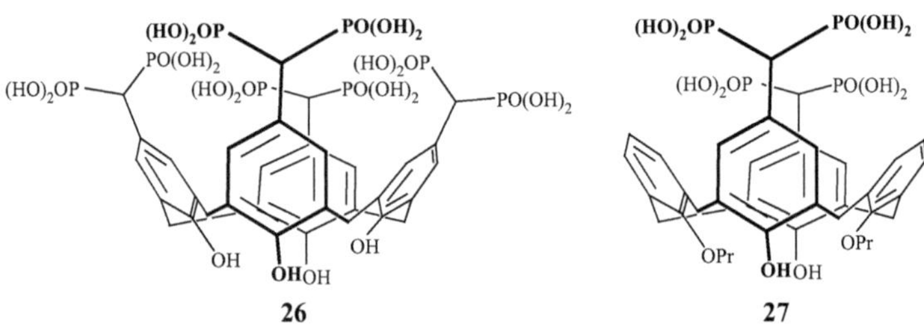

23a–c   24a–c

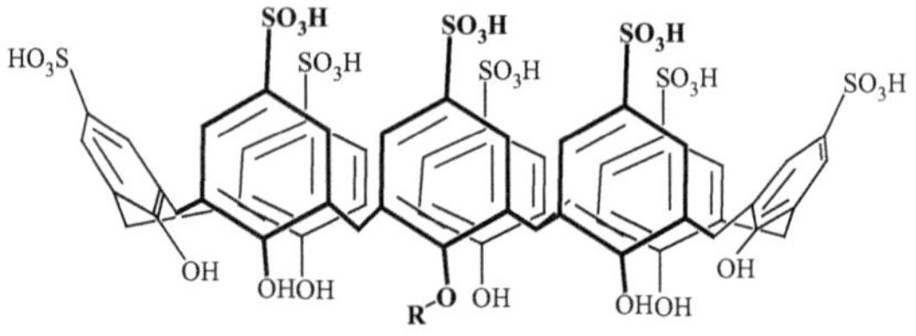

25a–c

a R = H; b R = CH$_2$COOH; c R = CH$_2$CH$_2$NH$_2$

Figure 6.10   Monosubstituted calix[*n*]arenes used in studies of anticoagulant effects by Coleman's group.[43]

26   27

Figure 6.11   Calixarene derivatives studied by Kalchenko and co-workers.[44]

confirmed by an HPLC assay, which showed a host–guest inclusion complex of **26** with the synthetic peptide Gly-Pro-Arg-Pro, an analogue of site 'A'. These experiments have also demonstrated that compound **26** is a specific inhibitor of fibrin polymerization and blood coagulation and could be used for the design of a new class of anti-thrombotic agents.

6.1.4   Insulin

The interactions between β-cyclodextrin and insulin have been studied by Wimmer and co-workers.[45] They studied the interaction between β-cyclo-dextrin and four non-carbohydrate-binding model proteins: ubiquitin, chy-motrypsin inhibitor 2 (CI2), S6 and insulin SerB9Asp by NMR spectroscopy and demonstrated that the interaction of β-cyclodextrin and these model pro-teins takes place at specific sites on the protein surface, and that solvent ac-cessibility of those sites is a necessary but not compelling condition for an interaction to occur. If this behaviour can be generalized, it might explain the wide range of different effects of cyclodextrins on different proteins: aggregation suppression (if residues responsible for aggregation are highly solvent-accessible), protection against degradation (if point of attack of a protease is sterically 'masked' by cyclodextrin), alteration of function (if residues involved in function are 'masked' by cyclodextrin). The exact effect of cyclodextrins on a given protein will always be related to the particular structure of this protein.

6.1.5 Signal Proteins—Histones

The post-translational modifications of proteins—for example, phosphorylation, acetylation and methylation—frequently induce biological activity by serving as on-switches for protein–protein interactions. Such localized and well defined sites for protein–protein binding collectively offer promise as targets for therapeutic applications and fundamental studies of chemical biology. In recent years many such modification sites have been discovered on the unstructured tails of proteins, and in particular the histones. These domains are unstructured in nature and do not present concave binding pockets, making them unsuitable for standard pharmaceutical agents.

Hof's group[46] has done a large amount of research to develop a family of sulfonato-calix[n]arene derivatives based supramolecular hosts **28a–e** that bind selectively and with high affinity to the trimethyllysine motifs of the histones (Figure 6.12).

They found that the parent sulfonatocalix[4]arene 2 is very well suited to bind trimethyllysine, displaying a K_d of 27.0 µM, with 70-fold selectivity over lysine, >100-fold selectivity over arginine, and even greater selectivity over all other known unmodified amino acids. NMR titration studies performed on more rigid biaryl hosts (**28a–e**) showed that all these derivatives were able to bond methylated lysine and that host **28a** has the strongest affinity for trimethyllysine (K_d of 15.6 µM) and best selectivity over unmethylated lysine (>150-fold).

Such residues are involved in gene regulation and oncogenesis. It is of great interest that they may be able to interact selectively with trimethyllysine sites, even in complex protein systems bearing a number of unmethylated lysine and arginine residues. They have further developed their system to generate a toolkit consisting of readily available dyes and calix[n]arene host molecules, which may be used to form dye-displacement sensors responding to a variety of cationic peptides. Using the data from only two or three of such sensors as a

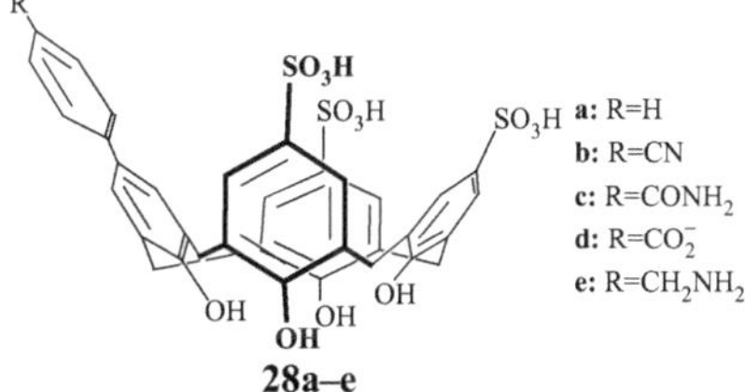

Figure 6.12	Sulfonatocalix[n]arene derivatives studied by Daze *et al.*[49] that bind trimethyllysine residues.

sensor array, fingerprints of data that discriminate robustly among many kinds of histone coding elements may be assembled; the sensors are claimed to be capable of discriminating among unmethylated, mono-, di-, and trimethylated lysines in a single histone tail sequence.[47]

In a parallel set of reports, Chini *et al.*[48] used various hydrophobic calix[*n*]arene derivatives to act as histone acetylase inhibitors.

6.1.6 Proteins Associated with Neurodegenerative Diseases

The pathology of many neurodegenerative diseases involves the formation of fibrillar plates, often termed amyloid fibres, as a result of conformational changes in the tertiary structures of proteins. The conformational shift from α-helical to β-sheet leads to exposure of hydrophobic surfaces on the proteins, followed by oligomerization and formation of protofilament aggregation into amyloid fibres. Finally, aggregation of the amyloid fibres into larger structures causes cell disruption and degeneration of the brain tissue. There exists a wide range of such neurodegenerative diseases,[49] and it can be seen that both intra- and extracellular proteins of widely differing nature are involved in such diseases. One general feature of the proteins involved is the presence of a proteoglycan or glycosylaminoglycan binding site, containing several basic amino acids.

Among these neurodegenerative diseases, those associated with the prion protein are, as of now, known to be infectious. The use of various para-sulfonato-calix[*n*]arenes for the detection of the pathogenic prion protein, PrP, has been studied by Coleman *et al.*[50] Given that the pathogenic mis-folded protein resists digestion by proteinase K, it was possible to remove all but this component in the analytical samples. Initially, Western Blot methods were used in the detection of PrP. There was a clear ring-size effect with respect to the macrocycle with increasing sensitivity of detection increasing from the ring size 4 to 6 to 8. More important, mono-substitution at the phenolic ring greatly enhanced the amplification of the Western Blot detection: up to 10 times as compared with the standard methods. Moreover it was possible to transfer the technology to ELISA-based detection, here reliable detection of the PrP protein was possible in human patients suffering from Creutzfeldt–Jacob disease. Fortunately the health crisis that might have arisen from contamination *via* blood transfusion has not occurred and the technology is not as yet in application.

6.1.7 Prostate-specific Antigen (PSA)

In 2003, Y. Lee and T. Kim developed a highly sensitive microarray protein chip coated with calix-crown derivatives having a bifunctional coupling property, allowing both efficient immobilization of capture proteins on solid matrices and making high-throughput analysis of protein–protein interactions possible.[51,52] For fabricating the protein microarray, they created a self-assembled monolayer (SAM) of calix[4]crown-5 derivatives **29** functionalized

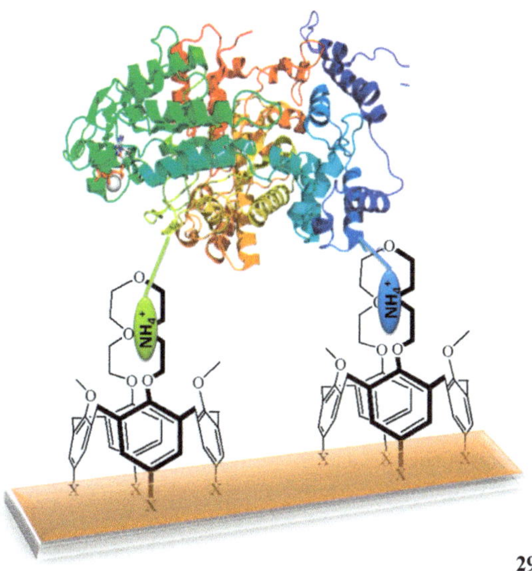

Figure 6.13 Immobilization of proteins on the slide glass modified with calix[4]-crown-5 derivatives, as described by Lee, Kim and co-workers.[52]

on the upper rim, either by thiols for grafting on the gold surface or by an aldehyde group for grafting on an amine modified glass slide. The proteins can thus be immobilized on the surface and have been used for the electrochemical immunosensing of glucose oxidase (GOx)-labelled C-reactive protein antigen (CRPAg) using the captured antibody monolayer immobilized on the calix[4]crown-5 surface (Figure 6.13).

6.2 Membrane Proteins

6.2.1 Transport Proteins

6.2.1.3 Surface Recognition of α-Chymotrypsin

By analogy with antibodies, which use changes in the sequence and conformation of six hypervariable loops in their protein structure for recognizing different antigens, Park and Hamilton designed the tetraloop peptide derivatives of calix[4]arene to allow the possible variability in the peptide units to be used for the selective recognition of different proteins.[53,54] Thus, using structural information for α-chymotrypsin, a selective receptor for this protein was designed and synthesized from calix[4]arene. It is a protein with surface arginine and lysine residues close to the entry of the active site; a calixarene was designed to interact with these residues and thus block access to the active site, acting as an inhibitor but not one directly bound to the active site (Figure 6.14).

These antibody mimics showed slow binding kinetics in an analogous manner to natural protein protease inhibitors. Complexation between the

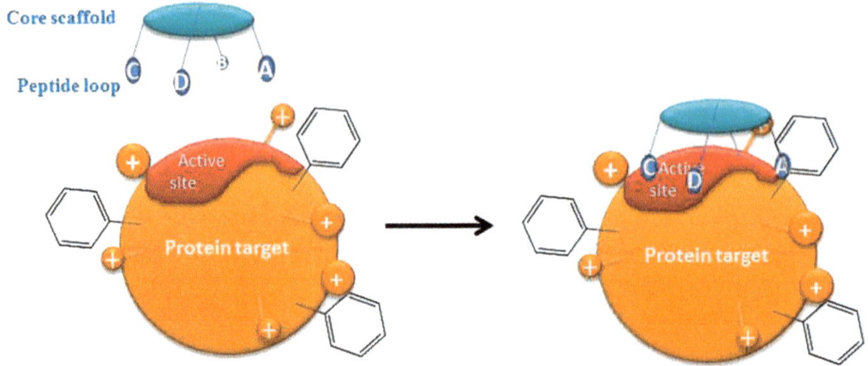

Figure 6.14 Artificial receptors based on 3-amino-5-aminomethylbenzoic acid for chymotrypsin surface recognition, described by Hamilton *et al.*[54]

receptor and α-chymotrypsin disrupted the interaction with proteinaceous inhibitors. These agents were particularly effective at blocking the soybean chymotrypsin–trypsin inhibitor complex.

6.2.2 Peripheral Proteins

6.2.2.1 Cytochrome c (Cyt c)

In the first report of protein extraction by calix[*n*]arenes, Oshima *et al.*[55] used a series of carboxylated *para*-tert-octyl calix[*n*]arene derivatives ($n = 4, 6, 8$) and demonstrated that compound **30** (Figure 6.15) showed high extraction capacity for the lysine-rich protein cytochrome *c,* a small haem protein found loosely associated with the inner membrane of the mitochondrion. Based on multi-electrostatic $COO^-\ldots NH_3^+$ interactions, up to 20 calixarenes bind to one molecule of Cyt *c*, which is in good agreement with the 19 lysine residues known to be present on the surface of this protein (Figure 6.16). They also described the successful quantitative extraction and recovery of Cyt *c* and lysosyme using the hexamer derivative.[56] Moreover, Cyt *c* complexed with **30** and dissolved in organic media such as chloroform, toluene or hexane, has been used as a catalyst for oxidative reactions.[57,58]

Crowley *et al.*[59] continued these studies and have recently reported the crystal structure of a protein–calixarene complex (Figure 6.17). The water-soluble *para*-sulfonatocalix[4]arene was shown to bind the lysine-rich cytochrome *c* at three different sites. They also obtained binding curves from NMR titrations which revealed an interaction process that involves multiple binding sites. Together, the data indicate a dynamic complex in which the calixarene explores the surface of cyt *c*, acting as an analogue of the electrical double layer found on colloids. The data also indicate that the calixarene is a mediator of protein–protein interactions, with potential applications in generating assemblies and promoting crystallization.

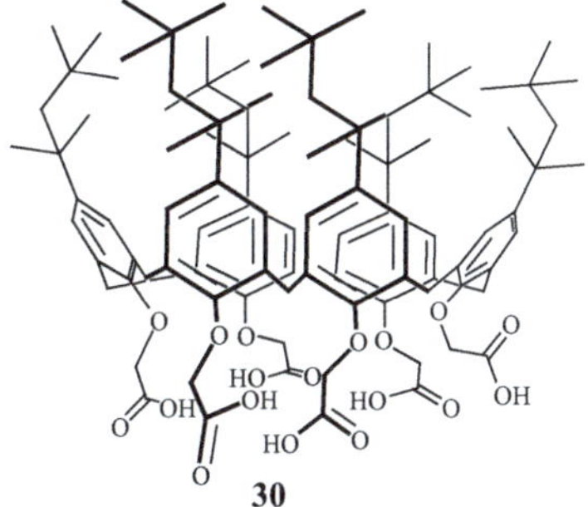

30

Figure 6.15 Molecular structure of the calix[6]arene carboxylic acid derivative used by Oshima.[56]

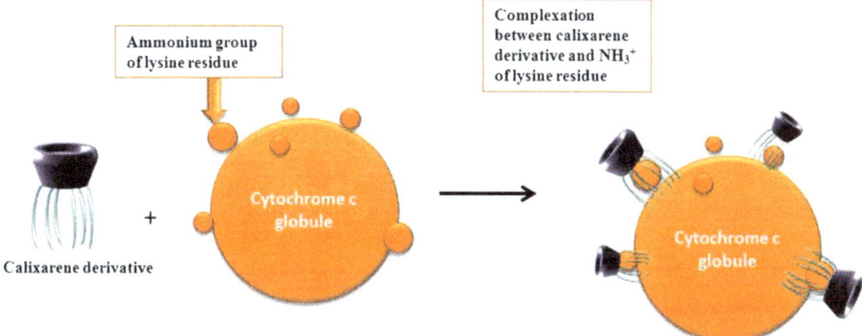

Figure 6.16 Schematic illustration of the cytochrome *c* extraction with calixarene derivatives described by Oshima *et al.*[57]

We have already seen that Hamilton and co-workers prepared a synthetic protein binding agent containing four peptide loops connected to a calix[4]-arene core (see Figure 6.14).[57,58] The loops contained negatively charged Gly-Asp-Gly-Asp, which complements the surface charge of cationic cyt *c*.[60,61] This allowed strong binding of the antibody mimic to cyt *c*, which disrupted its interaction with reducing agents. The binding of the receptor also disrupted the protein–protein interactions between cyt *c* and cytochrome *c* peroxidase.[62] The receptor competed effectively with cytochrome *c* peroxidase for the binding of cyt *c* by forming a 1:1 cyt *c*–receptor complex with a binding constant of around 108 M^{-1}.

Oshima *et al.*[63] have also studied the complexation between various crown ethers and cyt *c* in solution. They have demonstrated that the distribution of cyt *c* in Li$_2$SO$_4$/PEG aqueous biphasic system can be controlled by complexation with the crown ether, dicyclohexano-18-crown-6 (DCH18C6). The protein was quantitatively extracted into the PEG-rich phase in the presence of DCH18C6 and perchlorate ion (Figure 6.18). Of various crown ethers and their analogues that were investigated, only DCH18C6 was able to extract

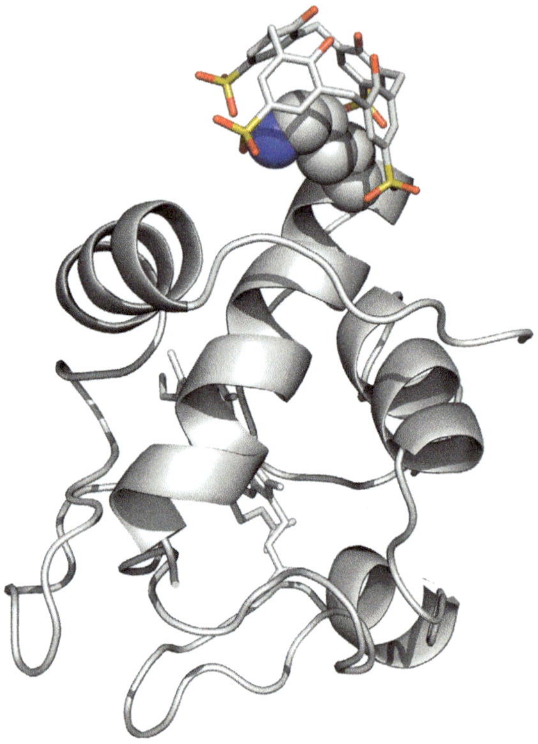

Figure 6.17 Structure of one of the cytochrome calix[4]arene sulfonate complexes present in the solid-state, as determined by Crowley and colleagues.[63] (Reproduced with permission from *Nature Chemistry*, 2012.)

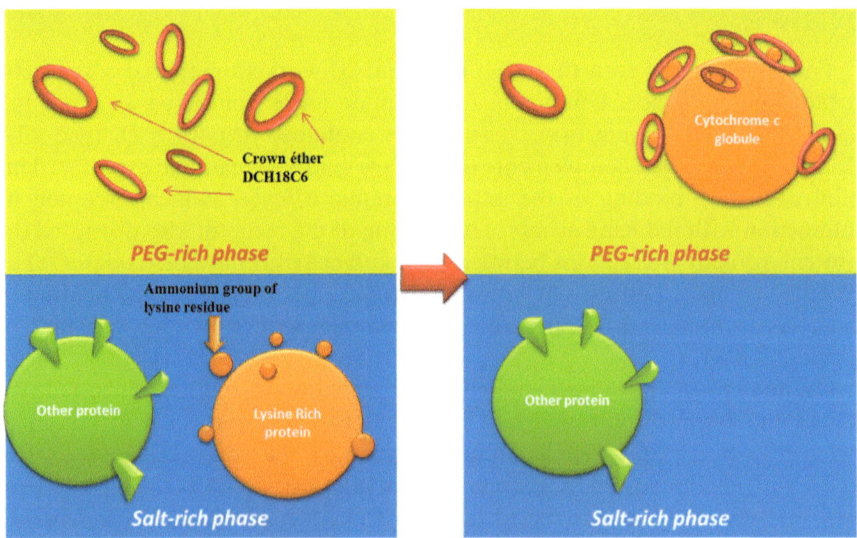

Figure 6.18 Complexation between various crown ethers and cytochrome *c* in solution studied by Oshima *et al.*[68] PEG = polyethylene glycol.

completely within 5 minutes. This lysine-rich protein was selectively extracted over other cationic proteins using DCH18C6.

6.2.2.2 Nucleotide Binding Domain 1 (NBD1) of Multidrug Resistance Protein MRP1

Recently, Baubichon-Cortay and co-workers published a paper dealing with the binding of calixarene derivatives on the NBD1 domain of MRP1.[64] The observed association constant is of the same order as that observed for ATP-Mg, the natural substrate for this protein, and is magnesium dependant.

6.2.2.3 ATP Binding Cassette Proteins

The use of amphiphilic calix[n]arenes for interaction with transmembrane proteins of the ATP class has been studied using fluorescence spectroscopy to probe the interaction of tris-carboxylatomethylene mono-alkoxycalix[4]arenes **31** with the protein.[65] This calixarene was specifically designed to form micelles and to use the known binding capacity of the anionic calix[n]arenes for cationic amino acids.[66,67] It was found that there was a selective partition of the extracted ABC protein BmrA into the aqueous media. For short alkyl chains the protein was found uniquely in the precipitate, whereas with long alkyl chains C10 and the above protein were located in the supernatant phase (Figure 6.19).

6.2.3 C-type Lectin Family

Multivalent non-covalent interactions are essential ingredients in the mediation of biological processes, as well as in the construction of complex structures for materials applications.[68] It is well known that sugar-binding lectin proteins bind to glycoproteins specifically through simultaneous multivalent interactions known as the 'glycoside cluster effect'. To date, many glycocalixarenes have been developed for specific binding to lectins.[69] Different spacer groups have been employed for grafting the sugar moieties to a calix[n]arene scaffold. Ungaro and co-workers for example prepared glycocalixarenes (**32,33**) using thiourea groups as spacers (Figure 6.20).[70–73]

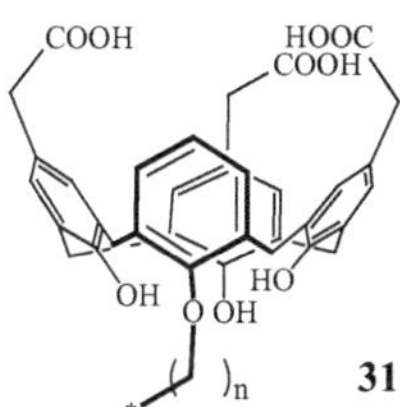

Figure 6.19 Carboxylated calix[4]arene derivatives studied by Coleman's group for NBD1 binding.[69]

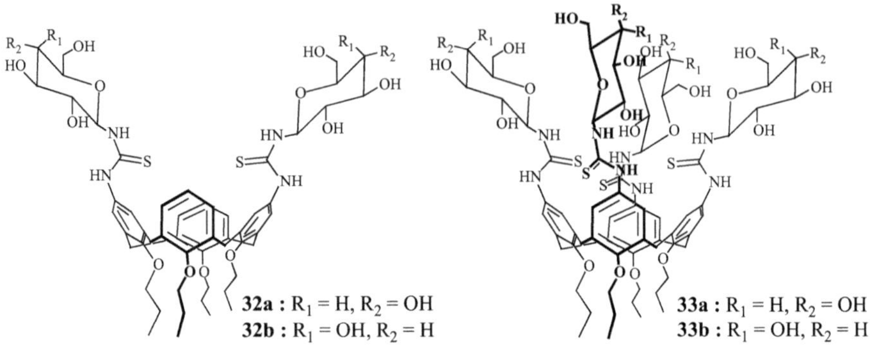

32a : R$_1$ = H, R$_2$ = OH
32b : R$_1$ = OH, R$_2$ = H

33a : R$_1$ = H, R$_2$ = OH
33b : R$_1$ = OH, R$_2$ = H

Figure 6.20 Calixarene derivatives used by Sansone *et al.*[73]

Turbidimetry experiments showed that tetraglucosyl and tetragalactosyl calixarene derivatives specifically bound to concanavalin A and peanut lectin (*Arachis hypogaea*), respectively. Consoli *et al.* synthesized *N*-acetyl-D-glucosamine clusters using amino-acid–calixarenes.[74] The glycosamino-acid–calixarenes obtained bound to lectin and increased inhibition of erythrocyte agglutination. Dondoni and co-workers also prepared glycoclusters based on calixarenes.[75,76]

Submillimolar concentrations of tetra- and octavalent sialoside clusters inhibited haemagglutination and viral infectivity mediated by BK and influenza A viruses.

Roy and co-workers prepared calixarene derivatives with sialic acid and *N*-acetylgalactosamine units on their lower rims.[77,78] Dendritic, water-soluble, carbohydrate-containing *p*-tert-butylcalix[4]arene derivatives were also synthesized in order to study their lectin-binding properties. It has been shown that the amphiphilic calix[4]arenes with 16 carbohydrate units displayed affinity for carbohydrate-binding proteins and polystyrene surfaces. Vidal and co-workers prepared seven topologically isomeric calix[4]arene glycoconjugates by the attachment of sugar moieties to a series of alkyne-derivatized calix[4]arene precursors.[79,80] They demonstrated that the trivalent conjugate exhibited higher affinity to a bacterial lectin than the monovalent conjugate.

In another paper, Bezouska *et al.*[81] reported that carboxylated calixarene derivatives (**34–36**) have a high affinity for CD69 leukocyte membrane receptor and act as specific antagonists providing complete protection against CD69-dependant apoptosis induced both by multivalent carbohydrate ligand and by antibody cross-linking (Figure 6.21). In the series tested, thiacalix[4]arene derivative **35** had the highest affinity for CD69 and proved to be the most specific inhibitor of CD69 identified so far in precipitation and cellular activation experiments. These researchers claim that this latest derivative may become an attractive tool in the protection of CD69 receptors *in vivo*, and thus protect CD69^{high} cytotoxic lymphocytes from apoptotic death and activation by the tumours.

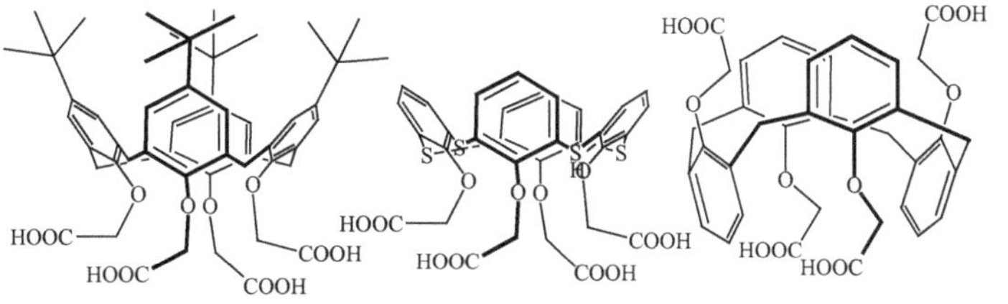

Figure 6.21   Carboxylated calixarene derivatives having affinity for the CD69 receptor of the leucocyte membrane, described by Bezouska *et al.*[81]

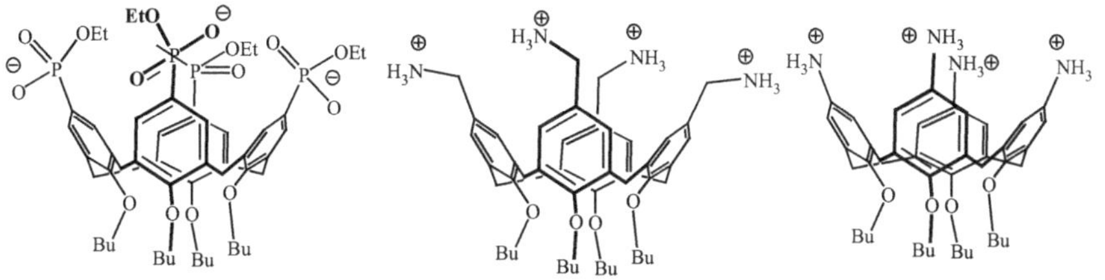

Figure 6.22   Calixarene derivatives used by Schrader.[17]

In a different approach, Schrader[17] constructed mixed monolayers of *para*-phosphonato-*O*-butyl-calix[4]arenes **37** and embedded these novel receptors into stearic acid monolayers at the air–water interface (Figure 6.22). Initial experiments showed excellent sensitivity towards lysine, argine, diarginine and triargine present in the subphase. Using shifts in the pressure/area diagrams, at protein concentrations of 10^{-8} M, weak to moderate effects were observed with acidic or neutral proteins such as ferritin and the Dps dodecamer—proteins containing basic domains, for example albumin and thrombin, led to larger effects. However, in the case of basic proteins such as histone H1 and chymotrypsin, cytochrome c and proteinase K, very large expansions in the apparent molecular areas were observed. The work was extended to cationic *para*-benzylammonium-calix[4]arenes **38** or *para*-aniliniumcalix[4]arenes **39** and here the selectivity was reversed with maximum sensitivity for the anionic proteins, including acylcarrier protein or ATPase. By careful choice a matrix for highly sensitive protein detection was constructed.

6.3   Conclusion

A wide range of protein-to-supramolecular-system interactions has been described in this chapter. It can be seen that the two options of using the supramolecular system—not only as a scaffold for coupling bioactive function, but also in the direct action of various molecules—have been widely used and that this field of research is expanding rapidly.

References

1. T. Kodadek, *Chem. Biol.*, 2001, **8**, 105–115.
2. T. Schrader and S. Koch, *Mol. Biosyst.*, 2007, **3**, 241–248.
3. M. W. Peczuh and A. D. Hamilton, *Chem. Rev.*, 2000, **100**, 2479–2494.
4. S. Jones and J. M. Thornton, *Proc. Nat. Acad. Sci. USA*, 1996, **93**, 13–20.
5. M. Luckey, *Membrane Structural Biology*, Cambridge University Press, Cambridge, 2008.
6. J. J. R. Fraústo da Silva and R. J. P. Williams, The biological chemistry of the elements, in *The Inorganic Chemistry of Life*, Oxford University Press, Oxford, 2nd edn, 2001.
7. D. A. Fulton and J. F. Stoddart, *Bioconj. Chem.*, 2001, **12**, 665–672.
8. J. W. Cornforth, P. D. Hart, G. A. Nicholls, R. J. Rees and J. A. Stock, Br., *J. Pharmacol. Chemother.*, 1955, **10**, 73–88.
9. E. Da Silva, A. N. Lazar and A. W. Coleman, *J. Drug Deliv. Sci. Technol.*, 2004, **14**, 3–20.
10. S. B. Nimse and T. Kim, *Chem. Soc. Rev.*, 2013, **42**, 366–386.
11. F. Perret, A. N. Lazar and A. W. Coleman, *Chem. Commun.*, 2006, (23), 2425–2438.
12. A. W. Coleman, F. Perret, A. Moussa, M. Dupin, Y. Guo and H. Perron, *Topics Curr. Chem.*, 2007, **277**, 31–88.
13. R. E. McGovern, H. Fernandes, A. R. Khan, N. P. Power and P. B. Crowley, *Nat. Chem.*, 2012, **4**, 527–533.
14. T. Oshima and Y. Baba, *J. Incl. Phenom. Macrocycl. Chem.*, 2012, **73**, 17–32.
15. L. Mutihac, C. Baltariu, A. Hotaranu and R. Mutihac, *Rev. Roum. Chim.*, 2006, **51**, 1041–1051.
16. S. Jebors, Y. Tauran, N. Aghajari, S. Boudebbouze, E. Maguin, R. Haser, A. W. Coleman and M. Rhimi, *Chem. Commun.*, 2011, **47**, 12 307–12 309.
17. R. Zadmard and T. Schrader, *J. Am. Chem. Soc.*, 2005, **127**, 904–915.
18. V. Martos, P. Castreno and J. Valero, *Curr. Opin. Chem. Biol.*, 2008, **12**, 698–706.
19. A. Casnati, F. Sansone and R. Ungaro, *Acc. Chem. Res.*, 2003, **36**, 246–254.
20. K. Yannakopoulou, *J. Drug. Deliv. Sci. Technol.*, 2012, **22**, 243–249.
21. T. Serno, R. Geidobler and G. Winter, *Adv. Drug. Deliv. Rev.*, 2011, **63**, 1086–1106.
22. S. Ohtake, Y. Kita and T. Arakawa, *Adv. Drug. Deliv. Rev.*, 2011, **63**, 1053–1073.
23. F. J. Otero-Espinar, A. Luzardo-Alvarez and J. Blanco-Mendez, *Mini-Rev. Med. Chem.*, 2010, **10**, 715–725.
24. Y. Sasaki and K. Akiyoshi, *Curr. Pharm. Biotech.*, 2010, **11**, 300–305.
25. D. A. Uhlenheuer, K. Petkau and L. Brunsveld, *Chem. Soc. Rev.*, 2010, **39**, 2817–2826.
26. T. J. Peters, *All About Albumin. Biochemistry, Genetics and Medical Applications*, Academic Press, New York, 1996.

27. C. C. Annarelli, L. Reyes, J. Fornazero, J. Bert, R. Cohen and A. W. Coleman, *Cryst. Eng.*, 2000, **3**, 173–177.
28. L. Memmi, A. Lazar, A. Brioude, V. Ball and A. W. Coleman, *Chem. Commun.*, 2001, 2474–2475.
29. E. Da Silva, C. F. Rousseau, I. Zanella-Cléon, M. Becchi and A. W. Coleman, *J. Incl. Phenom. Macro.*, 2006, **54**, 53–59.
30. J. Moubarak, E. Moreno, E. Diesis and A. W. Coleman, *Chem. J. Mold*, 2009, **4**, 94–100.
31. S. Kunsagi-Mate, S Lecomte, E. Eortmann, Kunsagi-Mate and B. Desbat, *J. Incl. Phenom. Macrocycl. Chem.*, 2010, **66**, 147–151.
32. S. Jebors, F. Fache, S. Balme, F. Devoge, M. Monachino, S. Cecillon and A. W. Coleman, *Org. Biomol. Chem.*, 2008, **6**, 319–329.
33. A. I Vovk, V. I. Kalchenko, S. A. Cherenok, V. P. Kukhar, O. V Muzychka and M. O. Lozynsky, *Org. Biomol. Chem.*, 2004, **2**, 3162–3166.
34. A. I. Vovk, L. A. Kononets, V. Y. Tanchuk, A. B. Drapailo, V. I. Kalchenko and V. P. Kukhar, *J. Incl. Phenom. Macroc. Chem.*, 2010, **66**, 271–277.
35. S. Cherenok, A. Vovk, I. Muravyova, A. Shivanyuk, V. Kukhar, J. Lipkowski and V. Kalchenko, *Org. Lett.*, 2006, **8**, 549–552.
36. T. Mecca, G. M. L. Consoli, C. Geraci and F. Cunsolo, *Bioorg. Med. Chem.*, 2004, **12**, 5057.
37. C. D. Gutsche and I. Alam, *Tetrahedron*, 1988, **44**, 4689–4694.
38. Z. Li, M. Wang, F. Wang, Z. Gu, G. Du, J. Wu and J. Chen, *Appl. Microbiol. Biotechnol.*, 2007, **77**, 245–255.
39. K. Haga, A. Nakamura and K. Yamane, *J. Jap. Soc. Biosci. Biotech. Agroc.*, 1995, **69**, 1041–1045.
40. K. Haga, R. Kanai, O. Sakamoto, M. Aoyagi, K. Harata and K. Yamane, *J. Biochem.*, 2003, **134**, 881–891.
41. C. Boutin, A. Stopin, F. Lenda, T. Brotin, J. P. Dutasta, N. Jamin, A. Sanson, Y. Boulard, F. Leteurtre and G. Huber, *et al.*, *Bioorg. Med. Chem.*, 2011, **19**, 4135–4143.
42. G. K. Seward, Y. Bai and N. S. Khan, *Chem. Sci.*, 2011, **2**, 1103–1110.
43. E. Da Silva, D. Ficheux and A. W. Coleman., *J. Incl. Phenom. Macroc. Chem.*, 2005, **52**, 201–206.
44. E. V. Lugovskoy, P. G. Gritsenko, T. A. Koshel, I. O. Koliesnik, S. O. Cherenok, O. I. Kalchenko, V. I. Kalchenko and S. V. Komisarenko, *FEBS Journal*, 2011, **278**, 1244–1251.
45. F. L. Aachmann, D. E. Otzen, K. L. Larsen and R. Wimmer, *Prot. Eng.*, 2003, **16**, 905–912.
46. K. D. Daze, T. Pinter, C. S. Beshara, A. Ibraheem, S. A. Minaker, M. C. F. Ma, R. J. M. Courtemanche, R. E. Campbell and F. Hof, *Chem. Sci.*, 2012, **3**, 2695–2699.
47. S. A. Minaker, K. D Daze, M. C. F. Ma and F. Hof, *J. Am. Chem. Soc.*, 2012, **134**, 11 674–11 680.

48. M. G. Chini, S. Terracciano, R. Riccio, G. Bifulco, R. Ciao, C. Gaeta, F. Troisi and P. Neri, *Org. Lett.*, 2010, **12**, 5382–5385.

49. J. Q. Trojanowski, M. Goedert, T. Iwatsubo and V. M.-Y. Lee, *Cell Death Diff.*, 1998, **5**, 832–845.

50. A. W. Coleman, F. Perret, S. Cecillon, A. Moussa, A. Martin, M. Dupin and H. Perron, *New J. Chem.*, 2007, **31**, 711–717.

51. Y. Lee, E. K. Lee, Y. W. Cho, T. Matsui, I.-C. Kang, T. Kim and M. H. Han, *Proteomics*, 2003, **3**, 2289–2304.

52. S. W. Oh, J. D. Moon, H. J. Lim, S. Y. Park, T. Kim, J. B. Park, M. H. Han, M. Snyder and E. Y. Choi, *FASEB J.*, 2005, **19**, 1335–1340.

53. H. S. Park, Q. Lin and A. D. Hamilton, *J. Am. Chem. Soc.*, 1999, **121**, 8–13.

54. H. S. Park, Q. Lin and A. D. Hamilton, *Proc. Nat. Acad. Sci. USA*, 2002, **99**, 5105–5109.

55. T. Oshima, M. Goto and S. Furusaki, *Biomacromolecules*, 2002, **3**, 438–450.

56. T. Oshima, H. Higuchi, K. Ohto, K. Inoue and M. Goto, *Langmuir*, 2005, **21**, 7280–7284.

57. T. Oshima, M. Sako, Y. Shikaze, K. Ohto, K. Inoue and Y. Baba, *Biochem. Eng. J.*, 2007, **35**, 66–70.

58. K. Shimojo, T. Oshima, H. Naganawa and M. Goto, *Biomacromolecules*, 2007, **8**, 3061–3066.

59. R. E. McGovern, H. Fernandes, A. R. Khan, N. P. Power and P. B. Crowley, *Nat. Chem.*, 2012, **4**, 527–533.

60. Y. Hamuro, M. C. Calama, H. S. Park and A. D. Hamilton, *Angew. Chem. Int. Ed.*, 1997, **36**, 2680–2692.

61. Q. Lin and A. D. Hamilton, *Comptes Rendus Chimie*, 2002, **5**, 441–444.

62. Y. Wei, G. L. McLendon, A. D. Hamilton, M. A. Case, C. B. Purring, Q. Lin, H. S. Park, C.-S. Lee and T. Yu, *Chem. Commun.*, 2001, (17), 1580–1581.

63. T. Oshima, A. Suetsugu and Y. Baba, *Analyt. Chim. Acta*, 2010, **674**, 211–219.

64. L. F. A. Nault, C. Girardot, A. Leydier, A. W. Coleman, T. Perrotton, S. Magnard and H. Baubichon-Cortay, *New. J. Chem.*, 2010, **34**, 1812–1822.

65. K. Suwinska, O. Shkurenko, C. Mbemba, A. Leydier, S. Jebors, A. W. Coleman, R. Matar and P. Falson, *New J. Chem.*, 2008, **32**, 1988–1998.

66. F. Perret and A. W. Coleman, *Chem. Commun.*, 2011, **47**, 7303–7319.

67. F. Perret, H. Peron, M. Dupin and A. W. Coleman, Calix-arenes as protein sensors, *in Topics in Current Chemistry, Creative Chemical Sensor Systems*, T. Schrader, Springer, Berlin, 2007, pp. 31–88.

68. J. D. Badjic, A. Nelson, S. J. Cantrill, W. B. Turnbull and J. F. Stoddart, *Acc. Chem. Res.*, 2005, **38**, 723–732.

69. L. Baldini, A. Casnati, F. Sansone and R. Ungaro, *Chem. Soc. Rev.*, 2007, **36**, 254–266.

70. F. Sansone, E. Chierici, A. Casnati and R. Ungaro, *Org. Biomol. Chem.*, 2003, **1**, 1802–1809.
71. A. Casnati, F. Sansone and R. Ungaro, *Acc. Chem. Res.*, 2003, **36**, 246–254.
72. S. André, F. Sansone, H. Kaltner, A. Casnati, J. Kopitz, H.-J. Gabius and R. Ungaro, *Chem. Biol. Chem.*, 2008, **9**, 1649–1661.
73. F. Sansone, L. Baldini, A. Casnati and R. Ungaro, *New J. Chem.*, 2010, **34**, 2715–2728.
74. G. M. L. Consoli, F. Cunsolo, C. Geraci and V. Sgarlata, *Org. Lett.*, 2004, **6**, 4163–4166.
75. A. Dondoni and A. Marra, *J. Org. Chem.*, 2006, **71**, 7546–7557.
76. A. Marra, L. Moni, D. Pazzi, A. Corallini, D. Bridi and A. Dondoni, *Org. Biomol. Chem.*, 2008, **6**, 1396–1409.
77. S. J. Meunier and R. Roy, *Tetrahedron Lett.*, 1996, **37**, 5469–5472.
78. R. Roy and J. M. Kim, *Angew. Chem. Int. Ed.*, 1999, **38**, 369–372.
79. S. Cecioni, R. Lalor, B. Blanchard, J.-P. Praly, A. Imberty, S. E. Matthews and S. Vidal, *Chem. Eur. J.*, 2009, **15**, 13 232–13 240.
80. S. Cecioni, S. Faure, U. Darbost, I. Bonnamour, H. Parrot-Lopez, O. Roy, C. Taillefumier, M. Wimmerov, J.-P. Praly, A. Imberty and S. Vidal, *Chem. Eur. J.*, 2011, **17**, 2146–2159.
81. B. Bezouska, R. Snajdrova, K. Krenek, M. Vancurova and A. Kadek, *et al.*, *Bioorg. Med. Chem.*, 2010, **18**, 1434–1440.

Cucurbiturils in Drug Delivery And For Biomedical Applications

NA'IL SALEH,*[a] INDRAJIT GHOSH[b] AND
WERNER M. NAU*[b]

[a] Department of Chemistry, College of Science, UAE University, P.O.
Box 15551, Al-Ain, United Arab Emirates; [b] School of Engineering and
Science, Jacobs University Bremen, Campus Ring 1, D-28759, Bremen,
Germany
*Email: n.saleh@uaeu.ac.ae; w.nau@jacobs-university.de

7.1   Introduction

Achieving the desired action of therapeutic/diagnostic agents in living systems
is a key challenge in pharmacology and medicine.[1–4] Over the past decades,
biomedical researchers have shifted their focus towards applications of artificial
nanoscale assemblies as biocompatible vehicles to improve the stability and
specificity of therapeutic and diagnostic agents. Such advanced formulation
approaches can have economic advantages over the development of new drug
candidates, and for the pharmaceutical industry, more efficient formulations
may have advantages from an intellectual-property protection point of view.

Among these nanoscaled assemblies, macrocycles,[1–4] which have the ability
to encapsulate a therapeutic agent non-covalently within their cavity, have
gained enormous attention in recent years. When compared to other drug
delivery systems such as dendrimers,[5] liposomes,[6] micelles,[7] carbon nanotubes,[8]
and hydrogels as well as polymers,[9] macrocycles frequently display advantages

Monographs in Supramolecular Chemistry No. 13
Supramolecular Systems in Biomedical Fields
Edited by Hans-Jörg Schneider
© The Royal Society of Chemistry 2013
Published by the Royal Society of Chemistry, www.rsc.org

in view of their thermal and chemical stability, their ability to self-assemble into different nanostructures, and their availability in various sizes, which have jointly contributed to the blossoming of macrocyclic drug vehicles in biomedical fields. Liposomes and micelles, for example, can readily embed hydrophobic drug molecules, but their limited stability (the micro-heterogeneous solutions cannot be stored for extended periods) and the lack of control on the rate of drug retention and release have greatly limited their broad use as drug delivery vehicles. Discrete supramolecular host molecules, and especially macrocyclic hosts, show guest binding both *in vitro* and *in vivo*; they can function both to capture drugs and to release them, the release kinetics can be tuned through the guest size and external stimuli, and they frequently show an astoundingly high biocompatibility (*i.e.*, biological inertness as well as low toxicity), thereby addressing and overcoming several challenges perceived for alternative drug delivery systems.

Among the different supramolecular host molecules, cyclodextrins (CDs)[1,2] have been most widely used for the formulation of therapeutic and diagnostic agents in pharmacology (see Chapter 5 by Ortiz Mellet *et al.*). However, their use in clinical conditions is generally limited to oral and topical drug delivery forms, since they can be nephrotoxic if they enter the body in a non-metabolized state.[10] In addition, the poor selectivity and low binding affinities ($K_{\text{binding}} < 10^4\ \text{M}^{-1}$)[11] of non-derivatized (natural) CDs towards guest molecules in general, and drug molecules in particular, requires excess concentrations of CDs to form CD•guest complexes quantitatively.

Several alternative macrocyclic host molecules have been suggested in the literature to have potential as drug delivery vehicles, for example derivatized calixarenes (see Chapter 6 by Perret and Coleman),[12–14] but they have yet to find actual pharmaceutical application. Recently, macrocycles known as cucurbiturils, along with derivatives and acyclic variants, have emerged as the next generation of macrocycles with clearly defined potential as drug vehicles. They are water-soluble, thermally as well as chemically stable, they show superior binding affinities with a large range of guest molecules, and preliminary toxicological studies are most promising.

Cucurbit[*n*]urils (CB*n*) ($n = 5$–10, with $n = 9$ not yet isolated)[15–20] are readily synthesized by a condensation reaction between glycoluril and formaldehyde (or paraformaldehyde) in strongly acidic media. The chemical structure of CB6 was solved by Freeman and others,[15] more than 70 years after its first synthesis in 1905.[21] In structural detail, CB homologues contain 5 to 10 glycoluril residues joined by two methylene bridges, which form two partially negatively charged hydrophilic carbonyl portals and an interior hydrophobic cavity with low polarity and polarizability (see Scheme 7.1 for chemical structures). Recently, the preparation[17,18,22] and purification[23] of CBs on a larger scale at economical cost has become possible, following different suggested procedures (*e.g.* bench-top,[17,18,22] microwave,[24] and a green method using alkylimidazolium) by different research groups.[25] In-depth studies from the Isaacs group have also revealed details of their formation mechanism during condensation.[26–27]

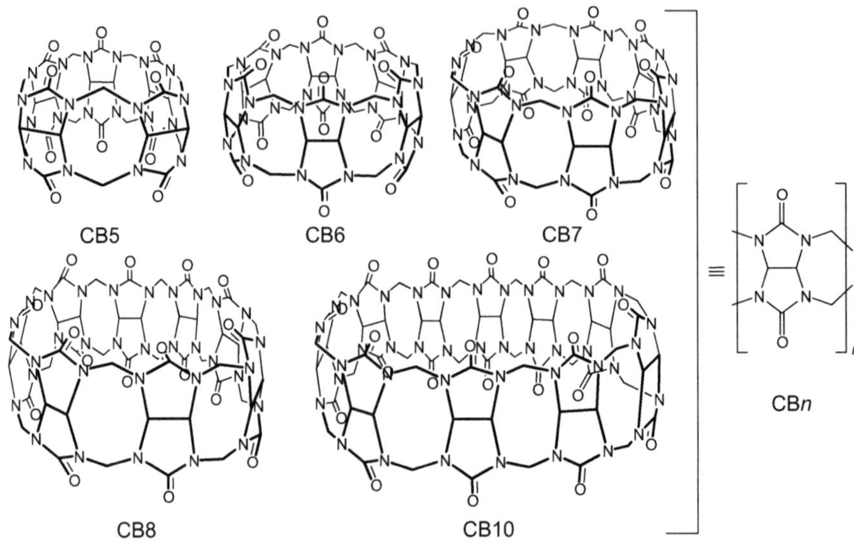

Scheme 7.1 Chemical structures of cucurbit[*n*]urils (CB*n*).

Although *in vivo* applications of CBs for medicinal therapeutics and diagnostics are emerging slowly,[3,4,10,28,29] the increasing number of publications on CB-based drug delivery furbishes evidence for the advent of these synthetic macrocycles as drug vehicles. In this chapter, we will discuss the binding of CBs with different drugs and biologically relevant molecules and summarize associated applications, for example, the amendment of a given drug's physical and texture properties (*e.g.*, the suppression of polymorphism), improvement of the drug's *in vivo* as well as thermal and chemical stabilities (required to allow long-term storage), the masking of unwanted taste, increase in the drug's solubility in water, enhancement of the ability to cross internal membranes, optimization of the drug's *in vivo* distribution and/or metabolism, the bioprotection of the delivered drug from binding to serum proteins in circulation tracks, the modification of the clearance mechanism, and the influence on the drug-pharmacokinetic effects. In addition, we will also allude to the *in vitro* and *in vivo* biocompatibility (cytotoxicity and toxicity of CBs and CB•guest complexes), the release of the drug molecules from the CB cavity, CB-based targeted drug delivery systems, the manufacturing and production of drug•CB complexes on an industrial scale (preformulation and formulation), all of which are relevant for drug delivery applications.

7.2 Factors Governing the Formation, Stability, and Properties of Cucurbituril•Guest Complexes

The host–guest chemistry of CBs has already been described in several review articles.[16,19] Binding of guest molecules with CBs is generally driven by composite supramolecular interactions. The hydrophobic effect as well as ion–dipole and dipole–dipole interactions have long been recognized to contribute to the binding of guests by CBs. The glycoluril core at the CB equator provides a hydrophobic void for the binding of neutral and hydrophobic molecules. Electrostatic interactions operate between cationic sites of the guests and the ureido oxygens at the CB portals. Recently, it has been reported[30,31] that the release of (high energy) water molecules from the cavity of CBs also plays a particularly important role for the binding of guest molecules. Finally, the size complementarity of the guests to the cavity volume of CBs is also an important denominator of the absolute binding constants of CBs.[30] Other things being equal, the binding constants become highest when the ratio of the guest volume and the inner cavity volume of the CB are close to 0.55.[30] (The 55% rule, originally established for capsular host–guest complexes.[32]) The binding constants of CBs with guest molecules are usually[4,28,30] in the range of 10^3–10^{10} M^{-1} with few engineered exceptions for which the binding constants become extremely high. For example, the binding constant of CB7 with ferrocene derivatives has been reported to reach up to 10^{15} M^{-1},[33–35] which is equivalent to that of the biotin–avidin pair, the strongest non-covalent interaction between two partners found in nature.

When guest molecules are non-covalently encapsulated inside the CB cavity, a modification of their physical and chemical properties always occurs due to an altered microenvironment, as well as confinement and isolation from the surrounding medium.[36] In addition, the host•guest complexes can also inherit beneficial properties of the host molecules. For example, when poorly water-soluble guest molecules are immersed in the cavity of CB7, their solubilities are enhanced owing to the higher solubility of CB7.[36–38] This phenomenon is often termed *"host-induced guest solubilization"* (also known for cyclodextrins). Moreover, for cationic guest molecules, the CBs' complexes can become more soluble than the uncomplexed hosts due to a higher guest solubility.[29,36,37,39] For example, the solubilities of CB8 complexes with cationic di- and trinuclear anticancer drugs are in the millimolar range, which is much higher in comparison to the intrinsic micromolar solubility of CB8.[39] There are only a few exceptions where the solubilities of CB*n*•cationic-guest complexes are lower than those of the free drug molecules. For example, complexes of CB6 and CB8 with some phenanthroline-containing platinum complexes are less soluble than the drugs themselves.[40]

Upon encapsulation inside the CB cavity, guest molecules can also take advantages of isolation or protection from the bulk solvent. Mohanty *et al.* have reported that CB7 can induce deaggregation and photostabilization of a fluorescent dye,[41] namely rhodamine 6G, which is commonly used in biotechnological applications such as fluorescence microscopy and fluorescence correlation spectroscopy. Recently, Biedermann *et al.* have also reported a fluorescence enhancement due to efficient deaggregation in the presence of CB8.[42]

In addition to the modification of the physical properties of guest molecules, owing to the presence of partially negative charged carbonyl portals, CBs also influence the chemical reactivities of guest molecules. In the simplest case, CBs modify the pK_a values of guest molecules and thereby also alter their chemical reactivities. CBs prefer binding to the protonated form of a guest molecule over its neutral form[43-46] and consequently shift the pK_a values of the conjugate acids of weakly basic guests (*e.g.*, amines) towards higher values. This phenomenon is known as *"host-assisted guest protonation"* or *"complexation-induced pK_a shifts"*.[44,47-50] The stabilization of the protonated species is mainly due to the formation of additional ion–dipole interactions between the cationic sites of the guest molecules and the carbonyl portals of the CBs. Shifts in the pK_a values allow the complexed guest molecules to remain protonated at pH values up to five units higher than the pK_a values of their free, uncomplexed forms; this has a direct influence on their chemical reactivities, their solubilities at higher pH values (an ionization effect which adds to the host-induced guest solubilization), and penetration abilities towards biological membranes, where required.[4,36,37,45,48,51-56] It is worth mentioning here that CDs are also known to affect the pK_a values of guest molecules but in the *"opposite"* direction (*i.e.*, negative pK_a shifts), because they tend to preferentially complex the neutral forms of guests.[49,57,58]

CBs can also influence the solid-state properties of the guest molecules. It is well known that pharmaceutical solids (drug substances and excipients) exist in different solid-state forms.[59] Changes in the physical state of a solid drug, during manufacturing or long-term storage, have been reported for many drug substances (*e.g.*, changes from an amorphous to a crystalline structure and *vice versa*, or from one crystal polymorph to another, or dehydration/hydration of the solid). As each solid-state structure of a certain drug is characterized by its own physical properties, including packing, thermodynamic, spectroscopic, kinetic, surface, and mechanical properties, such conversions between solid-state forms of a given drug substance should in turn affect its manufacturing reproducibility and cost, clinical efficacy, drug bioavailability, and even safety. Therefore, maintaining the physical stability of a pure drug substance during its processing, storage, and formulation in a usable dosage form can be a formidable task. As confirmed by differential scanning calorimetry (DSC) and thermogravimetric studies, CBs possess a very high thermal stability[60] because of the interactions of the outer methylene hydrogens at the bridging carbons with the carbonyl portals of neighbouring macrocycles found in their honeycomb-like crystal structure.[61] Therefore, formation of CB*n* host–guest complexes can impart significant thermal stability on many drugs in the solid-state. Beneficial effects of CBs on the thermal stability of the drug molecules will be exemplified in Section 7.4. It is worth mentioning here that X-ray powder diffraction (XRD), N_2-adsorption, and DSC confirmed the ability of CB7 to prevent interconversion of crystal polymorphs of some drugs, and to retain its amorphous structure[61] in the resulting complex, while having negligible effects on drug particle size and surface area.[62]

In an actual drug delivery process, the drug•host complex must pass through different biological media. There are five main fluids within the human body in which CBs and their CB•drug complexes may be dissolved and transported following administration: saliva, gastric and intestinal fluids, blood plasma, and nasal fluids.[29] The behaviour of the drug•CBn complexes can be very different in these media. For example, it has been observed that the solubilities of the drug•CBn complexes are higher in biological media than in water owing to the presence of cations or due to different pH. Note that inorganic metal ions, on one hand, bind to the carbonyl portals of CBs and, therefore, increase their solubilities[29,63,64] (as observed for cationic guest molecules) but at the same time decrease the effective binding constants of the guest molecules with CBs, which results in a displacement; the latter may or may not be desirable.[65–68] Solubility enhancement of the even-numbered CB homologues up to millimolar concentrations in salt-containing media (such as brine) is long known,[69] while metal ion-promoted decomplexation of CB•guest complexes has been subsequently scrutinized.[70–74]

7.3 Design of Cucurbituril Derivatives and Analogues with Improved Properties

Note that the use of even-numbered CBn homologues ($n = 6$, 8) as drug solubilizing agents is limited because of their low micromolar solubility in water. In general, the solubilities of the parent CBs (Scheme 7.1) are lower than those of many other host molecules of similar size, for example cyclodextrins and *p*-sulfonatocalixarenes, but their extraordinary binding affinities towards many guest molecules counterbalances this apparent disadvantage.[75,76] Nevertheless, to bypass this limitation, several functionalized CBs or acyclic CB analogues have recently been synthesized (Scheme 7.2).[26,77–82] Indeed, the reported modified CBs possess very high water solubility with respect to their parent structures. The solubilities of Me$_2$CB7 and CyCB7 in pure water are 264 mM and 181 mM, respectively.[83] The solubilities of the acyclic CBs are also very high and can be further enhanced by incorporating solubilizing sulfonato groups, as evidenced in recent reports.[38,84,85] Regarding the binding properties, all substituted CBs possess very similar recognition features for similar types of guest molecules, owing to the presence of the same recognition motifs (*i.e.*, a hydrophobic cavity laced with carbonyl portals).[30] However, due to their flexibility, the acyclic CBs are capable of binding an even wider range of guest molecules. For example, acyclic sulfonato CBs (CB-3 and CB-4 in Scheme 7.2) even bind neuromuscular blocking agents, which are too voluminous to bind inside the cavity of CB7 (see Section 7.4.8).[84] Note that due to their rigidity and barrel-shape, and in comparison to more flexible macrocycles—which in addition possess larger portals than their equator (cyclodextrins and calixarenes)— CBs have a particularly stringent size limitation for guest encapsulation (55% packing rule). Precedents have also been reported where the presence of aromatic substituents enhances the binding propensities of the substituted CBs towards aromatic guest molecules.[38] The binding affinities of fully cyclopentano-substituted CB6 or CB7 with dioxane and adamantylamine appear to be higher than those of the unsubstituted parents.[86] Considering the experimental trends, it can be generalized that the functionalization of CBs increases their solubilities

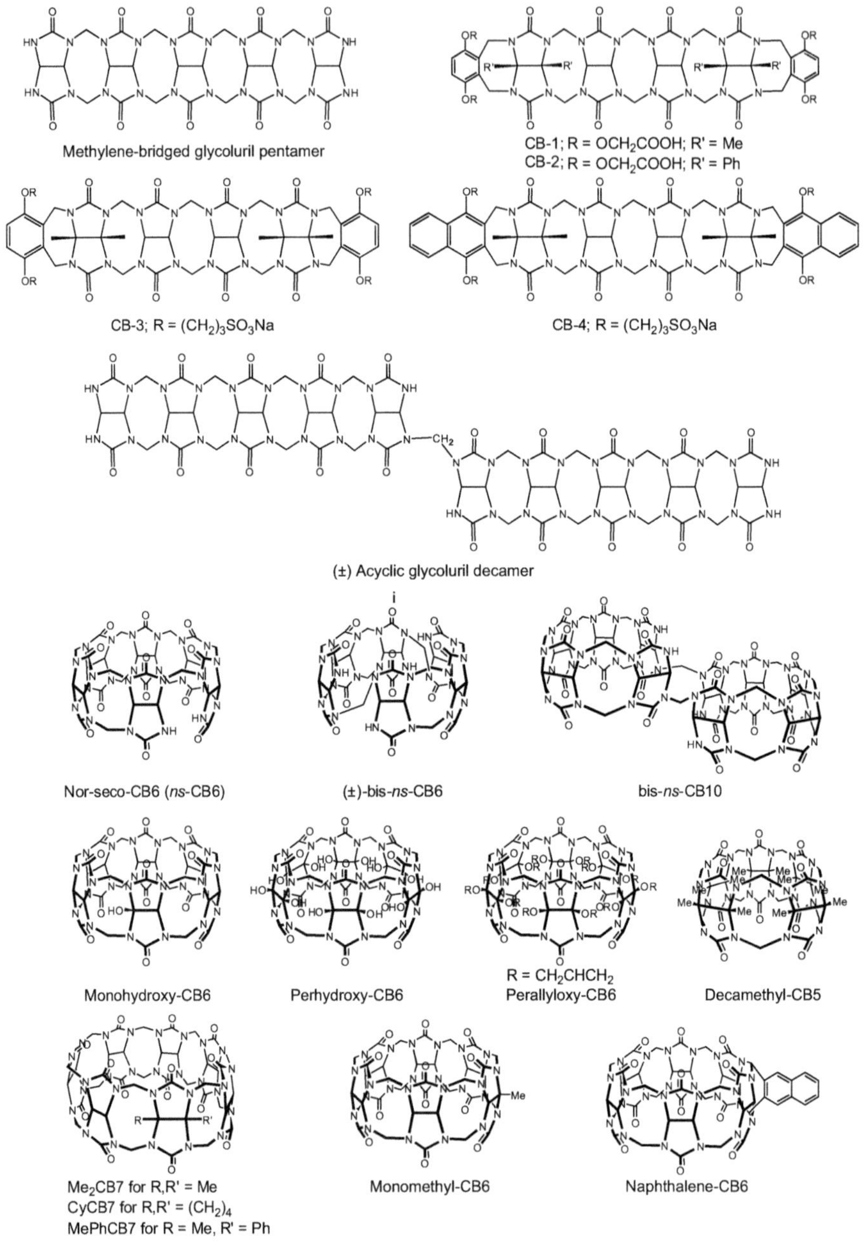

Methylene-bridged glycoluril pentamer

CB-1; R = OCH$_2$COOH; R' = Me
CB-2; R = OCH$_2$COOH; R' = Ph

CB-3; R = (CH$_2$)$_3$SO$_3$Na

CB-4; R = (CH$_2$)$_3$SO$_3$Na

(±) Acyclic glycoluril decamer

Nor-seco-CB6 (*ns*-CB6)

(±)-bis-*ns*-CB6

bis-*ns*-CB10

Monohydroxy-CB6

Perhydroxy-CB6

R = CH$_2$CHCH$_2$
Perallyloxy-CB6

Decamethyl-CB5

Me$_2$CB7 for R,R' = Me
CyCB7 for R,R' = (CH$_2$)$_4$
MePhCB7 for R = Me, R' = Ph

Monomethyl-CB6

Naphthalene-CB6

Scheme 7.2   Chemical structures of acyclic and functionalized cucurbit[*n*]urils.

and can improve their versatility in guest encapsulation, all of which presents important boundary conditions for drug encapsulation. Also noteworthy, the release of drug molecules from the acyclic CBs is faster;[38] again, this may or may not be desirable for the delivery of a particular drug.

7.4 Investigated Drug Molecules with Cucurbiturils as Macrocyclic Hosts

Encapsulation of drug molecules inside the cavity of CBs has been demonstrated to provide a range of distinct benefits. The present section encompasses a summary for most of the reports on CBs host–guest complexes with drugs, pharmaceutical agents, and other bioactive molecules. The corresponding structures discussed in this section are depicted in Scheme 7.3.

7.4.1 Anti-pathogenic Agents

Examples of the use of CBs to enhance the bioavailability of drugs have been demonstrated by studying the interactions of CB7 with benzimidazole-based drugs: albendazole, carbendazim, thiabendazole, and fuberidazole. These drug molecules possess significant solubility only in their protonated states. Solubilities of the neutral drug molecules are very poor. At pH 7.2, the reported solubilities of the drug molecules in water are 0.003, 0.160, 0.11, and 0.25 mM for albendazole, carbendazim, thiabendazole, and fuberidazole, respectively. Note that pK_a values of these drug molecules are in the range of 3.5–4.8, and therefore they are neutral at physiological pH, which is at least two units higher than the pK_a values of the drug molecules. As discussed in Section 7.2, CB7 binding increases the pK_a values of the conjugate acids of these drug molecules by 2–4 units and improves their solubilities by stabilizing the protonated forms at pH 7.2.[36] For albendazole, which is least soluble among the above-mentioned drugs, the solubility enhancement factor was 100 in the presence of 2 mM CB7. Not surprisingly, in another study, the solubility enhancement of albendazole was more pronounced when high concentrations of CBs were used and the solutions containing albendazole and CB homologues were prepared at low pH. The obtained solubilities for albendazole were 5.8 mM, 7.1 mM, and 2.7 mM with CB6, CB7, and CB8 respectively.[37] The most recent studies[83] show that the acyclic analogues of CBs and the CB7 derivative (Me$_2$CB7) can also increase the solubility of albendazole. For the acyclic analogue a 226-fold solubility enhancement was observed[38] and for Me$_2$CB7 the solubility of albendazole increased to 5.8 mM.

Upon encapsulation inside the cavity of CBs, a significant improvement in the photostability of several benzimidazole drugs has also been reported.[36] For example, fuberidazole and thiabendazole photobleached less effectively in the presence of CB7, with photostabilization factors amounting to 7 and 3, respectively. In addition, CB7 prevents the interconversion of crystal polymorphs of albendazole and retained the amorphous structure in the resulting complex.[62] The association of albendazole with CB7 also imparted an improved thermal stability on the drug. An improved photostability of Hoechst 34580, a common nuclear stain, has been reported upon encapsulation inside the cavity of CB7:[87] 40% of the dye molecules were photobleached under 2.5 h of white-light irradiation *versus* 7% in the presence of CB7. The enhanced photostability

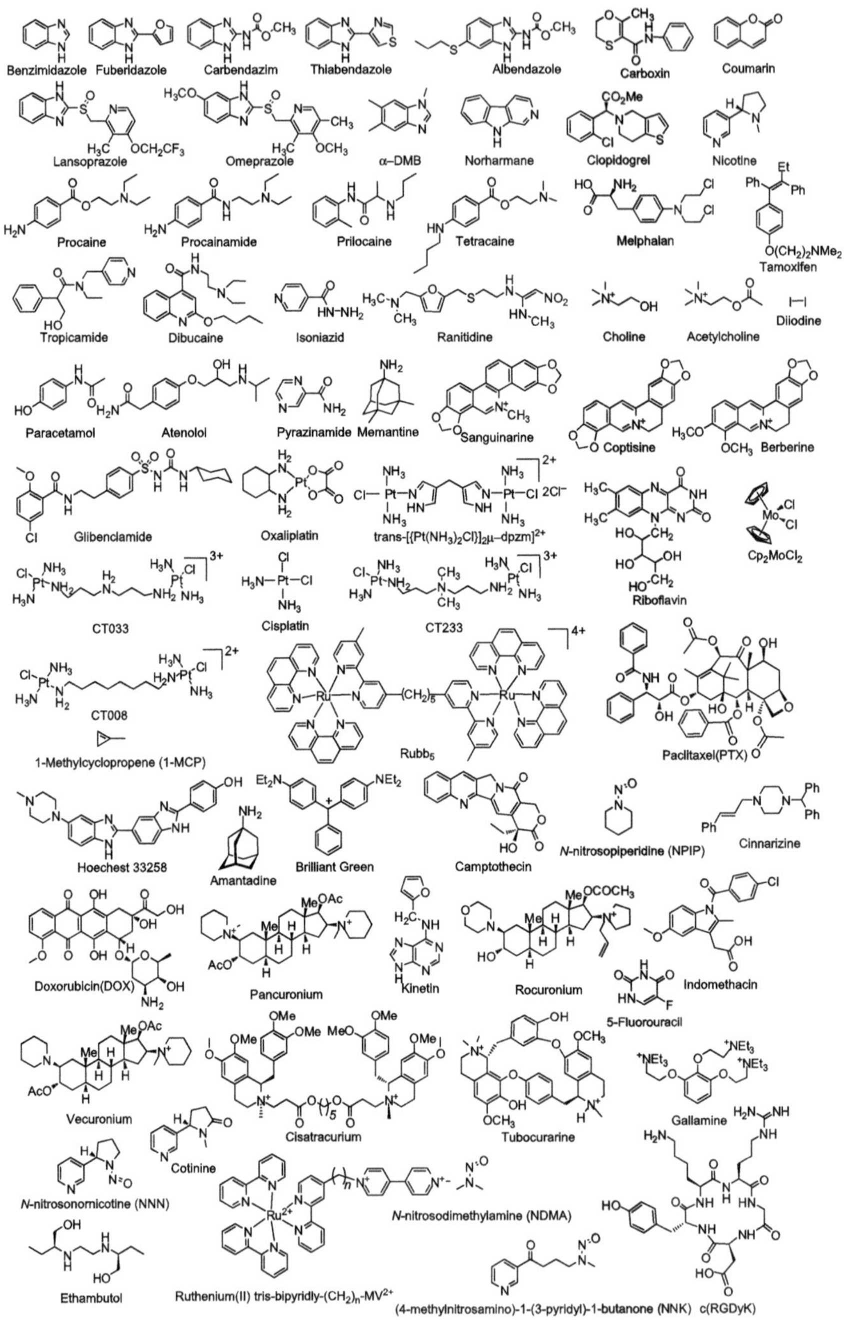

Benzimidazole   Fuberidazole   Carbendazim   Thiabendazole   Albendazole   Carboxin   Coumarin

Lansoprazole   Omeprazole   α–DMB   Norharmane   Clopidogrel   Nicotine

Procaine   Procainamide   Prilocaine   Tetracaine   Melphalan   Tamoxifen

Tropicamide   Dibucaine   Isoniazid   Ranitidine   Choline   Acetylcholine   Diiodine

Paracetamol   Atenolol   Pyrazinamide   Memantine   Sanguinarine   Coptisine   Berberine

Glibenclamide   Oxaliplatin   trans-[{Pt(NH$_3$)$_2$Cl}]$_2$μ–dpzm]$^{2+}$ Riboflavin   Cp$_2$MoCl$_2$

CT033   Cisplatin   CT233

CT008   Rubb$_5$ Paclitaxel(PTX)

1-Methylcyclopropene (1-MCP)

Hoechest 33258   Amantadine   Brilliant Green   Camptothecin   N-nitrosopiperidine (NPIP)   Cinnarizine

Doxorubicin(DOX)   Pancuronium   Kinetin   Rocuronium   Indomethacin   5-Fluorouracil

Vecuronium   Cisatracurium   Tubocurarine   Gallamine

N-nitrosonornicotine (NNN)   Cotinine   N-nitrosodimethylamine (NDMA)

Ethambutol   Ruthenium(II) tris-bipyridly-(CH$_2$)$_n$-MV^{2+} (4-methylnitrosamino)-1-(3-pyridyl)-1-butanone (NNK)   c(RGDyK)

Scheme 7.3   Chemical structures of the drug molecules and biologically active molecules discussed in this Section 7.4.

(factor of *ca.* 6) is due the encapsulation of Hoechst 34580 inside the cavity of CB7, which ensures, among other things, a protection of the excited state from water and oxygen.[88]

Stabilization of the active form of sanguinarine, a natural benzo[c]phenanthridine alkaloid, which exhibits anticancer, antimicrobial, and antifungal properties, was demonstrated upon binding with CB7.[56,89,90] CB7 preferentially binds the iminium form of the drug over its alkanolamine form and shifts the effective pK_a value of the iminium/alkanolamine equilibrium by *ca.* 3.6 units, from 7.2 to 10.8. It consequently expands the pH range in which the active iminium form is stable and improves the drug-activity near neutral pH. The effect of CB7 on the photooxidation of sanguinarine has also been demonstrated: The complexed drug was stabilized towards photoirradiation (photooxidation rate was slower by a factor of *ca.* 3) relative to the free drug. In this particular case, CB7 protects the drug molecule in its excited state from the attack of molecular oxygen.

Improved antifungal activity of a carboxin fungicide towards *Rhizoctonia solani* was achieved in the presence of CB8.[91] Mixing carboxin with CB8 in a 2:1 ratio improved the rate of growth inhibition at least by a factor of 3, and the use of CB-complexed fungicides has been considered to be an environmentally friendly approach. Although the authors did not mention any precise mechanism of action of the carboxin•CB8 complex, one possibility is that the macrocyclic complex actually binds more tightly to the biological targets (particularly proteins).[4]

In a separate study, it has been shown that the binding affinities of a drug molecule with the target site can be enhanced (with a concomitant increase in the target specificity) in the presence of a CB macrocycle.[92] In detail, the binding of the antimicrobial drug and fluorescent dye Brilliant Green (BG) with bovine serum albumin (BSA) as well as lysozyme was enhanced in the presence of CB7. Note that these studies provide a proof-of-principle for the formation of suprabiomolecular ternary complexes that might potentially be useful in the context of targeted drug delivery and drug specificity (see Section 7.6 for more examples).[92]

Interactions of additional drug molecules of this family with CBs have also been reported in the literature. For example, the fluorescence of berberine, an antimicrobial agent, was enhanced by a factor of 500 upon complexation with CB7, which encapsulates the dimethoxyisoquinoline portion of the guest. The reported binding constant of berberine with CB7 is 5-fold and 10 000-fold higher than those with β-cyclodextrin and *p*-sulfonatocalix[8]arene, respectively.[93,94] The antimicrobial alkaloid coptisine forms complexes with CBs as well, although no solution-NMR experiments have been reported to determine the mode of binding.[95] The tobacco alkaloid nicotine (used as insecticide) forms a weak inclusion complex with CB7, as demonstrated from a competitive fluorescence binding experiment using methylene blue as an indicator dye.[96] Collins, Day and co-workers investigated the interactions of the antibiotic drugs *trans*-[{PtCl(NH$_3$)$_2$}$_2$(μ-NH$_2$(CH$_2$)$_8$NH$_2$)]$^{2+}$ (CT008) and [{Ru(phen)$_2$}$_2$(μ-bb$_5$)]$^{4+}$ {phen = 1,10-phenanthroline; bb$_5$ = 1,5-bis[4(4'-methyl-2,2'-bipyridyl)]-pentane} (Rubb5) with CB10.[97]

7.4.2   Anti-neoplastic Agents

As is the case for the anti-pathogenic drugs, the solubility of the poorly soluble cytotoxic drug camptothecin (CPT)[38,98] was enhanced up to 0.4 and 47 µM in the presence of CB7 and CB8, respectively. Similarly, by using the acyclic sulfonato CBs (CB-3 and CB-4 in Scheme 7.2), large solubility enhancements have been reported for the antitumour drugs melphalan (with CB-3, factor 655), tamoxifen (with CB-3, factor 23), Paclitaxel (PTX) (with CB-3, factor of 2750), and CPT (with CB-4, factor 580).[38]

Examples of CB-assisted enhancements of the chemical stability of drugs through bioprotection have been reported in detail for platinum-based anti-cancer drugs.[39,40,99–105] Platinum-based drugs contain carrier am(m)ine ligands and labile ligands such as chlorides or carboxylate that are highly reactive towards thiol-containing plasma proteins and peptides (*e.g.*, glutathione), as well as the amino acids cysteine and methionine.[29] The thiol-containing groups of these proteins act as nucleophiles and replace the am(m)ine and labile ligands to convert the drugs into inactive metabolites. Full or partial encapsulation of the platinum drugs within the cavities of CBs can provide a steric and possibly electrostatic protection from such nucleophilic attacks. CBs bind the platinum drugs predominantly by encapsulating the hydrophobic components of their amine carrier ligands. This binding mode places the platinum metal centre near the junction of the macrocyclic cavity and the portals. Protection can thus be achieved even though the platinum metal centres are not completely immersed in the cavity of CBs. Accordingly, a significant decrease in degradation of CT008 and CT233 was observed in the presence of CB8, in which the linking ligand was found to fold closer to the metal.[105]

The rate of the reaction of a dinuclear platinum drug,[103] namely *trans*-((Pt(NH$_3$)$_2$Cl)$_2$µ-dpzm)$^{2+}$ (di-Pt), with guanosine was found to be reduced upon encapsulation inside the cavity of CB7. Furthermore, binding to CB7 and CB8 has been shown to slow down the reaction of platinum drugs derived from alkyl polyamines (CT008, CT033, and CT233)[105] with guanosine and cysteine by as much as 9-fold, without significantly affecting the drugs' cytotoxicity.[104]

Encapsulation of oxaliplatin anticancer drugs inside the cavity of CB7 (see Chapter 9 by Aldrich-Wright and colleagues),[101] through its cyclohexane component, has been shown to reduce the cytotoxic activity of the drug by 5–10-folds, depending on which of the five investigated cancer cell lines was used as target (A 549 human lung, SKOV-3 human ovarian, SKMEL-2 human melanoma, XF-498 human CNS, and HCT-15 human colon). The exact cause for this potentially undesirable efficacy modulation has not been clarified in detail; likely reasons are an effective decrease in the concentration of the active uncomplexed drug, or a different kinetics or mechanism of cellular uptake. The latter also depend upon whether the drug is more readily entering the cell in its complexed or uncomplexed form. In additional studies,[40] the encapsulation of a series of platinum phenanthroline-based DNA intercalator drugs with ancillary ligands (ethylene diamine or diaminocyclohexane) by CB6 to CB8 has led to mixed results, resulting in both positive and negative bioactivity.

The cytotoxicity of molybdocene dichloride (Cp_2MoCl_2)[106] in ovarian cancer (MCF-7 and 2008) cell lines increased upon encapsulation with CB7, with either increased solubility or improved membrane permeability being likely causes.

CBs not only protect platinum-based drugs from unwanted chemical reactions but can also stabilize them in the solid-state. For example, the solid oxaliplatin•CB7 complex experiences stabilization towards degradation. The colour of the free drug changes quickly, even in the dark, whereas the crystalline powder of the complex retained its original white color for more than one year.[101] Similar beneficial effects on the chemical stability are also observed in aqueous solution, where uncomplexed oxaliplatin was found to be only stable for 6 h at room temperature and for 24 h under refrigeration, while CB7-complexed oxaliplatin is stable even at high temperature and pressure.[101]

In an additional study, host–guest interaction-driven self-assembled nanostructures composed of diaminohexane-terminated gold nanoparticles $(AuNP-NH_2)$[107] capped by CB7 were shown to readily enter into MCF-7 cells as non-toxic endosomes. The *in vitro* antitumour activity of a CB8•fullerene[60] complex has also been tested and the result showed that the complex exhibited potent antitumour activity against HeLa cells.[108] The supramolecular CB8•fullerene[60] complex was simply prepared by grinding solid mixtures of macrocycle and guest in different mole ratios in a mixer mill.

7.4.3 Antagonist Agents

The effect of macrocyclic complexation on hydrolytic stability has been investigated for the histamine H_2-receptor antagonist ranitidine hydrochloride,[109] which is prone to decompose under humid conditions and at low pH. The stability of the drug in the presence of a slight excess of CB7 at pH 1.5 and 50 $^{\circ}$C was found to be more than two weeks with no decomposition compared to fast degradation within four days without CB7. The stability of the drug is (most) likely due to the prevention of nucleophilic attack by solvent molecules and steric hindrance towards the chemical reaction of an inactive and undesirable cyclic intermediate within the complex. A solubility enhancement of cinnarizine, a histamine antagonist, was observed upon binding with an acyclic analogue (CB-3 in Scheme 7.2) of CBs with a maximum enhancement factor of 354.[38]

The formation of a solid association complex of memantine, a NMDA glutamate antagonist,[110] with CB7 imparted an improved thermal stability accompanied by a large increase in the melting point of the drug•CB7 host–guest complex when compared to the free drug. The solid association complex of the selective β_1 receptor antagonist atenolol[110] with CB7 also showed increased thermal stability.

The use of CBs has also been suggested in the context of quality control of agricultural product harvesting induced by gaseous substances. Several reports on the encapsulation of gaseous substances in the cavities of CBs have emerged in the literature.[111–113] For example, it is well known that ethylene can profoundly affect the quality of harvested products, which may be beneficial or deleterious depending on the product, ripening stage, and desired use. Therefore, ethylene antagonists are of tremendous commercial value in the horticultural industry. 1-Methylcyclopropene (1-MCP),[114] which is used to prevent or slow down the ripening of fruits, needs to be stored in absorbents due to its gaseous and unstable characteristics, and CB6 was found to be a potent adsorbent for 1-MCP with about 4.5% uptake by weight. Furthermore, the release of 1-MCP from the complex has also been demonstrated in different solutions such as sodium bicarbonate, benzoic acid, and distilled water.

7.4.4 Enzyme Inhibitors

Cucurbit[7]uril (CB7) has been shown to mimic the activity of the enzyme monoamine oxidase when it encapsulates norharmane (NHM) (β-carboline) as a model substrate.[115] The encapsulation of NHM into CB7 moderately shifts the prototropic equilibrium of the protonated form (NHMH$^+$) to the base form (NHM), from pK_a 7.2 to 7.9. CB7 provides a binding pocket for the hydrophobic molecule, and the polar carbonyl portals offer an anchoring site to the positive charge of the cationic species (*i.e.*, NHMH$^+$), thereby shifting the pK_a value.

Saleh *et al.*[45] reported that CBs play an enzyme-mimetic role in bioactivation of prodrugs through complexation-induced pK_a shifts. Note that bioactivation is a process in which a prodrug, which is a pharmacological substance administered in an inactive (or significantly less active) form, is metabolized *in vivo* into its active metabolite. The effects of CB7 on the proton pump inhibitor drugs of the benzimidazole family were investigated, namely, lansoprazole and omeprazole, which are converted into their active forms by acid-promoted reactions. The activation takes place through an interesting reaction mechanism as depicted in Scheme 7.4. As can be seen, the pyridine ring of the prodrug (species I) is predominantly protonated (>99.9%) because it has a higher pK_a value than the benzimidazole residue (4.0 *vs.* 0.6). Nevertheless, for the free drug, the active form (*i.e.*, the cyclic sulfenamide) is formed through traces (<0.1%) of the structures with protonated benzimidazole residue and unprotonated pyridine residue (species II), since only this structure is capable of an intramolecular nucleophilic attack and subsequent dehydration. The active form of the drug molecule reacts with the cysteine residues of gastric (H$^+$–K$^+$)-ATPase, thereby inhibiting the enzyme, and reducing the production of gastric acid.

The challenge for the medicinal chemist, in order to accelerate activation, is to increase the basicity of the benzimidazole group while maintaining the basicity and nucleophilicity of the pyridine ring. Employment of CB7-induced pK_a shifts selectively increases the basicity of the benzimidazole moiety and consequently increases the population of species II and thereby accelerates the activation of the drug. The prodrug molecule, in its free state, is activated with an estimated half-life of about five minutes, whereas in the presence of CB7 it is

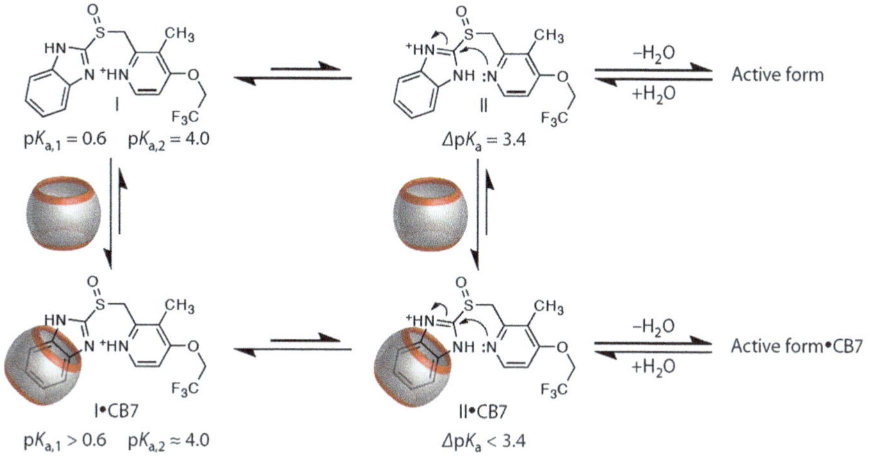

Scheme 7.4 Cucurbit[7]uril-assisted activation of lansoprazole, a prodrug. (Adapted with permission from Saleh *et al.*[45] Copyright 2008 Wiley-VCH Verlag GmbH & Co.)

activated within 20 seconds, which corresponds to a 15-fold faster activation. In addition to this effect of rapid activation, CB7 also prevents the active form of the drug from dimerization and degradation. In the absence of CB7, the cyclic sulfenamide derived from lansoprazole degrades rapidly at pH 2.9, with a half-life of about one hour. In the presence of 5 mM CB7, the half-life was more than three weeks, which corresponds to a stabilization factor of 500 or more.

Despite the above-mentioned enzyme-mimetic activities, CBs can also inhibit enzymatic reactions by complexing positively charged moieties (*e.g.*, guanidinium) in peptide substrates of proteases, which could interfere in biological processes.[116] Binding of CBs with amino acids, peptides, and proteins, and related applications are discussed in Section 7.9. Enzyme activities of trypsin and leucine aminopeptidase (LAP) towards certain peptide substrates were impeded in the presence of CB7.[116] The enzymatic activity remained unaltered in the presence of other substrates, and thus the inhibition was mainly due to the binding of CB7 to the substrate and not to the enzyme itself.

It is worth discussing here that CBs can also interfere or modulate enzymatic reactions by complexing inhibitors of enzymes.[117] Compounds consisting of an enzyme-binding group (benzenesulfonamide for bovine carbonic anhydrase or tacrine for acetylcholinesterase) and a CB7-binding group (1-adamantylamine, trimethylsilylmethyl amine, or hexylamine) were prepared, and the enzyme activity was measured in the presence of these compounds in their uncomplexed and complexed forms. In the case of carbonic anhydrase, the arylsulfonamide-containing compounds inhibited the enzyme activity, and the addition of CB7 restored the activity by sequestering the inhibitor from the active site. The inhibitor could be regenerated (released) when a strong competitor for CB7 was added to the solution.

7.4.5 Ocular Drugs

The bioavailability of ocular drugs is often limited by their permeability across the cornea as biological barrier. A case-study with tropicamide showed that CB7 reversibly encapsulates the drug molecule and at the same time increases the percentage of its cationic form at physiological pH due to host-assisted guest protonation; this remedies the otherwise required use of physiologically uncomfortable acidic conditions for dissolution of the drug.[51] Note that tropicamide is also known as an anticholinergic drug and is generally used as eye drops to achieve a mydriatic response (pupil dilation) in preparation for ophthalmological examination and surgery. Tropicamide is often formulated in aqueous solutions at pH between 4 and 5, below or very close to the pK_a value of the free drug (5.2–5.4), to enhance solubility and promote permeation across the negatively charged corneal epithelium. Topical application of such acidic solutions of the drug molecule results in significant, even if temporary, discomfort for patients. Both CB7 and CB8 encapsulate tropicamide and shift the pK_a by up to 2 units and thereby increase the degree of ionization by a factor of 30 at pH 7.

7.4.6 Vitamins and Hormones

The interaction of CBs with vitamins has been exemplified in a case study with the Co(III)-based coenzyme B12 (AdoCbl; Ado = 5′-deoxyadenosyl).[53] The pK_a of the α-axial ligand, 5,6-dimethylbenzimidazole nucleotide (α-DMB), is functionally important and, depending on the ligation state of the group through a shift in the protonation/ligation equilibrium, one differentiates "base-on" and "base-off" forms of the metal ion-centred coenzyme. In AdoCbl, a CB7-induced pK_a shift stabilizes the base-off form of the coenzyme even at higher pH and thereby mimics the biological function of AdoCbl-dependent enzymes (*e.g.*, mutases). Notably, the cavity of CB7 also provides a hydrophobic pocket for α-DMB, which is reminiscent of its biological recognition by methylmalonyl-CoA mutase.

Binding interactions of two other drugs of this family have also been studied with CBs. The formation of inclusion complexes of CB7 and methylated CB6 derivatives with kinetin[118] was confirmed in the solid-state and the host–guest complexation improved the solubility of kinetin in aqueous solution. The formation of a 1 : 1 riboflavin•CB7 complex in aqueous solution has also been reported.[119]

7.4.7 Anti-tuberculosis Agents

The protective effects of CBs towards unwanted chemical reactions of drug molecules are not limited to platinum complexes (Section 7.4.2). CB6 and CB7 prevent the acylation of the anti-tuberculosis drug isoniazid,[120] a reaction that is readily catalysed by acetyl coenzyme A (acetyl-CoA) in the human body, by lowering its nucleophilicity sufficiently, presumably due to steric hindrance towards reaction in the complex. It is interesting that for this drug molecule the resistance towards undergoing acylation was found to be independent of the nature of the acylating agent. The drug•CB7 complex was also found to be thermally more stable than the free drug, as shown by differential scanning calorimetry (DSC).[121]

The solid association complex of pyrazinamide and CB7 showed significantly improved thermal stability as well, again shown by DSC. The melting point of the host–guest complex is significantly higher than that of the free drug. Note that for most drug molecules, in their solid-state, heating the host–guest complexes to 100–200 °C—that is, beyond the melting point of the free drug— shows no melting of the samples until the cucurbituril itself begins to decompose at temperatures higher than 370 °C.[29] Elevated melting points of drug formulations are beneficial because a broader temperature range of the solid always reduces concerns related to processing and storage conditions.

7.4.8 Neurotransmitters and Neuromuscular Blockers

Interactions of CB7 with cationic and dicationic guest molecules bearing quaternary ammonium (and other "*onium*" groups) have been investigated frequently.[48,52,122] Choline and its derivatives,[123] along with phosphonium analogues, which possess trialkylonium head groups, form complexes with similar strengths as those of the corresponding tetraalkylonium cations.[48] The interaction of CB7 with succinylcholine, a potent nicotinic acetylcholine receptor blocker which is also used as a depolarizing neuromuscular relaxant, has been investigated.[52] Kim and co-workers reported that hexa(cyclohexyl)-CB6 binds to acetylcholine but not to choline.[122]

In a very promising study, Isaacs *et al.* investigated the recognition properties of acyclic sulfonato CBs (CB-3 and CB-4) with several neuromuscular blocking agents (NMBAs), namely rocuronium, vecuronium, pancuronium, gallamine, cisatracurium, and tubocurarine.[84] All the NMBAs formed inclusion type complexes with CB-3 with high binding constants (10^5–10^6 M^{-1}). Interestingly, the binding constants of the NMBAs (except tubocurarine) with the acyclic sulfonato derivative CB-4 were significantly higher (10^6–10^9 M^{-1}) than those for CB-3. The complexation/decomplexation *kinetics* of CB-4 and CB-3 showed also interesting differences. In detail, the studied guest molecules showed fast complexation/decomplexation kinetics on the NMR timescale with CB-3, while rocuronium, vecuronium, and cisatracurium showed slow complexation/decomplexation kinetics with CB-4. The other guest molecules showed fast or moderately fast complexation/decomplexation kinetics with CB-4. In the same study, the authors proposed that the acyclic sulfonato CBs have immense potential as reagents for the *reversal* of neuromuscular block after anesthesia. *In vivo* studies in rats showed that CB-4 has the ability to reverse the neuromuscular block induced by rocuronium. Notably, the binding affinity of CB-4 with rocuronium was found to be higher by a factor of 19 000 than that of Sugammadex, a γ-cyclodextrin derivative presently in clinical use as a reversal agent.

7.4.9   Local Anaesthetics

Cucurbit[7]uril (CB7) forms stable host–guest complexes with the local anaesthetics procaine, tetracaine, prilocaine, dibucaine, and procainamide in aqueous solution[54,124] as a result of ion–dipole interactions between the protonated ammonium groups of the guests and the carbonyl portals of CBs. The portion(s) of the guest molecules included in the host cavity depend on the pH of the solution and the relative sites of protonation. Small increases in the pK_a values of these drug molecules were observed in the presence of CB7.

7.4.10   Other Investigated Drug Molecules

Cucurbit[7]uril (CB7) binds to palmatine (an anti-inflammatory drug) and dehydrocorydaline.[123] The solubility of tolfenamic acid (another anti-inflammatory) was increased almost 37-fold through the formation of host–guest complexes with the acyclic sulfonato CB (CB-4 in Scheme 7.2). *o*-Carborane (a boron neutron capture therapeutic agent), which is water-insoluble, becomes soluble upon complexation with CB7.[125] Coumarin (an anticoagulant) was found to form stable inclusion complexes with CB7 and CB8 in aqueous solution.[126] Solid association complexes of paracetamol (an analgesic) and glibenclamide (an anti-diabetic drug) with CB7 afforded materials with beneficial thermal stabilities.[110] The solubilities of indomethacin (an anti-inflammatory) and clopidogrel (an anti-clotting drug) increased through the formation of host–guest complexes with acyclic sulfonato CBs (CB-4 and CB-3 in Scheme 7.2 for indomethacin and clopidogrel, respectively) by 56-fold and 1220-fold, respectively.[38] Very recently, small CBs have been shown to form inclusion complexes with molecular di-iodine (I_2), a well known wound antiseptic since more than 150 years. The binding constant with CB6 is sufficiently large (*ca.* 10^6 M^{-1})[127] to complex iodine efficiently in aqueous environments; the associated slow release of iodine in low concentrations could be highly desirable for antiseptic wound dressings.

7.5   Release of Drug Molecules from the Cavity of Cucurbiturils and Related Microheterogeneous Systems

7.5.1   Drug Release from Cucurbiturils

With the exception of a few systems that show very slow exchange kinetics (weeks to months),[65,97,128] host–guest complexes involving macrocycles display association and dissociation rate constants on the order of seconds or faster, which ensures a rapid dynamic equilibrium on the time scale typical for the action of common drugs. Nevertheless, it may be desirable to trigger the release of drugs by shifting the equilibrium towards the uncomplexed drug, depending on a specific stimulus, in a spatially and/or temporally controlled manner. The possibilities to achieve such a targeted drug delivery for CBs as macrocycles are discussed below.

7.5.1.1 Kinetics of Direct Drug Release by Dissolution

Guest molecules bind with CBs mainly by non-covalent, reversible interactions. Therefore, the free drug molecule will be available after administration, whichever route is taken (*e.g.*, intravenously or orally), through a dynamic chemical equilibrium, and the release of the administered encapsulated drug will thus occur "by itself", simply driven by dissolution of the drug complex and the associated dilution effect. Note that the degree of complexation always decreases with dilution. Indeed, upon dissolution, the drug albendazole was found to be released in seconds from the host cavity (in this case, CB7),[37] and even the release of dinuclear ruthenium complexes from the cavity of CB10 occurred within several hours.[97] Recently, it has been shown that the decomplexation process of the host–guest complex is faster for acyclic CB derivatives than for their cyclic analogues.[38] While a slow complexation process may not be ideal in the manufacturing processes, it may well be desirable to achieve sustained plasma concentrations of therapeutic agents over time (see Section 7.7 for the sustained release of cisplatin from its CB7 complex). It is also worth mentioning here that—for the kinetics of drug release—the absolute binding affinities (a thermodynamic measurement) of the guest molecules with CBs are not useful, as demonstrated by a saturation transfer experiment between CB8 and CB7 for the drug albendazole.[37] This is because of the tight carbonyl portal of CBs, which impose a constrictive binding for certain guests,[97] while other host–guest structures can be viewed as rotaxanes which display a mechanical barrier towards dethreading, for example, with the dinuclear ruthenium complex as guest (Rubb$_5$ in Scheme 7.3).[97]

7.5.1.2 Effect of Media and Additives on Drug Release

As mentioned before (see Section 7.2), inorganic cations bind to the portals of CBs and can therefore displace encapsulated guests, particularly when the guests are positively charged.[67] The effective binding constants of guest molecules are almost always smaller in the presence of salts.[65,68,73] The presence of inorganic cations in biological media therefore accelerates the release of drug molecules from the cavity of CBs and shifts the equilibrium towards decomplexation. Mohanty and co-workers demonstrated a proof-of-principle that a guest molecule, namely neutral red, can be effectively relocated from the cavity of CB7 into the hydrophobic cavity of bovine serum albumin (BSA) by addition of salts.[67] Thioflavin T (ThT), a dye molecule that is commonly used to diagnose amyloid fibrils,[129] forms 1 : 1 as well as 1 : 2 guest–host complexes with CB7. For the 1 : 1 complex, addition of the salt releases the guest molecule by decreasing the host–guest binding constant, whereas for the 1 : 2 complex, metal ions take part in forming a capsule-type ternary complex, a process which in this case could be directly followed through a characteristic fluorescence enhancement.[66,73,130] In a control experiment, when the competitor 1-adamantylamine (1-amantadine, which is actually also employed as an antiviral and antiparkinsonian drug) was added to the solution, the capsular complex was ruptured.[72] With CB8, ThT forms $1:2$ and $2:2$ host–guest complexes. The strong ion–dipole interactions provided by the carbonyl portals of CB8 support the stabilization of two π-stacked ThTs in both complexes. This study might have biological importance because binding of ThT with protein fibrils is mainly investigated by monitoring the changes in the characteristic fluorescence properties of ThT, which presumably can be modulated by CB7 and CB8 as co-additives.

The release of guest molecules can also be realized by targeted addition of a specific competitor. For example, the self-assembly resulting from host–guest interactions of diaminohexane-terminated gold nanoparticles (AuNP-NH$_2$) with CB7[107] was ruptured by using 1-adamantylamine as a strong competitor, which replaces AuNP-NH$_2$ from the cavity of CB7, as demonstrated by its increased cytotoxicity against MCF-7 cells.

An alternative release mechanism exploits the CB-induced pK_a shifts of the guest molecules. As the neutral forms of the drugs bind much more weakly with CBs than their corresponding protonated forms, a change in the pH of the medium from below pK_a' (the pK_a value of the complex) to above pK_a' effectively decreases the binding constants of the drugs, and the subsequent rapid release of the encapsulated drugs shifts the chemical equilibrium towards the uncomplexed guest (and host). Note that this concept can be advantageously employed at different stages of drug delivery. For example, drug•CB complexes with pK_a' values between 5 and 7 could be manufactured below pH 4 (because of the higher binding constants of the protonated forms and related higher degree of complexation), while the physiological pH itself would subsequently favour the efficient release of the drug molecules.

7.5.1.3 Drug Release by Photoinduced pH Jump

In addition to the influence of the pH of the biological medium, drug molecules can also be released by tuning the pH of the medium using photons as trigger and a suitable photobase as an auxiliary compound, as exemplified by Carvalho et al.[131] In the specific model system, the authors used Hoechst 33258 (a fluorescent DNA stain and antihelminthic drug) as guest, CB7 as host, and malachite green leucohydroxide (MGOH) as the photobase. The pK_a of Hoechst 33258 is 5.5 and the binding constant of the neutral form with CB7 is two orders of magnitude lower than that of the corresponding protonated form. Upon irradiation at 300 nm, MGOH generates OH$^-$ and increases the pH of the solution from 7 to 9, which effects the release of Hoechst 33258 from the host cavity (see Figure 7.1). When the experiment was performed in buffered solution, as a negative control, no drug release was observed.

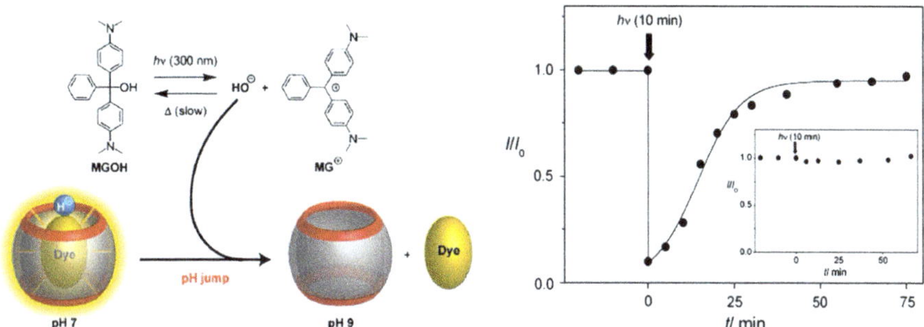

Figure 7.1	Release of a guest molecule from the cavity of CB7 upon photoinduced pH jump. The left part shows the schematic mechanism, while the graph on the right shows the decrease in fluorescence of the complexed guest, Hoechst 33258, upon irradiation of the photobase, the subsequent pH recovery, and—in the inset—the absence of fluorescence fluctuations as well as the corresponding decomplexation reaction in buffered solution. (Adapted with permission from Carvalho *et al.*[131] Copyright 2011 The Royal Society of Chemistry.)

7.5.2   Drug Release from Cucurbituril-Based Microheterogeneous Systems

In this section, we will discuss the use of CBs for the controlled release of drug molecules in different microheterogeneous systems, for example, hydrogels, surface-immobilized rotaxanes,[132] micellar copolymers, and nanocapsule polymers or vesicles.[133] Such microheterogeneous media have been tailored to release the drug molecules using different triggers, for example, pH, light, local heating, reductive cellular reactions, and competitors.

7.5.2.1   *Hydrogels*

The controlled release of the anticancer drug 5-fluorouracil was observed from CB6-containing alginate hydrogel beads.[134] Upon loading the drugs into the network structure of the hydrogel, the size of the beads increased from 2.5 mm to 3–4 mm with a loading capacity of 3.87–6.13 wt%. The slowest release with optimal loading (5.94 wt%) had a half-life ($t_{1/2}$) of 2.7 h.

7.5.2.2   *Micellar Copolymers*

Recently, Scherman and co-workers reported on the controlled release of guest molecules from a CB8-mediated micellar system. Double hydrophilic block copolymers are held together by CB8 (Scheme 7.5) to form a ternary complex, which readily self-assembles into micelle-like structures in aqueous media.[135] The supramolecular block copolymer assembly consists of poly(*N*-isopropylacrylamide) as a thermoresponsive block and poly(dimethylaminoethylmethacrylate) as the pH-responsive block.

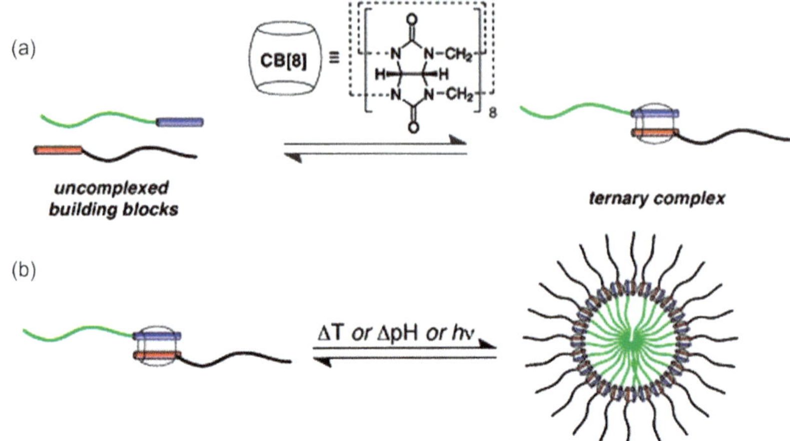

Scheme 7.5 (a) Cucurbituril CB8-assisted formation and (b) stimuli-induced disassembly of double hydrophilic block copolymers.
(Adapted with permission from Loh *et al.*[135] Copyright 2012 American Chemical Society.)

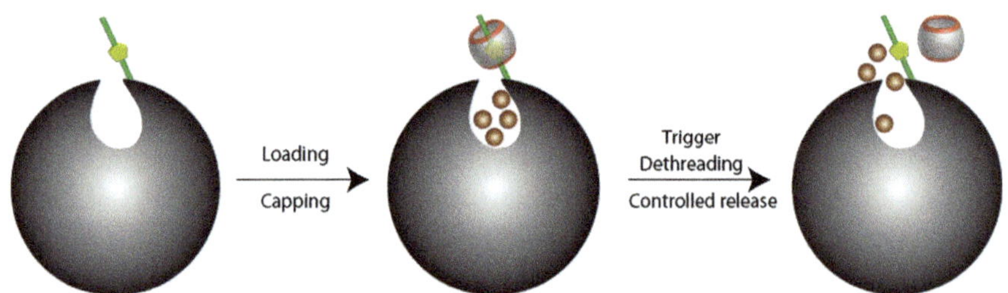

Scheme 7.6 Schematic illustration of the operational principle of cucurbituril-based supramolecular nanovalves for the controlled release of guest (drug) molecules.

The intracellular delivery of the antitumour drug doxorubicin, which was encapsulated inside the micellar structure, was demonstrated by the use of three stimuli/triggers: pH, temperature, and competitor. Disassembly of the micellar structure occurred upon changing the pH from 7 to 4, upon lowering the temperature from 37 °C to 15 °C, and upon addition of 1-adamantaneamine as competitive guest for CB8. It is worth mentioning that the release of doxorubicin from the micellar core to the cell nuclei of HeLa cells was also observed within a desirable time frame.

7.5.2.3 Surface-Immobilized Pseudorotaxanes

Zink and Stoddart explored supramolecular nanovalves (see Scheme 7.6) of the pseudorotaxane type, consisting of bisammonium stalks of 1,2,3-triazole derivatives and CB6, on the surface of mesoporous silica nanoparticles for the release of guest molecules in a controlled way in aqueous solution. The pores of the silica nanoparticles can be loaded with guest molecules (or with drug molecules in the case of drug delivery), and—depending on the pH of the solution—the pores can be unblocked or blocked by opening and closing the nanovalves to release or retain the preloaded guest molecules.[132,136] The pH-dependent binding of CB6 with the bisammonium stalks presents the operational principle of these nanovalves. At neutral and acidic pH, CB6 binds the protonated bisammonium stalks tightly due to ion–dipole interactions and, thereby, blocks the pores efficiently. At basic pH, deprotonation of the bisammonium stalks decreases the binding constants of the stacks with CB6 and unblocks the pores as a consequence of the dethreading of the CB6 molecules from the stacks. The system was investigated with the dye rhodamine B as a model guest, and the controlled release at basic pH was demonstrated in solution.

In a complementary study, Liu and Du[137] reported on the release of a calcein dye from a similar system consisting of 1,4-butanediammonium stacks. For this system, when the pH was adjusted to 10.5, deprotonation of the 1,4-butane-diamine stalks led to decomplexation of the rotaxane architecture to efficiently release the guest molecule from the pores. As an important extension of the versatility of the nanovalve design, the release of the guest molecules was alternatively achieved at neutral pH by employing competitors as additives. For example, when cetyltrimethylammonium bromide (CTAB) and 1,6-hexanediamine were added to the solution, the release of the guest molecules was also observed.

In another important study, the guest molecule calcein was also released by using an enzymatic reaction;[138] inspired by the supramolecular tandem assay principle (see Section 7.9.2),[139] lysine decarboxylase was used to produce cadaverine as product, which subsequently functioned as a competitor to open the nanovalves.

The concept of nanovalve-regulated drug release has also been investigated with other systems. For example, when the guest molecule was preloaded into the pores of a mesoporous silica matrix containing embedded AuNPs, the release of the guest molecule could be triggered by laser irradiation.[140] Laser irradiation with low intensity at the wavelength corresponding to the plasmon resonance of the AuNPs causes a local internal heating through dissipation of the photonic energy, which raises the local temperature above 60 °C to significantly decrease the ring-stalk binding and, thus, release the guest molecules. At too high lasing power, degradation of the matrix was observed. In this study, the authors mentioned that this light-sensitive nanostructure can increase the local temperature without significantly changing the bulk temperature, which could potentially be used for (spatially) controlled dual therapy involving the delivery of drug molecules to cells and necrosis through hyperthermia.

In a conceptually related way, supramolecular nanovalves with CB6 and *N*-(6-*N*-aminohexyl)aminomethyltriethoxysilane (condensed on the particle surface) have also been constructed on zinc-doped iron oxide (Fe_3O_4)

nanocrystals incorporated in mesoporous silica frameworks.[141] In this case, the guest molecules can be released by applying an oscillating magnetic field. When the local temperature rose due to magnetic oscillations, the ring-stalk binding decreased to affect the release of the guest molecule. The efficacy of this system was also tested *in vitro* on the breast cancer cell line MDA-MB-231 with doxorubicin as the entrapped drug.

7.6   Cucurbituril-Based Assemblies as Targeted Drug Delivery Systems

In addition to the controlled release of drug molecules, researchers have recently focused their interest on the synthesis of new derivatives of CBs or glycoluril-rich receptors for the delivery of guest molecules directly at the targeted site. In particular, functionalized CB6 derivatives have been employed (Scheme 7.7). The obvious advantage of targeted drug delivery is the reduction of side effects. An interesting example is the synthesis of an amphiphilic CB6 derivative that assembles into vesicles, the surface of which can be decorated *via* non-covalent interactions of alkylammonium-tagged guests with the free CB6 cavities.[133] The amphiphilic CB6 derivative was synthesized by reacting (allyloxy)$_{12}$-CB6 with 2-[2-(2-methoxyethoxy)-ethoxy]-ethanethiol. As a potential application, when the surface of the vesicles was decorated with a thiourea-linked α-mannose–spermidine conjugate and mixed with a solution containing concanavalin A—a lectin which shows specificity towards α-mannose—immediate aggregation was observed. As a negative control, when the vesicles were decorated with a galactose–spermidine conjugate, no aggregation was observed. The free ligand (*i.e.*, thiourea-linked α-mannose–spermidine) also did not aggregate with concanavalin A.

   Kim and co-workers also demonstrated that another CB6 derivative, (3-(6-hydroxyhexanethio)-propane-1-oxy)n-CB6, which can be synthesized by photoreaction of (allyloxy)$_{12}$-CB6 with 6-mercaptohexanol in methanol, can form spherical nanoparticles (CB6-NP) in water.[142] The entrapment of guest molecules inside the nanoparticle and the decoration of the nanoparticle surface were tested using Nile red as a model hydrophobic guest and a fluorescein-isothiocyanate–spermidine conjugate, respectively. The potential application of these nanoparticles in targeted drug delivery was shown by carrying out an *in vitro* study which involved decorating the nanoparticles non-covalently with a folate–spermidine conjugate on human ovarian carcinoma HeLa cells (as target cells) that had overexpressed folate receptors on their surface. The mechanism of cellular intake was confirmed to be folate-receptor mediated endocytosis. After endocytosis, Nile red was released, as monitored by confocal laser scanning microscopy. In continuation, the delivery of the antitumour drug PTX to HeLa cells was also demonstrated. The delivery of free PTX, CB6-NP•PTX, and the CB6-NP•PTX•folate–spermidine conjugate was monitored through the cell growth inhibition caused by the cytotoxicity of the delivered PTX. The cytotoxicity of the non-tagged and folate-tagged nanoparticles, as

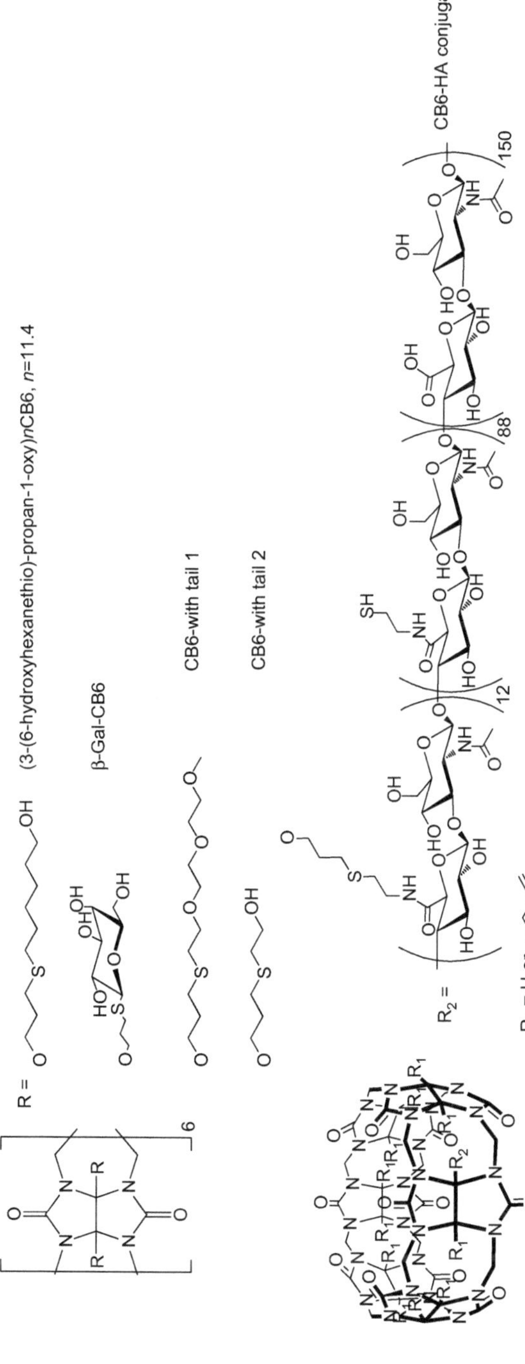

Scheme 7.7 Cucurbit[6]uril (CB6) derivatives that have been investigated for the purpose of targeted drug delivery.

manifested in cell growth inhibition, improved by a factor of 3.8 and 15.5 over that of free PTX, respectively. The increased cytotoxicity may be attributed to the facile internalization of the CB6-NP•PTX•folate–spermidine conjugate through receptor-mediated endocytosis.

A CB6-based galactose cluster could be delivered into HepG2 hepatocellular carcinoma cells *via* galactose-receptor mediated endocytosis.[143] Because the cells were known to overexpress galactose receptors, the galactose unit was covalently attached to the allyloxy group of (allyloxy)$_{12}$-CB6. In this study, the same fluorescein-isothiocyanate–spermine conjugate was used as a model drug and fluorescent probe. In a complementary study, the same CB6-galactose conjugate was utilized to non-covalently encapsulate dextran–spermine conjugates into hepatocytes containing asialoglycoprotein receptors. This model was used to demonstrate the possibility of having a receptor-mediated gene delivery system that is non-toxic and biocompatible.[144]

In another study, Kim *et al.*[145] described a polymeric nanocapsule that consists of disulfide-bridged CB6 and is sensitive to reduction. The CB6 derivative (allyloxy)$_{12}$-CB6 was employed in the synthetic route of these polymeric nanocapsules. When the nanocapsule was preloaded with carboxyfluorescein and treated with a reducing agent, namely dithiothreitol (DDT), the release of carboxyfluorescein was observed due to the rupture of the polymer nanocapsules. The potential application of this system in targeted drug delivery was illustrated by exploiting the CB6 cavity to embed a galactose–spermidine conjugate onto the surface and encapsulate carboxyfluorescein as targeting ligand and imaging probe inside the nanocapsule, respectively (Scheme 7.8). HepG2 hepatocellular carcinoma cells were used as the target. After incubation, a change in the fluorescence signal of carboxyfluorescein was observed inside the cells only for those polymeric nanocapsules that contained both the galactose–spermidine conjugate and carboxyfluorescein. No or very low fluorescence was observed for disulfide-bridged CB6 polymeric nanocapsules containing only carboxyfluorescein (this dye undergoes self-quenching when included in high concentration inside the nanocavities) or polymeric nanocapsules containing a dicarbon bridge instead of a disulfide bridge.

By using the same principle, controlled *in vitro* drug targeting of doxorubicin onto HeLa cells has also been reported.[142,146] An amphiphilic CB6 derivative with a disulfide bridge attached to its core was readily self-assembled into robust vesicles.[145,146] Receptor-mediated endocytosis was reported to be the mechanism of internalization of the vesicles into the targeted cells, and the release mechanism was assumed to be due to the reduction of disulfide bonds by cytoplasmic glutathione. The release of the entrapped doxorubicin was confirmed through its increased cytotoxicity.

In a recent study, a CB6-conjugated hyaluronate (CB6-HA) was synthesized and non-covalently decorated with a peptide–spermidine conjugate as a model for a drug that binds to and activates specifically the formyl peptide receptor (FPRL1), while fluorescein-isothiocyanate (FITC)–spermidine was used as the imaging probe (Figure 7.2).[147] Controlled drug targeting *in vitro* into B16F1 cells with HA receptors was demonstrated through the simultaneous

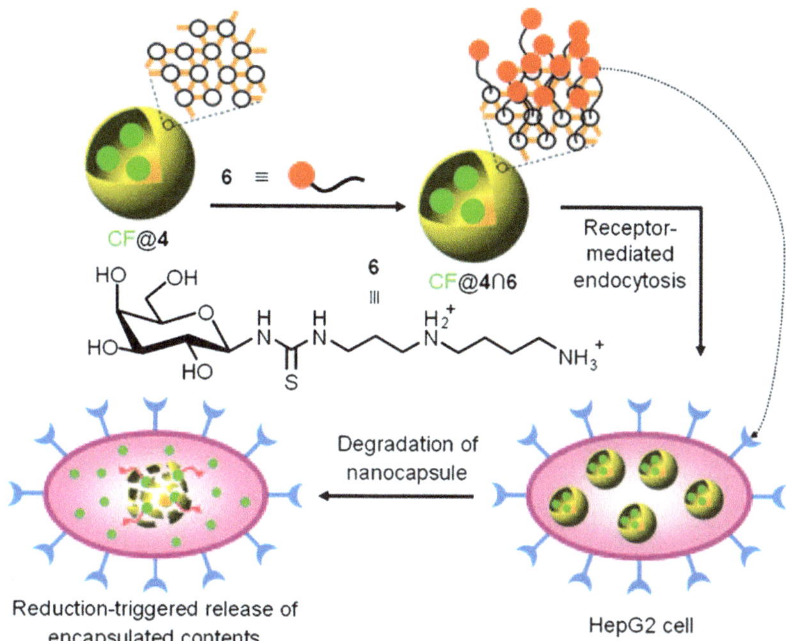

Scheme 7.8 Targeted uptake of decorated and loaded CB6-based nanoparticles into HePG2 cells and subsequent reduction-triggered release of guest molecules.
(Adapted with permission from Kim *et al.*[145] Copyright 2010 Wiley-VCH Verlag GmbH & Co.)

bioimaging of FITC–spermidine conjugated CB6-HA. The activation of the FPRL1 receptor results in increased intracellular Ca^{2+} levels, through which the delivery of the CB6-HA-peptide–spermidine conjugate could be demonstrated in FPRL1-expressing human breast adenocarcinoma (FPRL1/MCF-7) cells. The bright fluorescence signal of FLUO-3/AM served as indicator for enhanced Ca^{2+} concentrations. The stability of the system in biological media was also demonstrated *in vitro* as well as *in vivo*. In the animal experiment, male Sprague–Dawley rats were intravenously injected *via* their tail vein with a dose of 0.5 mg kg^{-1} body weight and the pharmacokinetics analysis of the collected blood samples was conducted by monitoring FITC fluorescence at 460 nm. A residence time of 3 days for the injected model drug was confirmed.

7.7 Studies of Cucurbiturils and their Drug Complexes *in Vitro* and *in Vivo*

Besides the large number of publications on the host–guest complexes of CBs with medicinally relevant compounds and their possible applications, the utilization of CBs as drug delivery vehicles requires detailed information on the *in vitro* and *in vivo* toxicological effects of free CBs and their drug complexes.

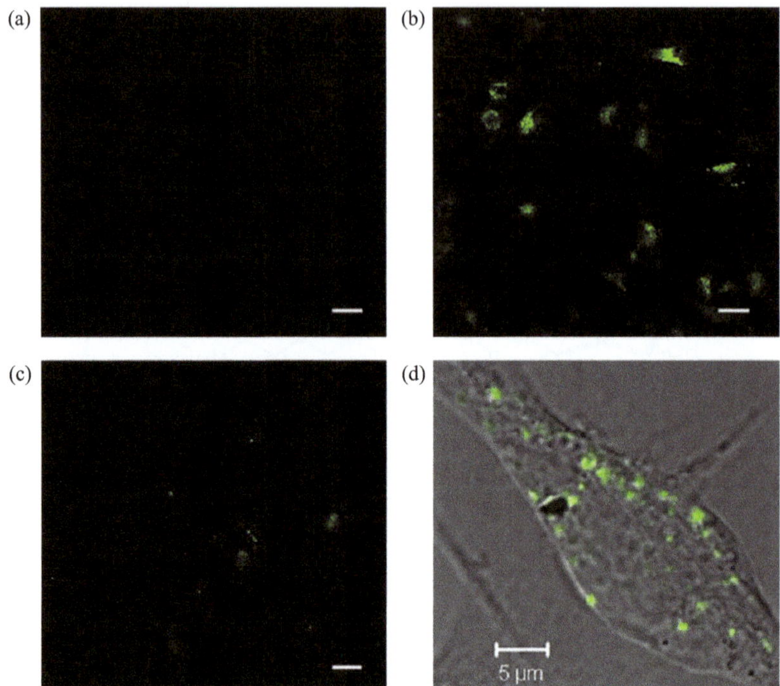

Figure 7.2 Confocal laser scanning microscopic images of B16F1 cells upon incubation with (a) fluorescein-isothiocyanate (FITC)–spermidine, (b) FITC–spermidine•CB6-HA, and (c) FITC–spermidine•CB6-HA in the presence of excess amount (1000-fold) of hyaluronic acid (HA). Part (d) shows the magnified image of B16F1 cells in the presence of FITC–spermidine•CB6-HA.
(Adapted with permission from Jung *et al.*[147] Copyright 2011 Elsevier Ltd.)

Even though the combined initial measurements demonstrate that CBs show little if not negligible cytotoxicity and toxicity, and that they can be formulated into dosage forms suitable for human drug administration, CBs are presently not in clinical use. In this section, we discuss the available *in vitro* and *in vivo* studies.

7.7.1 Penetration of Cucurbiturils and their Complexes into Cells

Cellular uptake of CB complexes has been demonstrated in several cell lines by using fluorescent tagging.[148,149] Visual confirmation of the uptake of CB complexes and their intracellular localization have accordingly been mainly conducted by fluorescence microscopy. Complexes of CB7 with FITC–spermine and Alexa Fluor 555-admantylpiperazine conjugates have been reported to be internalized by murine macrophage (RAW264.7) cancer cell lines and transported to lysosomes.[149] The complexes had intracellular stability for more than 2 hours. The uptake of CB7 complexes with acridine orange and pyronin Y was also demonstrated to occur in mouse muscle embryo cells (NIH/3T3).[148] In addition, the involvement of cellular receptors (see Section 7.6) in the cellular uptake of CB complexes has also been reported.[142–146] Although limited in number, these studies indicate that CB complexes (and intuitively CBs) are able to cross the cell membrane, which furbishes an important incentive to further investigate their biological stability and biocompatibility for the purpose of intracellular drug delivery.

7.7.2 Cytotoxicity and Toxicity of Cucurbiturils and their Functionalized Forms

The inherent cytotoxicity of a compound is the extent to which it kills cells or inhibits cell growth, while toxicity in general refers to the type and severity of a compound's side effects in living beings. Recently, several research groups including ours have evaluated the toxicity of CBs[149,150] and glycoluril oligomers[38] to demonstrate their biocompatibility. *In vitro* cell viability of CB7 using the 3-(4,5-dimethylthiazol-2-yl)-2,5-diphenyltetrazolium bromide (MTT) colorimetric assay in Chinese hamster ovary cells (CHO-K1) showed no significant cytotoxicity up to 1 mM concentration and 3 h incubation time; after 48 h incubation time an IC_{50} value of 0.53 mM was determined (Figure 7.3).[150] The low solubility (20 µM) of CB8 prevented accurate evaluation of its cytotoxicity in the biological media. In addition, no internal damage to the cells was observed in the cellular images upon administration of both CB7 and CB8 by using the MitoTracker Red CMXRos stain as probe. In the same study, CB7 and CB8 were also shown to be non-toxic *in vivo*.[150] The *in vivo* study was conducted through a maximal tolerated dose (MTD) experiment in female Balb/c mice. The CB7 macrocycle, intravenously administrated into the mice,

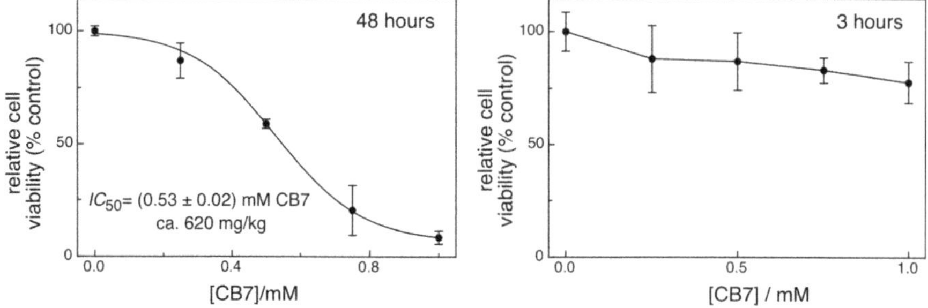

Figure 7.3 Relative cell viability of CHO-K1 cells plotted against the concentration of CB7 (*left*) and the incubation time determined by employing the MTT assay (see text) and monitoring formazan absorbance at 570 nm (*right*). (Adapted with permission from Uzunova *et al.*[150] Copyright 2010 The Royal Society of Chemistry.)

showed no significant acute toxicity as the mice tolerated a high dose of 200 mg kg^{-1} without sickness or weight loss. At doses of 200–250 mg kg^{-1}, and when administered as a fast injection, mice were found to go into a shock-like state, but recovered later. Therefore, the authors have estimated that the maximum tolerated dose of CB7 is 250 mg kg^{-1} when administered at a slow rate. A single orally administered dose of CB7 and CB8 as a mixture in equal proportions showed no toxicity up to 600 mg kg^{-1}.[150]

Toxicology and bioactivity have also been determined in another study for CB5, CB7, and three acyclic CB containers (pentamer and methyl and phenyl containing hexamer, see Scheme 7.2 for structures).[149] *In vitro* metabolic activity and cell cytotoxicity assays, such as cell viability (MTS) and cell lysis (adenylate kinase release) assays in human kidney (HEK 293), liver hepatocyte (HepG2), and murine macrophage (RAW264.7) cancer cell lines showed no cytotoxicity up to 1 mM concentrations of CB5 and CB7, as well as the other oligomers. Furthermore, haemolysis assays showed no significant increase in red blood cell lysis compared with the untreated samples, which suggests that the studied containers had good biocompatibility.

Recently, Isaacs and co-workers have also conducted cytotoxicity and toxicity studies of the new acyclic sulfonato CB containers, depicted in Scheme 7.2, to evaluate their biocompatibility.[38] *In vitro* cell viability (MTT) and cell lysis assays (adenylate kinase release) performed in human kidney (HEK 293), liver hepatocyte (HepG2), and monocyte (THP-1) cancer cell lines showed no cytotoxicity up to 10 mM concentrations. In this study, the authors selected kidney and liver cells because the drug molecules normally accumulate in these organs for processing and natural excretion and, consequently, exhibit their toxicity (mostly) at these locations. The selection of the THP-1 cell line was based on the fact that the obtained results could be used to determine the effects of the acyclic sulfonato CB containers (CB-3 and CB-4 in Scheme 7.2) towards immune cells. When haemolysis assays were performed with primary human red blood cells, no significant increase of red blood cell lysis was observed. The *in vivo* study was conducted through a maximal tolerated dose (MTD) experiment in Swiss Webster mice. The containers, intravenously administrated (bolus injection), showed no toxicity as the mice tolerated a high dose of 1230 mg kg^{-1} without sickness or weight loss. The mice appeared healthy, and the recovery rate of the treated mice was comparable with the control group of mice treated with phosphate-buffered saline (PBS) (Figure 7.4).

7.7.3　Activities of CB•Drug Complexes *in Vitro*

Note that although CBs are presently not recognized as being toxic, and whereas the use of their drug complexes could remedy several undesired side effects of the drug molecules (see Section 7.6), these agents should not decrease the activity of the drug molecules towards the targeted cell lines. Initial studies demonstrate that, in most cases, CBs do not alter the activities of the encapsulated drug molecules. In a case study, Day and Collins have evaluated the bioactivity of the albendazole•CB7 complex in human T-cell acute

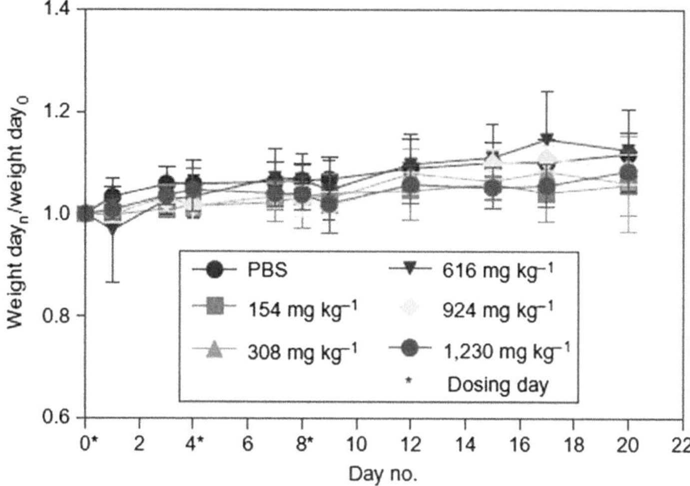

Figure 7.4 Results of the maximal tolerated dose (MTD) studies performed for the sulfonato derivative of the acyclic CB (CB-3 in Scheme 7.2) in Swiss Webster mice. The host molecule was dosed *via* the tail vein on different days (* indicates the dosage date).
(Adapted with permission from Ma *et al.*[38] Copyright 2012 Macmillan Publishers Limited.)

lymphoblastic leukaemia (CEM), ovarian (1A9), and colorectal (HT-29) cancer cell lines. The cytotoxicity of albendazole towards the targeted cell lines did not change upon encapsulation. The *in vitro* inhibition effects of an albendazole derivative also did not change towards HT-29 and human prostate cancer cells (PC-3).[37,55,151] In another study, captothecin encapsulated inside CB7 or CB8 showed comparable cytotoxic activity as the free drug in human lung (A549) and human leukaemia cells (K562).[98] Bioactivity assays in mycobacteria-infected (RAW264.7) macrophage cell lines showed that CB7-bound etham-butol[149] was as effective as its unbound form.

7.7.4   Activities of CB·Drug Complexes *in Vivo*

In vivo assays are pivotal for any medicinal application of CB·drug complexes. The *in vivo* cancer test against female Balb/c mice bearing human ovarian cancer was conducted for a CB7-encapsulated dinuclear platinum drug.[104] The result showed an increase in the drug's maximum tolerated dose from 0.1 to 0.45 mg kg⁻¹. When administered at 0.27 mg kg⁻¹, the drug·CB7 complex was just as active as the free drug at the equivalent dose. Wheate and co-workers investigated the activity of the cisplatin·CB7 complex for the treatment of ovarian cancer in mice.[152] Nude mice tolerated a dose of CB7 alone of 250 mg kg⁻¹ upon intraperitoneal injection and CB7 did not show any effect on the tumour growth rates of A2780 and A2780/cp70 xenografts or on the weight of the animals. Interestingly, the *in vivo* results showed that complexation of cisplatin with CB7 can beneficially reduce tumour growth although it was shown to have no effect on the *in vitro* cyctotoxicity of cisplatin.[39] The activity of the cisplatin•CB7 complex against A2780 tumours was similar (slightly increased growth delay) to that of the free drug, whereas the cisplatin•CB7 complex had a profound affect on the tumour growth of A2780/cp70 tumours with a 1.6-fold slower tumour-doubling time.[152]

To explain the increased *in vivo* activity of CB7-encapsulated cisplatin, Plumb *et al.*[152] measured the plasma concentrations of free and encapsulated cisplatin over time after injection. Analysis of the drug pharmacokinetics revealed that the total concentration of the cisplatin•CB7 complex over a period of 24 hours was significantly higher than that of free cisplatin administrated at an equivalent dose. The total area under the curve (AUC) of plasma platinum content *versus* time was significantly higher for the complex (28.8 h µg mL^{-1}) than for the free drug (16.3 h µg mL^{-1}). Thus, complexed cisplatin was retained in the circulation for a longer period of time than the free drug, which indirectly also supports the fact that CB7 protects the drug from unwanted chemical degradation. However, the peak drug serum concentration and uptake of the drug into different organs (tissue distribution in liver and kidney) were almost the same for the complexed and free drugs because the CB7 complex did not have any targeting ligand and CB7 is unable to exploit any enhanced permeability and retention effect.[152] Put together, the biological effects are of a purely pharmacokinetic, but not tumour targeting, nature. Previously, it had been proposed that the macrocyclic protection of the platinum drugs from degradation and reaction with thiol-containing groups could improve their drug activity.[152] Recently, Aldrich-Wright, Price, Wheate and colleagues[153] reported that the presence of metal ions decreases the diffusion of dinuclear platinum drugs (namely, *trans*-[{PtCl(NH$_3$)$_2$}$_2$$\mu$-dpzm]$^{2+}$, where dpzm is 4,4′-dipyrazolylmethane) upon complexation with CB7.

7.7.5 Other *in Vivo* Ramifications of Cucurbiturils

Despite the above-mentioned results, a great amount of detail regarding the behaviour of CBs and CB•guest complexes in the body remains underexplored: for example, their metabolism is not yet fully described,[151] the enzymes participating in their biotransformation as well as their distribution in the organs and tissues are unknown and their medicinal activity is far from being comprehensively explored. Noteworthy, unpublished data[151] have revealed that only 3.6% of the total CB was resorbed into the blood stream from the alimentary canal as measured by using ^{14}C-radiolabelled CB7 and CB8. Intravenous administration studies have also shown that clearance through urine has a mean half-life of 12.8 h. CB7 did *not* cross the blood–brain barrier, and accumulation in the liver and spleen was very low with respect to that in the kidneys. Putting the chemical and thermal stability of CB7 together with the relatively quick clearance through the urine (preceeded by higher activities in the kidneys), it may well be possible that CB7, in particular, is excreted without chemical modification.[151]

7.8 Pharmaceutical Formulation

The production of CB-based host–guest complexes as solid products involves initial mixing of the hosts and the guests at the required molar ratios in solution followed by isolation of the solid complexes, mostly by lyophilization,[37] ball-mill grinding,[108,154,155] or co-solvent processing.[125] The fact that the components are held together by non-covalent interactions requires some attention. In order to shift the dynamic chemical equilibrium between the free and the bound drug towards the latter in solution, one needs to calculate the concentration of the host molecule such that the drugs are mostly present in their complexed forms. Fortunately, CBs bind guest molecules with high binding affinity (see Section 7.2), which facilitates the preparation of solutions (and production of materials) with 99% and higher content of complexed drug even at relatively low (excess) CB concentrations.

Recently, the formulation of CBs into dosage forms suitable for clinical use has come into the focus.[29,155] CB6 and excipients could be compressed readily into stable and durable tablets for oral administration.[155] Wheate *et al.* mentioned that in this formulation, the tablets contained up to 50% microcrystalline CB6 (*w/w*) and the remainder was lactose (as diluent/bulking agent), Avicel (aids tablet compaction), talc and magnesium stearate (as lubricants and glidants), and Ac-Di-Sol (as disintegrant).[29] They have also demonstrated that these tablets exhibited suitable mechanical, packing, surface, and other processability properties, including tablet hardness, disintegration time in simulated gastric or intestinal fluid, and less than 2% weight loss on friability testing. The produced tablets also exhibited smooth surfaces without pitting or chipping upon compaction and they could be easily ejected from the die.[29] CB host–guest complexes have also been included in inserts for nasal delivery.[29] The compatibility of microcrystalline CB6 with various other excipients was also confirmed by DSC.[62,99,110,121]

A few recent reports[62,99,110,121] have also demonstrated (by using X-ray powder diffraction (XRD) and DSC techniques) the ability of CB7 to prevent interconversion of crystal polymorphs of some drugs and to retain the amorphous structure in the resulting CB7 complex.[62] N_2 adsorption studies confirmed that encapsulation of the drugs (*e.g.*, albendazole) inside CB7 did not affect the surface area and pore size distribution, which is desirable for pharmaceutical formulation.[62]

The selection of the preferred administration route also depends on the chemical properties of a particular CB•drug complex in different biological media. The salt and acid concentrations of these biological media mean that CB and their drug host–guest complexes can be solubilized to therapeutically relevant concentrations. Indeed, CB6 is soluble in simulated gastric fluid at concentrations up to 4 mM.[155] Furthermore, the varying aqueous solubility of the different CBs dictates which route of administration is preferable. For example, the high solubility of CB7 and sufficient solubility of CB6 in saline makes an intravenous administration viable, particularly for CB7, which is known to be excreted through urine.[151,152] The substituted water-soluble CB derivatives could similarly be suitable for intravenous administration. However, the low solubility of CB8 and its small enhancement by salts becomes an impediment to intravenous administration, especially when the drugs would need to be released from a potentially more soluble complex to produce free CB8. Nevertheless, the low solubility of CB8 in saline may be sufficient for oral dosage forms.

7.9 Other Biologically Relevant Applications Based on Cucurbiturils

The CBs have been widely utilized as synthetic receptors for molecular species with diverse biochemical functions. Studies on the molecular recognition of biological molecules by synthetic receptors are burgeoning as they can connect laboratory-based experimental observations on host–guest complexation phenomena with real-life effects not just limited to drug delivery, but extending also to diagnostic applications in medicinal and pharmaceutical fields. With equilibrium association constants up to 10^{15} M^{-1} in aqueous solution,[35] CBs are among the most promising class of receptors for targeting biological molecules with affinities and selectivities that are desirable or even necessary for several pharmacological applications *in vivo*.

7.9.1 Cucurbituril-Based Recognition of Biologically Important Species

In this subsection, the binding affinity of CBs with amino acids, peptides, and proteins is discussed. These compounds are relevant for CB-based drug delivery owing to their physiological omnipresence, and interactions with them could affect (1) the stabilities of the CB•drug complexes and (2) the (accelerated) release of the drug molecules from the cavity of CBs due to competitive binding. On the contrary, interactions with proteins could be potentially exploited for drug targeting. It is worth mentioning here that a recent review by Urbach and Ramilingham[156] already details several interactions of CBs with amino acids, peptides, and proteins.

7.9.1.1 *Amino Acids*

The first study on the interactions of amino acids with CBs was conducted using isothermal titration calorimetry (ITC).[157,158] The four amino acids Gly, Ala, Val, and Phe were found to form exclusion complexes with association to the portals of CB6, as evidenced by the small variance in the binding affinities of the amino acids, which had been selected to possess different size and hydrophobicity. In all cases, the binding was both enthalpically and entropically favoured. In a more recent study,[159] a CB6 analogue was found to bind Trp with one to two orders of magnitude higher affinity than to Phe and Tyr. Notably, no binding was observed for His, which is mainly due to the protonation of the side-chain which makes this amino acid more hydrophilic.

Danylyuk and Fedin[160] reported crystal structures of CB6 with Trp and its decarboxylation product tryptamine. The self-assembled structure indicated the formation of helical nanotubules in the case of tryptophan and stacked columns in the case of tryptamine through ion–dipole and stacking interactions between the indole rings and the glycoluril units of CB6. Interestingly, the formation of the solid-state complexes was found to be dependent on the type of salts used for solubilization of CB6, and the obtained results were explained in terms of the specific coordination mode of CB6 with the metal ions. Another CB6 analogue, bis-nor-seco-CB6, was found to bind stereoselectively to Phe.[161]

In comparison to CB6, CB7 binding with the aromatic amino acids is mainly of the inclusion type. The binding constants of CB7 with Phe in water have been reported to be 1.5×10^5 M^{-1} by competitive NMR titration[162] and 1.8×10^6 M^{-1} by ITC.[128] The binding affinities of CB7 for Trp, Tyr, Lys, Arg, Orn, and His (as well as the associated biogenic amines) were quantified by ITC and competitive fluorescent indicator displacement titrations.[139,163] Tao and co-workers studied the binding of CB7 with a series of amino acids by UV/Vis spectrophotometry.[164] They reported binding constants of CB7 with Phe, Tyr, and Trp of the order of 10^5 M^{-1} with 1 : 1 stoichiometries for CB7–amino-acid binding. In this study, much lower affinities of CB7 towards His, Glu, Met, Val, Leu, and Ala ($K_b < 500$ M^{-1}) were observed.

As the size of CB8 is quite large and the supramolecular interactions partly depend on the size complementarity of the guests with the host cavity, CB8 generally encapsulates two guest molecules to form *ternary* complexes. Frequently, an auxiliary guest is employed to encapsulate small guest molecules. Among others, methyl viologen (MV) often forms charge transfer complexes with small aromatic molecules. Therefore, a 1 : 1 complex of MV•CB8 has frequently been used to encapsulate aromatic guest molecules inside the cavity of CB8.[157,165] Charge-transfer interactions operate, but provide no decisive stabilization for the ternary complexes;[165] instead, the release of high-energy water presents the dominant driving force.[31] Urbach and co-workers reported a charge-transfer ternary complex of CB8 and MV and the amino acid Trp.[157] NMR measurements indicated that MV and the amino acid side-chain bind simultaneously within the cavity of CB8. The MV•CB8 complex was reported to bind Trp with 8- and 20-fold selectivity over Phe and Tyr, respectively. In contrast, no binding was observed for His[157] or the remaining 16 amino acids because of the absence of any aromatic group in their side chains.[158] Examples of other auxiliary guests have also been reported in the literature. The CB8 complexes with 2,7-dimethyldiazophenanthrenium (DPT)[166] and tetramethylbenzobis(imidazolium) (MBBI)[167] bind Trp with 10-fold higher and similar affinity, respectively, as the MV•CB8 complex. Note that the differential binding pattern of MV•CB8 towards aromatic *versus* non-aromatic amino acids is similar to that found for CB7 (see above), that is, the selectivity is comparable. The binding of CB8 with two equivalents of Trp and Phe[158] or with two equivalents of Tyr, His, or Leu showed that the side-chains bind within the cavity, and that the ammonium groups interact with the carbonyl oxygens of CB8, as evidenced by crystal structures.[168]

7.9.1.2   Peptides

The binding of CB6 with dipeptides and tripeptides was demonstrated early on by ITC.[169] Given the relatively small binding constants, the minimal variance with respect to size and sequence of the peptides, and the small size of CB6, it is likely that these peptides form exclusion complexes with interactions only at the portals of CB6, as already observed for amino acids. A recent electrospray ionization mass spectrometric (ESI-MS) study showed that CB6 (in the gas phase) has a high preference to bind the Lys residue in numerous peptides and, thereby, can form peptide•CB6 complexes.[170]

As can be projected from the affinity pattern observed for the natural amino acids, CB7 also binds with the aromatic residues in the peptides. Inoue, Kim and co-workers[171] demonstrated the binding of CB7 with peptides containing Phe at the N-terminus. The latter study focused on the ability of CB7 to discriminate diastereomers of dipeptides such as L-Phe–L-Pro *versus* L-Phe–D-Pro. Note that the cavity of CB7 is too small to bind more than one aromatic guest, and, therefore, 1 : 1 CB7•aromatic-residue complexes were observed.

The binding of CB7 with peptides containing Trp and Tyr residues was also the subject of another study.[172] In pure water, CB7 binds selectively to X-Gly *versus* Gly-X (where X = Phe, Tyr, Trp), and the selectivities are 2000–23 000-fold for the N-terminal aromatic peptides. This result is consistent with the prior work on the recognition of N-terminal aromatic peptides by CB8[173] and by the CB8•MV complex.[158] The strong and selective binding of Tyr-Lys *versus* Lys-Tyr by CB7 and the weak, yet selective binding of Lys-Tyr *versus* Tyr-Lys by CB6 was used to show that when these four molecules (2 hosts and 2 dipeptides) are mixed they "*self-sort*" into the CB7•Tyr-Lys and CB6•Lys-Tyr complexes. Studies on the binding of MV•CB8 with Trp[157] also revealed that the binary complex binds selectively to peptides containing an N-terminal Trp residue rather than to those with a C-terminal Trp, and the authors suggested this could be used in sequence-specific peptide recognition. This concept was successfully elaborated on an entire series of peptides.

7.9.1.3   Proteins

While supramolecular interactions of macrocyclic host molecules with proteins are most important for biological applications, the investigation of these interactions is challenging owing to the macromolecular nature and native structure of proteins, which generally limits the access of host molecules to the binding motifs exposed on the protein surface.[174,175] CB7, for example, binds to human insulin,[174] which has an N-terminal Phe residue on the B-chain (*i.e.*, PheB1, see Figure 7.5), which is exposed to the aqueous bulk and therefore accessible to macrocycles The crystal structure revealed that the last few residues of the B-chain unravelled, like a ball of string, to accommodate for the CB7 binding. These results also suggest that the terminus of a protein is well suited for targeting by a synthetic receptor because the terminal residue is a unique chemical epitope that cannot exist elsewhere on the protein, and because

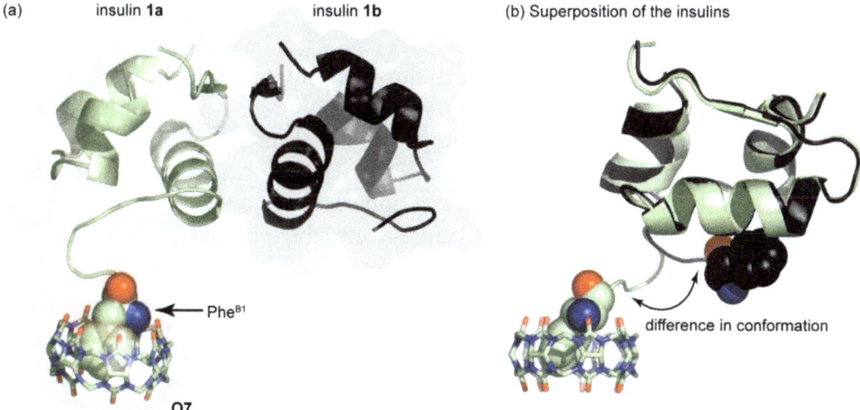

Figure 7.5 Crystal structure of the human insulin•CB7 complex (CB7 is labeled here as Q7). (a) Asymmetric unit of the crystal structure with two insulin molecules, human insulin **1a** and human insulin **1b** . As can be seen, cucurbit[7]uril (CB7) binds with the Phe aromatic residue at the B1 position. Composite part (b) shows the superposition of CB7-bound and free insulin to expose the CB7-induced partial unfolding at the N-terminus of the B-chain. (Adapted with permission from Chinai *et al.*[174] Copyright 2011 American Chemical Society.)

the terminus can unfold more easily than other sites on the protein in order to accommodate binding, and, for some proteins, potentially without loosing their biological activity. In another study, using a fluorescent indicator displacement assay, selective binding of CB7 with insulin within a larger series of other blood proteins and short peptides (all of which lacking an aromatic N-terminus) was reported,[175] which demonstrated the potential for applications such as continuous insulin monitoring.

The binding interactions of CB8 with proteins are expectedly very different from those for CB7. Recent investigations show that CB8 forms mainly ternary or higher-order complexes with proteins. For example, the fluorescence of the Trp residues on BSA can be moderately quenched (in the absence of macrocycle) by a sensitizer, namely 5,10,15,20-tetrakis(1-methyl-4-pyridinio)porphyrin (TPP).[176] Wang and co-workers reported that when CB8 was added to a solution mixture containing both BSA and TPP, the extent of quenching could be enhanced dramatically, due to the formation of a ternary complex *via* simultaneous inclusion of a pyridinium group of TPP and the indole side-chain of Trp in the cavity of CB8.[176] Such enhanced protein–sensitizer interactions have obvious implications for photodynamic therapy (see Chapter 10 by Schatz), if the sensitizer can be selectively brought closer or in contact with the target. A separate study showed that CB8 can be used to dimerize two protein molecules. When CB8 was added to a solution containing yellow fluorescent protein (YFP) and cyan fluorescent protein (CFP), each modified with an N-terminal recognition sequence for CB8 (Phy-Gly-Gly, FGG), hetero-FRET (Förster resonance energy transfer) became observable.

When CB8 was added to a solution containing only FGG-modified YFP, homo-FRET was observed.[177]

Another line of applications of CBs involves interactions with fluorescent dyes in the presence of proteins (lysozyme, BSA) under formation of ternary CB•protein•dye complexes[92] or in competition with proteins.[67] Thus, it could be demonstrated that the dye Brilliant Green (BG) forms a ternary complex with CB7 and proteins with positive cooperativity, and that drugs (using neutral red as a model) can be in principle released from CB7 upon addition of salts and effectively transferred to a protein (BSA). A study by Ghosh and Isaacs[117] also pointed to the involvement of ternary suprabiomolecular complexes between an enzyme, an inhibitor (which served as ditopic ligand), and CB7 (see also Section 7.4.4).

7.9.2 Application of Cucurbituril Host–Guest Complexes for Analytical and Diagnostic Bioassays

Cucurbituril host–guest complexes have also been employed for biologically relevant analytical and diagnostic applications. For example, they are used to monitor enzymatic reactions by the so-called supramolecular tandem assay principle (see Chapter 11 by Hennig and Dsouza *et al.*[178] for conceptual aspects). For these enzyme assays, a fluorescent dye is selected whose binding constant to a macrocycle in general, and to a CB in particular, lies in between the binding constant of the substrate and the product of the enzymatic reaction of interest. The dye and the macrocycle form the "reporter pair" for the tandem assay. For example, by using the Dapoxyl (a fluorescent dye)/CB7 reporter pair, the decarboxylation processes of different amino acids (Lys, Arg, His, Tyr, and Trp) to their corresponding biogenic amines (cadaverine, agmatine, histamine, tyramine, and tryptamine) can be conveniently monitored online and in homogeneous solution.[139,163,179] The binding constants of the biogenic amines with CB7 are up to three orders of magnitude higher than those of the corresponding amino acids, and the binding constant of Dapoxyl with CB7 falls in between. Therefore, in the presence of the amino acids, Dapoxyl remains complexed and strongly fluorescent; as the biogenic amines form in the progress of the enzymatic reaction, the dye is displaced from the CB7 cavity to result in a steady decrease in fluorescence until a plateau is reached and the conversion is complete. Note that the complexation and decomplexation process of the reporter molecule as well as the product with CB7 must be faster than the time scale of the enzymatic reaction in order for the assays to report directly on the enzyme kinetics, which is generally fulfilled. By coupling this principle with the intrinsic enantiospecificity of decarboxylases for L-amino acid substrates, multiparameter sensor arrays (for measuring concentrations of several amino acids in parallel) were designed that selectively signal the presence of a reactive pair of an L-amino acid and its corresponding decarboxylase.[163]

An interesting application of the sequence-selectivity of CBs for aromatic peptides has been demonstrated recently.[180] The metalloendopeptidase thermolysin selectively cleaves the amide bond at the nitrogen side of Phe residues in peptides, thus producing product peptides or intermediary peptides which contain an N-terminal Phe. CB7 binds weakly to non-terminal Phe but strongly to N-terminal Phe residues. The reporter pair in this assay was comprised of CB7 and acridine orange, which possesses a binding affinity that falls between the binding affinities of the substrate (Thr-Gly-Ala-Phe-Met–CONH$_2$) and product (Phe-Met–CONH$_2$) peptides. By measuring the rate of hydrolysis as a function of the substrate concentration, the sequence specificity (*e.g.*, Ser *vs.* Ala), the stereoselectivity (*e.g.*, L-Ala *vs.* D-Ala), and *endo- versus exo-*peptidase activity of the enzyme was determined according to the tandem assay principle. By using a known protease inhibitor, namely phosphoramidon, the assay was validated for the measurement of inhibitory constants, thereby illustrating potential for high-throughput screening.

In another study, Isaacs and co-workers showed that a napthalene-CB6 derivative can be used as a turn-on fluorescent sensor for biological amines, for example histamine.[82] Upon addition of Eu^{3+}, the intrinsic fluorescence of the napthalene-CB6 derivative quenched due to intra-complex resonance energy transfer from the naphthalene group to Eu^{3+}. Decomplexation of Eu^{3+} from the ureido carbonyl by histamine recovered the fluorescence. In a follow-up study, Anzenbacher, Isaacs and colleagues[181] demonstrated that the same sensing principle can also be used in cross-reactive array sensors, utilizing the complementary selectivities of two different derivatives of CBs (naphthalene-CB6 and CB-4, see Scheme 7.2), for the detection of cancer-associated nitrosamines (*e.g.*, *N*-nitrosonornicotine and 4-methylnitrosamino)-1-(3-pyridyl)-1-butanone). Note that host CB-4 also contains two naphthalene chromophores in its structure.

In other examples, CB7 has also been used to enhance the readout of enzyme-linked immunosorbent assays (ELISAs).[182] AuNPs have been shown to aggregate upon binding with CB7 on the Au surface. UV-Vis spectroscopy and dynamic light-scattering experiments demonstrated that urease, which produces ammonium ions upon reaction with urea, induces dissociation of the CB7-nanoparticle assemblies *via* the competition of ammonia for the binding to CB7. This activity was then applied to an ELISA for the detection of mouse immunoglobulin G.

Recently, Kim's group demonstrated that the moderately high binding affinity of CB6 with lysine residues in different peptides and proteins can potentially be used to probe the structures of proteins.[170] As an example, the authors studied CB6–ubiquitin complexes in the gas phase. Low-energy collision-induced dissociation (CID) of the CB6–ubiquitin complexes yielded different fragments. Further fragmentation of these ion peaks and related mass spectrometric analysis revealed details of the binding sites of CB6 with ubiquitin and enabled the team to deduce the protein structure in the solution phase, from which ionization takes place.

In another study, Kim and co-workers showed that thiol-containing glass surfaces can be decorated by perallyloxycucurbit[6]uril, where the functional groups at the CB equator act as anchoring units, retaining the CB6 cavity free to bind small molecules (or analytes).[77] Therefore, the entire system can act as a biochip for different sensing applications. The authors have demonstrated, in particular, that the free CB6 cavities are able to bind spermine as well as a spermine-tagged fluorophore (spermine–FITC).

7.9.3   Cucurbiturils for Bio-related Applications

An artificial non-covalent assembly for gene delivery, mimicking a virus-like particle, was established in the form of a ternary complex between PPI-DAB (G4) with CB6 and DNA, where PPI-DAB is a cationic dendrimer.[183] The formation of the ternary complex was based on the non-covalent and electrostatic interactions of PPI-DAB dendrimer with the macrocycle and DNA, respectively. The authors showed that the artificial assembly was able, in an efficient manner, to transfect mammalian cells (293 cells and Vero 76 cells) and the cytotoxicity of the PPI-DAB•CB6 complex in Vero 76 cells was found to be relatively low by employing the MTT assay.[183]

Peng and coworkers synthesized a series of MV-containing ruthenium *tris*-bipyridyl complexes, where the MV moiety was linked with a bipyridine group *via* carbon chains of different tether lengths (see Scheme 7.3 for structures), and demonstrated that the complexes can induce DNA photocleavage.[184] The complexes with longer carbon-chain linkage showed maximum photocleavage efficiency. CB8 binds with the MV moiety, hinders the back electron transfer process, and thereby increases the photocleavage efficiency.

Isaacs *et al.* demonstrated that glycoluril decamers of opposite handedness (see Scheme 7.2) can assemble together in water to display properties reminiscent of RNA folding.[185] The authors pointed out that metal ions play a crucial role for the stabilization of the double-helical structure. The formation of a helical structure was observed only in the presence of Li^+, Na^+, K^+, and Ca^{2+} ions and the others investigated (Cs^+, Be^{2+}, Mg^{2+}) did not induce the helical structure.

In another bio-related application of CBs, the selective capture of modified proteins was reported by exploiting the extraordinary binding constant of CB7 with ferrocenemethylammonium (AFc) ($K \sim 10^{12}$ M^{-1}).[186] Remarkably, CB7-immobilized beads were able to capture and isolate ferrocenylated plasma membrane proteins (PMPs) from a mixture of different proteins (Scheme 7.9).[186] The recovery of the proteins from the beads into solution was observed upon heating or treatment with a strong competitor.

The CB7–AFc-mediated capture of proteins has also been used to print well-aligned monolayers of proteins.[187] A single equivalent of ferrocenemethyl-ammonium was conjugated to yellow fluorescent protein. The protein–AFc conjugates were deposited onto a CB7 monolayer, which had been formed by spontaneous adsorption of CB7 onto Au and shown to produce a stable and densely packed layer of protein. The authors further demonstrated the compatibility of this approach with microcontact printing from a patterned poly(dimethylsiloxane) stamp. Interestingly, treatment of the surface with free

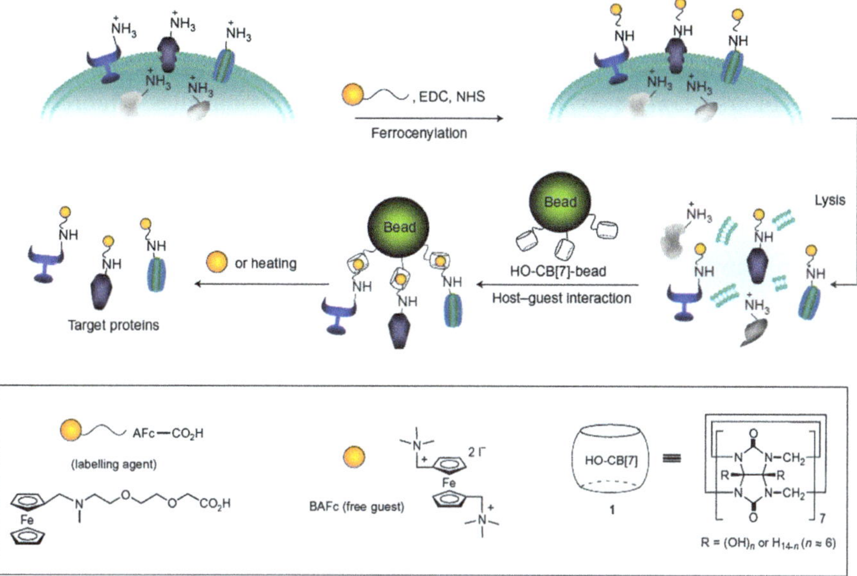

Scheme 7.9 Schematic representation of the use of a highly stable host–guest pair for isolation ("fishing") of plasma membrane proteins. 1-Ethyl-3-(3-dimethylaminopropyl) carbodiimide (EDC), *N*-hydroxysuccimidyl sepharose (NHS). (Adapted with permission from Lee *et al.*[186] Copyright 2011 Macmillan Publishers Limited.)

AFc ligand resulted in quantitative displacement of the protein, and, thus, printing was shown to be reversible.

The high stability of CB7–AFc or CB7-adamantylammonium complexes ($K = 10^{12}$–10^{15} M^{-1}) has also been utilized for the immobilization of proteins (*e.g.*, glucose oxidase as a model protein) on solid surfaces in three steps:[188] (1) anchoring CB7 units on an alkanethiolate mixed self-assembled monolayer (SAM) on Au, (2) attachment of AFc units to a target protein, and (3) immobilization of the "ferrocenylated" protein onto the CB7-attached SAM on Au. The surface immobilization of CB7 was achieved by an olefin cross-metathesis reaction between allyloxy-CB7 (see Section 7.5) and a vinyl-terminated SAM on gold.

The strong host–guest encapsulation of 1,6-diaminohexane (DAH)-conjugated hyaluronic acid (HA) with the cavity of CB6-cojugated HA resulted in the formation of a supramolecular hydrogel assembly of CB6/DAH-HA.[189] When these hydrogels were modified with the c(RGDyK) peptide (see Scheme 7.3 for the structure), the entrapped NHDF human fibroblast cells and NIH3T3 mouse fibroblast cells proliferated 5-fold within 14 days and 3 days, respectively, compared to the untreated hydrogels. *In vivo*, the hydrogel formed within minutes after subcutaneous injection of CB6-HA and DAH-HA solutions in nude mice. Histological analysis with haematoxylin and eosin (H&E) stains, which was performed in Balb/c mice 1 week after injection of the

CB6/DAH-HA hydrogel, showed negligible inflammation compared with normal mice,[189] thus confirming the biocompatibility of the synthesized hydrogels. Since the CB6-containing hydrogels could be readily stained with FITC or rhodamine B isothiocyanate (RBITC) conjugates, fluorescence imaging was used to monitor the localization of the hydrogel.

Recently, Kim and co-workers showed that transport of protons and alkali metal ions across lipid membranes is possible by using CB5 and CB6 as artificial ion channels.[122] The transmembrane transport of protons and Li^+, Cs^+, Rb^+, K^+ and Na^+ ions was demonstrated in the case of CB6-containing vesicles, while the transport of Li^+ and Na^+ was demonstrated in the case of CB5-containing vesicles.

7.10 Conclusions

Cucurbiturils have unique supramolecular recognition properties, which offer numerous advantages over the use of conventional macrocycles. These include up to femtomolar affinities, tunable complexation kinetics, pK_a shifts in favour of binding of the protonated forms of guests, and increased photochemical, chemical, and thermal stabilities of encapsulated guests. While these properties have long been recognized, their implementation in drug delivery systems has rapidly unfolded only during the past 10 years. Besides the use of discrete small-molecule formulations between cucurbiturils (as well as their acyclic analogues) and drug molecules, the use of functionalized cucurbiturils, which is presently actively under investigation for immobilization and bio-labelling, and the formation of hybrid structures between cucurbiturils and polymers will lead to future fields of investigation for medical therapeutics and diagnostics. Cucurbiturils already hold more prospects for drug delivery than calixarenes, and they are expected to rival cyclodextrins in many potential applications.

Acknowledgements

N.S. would like to acknowledge the Office of Research Support and Sponsored Projects (RSSP) at the United Arab Emirates University for their financial support of this project under grant number 21S041 within the framework of National Research Foundation funding program (NRF). I.G. and W.M.N. would like to thank the Deutsche Forschungsgemeinschaft (DFG, NA 686/5 and NA 686/6) for financial support.

References

1. K. Uekama, F. Hirayama and T. Irie, *Chem. Rev.*, 1998, **98**, 2045–2076.
2. J. Li and X. J. Loh, *Adv. Drug Deliv. Rev.*, 2008, **60**, 1000–1017.
3. J. W. Steed and P. A. Gale, *Supramolecular Chemistry: From Molecules to Nanomaterials*, John Wiley and Sons, 2012.
4. I. Ghosh and W. M. Nau, *Adv. Drug Deliv. Rev.*, 2012, **64**, 764–783.

5. S. Svenson and D. A. Tomalia, *Adv. Drug Deliv. Rev.*, 2005, **57**, 2106–2129.
6. A. Sharma and U. S. Sharma, *Int. J. Pharm.*, 1997, **154**, 123–140.
7. G. A. Husseini and W. G. Pitt, *Adv. Drug Deliv. Rev.*, 2008, **60**, 1137–1152.
8. L. Lacerda, A. Bianco, M. Prato and K. Kostarelos, *Adv. Drug Deliv. Rev.*, 2006, **58**, 1460–1470.
9. J. M. Dang and K. W. Leong, *Adv. Drug Delivery Rev.*, 2006, **58**, 487–499.
10. E. Shchepotina, E. Pashkina, E. Yakushenko and V. Kozlov, *Nanotechnol. Russ.*, 2011, **6**, 773–779.
11. M. V. Rekharsky and Y. Inoue, *Chem. Rev.*, 1998, **98**, 1875–1917.
12. E. Da Silva, A. N. Lazar and A. W. Coleman, *J. Drug Deliv. Sci. Technol.*, 2004, **14**, 3–20.
13. K. Wang, D. S. Guo, H. Q. Zhang, D. Li, X. L. Zheng and Y. Liu, *J. Med. Chem.*, 2009, **52**, 6402–6412.
14. G. S. Wang, H. Y. Zhang, D. Li, P. Y. Wang and Y. Liu, *Supramol. Chem.*, 2011, **23**, 441–446.
15. W. A. Freeman, W. L. Mock and N. Y. Shih, *J. Am. Chem. Soc.*, 1981, **103**, 7367–7368.
16. J. Lagona, P. Mukhopadhyay, S. Chakrabarti and L. Isaacs, *Angew. Chem. Int. Ed.*, 2005, **44**, 4844–4870.
17. J. Kim, I. S. Jung, S. Y. Kim, E. Lee, J. K. Kang, S. Sakamoto, K. Yamaguchi and K. Kim, *J. Am. Chem. Soc.*, 2000, **122**, 540–541.
18. A. Day, A. P. Arnold, R. J. Blanch and B. Snushall, *J. Org. Chem.*, 2001, **66**, 8094–8100.
19. E. Masson, X. X. Ling, R. Joseph, L. Kyeremeh-Mensah and X. Y. Lu, *RSC Adv.*, 2012, **2**, 1213–1247.
20. K. Kim, N. Selvapalam, Y. H. Ko, K. M. Park, D. Kim and J. Kim, *Chem. Soc. Rev.*, 2007, **36**, 267–279.
21. W. L. Mock and N. Y. Shih, *J. Org. Chem.*, 1986, **51**, 4440–4446.
22. C. Marquez, F. Huang and W. M. Nau, *IEEE Trans. Nanobiosci.*, 2004, **3**, 39–45.
23. S. Yi and A. E. Kaifer, *J. Org. Chem.*, 2011, **76**, 10275–10278.
24. N. J. Wheate, N. Patel and O. B. Sutcliffe, *Future Med. Chem.*, 2010, **2**, 231–236.
25. D. Z. Jiao, N. Zhao and O. A. Scherman, *Chem. Commun.*, 2010, **46**, 2007–2009.
26. L. Isaacs, *Isr. J. Chem.*, 2011, **51**, 578–591.
27. L. Isaacs, *Chem. Commun.*, 2009, 619–629.
28. D. H. Macartney, *Isr. J. Chem.*, 2011, **51**, 600–615.
29. S. Walker, R. Oun, F. J. McInnes and N. J. Wheate, *Isr. J. Chem.*, 2011, **51**, 616–624.
30. W. M. Nau, M. Florea and K. I. Assaf, *Isr. J. Chem.*, 2011, **51**, 559–577.
31. F. Biedermann, V. D. Uzunova, O. A. Scherman, W. M. Nau and A. De Simone, *J. Am. Chem. Soc.*, 2012, **134**, 15318–15323.
32. S. Mecozzi and J. Rebek, *Chem. Eur. J*, 1998, **4**, 1016–1022.

33. S. Moghaddam, C. Yang, M. Rekharsky, Y. H. Ko, K. Kim, Y. Inoue and M. K. Gilson, *J. Am. Chem. Soc.*, 2011, **133**, 3570–3581.
34. S. Moghaddam, Y. Inoue and M. K. Gilson, *J. Am. Chem. Soc.*, 2009, **131**, 4012–4021.
35. Y. H. Ko, I. Hwang, D. W. Lee and K. Kim, *Isr. J. Chem.*, 2011, **51**, 506–514.
36. A. L. Koner, I. Ghosh, N. Saleh and W. M. Nau, *Can. J. Chem.*, 2011, **89**, 139–147.
37. Y. J. Zhao, D. P. Buck, D. L. Morris, M. H. Pourgholami, A. I. Day and J. G. Collins, *Org. Biomol. Chem.*, 2008, **6**, 4509–4515.
38. D. Ma, G. Hettiarachchi, D. Nguyen, B. Zhang, J. B. Wittenberg, P. Y. Zavalij, V. Briken and L. Isaacs, *Nat. Chem.*, 2012, **4**, 503–510.
39. N. J. Wheate, D. P. Buck, A. I. Day and J. G. Collins, *Dalton Trans.*, 2006, 451–458.
40. S. Kemp, N. J. Wheate, S. Y. Wang, J. G. Collins, S. F. Ralph, A. I. Day, V. J. Higgins and J. R. Aldrich-Wright, *J. Biol. Inorg. Chem.*, 2007, **12**, 969–979.
41. J. Mohanty and W. M. Nau, *Angew. Chem. Int. Ed.*, 2005, **44**, 3750–3754.
42. F. Biedermann, E. Elmalem, I. Ghosh, W. M. Nau and O. A. Scherman, *Angew. Chem. Int. Ed.*, 2012, **51**, 7739–7743.
43. R. B. Wang, L. Yuan and D. H. Macartney, *Chem. Commun.*, 2005, 5867–5869.
44. V. Sindelar, S. Silvi and A. E. Kaifer, *Chem. Commun.*, 2006, 2185–2187.
45. N. Saleh, A. L. Koner and W. M. Nau, *Angew. Chem. Int. Ed.*, 2008, **47**, 5398–5401.
46. C. Marquez and W. M. Nau, *Angew. Chem. Int. Ed.*, 2001, **40**, 3155–3160.
47. A. D. St-Jacques, I. W. Wyman and D. H. Macartney, *Chem. Commun.*, 2008, 4936–4938.
48. I. W. Wyman and D. H. Macartney, *Org. Biomol. Chem.*, 2010, **8**, 253–260.
49. M. Shaikh, J. Mohanty, P. K. Singh, W. M. Nau and H. Pal, *Photochem. Photobiol. Sci.*, 2008, **7**, 408–414.
50. A. L. Koner and W. M. Nau, *Supramol. Chem.*, 2007, **19**, 55–66.
51. N. Saleh, M. A. Meetani, L. Al-Kaabi, I. Ghosh and W. M. Nau, *Supramol. Chem.*, 2011, **23**, 654–661.
52. I. W. Wyman and D. H. Macartney, *J. Org. Chem.*, 2009, **74**, 8031–8038.
53. R. B. Wang, B. C. MacGillivray and D. H. Macartney, *Dalton Trans.*, 2009, 3584–3589.
54. I. W. Wyman and D. H. Macartney, *Org. Biomol. Chem.*, 2010, **8**, 247–252.
55. Y. J. Zhao, M. H. Pourgholami, D. L. Morris, J. G. Collins and A. I. Day, *Org. Biomol. Chem.*, 2010, **8**, 3328–3337.
56. Z. Miskolczy, M. Megyesi, G. Tarkanyi, R. Mizsei and L. Biczok, *Org. Biomol. Chem.*, 2011, **9**, 1061–1070.
57. J. Mohanty, A. C. Bhasikuttan, W. M. Nau and H. Pal, *J. Phys. Chem. B*, 2006, **110**, 5132–5138.

58. X. Y. Zhang, G. Gramlich, X. J. Wang and W. M. Nau, *J. Am. Chem. Soc.*, 2002, **124**, 254–263.

59. N. Chieng, T. Rades and J. Aaltonen, *J. Pharm. Biomed. Anal.*, 2011, **55**, 618–644.

60. P. Germain, J. M. Letoffe, M. P. Merlin and H. J. Buschmann, *Thermochim. Acta*, 1998, **315**, 87–92.

61. D. Bardelang, K. A. Udachin, D. M. Leek, J. C. Margeson, G. Chan, C. I. Ratcliffe and J. A. Ripmeestert, *Cryst. Growth Des.*, 2011, **11**, 5598–5614.

62. N. Saleh, A. Khaleel, H. Al-Dmour, B. al-Hindawi and E. Yakushenko, *J. Therm. Anal. Calorim.*, 2013, **111**, 385–392.

63. Y. M. Jeon, H. Kim, D. Whang and K. Kim, *J. Am. Chem. Soc.*, 1996, **118**, 9790–9791.

64. D. Whang, J. Heo, J. H. Park and K. Kim, *Angew. Chem. Int. Ed.*, 1998, **37**, 78–80.

65. C. Marquez, R. R. Hudgins and W. M. Nau, *J. Am. Chem. Soc.*, 2004, **126**, 5806–5816.

66. S. D. Choudhury, J. Mohanty, H. Pal and A. C. Bhasikuttan, *J. Am. Chem. Soc.*, 2010, **132**, 1395–1401.

67. M. Shaikh, J. Mohanty, A. C. Bhasikuttan, V. D. Uzunova, W. M. Nau and H. Pal, *Chem. Commun.*, 2008, 3681–3683.

68. A. C. Bhasikuttan, H. Pal and J. Mohanty, *Chem. Commun.*, 2011, **47**, 9959–9971.

69. E. Meyer, *Inaugural-Dissertation, Heidelberg*, Germany, 1904.

70. T. C. S. Pace and C. Bohne, *Adv. Phys. Org. Chem.*, 2008, **vol. 42**, 167–223.

71. H. Tang, D. Fuentealba, Y. H. Ko, N. Selvapalam, K. Kim and C. Bohne, *J. Am. Chem. Soc.*, 2011, **133**, 20623–20633.

72. W. M. Nau, *Nat. Chem.*, 2010, **2**, 248–250.

73. A. C. Bhasikuttan, S. D. Choudhury, H. Pal and J. Mohanty, *Isr. J. Chem.*, 2011, **51**, 634–645.

74. W. Ong and A. E. Kaifer, *J. Org. Chem.*, 2004, **69**, 1383–1385.

75. J. W. Lee, S. Samal, N. Selvapalam, H. J. Kim and K. Kim, *Acc. Chem. Res.*, 2003, **36**, 621–630.

76. H. J. Buschmann, E. Cleve, K. Jansen, A. Wego and E. Schollmeyer, *Mat. Sci. Eng., C-Bio*, 2000, **14**, 35–39.

77. S. Y. Jon, N. Selvapalam, D. H. Oh, J. K. Kang, S. Y. Kim, Y. J. Jeon, J. W. Lee and K. Kim, *J. Am. Chem. Soc.*, 2003, **125**, 10186–10187.

78. N. Zhao, G. O. Lloyd and O. A. Scherman, *Chem. Commun.*, 2012, **48**, 3070–3072.

79. M. M. Ahmed, K. Koga, M. Fukudome, H. Sasaki and D. Q. Yuan, *Tetrahedron Lett.*, 2011, **52**, 4646–4649.

80. J. Z. Zhao, H. J. Kim, J. Oh, S. Y. Kim, J. W. Lee, S. Sakamoto, K. Yamaguchi and K. Kim, *Angew. Chem. Int. Ed.*, 2001, **40**, 4233–4235.

81. H. Isobe, S. Sato and E. Nakamura, *Org. Lett.*, 2002, **4**, 1287–1289.

82. D. Lucas, T. Minami, G. Iannuzzi, L. P. Cao, J. B. Wittenberg, P. Anzenbacher and L. Isaacs, *J. Am. Chem. Soc.*, 2011, **133**, 17966–17976.

83. B. Vinciguerra, L. P. Cao, J. R. Cannon, P. Y. Zavalij, C. Fenselau and L. Isaacs, *J. Am. Chem. Soc.*, 2012, **134**, 13133–13140.
84. D. Ma, B. Zhang, U. Hoffmann, M. G. Sundrup, M. Eikermann and L. Isaacs, *Angew. Chem. Int. Ed.*, 2012, **51**, 11358–11362.
85. C. Shen, D. Ma, B. Meany, L. Isaacs and Y. H. Wang, *J. Am. Chem. Soc.*, 2012, **134**, 7254–7257.
86. F. Wu, L. H. Wu, X. Xiao, Y. Q. Zhang, S. F. Xue, Z. Tao and A. I. Day, *J. Org. Chem.*, 2012, **77**, 606–611.
87. W. H. Lei, Q. X. Zhou, G. Y. Jiang, Y. J. Hou, B. W. Zhang, X. X. Cheng and X. S. Wang, *ChemPhysChem*, 2011, **12**, 2933–2940.
88. C. Marquez and W. M. Nau, *Angew. Chem. Int. Ed.*, 2001, **40**, 4387–4390.
89. C. F. Li, L. M. Du and H. M. Zhang, *Spectroc. Acta Pt. A-Molec. Biomolec. Spectr.*, 2010, **75**, 912–917.
90. L. Biczok, V. Wintgens, Z. Miskolczy and M. Megyesi, *Isr. J. Chem.*, 2011, **51**, 625–633.
91. H. Liu, X. Wu, Y. Huang, J. He, S. F. Xue, Z. Tao, Q. J. Zhu and G. Wei, *J. Inclusion Phenom. Macrocyclic Chem.*, 2011, **71**, 583–587.
92. A. C. Bhasikuttan, J. Mohanty, W. M. Nau and H. Pal, *Angew. Chem. Int. Ed.*, 2007, **46**, 4120–4122.
93. M. Megyesi, L. Biczok and I. Jablonkai, *J. Phys. Chem. C*, 2008, **112**, 3410–3416.
94. N. Dong, L. N. Cheng, X. L. Wang, Q. Li, C. Y. Dai and Z. Tao, *Talanta*, 2011, **84**, 684–689.
95. C. F. Li, L. M. Du, W. Y. Wu and A. Z. Sheng, *Talanta*, 2010, **80**, 1939–1944.
96. Y. Y. Zhou, H. P. Yu, L. Zhang, H. W. Xu, L. A. Wu, J. Y. Sun and L. Wang, *Microchim. Acta*, 2009, **164**, 63–68.
97. M. J. Pisani, Y. J. Zhao, L. Wallace, C. E. Woodward, F. R. Keene, A. I. Day and J. G. Collins, *Dalton Trans.*, 2010, **39**, 2078–2086.
98. N. Dong, S. F. Xue, Q. J. Zhu, Z. Tao, Y. Zhao and L. X. Yang, *Supramol. Chem.*, 2008, **20**, 659–665.
99. A. R. Kennedy, A. J. Florence, F. J. McInnes and N. J. Wheate, *Dalton Trans.*, 2009, 7695–7700.
100. M. S. Bali, D. P. Buck, A. J. Coe, A. I. Day and J. G. Collins, *Dalton Trans.*, 2006, 5337–5344.
101. Y. J. Jeon, S. Y. Kim, Y. H. Ko, S. Sakamoto, K. Yamaguchi and K. Kim, *Org. Biomol. Chem.*, 2005, **3**, 2122–2125.
102. N. J. Wheate, R. I. Taleb, A. M. Krause-Heuer, R. L. Cook, S. Wang, V. J. Higgins and J. R. Aldrich-Wright, *Dalton Trans.*, 2007, 5055–5064.
103. N. J. Wheate, A. I. Day, R. J. Blanch, A. P. Arnold, C. Cullinane and J. G. Collins, *Chem. Commun.*, 2004, **10**, 1424–1425.
104. N. J. Wheate, *J. Inorg. Biochem.*, 2008, **102**, 2060–2066.
105. Y. J. Zhao, M. S. Bali, C. Cullinane, A. I. Day and J. G. Collins, *Dalton Trans.*, 2009, 5190–5198.
106. D. P. Buck, P. M. Abeysinghe, C. Cullinane, A. I. Day, J. G. Collins and M. M. Harding, *Dalton Trans.*, 2008, 2328–2334.

107. C. Kim, S. S. Agasti, Z. J. Zhu, L. Isaacs and V. M. Rotello, *Nat. Chem.*, 2010, **2**, 962–966.
108. G. C. Jiang and G. T. Li, *J. Photochem. Photobiol. B-Biol.*, 2006, **85**, 223–227.
109. R. B. Wang and D. H. Macartney, *Org. Biomol. Chem.*, 2008, **6**, 1955–1960.
110. F. J. McInnes, N. G. Anthony, A. R. Kennedy and N. J. Wheate, *Org. Biomol. Chem.*, 2010, **8**, 765–773.
111. A. I. Day, A. P. Arnold and R. J. Blanch, *US Patent 2003/0140787A1*, 2003.
112. B. S. Kim, Y. H. Ko, Y. Kim, H. J. Lee, N. Selvapalam, H. C. Lee and K. Kim, *Chem. Commun.*, 2008, 2756–2758.
113. M. Florea and W. M. Nau, *Angew. Chem. Int. Ed.*, 2011, **50**, 9338–9342.
114. Q. Zhang, Z. Zhen, H. Jiang, X. G. Li and J. A. Liu, *J. Agric. Food Chem.*, 2011, **59**, 10539–10545.
115. R. B. Wang, I. W. Wyman, S. H. Wang and D. H. Macartney, *J. Inclusion Phenom. Macrocyclic Chem.*, 2009, **64**, 233–237.
116. A. Hennig, G. Ghale and W. M. Nau, *Chem. Commun.*, 2007, 1614–1616.
117. S. Ghosh and L. Isaacs, *J. Am. Chem. Soc.*, 2010, **132**, 4445–4454.
118. Y. Huang, S. F. Xue, Z. Tao, Q. J. Zhu, H. Zhang, J. X. Lin and D. H. Yu, *J. Inclusion Phenom. Macrocyclic Chem.*, 2008, **61**, 171–177.
119. Y. Y. Zhou, J. Y. Sun, H. P. Yu, L. Wu and L. Wang, *Supramol. Chem.*, 2009, **21**, 495–501.
120. H. Cong, C. R. Li, S. F. Xue, Z. Tao, Q. J. Zhu and G. Wei, *Org. Biomol. Chem.*, 2011, **9**, 1041–1046.
121. N. J. Wheate, V. Vora, N. G. Anthony and F. J. McInnes, *J. Inclusion Phenom. Macrocyclic Chem.*, 2010, **68**, 359–367.
122. Y. J. Jeon, H. Kim, S. Jon, N. Selvapalam, D. H. Oh, I. Seo, C. S. Park, S. R. Jung, D. S. Koh and K. Kim, *J. Am. Chem. Soc.*, 2004, **126**, 15 944–15 945.
123. C. J. Li, J. Li and X. S. Jia, *Org. Biomol. Chem.*, 2009, **7**, 2699–2703.
124. D. Ma, R. Glassenberg, S. Ghosh, P. Y. Zavalij and L. Isaacs, *Supramol. Chem.*, 2012, **24**, 325–332.
125. R. J. Blanch, A. J. Sleeman, T. J. White, A. P. Arnold and A. I. Day, *Nano. Lett.*, 2002, **2**, 147–149.
126. R. B. Wang, D. Bardelang, M. Waite, K. A. Udachin, D. M. Leek, K. Yu, C. I. Ratcliffe and J. A. Ripmeester, *Org. Biomol. Chem.*, 2009, **7**, 2435–2439.
127. H. S. El-Sheshtawy, B. S. Bassil, K. I. Assaf, U. Kortz and W. M. Nau, *J. Am. Chem. Soc.*, 2012, **134**, 19935–19941.
128. M. V. Rekharsky, T. Mori, C. Yang, Y. H. Ko, N. Selvapalam, H. Kim, D. Sobransingh, A. E. Kaifer, S. M. Liu, L. Isaacs, W. Chen, S. Moghaddam, M. K. Gilson, K. M. Kim and Y. Inoue, *Proc. Natl. Acad. Sci. USA*, 2007, **104**, 20 737–20 742.
129. R. Khurana, C. Coleman, C. Ionescu-Zanetti, S. A. Carter, V. Krishna, R. K. Grover, R. Roy and S. Singh, *J. Struct. Biol.*, 2005, **151**, 229–238.

130. S. D. Choudhury, J. Mohanty, H. P. Upadhyaya, A. C. Bhasikuttan and H. Pal, *J. Phys. Chem. B*, 2009, **113**, 1891–1898.
131. C. P. Carvalho, V. D. Uzunova, J. P. Da Silva, W. M. Nau and U. Pischel, *Chem. Commun.*, 2011, **47**, 8793–8795.
132. S. Angelos, Y. W. Yang, K. Patel, J. F. Stoddart and J. I. Zink, *Angew. Chem. Int. Ed.*, 2008, **47**, 2222–2226.
133. H. K. Lee, K. M. Park, Y. J. Jeon, D. Kim, D. H. Oh, H. S. Kim, C. K. Park and K. Kim, *J. Am. Chem. Soc.*, 2005, **127**, 5006–5007.
134. X. L. Huang, Y. B. Tan, Q. F. Zhou and Y. X. Wang, *E-Polymers*, 2008, 11.
135. X. J. Loh, J. del Barrio, P. P. C. Toh, T. C. Lee, D. Z. Jiao, U. Rauwald, E. A. Appel and O. A. Scherman, *Biomacromolecules*, 2012, **13**, 84–91.
136. S. Angelos, N. M. Khashab, Y. W. Yang, A. Trabolsi, H. A. Khatib, J. F. Stoddart and J. I. Zink, *J. Am. Chem. Soc.*, 2009, **131**, 12 912–12 914.
137. J. S. Liu and X. Z. Du, *J. Mater. Chem.*, 2010, **20**, 3642–3649.
138. J. S. Liu, X. Z. Du and X. F. Zhang, *Chem. Eur. J.*, 2011, **17**, 810–815.
139. A. Hennig, H. Bakirci and W. M. Nau, *Nat. Methods*, 2007, **4**, 629–632.
140. J. Croissant and J. I. Zink, *J. Am. Chem. Soc.*, 2012, **134**, 7628–7631.
141. C. R. Thomas, D. P. Ferris, J. H. Lee, E. Choi, M. H. Cho, E. S. Kim, J. F. Stoddart, J. S. Shin, J. Cheon and J. I. Zink, *J. Am. Chem. Soc.*, 2010, **132**, 10623–10625.
142. K. M. Park, K. Suh, H. Jung, D. W. Lee, Y. Ahn, J. Kim, K. Baek and K. Kim, *Chem. Commun.*, 2009, 71–73.
143. J. Kim, Y. Ahn, K. M. Park, Y. Kim, Y. H. Ko, D. H. Oh and K. Kim, *Angew. Chem. Int. Ed.*, 2007, **46**, 7393–7395.
144. S. K. Kim, K. M. Park, K. Singha, J. Kim, Y. Ahn, K. Kim and W. J. Kim, *Chem. Commun.*, 2010, **46**, 692–694.
145. E. Kim, D. Kim, H. Jung, J. Lee, S. Paul, N. Selvapalam, Y. Yang, N. Lim, C. G. Park and K. Kim, *Angew. Chem. Int. Ed.*, 2010, **49**, 4405–4408.
146. K. M. Park, D. W. Lee, B. Sarkar, H. Jung, J. Kim, Y. H. Ko, K. E. Lee, H. Jeon and K. Kim, *Small*, 2010, **6**, 1430–1441.
147. H. Jung, K. M. Park, J. A. Yang, E. J. Oh, D. W. Lee, K. Park, S. H. Ryu, S. K. Hahn and K. Kim, *Biomaterials*, 2011, **32**, 7687–7694.
148. P. Montes-Navajas, M. Gonzalez-Bejar, J. C. Scaiano and H. Garcia, *Photochem. Photobiol. Sci.*, 2009, **8**, 1743–1747.
149. G. Hettiarachchi, D. Nguyen, J. Wu, D. Lucas, D. Ma, L. Isaacs and V. Briken, *PLoS One*, 2010, **5**, e10514.
150. V. D. Uzunova, C. Cullinane, K. Brix, W. M. Nau and A. I. Day, *Org. Biomol. Chem.*, 2010, **8**, 2037–2042.
151. A. I. Day and J. G. Collins, in *Supramolecular Chemistry: From Molecules to Nanomaterials*, P. A. Gale and J. W. Steed, 2012, vol. 3 , pp. 983-1000.
152. J. A. Plumb, B. Venugopal, R. Oun, N. Gomez-Roman, Y. Kawazoe, N. S. Venkataramanan and N. J. Wheate, *Metallomics*, 2012, **4**, 561–567.
153. N. J. Wheate, P. G. A. Kumar, A. M. Torres, J. R. Aldrich-Wright and W. S. Price, *J. Phys. Chem. B*, 2008, **112**, 2311–2314.

154. F. Constabel and K. E. Geckeler, *Tetrahydron lett.*, 2004, **45**, 2071–2073.
155. S. Walker, R. Kaur, F. J. McInnes and N. J. Wheate, *Mol. Pharm.*, 2010, **7**, 2166–2172.
156. A. R. Urbach and V. Ramalingam, *Isr. J. Chem.*, 2011, **51**, 664–678.
157. M. E. Bush, N. D. Bouley and A. R. Urbach, *J. Am. Chem. Soc.*, 2005, **127**, 14 511–14 517.
158. P. Rajgariah and A. R. Urbach, *J. Inclusion Phenom. Macrocyclic Chem.*, 2008, **62**, 251–254.
159. J. Lagona, B. D. Wagner and L. Isaacs, *J. Org. Chem.*, 2006, **71**, 1181–1190.
160. O. Danylyuk and V. P. Fedin, *Cryst. Growth Des.*, 2012, **12**, 550–555.
161. W. H. Huang, P. Y. Zavalij and L. Isaacs, *Angew. Chem. Int. Ed.*, 2007, **46**, 7425–7427.
162. S. M. Liu, C. Ruspic, P. Mukhopadhyay, S. Chakrabarti, P. Y. Zavalij and L. Isaacs, *J. Am. Chem. Soc.*, 2005, **127**, 15959–15967.
163. D. M. Bailey, A. Hennig, V. D. Uzunova and W. M. Nau, *Chem. Eur. J.*, 2008, **14**, 6069–6077.
164. H. Cong, L. L. Tao, Y. H. Yu, F. Yang, Y. Du, S. F. Xue and Z. Tao, *Acta Chimica Sinica*, 2006, **64**, 989–996.
165. F. Biedermann and O. A. Scherman, *J. Phys. Chem. B*, 2012, **116**, 2842–2849.
166. Y. H. Ling, W. Wang and A. E. Kaifer, *Chem. Commun.*, 2007, 610–612.
167. F. Biedermann, U. Rauwald, M. Cziferszky, K. A. Williams, L. D. Gann, B. Y. Guo, A. R. Urbach, C. W. Bielawski and O. A. Scherman, *Chem. Eur. J.*, 2010, **16**, 13716–13722.
168. J. M. Yi, Y. Q. Zhang, H. Cong, S. F. Xue and Z. Tao, *J. Mol. Struct.*, 2009, **933**, 112–117.
169. H. J. Buschmann, E. Schollmeyer and L. Mutihac, *Thermochim. Acta*, 2003, **399**, 203–208.
170. S. W. Heo, T. S. Choi, K. M. Park, Y. H. Ko, S. B. Kim, K. Kim and H. I. Kim, *Anal. Chem.*, 2011, **83**, 7916–7923.
171. M. V. Rekharsky, H. Yamamura, C. Inoue, M. Kawai, I. Osaka, R. Arakawa, K. Shiba, A. Sato, Y. H. Ko, N. Selvapalam, K. Kim and Y. Inoue, *J. Am. Chem. Soc.*, 2006, **128**, 14871–14880.
172. M. V. Rekharsky, H. Yamamura, Y. H. Ko, N. Selvapalam, K. Kim and Y. Inoue, *Chem. Commun.*, 2008, 2236–2238.
173. L. M. Heitmann, A. B. Taylor, P. J. Hart and A. R. Urbach, *J. Am. Chem. Soc.*, 2006, **128**, 12574–12581.
174. J. M. Chinai, A. B. Taylor, L. M. Ryno, N. D. Hargreaves, C. A. Morris, P. J. Hart and A. R. Urbach, *J. Am. Chem. Soc.*, 2011, **133**, 8810–8813.
175. K. Sladkova, J. Houska and J. Havel, *Rapid Commun. Mass Spectrom.*, 2009, **23**, 3114–3118.
176. W. H. Lei, G. Y. Jiang, Q. X. Zhou, B. W. Zhang and X. S. Wang, *Phys. Chem. Chem. Phys.*, 2010, **12**, 13255–13260.
177. H. D. Nguyen, D. T. Dang, J. L. J. van Dongen and L. Brunsveld, *Angew. Chem. Int. Ed.*, 2010, **49**, 895–898.

178. R. N. Dsouza, A. Hennig and W. M. Nau, *Chem. Eur. J.*, 2012, **18**, 3444–3459.
179. W. M. Nau, G. Ghale, A. Hennig, H. Bakirci and D. M. Bailey, *J. Am. Chem. Soc.*, 2009, **131**, 11558–11570.
180. G. Ghale, V. Ramalingam, A. R. Urbach and W. M. Nau, *J. Am. Chem. Soc.*, 2011, **133**, 7528–7535.
181. T. Minami, N. A. Esipenko, B. Zhang, L. Isaacs, R. Nishiyabu, Y. Kubo and P. Anzenbacher, *J. Am. Chem. Soc.*, 2012, **134**, 20021–20024.
182. R. de la Rica and A. H. Velders, *Small*, 2011, **7**, 66–69.
183. Y. B. Lim, T. Kim, J. W. Lee, S. M. Kim, H. J. Kim, K. Kim and J. S. Park, *Bioconjugate Chem.*, 2002, **13**, 1181–1185.
184. S. G. Sun, Y. X. He, Z. G. Yang, Y. Pang, F. Y. Liu, J. L. Fan, L. C. Sun and X. J. Peng, *Dalton Trans.*, 2010, **39**, 4411–4416.
185. W. H. Huang, P. Y. Zavalij and L. Isaacs, *Org. Lett.*, 2009, **11**, 3918–3921.
186. D. W. Lee, K. M. Park, M. Banerjee, S. H. Ha, T. Lee, K. Suh, S. Paul, H. Jung, J. Kim, N. Selvapalam, S. H. Ryu and K. Kim, *Nat. Chem.*, 2011, **3**, 154–159.
187. J. F. Young, H. D. Nguyen, L. T. Yang, J. Huskens, P. Jonkheijm and L. Brunsveld, *ChemBioChem*, 2010, **11**, 180–183.
188. I. Hwang, K. Baek, M. Jung, Y. Kim, K. M. Park, D. W. Lee, N. Selvapalam and K. Kim, *J. Am. Chem. Soc.*, 2007, **129**, 4170–4171.
189. K. M. Park, J. A. Yang, H. Jung, J. Yeom, J. S. Park, K. H. Park, A. S. Hoffman, S. K. Hahn and K. Kim, *ACS Nano*, 2012, **6**, 2960–2968.

Nucleic Acids as Supramolecular Targets

ENRIQUE GARCÍA-ESPAÑA,*[a] IVO PIANTANIDA*[b] AND HANS-JÖRG SCHNEIDER*[c]

[a] Universitat de València, Instituto de Ciencia Molecular, 46980 Paterna (Valencia), Spain; [b] Rudjer Boskovic Institute HR-10002 Zagreb; [c] Universität des Saarlandes FR Organische Chemie, Saarbrücken, Germany
*Email: Enrique.Garcia-es@uv.es; pianta@irb.hr; ch12hs@rz.uni-sb.de

8.1 Introduction

Selective non-covalent interactions with DNA and RNA, including replication and transcription, are of central importance in life, and in themselves can be considered as supramolecular chemistry *in vivo*. Many drugs base their activity on interaction with DNA/RNA by interfering with natural and malign processes.[1] Nucleic acids are obvious targets for the exploration of supramolecular complexions, in particular as nucleic acids in contrast to proteins have regular and repeating structures. Owing to their anionic phosphate backbone, the obvious prerequisite for most ligands is the presence of cationic sites, usually in the form of protonated polyamines. Polyamines themselves are involved in many biological functions, linked with cell growth, survival and proliferation, as discussed in several books[2,3] and reviews.[4]

Cellular DNA is mostly present in double-stranded form and non-covalent interactions with smaller ligands occur preferentially in the DNA minor groove,[1,3,5] less often in the major groove,[6] or by intercalation between adjacent

Monographs in Supramolecular Chemistry No. 13
Supramolecular Systems in Biomedical Fields
Edited by Hans-Jörg Schneider
© The Royal Society of Chemistry 2013
Published by the Royal Society of Chemistry, www.rsc.org

base pairs[3] and through metal coordination to the bases;[7] examples focused on the anti-tumor platinum derivatives are discussed in Chapter 9 of this book. Selective targeting of RNA has gained particular attention following the growing importance of retroviruses, including antibiotics targeting RNA.[1,3]

The present chapter intends to highlight the interactions of nucleic acids with synthetic ligands, which were developed in the framework of supramolecular chemistry. These studies have allowed exploration of the underlying fundamental binding mechanisms, most often on the basis of *in vitro* measurements with model systems: synthetic polynucleotides, such as DNA alternating-(dAdT-dAdT; dGdC-dGdC) or homo- (dA-dT; dG-dC), and corresponding RNA sequences. A better understanding of small molecule binding modes to DNA/RNA will help in development of antitumor drugs[8] and markers for biomedical studies,[9] which base their activity on polynucleotide targeting.

To demonstrate the applications of supramolecular concepts on the research of small DNA-binding molecules we have chosen as a starting point the biologically very common family of polycationic aliphatic amines, which ensure that all genomes of eukaryotic organisms are tightly packaged into chromatin. A huge variety of DNA/RNA-targeting synthetic polyamine-based analogues exhibited in many cases improved and/or altered DNA–RNA interactions. Structural variations, in particular of macrocyclic amines, have led to unexpected activities such as base flipping, and to several new prototypes of high biological activity, holding promise for a variety of therapeutic applications. Increasingly important is the possible use of polyamines as non-viral vectors for gene therapy;[10] lipophilic or amphiphilic compounds such as polyethylene imine are particularly promising for transfection.[11]

Many host systems can be equipped with attached cationic sites to increase associations with nucleic acids, in particular with the aim of either targeting with drugs bound inside the host cavity, or for use as non-viral carriers for the transfection of bound nucleic acids. Cyclodextrins with amino group substituents are very efficient DNA binders.[12] For an extensive discussion of cyclodextrin-based systems we refer to Chapter 5 of this book. Intriguing new possibilities to control nucleic acids functions and to develop new drugs are found with supramolecular metal complexes, which also allow selective hydrolysis of natural nucleic acids. DNA-based nanotechnology can lead to exciting new medicinal and biotechnological applications. The present chapter cannot be comprehensive, but it is hoped that along with the leading references cited the reader will gain insight into a rapidly developing field of biomedical supramolecular chemistry addressing nucleic acids, with promising therapeutic and diagnostic or sensing applications.

8.2 General Binding Modes with Nucleic Acids

This section intends to provide the less-informed reader with some basic information and references on the principles which rule both formation of the relevant conformations of nucleic acids and their non-covalent binding modes with smaller, often synthetic, ligands.[3] Covalent reactions such as those with alkylating agents or with many cisplatin drugs (see Chapter 9) are generally irreversible; they inhibit transcription, translation, and replication of DNA and are often used as anticancer drugs.

8.2.1 Groove Binding

The physiologically most relevant B polymorph of DNA bears base pairs approximately orthogonal to the helix and has a major groove width of 11.6 Å and depth of 8.5 Å, with multiple interaction sites and a relatively strong binding to small ligands, including bulky molecules.[13] The minor groove, with a width of 6.0 Å, is much smaller but has a depth of 8.2 Å—still relatively deep—and exhibits a higher density of negative charge, particularly in AT-rich sequences. Many small molecules, such as distamycin, netropsin or Hoechst 33528, for example, bind to the minor groove, primarily through a set of hydrogen bonds with base pairs and the sugar parts of the DNA backbone; the molecules also exhibit additional interactions by ion pairing between positively charged nitrogen centers in the drugs, and the DNA rim phosphates.[3,5] The binding is sterically controlled by a convex shape of ligands, which match the bending of the DNA groove. The often-found selectivity to AT as opposed to GC pairs is due to steric hindrance of guanine amino groups in the minor groove (Figure 8.1).[13] An example with distamycin is shown in Figure 8.2.[14] It contains a positively charged amidino group and three pyrrole rings linked by amide bonds; their amide N-H protons are involved in bifurcated hydrogen bonds to N3 of adenines and O2 of thymines.

Folded RNAs exhibit more divergent structures than DNAs, including helical duplexes, bulges, loops, knots and triple-strands.[15] Targeting of RNA with small ligands has become important due to the increasingly recognized relevance of retroviruses and the key role of RNA in gene replication and expression.,[16] Ligands with cationic charge centers in a scaffold of limited flexibility bind preferentially to RNA, most likely within the major groove; for instance, RNA-selective aminoglycoside antibiotics contain several

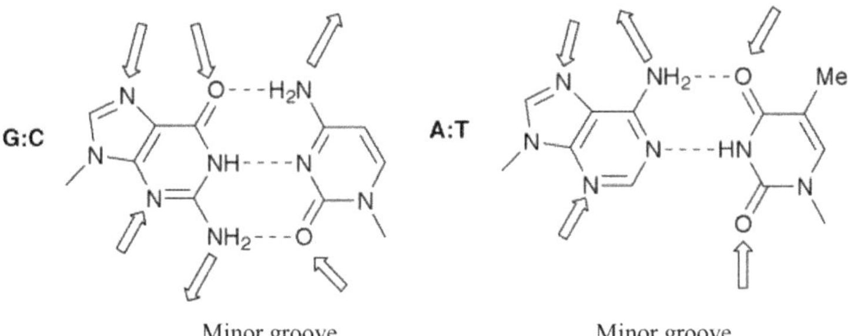

Figure 8.1 Potential hydrogen bond acceptor and donor sites at the edges of A:T and G:C Watson–Crick base pairs; directionality of hydrogen bonds indicated by the arrows.
Reproduced by permission of The Royal Society of Chemistry, ref. 13a.

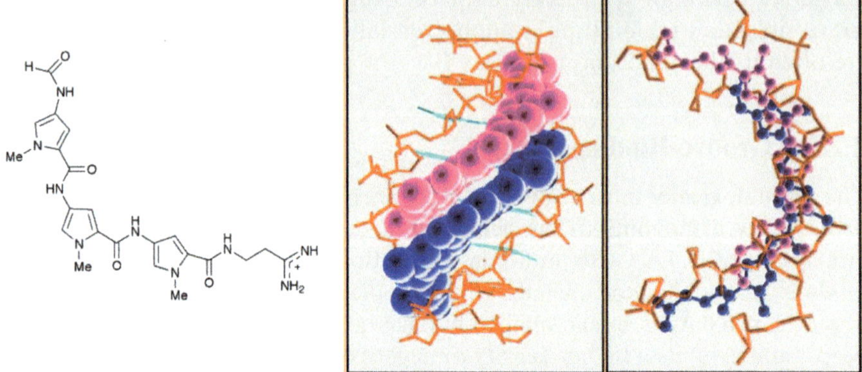

Figure 8.2   A 2 : 1 distamycin · complex with the DNA model octanucleotide d(ICA-TATIC)$_2$, in an expanded minor goove in order to accommodate two drugs side-by-side ("thickness" of one drug: 3.4 Å). The drugs are shown as a space-filled model, the DNA as a stick model; all bases are omitted for clarity.
(Reprinted with permission from Elsevier, ref. 14a.)

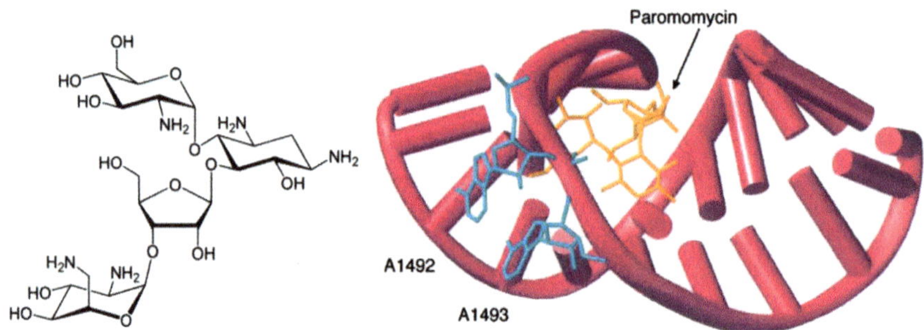

Figure 8.3   RNA binding, NMR-derived structure: A-region, 16S rRNA; the antibiotic paromomycin inside major groove, adenines A1492 and A1493 (blue) in minor groove.
(Reproduced with permission from VanLoock *et al.*[17] *J. Mol. Biol.* 1999, **285**, 2069.)

amino-sugars with ammonium centers, and exhibit polar contacts with RNA backbone and nucleobases in the major groove (Figure 8.3).[17]

8.2.2   Intercalation

Planar aromatic systems, like 9-aminoacridine or ethidium bromide, preferably inserted between base pairs.[3] They severely deform the DNA double-helix, leading to partial unwinding and helix elongation (Figure 8.4). Intercalators interfere in this way with replication and transcription, and can therefore act as

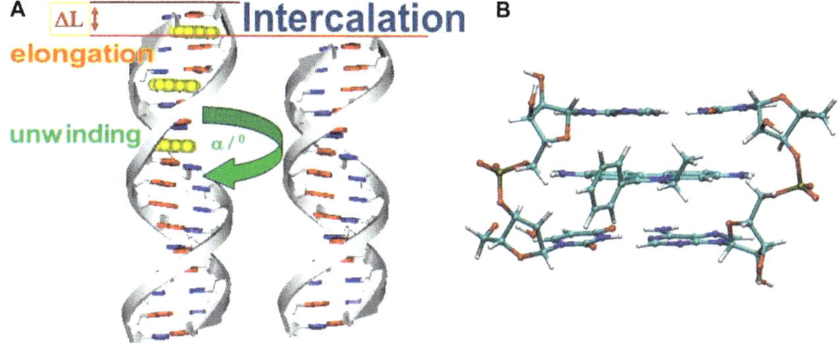

Figure 8.4 (A) *Left*: Planar aromatic moiety (yellow) inserted between every second set of base pairs (neighbor exclusion principle) causes deformation of double-stranded (ds) helix compared to free DNA/RNA (*right*). (B) X-ray structure of an intercalator–ds-dinucleotide complex. ΔL = length increase.

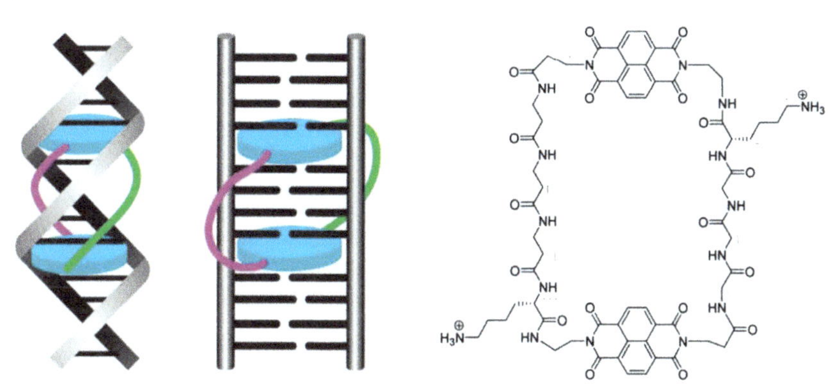

Figure 8.5 Bis-intercalation with a bis-naphthalene diimide.
(Reprinted with permission from Chu *et al.*[21] *J. Am. Chem. Soc.*, 2009, **131**, 3499. Copyright 2009 American Chemical Society.)

antitumor agents, but also as mutagenic compounds. Square-planar metal complexes are also very effective intercalators;[18] corresponding platinum complexes are discussed in Chapter 9. Usually there is no base or sequence selectivity, in line with a predominantly dispersive binding mechanism.[19] The affinity of intercalators depends primarily on the size of the aromatic surface, to a minor extent influenced by the presence of heteroatoms within the arenes.[20]

8.2.3 Bis-intercalation

A simultaneous insertion of two aryl units can occur, with subsequent enhanced affinity, if the chain connecting the aryl units is long enough; otherwise the so-called next-neighbor exclusion prohibits simultaneous insertion between closer nucleobases which would lead to too heavy helix distortions. A specific

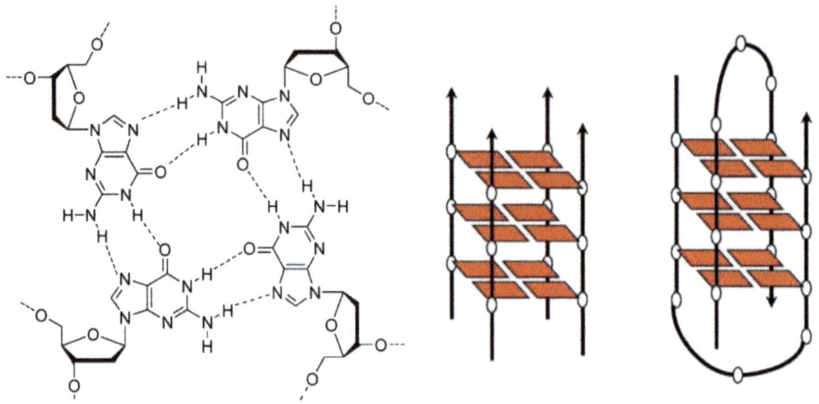

Figure 8.6 Quadruplex formation from four guanines: (A) tetramolecular with parallel strands; (B) bimolecular with diagonal loops.
Reproduced by permission of The Royal Society of Chemistry, ref. 22a.

case of threading bis-intercalation (Figure 8.5) demonstrates the dynamic nature of ds-DNA/RNA, whereby continuous opening and closing of the double strand (ds) allows insertion of cyclic or sterically hindered molecules in otherwise impossible ways.[3,21]

8.2.4 Quadruplexes

Quadruplexes are found for instance, in telomers of DNA and consist most often of four strands of stacked guanine (G) tetrads held together by Hoogsteen hydrogen bonds (Figure 8.6).[22] Suitable ligands can stabilize the quadruplex structure, inhibiting in that way telomerase activity; this can result in a number of biologically relevant effects on which some anticancer strategies are based (see Section 8.6).

8.3 Biogenic Polyamine Interactions with DNA/RNA

The biogenic polyamines spermine ($H_2N(CH_2)_3NH(CH_2)_4NH(CH_2)_3NH_2$), putrescine ($NH_2(CH_2)_4NH_2$) and spermidine ($NH_2(CH_2)_3NH(CH_2)_4NH_2$) are part of the polyamine metabolic pathway and have been studied in much detail in combination with nucleic acids. Generally, there seems to be no single way of interaction.[4a] Crystallographic analyses of complexes formed between biogenic polyamines and double-stranded DNA oligonucleotides show the expected electrostatic interactions with the phosphate groups, but also some contacts with the base residues. NMR analyses in solution, however, show that spermine has a high mobility in DNA complexation, still allowing some base selectivity. Spermine was found to induce a B-to-Z conformational transition on GC-rich oligonucleotides, with a lower affinity for AT-rich regions.[23] At variance to the above-mentioned is the finding that spermine can differentiate between DNA sequences, with preferential stabilization of A-tracts.[24] Comparative studies of

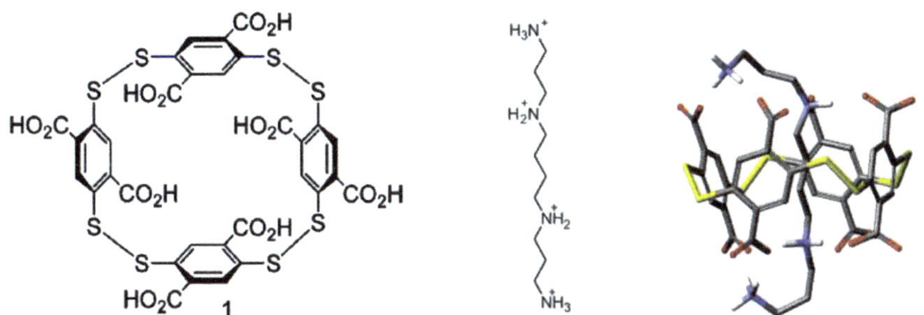

Figure 8.7 A macrocyclic receptor (**1**) which binds spermine inside the cavity. (Reprinted with permission from *J. Am. Chem. Soc.*, 2006, **128**, 10 253. Copyright 2006 American Chemical Society.)

different oligoamines showed that at low polycation concentrations, putrescine binds preferently at the minor and major grooves of ds-DNA, whereas spermine and spermidine bind to the major groove. At high polycation concentrations, putrescine interaction with the bases is weak, whereas strong base binding occurred for spermidine in the major and minor grooves of DNA duplex, at variance to major groove binding preferred by spermine cations.[25] Computational approaches suggested that the binding of spermine in the DNA major groove is limited to GC-rich regions, with a preferential minor groove binding in alternating AT sequences.

An intriguing approach to control binding of natural polyamine was demonstrated with the macrocyclic receptor **1**; the multiple negative charges bind spermine, a possible target for cancer treatment, inside the cavity (Figure 8.7) with a dissociation constant of 22 nM, an affinity higher than that of free spermine with DNA.[26] The ability of the receptor to remove DNA-bound spermine was confirmed by demonstration of the change of left-handed DNA conformation as induced by spermine back to the right-handed form upon addition of the receptor.

8.4 Synthetic Linear Polyamines

Investigations with synthetic linear polyamines provide a better understanding of the interaction with DNA/RNA, based on different nature, number and various locations of the positively charged centers. The correlation of many polyamine affinities, measured by the fluorescence competition assay with ethidium bromide, for example, shows a linear dependence on the number of possible salt bridges. Three biogenic amines, *viz.* putrescine (**2**), spermidine (**5**) and spermine (**9**), also follow this correlation (Figures 8.8 and 8.9).[27] The same correlation is observed with several hundreds of organic and inorganic ion pairs as expected, dependent on the ionic strength of the medium, with a surprisingly small dependence on the nature of the involved anions and cations.[28] The interaction between polyamines and calf thymus DNA exhibits, remarkably, the same correlation, yielding again a constant increment of $5 \pm 1 \, \text{kJ} \, \text{mol}^{-1}$ per

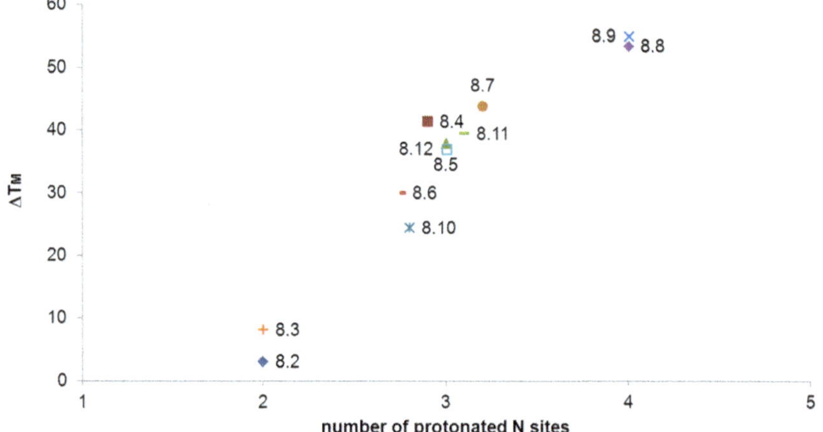

Figure 8.8 Selected polyamines, including putrescine (**2, 8.2**), spermidine (**5, 8.5**) and spermine (**9, 8.9**); compounds **2** to **12** refer to **8.2** to **8.12**, etc.

Figure 8.9 Plot of the ΔTm values versus the number of positive charges at pH = 6.25 for the interaction of a series of acyclic polyamines with polyA-polyU.

single salt bridge. Notably, permethylated analogues of spermine as well as of most macrocyclic amines follow the same correlation lines for DNA association, which allows prediction of polyamine affinities by applying the same free energy increments.[29] This does not exclude limited base selectivity, which requires only smaller free energy contributions in comparison to the stronger electrostatic interaction determining the total affinity.[30]

Synthetic modifications include also α-methylated polyamine analogues, which induce DNA condensation and sterical protection of cellular DNA from oxidative damage;[31] they are ubiquitous cellular cations and exert multiple biological functions. Very close polyamine analogues of putrescine, spermidine, and spermine mimic these biogenic polyamines at micromolar level but are unable to substitute for the natural polyamines in maintaining cell proliferation, which would be suggestive of possible biomedical application.[32]

8.5 Polyamine–Aryl Conjugates

The attachment of aromatic moieties, in particular heterocycles, to polyamines allows additional interactions in form of intercalation and hydrogen bonding. Biomedical implications of polyamine–aryl conjugates often point to more efficient cellular uptake, and consequently to increased biological activity in relation to alkyl amines. For instance, a series based on naphtalimido derivatives of the biogenic polyamines spermidine (**13**), spermine (**14**) and oxaspermine (**15**) (Figure 8.10) showed strong DNA binding and high *in vitro* cytotoxicity due to their preferential localization inside the nucleus as evidenced by fluorescence microscopy.[33]

In the same line, the drug F14512 (**17**; Figure 8.11), which is an etoposide derivative that contains a spermine in place of the C4 glycosidic moiety, turned out to be a more efficient topoisomerase II poison than etoposide (**16**).[34] The drug was designed to exploit the **polyamine transport system** (PTS) which is upregulated in several types of cancers. The presence of the spermine moiety enhances the binding affinity of the drug for DNA with respect to etoposide. The spermine–drug core linkage is critical for these attributes, yielding tighter binding and an increased stability of the ternary topoisomerase II–drug–DNA complex.[34]

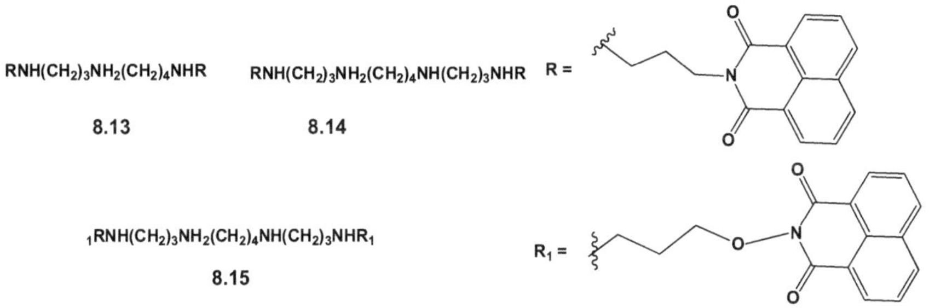

Figure 8.10 Naphthalimido derivatives of biogenic polyamines, including spermidine (**13**) and spermine (**14**).

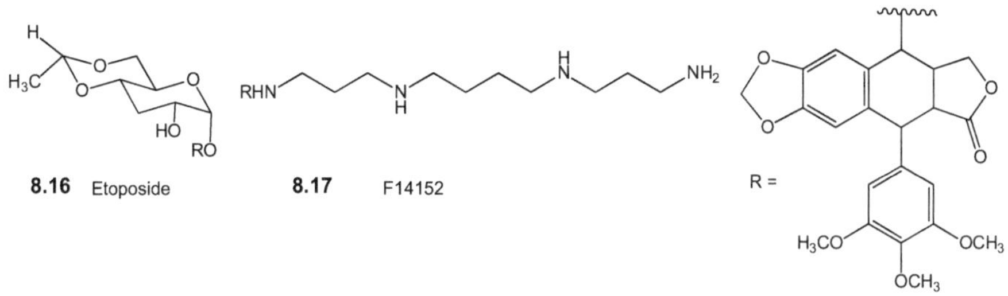

Figure 8.11 Etoposide (**16**) and the drug derivative F14152 (**17**).

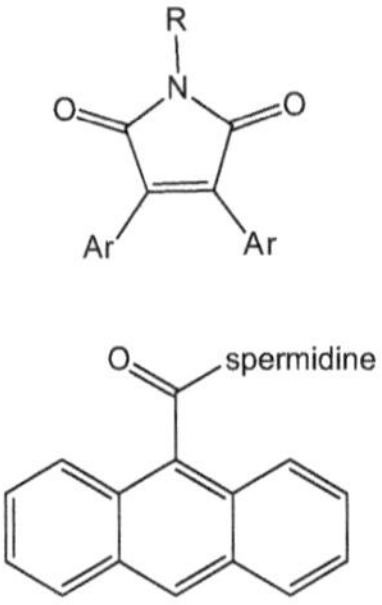

	Ar	R		Ar	R
8.18	3-indolyl	spermine	**8.22**	1-pyrenyl	spermine
8.19	phenyl	spermine	**8.23**	3-indolyl	spermidine
8.20	2-naphthyl	spermine	**8.24**	3-indolyl	tren
8.21	1-naphthyl	spermine	**8.25**	3-indolyl	en

8.26

Figure 8.12 Bisaryl ligands **18** to **25** and the reference compound **26**.

The hydrophilic/lipophilic balance has been shown to be important for the antiproliferative activity of open lactone camptothecin–polyamine conjugates.[35] Furthermore, the attachment of spermine chains to well-known DNA intercalators such as acridine, adds several specific characteristics, for example, selectivity towards alternating purine–pyrimidine oligodeoxyribonucleotides and spermine-induced aggregated structures for DNA packaging.[36] Another intriguing mode of large DNA aggregation is induced by pteridine–polyamine conjugates which, instead of the more common continuous DNA-shrinking process, caused a peculiar combination of discontinuous processes, resulting in specific formation of beads-on-a-chain structures on giant DNA.[37] Comparison of acridine[36] and pteridine–polyamine[37] conjugates with DNA aggregation systems caused by simple polyamines made evident that different kinds of interaction contribute to the DNA-folding process.

Spermine–bisaryl conjugates (Figure 8.12) were shown to be potent inducers of B- to Z-DNA transitions.[38] Among the tested ligands, only **20** showed a higher binding affinity for Z-DNA than for B-DNA. In this case, the bisaryl-maleimide part increases the binding affinity to Z-DNA by 20–30 times compared with spermine.[23] Thermodynamic analysis suggested that although the initial binding of **20** occurs through electrostatic interactions of the spermine-tail with DNA, the sterical features of the bisaryl-maleimide part give the major contribution to the B–Z transition. Such transition was shown to be both enthalpically and entropically favorable, in contrast to the B–Z transitions induced by $[Co(NH_3)_6]^{3+}$ or Na^+, which are only entropy-driven processes.

Many polyamine-linked bis-intercalators have been extensively studied and described in several reviews and books;[3,39] in general the polycationic linker gives the major contribution to the affinity of the compound towards DNA/RNA. The best-known systems are the cyanine-dimers TOTO (Thiazole Orange dimer) and the YOYO (Oxazole Yellow dimer) dye families, revolutionary DNA/RNA fluorimetric markers with sensitivity comparable to radioactive probes.[40]

The combination of aliphatic amine linkers with two phenanthridines acting as protonable DNA-intercalators, yielded pH modulated affinity of bis-phenanthridine towards DNA/RNA as well as pH-dependent specific fluorimetric response for poly G.[41] Intercalator–nucleobase conjugates were

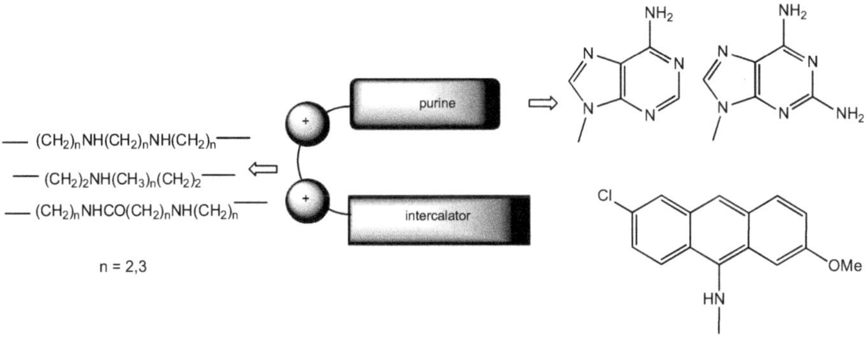

purine

intercalator

— (CH₂)ₙNH(CH₂)ₙNH(CH₂)ₙ—

— (CH₂)₂NH(CH₃)ₙ(CH₂)₂—

— (CH₂)ₙNHCO(CH₂)ₙNH(CH₂)ₙ—

n = 2,3

NH₂   NH₂   NH₂   Cl   OMe   HN

Figure 8.13 Example of intercalator–nucleobase conjugates designed for the DNA-abasic site recognition.[42,43]

designed to bind selectively to abasic sites of ds-DNA complementary to nucleobases of the conjugate.[42] One of the most extensively studied groups of such compounds is characterized by oligoamine linkers between the intercalator and the nucleobase (Figure 8.13), in which aliphatic amines have a dual role: when positively charged by protonation they form additional electrostatic interactions with DNA phosphates, while in neutral form at basic conditions they catalyze β-elimination of the DNA backbone.

Very recently, chiral perylene bisimide (PBI)–spermine conjugates were designed, which preferentially stack in the helical arrangements induced by the presence of a chiral bias. Such PBI–spermine conjugates formed dimeric aggregates within DNA/RNA grooves, revealing a spectroscopic response highly sensitive to the properties of the binding site (width, depth and binding motif of the groove). Such self-adjusting PBI dimers giving selective induced circular dichroism (CD) signals offer new opportunities for the dynamic sensing of DNA/RNA secondary structure.[44]

8.6 Interactions of Polyamine and Polyamine-Conjugates with DNA G-Quadruplexes

A large group of polyamine–aryl conjugates that use polyamines to control supramolecular organization are ligands targeting G-quadruplexes. These ligands have aromatic surfaces large enough to take advantage of stacking favorably with the large quadruplex-guanines, in comparison to smaller aromatic surfaces of base pairs in ds-DNA/RNA. Cationic side-chains, which sterically prevent intercalation into ds-DNA/RNA but do not hamper stacking on top or bottom of G-4 systems, also have beneficial impact on G-quadruplex targeting. Systematic studies with PBI–polyamine conjugates, named POL (**29** to **32**; Figure 8.14), revealed the crucial impact of polyamine side-chain charge density on the efficacy of the compounds to induce G-quadruplex structures, based on the sterically controlled interaction between cationic side-chains and phosphate backbone. Moreover, selectivity of some PBI–polyamine conjugates towards

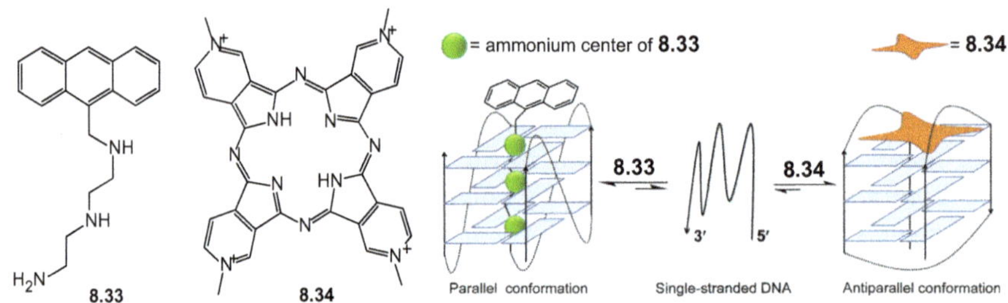

Figure 8.14 Representatives of the POL family of polyamine G-quadruplex targeting ligands.

Figure 8.15 Switching between parallel and antiparallel folding forms of telomeric G-quadruplex induced by ligands **33** and **34**.
(Adapted from Rodriguez *et al.*[46] with permission from Wiley VCH.)

G-quadruplex over ds-DNAs was also observed,[45] and attributed to aggregation of PBI dyes stacked upon each other within the ds-DNA groove, while the cationic chains for sterical reasons only partially interacted with DNA phosphates, as distinct from much better contact between polyamines and phosphates in the G-quadruplex complex.

The anthracene–polyamine conjugate (**33**) shown in Figure 8.15 revealed a unique mode of interaction with G-quadruplex, whereby the polycationic amine was not externally bound to DNA phosphates but instead was inserted through the central quadruplex channel, in that way mimicking K^+ ions.[46] Interestingly, the strength of metal ions located in the central channel of the G-quadruplex was estimated to decrease in the order $Li^+ > Na^+ > K^+$ reflecting the strength of cation–π effect in the quadruplex interior. Moreover, the authors suggested on the basis of the CD spectra the existence of a dynamic process of ligand-driven conformational switching of telomeric DNA quadruplex (telo24) between parallel or antiparallel folding forms as a result of the addition of the anthracene polyamine **33** or of the porphyrinor **34**.

Further applications of perylene-1,3-diaminopropane or spermine conjugates took advantage of the ability of aliphatic amines to react as nucleophiles with proximate guanines in the presence of one-electron oxidants to yield the oxodG sites, thus revealing novel opportunities in designing potential telomerase inhibitors or for probing the G-quadruplex topologies.[47]

A remarkable affinity to quadruplexes with a binding strength of about seven orders of magnitude was observed with the macrocycle BOQ1 (**35**) shown in Figure 8.16,[48] the affinity for duplex DNA was about 10 times less. In addition, the macrocycle was more selective than the monomeric control compound MOQ2 (**36**), which is not able to discriminate between double-stranded and quadruplex DNA. While fluorometric titrations supported the strong interaction of BOQ1 with G-quadruplex, competition dialysis experiments with a panel of different DNA structures confirmed the specificity of BOQ1 (**35**) for G-quadruplex DNA. The ΔTm value obtained by fluorescence resonance energy transfer (FRET) ($\Delta Tm = +28\,^{\circ}C$) was one of the highest values obtained for a G-quadruplex ligand. The macrocycle was shown also to induce a strong inhibition of telomerase ($IC_{50} = 0.13\,\mu M$).

In general metal-free planar heteroaromatic molecules have been proposed to interact with G-quartets. However, there are other features of G-quadruplexes that can be advantageously used to design quadruplex binders. For instance, interaction with phosphate backbone or with bases outside the quartets, as well as replacement of the alkali cations located within the central channel of the quadruplex by other metal ions or charged organic molecules are factors that can be exploited for this purpose.[49] In this respect, planar metal complexes can be interesting quadruplex binders.[50] First, the electron-withdrawing properties of metal centers reduces the electron density on coordinated aromatic systems affording electron-poor π systems, which are expected to stack more efficiently with the end G-quartets. On the other hand, metal ions can be placed at the center of the G-quartets with the alkali ions stabilizing by the cation–π effect the quadruplex structure. Apart from these features, the functionalization of the planar complexes with lateral chains that bear additional positive charges can be a good strategy to improve the binding through electrostatic interactions with the negative backbone in the loops or grooves of the quadruplex. For this purpose several strategies have been followed. The first one uses metalloporphyrin compounds with positively charged substituents at the meso positions. Such substituents can be polyamines as those depicted in Figure 8.17a.

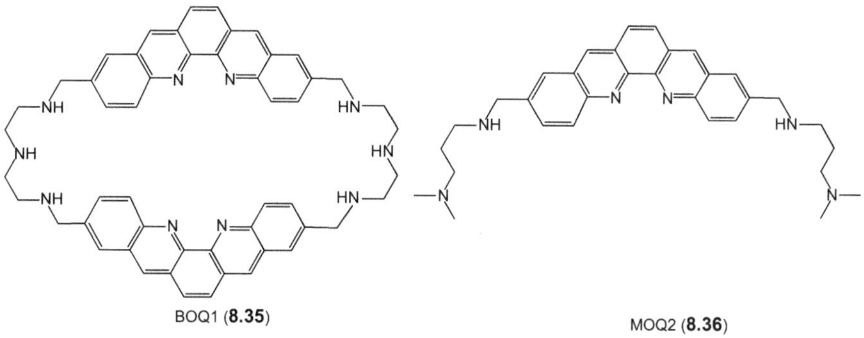

Figure 8.16 Polyazamacrocycle BOQ1 (**35**) and its related open-chain counterpart MOQ2 (**36**).

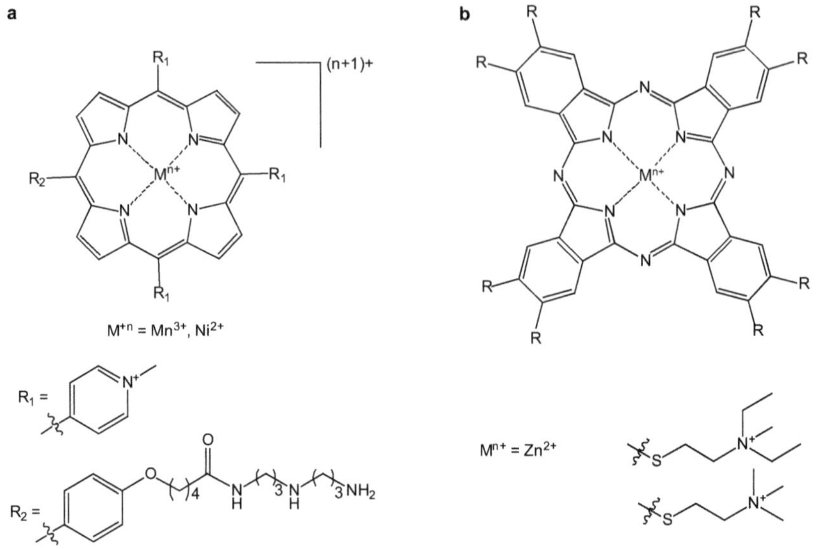

Figure 8.17   (a) Polyamine-modified metalloporphyrins and (b) phtalocyanine metal complexes.

Metallophtalocyanines have a more extended aromatic system which matches better the electronic and stereochemical requirements of the G-tetrads. Zn^{2+} and Ni^{2+} complexes of phtalocyanines showed improved binding affinities to the h-telo quadruplex relative to the simple porphyrin complex.

The phtalocyanines are generally functionalized with either four or eight side arms (Figure 8.17b) containing quaternary ammonium groups. In general, the larger the number of positive charges, the larger is the binding affinity for the G-quadruplexes and the greater the inhibition of the telomerase. Another class of planar macrocyclic rings used to achieve large affinities for G-quadruplex structures is trianionic corrole ligands, which are very effective for stabilizing high oxidation states of metal complexes and may provide additional geometric and electronic features. Compounds of this class, which have been designed to target h-telo and the G-rich sequences from the *c-myc* gene promoter region, are shown in Figure 8.18.

8.7 Selection between DNA and RNA Grooves/ Cyclophanes Interacting with DNA and RNA/ Unfolding of Double Strands

The deep major groove of double-stranded RNA with its large negative potential is the preferred binding site for many polyamines, including most aminoglycosides; for example, the well known RNA-binder antibiotic streptomycin increased the melting point of RNA model polyA-polyU by 6.9 °C, as compared to

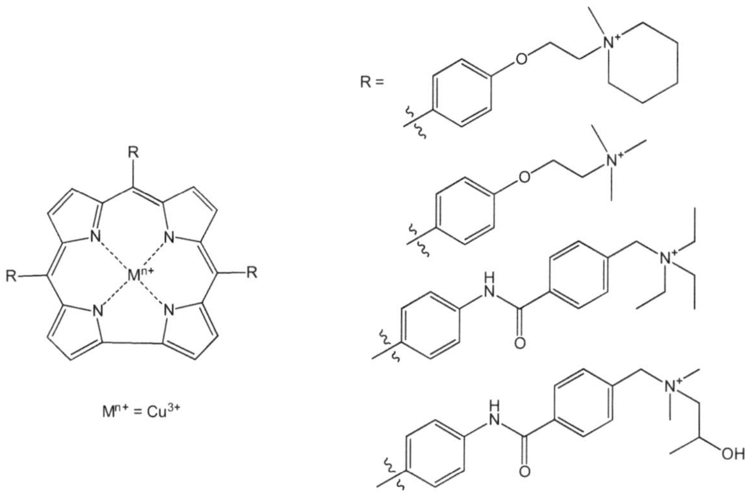

Figure 8.18 Corrole complexes.

0.3 °C with DNA-model poly(dA)-poly(dT); the antibiotic neomycin exhibits distinctly larger differences (see Figure 8.19). The affinities of antibiotic aminoglycosides depend mostly on the number of positively charged nitrogen centers, with additional weaker binding contributions like stacking and hydrogen bonding. The examples **37** to **49** in Figure 8.19 demonstrate that synthetic polyamines allow very distinct selectivity changes, surpassing those of natural binders. Thus, the melting point difference between the RNA and DNA models with ligand **45** amounts to $\Delta Tm = 42\,°C$, due to the large charge accumulation within a small space. Ligand **49** has been designed to exhibit the same large size as the binding part of the zinc finger protein, which is accommodated in the wider DNA groove; in consequence, a large affinity to ds-DNA is observed.[51]

Macrocyclic amines are able to distinguish between ds-RNA and ds-DNA, depending primarily on the number and position of arenes within the ligand. Thus, the macrocycle **38** shows a large affinity to the RNA model, in contrast to ligands such as **40** which bear several benzene units and exhibit an unusually large preference for DNA (Figure 8.19). In the same line the oxaaza macrocycle **39** which has no aryl groups shows preference for RNA.[52] However, in general, macrocycles are less efficient in increasing the ΔTm than are acyclic polyamines with the same number of positive charges in their structure. The azacyclophane **40**, bearing four benzene rings in rigid frameworks exhibits affinities which depend significantly on the ring size. The smaller macrocycles CP33 (**41**), CP44 (**42**) and CP55 (**43**) show, as do other related macrocycles, a moderate preference for DNA, whereas the larger CP66 (**44**) leads to a dramatic destabilization of ds-RNA.[53] The phosphates in the RNA groove match the cationic ammonium centers of the macrocycle (**44**) less efficiently than those in the larger B-DNA groove, as visible in modeling studies. Instead of the normal groove binding, the smaller RNA groove prefers the CP66 (**44**) ring by an intracavity inclusion of a nucleobase with concomitant base flipping, resulting in

Melting point changes, ΔT_m(°C)

		on DNA	on RNA
8.37	Neomycin	11	33
8.9	Spermine $H_2N(CH_2)_3NH(CH_2)_4NH(CH_2)_3NH_2$	17	40
8.38	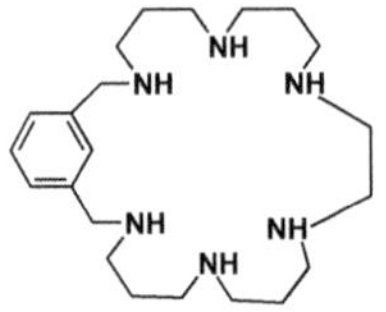	3	52
8.39	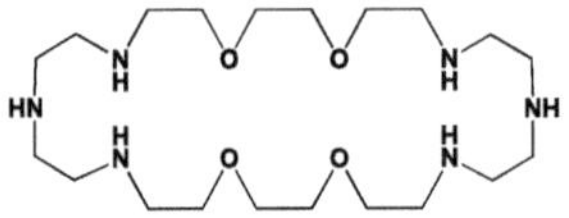	11.1	28.8
8.40	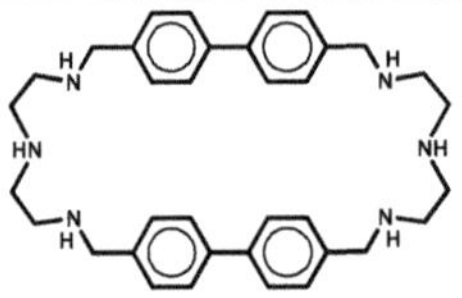	24	−6
8.41	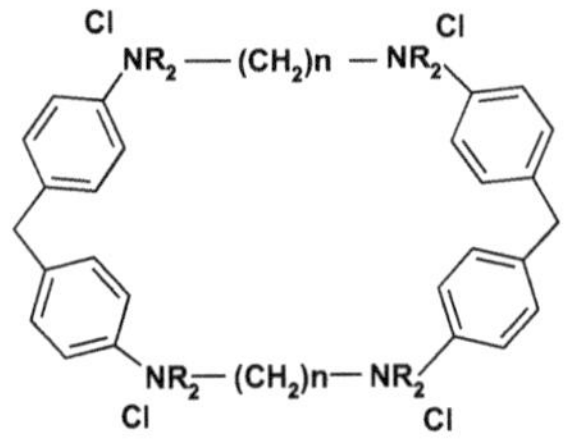 CP33 (n=3)	30	27
8.42	CP44 (n=4)	36	14
8.43	CP55 (n=5)	28	6
8.44	CP66 (n=6)	27	−6
8.45	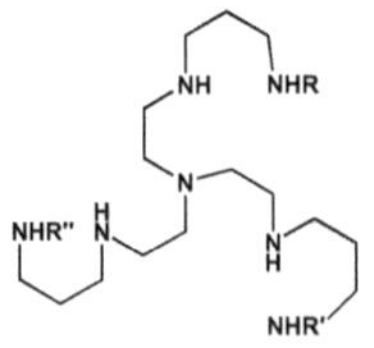	1	43

Figure 8.19

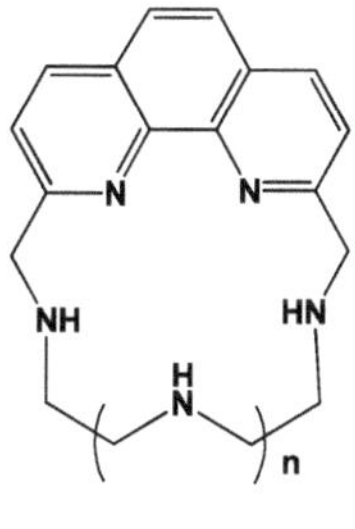

8.46	a (n=1)	12	9
8.47	b (n=2)	23	13
8.48	c (n=3)	27	14
8.49		−2	14

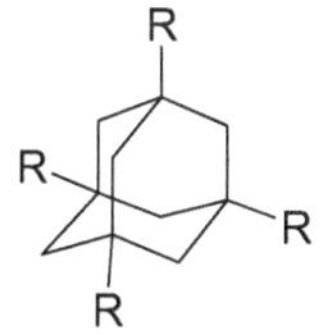

R = OOC– CH$_2$$^+NMe_2$—

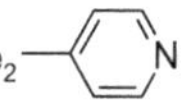

Figure 8.19 Examples of RNA/DNA-selective polyamines; in **8.41** to **8.45** : R = Me. Melting point changes, ΔTm, with polyA-polyU as RNA model and with poly(dA)-poly(dT) as DNA model; charges at N atoms omitted. Measured usually at ligand-to-nucleic-acid ratio $r = 0.3$. Source: from refs. 27, 51 and 52.

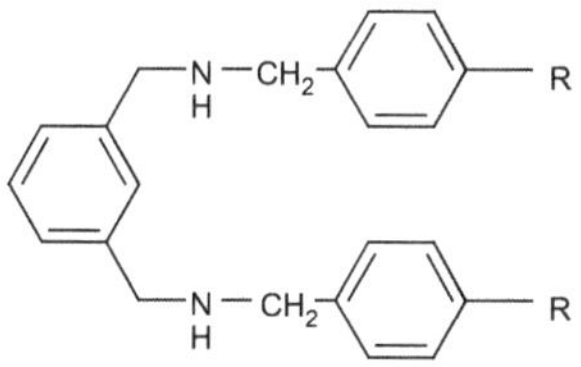

Figure 8.20 Ligand for unwinding of double strands by intercalation with a single strand (R = Cl,Br, H, etc.)[20]

unwinding of the RNA helix. Circular dichroism (CD) spectra show with CP66 (**44**) and RNA, but not with DNA denaturation. NMR-spectroscopy supports that nucleobase inclusion occurs only with the RNA model. Such base-flipping has been known until now to be controlled by corresponding enzymes, including some methyltransferases, glycosylases, and photolyases.[54] Similarly, macrocycles **46** to **48** containing a phenanthroline unit show small ΔTm value differences between DNA and RNA, or even preference for DNA, suggesting intercalation of the phenanthroline ring.[52]

Unwinding of duplexes can also be achieved with cleft-like amines such as those in Figure 8.20, which are designed to intercalate only with single-stranded nucleic acids. Intercalation in double strands requires the presence of larger aromatic units such as purine, naphthalene, etc.[20] The ligands (Figure 8.20), performing as artificial helicases, bear only phenyl groups and can therefore complex only single strands; they cause lowering of the calf-thymus melting by for instance up to 28 °C (with R = Cl), depending on the ligand-to-nucleic-acid ratio.[55] Such ligands can be the basis of potential antiviral and anticancer drugs and can also be useful for the polymerase chain reaction, in which dissociation of folded nucleic acid is a necessary step, usually requiring the use of heat-resistant enzymes.

8.8   Interaction of Polyamine Calixarenes with DNA/RNA

Calixarenes bearing amino groups,[56] which are also discussed in Chapter 6, have the advantage that they interact selectively with the nucleic acid backbone by virtue of their relatively rigid (from a conformational viewpoint), pre-organized cationic substituents at the calix rim; they can be equipped with additional groups (*e.g.* hydrogen bonding moieties) for the sake of selective DNA recognition or with lipophilic groups with the aim of self-assembly towards larger aggregates.[57]

The cone-shaped calix[4]arene (**50**) and the calix[4]1,3-alternate (**51**) form bearing four permethylammonium groups at the upper rim (Figure 8.21)[58] bind nucleotides and nucleosides with differences of $5 \pm 1\,\mathrm{kJ\,mol^{-1}}$ for one salt bridge, indicating a predominant ion-pairing mechanism. Larger macrocycles show only a moderate increase in affinity, but at the same time a moderately increased nucleobase discrimination. The role of calixarenes in most applications is not based on typical intracavity host–guest complexation, but rests on the possibility to arrange both the cationic polyamine groups with the calix cone in a narrow space on one side of the macrocycle, and other kinds of group, *e.g.* lipophilic chains, on the opposite side. Association with double-stranded nucleic acids exhibits again larger binding with the four cationic groups in the

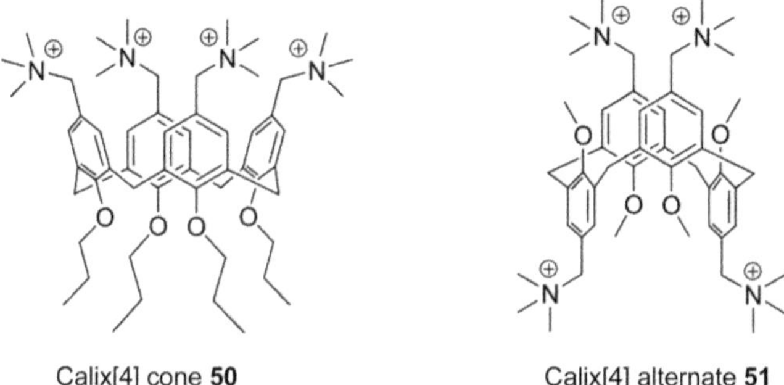

Figure 8.21   Calix[4]arenes bearing quaternary amine side-chains.

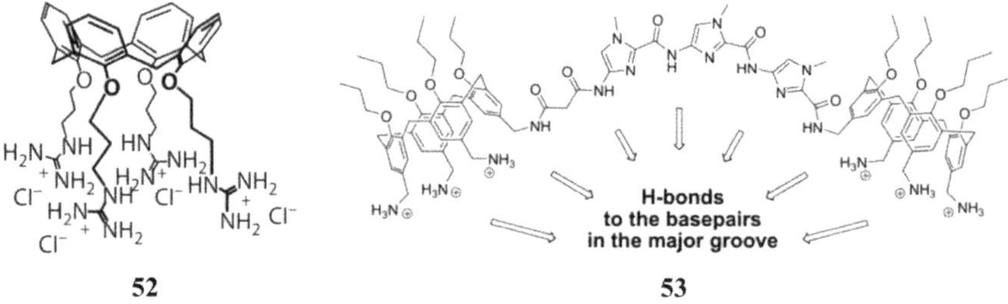

Figure 8.22 Calix[4]arenes bearing guanidinium groups (**52**); double amine calix[4]-arenes interconnected by a *N*-methyl imidazole linker (**53**).

cone form, **50**, with a distinct and preference for the RNA model polyA-polyU. Both the calix[4] alternate form, **51,** and the larger calixes show smaller affinities to the nucleic acid duplexes.

Calixarenes substituted with guanidinium groups instead of amino groups also bind efficiently to nucleic acids, but have the advantage of permanent positive charges over a larger pH range (Figure 8.22).[56] The guanidinium-calixarene **52** is a highly efficient and remarkably non-toxic vector if used in combination with a helper lipid, such as 1,2-dioleoyl phosphatidylethanolamine.[56b]

Calixarene dimers with protonated amino groups at the upper rim connected over a suitable flexible bridging unit are promising candidates for selective multivalent associations with nucleic acids. The *N*-methyl pyrrole and *N*-methyl imidazole units in the bridge of **53** were designed to bind into the major groove of DNA, with hydrogen bonds towards the base pairs by their amide protons and the imidazole nitrogen lone pairs. Indeed, a distinct preference for double-stranded polydG-polydC duplexes was found, with a nanomolar affinity.[57]

Multicalixarenes such as **54** (Figure 8.23) may also be used as vectors (see Section 8.9) but only when R1 is a fully protonated aliphatic amine at pH 7 (glycine residue with an NH_3^+ group).[59] Amphiphilic cationic calixarenes (**55**) self-assemble at very low critical micelle concentrations, and expose their cationic sites to the micelle surface. With DNA, single plasmid DNA nanoparticles are formed which exhibit promising transfection, condensing DNA into small nanoparticles of about 50 nm diameter; they have at the same time a low cytotoxicity.[60]

Dimeric calixarenes decorated with various aliphatic, aromatic amines and guanidines also showed quite rare binding to the ds-DNA major groove.[61] The molecular modeling designed on the basis of experimental evidence for 1:1 complexes between calixarene dimers and double-stranded oligonucleotides (20 base pairs) revealed that the ligands' ammonium and guanidinium moieties form a network of hydrogen bonds with acceptors at the Hoogsteen sites of the DNA base pairs, while the butoxy tails point away from the DNA molecule. While shorter ammonium dimers bridge up to six base pairs and form hydrogen bonds with their nucleic bases, the extended cationic fingers of the guanidinium dimers bridge up to nine base pairs. The latter group of compounds was even

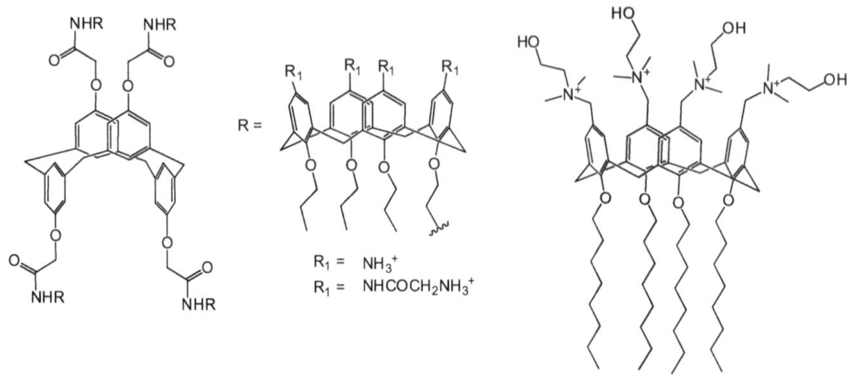

8.54

8.55

Figure 8.23 Multicalixarenes (**54**), self-assembling amphiphilic calixarenes (**55**).

able to stretch over the complete width of the major groove to effect interactions with the opposite phosphodiester backbone anions; these additional interactions explain their superior DNA affinities. Intriguingly, the 4-carbon alkyl bridge between both calix[4]arene heads is located at approximately 3–4 Å distance from the major groove floor, allowing additional nucleobase recognition. Moreover, additional ammonium substituents attached to the alkyl bridge are predicted to further establish favorable hydrogen bond contacts to DNA bases and phosphates.

8.9 Supramolecular Gene Delivery Systems

The simplest polyamine-based concept of selective delivery of drugs to tumor cells is based on the polyamine transport system (PTS), which is an energy-dependent machinery frequently overexpressed in cancer cells with a high demand for polyamines. The high specific activity of the PTS in tumor cells is thought to be associated with the inability of biosynthetic enzymes to provide sufficient levels of polyamines to sustain rapid cell division. These bioproduction constraints are partially offset by scavenging polyamines from exogenous sources and many tumor types have been shown to contain high levels of polyamines resulting from an active PTS importing exogenous polyamines. An active PTS has been characterized in a large number of tumor cell lines.[62,63]

A good example of a selectively delivered polyamine-containing drug to cancer cells consisted of epipodophyllotoxin core-targeting topoisomerase II with a spermine moiety introduced as a cell delivery vector (see Figure 8.11 in section 8.4).[34] The polyamine tail actually had multiple functions: increased water solubility of otherwise hydrophobic compound; increased affinity toward DNA by electrostatic interactions; and facilitated selective uptake by tumor cells via the PTS.[63]

8.9.1 Dendron-Based Polyamine Structures

Among dendrons with different amine surface groups, spermine derivatives were shown to be the most effective DNA binders, while the *N*,*N*-di-(3-aminopropyl)-*N*-(methyl)amine (DAPMA)-functionalized dendrons were, although still modest, the most effective systems for gene delivery.[64] Further research in that field led to molecular dynamics analysis of dendritic molecules with spermine surface groups and their interactions with double-helical DNA, aiming to a better understanding of DNA-cell delivery. The proposed self-assembly model explained why the first generation of dendrons (G1) is more effective in binding DNA than the second generation analogue (G2, Figure 8.24), due to the compensation effect of the latter in which some spermine chains "sacrifice" themselves and screen the complex, which enables the other spermine residues to bind more effectively to DNA.[65]

However, the extremely high affinity of dendritic polyamines towards DNA may be a drawback for *in vivo* applications, for example in gene therapy, since such almost irreversible complexes hamper the transfection of DNA in cells. One of the efficient ways to circumvent this problem is the design of degradable vectors, in which degradation of a multivalent array into smaller units, for instance individual monovalent ligands, causes dramatic decrease of affinity, meaning that the ligands can no longer intervene in the biological pathways. In line with this hypothesis new dendron–spermine analogues have been prepared, and efficient spontaneous degradation of multivalent interactions demonstrated under biological conditions (Figure 8.25), including the time scale relevant for the successful delivery to DNA before degradation takes place.[66]

The "combinatorial library approach" to gene delivery relies on supramolecular self-assembly of individual components by random incorporation of various hydrophobic and hydrophilic domains to yield candidates with the appropriate balance of DNA binding strength and endosomolytic properties. The here described system included a DNA binding agent (a dioleylspermine derivative), combined with membrane permeation (oligoarginine) or targeting agent (folic acid). A combination of electrostatic attraction and the

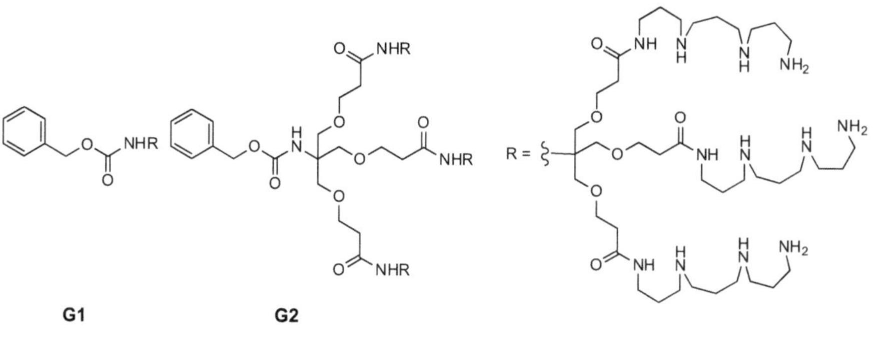

Figure 8.24 G1 and G2 dendrons with surface spermidine ligands and a model for a G-dendron bound to DNA.

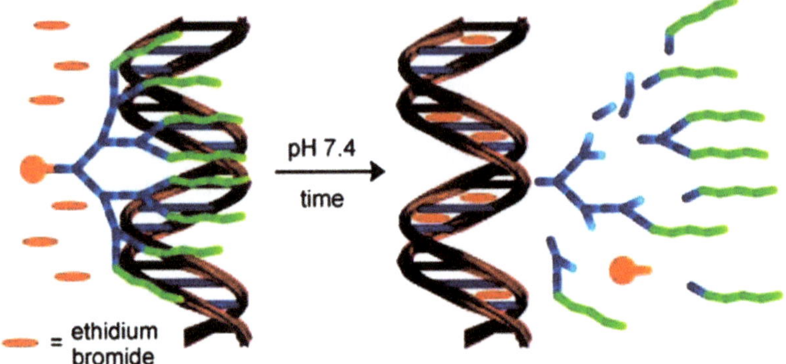

Figure 8.25 Experiment designed to assay the ability of a dendron to release from DNA on degradation. Ethidium bromide is used as a fluorescent probe of the availability of "free" DNA in solution.
(Reprinted from Welsh *et al.*[66] with permission from Wiley & Co.)

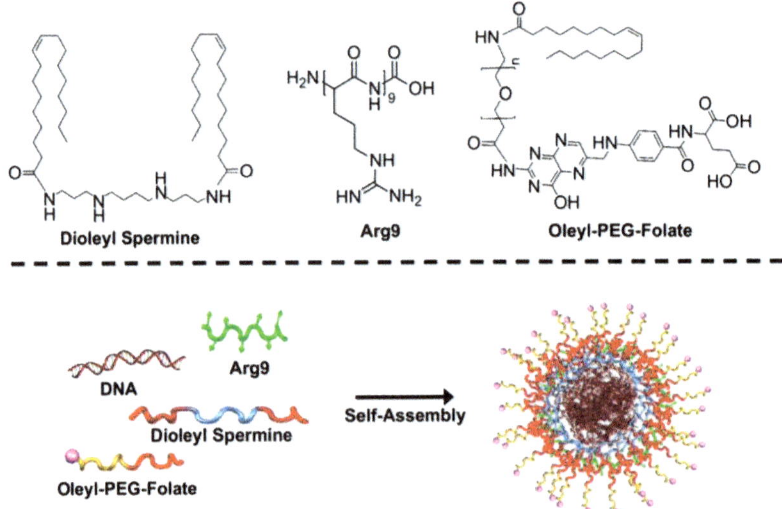

Figure 8.26 Schematic representation of the supramolecular self-assembly of individual vector components with plasmid DNA.[67] PEG = polyethyleneglycol.
(Reprinted with permission from Gabrielson and Cheng[67], Elsevier)

hydrophobic effect is used to bring the individual groups together (Figure 8.26) to form nanoscale complexes with DNA.[67] The resulting cell-specific formulations, with superior transfection efficiency to lipofectamine 2000, have greater transfection efficiency than the individual components.

Another approach to polyamine-based supramolecular non-viral DNA delivery relied on cationic polymers based on polyamines, in which the charge density of ammonium ionenes is simply manipulated with monomer selection; others have shown that charge density influences transfection efficiencies and cytotoxicity.[68] Unlike other cationic vectors for gene delivery, aliphatic ionenes containing quaternary amines provide permanent charge density, and endosomal pH changes do not alter charge density in these strong polyelectrolytes.[69] Cationic polymers incorporating aliphatic polyamines can also efficiently transport RNAs as demonstrated by self-assembled PEG-based block catiomer acting as carrier of siRNA, additionally equipped with amine-protonation equilibrium designed to enhance intracellular gene silencing.[70]

8.10   DNA Cleavage by Metal-free Polyamines

While metal complexes of polyamines ligands have been widely used as nuclease mimetics (see section 8.12), the number of studies performed with metal-free cleaving agents is much more limited. However, after resolving the structure of the metal-free active site of staphylococcal nuclease, which imparts a 10^{16}-fold rate enhancement of DNA hydrolysis,[71] several groups, pioneered by the work of Göbel[72] and Breslow,[73] focused their interest on this topic. The use of metal-free ligands can be interesting for many applications in view of the generally higher toxicity of metal ions in biological systems. Regarding polyamine ligands there are several studies mostly involving small tri- or tetra-azamacrocyclic ligands. One of the first relevant examples of this chemistry concerns the enhanced photocleavage of plasmid pBR322 supercoiled DNA by sapphyrin ligands appended with polyamine chains (Figure 8.27).[74] Molecules with DNA photocleavage properties may be useful in photodynamic therapy (see Chapter 15).

Regarding this point, relatively recent studies showed DNA photocatalytic cleaving activity of porphyrin units appended with spermidine or spermine units.[75] The oxidative photocatalytic activity was found to be dependent on the attached polyamine (Figure 8.28).

Photocleavage of HIV-DNA by quinacridine derivatives triggered by triplex formation was shown to occur with the ligands shown in Figure 8.29.[76] Such

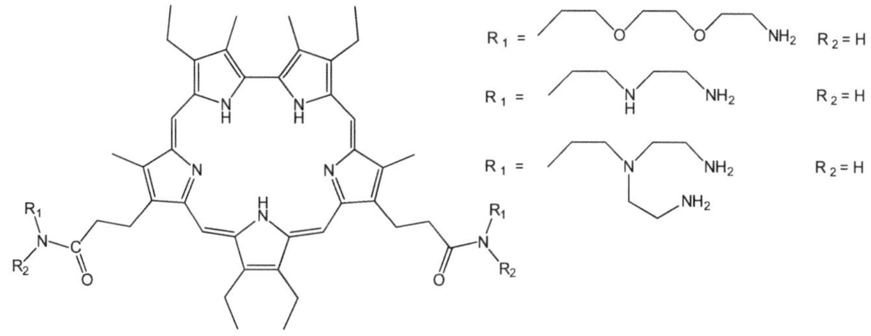

Figure 8.27   Sapphyrin compounds decorated with polyamine chains.

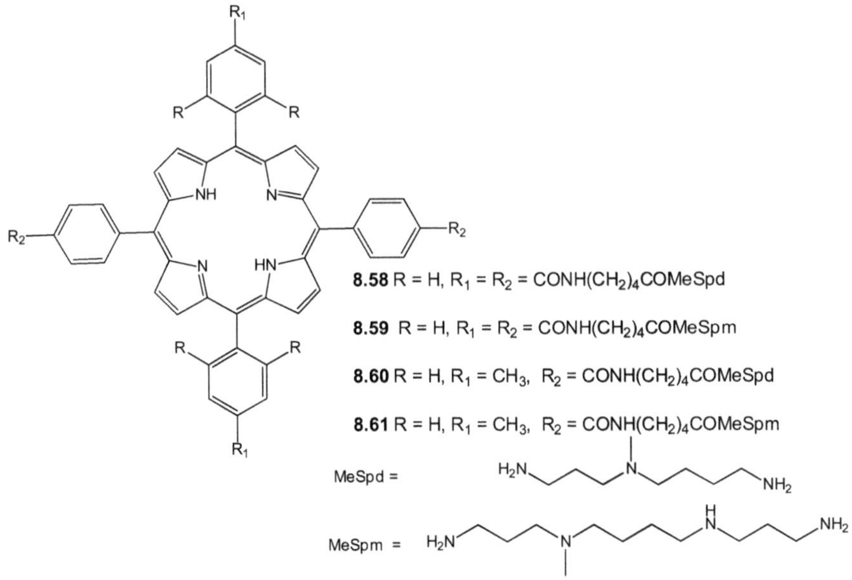
R1
R
R
NH
N
HN
N
R2
R2
R
R
R1
8.58 R = H, R1 = R2 = CONH(CH2)4COMeSpd
8.59 R = H, R1 = R2 = CONH(CH2)4COMeSpm
8.60 R = H, R1 = CH3, R2 = CONH(CH2)4COMeSpd
8.61 R = H, R1 = CH3, R2 = CONH(CH2)4COMeSpm
MeSpd =
H2N
N
NH2
MeSpm =
H2N
N
H
N
NH2

Figure 8.28 Polyamine substituted porphyrins revealing photoactive DNA cleavage. Spm, spermine; Spd, spermidine.

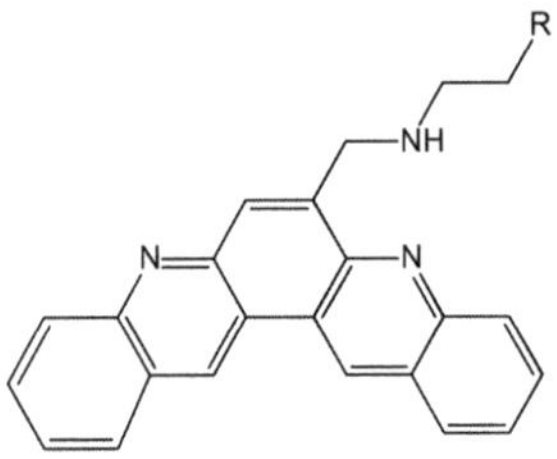
R
NH
N
N

PQ1 R = CH_3 **(8.62)**

PQ2 R = $CH_2N(CH_3)_2$ **(8.63)**

PQ3, R = NC_5H_{10} (piperidinyl) **(8.64)**

Figure 8.29 General formula of aminoquinacridine derivatives.

ligands have been shown to stabilize triple-helical DNA, exhibiting photoactive properties, which makes them suitable as DNA cleavage agents. The compound PQ3 (**64**), Figure 8.29, turned out to be the most active compound of the series, cleaving efficiently ds-DNA in the vicinity of the HIV polypurine tract when this sequence had formed a triplex with a 16-mer pyrimidine triplex-forming oligonucleotide. Anthracene derivatives appended to cyclen moieties through a peptide skeleton have also revealed interesting behavior in their interaction with DNA, including photocleavage (Figure 8.30).[77]

By comparing the binding constants, viscometry results and DNA Tm measurements, the authors of this study concluded that while **65** interacts with DNA through groove binding, **66** binds to DNA with a multiple binding mode that involves groove binding and π-stacking. The binding constant of compound **66** to calf-thymus- (ct) DNA is 100-fold higher than that for **65** and the

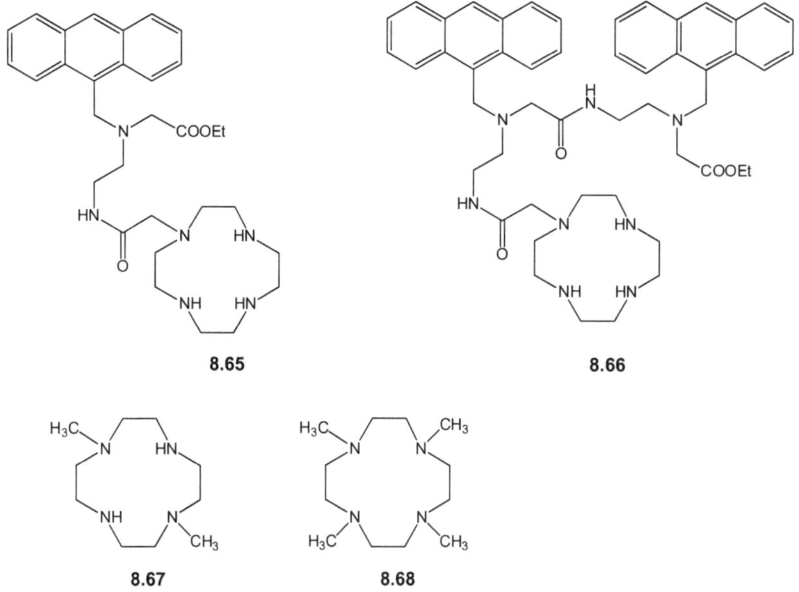

Figure 8.30 Cyclen and cyclen-anthracene derivatives.

compound is also much more efficient in DNA photocleavage. It is interesting that the monoanthryl compound **65** has sequence selectivity with a 36 times higher affinity for GC than for AT sequences.

Apart from photocatalytic cleavage, a number of polyamine conjugates operate by a hydrolytic mechanism. Two examples are *trans* dimethylated (**67**) and tetramethylated (**68**) cyclens which are presented in Figure 8.30.[78] In particular, and quite surprisingly, *trans* dimethylated cyclen was able to hydrolyze ds-DNA at physiological conditions (37 °C, pH = 7.2). Several experiments showed that DNA cleavage takes place by a hydrolytic pathway close to that of the natural enzyme.

Another example of functionalized cyclen moieties acting as DNA catalyst are urea bridged cyclen, peptide nucleic acid (PNA) conjugates (**69** to **71**, Figure 8.31). The bicyclen urea bridged compound was more effective in cleaving DNA than the monocyclen one, again with a hydrolytic mechanism. Polyacrylamide electrophoresis (PAGE) and MALDI-TOFF analysis indicated a site-selective hydrolysis.

A further example with a surprisingly high catalytic efficiency of the free-ligand compared with its Zn(II) complex was the 1,4,7-triazacyclononane derivative **72** (Figure 8.32) containing lateral hydroxyethyl and guanidiniethyl side-chains.[79]

DNA cleavage with **72** reaches a 10^6-fold rate enhancement and exhibits a much higher efficiency than the corresponding non-guanidinium *N*-(2-hydroxyethyl)-1,4,7-triazacyclononane and the parent 1,4,7-triazacyclonane (TACN). Compounds **73** and **74** combine TACN with an anthraquinone unit.[80]

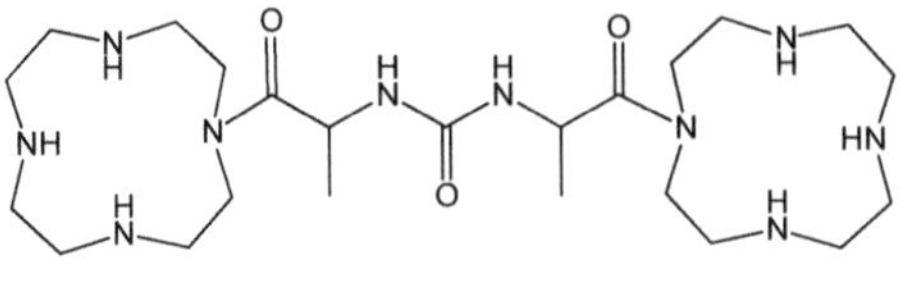

8.69

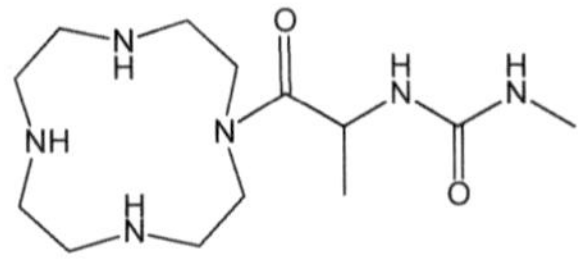

8.70

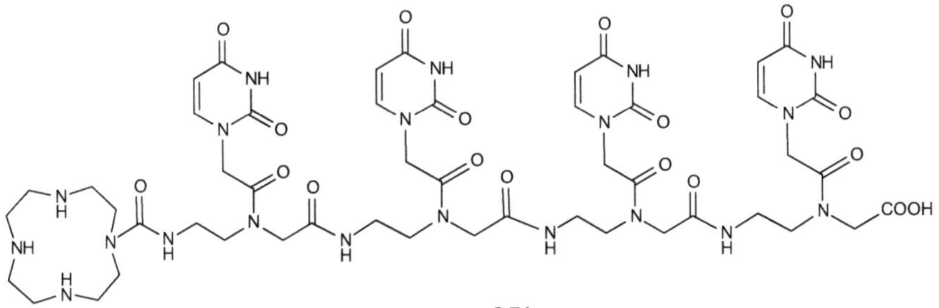

8.71

Figure 8.31 Peptide nucleic acid cyclen conjugates.

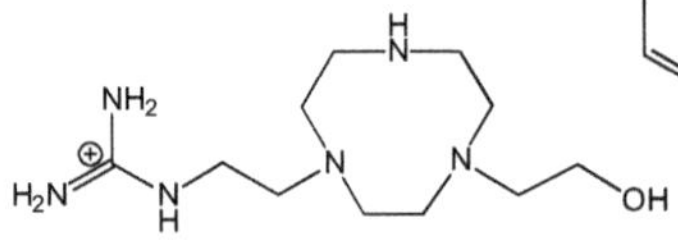

8.72

n = 1 **8.73**
n = 2 **8.74**

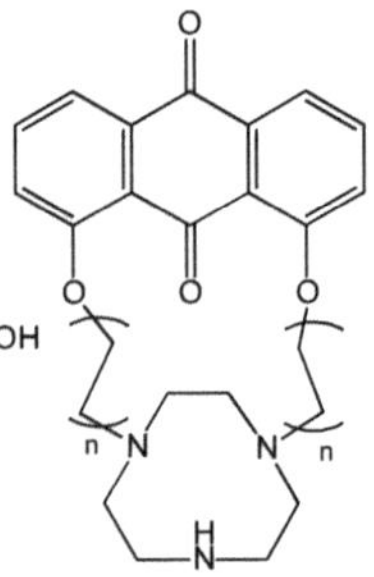

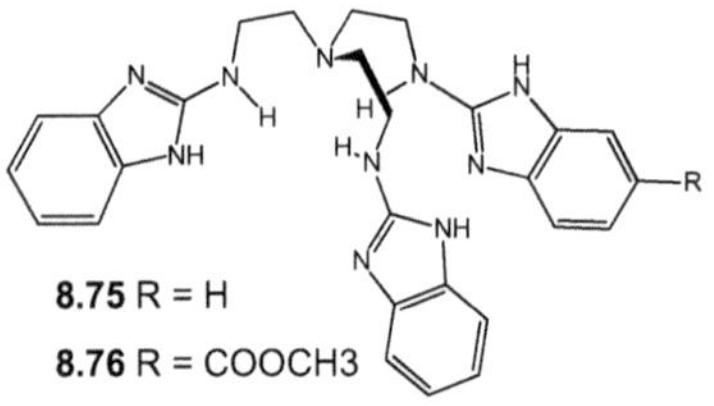

8.75 R = H
8.76 R = COOCH3

Figure 8.32 Triazacyclononane and tren derivatives with DNA cleavage activity.

Fluorescence and CD spectra suggested an intercalative binding mode with the highest binding ability due to **74** which has longer linkers between anthraquinone and TACN. The apparent first-order rate constants of **73** and **74** for DNA cleavage indicated *ca.* 50- and 80-fold rate accelerations with respect to TACN. Such rate enhancements were ascribed to a cooperative effect due to intercalation.

Tris(2-aminobenzimidazoles) **75** and **76** proved to be among the most efficient metal-free catalysts for RNA cleavage (Figure 8.32, *right*).[81] At 1 mM concentration, the half-life of RNA compares well with those achieved by many metal complexes. As shown by fluorescence correlation spectroscopy, the active species seems to be aggregates of substrate and catalyst leading to denaturation of the RNA secondary structure. Compound **76** was shown to be active also in the non-aggregated state at very low concentration.

8.11 Binding Modulation by pH, Metal Ions and Other Effectors

Ligands interacting with nucleic acids as a function of pH are of interest—in view of the lower pH in cancer tissues, for example. The binding of aminoglycosides is known to depend on pH.[82] Ru(II) polypyridyl complexes can associate with DNA under photocontrol and as a function of pH.[83] Other pH-driven switches based on DNA have been reported.[84] Several diseases are characterized by abnormal metal ion levels: elevated Fe and likely also Cu levels occur in the brain of patients with Alzheimer's disease.[85] Interactions which depend on the presence of metal ions such as zinc are of interest for antisense agents, which are specific for cells containing elevated amounts of, for instance, Zn(II) ions in nerve, sperm, and some cancer cells; also malaria parasites infecting red blood cells are known to accumulate Zn(II).[86]

Metal-dependent associations can materialize with ligands which change their conformation due to allosteric effects. As illustrated in Figure 8.33 the presence of Cu^{2+} ions with a binaphthyl polyamine, for example, leads to dramatic variations of both DNA and RNA duplex stabilities, as only in absence of the metal is a bisintercalation by the ligand possible.[87]

Ligands which exhibit pH-controlled interactions are compounds such as the bis-phenanthridines mentioned already in Section 8.5.[41] Recently it has been shown that conformational reorganization of scorpiand-like azamacrocycles

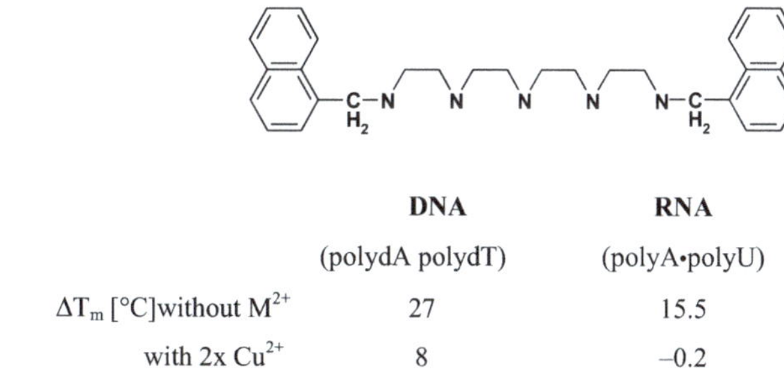

	DNA	RNA
	(polydA polydT)	(polyA•polyU)
ΔT_m [°C] without M^{2+}	27	15.5
with 2x Cu^{2+}	8	−0.2

(Cu^{2+} alone negligible effects at $\gamma = 0.1$)

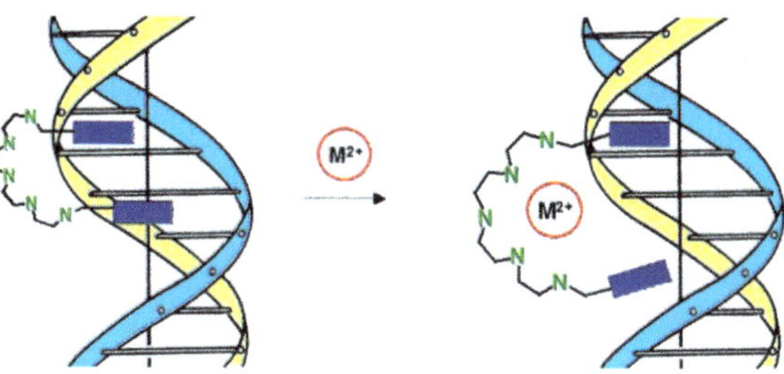

Figure 8.33 Allosteric control of association with duplexes by metal ions; melting point changes with and without metal ion.
(Reprinted with permission from Lomadze *et al.*[87] Elsevier)

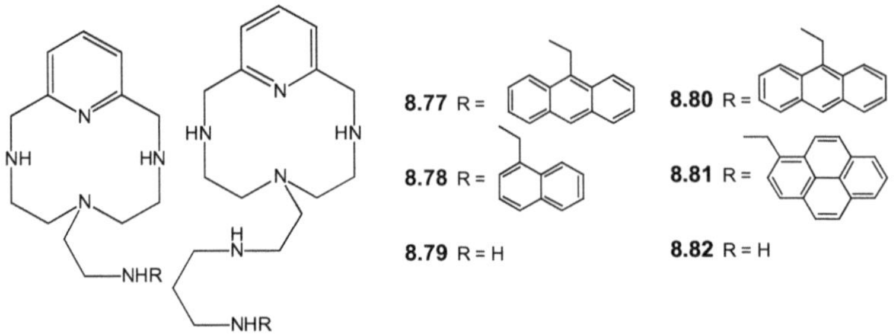

Figure 8.34   Representation of a scorpiand and related ligands.

brought about by pH changes or by metal coordination can have significant effects in the modulation of ct-DNA binding and in cell viability.[88] Macrocycles **77**, **80** and **81** (Figure 8.34), consisting of a macrocyclic pyridinophane core appended with a polyamine arm containing the classical intercalative units anthracene and pyrene, change from a closed to an open conformation as a function of pH; when all secondary amino groups of the compounds are protonated, electrostatic repulsion causes the ligand to adopt an open conformation to minimize the electrostatic affect. However, as soon as one secondary amine deprotonates the compounds adopt a closed conformation with the condensed aromatic ring of the tail π-stacked with the pyridine ring. Coordination of Cu(II) and Zn(II) to **77**, **80** and **81** leads also to closed conformations, since these metal ions need the nitrogen atoms in the tail to complete their first coordination spheres. Crystal structures show that the metal ions are completely embedded by the ligand and cannot interact with other molecules unless dissociation occurs. The pH at which the reorganization occurs can be monitored following the changes in the UV spectra. As expected, metal ions, particularly Cu^{2+}, cause the molecules to close up at lower pH values.

UV/Vis spectroscopy, steady-state fluorescence, circular dichroism, ethidium bromide displacement assays, and viscosity measurements have proved that the adoption of the closed conformation by metal complexes leads to a less effective intercalation of the compounds to ct-DNA in solution. Cell viability assays in cellular cultures of human bladder cancer cell lines T24, 253J and UMUC-3 showed that **77**, **80** and **81** were *ca.* 20 times more cytotoxic than **78**, **79** and **82**, which are analogous to the former triad but lack intercalating units, suggesting that these units mediate the cytotoxic effects in the cells. While the half-maximal inhibitory concentrations (IC_{50}'s) for **77**, **80** and **81** were in the micromolar range, the IC_{50}'s of **78**, **79** and **82** were above 100 μM, except for **78** and **79** in 253J cells. The metal complexes were also less toxic than the free ligands, cytotoxicity being not observable below 100 μM. Cells were almost 100% viable when incubated with metal complexes for 48 h, while they were less than 20% viable when incubated with uncoordinated ligands. Moreover,

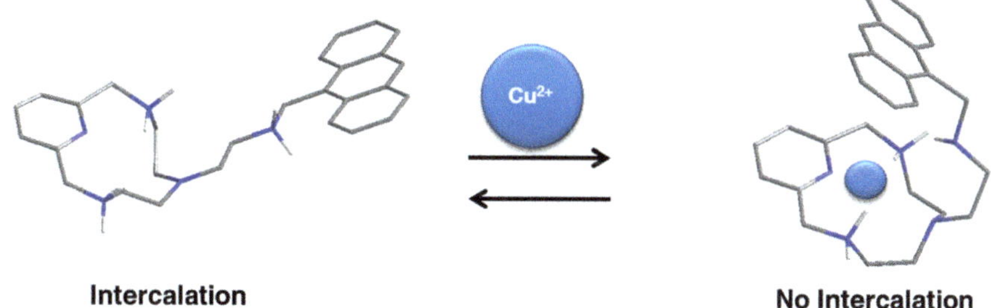

Figure 8.35 Scheme of the metal-ion driven "on–off" switch due to the intercalative binding mode in scorpiand-like compounds.

metal ions blocked the cytotoxicity of **77**, **80** and **81** further when added on top of the culture medium of cells that had been already treated for 24 h with these ligands.

Cellular damage activates the ATM kinase, which in turn phosphorylates the p53 protein at the Ser15 residue.[89] Interestingly, ligand **80,** but not its Zn(II) complex, induced the cellular DNA damage response, as indicated by an increase in the phosphorylation of the Ser15 residue of the p53 protein. All these results suggest that the interaction of the metals with **77**, **80** and **81** and the concomitant conformational change of the ligand (Figure 8.35) modulate, at least in part, their cytotoxic effects on the cells.

Peptide nucleic acids (PNAs) with a Zn(II) chelating unit (L) conjugated through aromatic rings (Figure. 8.31)[90] showed that binding of Zn^{2+} to the chelating units enhances the affinity of the PNA probes to single-stranded DNA significantly, as reflected by increases of Tm of at most *ca.* 12 °C. The ligands hybridized to DNA within a few minutes at 22 °C.

A modulation of DNA binding by an external effector molecule was observed in a system that combined the potent intercalating DNA properties of anthrylamines with the known drug-binding capability of cyclodextrin (see Figure 8.36).[91] Examination by ^{1}H NMR and circular dichroism (CD) spectroscopies revealed that addition of 1-adamantol, a strong intracavity binder of cyclodextrins, removes the anthryl unit out of the β-cyclodextrin cavity. The release of the anthryl unit permits its intercalation into ds-DNA, as indicated by NMR.

Adverse effects of metal coordination in ligand binding to G-quadruplexes have been noted in Cu(II) complexes of the bisquinolinium tweezer-like ligand **83** (Figure 8.37). The ligand alone has a closed arrangement due to formation of intramolecular hydrogen bonds between the amide NH groups and the nitrogen atom of the central pyridine ring. Cu(II) coordination to the carbonyl amide groups and pyridine disrupts the internal hydrogen bonding, yielding a more open structure (Figure 8.37). Based on the CD spectra, it seems that while unmetalated **83** strongly binds to the G-quadruplex, the Cu(II)-induced open structure significantly weakens this binding.[92]

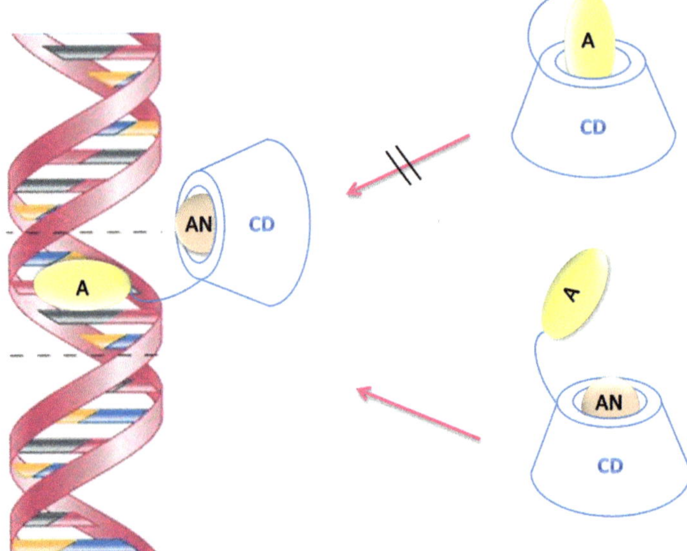

Figure 8.36   Adamantol (**AN**) as an effector in the intercalative binding mode of an anthracene unit (**A**).

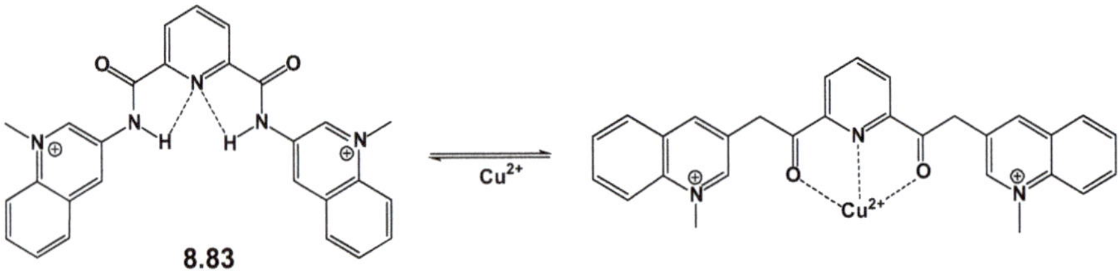

8.83

Figure 8.37   Coordination of Cu^{2+} to ligand **83** disrupts hydrogen bonds and weakens the binding to G-quadruplexes.

8.12   Selective DNA/RNA Cleavage with Metal Complexes

Notwithstanding the actual upsurge of interest in metal-free artificial nucleases discussed in Section 8.10, most of the studies on this topic have been carried out with metal complexes.[93] Biological nucleases typically include Mg(II), Ca(II), Zn(II), Mn(II) and Fe(II)/Fe(III) metal centers within their active sites, as shown by alkaline phosphatase as an example (Figure 8.38).[94] The role of the different metal ions and the underlying mechanisms of phosphate ester hydrolysis have been aptly discussed in many reviews[93] and need not be repeated here.

Natural enzymes have evolved to levels of sophistication that are almost unreachable by abiotic catalysts. However, the implementation of different

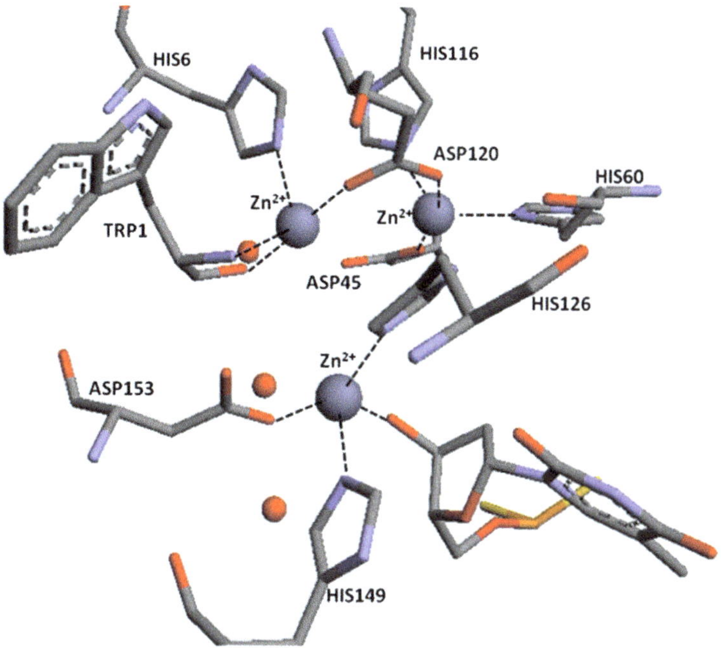

Figure 8.38 The active site of alkaline phosphatase.
(Adapted from Kim and Wyckoff,[94] DOI:10.2210/pdb1alk/pdb.)

metal ions in supramolecular complexes allows development of artificial DNA or RNA cleaving catalysts; attachment of suitable ligands to such complexes can enhance their affinity towards nucleic acids, and/or can even allow sequence selectivity. Hydrolysis of DNA proceeds extremely slowly with a half-life of about 37 million years.[95] The extraordinary inertness of DNA has made supercoiled plasmid DNA a frequently used target. A single scission of plasmid DNA transforms the supercoiled form (form I) to a circular form (nicked form II); a second scission gives rise to a linear form (form III). Form II can be enzymatically re-ligated to give back supercoiled DNA (form I), which is proof-of-concept for a clean hydrolytic mechanism and serves to discard oxidative mechanisms. RNA hydrolysis is much faster than that of DNA owing to the action of the 2′-OH ribose function, which can act as an internal nucleophile; the nucleophilic attack is also assisted by general base catalysis, for which metal ions, in particular Mg(II) ions, play an essential role.[96] Large gains in reactivity may in future be expected from a change to a less water-containing environment; for instance, the cleavage of RNA models by dinuclear Zn(II) complexes is increased by up to 10^{17} in ethanol, whereas in water, little acceleration is observed.[97] Implementation of DNA recognition ligands leads to artificial restriction enzymes, for which already a large number of reviews are available.[93] In the next paragraphs, we will include a few examples of the advances made with synthetic metallonucleases in recent years; in the first part we will discuss metallonucleases based on first-row transition and postransition metal ions, particularly Zn(II), and in the second part we will deal with lanthanide metal complexes.

8.12.1 Complexes of First-row Transition Metal Ions and Zinc(II)

Several metallo complexes have been developed, which can selectively cleave even large genomic ds-DNA molecules. This happens mostly *via* the hydrolysis of DNA phosphodiester links, in a similar way as natural nucleases do, and is an important step towards gene manipulation. Metallonucleases for RNA that are analogues of natural ribozymes have also been reported.[93g] They can promote the selective inhibition of protein expression in cell culture and thus have potential for therapeutic application.

A very efficient system for DNA cleavage is the binuclear Fe(III) complex with the ligand shown in Figure 8.39.[98] At $100\,\mu M$ the complex hydrolyses supercoiled DNA at pH 7 and $37\,^\circ C$ with $k = 2.1 \times 10^{-3}\,s^{-1}$, which corresponds to a half-life of just 5 min, one of the best values so far reported. The cleavage does not depend on O_2 and the linear form can be re-ligated almost quantitatively, supporting a hydrolytic mechanism. The presence of an oxo group bridging the Fe(III) centers has been postulated to play a key role in the observed rate enhancement.

There are several enzymes for phosphoester hydrolysis that have two or even three Zn(II) metal ions in their active sites. Therefore, different abiotic ligands which are able to incorporate either two or three Zn(II) ions with coordinatively unsaturated sites have been designed and prepared during recent years. Complexes $[Zn_2(\mathbf{84})]^{3+}$ and $[Zn_3(\mathbf{85})]^{6+}$ shown in Figure 8.40 are representative examples.[99]

The maximum rate was observed for the trinuclear complex, which catalyzed the hydrolysis much more efficiently than the binuclear complex.

Hirota and co-workers observed that photoisomerization of an azobenzene moiety connecting two tridentate Zn^{2+} coordinating units had a strong influence on DNA hydrolysis[100] (Figure 8.41). Although both the *trans-* and *cis-*isomers

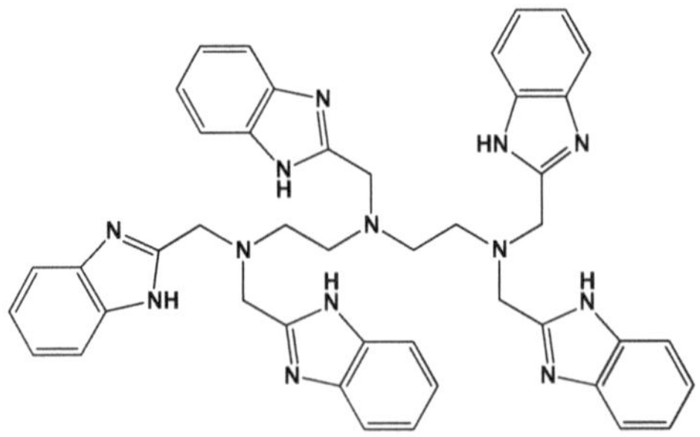

Figure 8.39 Benzylimidazole derivative of diethylenetriamine.

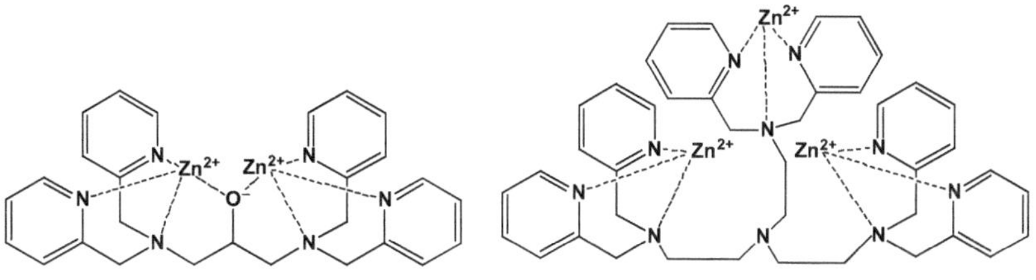

$[Zn_2(8.84)]^{3+}$

$[Zn_3(8.85)]^{6+}$

Figure 8.40 Bi- and trinuclear Zn(II) complexes.

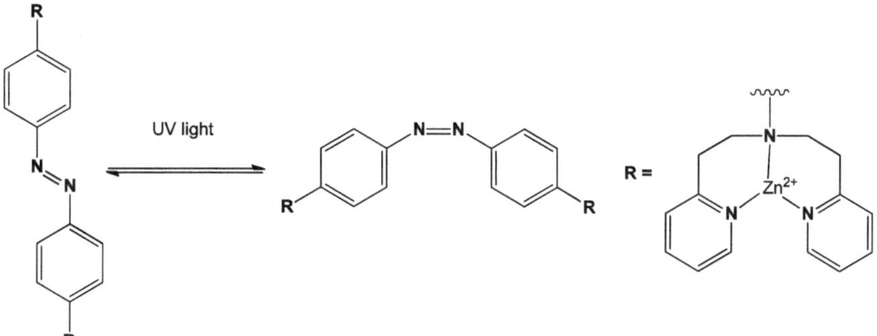

Figure 8.41 Photoisomerization "on–off" switch of DNA cleavage activity.

bind to DNA with similar strengths, only the *cis*-isomer with the right disposition and separation of metal centers is able to cleave DNA.

An allosteric and sequence-selective ligand which cleaves ds-DNA by an oxidative radical mechanism was described by Dervan in 1987. It contains two oligo-pyrrolecarboxamide (netropsin) moieties linked by a polyether, and an iron-chelating EDTA unit. With a 517 base-pair restriction fragment that could be obtained in the presence of dioxygen, a reductant and Fe(II) ions preferred cleavage at AT-rich binding sites, particularly if Sr(II) or Ba(II) ions were added; these latter coordinate to the polyether chain and thus change the ligand conformation to a convex shape.[101]

Highly selective RNA hydrolysis was achieved with a dinuclear zinc complex tethered through a hexamethylene chain *via* a phosphoramidite TPBA to the 5′ ends of a DNA sequence (Figure 8.42).[102] The DNA portion in the TPBA-DNA oligomer was complementary, by Watson–Crick type base pairs AU, AT and GC, to the U4–C23 part in the target RNA sequence; as a consequence, selective Zn(II)-catalyzed scission could occur at the 3′ side of a cytosine residue (C24, marked by arrow) in the RNA.

One new approach is the grafting of metal complexes on the surface of nanoparticles. For example, the Zn^{2+} complexes of bis(2-amino-pyridinyl-6-methyl)amine (BAPA) assembled on the surface of gold nanoparticles through

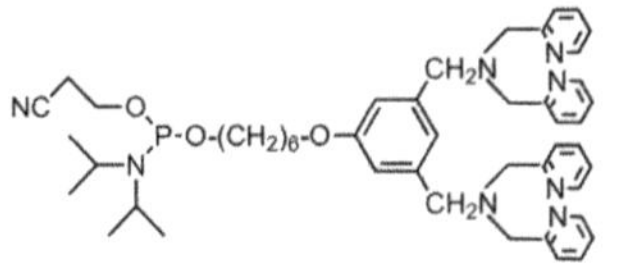

phosphoramidite TPBA linker

1 10 20 30

5' AUC UGG AGG UCC UGU GUU CGA UCC ACA GAA UUC GUC 3' RNA

3' ACC TCC AGG ACA CAA GCT AG**X** 5' TPBA-DNA

Figure 8.42 Selective RNA hydrolysis based on nucleobase complementarity; using a DNA-like oligomer tethered with a phosphoramidite (TPDA, *N,N,N′,N′*-tetra-kis(2-pyridylmethyl)-3,5-bis(aminomethyl)benzene) linker bearing Zn(II) ions. X is the residue carrying TPBA; the arrow indicates the site of selective scission.[102]

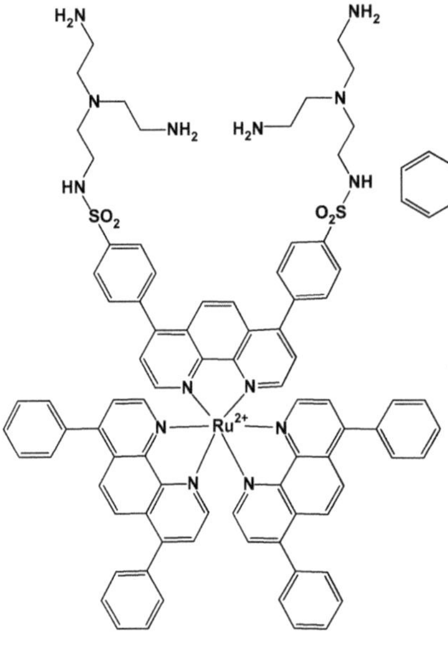

8.86

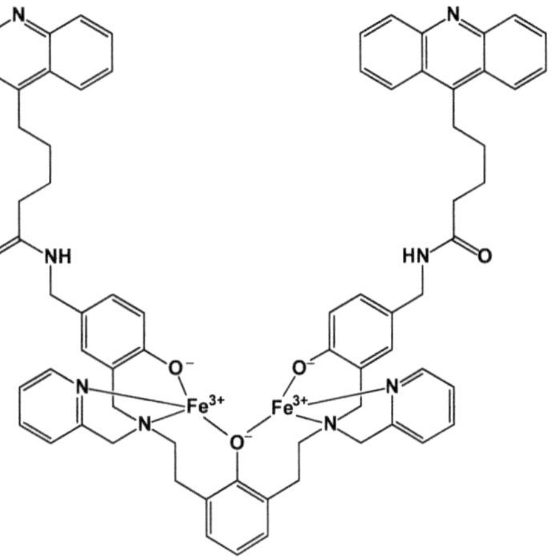

8.87

Figure 8.43 Ru(II) bipyridyl complexes decorated with tren amines.

thiol linkages have proved to display interesting behavior in DNA cleavage.[103] This system, although with moderate reactivity ($k = 2 \times 10^{-6}$ s^{-1} at 37 °C and pH 7, [Zn(II)] = 15 μM) promotes direct double-strand scission with conversion of the supercoiled form into the linear form.

Alternative strategies to obtain efficient catalysis consist in conjugating the metal complexes to molecules that can reinforce the binding to nucleic acids. This idea was already advanced by J. K: Barton and co-workers in 1987, who decorated a ruthenium intercalating complex with polyamine-charged side

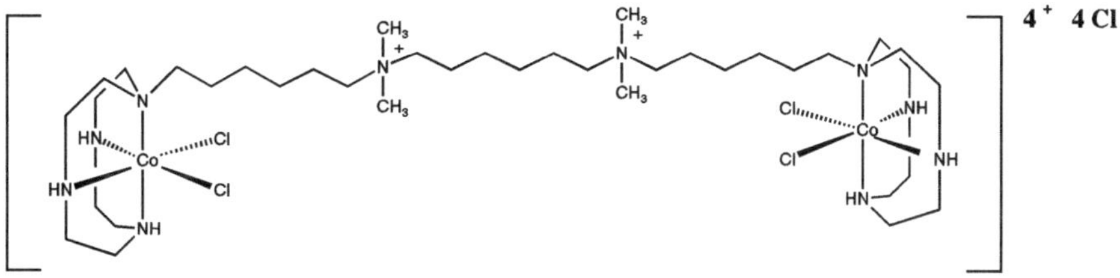

Figure 8.44 Co(III)-cyclen complexes connected by peralkylated amines.

arms to facilitate the binding to DNA (**86**, Figure 8.43).[104] The complexation of the non-redox metal ions Zn(II), Cd(II) and Pb(II) in the arms of compound **86** led to DNA cleavage with a moderate efficiency, which reached 40%. With Cu(II) or Co(II) metal ions, significantly enhanced cleavage was observed, but only a small re-ligation level, suggesting a partial redox mechanism.

Combination of polynuclear metal ions and intercalating moieties has been realized with compound **87** (Figure 8.43), in which a binucleating ligand for Fe(III) was connected with acridine units.[105] This compound has $k = 6.6 \times 10^{-3}$ s^{-1} at pH 7 and 37 °C for a 56 µM concentration of the complex in the hydrolysis of supercoiled pBR322 DNA, which implies an impressive half-life of less than two minutes. The two acridine moieties seem to play an important role due to their intercalation with DNA; a 300-fold increase of rate compared to the complex without acridine moieties could be obtained, which also was much greater for the compound with two acridine moieties than with just one.

Association with supramolecular metal complexes bearing multiple positive charges leads to significant enhancement in Michaelis–Menten K_M values by ion pairing with the negatively charged DNA rim; the presence of two metal centers can increase the catalytic k_{cat} constants. The binuclear Co(III) complex shown in Figure 8.44 was found to cleave supercoiled plasmid DNA at 37 °C with a 10 million-fold rate enhancement in presence of 0.05 M catalyst—in only 2 hours one could observe an almost complete cleavage without any products from radical reaction.[106] Saturation kinetics showed that the K_M value increases with the length of the spacer, from 1.0×10^3 M^{-1} in the case of [Co(cyclen)]$^{3+}$ to 5.5×10^3 M^{-1} for the complex with a C6 spacer; in contrast, the k_{cat} values were rather constant. For possible biotechnological applications, hydrolytic cleavage with intact oligonucleotide ends is essential, and has been demonstrated also by re-ligation experiments with other metal complex catalysts.[107]

A related approach has been used by Mao and co-workers, who introduced quaternized amines into the exterior of bis(bipyridine)zinc(II) complexes (Figure 8.45).[108] Hydroxylated [ZnL$_2$(OH)]$^{3+}$ seems to be the active species in the catalysis: the metal-bound hydroxide group would be performing the nucleophilic attack while the alkylated ammonium groups could bind to the next phosphoryl groups in DNA.

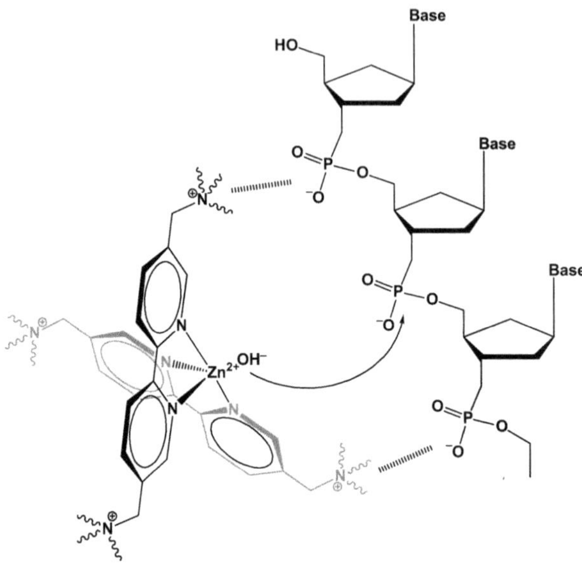

Figure 8.45 Scheme of the interaction with DNA of a tris(bipyridine) zinc(II) complex decorated with quaternized amines.

Another strategy developed by Kimura and co-workers to strengthen the binding and improve the catalytic ability consisted in appending nucleobases to coordinatively unsaturated macrocyclic zinc(II) complexes.[109] The zinc-cyclen complex $[Zn(\mathbf{8.88})]^{2+}$ (Figure 8.46) with pendant acridine groups shows affinities with nucleosides of log $K = 7.2$ for thymine dT and log $K = 6.9$ for uracil, consistent with the basicities of the conjugate nucleobase at the N(3) position. The Zn(II)-cyclen complex lacking the acridine moiety was found to be about 100 times weaker, the difference of $\Delta\Delta G = 10$ kJ/mol being due to stacking with the nucleobase. Interaction of the Zn(II)-cyclen–acridine complex with poly(dT) occurs with T deprotonation and stacking with $K_{app} = 3.6 \times 10^4 \, M^{-1}$. Studies with double-stranded poly(A).poly(U) indicated that all Zn(II)-cyclen complexes penetrate the core of the helix with binding to the N(3)-deprotonated uracil bases, with subsequent selective denaturation of the double-stranded nucleic acid.

DNAase footprinting experiments showed that the Zn(II)-cyclen acridinyl unit binds only to the thymine groups, and disintegrates the A–T link so that the separated adenine partners are more exposed to the nuclease action. The complex is also able to inhibit the binding of human TATA binding protein to the so-called TATA box, which is an AT-rich DNA sequence located at 25–30 bases upstream from the transcriptional start sites and is an essential element of the promoter for eukaryotic RNA polymerase, playing a key role in regulating the overall level of transcription. The concentration required for 50% inhibition (IC$_{50}$) of the TBP–DNA complex formation was 15 µM; related zinc complexes worked even at 0.4 µM. The complex $[Zn_2(\mathbf{8.89})]^{4+}$ was shown to

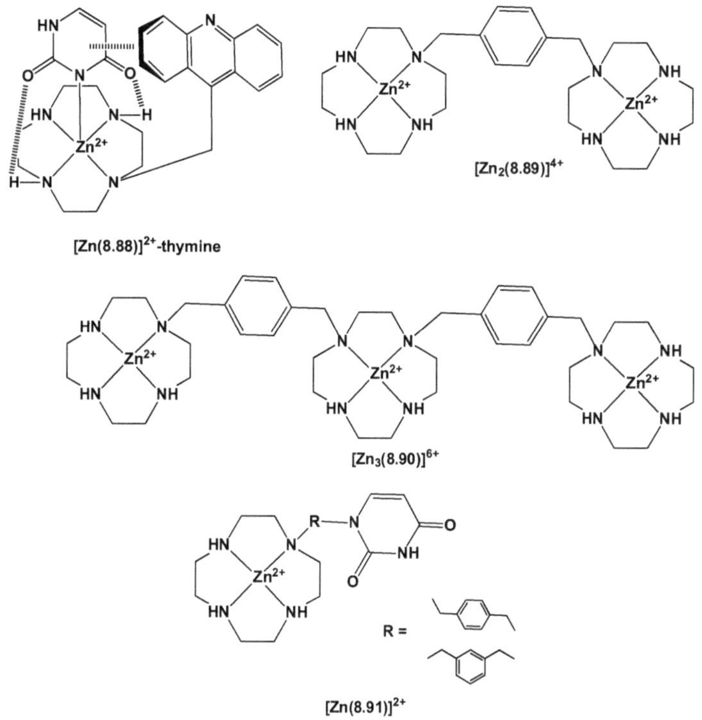

[Zn(8.88)]$^{2+}$-thymine

[Zn$_2$(8.89)]$^{4+}$

[Zn$_3$(8.90)]$^{6+}$

R =

[Zn(8.91)]$^{2+}$

Figure 8.46 Zn(II)-cyclen monomers and oligomers.

hold promise for prevention of formation of complexes of key viral RNA with proteins; this plays a role in transcription of the HIV-1 genome which is facilitated by a HIV-1 regulatory protein. With the tris Zn(II)-cyclen derivative [Zn$_3$(**8.90**)]$^{6+}$ the K_d value for the TAR Tat complexation and the IC$_{50}$ value was as low as 15 nM, which may offer as new strategy to fight AIDS. In comparison, aminoglycoside antibiotics such neomycin show inhibition values for TAR RNA-Tat protein complex formation of IC$_{50}$ around 1 μM.

A recent example, in which uracil moieties are linked to Zn(II)-cyclen complexes ([Zn(**8.91**)]$^{2+}$ through aryl spacers, is shown in Figure 8.46. The results showed that these complexes are able to dramatically accelerate plasmid DNA cleavage at very low concentration of complex (4 μM).[110]

8.12.2 Lanthanide Complexes

The hydrolytic cleavage of DNA or RNA is efficiently catalyzed also by lanthanide ions. Supramolecular complexes with suitable ligands (Figure 8.47) of these highly charged ions have led to spectacular results with respect to catalytic efficiency as well as selectivity.[93] DNA hydrolysis as well as RNA hydrolysis involve Lewis acid catalysis by metal ion or by metal-bound water, and base catalysis by metal-bound hydroxide. Saturation kinetics with plasmid DNA

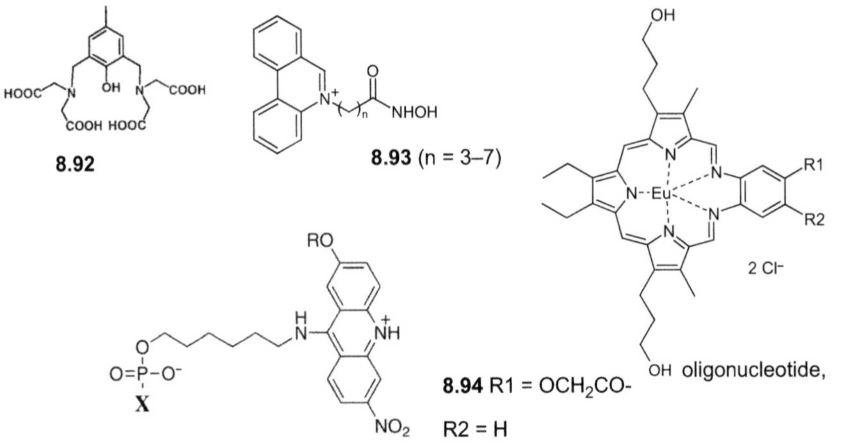

8.92

8.93 (n = 3–7)

8.94 R1 = OCH$_2$CO-

R2 = H

2 Cl$^-$

OH oligonucleotide,

8.95 R = Me, Et, iProp

3' d(ACT CTG CTA CTG ACC TAG **X**) 5'

Figure 8.47 Complexes and ligands used for lanthanide-induced hydrolyses.

show little dependence of the k_{cat} values on the metal ion, in contrast to the K_M values, in line with spectroscopically observed lanthanide-induced conformational changes.[111] For RNA cleavage the last three lanthanide ions Tm(III), Yb(III), Lu(III) as well as Ce(III) ions are very efficient. The high charge density in Ce(IV) makes it superior for DNA cleavage; pentacoordinated cerium(IV) complexes were found to accelerate DNA hydrolysis more than 10^{11} fold.[93c–e] An optimal distance between the lanthanide ions in binuclear complexes is critical, as shown by different lanthanide complexes exhibiting order of magnitude efficiency differences.[112] Cofactors including functions such as polyhydroxy groups, intercalators,[113a] or amino acids such as serine, histidine, and aspartic acid[113b] enhance the hydrolysis of plasmid DNA by lanthanide ions considerably. Religation experiments confirmed hydrolysis with intact ends.[107]

Dicerium complexes such as **92** hydrolyze plasmid DNA with linear fragments stemming not from random single-strand breaks, and cleave with high regioselectivity (more than 90% 5'-OPO$_3$ and 3'-OH ends).[114] The hydroxamic acids **93** are connected to a phenanthridinium group as stacking unit and found to cleave plasmid DNA with Fe^{2+}, VO^{2+} and lanthanide(III) ions.[115] Optimal results were observed with a chain length $n = 5$; the reaction rate increased with higher pH, indicating that the active forms are hydroxo complexes. The europium(III) texaphyrin **94**, conjugated to a synthetic oligonucleotide complementary in sequence to a synthetic RNA 30-mer as a substrate, was employed for sequence-selective RNA cleavage.[116] Such complexes were also conjugated with antisense DNA fragments to achieve a ribozyme mimic.[117] Site-selective hydrolyses were achieved with acridine-DNA conjugates such as **95**, which were hybridized to complementary RNA in combination with

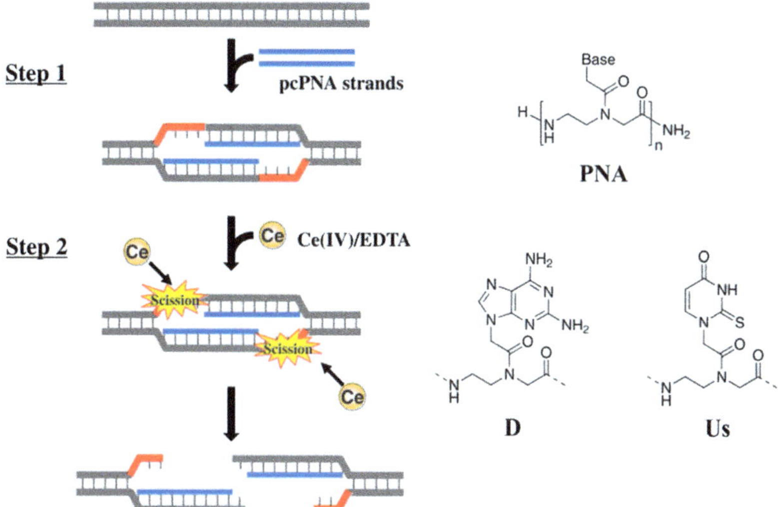

Figure 8.48 Selective cleavage of ds-DNA by base pairing with peptide nucleic acids (PNAs) which invade the ds-DNA, forming ss-DNA portions (red lines. step 1); after this the phosphodiester linkages in the single stranded parts are hydrolyzed by Ce(IV)/EDTA. Structures of PNA and pseudo-complementary bases D and Us are shown on the right-hand side; the symbol U is used for 2-thiouracil in the sequences.
(Reproduced from Komiyama and Sumaoka,[121] with permission by the Chemical Society of Japan.)

another unmodified DNA that hybridizes to the adjacent fraction of the RNA; the conjugates were used as site-selective RNA activators for site-selective RNA cleavage catalyzed by Lu(III).[118]

Conjugation of groove binders with metal complexes holds promise not only for enhanced catalytic efficiency, but also for the development of sequence-selective cleavage. The Fe(III)-complex of a hydroxamic acid containing two netropsin units was shown to lead to significantly enhanced DNA cleavage.[119] Distamycin-linked hydroxamic acids combined with vanadyl ion and hydrogen peroxide were shown in experiments with an end-labeled DNA fragment to exhibit either specific oxidative cleavage flanking 8 and 10 AT sites, or preferential cleavage of 5′-end-labeled strand at the AT sites, depending on the spacer.[120]

Even large genomic DNAs have recently become a possible target. With natural restriction enzymes bearing only a 6-base-pair recognition sequence, the cleavage occurs at many sites. In contrast, artificial catalysts can be equipped with many more recognition sites, and therefore are particularly promising, potentially even with very long DNAs such as in the human genome with its 3×10^9 base pairs. For example, one can combine peptide nucleic acids (PNAs) with a complementary Ce(IV)/EDTA complex, as shown on Figure 8.48.[121] The target site in the unfolded DNA part is located by two strands of PNA and selectively hydrolyzed by a Ce(IV)/EDTA complex, with remarkably little unspecific cleavage.

8.13 DNA Bionanotechnology

Finally we want to make the reader at least aware of the fascinating biomedical applications based on the unique regular hydrogen bonding and self-assembly possibilities of nucleic acids, first developed by N. C. Seeman.[122] A full discussion of this most actively pursued research field is beyond the scope of the present chapter, instead we refer to recent reviews[123,124] and briefly explain some fundamental supramolecular features.

Figure 8.49 illustrates the principles used for the construction of large two-dimensional nanostructures.[122] Sequences of ss-DNA are designed which do not allow duplex formation, for example that of a four-arm branched junction, which contains the maximum number of possible Watson–Crick base pairs (Figure 8.49a). If these units bear sticky ends with unpaired bases (Figure 8.49b) they can combine to form even larger assemblies (Figure 8.49c). In this way one can obtain large two- or three-dimensional nanoparticles.

The so-called Origami-technique uses long ss-DNA as a scaffold, usually the 7.3 kilobase genome of a bacteriophage, which is mixed with an excess of smaller complementary strands. These smaller strands or "staples", with 32 bases for example, are complementary by base-pairing to distinct segments of the long ss-DNA (Figure 8.50). In this way 2D, and recently 3D, DNA nanostructures can be obtained, including stars, squares, rectangles, triangles and even smiley faces.[123b,c] DNA-block polymers are accessible from coupling of oligodeoxyribonucleotides onto polymers.[123d] Intriguing hybrid DNA–protein or DNA–nanoparticle systems have also been developed. All these supramolecular assemblies open new ways to the encapsulation and delivery of drugs, DNA detection and gene therapy.[125] In DNA-controlled synthesis reactants or even enyzmes can be brought into close proximity by hybridization to a complementary DNA strand.[123e]

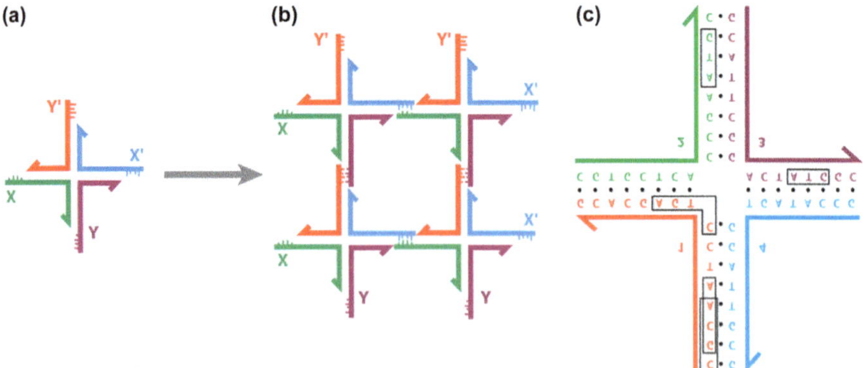

Figure 8.49 Formation of four-armed junction assemblies from ss-DNA strands, based on optimal number of possible Watson–Crick base pairs (see text). (Reprinted with permission from Seeman.[122] *Annu. Rev. Biochem.*, Confirmation Nr 11076647.)

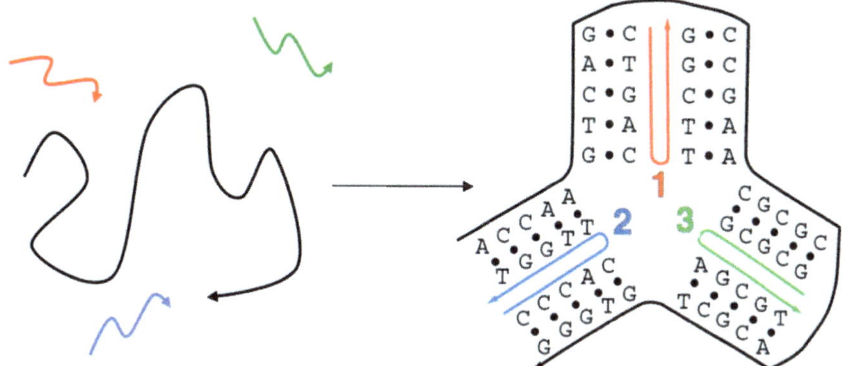

Figure 8.50 The Origami-technique: folding of a large ss-DNA by adding small nucleic acids staples (red, green, blue) (see also text). (Reprinted with permission from Schnitzler and Herrmann,[123d] *Acc. Chem. Res.*, 2012, **45**, 1419. Copyright 2012 American Chemical Society.)

Acknowledgements

Financial support by the Spanish Ministerio de Economía y Competitividad and FEDER funds of the E. U. (Projects CONSOLIDER INGENIO 2010 CSD2010-00065), Generalitat Valenciana (PROMETEO 2011/008) is gratefully acknowledged by E. G. E. We thank Estefanía Delgado-Pinar, Javier Pitarch Jarque and Mario Inclán for their careful reading of the chapter and help with the production of some of the figures. I.P. thanks the Ministry of Science, Education and Sport of the Republic of Croatia for financial support (grant no. 098-0982914-2918) during this work.

References

1. *The Organic Chemistry of Drug Design and Drug Action*, ed. R. B. Silverman, Elsevier/Academic Press, New York, 2004.
2. (a) S. S. Cohen, *A Guide to the Polyamines*. 1998, Oxford University Press, New York; (b) *Polyamine Drug Discovery*, ed. P. M. Woster and R. A. Casero, Jr., Royal Society of Chemistry, 2011; (c) Y. Liu and W. D. Wilson, *Methods Mol. Biol.*, 2010, **613**, 1.
3. *DNA and RNA Binders, from small molecules to drugs*, ed. M. Demeunynck, C. Bailly and W. D. Wilson, Wiley-VCH, Weinheim, 2002; *DNA and RNA structure. Nucleic Acids in Chemistry and Biology*, ed. G. M. Blackburn and M. J. Gait, 1996, Oxford University Press, New York, 1996, 2nd edn.
4. (a) M. A. Medina, J. L. Urdiales, C. Rodriguez-Caso, F. J. Ramirez and F. Sanchez-Jimenez, *Crit. Rev. Biochem. Mol.*, 2003, **38**, 23; (b) R. A. Casero and P. M. Woster, *J. Med. Chem.*, 2009, **52**, 4551; (c) F. Liang, S. H. Wan, Z. Li, X. Q. Xiong, L. Yang, X. Zhou and C. Wu,

Curr. Med. Chem., 2006, **13**, 711; (d) U. Bachrach, *Curr. Prot. Peptide Sci.*, 2005, **6**, 559.

5. P. D. Dervan and B. S. Edelson, *Curr. Opin. Struc. Biol.*, 2003, **13**, 284.

6. M. J. Hannon, *Chem. Soc. Rev.*, 2007, **36**, 280.

7. (a) A. D. Richards and A. Rodger, *Chem. Soc. Rev.*, 2007, **36**, 471; (b) B. M. Zeglis, V. C. Pierre and J. K. Barton, *Chem. Commun.*, 2007, **00**, 4565; (c) K. L. Haas and K. J. Franz, *Chem. Rev.*, 2009, **109**, 4921; (d) P. C. A. Bruijnincx and P. J. Sadler, *Curr. Opin. Chem. Biol.*, 2008, **12**, 197.

8. A. M. Havelka, M. Berndtsson, M. H. Olofsson, M. C. Shoshan and S. Linder, *Mini Rev. Med. Chem.*, 2007, **7**, 1035.

9. A. Hillisch, M. Lorenz and S. Diekmann, *Curr. Opin. Struc. Biol.*, 2001, **11**, 201.

10. M. A. Mintzer and E. E. Simanek, *Chem. Rev.*, 2009, **109**, 259.

11. (a) B. Demeneix and J. P. Behr, *Non-Viral Vectors for Gene Therapy*, 2005, 2nd edn: Part 1, 217–230; (b) P. Midoux, G. Breuzard, J. P. Gomezand and C. Pichon, *Curr. Gene Ther.*, 2008, **8**, 335; (c) R. Srinivas, S. Samanta and A. Chaudhuri, *Chem. Soc. Rev.*, 2009, **38**, 3326.

12. A. V. Eliseev and H.-J. Schneider, *J. Am. Soc. Chem.*, 1994, **116**, 6081.

13. (a) S. Neidle, *Nat. Prod. Rep.*, 2001, **18**, 291; (b) G. S. Khan, A. Shah, Z.-U. Rehman and D. Barker, *J. Photochem. Photobiol. B*, 2012, **115**, 105.

14. (a) X. Chen, B. Ramakrishnan and M. J. Sundaralingam, *J. Mol. Biol.*, 1997, **267**, 1157; (b) M. Coll, C. A. Frederick, A. H. J. Wang and A. Rich, *Proc. Natl. Acad. Sci. USA*, 1987, **84**, 8385.

15. S. E. Butcher and A. M. Pyle, *Acc. Chem. Res.*, 2011, **44**, 1302.

16. (a) J. Gallego and G. Varani, *Acc. Chem. Res.*, 2001, **34**, 836; (b) S. Fulle and H. Gohlke, *J. Mol. Recog.*, 2010, **23**, 220; (c) Ke Li, M. Fernandez-Saiz, C. T. Rigl, A. Kumar, K. G. Ragunathna, A. W. C. McConnaughie, D. W. Boykin, H.-J. Schneider and W. D. Wilson, *Bioorg. Med. Chem.*, 1997, 1157.

17. M. S. VanLoock, T. R. Easterwood and S. C. Harvey, *J. Mol. Biol.*, 1999, **285**, 2069.

18. (a) H.-K. Liu and P. J. Sadler, *Acc. Chem. Res.*, 2011, **44**, 349; (b) T. Biver, F. Secco and M. Venturini, *Coord. Chem. Rev.*, 2008, **252**, 1163.

19. D. Reha, M. Kabelac, F. Ryjacek, J. Sponer, J. E. Sponer, M. Elstner, S. Suhai and P. Hobza, *J Am. Chem. Soc.*, 2002, **124**, 3366.

20. H.-J. Schneider and J. Sartorius, *J. Chem. Soc., Perkin Trans*, 1997, **2**, 2319.

21. Y. Chu, D. W. Hoffman and B. L. Iverson, *J. Am. Chem. Soc.*, 2009, **131**, 3499.

22. (a) P. Murat, Y. Singh and E. Defrancq, *Chem. Soc. Rev.*, 2011, **40**, 5293; (b) M. L. Bochman, K. Paeschke and V. A. Zakian, *Nat. Rev. Genet.*, 2012, **13**, 770.

23. A. Parkinson, M. Hawken, M. Hall, K. J. Sanders and A. Rodger, *Phys Chem Chem Phys*, 2000, **2**, 5469.

24. M. M. Patel and T. J. Anchordoquy, *Biophys. Chem.*, 2006, **122**, 5.
25. A. A. Ouameur and H. A. Tajmir-Riahi, *J. Biol. Chem.*, 2004, **279**, 42041.
26. L. Vial, R. F. Ludlow, J. Leclaire, R. Perez-Fernandez and S. Otto, *J. Am. Chem. Soc.*, 2006, **128**, 10253.
27. D. K. Chand, H.-J. Schneider, J. A. Aguilar, F. Escartí and E. García-España, *Inorg. Chim. Acta*, 2001, **316**, 71.
28. H.-J. Schneider, *Angew. Chem., Int. Ed. Engl.*, 2009, **48**, 3924.
29. H.-J. Schneider and T. Blatter, *Angew. Chem., Int. Ed. Engl.*, 1992, **31**, 1207.
30. H.-J. Schneider and A. Yatsimirsky, *Chem. Soc. Rev.*, 2008, 263.
31. I. Nayvelt, M. T. Hyvonen, L. Alhonen, I. Pandya, T. Thomas, A. R. Khomutov, J. Vepsalainen, R. Patel, T. A. Keinanen and T. J. Thomas, *Biomacromolecules*, 2010, **11**, 97.
32. C. N. N'soukpoe-Kossi, A. A. Ouameur, T. Thomas, A. Shirahata, T. J. Thomas and H. A. Tajmir-Riahi, *Biomacromolecules*, 2008, **9**, 2712.
33. V. Pavlov, P. K. T. Lin and V. Rodilla, *Chem.-Biol. Interact.*, 2001, **137**, 15.
34. A. C. Gentry, S. L. Pitts, M. J. Jablonsky, C. Bailly, D. E. Graves and N. Osheroff, *Biochemistry*, 2011, **50**, 3240.
35. C. Samor, A. Guerrini, G. Varchi, G. L. Beretta, G. Fontana, E. Bornbardelli, N. Carenini, F. Zunino, C. Bertucci, J. Fiori and A. Battaglia, *Bioconjugate Chem.*, 2008, **19**, 2270.
36. L. Perez-Flores, A. J. Ruiz-Chica, J. G. Delcros, F. M. Sanchez-Jimenez and F. J. Ramirez, *Spectrochim. Acta A*, 2008, **69**, 1089.
37. N. Chen, A. A. Zinchenko, S. Murata and K. Yoshikawa, *J. Am. Chem. Soc.*, 2005, **127**, 10910.
38. I. Doi, G. Tsuji, K. Kawakami, O. Nakagawa, Y. Taniguchi and S. Sasaki, *Chem. Eur. J.*, 2010, **16**, 11993.
39. L.-M. Tumir and I. Piantanida, *Mini-Rev. Med. Chem.*, 2010, **10**, 299.
40. (a) A. N. Glazer and H. S. Rye, *Nature*, 1992, **359**, 859; (b) H. S. Rye, S. Yue, M. A. Quesada, R. P. Haugland, R. A. Mathies and A. N. Glazer, *Method Enzymol.*, 1993, **217**, 414.
41. G. Malojčić, I. Piantanida, M. Marinić, M. Žinić, M. Marjanović, M. Kralj, K. Pavelić and H. J. Schneider, *Org. Biomol. Chem.*, 2005, **3**, 4373.
42. J. Lhomme, J. F. Constant and M. Demeunynck, *Biopolymers*, 1999, **52**, 65.
43. K. Alarcon, M. Demeunynck, J. Lhomme, D. Carrez and A. Croisy, *Bioorg. Med. Chem. Lett.*, 2001, **11**, 1855.
44. T. H. Rehm, M. Radić Stojković, S. Rehm, M. Škugor, I. Piantanida and F. Würthner, *Chem. Sci.*, 2012, **3**, 3393.
45. (a) E. Micheli, C. M. Lombardo, D. D'Ambrosio, M. Franceschin, S. Neidle and M. Savino, *Bioorg. Med. Chem. Lett.*, 2009, **19**, 3903; (b) M. Franceschin, C. M. Lombardo, E. Pascucci, D. D'Ambrosio, E. Micheli, A. Bianco, G. Ortaggi and M. Savino, *Bioorg. Med. Chem.*, 2008, **16**, 2292.
46. R. Rodriguez, G. D. Pantos, D. P. N. Goncalves, J. K. M. Sanders and S. Balasubramanian, *Angew. Chem. Int. Ed.*, 2007, **46**, 5405.

47. P. Chitranshi and L. Xue, *Bioorg. Med. Chem. Lett.*, 2011, **21**, 6357.

48. M.-P. Teulade-Fichou, C. Carrasco, L. Guittat, C. Vailly, P. Alberti, J.-L. Mergny, A. David, J.-M. Lehn and W. D. Wilson, *J. Am. Chem. Soc.*, 2003, **125**, 4732.

49. (a) J. T. Davies, *Angew. Chem. Int. Ed.*, 2004, **43**, 668; (b) T. Ou, W-j. Lu, J.-H. Tan, Z.-S. Huang, K:-Y. Wong and L.-Q. Gu, *ChemMedChem*, 2008, **3**, 690; (c) S. Neidle, *Curr. Opin. Struct. Biol.*, 2009, **19**, 239; (d) S. Balasubramanian and S. Neidle, *Curr. Opin. Chem. Biol.*, 2009, **13**, 345; (e) E. Gavathiotis, R. A. Heald, M. F. G. Stevens and M: S. Searle, *J. Mol. Biol.*, 2003, **334**, 25.

50. S. N. Georgiadis, N. H. Abd Karim, K. Suntharalingam and R. Vilar, *Angew. Chem. Int. Ed.*, 2010, **49**, 4020.

51. N. Lomadze and H.-J. Schneider, *Tetrahedron Lett.*, 2002, **43**, 4403.

52. D. K. Chand, H.-J. Schneider, A. Bencini, A. Bianchi, C. Giorgi, S. Ciattini and B. Valtancoli, *Chem. Eur. J.*, 2000, **6**, 4001.

53. (a) M. Fernandez-Saiz, H.-J. Schneider and W. D. Wilson, *J. Am. Chem. Soc.*, 1996, **118**, 4739; (b) M. Fernandez-Saiz, F. Werner, T. M. Davis, H.-J. Schneider and W. D. Wilson, *Eur. J. Org. Chem.*, 2002, 1077.

54. R. J. Roberts, *Cell*, 1995, **82**, 9, and references cited therein.

55. A. Ali, M. Gasiorek and H.-J. Schneider, *Angew. Chem., Int. Ed. Engl.*, 1998, **37**, 3016.

56. (a) L. Baldini, A. Casnati, F. Sansone and R. Ungaro, *Chem. Soc. Rev.*, 2007, **36**, 254; (b) F. Sansone, L. Baldini, A. Casnati and R. Ungaro, *New J. Chem.*, 2010, **34**, 2715.

57. Review see: M. S. Peters, M. Li and T. Schrader, *Nat. Prod. Commun.*, 2012, **7**, 409.

58. Y. Shi and H.-J. Schneider, *J. Perkin Trans.*, 1999, **2**, 1797.

59. R. Lalor, J. L. DiGesso, A. Mueller and S. E. Matthews, *Chem. Commun.*, 2007, 4907.

60. R. V. Rodik, A. S. Klymchenko, N. Jain, S. I. Miroshnichenko, L. Richert, V. I. Kalchenko and Y. Mely, *Chem. Eur. J.*, 2011, **17**, 5526.

61. W. B. Hu, C. Blecking, M. Kralj, L. Šuman, I. Piantanida and T. Schrader, *Chem. Eur. J.*, 2012, **18**, 3589.

62. P. M. Cullis, R. E. Green, L. Merson-Davies and N. Travis, *Chem. Biol.*, 1999, **6**, 717.

63. J. M. Barret, A. Kruczynski, S. Vispe, J. P. Annereau, V. Brel, Y. Guminski, J. G. Delcros, A. Lansiaux, N. Guilbaud, T. Imbert and C. Bailly, *Cancer Res.*, 2008, **68**, 9845.

64. S. P. Jones, G. M. Pavan, A. Danani, S. Pricl and D. K. Smith, *Chem. Eur. J.*, 2010, **16**, 4519.

65. (a) G. M. Pavan, A. Danani, S. Pricl and D. K. Smith, *J. Am. Chem. Soc.*, 2009, **131**, 9686; (b) P. Posocco, S. Pricl, S. Jones, A. Barnard and D. K. Smith, *Chem. Sci.*, 2010, **1**, 393.

66. D. J. Welsh, S. P. Jones and D. K. Smith, *Angew. Chem. Int. Ed.*, 2009, **48**, 4047.

67. N. P. Gabrielson and J. J. Cheng, *Biomaterials*, 2010, **31**, 9117.

68. A. N. Zelikin, D. Putnam, P. Shastri, R. Langer and V. A. Izumrudov, *Bioconjugate Chem.*, 2002, **13**, 548.

69. A. Akinc, M. Thomas, A. M. Klibanov and R. Langer, *J. Gene Med.*, 2005, **7**, 657.

70. K. Itaka, N. Kanayama, N. Nishiyama, W. D. Jang, Y. Yamasaki, K. Nakamura, H. Kawaguchi and K. Kataoka, *J. Am. Chem. Soc.*, 2004, **126**, 13612.

71. F. A. Cotton, E. E. Hezzen Jr. and M. J. Legg, *Proc. Natl. Acad. Sci. USA*, 1979, **76**, 2551.

72. M. W. Göbel, J. W. Batts and G. Dürner, *Angew. Chem. Int. Ed.*, 1992, **31**, 207.

73. E. V. Anslyn and R. Breslow, *J. Am. Chem. Soc.*, 1989, **111**, 5972.

74. J. L. Sessler, A. Andrievsky, P. I. Sansom and V. Kral, *Biorg. Med. Chem. Lett.*, 1997, **11**, 1433.

75. G. García, V. Sararzy, V. Sol, C. Le Morvan, R. Granet and S. Alves, *Bioorg. Med. Chem.*, 2009, **17**, 767.

76. M.-P. Teulade-Fichou, D. Perrin, A. Boutorine, D. Polverrari, J.-P. Vigneron, J.-M. Lehn, J.-S. Sun, T. Garestier and C. Hélène, *J. Am. Chem. Soc.*, 2001, **123**, 9283.

77. Y. Huang, Y. Zhang, J. Zhang, D.-W. Zhang, Q.-S. Lu, J.-L. liu, S.-Y. Chen, H.-H. Lin and X.-Q. Yu, *Org. Biomol. Chem.*, 2009, **7**, 2278.

78. S. H. Wan, F. Liang, X. Q. Xiong, L. Yang, X. J. Wu, P. Wang, X. Zhou and C. T. Wu, *Bioorg. Med. Chem. Lett.*, 2006, **16**, 2804.

79. X. Sheng, X.-M. Lu, J.-J. Zhang, Y.-T. Chen, G.-Y. Lu, Y. Shao, F. Liu and Q. Xiu, *J. Org. Chem.*, 2007, **72**, 1799.

80. W. Xu, X. Yang, L. Yang, Z.-L. Jia, F. Liu and G.-Y. Lu, *New J. Chem.*, 2010, **34**, 2654.

81. U. Scheffer, A. Strick, V. Ludwig, S. Peter, E. Kalden and M. W. Göbel, *J. Am. Chem. Soc.*, 2005, **127**, 2211.

82. F. Freire, I. Cuesta, F. Corzana, J. Revuelta, C. González, M. Hricovini, A. Bastida, J. Jiménez-Barbero and J. L. Asensio, *Chem. Commun.*, 2007, 174.

83. Q. Li, D. Sun, Y. Zhou, D. Liu, Q. Zhang and J. Liu, *Inorg. Chem. Commun.*, 2012, **20**, 142.

84. (a) L. F. Chen, J. C. Di, C. Y. Cao, Y. Zhao, Y. Ma, J. Luo, Y. Q. Wen, W. G. Song, Y. L. Song and L. Jiang, *Chem. Commun.*, 2011, **47**, 2850; (b) A. Lee and M. C. DeRosa, *Chem. Commun.*, 2010, **46**, 418.

85. M. Loef and H. Walach, *Br. J. Nutr.*, 2012, **107**, 7.

86. (a) C. J. Frederickson, *Int. Rev. Neurobiol.*, 1989, **31**, 145; (b) M. P. Cuajungco and G. J. Lees, *Neurobiol. Dis.*, 1997, **4**, 137; (c) C. J. Frederickson, S. W. Suh, D. Silva, C. J. Frederickson and R. B. Thompson, *J. Nutr.*, 2000, **42**, 877; (d) P. D. Zalewski, X. Jian, L. L. Soon, W. G. Breed, R. F. Seamark, S. F. Lincoln, A. D. Ward and F. Z. Sun, *Reprod. Fertil. Dev.*, 1996, **8**, 1097; (e) E. J. Margalioth, J. G. Schenker and M. Chevion, *Cancer*, 1983, **52**, 868.

87. N. Lomadze, E. Gogritchiani, H.-J. Schneider, M. T. Albelda, J. Aguilar, E. García-España and S. V. Luis, *Tetrahedron Lett.*, 2002, **43**, 7801.

88. M. Inclán, M. T. Albelda, J. C. Frías, S. Blasco, B. Verdejo, C. Serena, C. Salat-Canela, M. L. Díaz, A. García-España and E. García-España, *J. Am. Chem. Soc.*, 2012, **134**, 9644.

89. S. Y. Shieh, M. Ikeda, Y. Taya and C. Prives, *Cell*, 1997, **91**, 325.

90. A. Mohkir, R. Krämer and H. Wolf, *J. Am. Chem. Soc.*, 2004, **126**, 6208.

91. T. Ikeda, K. Yoshida and H. J. Schneider, *J. Am. Chem. Soc.*, 1995, **117**, 1453.

92. D. Monchaud, P. Yang, L. Lacroix, M.-P. Teulade-Fichou and J.-L. Mergny, *Angew. Chem. Int. Ed.*, 2008, **47**, 4858.

93. Recent reviews: (a) F. Mancin, P. Scrimin, P. Tecilla and U. Tonellato, *Chem. Commun.*, 2005, 254; (b) F. Mancin, P. Scrimin and P. Tecilla, *Chem. Commun.*, 2012, **48**, 5545; (c) H. Katada and M. Komiyama, *Curr. Gene Therapy*, 2011, **11**, 38; (d) H. Katada and M. Komiyama, *Chem-BioChem*, 2009, **10**, 1279; (e) Y. Aiba, J. Sumaoka and M. Komiyama, *Chem. Soc. Rev.*, 2011, **40**, 5657; (f) R. Ott and R. Krämer, *Appl. Microbiol. Biotechnol.*, 1999, **52**, 761; (g) J. R. Morrow and O. Iranzo, *Curr. Opin. Chem. Biol.*, 2004, **8**, 192; (h) H.-J. Schneider and A. K. Yatsimirsky, in *The Lanthanides and Their Interrelations With Biosystems*, ed. H. Sigel, Marcel Dekker, New York, 2003, pp. 369–462. *Metal ions in biological systems*, vol. 40; (i) M. Komiyama, *Ibid.*,863; (j) S. J. Franklin, *Curr. Opin. Chem. Biol*, 2001, **5**, 201; (k) C. M. Dupureur, *Curr. Opin. Chem. Biol*, 2008, **12**, 250.

94. E. E. Kim and H. W. Wyckoff., *J. Mol. Biol.*, 1991, **218**, 449.

95. G. K. Schroeder, C. Lad, P. Wyman, N. H. Willimas and R. Wolfenden, *Proc. Natl. Acad. Sci. USA*, 2006, **103**, 4052.

96. (a) J. Schnabl and R. K. O. Sigel, *Curr. Opin. Chem. Biol.*, 2010, **14**, 269; (b) D. O. Corona-Martínez, P. Gomez-Tagle and A. K. Yatsimirsky, *J. Org. Chem.*, 2012, **77**, 9110, and references cited therein.

97. C. T. Liu, A. A. Neverov and R. S. Brown, *J. Am. Chem. Soc.*, 2008, **130**, 16711.

98. C. Liu, S. Yu, D. Li, Z. Liao, X. Sun and H. Xu, *Inorg. Chem.*, 2002, **41**, 913.

99. M. Yashiro, A. Ishikubo and M. Komiyama, *Chem. Commun.*, 1997, 83.

100. A. Panja, T. Matsuo, S. Nagao and S. Hirota, *Inorg. Chem.*, 2011, **50**, 11437.

101. J. H. Griffin and P. B. Dervan, *J. Am. Chem. Soc.*, 1987, **109**, 6840.

102. S. Matsuda, A. Ishikubo, A. Kuzuya, M. Yashiro and M. Komiyama, *Angew. Chem., Int. Ed.*, 1998, **37**, 3284.

103. R. Bonomi, F. Selvestrel, V. Lombardo, C. Sissi, S. Polizzi, F. Mancin, U. Tonellato and P. Scrimin, *J. Am. Chem. Soc.*, 2008, **130**, 3823.

104. L. A. Basile, A. L. Raphael and J. K. Barton, *J. Am. Chem. Soc.*, 1987, **109**, 7550.

105. X. Chen, J. Fan, X. Peng, J. Wang, S. Sun, R. Zhang, T. Wu, F. Zhang, J. Liu, F. Wang and S. Ma, *Biorg. Med. Chem. Lett.*, 2009, **19**, 4139.
106. R. Hettich and H.-J. Schneider, *J. Am. Chem. Soc.*, 1997, **119**, 5638.
107. (a) R. Hettich and H.-J. Schneider, *J. Perkin Trans.*, 1997, **2**, 2069; (b) J. Sumaoka, Y. Azuma and M. Komiyama, *Chem. Eur. J.*, 1998, **4**, 205.
108. Y. An, Y.-Y. Lin, H. Wang, H.-Z. Sun, M.-L. Tong, L.-N. Ji and Z.-W. Mao, *Dalton Trans.*, 2007, 1250.
109. S. Aoki and E. Kimura, *Chem. Rev.*, 2004, **104**, 769.
110. C.-Q. Xia, N. Jiang, X. Peng, J. Zhang, S.-Y. Chen, H.-H. Lin, X.-Y. Tan, Y. Yue and X.-Q. Yu, *Biorg. Med. Chem. Lett.*, 2006, **14**, 5756.
111. A. Roigk, R. Hettich and H.-J. Schneider, *Inorg. Chemistry*, 1998, **37**, 751.
112. K. G. Ragunathan and H.-J. Schneider, *Angew. Chem. Int. Ed. Engl.*, 1996, **35**, 1219.
113. (a) J. Rammo, R. Hettich, A. Roigk and H. J. Schneider, *Chem. Commun.*, 1996, 105; (b) A. Roigk and H. J. Schneider, *Chem. Eur. J.*, 2001, 205.
114. M. E. Branum, A. K. Tipton, S. Zhu and L. Que, Jr., *J. Am. Chem. Soc.*, 2001, **123**, 1898.
115. S. Hashimoto and Y. Nakamura, *J. Perkin Trans.*, 1996, **1**, 2623.
116. D. Magda, R. A. Miller, J. L. Sessler and B. L. Iverson, *J. Am. Chem. Soc.*, 1994, **116**, 7439.
117. D. Magda, M. Wright, S. Crofts, A. Lin and J. L. Sessler, *J. Am. Chem. Soc.*, 1997, **119**, 6947.
118. A. Kuzuya, Y. Shi, K. Tanaka, K. Machida and M. Komiyama, *Chem. Lett.*, 2009, **38**, 432.
119. S. Hashimoto, K. Itai, Y. Takeuchi and Y. Nakamura, *Heterocycl. Commun.*, 1997, **3**, 307.
120. S. Hashimoto, T. Inui and Y. Nakamura, *Chem. Pharm. Bull.*, 2000, **48**, 603.
121. M. Komiyama and J. Sumaoka, *Bull. Chem. Soc. Jpn*, 2012, **85**, 533; see also T. Lönnberg, Y. Aiba, Y. Hamano, Y. Miyajima, J .Sumaoka and M. Komiyama, *Chem. Eur. J.,* 2010, **16**, 855.
122. N. C. Seeman, *Annu. Rev. Biochem.*, 2010, **79**, 65.
123. (a) R. F. Service, *Science*, 2011, **332**, 1140; (b) B. Sacca and C. M. Niemeyer, *Angew. Chem. Int. Ed.*, 2012, **51**, 58; (c) R. M. Zadegan and M. L. Norton, *Int. J. Molecular Sci.*, 2012, **13**, 7149; (d) T. Schnitzler and A. Herrmann, *Acc. Chem. Res.*, 2012, **45**, 1419; (e) F. C. Simmel, *Curr. Opin. Biotechnol.*, 2012, **23**, 516.
124. (a) G. Mayer, *Angew. Chem. Int. Ed.*, 2009, **48**, 2672; (b) G. Mayer, S. Lennarz, F. Rohrbach and F. Tolle, *Angew. Chem. Int. Ed.*, 2011, **50**, 12400.
125. (a) O. I. Wilner and I. Willner, *Chem. Rev.*, 2012, **112**, 2528; (b) H. An and B. Jin, *Biotechnol. Adv.*, 2012, **30**, 1721.

CHAPTER 9

Biomolecular Interactions of Platinum Complexes

BENJAMIN W. HARPER, FENG LI, RHYS BEARD,
K. BENJAMIN GARBUTCHEON-SINGH, NEVILLE S. NG
AND JANICE R. ALDRICH-WRIGHT*

Nanoscale Organisation and Dynamics Group, School of Science and
Health, University of Western Sydney, Locked Bag 2579, Penrith South DC,
NSW, 1797, Australia
*Email: j.aldrich-wright@uws.edu.au

9.1 Introduction

The *in vitro* and *in vivo* effects of platinum complexes have been well reported. However, it is their interactions with biologically relevant structures and biomacromolecules such as DNA and proteins that can determine their therapeutic effects. The pharmacological properties of platinum complexes are dependent on binding interactions, and the investigation of such interactions is essential to gain insight into the mechanisms of action. Interactions with DNA can inhibit cellular processes such as replication and transcription, which are vital for the proliferation of cells.[1,2] By investigating how platinum complexes achieved these different interactions we may be able to determine the relationships between the structural characteristics, biophysical properties and biological activity in order to design more effective anticancer drugs.

Monographs in Supramolecular Chemistry No. 13
Supramolecular Systems in Biomedical Fields
Edited by Hans-Jörg Schneider
© The Royal Society of Chemistry 2013
Published by the Royal Society of Chemistry, www.rsc.org

9.2 *cis*-Diamminedichloro Platinum(II)—Cisplatin

Since the opportune rediscovery of Peyrone's chloride, cisplatin (**1**, $[Pt(NH_3)_2Cl_2]$, see Figure 9.2) by Rosenberg in 1965,[3,4] platinum complexes have been widely used for the treatment of various carcinomas. The clinical success of cisplatin against a number of cancer types, including testicular, breast, ovarian, melanoma, lymphoma, myelomas, bladder, lung and neck, accounts for the Food and Drug Administration's approval of the drug and its application.[5,6]

Cisplatin interacts with a wide range of biomolecules, including proteins, but it is generally accepted that its anticancer properties result from coordinate binding to nuclear DNA.[7] Cisplatin is administered intravenously, and remains as a chloride due to the high chloride ion concentration in the blood stream (~ 100 mM).[8,9] Intracellular uptake of cisplatin depends on a number of factors including the concentration of sodium and potassium ions, pH, and the high polarity of cisplatin which impedes the rate of cellular uptake.[10] Previously, it was proposed that cisplatin entered the cell *via* passive diffusion;[11] however, more recent evidence has implicated transport proteins such as protein copper transporter-1 protein (CTR1) in cisplatin influx.[12,13] Once inside the cell, cisplatin encounters a significantly lower chloride concentration (4–20 mM) which facilitates hydrolysis of one or both chloride ligands, forming $[Pt(NH_3)_2Cl(H_2O)]^+$ or $[Pt(NH_3)_2(H_2O)_2]^{2+}$, respectively.[14–16] Interactions between cisplatin and DNA occur through coordinative/covalent binding. Coordinative binding with DNA is characterised by the formation of adducts, with the N7 position on the imidazole rings of guanine (G) and to a lesser extent, adenine (A) of DNA, which is irreversible and non-specific.[17]

Cisplatin can form monofunctional (2%) adducts and various types of bifunctional adducts (Figure 9.1). The most common adducts, 1,2-(GpG) and 1,2-(ApG) intra-strand crosslinks, are formed between adjacent base pairs on the same DNA strand and represent 60–65% and 20–25% of the adduct formed, respectively. Adducts that are less frequently formed are (i) 1,3-GpXpG intra-strand crosslinks (2%), where a base pair separates the two guanine bases bound to platinum, (ii) monofunctional guanine adducts (2%), (iii) G–G (2%) inter-strand crosslinks, where platinum adduct forms between two guanines on two different strands of DNA[10] as well as (iv) protein–DNA crosslinks.[15,18,19] The formation of these crosslinks can cause three-dimensional changes in the structure of DNA such as localised bending and unwinding.[20] These distortions can obstruct, or inhibit the cellular processes involving DNA[20] replication and transcription preventing proliferation of cancer cells,[21–23] and in some cases lead to programmed cell death (apoptosis).[24]

Some of the biophysical properties limit the efficacy of cisplatin. Low solubility in aqueous solutions, and therefore the low solubility in the blood, leads to protracted intravenous drug administrations with hydration and diuresis required to minimise nephrotoxicity.[16,22,25] Side-effects such as neurotoxicity (nervous system damage), myelosuppression (reduction in bone marrow activity), ototoxicity (hearing damage), and nephrotoxicity (impaired kidney function) significantly limit patient doses.[26–29] Cisplatin resistance is either

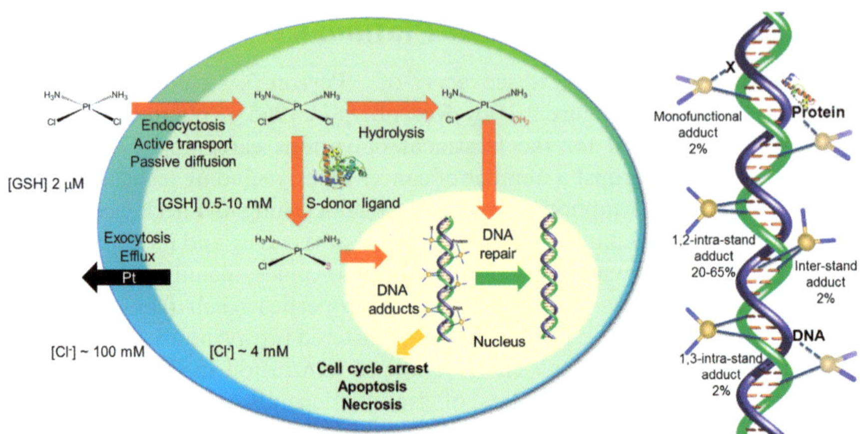

Figure 9.1 Due to the chloride concentration outside the cell cisplatin remains mostly unreacted. Once it has diffused into the cell cisplatin hydrolyses and can react with intracellular sulfur-containing proteins, cellular DNA or RNA. Once in the nucleus, cisplatin binds to DNA through covalent bonds (irreversible) with the purine bases of DNA, and if not repaired is able to prevent DNA transcription and replication and ultimately cause cell cycle arrest and cell death.[6,34,35] Some of the adducts formed between cisplatin and DNA are shown. (DNA double helix models were generalised by the Persistence of Vision Raytracer [POV-Ray] program.[36])

intrinsic or acquired[16,30] and can be attributed to decreased membrane transport of the drug, increased cytoplasmic detoxification, increased DNA repair and increased DNA tolerance to damage. Cisplatin deactivation by S-donor ligands found in, for example, glutathione and metallothionein, constitutes another problem to be overcome.[5,31–33]

9.3 Cisplatin Derivatives

In efforts to overcome the clinical disadvantages of cisplatin and produce a new compound which would demonstrate improved properties, thousands of platinum-based complexes have been synthesised. Over the past 40 years, 33 platinum-based drugs have entered clinical trials, with only *cis*-diamine(1,1-cyclobutanedicarboxylato)platinum(II) (carboplatin, **2**, 1993) and [(1*R*,2*R*)-diaminocyclohexane (ethanedioato-*O,O′*)]platinum(II) (oxaliplatin, **3**, 2002) achieving international therapeutic approval.[33] Three others have achieved approval in individual countries: diammine[(hydroxy-κO)acetato(2-)-κO]platinum(II) (nedaplatin, **4**, Japan); [SP-4-2-[4*R*-(2*a*,4*a*,5*b*)]]-[2-(1-methylethyl)-1,3-dioxolane-4,5-dimethanamine-*N,N′*][propanedioato(2-)-*O,O′*]-platinum(II) (heptaplatin, **5**, South Korea); and [(1*R*,2*R*)-2-(aminomethyl)cyclobutyl]methanamine-2-hydroxypropanoic acid platinum(II) (lobaplatin, **6**, China). Currently, four platinum drugs are in various stages of clinical trials: (OC-6-43)-*bis*(acetato)amminedichloro(cyclohexylamine) platinum(IV) (satraplatin, **7**); azane-(2-methylpyridine) platinum(II) dichloride (picoplatin, **8**);

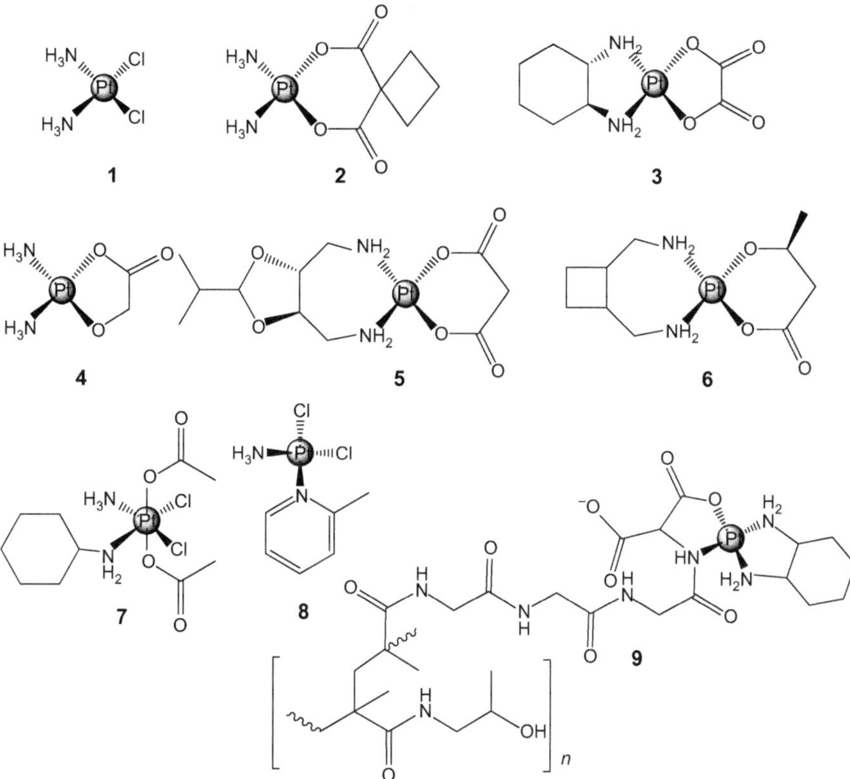

Figure 9.2 Currently approved platinum-based chemotherapeutic drugs: **1**, cisplatin; **2**, carboplatin; **3**, oxaliplatin; **4**, nedaplatin; **5**, heptaplatin and **6**, loba-platin. Current platinum-based drugs in clinical trials: **7**, Satraplatin; **8**, picoplatin and **9**, ProLindac™.

ProLindac™ a diaminocyclohexane platinum hydroxypropylmethacrylamide polymer prodrug (**9**); and lipoplatin, a liposomal derivative of cisplatin.

One of the first successful cisplatin derivatives used in the clinic was carbo-platin (**2**, Figure 9.2), because it does not induce nephrotoxicity and demon-strated a much lower toxicity profile than cisplatin.[21,37] The reason for this improved efficacy is that carboplatin incorporates the more stable cyclobutane dicarboxylate as a leaving group, which effectively reduces the rate of hy-drolysis.[10,38] As a result, carboplatin can be administered at much higher doses; however, this advantage is limited by myelosuppression, in particular throm-bocytopenia. A 20–40% increase in concentration is required to produce an equivalent efficacy compared with cisplatin, and the rate of adduct formation by carboplatin is 10-fold slower.[39] It also demonstrates cross-resistance with cisplatin as a consequence of the similar mechanism of action and as such is ineffective against many cancers that display an intrinsic or induced cisplatin resistance.[40] Despite this, carboplatin remains a vital component in many chemotherapy protocols.[41,42]

Further developments produced substitution of the amine ligands with mono- or bi-dentate ligands, evoking changes in the steric, electronic and basicity effects.[43] Two examples of the resulting platinum(II) complexes are *cis*-dichloro(*trans*-1R,2R-diaminocyclohexane)platinum(II) and *cis*-dichloro(1,2-diaminobenzene)platinum(II), which both showed a better therapeutic index than cisplatin in mice implanted with Sarcoma 180, and comparable cytotoxicity towards PC6 tumours in mice and L1210 tumour cells.[43–45] However, complexes containing 1R,2R-diaminocyclohexane and 1,2-diaminobenzene produced compounds with poor solubility in water and as a consequence slurries were required to test the complexes.[43] Furthermore, poor solubility in the blood limited clinical use.

In third generation cisplatin analogues, both the leaving and the amines groups have been replaced; for example, oxaliplatin (**3**, Figure 9.2) contains two bidentate ligands, a 1R,2R-diaminocyclohexane and an oxalate. This produces a compound that is water-soluble, is not cross-resistant with cisplatin,[46,47] has lower nephrotoxicity and myelosuppression compared to cisplatin, has less haematological toxicity than carboplatin at therapeutic doses[48–50] and demonstrates a wider spectrum of activity than cisplatin or carboplatin.[51,52] It has been widely used in the treatment of ovarian and colorectal cancers after cisplatin or carboplatin has failed, or when the cancer has recurred.[23,30] In particular it is used in conjunction with 5-fluorouracil to treat human colorectal cancer which is resistant to both cisplatin and carboplatin.[53] These properties are attributed to the 1R,2R-diaminocyclohexane ligand.[54] Its clinical use, however, has diminished owing to its significant neurotoxicity.[38,52,55] The other chiral analogues (*RR*, *SS*, *RS*) of 1,2-diaminocyclohexane platinum(II) complexes are still being investigated due to their activity in cisplatin-resistant cells, but none has shown significant therapeutic advantage over cisplatin or carboplatin.[56–58]

9.3.1 Initially Proposed Structure–Activity Relationships

Of the thousands of platinum-based complexes that have been designed, synthesised and examined for their antitumour activity,[37] only 33 compounds have entered clinical trials and all but a few are cisplatin analogues. From this information Cleare and Hoeschele[59,60] initially proposed that a number of structural requirements for platinum(II) complexes were necessary for activity against tumours. These structural requirements describe square–planar platinum(II) complexes with the general formula *cis*-[PtX$_2$(NH$_2$R)$_2$], where NH$_2$R is an inert amine (such as NH$_3$ or ligands containing stable primary or secondary amine groups) and X is two anionic leaving groups with intermediate binding strength to platinum, a weak *trans* effect[37,59,60] (to facilitate coordination to the DNA), and an overall neutral charge.[48,61] The *trans* effect is defined by the influence upon substitution in square–planar complexes of ligands, *trans* to the position of substitution. The intensity the *trans* effect is measured by the increase in rate of substitution and follows this order: F$^-$, H$_2$O, OH$^-$ <NH$_3$ <py< Cl$^-$ <Br$^-$ <I$^-$, SCN$^-$, NO$_2$$^-$, SC(NH$_2$)$_2$, Ph$^-$ <SO$_3$$^{2-}$ <PR$_3$,

AsR_3, SR_2, CH_3^- <H^-, NO, CO, CN^-, C_2H_4. These ligands also exert their *trans* effect in the same order upon bond lengths and stability.

It is estimated that $\leq 1\%$ of the administered cisplatin reaches the intended target, namely DNA.[31] In addition, clinical use is limited by severe side-effects, which often results in the drug being administered at suboptimal doses and affords tumours the opportunity to develop resistance. Tumour resistance may occur in many forms including reduced uptake or increased drug efflux by tumour cells; improved repair; tolerance due to the up-regulation of DNA repair and tolerance pathways; and degradation by intracellular thiols due to increased glutathione levels (as high as 10 mM) in drug-resistant cells.[32,33]

The subsequent design of platinum complexes, to overcome these drawbacks, has seen a departure from "the Cleare and Hoeschele rules".[37] Researchers have developed compounds that break these structural conventions and demonstrate superior cytotoxicity than cisplatin.[9,10,62,63] Many of these platinum complexes do not simply coordinately bind to DNA, but interact through different mechanisms and a combination of mechanisms to evoke different, and more effective, responses.[64]

9.4 Types of DNA Binding Interactions

DNA offers many potential sites for molecules to bind owing to its size and complexity. Cisplatin and its structural analogues are thought to produce their therapeutic effect by coordinating to the N7 of base pairs (guanine or adenine); however, this is not the only way in which platinum complexes can interact with DNA. Positively charged compounds are able to bind to DNA (negatively charged by virtue of its sugar-phosphate backbone) through two main modes of interaction: irreversible covalent/coordinative binding; or reversible intermolecular associations, which can then be further divided into electrostatic binding, groove binding and intercalation. Metal complexes may exhibit a preference for a particular binding mode or nucleotide sequence, depending on the size and shape of the molecule.

9.4.1 Irreversible Covalent/Coordinative Binding

Cisplatin and many of its analogues bind coordinately to DNA.[65] This requires that the leaving groups be replaced by water to give a reactive, positively charged species. This charged species is electrostatically attracted to the DNA, before coordinate binding. Examples of crystal structures of platinum complexes that coordinately bind to DNA are shown in Figure 9.3. A number of multinuclear platinum complexes has also been developed,[66] some of which have shown activity in both cisplatin-sensitive and cisplatin-resistant cell lines.[24] The most well-studied complexes are the bi-([{*trans*-PtCl(NH$_3$)$_2$}$_2$(H$_2$N(CH$_2$)$_4$NH$_2$)]$^{2+}$ 1,1/t,t) and trinuclear (BBR3464)[66] with *trans* geometry that covalently binds to DNA (Figure 9.4).[67]

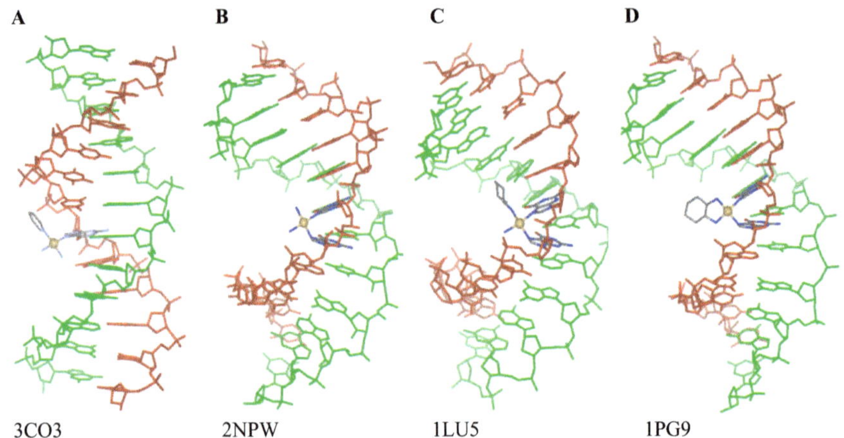

Figure 9.3   Representations from the X-ray crystal structures showing (A) a mono DNA adduct with [Pt(ammine)$_2$(cyclohexylamine)]$^+$ (PDB 3CO3);[68] (B) a 1,2-intra-strand DNA adduct with cisplatin (PDB 2NPW);[69] (C) an 1,2-intra-strand DNA adduct with [Pt(ammine)(cyclohexylamine)]$^{2+}$ (PDB 1LU5);[70] and (B) a 1,2-intra-strand DNA adduct with oxaliplatin (PDB 1PG9).[71] Hydrogen atoms and solvent molecules have been omitted for clarity.

9.4.2   Reversible Intermolecular Associations

Reversible intermolecular associations such as electrostatic binding occur due to the interaction between water and cations, such as Na$^+$ and the polynuclear platinum complex TriplatinNC (Figure 9.4) with the negatively charged phosphate backbone at the exterior surface of the DNA helix. The reduction in net charge of the DNA and the removal of hydration can produce conformational changes in the DNA.

9.4.3   Groove Binders

Groove binding involves matching the contour of the curved major or minor grooves of DNA with molecules that bend to fit into the curvature.[74] These molecules usually consist of a number of aromatic rings attached by flexible amide linkers that allow them to twist in order to make weaker intermolecular forces with the floor of the groove. A comprehensive description of these interactions is provided in Chapter 8 of this book.

Experiments have confirmed that the sequence selectivity of these small groove binders can be expanded if the molecules are linked by platinum(II), for example, bis-netropsin and bis-distamycin (Figure 9.5).[75] These larger molecules still form cooperative associations and show considerable affinity for the sequence, d(TTTTAAAA)$_2$, even when linked by platinum. However, the sequence specificity of these compounds is limited to a few sequences of DNA and they must form dimers in solution to bind with any selectivity to DNA.

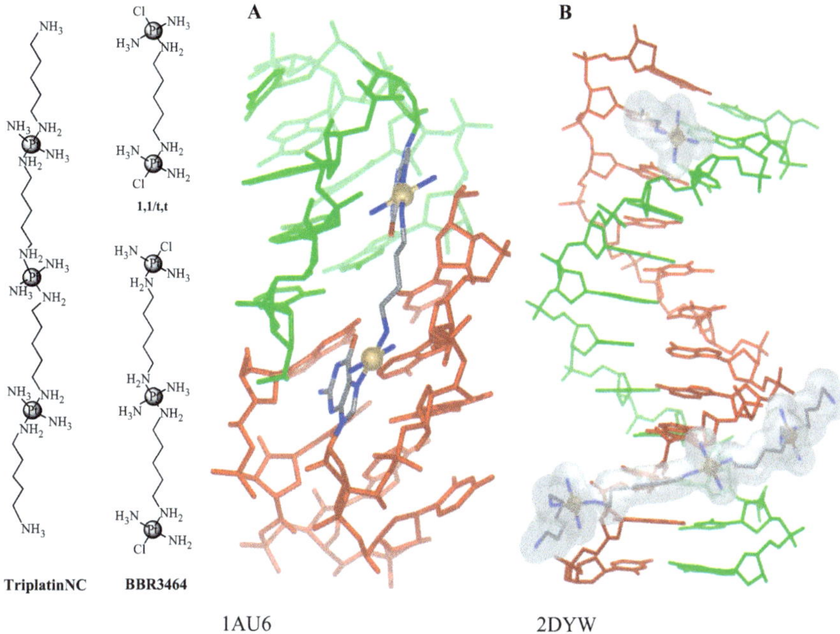

Figure 9.4 Molecular models of the complexes as well as representations from the X-ray crystal structures showing (A) a bisplatinum compound, 1,1/t,t cross-linked to a GpC site in DNA (PDB 1AU6).[72] (B) Electrostatic DNA binding by a polynuclear platinum complex, TriplatinNC, in semi-transparent space-filling representation (PDB 2DYW).[73] Hydrogen atoms and solvent molecules have been omitted for clarity.

Brabec and co-workers[76] have attached platinum to the end of distamycin in an effort to harness the sequence-selective attributes of this groove binder.

The effect of attaching platinum to the terminal end of a pyrrole–pyrrole–imidazole polyamide (PyPyIm-Pt, DJ1953-2) would determine if the resulting complex would bind sequence-specifically in the minor groove and then bind covalently to DNA. The platinum polyamide, DJ-1953-2,[77] formed a dimer (as shown in Figure 9.6), initially groove-bound to the target sequence of DNA, d(CATTGTCAGAC)₂, with greater affinity than the mismatched sequences and subsequently covalently bound. Complexes that could groove and coordinately bind include, [{*trans*-PtCl(NH₃)₂}(N₂H(CH₂)₄NH₂)]$^{2+}$ (1,1/t,t) (see Figure 9.4), cisplatin-distamycin and the parallel dimer of DJ1953-2.

9.4.4 Intercalation

Intercalation is reversible, and is stabilised by a combination of electrostatic, hydrogen bonding, entropic, van der Waals and hydrophobic interactions. As a consequence of intercalation, DNA is lengthened, stiffened and unwound as well as losing regular helical structure[49,78] but this effect is dependent upon the "depth of penetration."[50,79,80] Intercalation anchors the metal complex with a

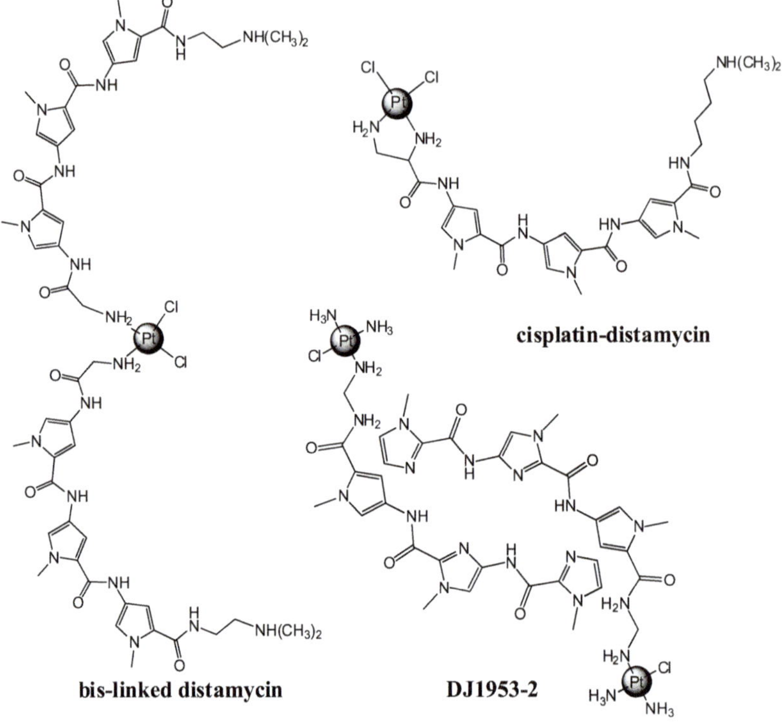

Figure 9.5   Examples potential groove binders as well as molecules that could bind by a combination of groove and coordinate binding.

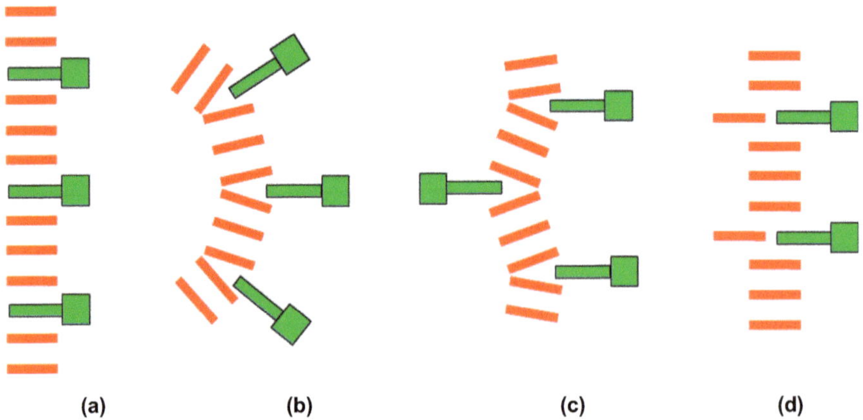

Figure 9.6   A schematic showing the different forms of intercalation. The base pairs are in red, and the intercalators are in green (a) intercalation; (b) and (c) semi-intercalation; and (d) quasi-intercalation.

distinct orientation within the intercalation site, allowing functional groups to interact with the DNA grooves. (2,2′:6′,2″-Terpyridine) platinum(II) chloride, [Pt(terpy)Cl]$^{+}$ and 2-hydroxyethanethiolato (2,2′:6′,2″-terpyridine)

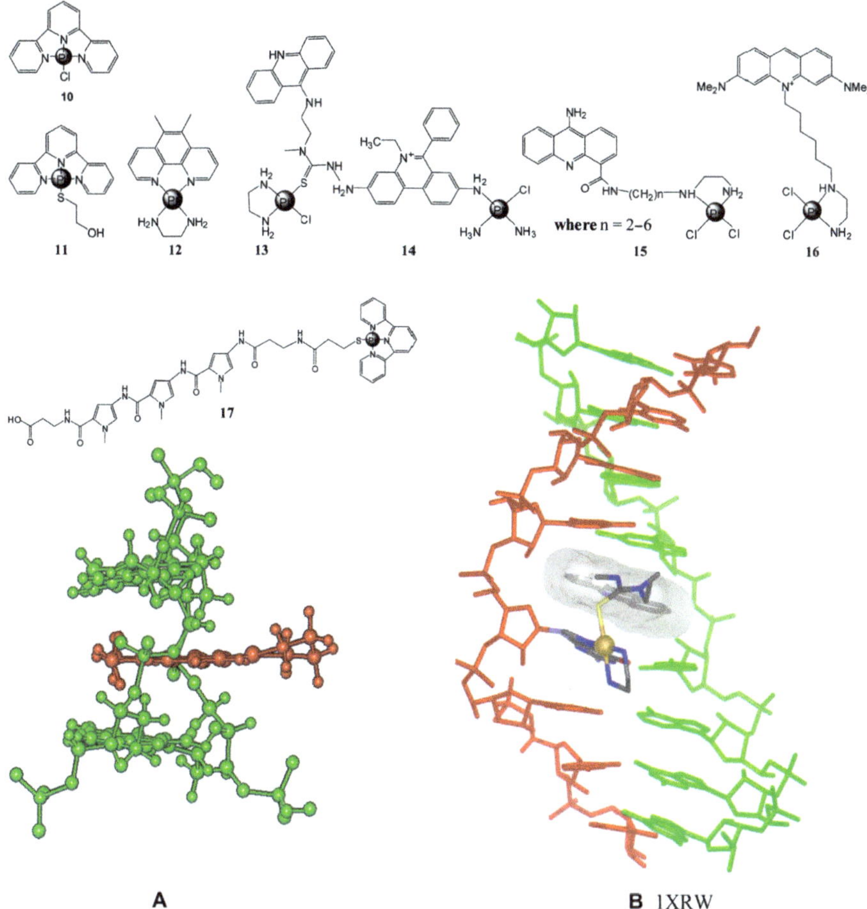

A **B** 1XRW

Figure 9.7 Some examples of platinum(II) complexes that intercalate. (A) A molecular model of [Pt(en)(56Me$_2$phen)]$^{2+}$ (red) intercalated from the minor groove. For simplicity, only the T2 and C3 (green) base pairs have been shown, since they constitute the actual binding site.[63] (B) A representation from the X-ray crystal structures showing PT-ACRAMTU both coordinately binding and intercalating into DNA (PDB 1XRW),[97] acridine in semi-transparent space-filling representation. Hydrogen atoms and solvent molecules have been omitted for clarity.

platinum(II), ([Pt(terpy)HET]$^{2+}$ (Figure 9.7, **10** and **11**, respectively)[81,82] both intercalate into the DNA base pair stack.[83,84] Other fully coordinated square planar platinum(II) complexes in the form of [Pt(A$_L$)(I$_L$)]$^{2+}$ where, for example, A$_L$ could be 1,2-diaminoethane or 1,2-diaminocyclohexane and I$_L$ is 1,10-phenanthroline (phen) or 5,6-tetramethyl-1,10-phenanthroline (56Me$_2$phen),[85,86] have shown activity in cultured L1210 murine leukaemia cells and bind DNA by intercalation (Figure 9.7, **12**).[64,87,88]

This type of interaction is not only limited to square planar platinum(II) complexes as octahedral platinum(IV) complexes, but have also been shown to intercalate into DNA. The coordination geometry of square planar complexes allows for deeper intercalator insertion than octahedral geometry.[87,88] Platinum(II) complexes containing intercalating ligands stack between the base pairs of DNA,[89,90] depending on the choice of the ancillary ligands which may insert beyond the platinum(II) centre.[93,94] In the case of octahedral complexes the disposition of the ancillary ligands in relation to the intercalating ligand produces steric interactions with the DNA preventing full insertion.[87,88] The presence of an intercalator inserted between two base pairs excludes the neighbouring intercalation sites from being occupied.[49,78,94] This phenomenon is referred to as the Nearest Neighbour Exclusion Model.[94–96] It results in periodic intervals of 10.2Å between two drugs binding as it spans three binding sites.

Examples of complexes that can interact through intercalation and some other binding forms have been reported.[97,118,119] Complexes that can intercalate and coordinately bind, including complexes that link or tether cisplatin[98,99] such as PT-ACRAMTU ([(PtCl(en)(ACRAMTU-S)](NO$_3$)$_2$, where en = 1,2-diaminoethane and ACRAMTU =1-[2-(acridin-9-ylamino)ethyl]-1,3-dimethylthiourea)]; **13**). PT-ACRAMTU forms adducts to guanine-N7, selectively at 5′-CG sites in the major groove ~80% of the time (Figure 9.7, **13**, 1XRW).[97]

9.5 Multinuclear Platinum(II) Compounds

While the design of cisplatin derivatives for improved anticancer activity has been extensively studied since the early 1980s, there has also been much interest in the development of multinuclear platinum(II) complexes that offer a similar covalent-binding mechanism of action which produces a range of different DNA adducts. Multinuclear drugs offer advantages over mononuclear platinum drugs; these include high cationic charge, facilitating DNA attraction, as well as the capability to produce new adducts offering low resistance factors. In the 1990s there was a large degree of interest surrounding multinuclear platinum complexes, with over 50 complexes synthesised and biologically assessed.[100] A summary of a selection of these complexes can be seen in Figure 9.8. Farrell's research group has played a leading role in the design and synthesis of multinuclear platinum complexes with initial complexes designed around a cisplatin motif.[101–103] The linking of two cisplatin centres by a flexible diamminoalkane tether produced promising results, with cytotoxicity against cisplatin-resistant cell lines and the capability of forming distinctively different, long range intra- and inter-strand DNA adducts.[104]

Inspired by the promising results of Farrell[101,102,104] a variety of binuclear cisplatin-based complexes were synthesised by both Broomhead's and Komeda's teams with the incorporation of aromatic tethers. Broomhead and his co-workers developed cisplatin complexes linked by a flexible aromatic 4,4′-dipyrazolylmethane tether to produce new platinum-DNA adducts.[105] While the novel DNA interactions were not fully characterised, covalent adducts as well as minor groove binding were proposed, with cytotoxicity comparable to cisplatin in both cisplatin-sensitive and cisplatin-resistant cell lines.[106,107]

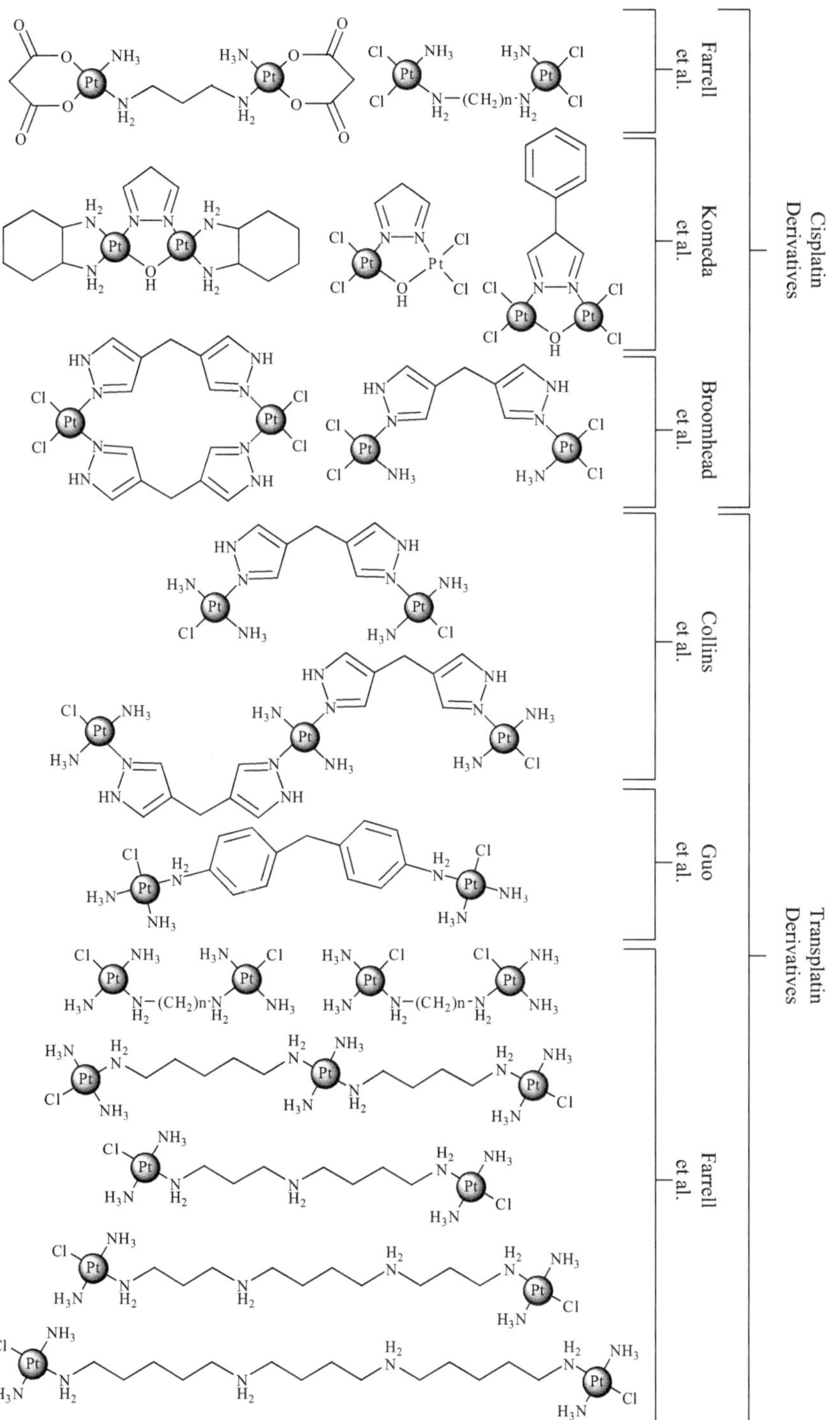

Figure 9.8 A selection of multinuclear *cis*- and *trans*-platinum complexes previously synthesised.

Collin's research team synthesised structurally similar transplatin complexes informed by Broomhead's work, which incorporated a single 4,4-dipyrazolylmethane group for a binuclear-platinum complex or two 4,4′-dipyrazolylmethane tethers for the trinuclear-platinum complex. The resultant transplatin complexes showed significant cellular uptake and accumulation, as well as significant activity against murine leukaemia L1210, human ovarian cancer cell line 2008, and the respective cisplatin resistant cell lines. *In vitro* DNA binding showed that both the di- and trinuclear complexes were efficient inter-strand cross-linking agents, while the dinuclear complex formed a greater number of inter-strand links. Molecular modelling suggested that both complexes are incapable of forming the 1,2-intra-strand adduct due to the rigidity of the aromatic linkers.[108,109] A similar monocovalent dinuclear-platinum complex with an aromatic 4,4′-methylenedianiline tether was synthesised by Guo's research group. This complex showed much higher cytotoxicity than cisplatin at low concentrations against both the murine leukaemia P-388 and human non-small-cell lung cancer A-549 cell lines. Interactions of this complex with DNA using circular dichroism, indicated a transformation of the tertiary structure of the DNA from B to Z form.[110,111]

In a slightly different approach, research by Komeda *et al.*[112–114] produced azole-bridged dinuclear transplatinum complexes with the aim of producing 1,2-intra-strand DNA cross-links like cisplatin but with minimal distortion of the DNA scaffold, in order to alter the toxicity profile of the complex. The formation of a sterically distinct adduct with minimal distortion of the DNA backbone was shown to be feasible in an NMR binding study[114] while increased cytotoxicity was also shown, with the results proposed to be due to mitigation of the DNA-damage repair mechanism in cisplatin resistant cell lines.[114,115]

Todate the most successful complexes have come from Farrell's research group which has synthesised a wide range of bi- and trinuclear *trans*-platinum complexes (see Figure 9.4), providing clinically promising results. DNA binding studies have revealed a marked increase in inter-strand cross-linking of about 40-fold over cisplatin.[116] As well as the increase in inter-strand binding, the kinetics for the formation of both covalent bonds to form a 1,4-GG cross-link have been shown to be 10 to 55-fold faster than the 1,2-GG intra-strand cross-links formed by cisplatin.[117] Due to the length of these complexes the formation of both intramolecular inter-strand cross-links and inter-duplex cross-links were observed, in conjunction with a wider spectrum of cytotoxicity.[116]

The clinically promising multinuclear platinum complex [{*trans*-PtCl(NH$_3$)$_2$}$_2$μ-(*trans*-Pt(NH$_3$)$_2$(H$_2$N(CH$_2$)$_6$NH$_2$)$_2$)]$^{4+}$ (BBR3464) has been shown to effectively bind to DNA causing long-range inter- and intra-strand DNA adducts of up to 6 bases apart as illustrated in Figure 9.4. This long-range adduct formation has been shown to efficiently unwind DNA, as well as transform B-DNA to its Z-form.[118–120] While the binding of both bi- and tri-nuclear complexes is similar, the trinuclear complex has a greater affinity for DNA due to the increased cationic charge on the complex by addition of a biologically inactive central platinum monomer in the tether.[121,122] Extensive cellular studies have been undertaken using BBR3464, with results showing a highly cytotoxic profile with activity at least 20 fold greater than cisplatin against a range of human cell lines both sensitive and resistant to cisplatin.[120,123,124] The high cellular potency of BBR3464, particularly in cisplatin-resistant cell lines and tumour models, is attributed to the novel and irreversible interaction with DNA as well as a correlation with the increase in cellular uptake and accumulation.[120,123,125–127]

While the overall cytotoxicity is important, the lack of cross-resistance and the novel DNA adducts when compared to cisplatin offers a promising route for future drug design. Factors that contribute to this activity include reduced deactivation by intracellular thiols,[128,129] cell cycle arrest,[130] activation of p-53 independent cell death[126] as well as a reduction in the expression of the DNA mis-match repair protein PMS2.[126,127] While the preclinical trials and cellular studies of this series of drugs seemed promising, the clinical trials have not been as positive. Initial phase I trials showed high neutropenia and gastrointestinal toxicity contributed to the prolonged half-life and platinum accumulation.[131] Successful variation of the treatment regime was undertaken leading to phase ii trials in patients with gastric or gastro-oesophageal adenocarcinoma,[132] advanced ovarian cancer,[133] non-small cell lung cancer[134] and sensitive or refractory small cell lung cancer.[135] The lack of activity in all phase II trials to date does not support further evaluation of BBR3464 against any cancers tested thus far, with no further new trials reported.

9.6 Terpyridineplatinum(II)-Based Intercalators

9.6.1 Monointercalators

The first terpyridineplatinum(II) compound produced, $[Pt(terpy)Cl]^+$ (**10**, Figure 9.7) was shown to bind to calf thymus(ct)-DNA.[81] Binding studies revealed that covalent binding (platination) and intercalation both played a role.[136,137] It was determined that **10** initially intercalates *via* the aromatic terpyridine (terpy) ligand, inserting between the base pairs, leading to modifications in the structure and properties of the ct-DNA.[136,138,139] Complex **10** subsequently forms covalent bonds to base pairs in DNA *via* the loss of the labile chloride ligand (a good leaving group) allowing aquation to $[Pt(terpy)H_2O]^{2+}$ which readily forms covalent bonds to base pairs of DNA.[81,137] In order to determine the effect of intercalation alone, over the well-studied chemical interaction of covalent binding to DNA, substitution of the chloride ion with a thermodynamically stable ligand with a much slower rate of substitution was investigated.

The substitution of the reactive chloride ion in **10** produced **11** (Figure 9.7) which was the prototype for a family of terpyridineplatinum(II) intercalating complexes. Since then, various functional groups have been used to synthesise a range of analogous complexes. The properties of the resulting complexes can be tailored by the specific choice of the group coordinated to the $[Pt(terpy)]^{2+}$ moiety affecting the structure–activity relationships, self-stacking,[140] photophysical properties and interaction with biomolecules including proteins and DNA.

9.6.2   Ligand Substitution and Attached Groups

The potential of terpyridineplatinum(II) complexes as anticancer drugs was indicated by the reported cytotoxicity of several complexes against human ovarian carcinoma cell lines and cell lines resistant to cisplatin.[141–144] Linear dichroism studies and unwinding ligation studies showed that the complexes bound to ct-DNA by intercalation, whereas circular dichroism and fluorescence spectroscopy (ethidium bromide displacement) experiments were used to determine the binding constant.[144,145] In addition to intercalation, $[Pt(terpy)]^{2+}$ complexes also form covalent bonds to DNA base pairs by nucleophilic substitution of the fourth coordinate ligand.[141,146] Terpyridineplatinum(II) complexes have been shown to bind preferentially to GC-residues in nucleotides such as quadruplex DNA at the end of telomeres, giving a potential target for selective drugs to interact. Platination is not limited to G-residues in DNA, and ligand substitution of the functional groups of thiols and the imidazoles on proteins can also occur to produce cell death.[143,147] The different mode of action compared to cisplatin may explain the biological activity and indicates that a new class of platinum antitumour agents has been found with little to no cross-resistance to cisplatin.

Terpyridineplatinum(II) complexes with chloro, hydroxy and picoline ligands coordinated at the fourth position are particularly susceptible to nucleophilic substitution.[144] The rigid tridentate geometric arrangement of the $2,2':6',2''$-terpyridine ligand forces the middle pyridine to form a shorter bond to the platinum, this in turn forces the ligand that is *trans* to the middle pyridine to have a slightly weaker and longer bond making it much more susceptible to substitution.[145] The use of sulfur ligands results in relatively inert complexes due to the high affinity of sulfur for platinum(II),[136,143] increasing the stability of the complexes and preventing nucleophilic substitution. Interactions of these stable complexes with the DNA can therefore be attributed to intercalation alone and not platination.

A series of terpyridineplatinum complexes with various fourth coordinate ligands (Figure 9.9) have been prepared to investigate the effect of the ligand on anticancer activity. Lowe's group derived a range of compounds to assess the substitution of the fourth coordinate ligand. Derivatives of functional groups on the pyridine bound to complex **18** (R = 4-Me-C_5H_4N, 4-Br-C_5H_4N, 4-MeCO-C_5H_4N, 4-Me$_2$N-C_5H_4N, 2-F-C_5H_4N, 3-F-C_5H_4N, thiazole or imidazole) were found to be more effective than carboplatin but not cisplatin in the human ovarian cancer cell lines including CHI, CHIcisR, CHIdoxR, A2780, A2780cisR and SKOV3, although substituted functional pyridines had little effect on the cytotoxicity of the complex.[144] Replacing the pyridine ligand with a readily labile group such as H_2O or MeCN (**19**) did not improve cytotoxicity, also suggesting that DNA platination is not critical for high cytotoxicity.[148,149] The effect of substituents at the $4'$-position was evaluated with derivatives of complex **18** against the human ovarian cell lines (X = N(CH_2CH_2OH)$_2$, NMe(CH_2-CH_2OH), Cl, Br, OMe, 4-Me-C_6H_4, 4-Br-C_6H_4, N(CH_2CH_2)$_2$, NHNH$_2$, NMeNH$_2$, 4-OH, 4-NH$_2$, NHCOCH$_3$ shown in Figure 9.9). Complexes of this

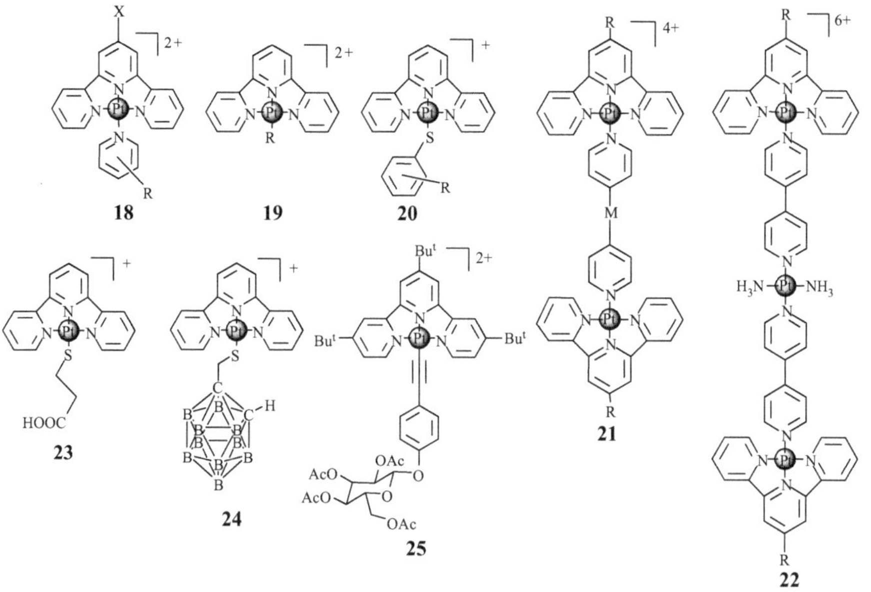

Figure 9.9 The chemical structures of a range of platinum(II) terpyridine complexes.

structural type showed a significant loss of activity due to the large electron-donating substituents in the 4′-position, whereas the introduction of a 4′-chloro group increased activity.[144] Complex **18** (where X = H) was shown to intercalate strongly with DNA with a binding constant in the order of 2×10^7 M^{-1}, more than two orders of magnitude greater than ethidium bromide and complex **10** (see Figure 9.4).[141,147] Substituting the fourth coordinate ligand with an aromatic thiol was evaluated by McFadyen *et al.*[150] with derivatives of complex **20** (R = 4-H, 2-OCH$_3$, 3-OCH$_3$, 4-OCH$_3$, 4-NO$_2$, 4-F, 4-Cl, 4-Br, 4-CH$_3$, 4-NH$_3$). This produced active compounds which exert their effects rapidly with widespread cell lyses revealed under microscopic examination.

Lowe and his co-workers synthesised a large range of bi- and trinuclear terpyridineplatinum(II) bisintercalators such as **21** and **22**, with rigid and flexible linkers.[144,151] Bisintercalating compounds may have higher affinities due to their increased charge and ability to span a greater number of DNA base pairs unobtainable by monointercalators.[151] By spanning a larger number of DNA base pairs, bisintercalators may elicit more information and target specific sequences of DNA.[152] This coupled with the potential for more effective DNA–drug associations due to increased affinity should result in clinically relevant cytotoxicity profiles. The cytotoxicity of the complexes was reported to depend on three variables: linker length, charge density, and counter ions.[144] Complexes with shorter linkers displayed higher activity than those with longer linkers. For example, the activity of complex with the generic structure of **21** was improved when M = no linker or *trans*-CH = CH- over butadiyne, 1,4-diethynylbenzene or 4,4′-diethynylbipyridine.[144] The more highly charged complexes were also more cytotoxic.[144,151] The high charge density and greater electrostatic stress of the shorter linked complexes was believed to increase DNA interactions, which would lead to their greater cytotoxicity.[144] Moreover, the tetrafluoroborate salts of the complexes were more active than their corresponding water-soluble nitrate salts. All of the bisintercalators displayed little or no cisplatin or carboplatin cross-resistance in matched cell lines.[144] A study of complex **21** (where M = *trans*-CH = CH-) with the oligonucleotide d(CGTACG)$_2$ revealed that the complex could undergo ligand substitution by the oligonucleotide.[144] Nucleophilic substitution by nucleobases is well established with cisplatin; however, the degree of lability of the linking chain in complex **21** (where M = *trans*-CH = CH-) was unexpected.[144] In this study the linking group was slowly displaced, producing a reactive electrophilic platinum species in solution that formed an irreversible coordinate covalent adduct through the N7 position of the terminal G-residue.[141,144] These complexes inhibit thioredoxin reductase, a key component in the metabolism of human tumours.[143,153,154]

It was proposed that these complexes act on both DNA and thioredoxin reductase (Figure 9.10), producing an increased activity when compared to cisplatin.[143,151] They have also been reported to inhibit cysteine proteases by binding to the active site cysteine (as shown in Figure 9.10).[153,154] It was also proposed that transport of these complexes by human serum albumin might be a reason for their higher activity.[151,155]

While the activity and cytotoxicity of a range of terpyridineplatinum(II) complexes seems promising, trials *in vivo* have not been reported. Functionalisation of these complexes to make them more selective through a prodrug approach offers some advantages, and as a result there have been a number of different research approaches. The incorporation of a terpyridineplatinum(II)

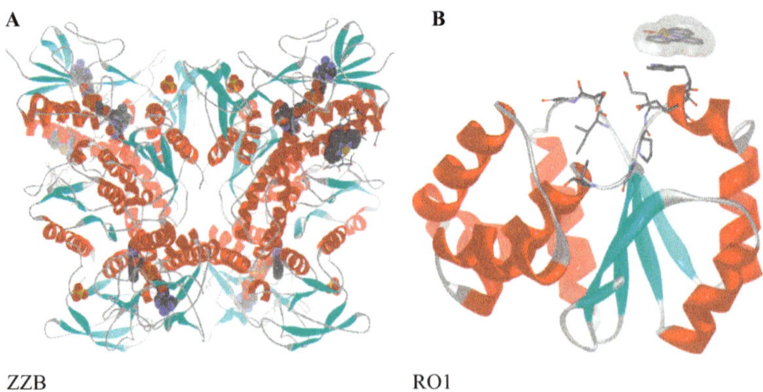

Figure 9.10   Representations from X-ray crystal structures depicting the interaction of [Pt(terpy)(HET)]$^+$ with (A) mammalian topoisomerases and human thioredoxin reductase 1(PDB 2ZZB)[154] and (B) cysteine proteases (PDB 3RO1).[153] Hydrogen atoms and solvent molecules have been omitted for clarity.

unit into a sequence selective polyamide was explored using (3-mercaptopropionic) terpyridineplatinum(II) [Pt(terpy)(MPA)]$^{2+}$ and solid phase synthesis. The aim was to produce complexes similar to **17** that exhibit DNA sequence specificity in order to mitigate toxicity and side-effects typically associated with platinum(II)-based drugs.[156,157]

Terpyridineplatinum(II) complexes containing *bis*-(thioalkyl)-dicarba-*closo*-dodecaborane (**9**) (carborane) ligands, such as complex **24**, are synergistic intercalators.[158] They combine the effects of metallointercalators with carborane cages, which are used in radiation therapy. Although high cytotoxicity in the absence of thermal radiation is not expected, complex **24** is cytotoxic in the L1210 and ovarian cancer cell lines.[158,159] It shows comparable cytotoxicity to cisplatin in the L1210 cell line and eight times greater cytotoxicity than cisplatin in the resistant cell line L1210cisR.[158]

Aryl and alkynyl terpyridineplatinum complexes have received much attention in recent times partly due to the stability of the strong Pt–C bond which slows hydrolysis and the luminescent properties of terpyridineplatinum(II) complexes with a range of acetylide ligands which exhibit long-lived emission states.[160–162] A series of these complexes has been synthesised, two of which are shown in Figure 9.9. In more recent studies by Che and co-workers, terpyridineplatinum(II) complexes with glycosylated acetylide and arylacetylide ligands have been shown to bind to DNA. They were shown to be luminescent and therefore have useful properties to be considered as biological probes. The results of cell line testing against five human cell carcinoma lines showed potency up to 100 times greater than cisplatin at killing human cancer cell lines.[163] The novel ideas and innovations to this field offer constant improvement to current drug design.

9.7 Platinum(II) Metallointercalators

Lippard and co-workers expanded upon their initial DNA binding experiments using platinum complexes [Pt(A$_L$)(I$_L$)]$^{2+}$ containing planar aromatic moieties (I$_L$) such as phen and 2,2′-bipyridine (bpy) and non-aromatic (A$_L$) bidentate 1,2-diaminoethane.[164] To establish the binding mechanism, ethidium bromide displacement assays were undertaken with increasing concentrations of the metal complexes. It was determined that compounds containing fused, aromatic moieties, such as phen and bpy, were capable of competitive ethidium bromide binding, as a decreased slope with no change in the intercept was observed.[164] This behaviour was indicative of intercalation and was further verified by the unwinding of DNA, determined by the sedimentation velocity and gel electrophoresis methods. Evidence of intercalation was further reinforced by X-ray crystallography of ct-DNA containing [Pt(phen)(en)]$^{2+}$ and [Pt(bpy)(en)]$^{2+}$.[165]

While the binding mechanisms for metallointercalators were under review, the activity of the compounds against tumour cells had only been presumed and not confirmed.[86] McFadyen *et al.*[85] determined the biological properties of platinum(II) metallointercalators including complexes of the type [Pt(A$_L$)(I$_L$)]$^{2+}$, where A$_L$ was 1,2-diaminoethane and I$_L$ was bpy, phen or

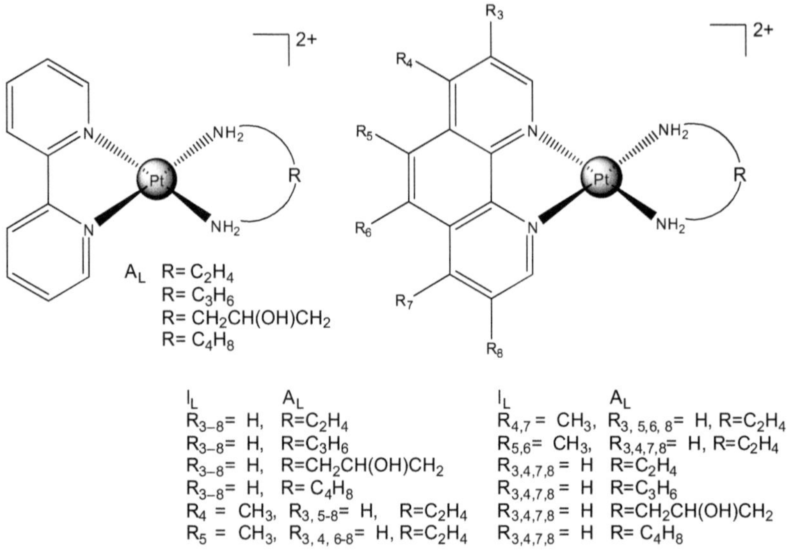

Figure 9.11 Examples of some of the metallointercalators investigated.

3,4,7,8-tetramethyl-1,10-phenanthroline (Figure 9.11). These complexes were shown to demonstrate a broad range of *in vitro* activities against L1210 leukaemia cell lines but no antitumour activity *in vivo* in mice.[166] Brodie *et al.*[63] expanded the series to include more methylated phenanthroline ligands such as, 4-methyl-1,10-phenanthroline, 5-methyl-1,10-phenanthroline, 4,7-dimethyl-1,10-phenanthroline or 5,6-dimethyl-1,10-phenanthroline. The observed IC_{50} values against L1210 cells ranged from 1.7 ± 0.3 to $>50\,\mu M$ with the most active compound being [Pt(en)(56Me$_2$phen)]Cl$_2$.[63] It was also reported that the [Pt(en)(phen)]Cl$_2$ was the most effective intercalator but did not exert the highest toxicity.[63,167]

Altering the intercalating ligand (I_L) by attaching various electron-withdrawing groups on the ligand was employed to further define the cytotoxicity, and also introduce chiral ancillary ligands (A_L) such as 1,2-diaminocyclohexane (*SS*-dach, *RR*-dach and *RS*-dach).[168] The effect of chirality was of particular interest because the established anticancer drug oxaliplatin incorporates *RR*-dach.[169] The most significant effect was the observed cytotoxicity when chiral ancillary ligands were incorporated with the order or activity *SS*>*RR*>en. Against L1210 murine leukaemia cells, the most cytotoxic complex identified was (5,6-dimethyl-1,10-phenanthroline) (1*S*,2*S*-diaminocyclohexane) platinum(II) (56MESS) (Figure 9.11), having an IC_{50} of $0.009 \pm 0.002\,\mu M$,[170] which is 5–10 times more cytotoxic against L1210 cancer cells than cisplatin.[168] Complex 56MESS has outperformed cisplatin in testing against several human cell lines (Figure 9.12).[171] Experiments which examined the IC_{50} values in human colon HT29, ovarian IGROV1 and rat colon PROb cell lines for 18 metallointercalators, where $A_L =$ en, *SS*-dach or *RR*-dach, showed that activity was sustained and increased if incubated for 72 h.[171]

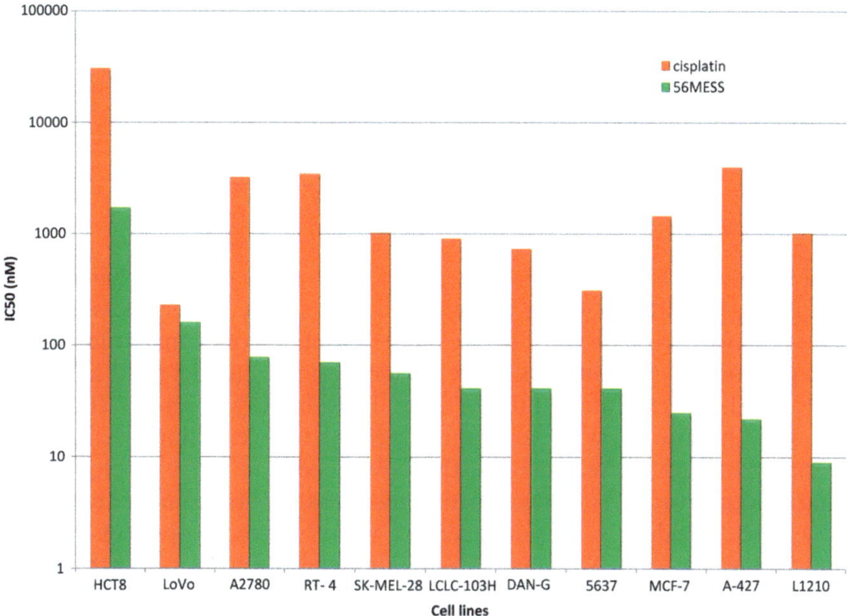

Figure 9.12   Comparison of the IC_{50} values of cisplatin and 56MESS.

The promising *in vitro* studies led to *in vivo* studies of the drug. This revealed a different story for 56MESS, which showed very little anticancer activity against BD-IX rats bearing peritoneal carcinomatosis, independent of the method of administration.[171] The results also showed the drug displayed nephrotoxicity at pharmacological doses, whereas cisplatin displayed toxicity but also completely eradicated the tumour. The failure of the compound *in vivo* was hypothesised to be due to its deactivation by intercellular compounds such as glutathione or the metallothionein family of proteins.[171] However, the significance of deactivation of 56MESS by glutathione is unlikely, given that NMR experiments have demonstrated that the *in vitro* degradation of 56MESS ($t_{\frac{1}{2}} = 68$ h) was significantly lower than cisplatin ($t_{\frac{1}{2}} = 0.5$ h).[172]

Further studies have indicated that changing the chiral ancillary ligand to 1,2-diaminocyclopentane[173] or 2,3-diaminobutane[174,175] results in the *RR* form being the most cytotoxic enantiomer, which is the opposite stereochemistry to the previously tested compounds. The most cytotoxic compounds were (5,6-dimethyl-1,10-phenanthroline)(1*R*,2*R*-diaminocyclopentane)platinum(II)[173] and 5,6-dimethyl-1,10-phenanthroline)(2*R*,3*R*-diaminobutane)platinum(II) (Figure 9.11),[175] with IC_{50} values of $0.17 \pm 0.04\,\mu M$ and $0.48 \pm 0.02\,\mu M$, respectively.

Platinum(II) metallointercalators offer potential as anticancer agents as evident by *in vitro* cancer cell line cytotoxicity, nonetheless inconsistencies may be encountered upon transition from *in vitro* studies to an *in vivo* tumour model. Many reasons have been suggested for these differences but the most common is that the drug is unable to localise within the target cells. Various modifications can be carried out to increase the selective anticancer cytotoxicity of the drug and thus reduce host toxicity, including nanorod formation,[176] encapsulation[177,178] and conversion of platinum(II) complexes into platinum(IV) complexes to decrease the activity of the complexes until activated in the cell, to increase the selectivity of the drug towards the cancer cells.[179]

9.8 Platinum(IV) Compounds

While there has been unquestionable clinical success from using platinum(II) complexes, the drawbacks and severe side-effects have driven the search for drugs of increased efficacy. Platinum(IV) complexes offer many advantages over platinum(II), including a lower toxicity profile, and increased kinetic inertness and reduced intercellular activity. Platinum(IV) drugs are sufficiently stable and inert to allow circulation through the body enabling localisation to the desired target where, typically, they are reduced to active cytotoxic platinum(II) derivatives.[180,181] The relative inertness of platinum(IV) complexes reduces the incidences of side-reactions with *in vivo* components, namely intracellular or extracellular sulfur-containing proteins (*e.g.* glutathione or metallothionein) that typically deactivate platinum(II) complexes.[182,183]

Platinum(IV) complexes of the general formula *cis*-$[PtX_2Y_2(NH_2R)_2]$ where X is the equatorial leaving groups, Y is the axial ligands and NH_2R is an inert amine—exert their antitumour effects, like platinum(II) complexes, through coordinative binding to DNA. There is persuasive evidence to suggest that octahedral platinum (IV) complexes (see Figure 9.2) are rapidly reduced to platinum(II) derivatives *in vivo*[62] and form mono- and bifunctional adducts, both intra- and inter-strand.[32,62,184–187]

Platinum(IV) drugs have shown promising preclinical success and as a result have entered clinical trials. A number of platinum(IV) complexes lack cross-resistance with cisplatin, exhibit lower toxicity and are suitable for oral administration.[37,62] Some of the most promising drugs, which are shown in Figure 9.13 and previously in Figure 9.2, include tetraplatin also known as ormaplatin (**26**), iproplatin (**27**), diamminedichlorodihydroxy platinum(IV) (oxoplatin, **28**) and satraplatin (**7**). Satraplatin (**7**, JM216, Figure 9.2) is the most successful of the platinum(IV) complexes; however, *cis*-dichlorobis(isopropylamine)trans-dihydroxyplatinum(IV) (iproplatin, JM-9, **27**) and tetraplatin (**26**)[188] are also undergoing clinical development. Although transplatin

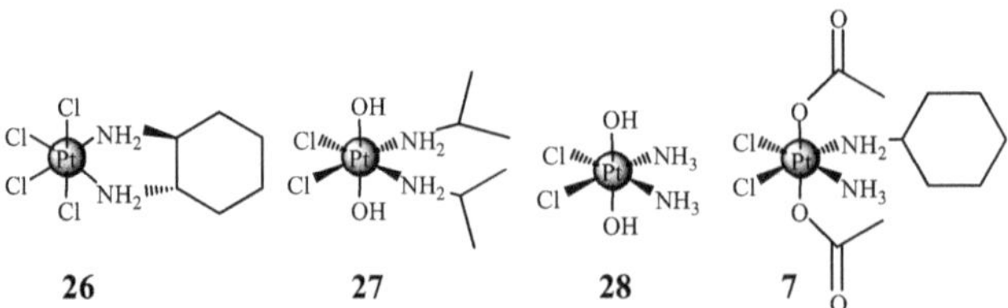

Figure 9.13   The chemical structures of platinum(IV) complexes tetraplatin or ormaplatin (**26**), iproplatin (**27**), oxoplatin (**28**) and satraplatin (**7**).

was found to be inactive against tumours, substitution of the am(m)ine ligands has produced a number of active antitumour *trans*-coordinated complexes.[189] The *trans*-platinum(IV) complex, JM335 (**9**), has demonstrated higher antitumour activity than its *cis*-isomer, JM149, in human ovarian carcinoma cells.[22]

Ormaplatin showed severe neurotoxicity as well as myelosuppression in phase I trials, with no phase II trials yet reported.[190,191] Iproplatin (**27**) also experienced limited success in phase II trials with results on par with cisplatin.[192–194] Oxoplatin (**28**) and satraplatin (**7**) offered much more promising results. Oxoplatin is relatively inert under mild conditions and therefore can be administered orally. Oxoplatin is reduced to cisplatin by intracellular reducing agents, as well as by ascorbic acid and hydrochloric acid, as evident where exposure to 0.1 M HCl resulted in a two-fold increase in activity.[195] Reduction/ activation under acidic conditions is promising as it can be utilised in the acidic microenvironment of solid tumours. Satraplatin was the first orally active platinum analogue and has recently completed phase III trials with very promising outcomes including an increase in progression-free survival, a decrease in pain response, and encouraging results for prostate-specific antigen response.[196] Satraplatin has shown similar anticancer activity to carboplatin but lacks the nephrotoxicity, neurotoxicity and ototoxicity, with patient responses indicating that it is much better tolerated than cisplatin.[196]

While there appear to be successful platinum(IV) prodrugs like satraplatin, the search continues for new drugs based on this prodrug model. The addition of tags capable of therapeutic targeting and activation is being explored. One method for active drug release under current research is that of photoactivatable platinum complexes. This method employs a platinum(IV) prodrug with light-sensitive ligand(s) that can be photoactivated at the site of the tumour to release active antitumour agents (Figure 9.14). This strategy has been advanced with the development of fibre optic and laser technologies that can focus irradiation with a defined wavelength and intensity to identified targets such as internal organs.[197] The platinum-photoactivatable classes developed so far are typically based on diiodo-platinum(IV) and diazido-platinum(IV) diammine complexes, which appear to have a unique photoactivation ability based on the oxidation state of the platinum centre and metal-to-ligand charge transitions.[197,198]

While platinum(IV) drugs capable of triggered release appear to offer a promising pathway for therapeutics, the addition of functional groups capable of

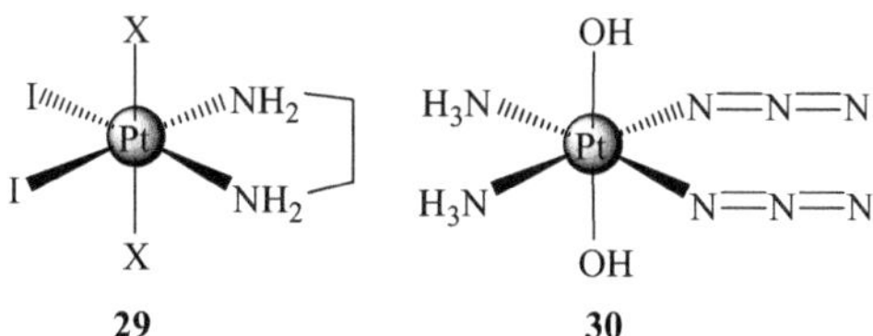

29　　　　**30**

Figure 9.14　The general structure of photoactivatable complexes based on diiodo-(**29**) and diazido-platinum(IV) (**30**) complexes.

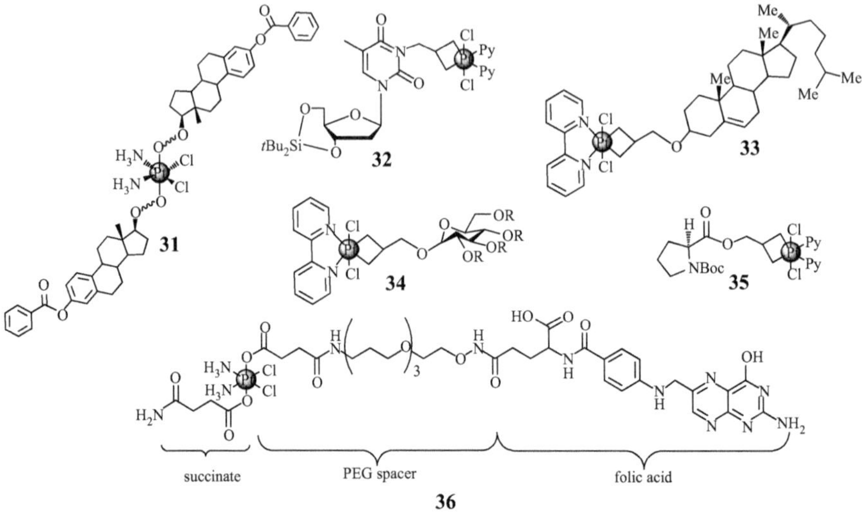

Figure 9.15 The chemical structure of a range of biologically functionalised platinum(IV) complexes.

biological targeting are receiving increased attention. One such targeting approach has included the attachment of cell-sensitising oestradiol units to a platinum(IV) compound (**31**) as seen in Figure 9.15.[199] The use of oestrogen in drug targeting and delivery has two main advantages: oestrogen receptors are overexpressed on the surface of many cancer cells, and, oestrogen induces overexpression of the human gene high-mobility group box 1 (HMGB1) which can prevent repair of cisplatin–DNA adducts.[7,200,201] As such, the use of compounds like oestrogen, which target as well as sensitise platinum(IV) prodrugs, are of great interest. In this case, the steroid is released from the platinum complex upon reduction and before DNA binding,[201] releasing both cisplatin and estrogen. The proposed mechanism of action of these compounds suggests that when the drug–steroid complex enters the cell, the complex is reduced releasing cisplatin and two equivalents of modified oestrogen. Hydrolysis of the modified oestrogen and subsequent oestrogen receptor binding may lead to HMGB1 upregulation, resulting in shielding repair of the cisplatin–DNA adducts.[199,200]

Recently, platinacyclobutanes bearing biological components have been synthesised by Stocker and Hobard[202] (Figure 9.15), where platinum(IV) complexes containing thymidine (**32**), cholesterol (**33**), glucose (**34**) and proline (**35**) were coupled to a cyclopropylmethanol, which is reduced to the active platinum(II) derivative. These components were designed to both increase water solubility and increase uptake in cancerous tissues. Dhar *et al.*[181,203] synthesised a platinum(IV) complex (**36**), with the main objective of targeting, protecting and delivering cisplatin (upon intracellular reduction) to the cancerous cell. The platinum(IV) complex designed for this purpose contains succinate as one its axial ligands for attachment to the amine-functionalised single-wall nanotube and a folic acid derivative as the other axial ligand, separated from the platinum centre by a polyethylene glycol (PEG) spacer, making the compound more water-soluble and biocompatible. While there has been limited clinical trials of targeting platinum(IV) complexes, the recent advances in prodrug design offer immense potential for future chemotherapeutic treatment.

9.9   Other Metals

Platinum has been the focus of this chapter mainly owing to its demonstrated anticancer activity; however, many other transition metal complexes interact with nucleic acids. As early as the 1950s Dwyer recognised that metal complexes could provide structural information about nucleic acids.[204–208] This insight is still relevant today, with the on-going investigations into the interactions of transition metal complexes with macromolecules within the cell. Some transition metals and their interactions are briefly introduced here; lanthanide, zinc and copper complexes are discussed in detail in Chapter 8 with respect to their cleavage of nucleic acids and interactions with quadruplex DNA.

9.9.1   Ruthenium, Iridium, Osmium, Rhodium and Iron complexes

Ruthenium complexes have dominated many of these investigations because of their synthetic versatility and characteristic spectroscopy.[209–211] Their potential as probes for nucleic acids has been demonstrated by several researchers in the last decade with the binding interactions of $[Ru(phen)_2dppz]^{2+}$ (where dppz is dipyrido[3,2-a:2′3′-c]phenazine) being a frequent example. In solution it exhibits enhanced luminescence in the presence of DNA, and because of this it is known as a "light-switch" complex. The potential of this complex could not be fully realised without detailed information about the binding interaction. Recent X-ray structural studies[212] have confirmed previous NMR studies which showed that these complexes bind from the minor groove[213] with some sequence specificity.

Simple ruthenium arene complexes of the type $[(\eta^6\text{-arene})Ru(en)(Cl)]^+$ (where en = 1,2-diaminoethane) displayed anticancer activity (in A2780 human ovarian cancer cell line) which improved as the size of the arene increased (benzene < p-cymene < biphenyl < dihydroantracene < tetrahydroanthracene).[214,215] They are water-soluble and the metal–ligand bonds formed are relatively inert towards displacement under physiological conditions.[215] This activity is accomplished by targeting guanine bases of DNA oligomers to form adducts.[216,217] X-ray studies have shown that the arene moiety stacks between the DNA base pairs and the –NH– of the en hydrogen bonds with the oxygen on the guanine base (Figure 9.16).[216] For a more comprehensive description of this class of complexes, any of the reviews by Sadler,[218,219] Therrien[220] or Dyson[221–223] are excellent places to start.

Octahedral ruthenium, iridium and osmium complexes that mimic the structure of staurosporine, the target of the BRAF serine/threonine kinase,

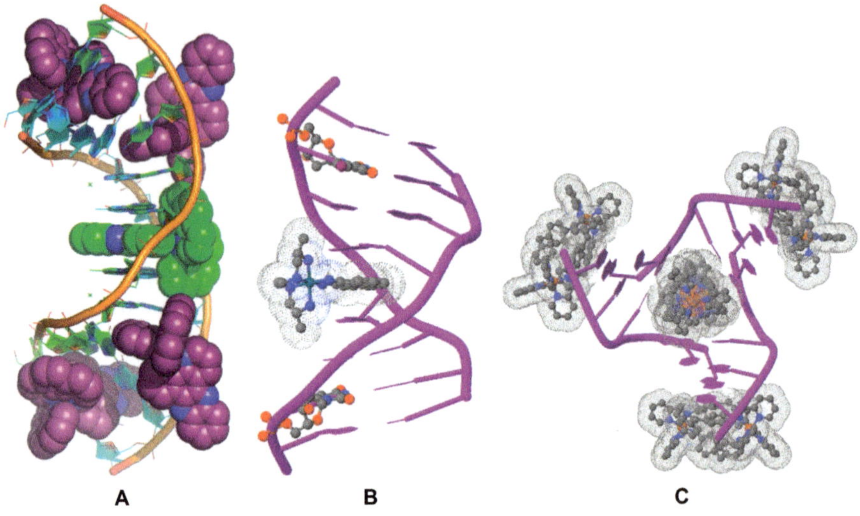

Figure 9.16 Images generated from X-ray data of (A) Λ-[Ru(phen)₂dppz)]²⁺ where the dppz intercalates symmetrically and perpendicularly from the minor groove at the central TA/TA step of the d(CCGGTACCGG)₂ duplex. A second ruthenium complex intercalates through the combination of a shallower angled intercalation at the C1C2/G9G10 site at the end of the duplex, and semi-intercalation at the G3G4 step of an adjacent duplex.[212] (B) [Rh(Me₂trien)(phi)]³⁺ intercalating selectively into 5′-TGCA-3′,[224] and (C) [Fe₂L₃]⁴⁺ fitting perfectly into the central hydrophobic cavity of a DNA three way junction.[225,226]

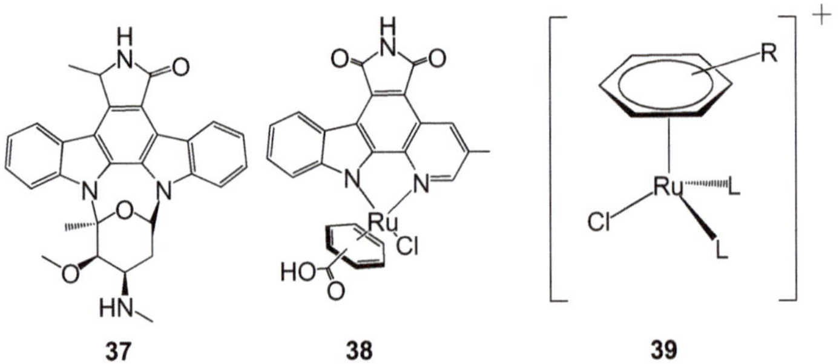

Figure 9.17 The chemical structures of staurosporine (**37**), the organoruthenium compound CS292 (**38**)[227] and an Ru(II)-arene (**39**), where **L** represents either a mono- or bidentate ligand and arene **R** could be any aromatic cyclic group.

have been reported to inhibit important cellular enzymes or have direct anticancer activity. Libraries of these complexes have been synthesised and CS292 (**38**, Figure 9.17), a ruthenium based organometallic kinase inhibitor, has been reported to bind effectively with high binding affinity to BRAF.[227–233]

Selective interactions with DNA have been demonstrated by [Rh(R,R-Me$_2$trien)(phi)]$^{3+}$, Δ-[Rh(bpy)$_2$(chrysi)]$^{3+}$ and [Rh(Me$_2$trien)(phi)]$^{3+}$ (where R,R-Me$_2$trien = $2R,9R$-2,9-diamino-4,7-diazadecane, phi = 9,10-diaminophenanthrene, bpy = 2,2′-bipyridine, chrysi = 5,6-chrysenequinone diimine). These Rh complexes were designed specifically to intercalate into the 5′-TGCA-3′ sequences of DNA and have produced a high-resolution view of an octahedral metal complex intercalated into the base pairs of DNA.[224] A new mode of recognition, at three-way DNA junction, has been demonstrated by an iron helicate complex, [Fe$_2$L$_3$]^4 L = ($NE,N′E$)-4,4′-methylenebis(N-(pyridin-2-ylmethylene)aniline)) bound perfectly into the central hydrophobic cavity.[224–226,234,235]

9.9.2 Copper Complexes

Since the initial work of Sigman and co-workers in the 1970s, there has been substantial interest in copper complexes of 1,10-phenanthroline and its derivatives. These compounds bind to DNA, but not by intercalation, as is often the case for compounds containing phenanthroline. *Bis*(1,10-phenanthroline)copper(II) complexes (**40**, Figure 9.18) are well known for their ability to cleave DNA, particularly in the presence of hydrogen peroxide.[236] The mechanism of action of these compounds is still under examination; however, it is believed that the [Cu(phen)$_2$]$^{2+}$ complex is reduced to [Cu(phen)$_2$]$^{+}$, which then binds to DNA non-covalently at the minor groove. The DNA interaction is then oxidised in the presence of dihydrogen peroxide, which then goes on to perform an oxidative attack, leading to DNA cleavage.[237]

9.9.3 Titanium and Vanadium Complexes

Titanium complexes have been shown to possess a wide spectrum of anti-tumour properties. While the mechanism of action has not been determined, it is different to that of cisplatin. It has been proposed that titanium complexes inhibit the cell cycle by interacting with DNA but the overall mechanism is likely to be complex, involving transport and delivery into cancer cells by biomacromolecules.[239–241] Vanadocene(IV) dichloride (Figure 9.19)

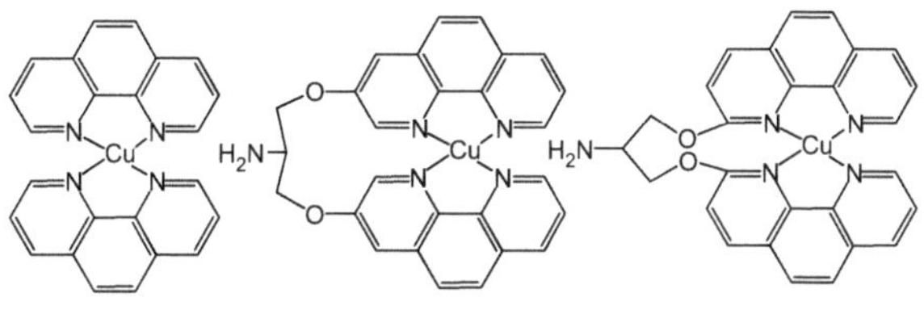

40 **41** **42**

Figure 9.18 *Bis*-(1,10-phenanthroline)copper(II) complexes where the addition of a serinol bridge, known as "Clip", joins two phenanthroline ligands together *via* either position 3 (**41**) or 2 (**42**) to increases the binding affinity. The "clipped" complexes are more efficient nucleases than the parent compound.[238]

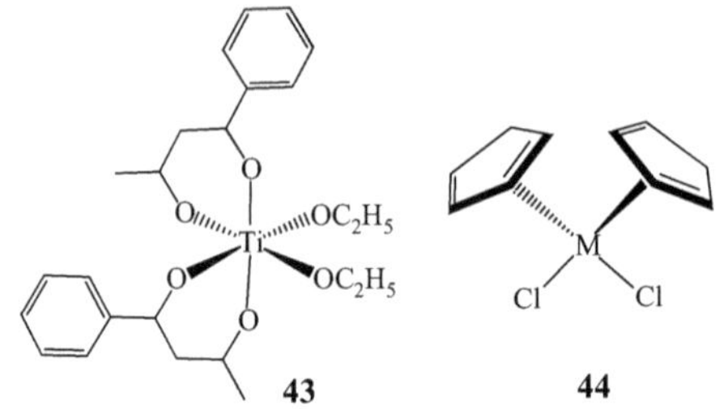

Figure 9.19 Titanium complexes butotitane (**43**) and metalloocene dichloride (**44**), where M = Ti, V.

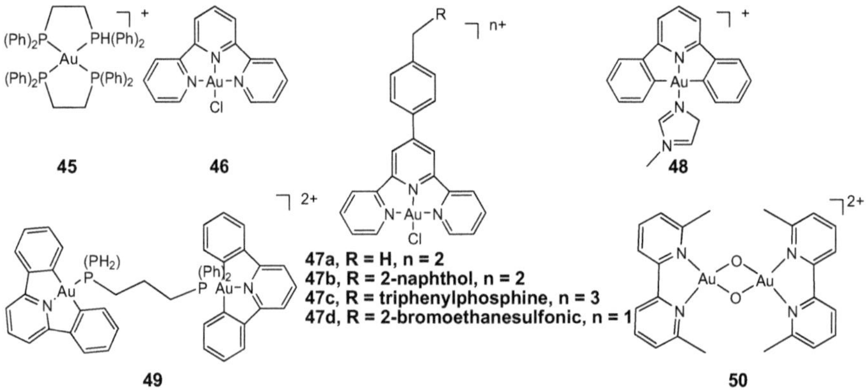

Figure 9.20 The structure of some gold(I) and (III) complexes for $[Au(dppe)_2]^+$ (**45**), $[Au(terpy)Cl]^{2+}$ (**46**), $[Au(4-(4R-phenyl)terpy)Cl]^+$ (**47**), $[Au(2,6-diphenylpyridine)(meim-1)]^+$ (**48**), $[\{Au(2,6-diphenylpyridine)\}_2\mu-1,2-bis(diphenylphosphino)\ propane]^{2+}$ (**49**) and $[\{Au(6,6-dimethyl-2,2-bipyridine)\}_2(\mu-O)_2]^{2+}$ (**50**).

demonstrates low toxicity in humans and effective anticancer activity, and like titanocene dichloride, it accumulates in the nucleus,[242,243] distorts the secondary structure of DNA,[244] and inhibits DNA and RNA synthesis.[245,246]

9.9.4 Gold Complexes

Recently, gold complexes have been reported to elicit their antiproliferative potency as a consequence of their interactions with mitochondrial DNA.[247] $[Au(dppe)_2]^+$ is lipophilic (**45**, Figure 9.20) and induces mitochondrial membrane permeability by non-selective concentration. Ligands, such as 1,10-phenanthroline, 2,2':6',2''-terpyridine and 2,6-diphenylpyridine) have been used to modulate their reduction potential and fast hydrolysis rate under physiological conditions. Charged, planar complexes such as $[Au(terpy)Cl]Cl_2$ (**46**, **47a-d**, **48**

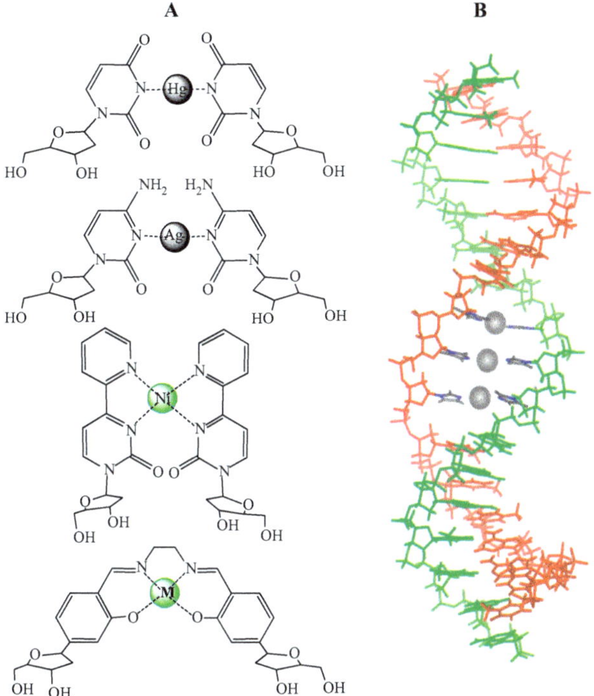

Figure 9.21 (A) The structure of selected metal-mediated base pairs of natural nucleic acids and designed synthetic nucleosides. (B) Structure of the duplex containing the Ag-mediated imidazole base pairs in its centre.[254]

and **49,** Figure 9.20) have the potential to intercalate into DNA. Some gold(III) terpyridine derivatives have been reported to show a correlation between DNA binding affinity and cytotoxicity, which suggests that the structure can be further modulated to improve activity.[248,249] The cytotoxicity as well as DNA binding affinity are changed significantly with different ligand substitutions to either the monodentate ligand, Cl or the 4-position of the terpy.[250] As an example of structure influence, [Au(2,6-diphenylpyridine)(meim-1)]$^{+}$ has been demonstrated to enhance G-quadruplex assembly.[248,250]

9.9.5 Metal Mediated Base Pair Formation

Metal ions can interact with nucleic acids in another profoundly different way, replacing the hydrogen bonds in Watson–Crick type base pairs by metal–ligand interactions inside the double helix, forming a metal-mediated base pair.[251–260] Certain transition metals, namely Ag(I),[252,261] Hg(II),[252,262,263] Zn(II),[264] Co(II),[264] Ni(II),[264] Cu(II),[265–267] VO(II),[266] Mn(III)[266] and Fe(III)[266], can coordinately bind either to natural nucleic acids or designed synthetic nucleosides (Figure 9.21). Incorporation of metal ions into DNA strands has significant impact for designing new nanomaterials such as DNA wires.[264] A number of recent reviews provide a great start if more information about these fascinating interactions with nucleic acids is to be surveyed.[251–253,256–259,261,268]

9.10 Conclusion

Since the initial structure–activity relationship postulations of Rosenberg, Cleare and Hoeschele, numerous platinum anticancer compounds have been designed that both comply with and challenge the early convention. Many, if not virtually all, such complexes primarily exhibit DNA-binding properties, and more are being discovered that can affect their biological effects through various interactions with the same biomacromolecular target. Strategies that investigate innovative designs of the platinum anticancer compounds, in order to imbue *in vivo* stability, provide effective, selective delivery, while increasing efficacy markedly increases the prospect of success for the future developments of anticancer drugs. There are questions that still need to be answered, as structure–activity relationships don't successfully predict the cytotoxicity with respect to DNA binding, or the amount of DNA distortion. If coordinate binding is not essential for cytotoxicity, then what are the "new" rules of design?

Clearly interactions with nucleic acids are not limited to platinum. Many other metals complexes show affinity and activity. The interactions of these metal complexes are diverse and provide scope for design of diagnostic probes and medicinal agents. Metal-mediated DNA structures will provide the "building blocks" for DNA-based nanotechnology.

Acknowledgements

The authors would like to thank the University of Western Sydney for financial support through internal research grants. B. W. J. Harper, K. B. Garbutcheon-Singh and N. S. Ng were supported by an Australian Postgraduate Award from the University of Western Sydney.

References

1. M. J. Waring, *Drugs Exp. Clin. Res.*, 1986, **12**, 441–453.
2. M. J. Waring, *Annu. Rev. Biochem.*, 1981, **50**, 159–192.
3. B. Rosenberg, L. Van Camp and T. Krigas, *Nature*, 1965, **205**, 698–699.
4. K. R. Barnes, A. Kutikov and S. J. Lippard, *Chem. Biol.*, 2004, **11**, 557–564.
5. R. A. Alderden, H. R. Mellor, S. Modok, T. W. Hambley and R. Callaghan, *Biochem. Pharmacol.*, 2006, **71**, 1136–1145.
6. I. Kostova, *Recent Pat. Anti-Cancer Drug Discov.*, 2006, **1**, 1–22.
7. J. C. Huang, D. B. Zamble, J. T. Reardon, S. J. Lippard and A. Sancar, *Proc. Natl Acad. Sci. USA*, 1994, **91**, 10 394–10 398.

8. M. A. Fuertes, J. Castilla, C. Alonso and J. M. Perez, *Curr. Med. Chem.*, 2003, **10**, 257–266.

9. A. S. Abu-Surrah and M. Kettunen, *Curr. Med. Chem.*, 2006, **13**, 1337–1357.

10. L. Kelland, *Nat. Rev. Cancer*, 2007, **7**, 573–584.

11. D. P. Gately and S. B. Howell, *Br. J. Cancer*, 1993, **67**, 1171–1176.

12. S. Ishida, J. Lee, D. J. Thiele and I. Herskowitz, *Proc. Natl Acad. Sci. USA*, 2002, **99**, 14298–14302.

13. K. Katano, A. Kondo, R. Safaei, A. Holzer, G. Samimi, M. Mishima, Y.-M. Kuo, M. Rochdi and S. B. Howell, *Cancer Res.*, 2002, **62**, 6559–6565.

14. S. M. Cohen and S. J. Lippard, in *Prog. Nucleic Acid Res. Mol. Biol.*, Academic Press, San Diego, 2001, vol. 67, pp. 93-130.

15. M. S. Davies, S. J. Berners-Price and T. W. Hambley, *J. Inorg. Biochem.*, 2000, **79**, 167–172.

16. E. R. Jamieson and S. J. Lippard, *Chem. Rev.*, 1999, **99**, 2467–2498.

17. A. M. Fichtinger-Schepman, J. L. van der Veer, J. H. den Hartog, P. H. Lohman and J. Reedijk, *Biochemistry*, 1985, **24**, 707–713.

18. V. Calderone, A. Casini, S. Mangani, L. Messori and P. L. Orioli, *Angew. Chem. Int. Ed.*, 2006, **45**, 1267–1269.

19. Q. Liu, I. A. Kriksunov, R. Graeff, C. Munshi, H. C. Lee and Q. Hao, *J. Biol. Chem.*, 2006, **281**, 32861–32869.

20. K. W. Jennette, S. J. Lippard, G. A. Vassiliades and W. R. Bauer, *Proc. Natl Acad. Sci. USA*, 1974, **71**, 3839–3843.

21. J. Reedijk, *Proc. Natl Acad. Sci. USA*, 2003, **100**, 3611–3616.

22. L. R. Kelland, C. F. J. Barnard, K. J. Mellish, M. Jones, P. M. Goddard, M. Valenti, A. Bryant, B. A. Murrer and K. R. Harrap, *Cancer Res.*, 1994, **54**, 5618–5622.

23. P. Kopf-Maier and H. Kopf, *Chem. Rev.*, 1987, **87**, 1137–1152.

24. M. A. Fuertes, J. Castilla, C. Alonso and J. M. Perez, *Curr. Med. Chem.: Anti-Cancer Agents*, 2002, **2**, 539–551.

25. N. P. Farrell, *Semin. Oncol.*, 2004, **31**, 1–9.

26. B. A. Chabner, C. J. Allegra, G. A. Curt and P. Calabresi, in *Goodman and Gilman's The Pharmacological Basis of Therapeutics*, ed. J. G. Hardman, L. E. Limbird and A. G. Gilman, McGraw-Hill, New York, 9th Int. edn, 1996.

27. D. Hayes, E. Cvitkovic, R. Golbey, E. Scheiner and I. H. Krakoff, *Proc. Am. Assoc. Cancer Res.*, 1976, **17**, 169.

28. P. J. Loehrer and L. H. Einhorn, *Ann. Intern. Med.*, 1984, **100**, 704–713.

29. N. J. Wheate, R. I. Taleb, A. M. Krause-Heuer, R. L. Cook, S. Wang, V. J. Higgins and J. R. Aldrich-Wright, *Dalton Trans.*, 2007, 5055–5064.

30. L. R. Kelland, *Drugs Exp. Clin. Res.*, 2000, **59**, 1–8.

31. N. Graf and S. J. Lippard, *Adv. Drug Deliv. Rev.*, 2012, **64**, 99–1004.

32. M. D. Hall, H. R. Mellor, R. Callaghan and T. W. Hambley, *J. Med. Chem.*, 2007, **50**, 3403–3411.

33. N. J. Wheate, S. Walker, G. E. Craig and R. Oun, *Dalton Trans.*, 2010, **39**, 8113–8127.

34. Z. D. Bugarcic, J. Bogojeski, B. Petrovic, S. Hochreuther and R. van Eldik, *Dalton Trans.*, 2012, **41**, 12 329–12 345.

35. G. Han, N. S. Chari, A. Verma, R. Hong, C. T. Martin and V. M. Rotello, *Bioconjug. Chem.*, 2005, **16**, 1356–1359.

36. M. Milev, O. M. Nickel and A. Dominik, University of Applied Sciences Giessen, Website, [Online], 2010.

37. E. Wong and C. M. Giandomenico, *Chem. Rev.*, 1999, **99**, 2451–2466.

38. T. W. Hambley, *Coord. Chem. Rev.*, 1997, **166**, 181–223.

39. R. J. Knox, F. Friedlos, D. A. Lydall and J. J. Roberts, *Cancer Res.*, 1986, **46**, 1972–1979.

40. M. E. Gore, I. Fryatt, E. Wiltshaw, T. Dawson, B. A. Robinson and A. H. Calvert, *Br. J. Cancer*, 1989, **60**, 767–769.

41. A. E. Taylor, E. Wiltshaw, M. E. Gore, I. Fryatt and C. Fisher, *J. Clin. Oncol.*, 1994, **12**, 2066–2070.

42. D. S. Alberts, S. Green, E. V. Hannigan, R. O'Toole, D. Stock-Novack, P. Anderson, E. A. Surwit, V. K. Malvlya, W. A. Nahhas and C. J. Jolles, *J. Clin. Oncol.*, 1992, **10**, 706–717.

43. M. J. Cleare and J. D. Hoeschele, *Bioinorg. Chem.*, 1973, **2**, 187–210.

44. T. A. Connors, M. Jones, W. C. Ross, P. D. Braddock, A. R. Khokhar and M. L. Tobe, *Chem-Biol. Interact.*, 1972, **5**, 415–424.

45. G. R. Gale, E. M. Walker, Jr., L. M. Atkins, A. B. Smith and S. J. Meischen, *Res. Commun. Chem. Pathol. Pharmacol.*, 1974, **7**, 529–538.

46. T. Tashiro, Y. Kawada, Y. Sakurai and Y. Kidani, *Biomed. Pharmacother.*, 1989, **43**, 251–260.

47. A. J. Kraker and C. W. Moore, *Cancer Res.*, 1988, **48**, 9–13.

48. A. Pasini and F. Zunino, *Angew. Chem. Int. Ed.*, 1987, **26**, 615–624.

49. W. Saenger, *Principles of Nucleic Acid Structure*, Springer-Verlag, New York, 1984.

50. U. Pindur, M. Haber and K. Sattler, *J. Chem. Educ.*, 1993, **70**, 263–272.

51. J. L. Misset, H. Bleiberg, W. Sutherland, M. Bekradda and E. Cvitkovic, *Crit. Rev. Oncol. Hematol.*, 2000, **35**, 75–93.

52. F. Levi, G. Metzger, C. Massari and G. Milano, *Clin. Pharmacokinet.*, 2000, **38**, 1–21.

53. J. L. Misset, *Br. J. Cancer*, 1998, **77**, 4–7.

54. J. H. Burchenal, K. Kalaher, K. Dew, L. Lokys and G. Gale, *Biochimie*, 1978, **60**, 961–965.

55. L. Kelland, *Nat. Rev. Cancer*, 2007, **7**, 573–584.

56. T. Dragovich, D. Mendelson, S. Kurtin, K. Richardson, D. Von Hoff and A. Hoos, *Cancer Chemother. Pharmacol.*, 2006, **58**, 759–764.

57. J. R. Rice, J. L. Gerberich, D. P. Nowotnik and S. B. Howell, *Clin. Cancer Res.*, 2006, **12**, 2248–2254.

58. M. P. Hacker, A. R. Khokhar, I. H. Krakoff, D. B. Brown and J. J. McCormack, *Cancer Res.*, 1986, **46**, 6250–6254.

59. M. J. Cleare and J. D. Hoeschele, *Platinum Metals Rev.*, 1973, **17**, 2–13.
60. M. J. Cleare and J. D. Hoeschele, *Bioinorg. Chem.*, 1973, **2**, 187–210.
61. J. Reedijk, *Pure Appl. Chem.*, 1987, **59**, 181–192.
62. M. D. Hall and T. W. Hambley, *Coord. Chem. Rev.*, 2002, **232**, 49–67.
63. D. Wang and S. J. Lippard, *Nat. Rev. Drug Discov.*, 2005, **4**, 307–320.
64. C. R. Brodie, J. G. Collins and J. R. Aldrich-Wright, *Dalton Trans.*, 2004, 1145–1152.
65. H. Huang, L. Zhu, B. R. Reid, G. P. Drobny and P. B. Hopkins, *Science*, 1995, **270**, 1842–1845.
66. N. Farrell, in *Metal Ions in Biological Systems*, A. Sigel and H. Sigel, Marcel Dekker, New York, 2004, vol. 42 .
67. C. Manzotti, G. Pratesi, E. Menta, R. Di Domenico, E. Cavalletti, H. H. Fiebig, L. R. Kelland, N. Farrell, D. Polizzi, R. Supino, G. Pezzoni and F. Zunino, *Clin. Cancer Res.*, 2000, **6**, 2626–2634.
68. K. S. Lovejoy, R. C. Todd, S. Zhang, M. S. McCormick, J. A. D'Aquino, J. T. Reardon, A. Sancar, K. M. Giacomini and S. J. Lippard, *Proc. Natl Acad. Sci. USA*, 2008, **105**, 8902–8907.
69. Y. Wu, D. Bhattacharyya, C. L. King, I. Baskerville-Abraham, S.-H. Huh, G. Boysen, J. A. Swenberg, B. Temple, S. L. Campbell and S. G. Chaney, *Biochemistry*, 2007, **46**, 6477–6487.
70. A. P. Silverman, W. Bu, S. M. Cohen and S. J. Lippard, *J. Biol. Chem.*, 2002, **277**, 49743–49749.
71. Y. Wu, P. Pradhan, J. Havener, G. Boysen, J. A. Swenberg, S. L. Campbell and S. G. Chaney, *J. Mol. Biol.*, 2004, **341**, 1251–1269.
72. D. Yang, S. S. G. E. van Boom, J. Reedijk, J. H. van Boom, N. Farrell and A. H.-J. Wang, *Nat. Struct. Mol. Biol.*, 1995, **2**, 577–586.
73. S. Komeda, T. Moulaei, K. K. Woods, M. Chikuma, N. P. Farrell and L. D. Williams, *J. Am. Chem. Soc.*, 2006, **128**, 16092–16103.
74. S. K. Kim and B. Norden, *FEBS Lett.*, 1993, **315**, 61–64.
75. S. L. Grokhovsky, A. N. Surovaya, G. Burckhardt, V. F. Pismensky, B. K. Chernov, C. Zimmer and G. V. Gursky, *FEBS Lett.*, 1998, **439**, 346–350.
76. H. Loskotová and V. Brabec, *Eur. J. Biochem.*, 1999, **266**, 392–402.
77. D. Jaramillo, N. J. Wheate, S. F. Ralph, W. A. Howard, Y. Tor and J. R. Aldrich-Wright, *Inorg. Chem.*, 2006, **45**, 6004–6013.
78. W. D. Wilson and R. L. Jones, *Adv. Pharmacol. Chemother.*, 1981, **18**, 177–222.
79. M. H. Werner, A. M. Gronenborn and G. M. Clore, *Science*, 1996, **271**, 778–784.
80. P. Lincoln and B. Norden, *J. Phys. Chem. B*, 1998, **102**, 9583–9594.
81. K. W. Jennette, S. J. Lippard, G. A. Vassiliades and W. R. Bauer, *Proc. Natl Acad. Sci. USA*, 1974, **71**, 3839–3843.
82. A. H.-J. Wang, J. Nathans, G. V. D. Marel, J. H. V. Boom and A. Rich, *Nature*, 1978, **276**, 471.
83. K. W. Jennette, S. J. Lippard, G. A. Vassiliades and W. R. Bauer, *Proc. Natl Acad. Sci. USA*, 1974, **71**, 3839–3843.

84. P. J. Bond, R. Langridge, K. W. Jennette and S. J. Lippard, *Proc. Natl Acad. Sci. USA*, 1975, **72**, 4825–4829.
85. D. W. McFadyen, L. P. G. Wakelin, I. A. G. Roos and V. A. Leopold, *J. Am. Chem. Soc.*, 1985, **28**, 1113–1116.
86. W. B. Pratt, in *The Anticancer Drugs*, W. B. Pratt and R. W. Ruddon, Oxford University Press, New York, 1979.
87. M. Cusumano, M. L. D. Petro, A. Giannetto, F. Nicolo and E. Rotondo, *Inorg. Chem.*, 1998, **37**, 563–568.
88. D. M. Fisher, Honours Thesis, University of Sydney, 2000.
89. M. Howe-Grant, K. C. Wu, W. R. Bauer and S. J. Lippard, *Biochemistry*, 1976, **15**, 4339–4346.
90. S. J. Lippard, *Acc. Chem. Res.*, 1978, **11**, 211–217.
91. A. H. J. Wang, J. Nathans, G. van der Marel, J. H. van Boom and A. Rich, *Nature*, 1978, **276**, 471–474.
92. J. K. Barton, *Science*, 1986, **15**, 8.
93. J. K. Barton, *Science*, 1986, **233**, 727–734.
94. G. M. Blackburn and M. J. Gait, *Nucleic Acids in Chemistry and Biology*, Oxford University Press, New York, 2nd edn, 1996.
95. W. Bauer and J. Vinograd, *J. Mol. Biol.*, 1970, **47**, 419–435.
96. D. M. Crothers, *Biopolymers*, 1968, **6**, 575–584.
97. H. Baruah, M. W. Wright and U. Bierbach, *Biochemistry*, 2005, **44**, 6059–6070.
98. W. I. Sundquist, D. P. Bancroft, L. Chassot and S. J. Lippard, *J. Am. Chem. Soc.*, 1988, **110**, 8559–8560.
99. R. J. Holmes, M. J. McKeage, V. Murray, W. A. Denny and W. D. McFadyen, *J. Inorg. Biochem.*, 2001, **85**, 209–217.
100. N. J. Wheate and J. G. Collins, *Coord. Chem. Rev.*, 2003, **241**, 133–145.
101. N. Farrell and Y. Qu, *Inorg. Chem.*, 1989, **28**, 3416–3420.
102. N. P. Farrell, S. G. De Almeida and K. A. Skov, *J. Am. Chem. Soc.*, 1988, **110**, 5018–5019.
103. M. E. Oehlsen, A. Hegmans, Y. Qu and N. Farrell, *Inorg. Chem.*, 2005, **44**, 3004–3006.
104. N. Farrell, Y. Qu and M. P. Hacker, *J. Med. Chem.*, 1990, **33**, 2179–2184.
105. J. A. Broomhead, L. M. Rendina and M. Sterns, *Inorg. Chem.*, 1992, **31**, 1880–1889.
106. J. A. Broomhead and M. J. Lynch, *Inorg. Chim. Acta*, 1995, **240**, 13–17.
107. J. A. Broomhead, L. M. Rendina and L. K. Webster, *J. Inorg. Biochem.*, 1993, **49**, 221–234.
108. N. J. Wheate, C. Cullinane and L. K. Webster, *Anticancer Drug Design*, 2001, **16**, 91–98.
109. N. J. Wheate and J. G. Collins, *Curr. Med. Chem.*, 2005, **5**, 267–279.
110. D. Fan, X. Yang, X. Wang, L. Zhang and Z. Guo, *J. Biol. Inorg. Chem.*, 2007, **12**, 655–665.
111. S.-X. Guo, D. N. Mason, S. A. Turland, E. T. Lawrenz, L. C. Kelly, G. D. Fallon, B. M. Gatehouse, A. M. Bond, G. B. Deacon, A. R. Battle,

T. W. Hambley, S. Rainone, L. K. Webster and C. Cullinane, *J. Inorg. Biochem.*, 2012, **115**, 226–239.

112. S. Komeda, M. Lutz, A. L. Spek, M. Chikuma and J. Reedijk, *Inorg. Chem.*, 2000, **39**, 4230–4236.

113. S. Komeda, M. Lutz, A. L. Spek, Y. Yamanaka, T. Sato, M. Chikuma and J. Reedijk, *J. Am. Chem. Soc.*, 2002, **124**, 4738–4746.

114. S. Komeda, H. Ohishi, H. Yamane, M. Harikawa, K. Sakaguchi and M. Chikuma, *Dalton Trans.*, 1999, 2959–2962.

115. S. Komeda, G. V. Kalayda, M. Lutz, A. L. Spek, Y. Yamanaka, T. Sato, M. Chikuma and J. Reedijk, *J. Med. Chem.*, 2003, **46**, 1210–1219.

116. T. Muchova, S. M. Quintal, N. P. Farrell, V. Brabec and J. Kasparkova, *J. Biol. Inorg. Chem.*, 2012, **17**, 239–245.

117. J. W. Cox, S. J. Berners-Price, M. S. Davies, Y. Qu and N. Farrell, *J. Am. Chem. Soc.*, 2001, **123**, 1316–1326.

118. Y. Qu, N. Farrell, J. Kasparkova and V. Brabec, *J. Inorg. Biochem.*, 1997, **67**, 174–174.

119. N. Farrell, T. G. Appleton, Y. Qu, J. D. Roberts, A. P. S. Fontes, K. A. Skov, P. Wu and Y. Zou, *Biochemistry*, 1995, **34**, 15 480–15 486.

120. A. L. Harris, X. Yang, A. Hegmans, L. Povirk, J. J. Ryan, L. Kelland and N. P. Farrell, *Inorg. Chem.*, 2005, **44**, 9598–9600.

121. T. D. McGregor, A. Hegmans, J. Kašpárková, K. Neplechova, O. Novakova, H. Peňazová, O. Vrana, V. Brabec and N. Farrell, *J. Biol. Inorg. Chem.*, 2002, **7**, 397–404.

122. A. L. Harris, J. J. Ryan and N. Farrell, *Mol. Pharmacol.*, 2006, **69**, 666–672.

123. J. D. Roberts, J. Peroutka and N. Farrell, *J. Inorg. Biochem.*, 1999, **77**, 51–57.

124. C. Manzotti, G. Pratesi, E. Menta, R. Di Domenico, E. Cavalletti, H. H. Fiebig, L. R. Kelland, N. Farrell, D. Polizzi and R. Supino, *Clin. Cancer Res.*, 2000, **6**, 2626–2634.

125. T. Servidei, C. Ferlini, A. Riccardi, D. Meco, G. Scambia, G. Segni, C. Manzotti and R. Riccardi, *Eur. J. Cancer*, 2001, **37**, 930–938.

126. G. Pratesi, P. Perego, D. Polizzi, S. Righetti, R. Supino, C. Caserini, C. Manzotti, F. Giuliani, G. Pezzoni and S. Tognella, *Br. J. Cancer*, 1999, **80**, 1912.

127. P. Perego, L. Gatti, C. Caserini, R. Supino, D. Colangelo, R. Leone, S. Spinelli, N. Farrell and F. Zunino, *J. Inorg. Biochem.*, 1999, **77**, 59.

128. B. A. J. Jansen, J. Brouwer and J. Reedijk, *J. Inorg. Biochem.*, 2002, **89**, 197–202.

129. M. E. Oehlsen, A. Hegmans, Y. Qu and N. Farrell, *J. Biol. Inorg. Chem.*, 2005, **10**, 433–442.

130. C. Billecke, S. Finniss, L. Tahash, C. Miller, T. Mikkelsen, N. P. Farrell and O. Bögler, *Neuro-oncol.*, 2006, **8**, 215–226.

131. C. Sessa, G. Capri, L. Gianni, F. Peccatori, G. Grasselli, J. Bauer, M. Zucchetti, L. Vigano, A. Gatti and C. Minoia, *Ann. Oncol.*, 2000, **11**, 977–983.

132. D. Jodrell, T. Evans, W. Steward, D. Cameron, J. Prendiville, C. Aschele, C. Noberasco, M. Lind, J. Carmichael and N. Dobbs, *Eur. J. Cancer*, 2004, **40**, 1872–1877.

133. A. Calvert, H. Thomas, N. Colombo, M. Gore, H. Earl, L. Sena, G. Camboni, P. Liati and C. Sessa, *Eur. J. Cancer*, 2001, **37**, 260–260.

134. G. Scagliotti, S. Novello, L. Crinò, F. De Marinis, M. Tonato, C. Noberasco, G. Selvaggi, F. Massoni, B. Gatti and G. Camboni, *Lung Cancer*, 2003, **41**, s223.

135. T. A. Hensing, N. H. Hanna, H. H. Gillenwater, M. G. Camboni, C. Allievi and M. A. Socinski, *Anti-cancer drug*, 2006, **17**, 697–704.

136. K. W. Jennette, J. T. Gill, J. A. Sadownick and S. J. Lippard, *J. Am. Chem. Soc.*, 1976, **98**, 6159–6168.

137. W. D. McFadyen, L. P. G. Wakelin, I. A. G. Roos and B. L. Hillcoat, *Biochem. J.*, 1987, **242**, 177–183.

138. C. Yu, H.-y. Chan, W. K. and V. W. Yam, *Proc. Natl Acad. Sci. USA*, 2006, **103**, 19652–19657.

139. W. D. McFadyen, L. P. G. Wakelin, I. A. G. Roos and B. L. Hillcoat, *Biochem. J.*, 1986, **238**, 757–763.

140. R. D. Gillard and A. Sengül, *Transition Metal Chemistry (London)*, 2001, **26**, 339–344.

141. A. McCoubrey, H. C. Latham, P. R. Cook, A. Rodger and G. Lowe, *FEBS Lett.*, 1996, **380**, 73–78.

142. G. Lowe, A. S. Droz, T. Vilaivan, G. W. Weaver, J. J. Park, J. M. Pratt, L. Tweedale and L. R. Kelland, *J. Med. Chem.*, 1999, **42**, 3167–3174.

143. K. Becker, C. Herold-Mende, J. J. Park, G. Lowe and R. H. Schirmer, *J. Med. Chem.*, 2001, **44**, 2784–2792.

144. G. Lowe, A. S. Droz, T. Vilaivan, G. W. Weaver, J. J. Park, J. M. Pratt, L. Tweedale and L. R. Kelland, *J. Med. Chem.*, 1999, **42**, 3167–3174.

145. G. Lowe, A. Droz, T. Vilaivan, G. W. Weaver, L. Tweedale, J. M. Pratt, P. Rock, V. Yardley and S. L. Croft, *J. Med. Chem.*, 1999, **42**, 999–1006.

146. S. Wee, R. A. J. O'Hair and W. D. McFadyen, *Rapid Commun. Mass Spectrom.*, 2005, **19**, 1797–1805.

147. A. McCoubrey, H. C. Latham, P. R. Cook, A. Rodger and G. Lowe, *FEBS Lett.*, 1996, **380**, 73–78.

148. M. Howe-Grant, K. C. Wu, W. R. Bauer and S. J. Lippard, *Biochemistry*, 1976, **15**, 4339–4346.

149. C. S. Peyratout, T. K. Aldridge, D. K. Crites and D. R. McMillin, *Inorg. Chem.*, 1995, **34**, 4484–4489.

150. W. D. McFadyen, L. P. G. Wakelin, I. A. G. Roos and V. A. Leopold, *J. Medicin. Chem.*, 1985, **28**, 1113–1116.

151. N. J. Wheate, C. R. Brodie, J. G. Collins, S. Kemp and J. R. Aldrich-Wright, *Mini-Rev. Medicin. Chem.*, 2007, **7**, 627–648.

152. J. G. Collins and N. J. Wheate, *J. Inorg. Biochem.*, 2004, **98**, 1578–1584.

153. Y.-C. Lo, T.-P. Ko, N.-C. Wang and A. H.-J. Wang, *J. Biomol. Struct. Dyn.*, 2011, **29**, 267–282.
154. Y.-C. Lo, T.-P. Ko, W.-C. Su, T.-L. Su and A. H. J. Wang, *J. Inorg. Biochem.*, 2009, **103**, 1082–1092.
155. S. Bonse, J. M. Richards, S. A. Ross, G. Lowe and L. Krauth-Siegel, *J. Med. Chem.*, 2000, **43**, 4812–4821.
156. M. van Holst, D. Le Pevelen and J. Aldrich-Wright, *Eur. J. Inorg. Chem.*, 2008, 4608–4615.
157. M. van Holst, D. Le Pevelen and J. Aldrich-Wright, *Eur. J. Inorg. Chem.*, 2009, 691.
158. S. L. Woodhouse, E. J. Ziolkowski and L. M. Rendina, *Dalton Trans.*, 2005, **17**, 2827–2829.
159. J. A. Todd, P. Turner, E. J. Ziolkowski and L. M. Rendina, *Inorg. Chem.*, 2005, **44**, 6401–6408.
160. V. W. Yam, K. Chan, H-Y. K. Wong, M-C. and N. Zhu, *Chemistry – Eur. J.*, 2005, **11**, 4535–4543.
161. V. W. Yam, C.-H. Tao, L. Zhang, K. Wong, M-C and K.-K. Cheung, *Organometallics*, 2001, **20**, 453–459.
162. V. W.-W. Yam, R. P.-L. Tang, K. M.-C. Wong, X.-X. Lu, K.-K. Cheung and N. Zhu, *Chemistry – Eur. J.*, 2002, **8**, 4066–4076.
163. D.-L. Ma, T. Y.-T. Shum, F. Zhang, C.-M. Che and M. Yang, *Chem. Commun.*, 2005, **37**, 4675–4677.
164. M. Howe-Grant, K. C. Wu, W. R. Bauer and S. J. Lippard, *Biochemistry*, 1976, **15**, 4339–4346.
165. S. J. Lippard, P. J. Bond, K. C. Wu and W. R. Bauer, *Science*, 1976, **194**, 726–728.
166. W. D. McFadyen, L. P. Wakelin, I. A. Roos and V. A. Leopold, *J. Med. Chem.*, 1985, **28**, 1113–1116.
167. K. J. Davis, J. A. Carrall, B. Lai, J. R. Aldrich-Wright, S. F. Ralph and C. T. Dillon, *Dalton Trans.*, 2012, **41**, 9417–9426.
168. S. Kemp, N. J. Wheate, D. P. Buck, M. Nikac, J. G. Collins and J. R. Aldrich-Wright, *J. Inorg. Biochem.*, 2007, **101**, 1049–1058.
169. C. R. Culy, D. Clemett and L. R. Wiseman, *Drugs*, 2000, **60**, 895–924.
170. D. M. Fisher, R. R. Fenton and J. R. Aldrich-Wright, *Chem. Commun.*, 2008, 5613–5615.
171. J. Moretto, B. Chauffert, F. Ghiringhelli, J. R. Aldrich-Wright and F. Bouyer, *Invest. New Drugs*, 2011, **29**, 1164–1176.
172. K. B. Garbutcheon-Singh, M. P. Grant, B. W. Harper, A. M. Krause-Heuer, M. Manohar, N. Orkey and J. R. Aldrich-Wright, *Curr. Top. Med. Chem.*, 2011, **11**, 521–542.
173. K. B. Garbutcheon-Singh, P. Leverett, S. Myers and J. R. Aldrich-Wright, *Dalton Trans.*, 2013, **42**, 918–926.
174. G. H. Bulluss, Honours Thesis, University of Sydney, 1999.
175. M. Manohar, Dissertation, University of Western Sydney, 2010.
176. A. M. Krause-Heuer, N. J. Wheate, W. S. Price and J. Aldrich-Wright, *Chem. Commun.*, 2009, 1210–1212.

177. S. Kemp, N. J. Wheate, S. Wang, J. G. Collins, S. F. Ralph, A. I. Day, V. J. Higgins and J. R. Aldrich-Wright, *J. Biol. Inorg. Chem.*, 2007, **12**, 969–979.
178. V. B. Jadhav, Y. J. Jun, J. H. Song, M. K. Park, J. H. Oh, S. W. Chae, I.-S. Kim, S.-J. Choi, H. J. Lee and Y. S. Sohn, *J. Control. Release*, 2010, **147**, 144–150.
179. L. J. Wilkoff, E. A. Dulmadge, M. W. Trader, S. D. Harrison, Jr. and D. P. Griswold, Jr., *Cancer Chemother. Pharmacol.*, 1987, **20**, 96–100.
180. M. J. Ferguson, F. Y. Ahmed and J. Cassidy, *Drug Resist. Update*, 2001, **4**, 225–232.
181. S. Dhar, F. X. Gu, R. Langer, O. C. Farokhzad and S. J. Lippard, *Proc. Natl Acad. Sci. USA*, 2008, **105**, 17 356–17 361.
182. D. Screnci, M. McKeage, P. Galettis, T. Hambley, B. Palmer and B. Baguley, *Br. J. Cancer*, 2000, **82**, 966–972.
183. B. Desoize and C. Madoulet, *Crit. Rev. Oncol. Hematol.*, 2002, **42**, 317–325.
184. S. Choi, C. Filotto, M. Bisanzo, S. Delaney, D. Lagasee, J. L. Whitworth, A. Jusko, C. Li, N. A. Wood, J. Willingham, A. Schwenker and K. Spaulding, *Inorg. Chem.*, 1998, **37**, 2500–2504.
185. L. Ellis, H. Er and T. Hambley, *Aust. J. Chem.*, 1995, **48**, 793–806.
186. J. F. Vollano, S. Al-Baker, J. C. Dabrowiak and J. E. Schurig, *J. Med. Chem.*, 1987, **30**, 716–719.
187. J. J. Wilson and S. J. Lippard, *Inorg. Chem.*, 2011, **50**, 3103–3115.
188. D. Lebwohl and R. Canetta, *Eur. J. Cancer*, 1998, **34**, 1522–1534.
189. J. M. Perez, M. A. Fuertes, C. Alonso and C. Navarro-Ranninger, *Crit. Rev. Oncol. Hematol.*, 2000, **35**, 109–120.
190. R. J. Schilder, F. P. LaCreta, R. P. Perez, S. W. Johnson, J. M. Brennan, A. Rogatko, S. Nash, C. McAleer, T. C. Hamilton and D. Roby, *Cancer Res.*, 1994, **54**, 709–717.
191. T. J. O'Rourke, G. R. Weiss, P. New, H. Burris, G. Rodriguez, J. Eckhardt, J. Hardy, J. G. Kuhn, S. Fields and G. M. Clark, *Anti-cancer Drug.*, 1994, **5**, 520–526.
192. L. Pendyala, J. W. Cowens, G. B. Chheda, S. P. Dutta and P. J. Creaven, *Cancer Res.*, 1988, **48**, 3533–3536.
193. J. Vermorken, S. Gundersen, M. Clavel, J. Smyth, P. Dodion, J. Renard and S. Kaye, *Ann. Oncol.*, 1993, **4**, 303–306.
194. H. Anderson, J. Wagstaff, D. Crowther, R. Swindell, M. J. Lind, J. McGregor, M. Timms, D. Brown and P. Palmer, *Eur. J. Cancer Clin. Oncol.*, 1988, **24**, 1471–1479.
195. U. Olszewski, F. Ach, E. Ulsperger, G. Baumgartner, R. Zeillinger, P. Bednarski, G. Hamilton and M. Coluccia, *Metal Based Drugs*, 2009, **2009**, 42.
196. H. Choy, C. Park and M. Yao, *Clin. Cancer Res.*, 2008, **14**, 1633–1638.
197. P. J. Bednarski, F. S. Mackay and P. J. Sadler, *Anti-Cancer Agents Med. Chem.*, 2007, **7**, 75–93.
198. P. J. Bednarski, R. Grünert, M. Zielzki, A. Wellner, F. S. Mackay and P. J. Sadler, *Chem. Biol.*, 2006, **13**, 61–67.

199. K. R. Barnes, A. Kutikov and S. J. Lippard, *Chem. Biol.*, 2004, **11**, 557–564.
200. K. Chau, H. Lam and K. Lee, *Exp. Cell Res.*, 1998, **241**, 269–272.
201. Q. He, C. H. Liang and S. J. Lippard, *Proc. Natl Acad. Sci. USA*, 2000, **97**, 5768–5772.
202. B. L. Stocker and J. O. Hoberg, *Organometallics*, 2006, **25**, 4537–4541.
203. S. Dhar, Z. Liu, J. Thomale, H. Dai and S. J. Lippard, *J. Am. Chem. Soc.*, 2008, **130**, 11467–11476.
204. F. P. Dwyer and E. C. Gyarfas, *J. Proc. R. Soc. NSW*, 1950, **83**, 170–173.
205. F. P. Dwyer, E. C. Gyarfas and M. F. O'Dwyer, *Nature*, 1951, **167**, 1036.
206. B. Bosnich and F. P. Dwyer, *Aust. J. Chem.*, 1966, **19**, 2229–2233.
207. D. P. Mellor, *Proc. Roy. Aust. Chem. Inst.*, 1970, **37**, 199–208.
208. N. L. Kilah and E. Meggers, *Aust. J. Chem.*, 2012, **65**, 1325–1332.
209. M. R. Gill and J. A. Thomas, *Chem. Soc. Rev.*, 2012, **41**, 3179–3192.
210. A. W. McKinley, P. Lincoln and E. M. Tuite, *Coord. Chem. Rev.*, 2011, **255**, 2676–2692.
211. M.-Y. Wei, L.-H. Guo and P. Famouri, *Microchim. Acta*, 2011, **172**, 247–260.
212. H. Niyazi, J. P. Hall, K. O'Sullivan, G. Winter, T. Sorensen, J. M. Kelly and C. J. Cardin, *Nat. Chem.*, 2012, **4**, 621–628.
213. A. Greguric, I. D. Greguric, T. W. Hambley, J. R. Aldrich-Wright and J. G. Collins, *Dalton Trans.*, 2002, 849–855.
214. S. J. Dougan, M. Melchart, A. Habtemariam, S. Parsons and P. J. Sadler, *Inorg. Chem.*, 2006, **45**, 10 882–10 894.
215. Y. K. Yan, M. Melchart, A. Habtemariam and P. J. Sadler, *Chem. Commun.*, 2005, 4764–4776.
216. H. Chen, J. A. Parkinson, S. Parsons, R. A. Coxall, R. O. Gould and P. J. Sadler, *J. Am. Chem. Soc.*, 2002, **124**, 3064–3082.
217. R. E. Aird, J. Cummings, A. A. Ritchie, M. Muir, R. E. Morris, H. Chen, P. J. Sadler and D. I. Jodrell, *Br. J. Cancer*, 2002, **86**, 1652–1657.
218. S. J. Dougan and P. J. Sadler, *CHIMIA Int. J. Chem.*, 2007, **61**, 704–715.
219. M. Melchart and P. J. Sadler, in *Bioorganometallics*, Wiley-VCH Verlag, Weiheim FRG, 2006, pp. 39–64.
220. B. Therrien, *Eur. J. Inorg. Chem.*, 2009, **2009**, 2445–2453.
221. M. Hanif, A. A. Nazarov, A. Legin, M. Groessl, V. B. Arion, M. A. Jakupec, Y. O. Tsybin, P. J. Dyson, B. K. Keppler and C. G. Hartinger, *Chem. Commun.*, 2012, **48**, 1475–1477.
222. K. J. Kilpin, C. M. Clavel, F. Edafe and P. J. Dyson, *Organometallics*, 2012, **31**, 7031–7039.
223. A. A. Nazarov, J. Risse, W. H. Ang, F. Schmitt, O. Zava, A. Ruggi, M. Groessl, R. Scopelitti, L. Juillerat-Jeanneret, C. G. Hartinger and P. J. Dyson, *Inorg. Chem.*, 2012, **51**, 3633–3639.
224. C. L. Kielkopf, K. E. Erkkila, B. P. Hudson, J. K. Barton and D. C. Rees, *Nat. Struct. Mol. Biol.*, 2000, **7**, 117–121.

225. D. Boer, J. Kerckhoffs, Y. Parajo, M. Pascu, I. Usón, P. Lincoln, M. Hannon and M. Coll, *Angew. Chemie, Int. Ed.*, **49**, 2336–2339.
226. A. Oleksi, A. G. Blanco, R. Boer, I. Usón, J. Aymamí, A. Rodger, M. J. Hannon and M. Coll, *Angew. Chemie, Int. Ed.*, 2006, **45**, 1227–1231.
227. X. Xie, S. P. Mulcahy and E. Meggers, *Inorg. Chem.*, 2009, **48**, 1053–1061.
228. G. E. Atilla-Gokcumen, C. L. Di and E. Meggers, *J. Biol. Inorg. Chem.*, 2011, **16**, 45–50.
229. G. E. Atilla-Gokcumen, N. Pagano, C. Streu, J. Maksimoska, P. Filippakopoulos, S. Knapp and E. Meggers, *ChemBioChem*, 2008, **9**, 2933–2936.
230. S. Dieckmann, R. Riedel, K. Harms and E. Meggers, *Eur. J. Inorg. Chem.*, **2012**, 813–821.
231. D. S. Williams, P. J. Carroll and E. Meggers, *Inorg. Chem.*, 2007, **46**, 2944–2946.
232. S. Mollin, S. Blanck, K. Harms and E. Meggers, *Inorg. Chim. Acta*, 2012, **393**, 261–268.
233. P. K. Sasmal, C. N. Streu and E. Meggers, *Chem. Commun.*, 2013, **49**, 1581–1587.
234. M. J. Hannon, V. Moreno, M. J. Prieto, E. Molderheim, E. Sletten, I. Meistermann, C. J. Isaac, K. J. Sanders and A. Rodger, *Angew. Chemie, Int. Ed.*, 2001, **40**, 879–884.
235. B. M. Zeglis, V. R. C. Pierre, J. T. Kaiser and J. K. Barton, *Biochemistry*, 2009, **48**, 4247–4253.
236. D. S. Sigman, D. R. Graham, V. D'Aurora and A. M. Stern, *J. Biol. Chem.*, 1979, **254**, 12269–12272.
237. M. Pitié, C. J. Burrows and B. Meunier, *Nucleic Acids Res.*, 2000, **28**, 4856–4864.
238. A. Robertazzi, A. V. Vargiu, A. Magistrato, P. Ruggerone, P. Carloni, P. d. Hoog and J. Reedijk, *J. Phys. Chem. B*, 2009, **113**, 10 881–10 890.
239. E. Melendez, *Crit. Rev. Oncol. Hematol.*, 2002, **42**, 309–315.
240. B. K. Keppler and M. Hartmann, *Metal-Based Drugs*, 1994, **1**, 145–150.
241. F. Caruso, M. Rossi, C. Opazo and C. Pettinari, *Bioinorg. Chem. Appl.*, 2005, **3**, 317–329.
242. P. Köpf-Masier and D. Krahl, *Chem.-Biol. Interact.*, 1983, **44**, 317–328.
243. P. Köpf-Masier and D. Krahl, *Naturwissenschaften*, 1981, **68**, 273–274.
244. P. Köpf-Masier, *J. Struct. Biol.*, 1990, **105**, 34–45.
245. P. Köpf-Masier, H. Köpf and W. Wagner, *Naturwissenschaften*, 1981, **68**, 272–273.
246. P. Köpf-Masier and H. Köpf, *Naturwissenschaften*, 1980, **67**, 415–416.
247. P. J. Barnard, M. V. Baker, S. J. Berners-Price and D. A. Day, *J. Inorg. Biochem.*, 2004, **98**, 1642–1647.
248. A. Casini, G. Hartinger Christian, C. Gabbiani, E. Mini, J. Dyson Paul, K. Keppler Bernhard and L. Messori, *J. Inorg. Biochem.*, 2008, **102**, 564–575.
249. P. F. Shi, Q. Jiang, Y. M. Zhao, Y. M. Zhang, J. Lin, L. P. Lin, J. Ding and Z. J. Guo, *J. Inorg. Biochem.*, 2006, **11**, 745–752.

250. R. Wai-Yin Sun, D.-L. Ma, E. L.-M. Wong and C.-M. Che, *Dalton Trans.*, 2007, 4884–4892.
251. J. Müller, *Eur. J. Inorg. Chem.*, 2008, **2008**, 3749–3763.
252. A. Ono, H. Torigoe, Y. Tanaka and I. Okamoto, *Chem. Soc. Rev.*, 2011, **40**, 5855–5866.
253. Y. Takezawa and M. Shionoya, *Acc. Chem. Res.*, 2012, **45**, 2066–2076.
254. S. Johannsen, N. Megger, D. Boehme, R. K. O. Sigel and J. Mueller, *Nat. Chem.*, 2010, **2**, 229–234.
255. J. Mueller, E. Freisinger, P. Lax, D. A. Megger and F.-A. Polonius, *Inorg. Chim. Acta*, 2007, **360**, 255–263.
256. K. Seubert, D. Boehme, J. Koesters, W.-Z. Shen, E. Freisinger and J. Mueller, *Z. Anorg. Allg. Chem.*, 2012, **638**, 1761–1767.
257. G. H. Clever, C. Kaul and T. Carell, *Angew. Chemie, Int. Ed.*, 2007, **46**, 6226–6236.
258. G. H. Clever, K. Polborn and T. Carell, *Angew. Chemie, Int. Ed.*, 2005, **44**, 7204–7208.
259. G. H. Clever and M. Shionoya, *Coord. Chem. Rev.*, 2010, **254**, 2391–2402.
260. K. Tanaka, G. H. Clever, Y. Takezawa, Y. Yamada, C. Kaul, M. Shionoya and T. Carell, *Nat. Nanotechnol.*, 2006, **1**, 190–194.
261. D. A. Megger, G. C. Fonseca, F. M. Bickelhaupt and J. Muller, *J. Inorg. Biochem.*, 2011, **105**, 1398–1404.
262. Y. Miyake, H. Togashi, M. Tashiro, H. Yamaguchi, S. Oda, M. Kudo, Y. Tanaka, Y. Kondo, R. Sawa, T. Fujimoto, T. Machinami and A. Ono, *J. Am. Chem. Soc.*, 2006, **128**, 2172–2173.
263. Y. Tanaka, S. Oda, H. Yamaguchi, Y. Kondo, C. Kojima and A. Ono, *J. Am. Chem. Soc.*, 2006, **129**, 244–245.
264. P. Aich, S. L. Labiuk, L. W. Tari, L. J. T. Delbaere, W. J. Roesler, K. J. Falk, R. P. Steer and J. S. Lee, *J. Mol. Biol.*, 1999, **294**, 477–485.
265. L. Zhang, A. Peritz and E. Meggers, *J. Am. Chem. Soc.*, 2005, **127**, 4174–4175.
266. G. H. Clever, Y. Söltl, H. Burks, W. Spahl and T. Carell, *Chem. – Eur. J.*, 2006, **12**, 8708–8718.
267. H. Weizman and Y. Tor, *J. Am. Chem. Soc.*, 2001, **123**, 3375–3376.
268. S. Johannsen, N. Megger, D. Böhme, K. O. SigelRoland and J. Müller, *Nat. Chem.*, 2010, **2**, 229–234.

Supramolecular Metal Complexes for Imaging and Radiotherapy

JÜRGEN SCHATZ*[a] AND DANIEL SCHÜHLE[b]

[a] Organic Chemistry 1, Department of Chemistry and Pharmacy, University of Erlangen, Henkestraße 42, 91054 Erlangen, Germany; [b] Heimerle + Meule GmbH, Dennigstrasse 16, 75179 Pforzheim, Germany
*Email: juergen.schatz@fau.de

10.1 Introduction

Metal ions play a vital role in biological and/or biochemical systems. When it comes to using artificial, non-natural metal-based systems, it is important to provide the "correct" chemical environment for metal ions, mostly a suitable coordination sphere, due to the potentially significant toxicity of aquo ions. Supramolecular entities such as crown or lariat ethers, cryptands, carcerands, and others are used to encapsulate the necessary metal ion to modify the chemical and biological properties of the metal center (Figure 10.1A). Here, the borderline between "classic" and "supramolecular" coordination chemistry is fuzzy. Therefore, in the following compilation we will give some examples for "classic" coordination compounds along with selected examples for supramolecular systems that are formed using non-covalent bonds. Figure 10.1 reveals two other setups that are often found for supramolecular metal complexes in a biological context. Arrangement B shows a supramolecular platform that defines number and relative spatial positions of metal centers.

Monographs in Supramolecular Chemistry No. 13
Supramolecular Systems in Biomedical Fields
Edited by Hans-Jörg Schneider
© The Royal Society of Chemistry 2013
Published by the Royal Society of Chemistry, www.rsc.org

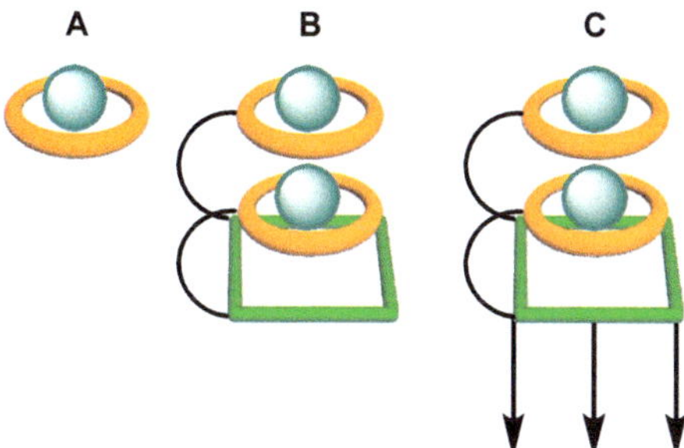

Figure 10.1 Supramolecular coordination compounds in various complexes. (A) Binding of a metal ion (light blue ball) by a multidentate entity (golden circle) resulting in a stable metal complex. (B) Attachment of two or more of such complexes to a supramolecular platform (green rectangle). (C) Targeting vectors (black arrows) control biodistribution of supramolecular metal complexes.

This is helpful if the relative concentration of metal ions per molecule or co-operative interaction between metal centers is an important feature (Figure 10.1B). This approach can be further expanded when additional targeting vectors (Figure 10.1C, symbolized by arrows) are introduced that ensure defined, tailor-made interactions with biological targets resulting in a higher biological impact.

In today's supramolecular chemistry, nanoparticles or metal clusters can play an important role. Their use in molecular diagnostics and therapeutics has been reviewed recently[1] and is therefore not included in this compilation, which is focused on molecular complexes with applications in imaging and radiotherapy. Additional details can also be found in Chapter 15 of the present book. General aspects of coordination chemistry in medicine can be found elsewhere.[2]

10.2 Supramolecular Chemistry in Imaging

Modern medicinal diagnoses strongly rely on imaging methods used to study biological (mis-)function down to cellular level. In clinical routine, diverse imaging methodologies such as magnetic resonance imaging (MRI, *cf.* Section 10.2.1), computed tomography (CT), positron emission tomography (PET), and single photon emission tomography (SPECT) have found widespread application. In addition, spectroscopic properties (UV/Vis, fluorescence, phosphorescence, luminescence, ultrasound) are used to gain insight into biological systems.

All of these imaging modalities either require or can benefit from the use of contrast agents that can be based on metal complexes. In this chapter general chemical aspects of these contrast agents are discussed and the interested reader is referred to more specialized literature such as the Chapter 15 in this book.

10.2.1 Magnetic Resonance Imaging

Magnetic resonance imaging (MRI) is one of the most powerful tools in medical imaging. Here, the resonance of water protons is observed and the image reflects the biological environment. After aligning proton spins in a static magnetic field, radiofrequencies are used to probe the proton spins, thus generating information from the environment of the spins.[3] Different intensities in images originate either from differences in proton density or in the longitudinal (T_1) or transversal (T_2) relaxation time of water protons. While this technique provides endogenous contrast, about one-third of all MRI scans worldwide is performed using contrast agents to obtain information that is not accessible without these artificial compounds. All clinically approved MRI contrast agents affect the ability of para- or ferromagnetic to enhance the relaxation times of water. Most approved contrast agents contain gadolinium(III), which shortens T_1 of surrounding water molecules very efficiently and leads to a brightening in MR images of areas that contain the agent.[4] The ability of shortening T_1 is expressed using the term "longitudinal relaxivity", or often just relaxivity (r_1). The higher the relaxivity (given in $mM^{-1}s^{-1}$) of a paramagnetic complex, the more efficiently it shortens T_1, thus the more contrast it generates or the lower dose of contrast agent can be used to achieved the same contrast.

The contrast agent $[Gd(H_2O)]^{3+}$ itself is very efficient; however, the high toxicity of the metal ion prevents its use in "real" medical applications. Supramolecular chemistry can help to create a coordination sphere in such a way that the toxicity problem can be effectively overcome. It is essential for translation from preclinical studies to clinical settings that the complexes are thermodynamically highly stable and show very high kinetic inertness.[5] Following this premise, many gadolinium(III) chelates are used in clinical routine (*cf.* Figure 10.2 for selected examples and Table 10.1 for selected properties).

As can be seen in Figure 10.3, common open-chain or macrocyclic ligands for Gd(III) complexation are hepta- or octadentate, and are oligoaminocarboxylates or derivatives thereof. In most complexes of this type, Gd(III) has a coordination number of 9, leaving one or two coordination sites for water, respectively. The mechanism of how paramagnetic compounds shorten relaxation times is well understood but is beyond the scope of this chapter. Details can be found in excellent reviews.[32,33] Briefly, coordinated water molecules experience a very efficient relaxation due to the paramagnetic ion. This relaxation effect is transduced to the bulk by rapid exchange of coordinated water molecules with the bulk. In the end, these complexes "catalyze" relaxation effects of surrounding nuclei.

As MRI is a very important diagnostic tool, there is considerable need for contrast agents having higher relaxivity. Several approaches can be followed to this end: since water exchange between coordinated and bulk water is essential to 'spread' the influence of Gd(III) within the surrounding water, it is not surprising that the average water residence time of water coordinated to the metal ion, τ_M, has a strong influence on relaxivity. It can be optimized, for

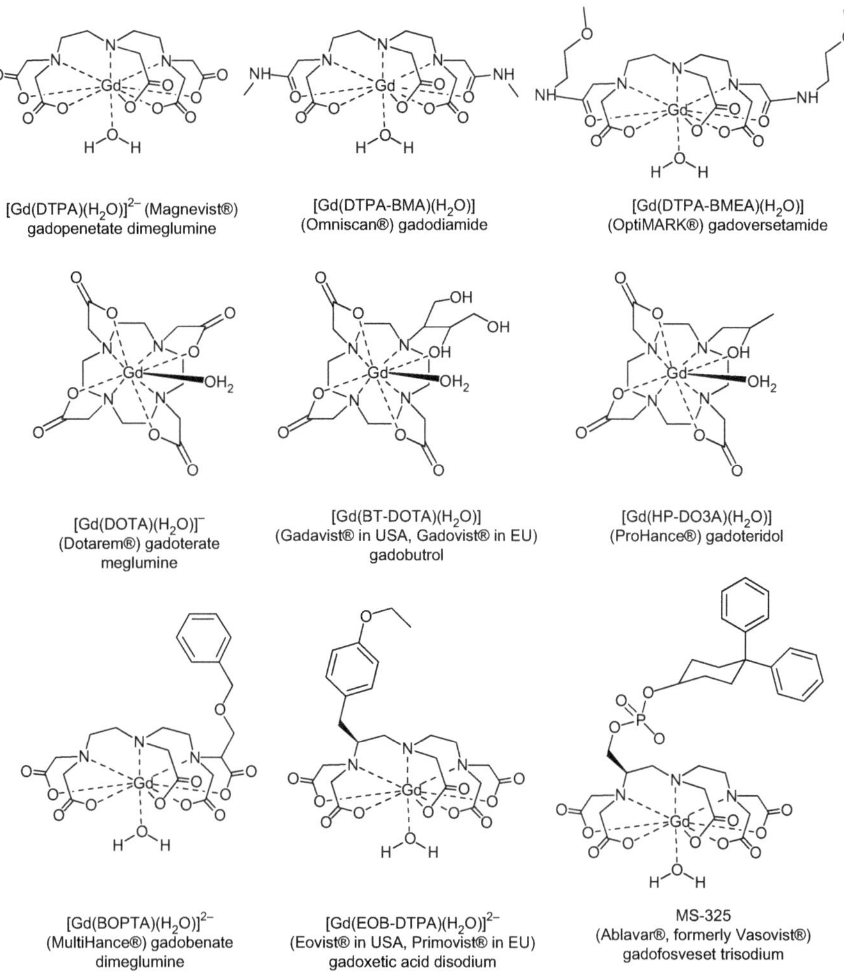

Figure 10.2 Structures and names of clinically approved MRI contrast agents (counterion not shown). Chemical name (*e.g.* [Gd(DTPA)(H$_2$O)]$^{2-}$), trade name (*e.g.* Magnevist®) and generic name (*e.g.* gadopenetate dimeglumine) are given.

example, by choosing donor atoms within ligands in a suitable way. However, from Tables 10.1 and 10.2 it is apparent that relaxivity for all chelates is rather low, with the exception of those approved agents that bind to plasma proteins, thanks to hydrophobic interactions with human serum albumin (HSA). As an example, the relaxivity of MS-325 is increased from 5.8 to 27.7 mM^{-1} s^{-1}. This is due to a very efficient increase in the molecular tumbling time (τ_R) of the **Gd-H** vector. Only when τ_R is optimized are the full effects of different τ_M values reflected in relaxivity.[34] In other words, large, slowly tumbling systems are required for more efficient contrast agents. Since synthesis of such agents

Table 10.1 Relaxivities ($mM^{-1} s^{-1}$) of some clinically approved MRI contrast agents in water and human plasma at 37 °C and 0.47, 1.5 and 3 tesla (T).[6,7]

Compound	1.5 T H$_2$O		0.47 T plasma[a]		1.5 T plasma		3 T plasma	
	r$_1$	r$_2$	r$_1$	r$_2$	r$_1$	r$_2$	r$_1$	r$_2$
[Gd(DTPA)(H$_2$O)]$^{2-}$	3.3	3.9	3.8	4.1	4.1	4.6	3.7	5.2
[Gd(DTPA-BMA)(H$_2$O)]	3.3	3.6	4.4	4.6	4.3	5.2	4.0	5.6
[Gd(DTPA-BMEA)(H$_2$O)]	3.8	4.2	5.7	6.6	4.7	5.2	4.5	5.9
[Gd(EOB-DTPA)(H$_2$O)]$^{2-}$	4.7	5.1	8.7	13	6.9	8.7	6.2	11
[Gd(BOPTA)(H$_2$O)]$^{2-}$	4.0	4.3	9.2	13	6.3	8.7	5.5	11
MS-325[8]	5.8	6.7	47.2	57.6	27.7	72.6	9.9	73.0
[Gd(HP-DO3A)(H$_2$O)]	2.9	3.2	4.8	6.1	4.1	5.0	3.7	5.7
[Gd(BT-DO3A)(H$_2$O)]	3.3	3.9	6.1	7.4	5.2	6.1	5.0	7.1
[Gd(DOTA)(H$_2$O)]$^{-}$	2.9	3.2	4.3	5.5	3.6	4.3	3.5	4.9

[a]At 40 °C.

is tedious and time consuming, supramolecular chemistry offers the unique possibility to create large structures—"just" by non-covalent assembly. Furthermore, contrast agents which combine multiple gadolinium ions in one molecular entity lead to superior, often multiplied magnetic, properties.[33] Supramolecular chemistry is therefore an ideal tool for both approaches, that is, control of molecular mobility and water exchange as well as number of gadolinium centers within the assembly.

For a more detailed understanding of the physical and coordination chemistry of gadolinium-based MRI contrast agents, we refer to several excellent reviews explaining their coordination chemistry[35–46] or medical application.[47–57]

In the following paragraphs, we will highlight the use of some supramolecular platforms in this field and also give examples of how self-assembly can be used to give information about physiological abnormalities.

Dynamic covalent chemistry can be used to attach probes to biological targets *in vivo*. A nice example (Figure 10.4) uses the overexpressed sialic acid on the surface of tumor cells as a target.[58] Chemical reaction of a phenylboronic acid with sialic acid diols overexpressed by the tumor cells results in the formation of a cyclic boronic ester that is stable enough to enrich the probe at the tumor site and to image it. Additional ionic interactions between the benzyl ammonium group and the carboxylic acid of the sialic acid intensifies the non-covalent interaction and leads to some specificity over other diol-containing carbohydrates. It should be added that, in principle, every contrast agent that contains a targeting group functions on supramolecular chemistry principles since, in most cases, only supramolecular interactions or dynamic bond formation leads to targeting.

In the following paragraphs, classes of supramolecular platforms are presented along with selected examples of how they can be used to enhance MRI contrast agents.

Table 10.2 Relaxivity of a selection of MRI contrast agents (names of ligands are given; relaxivities are in respect of the corresponding Gd(III) complexes).

Compound	r_1 $(mM^{-1}s^{-1})$	H freq. (MHz)	Temp. (°C)	pH	Ref. no.
MP-2269	6.2	20	40		9
B-21326/7	6.78	20	39	7.0	10
DTPA-L1	3.7	20			11
DOTMA	3.8	20	25		12
DOTA-pNB	5.4	20	25		13
DOTA(BOM)	5.4	20	25	7.4	14
cis-DOTA(BOM)$_2$	6.8	20	25	7.4	14
trans-DOTA(BOM)$_2$	6.5	20	25	7.4	14
DOTA(BOM)$_3$	7.5	20	25	7.4	14
DOTA(BOM)	5.4	20	25	7.4	15
cis-DOTA(BOM)$_2$	5.7	20	25	7.4	15
trans-DOTA(BOM)$_2$	5.8	20	25	7.4	15
DOTA(BOM)$_3$	6.7	20	25	7.4	15
DTPA-MpNPA	5.08	20	25		16
DTPA-BMA	4.58	20	5		17
DTPA-BMA	4.39	20	25		17
DTPA-BMA	3.96	20	35		17
DTPA-BPA	4.66	20	25		16
DTPA-BpAPA	4.12	20	25		16
DTPA-BpNPA	3.78	20	25		16
DTPA-BpTFPA	3.71	20	25		16
DTPA-B*i*BA	4.27	25			18
DTPA-BMMEA	4.1	20	40		19
DTPA-BHMEA	4.2	20	40		19
DTPA-BBA	4.08	20	37	7	20
DTPA-cis-BAM	3.2	20	40		21
DTPA-PenAM	3.6	20	40		21
DTPA-OAM	4.2	20	40		21
16-DTPA-PN	2.8	64	24		22
16-DTPA-PN	3.7	250	24		22
17-DTPA-BN	2.5	64	24		22
17-DTPA-BN	3.4	250	24		22
16-DTPA-PN-OH	3.5	250	23		23
DO3MA	4.4	20	40		24
DO3A	4.8	20	40		24
DO3A-L2	4.49	20	39	7.3	25
DO3A-L1	4.03	20	39	7.3	25
DO3A-L4	5.19	20	39	7.3	25
DO3A-L3	4.33	20	39	7.3	25
DOTPMB	2.8				26
BPO4A	1.71	40	25	7	27
BPO4A	1.66	40	25	12	27
DOTPMe	2.09	64	25		28
DOTMP-MBBzA	3.09	20	25		29
DOTMP-MPA	3.08	20	25		29
CF$_3$CH$_2$PO$_3$DOTA	2.5	40	25	6.8	30
DOTEP	5.1	40	25	7	31

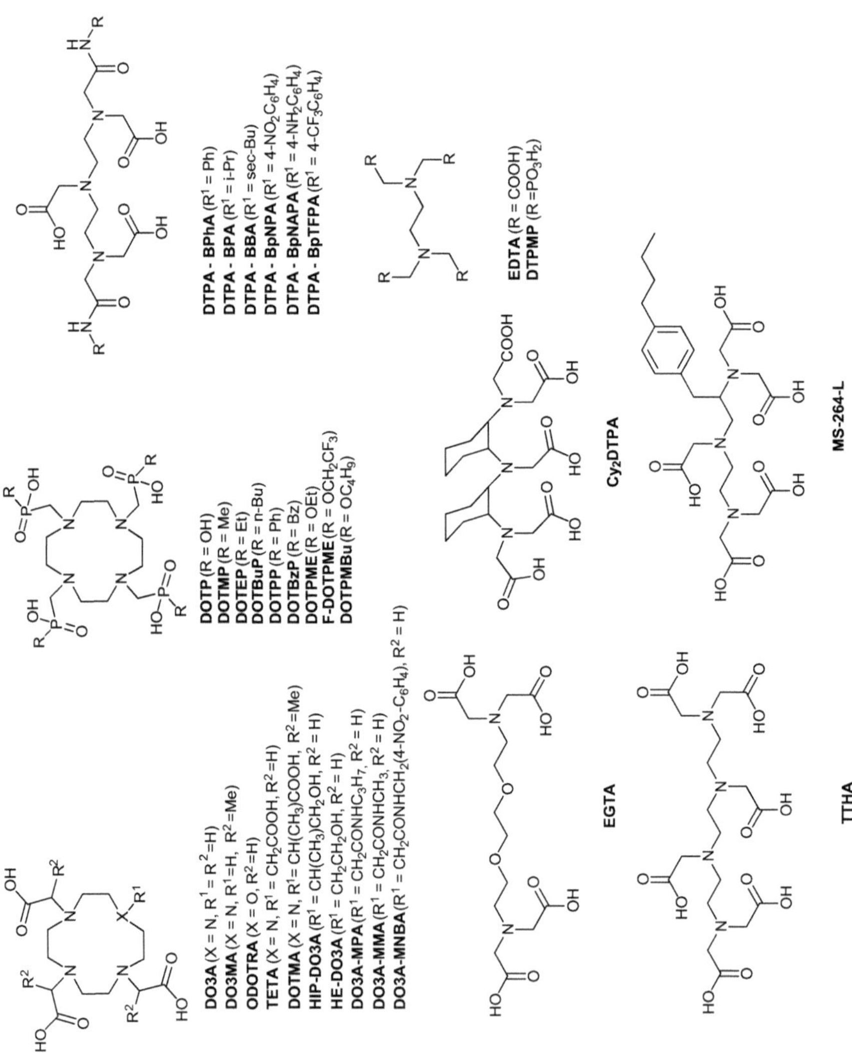

Figure 10.3 Selected ligands for Gd(III)-based MRI contrast agents.

10.2.1.1　Cyclodextrins

Cyclodextrins are easily available, water-soluble, toroidal-shaped macrocycles that exhibit a hydrophobic cavity formed by 6, 7 or 8 glucopyranoside units (α-, β-, γ-cyclodextrin, respectively). Additional information about cyclodextrins can be found in Chapter 5 of this book. Thanks to their commercial availability and physicochemical properties, cyclodextrins (CDs) are ideal supramolecular molecules and it is obvious to use such macrocycles as core components for MRI contrast agents. Here, metal complexes have to be linked to the sugar core using covalent bonds. A modular approach is desirable to allow easy synthetic access of functionalized molecules. Therefore, click chemistry[59–66] or simple (thio)urea bond formation[67,68] can be used as efficient synthetic strategy to covalently link complexes to the CD skeleton. As an example, heptamer **1** (Figure 10.5) has a good relaxivity of $43.4\,\text{mM}^{-1}\,\text{s}^{-1}$ ($6.2\,\text{mM}^{-1}\,\text{s}^{-1}$ per gadolinium) at $9.4\,\text{T}$.[69]

Commonly, β-CDs are used as the core molecule. Using click reactions, up to 7 Gd(III) complexes can be attached to the core. The hydrophobic interior provides an additional binding site for hydrophobic moieties of biologically interesting targets.

In a systematic study investigating the influence of number and arrangement of DOTA-type Gd(III) complexes on relaxivity, several contrast agents were prepared based on either the benzene or cyclodextrin skeleton.[60] Benzene based trimer **2** and hexamer **3** as well as cyclodextrin heptamer **4** exhibited relaxivities ($60\,\text{MHz}$, $37\,^{\circ}\text{C}$) of 18, 66, and $84\,\text{mM}^{-1}\,\text{s}^{-1}$, respectively (Figure 10.5). Taking the relaxivity ($3.2\,\text{mM}^{-1}\,\text{s}^{-1}$) of the monomeric Gd(III) unit into account, it became clear that a high local density of signaling units attached to a core molecule is beneficial to contrast agent properties when the large system is rigid. Reasons are that τ_R increases with molecular weight and that the restricted internal motion, which can be achieved by steric or coulomb repulsion, for example, slows down the tumbling of the **Gd-H** vector even further.

Subtle modification of the coordination sphere to optimize τ_M can further enhance the properties. Exploiting a phosphonate based DOTA binding pocket for Gd(III) bound to a cyclodextrin core by thiourea linkers (**5**, Figure 10.5), relaxivities well over $100\,\text{mM}^{-1}\text{s}^{-1}$ or >15 per Gd(III) ion ($20\,\text{MHz}$, $25\,^{\circ}\text{C}$) can be observed.[67,68]

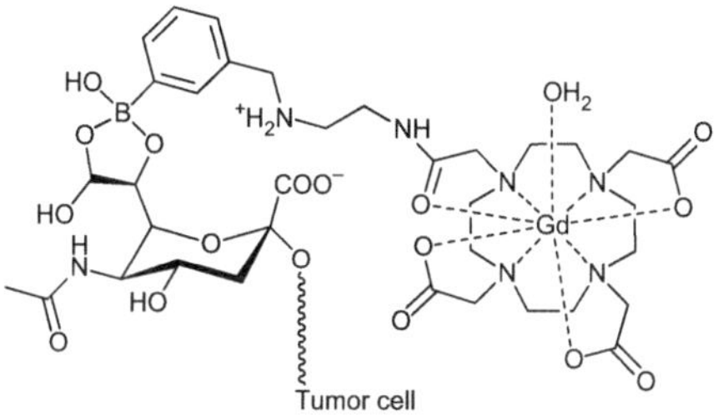

Figure 10.4　A phenylboronic acid-based Gd-MRI contrast agent that is bound to sialic acid-rich tumor surfaces.[58]

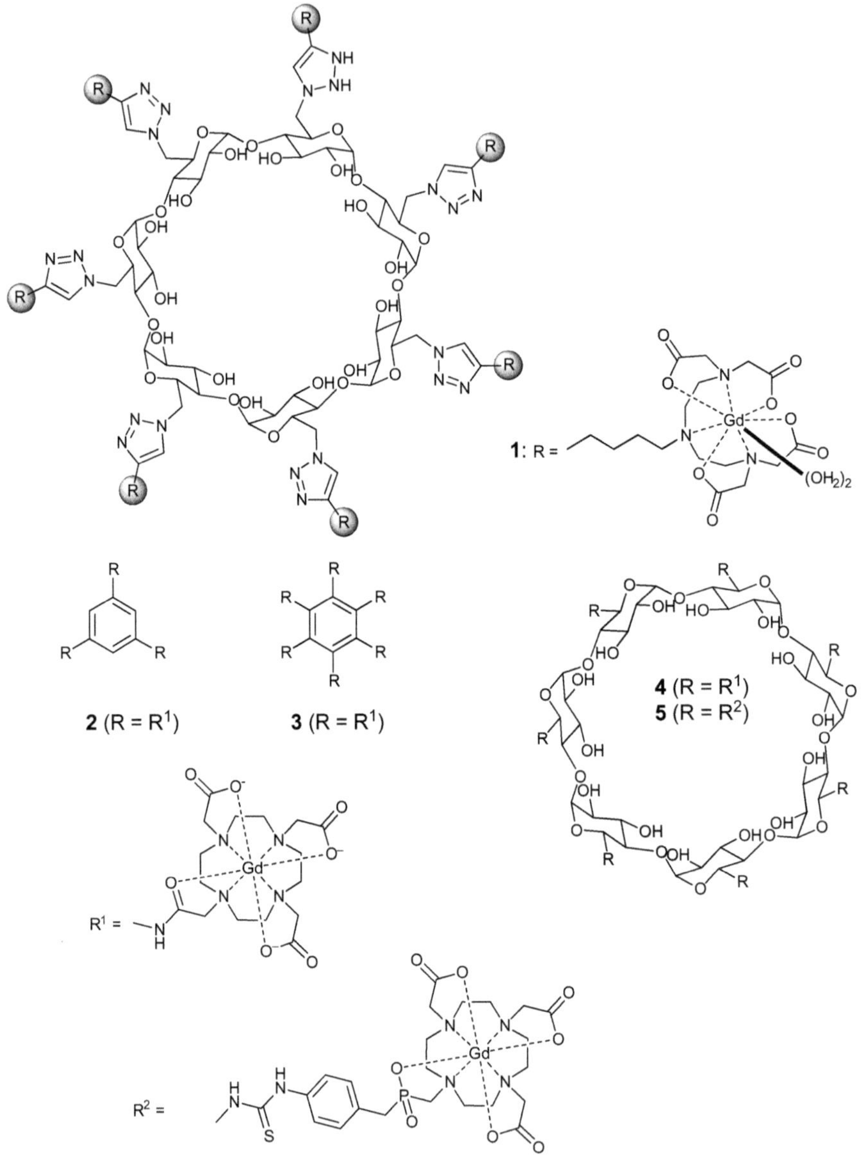

Figure 10.5 Benzene- and Cyclodextrin-based MRI contrast agents.[69]

10.2.1.2 Calixarenes

Calix[*n*]arenes, cyclic oligomers of phenols and formaldehyde, are an excellent class of supramolecular molecules. Both, the upper (wide) or lower (narrow) rim of the calixarenes provide n-positions to attach metal complexes. In addition, site-selective functionalization is possible, making calixarenes ideal platforms that allow a high local density of signaling units accompanied by potential targeting vectors attached at the opposite side of the macrocyclic skeleton.[70,71] Additional information about cyclodextrins can be found in Chapter 6 of this book.

An early example of a calixarene-based MRI agent was the calix[4]arene tetraamide **6**.[72] In this case, the calixarene core was used to prearrange donor groups for Gd(III) coordination (Figure 10.6). Relaxivity is reasonable (dimethyl sufoxide/water 9 : 1, 20 °C, 400 MHz, 3.40 mM^{-1} s^{-1}) but the stability of the Gd(III) complex is far too low.

An octa-coordination sphere around the gadolinium ion can be formed by adding four ancillary nitrogen ligands, which leads to an increase of the stability constant by an order of magnitude ($K_{ass} = 2 \times 10^5$ M^{-1}). Calix[4]arene **7** interacts with HSA as a potential biological target proving the "targeting vector" principle (*cf.* Figure 10.1).

Similarly, imidodiacetic acid calixarene **8** forms stable complexes with Gd(III) ($K_{ass} = 2.4 \times 10^4$ M^{-1}) with reasonable relaxivity (9.6 s^{-1} mM^{-1}, 20 MHz, 25 °C). The HSA-adduct is rather stable ($K_A = 2.4 \times 10^4$ M^{-1}) and has a relaxivity as high as 60 s^{-1} mM^{-1}.[73]

The water-soluble calix[8]arene **9** does form reasonably stable complexes with 8 Gd(III) ions ($K_{ass} = 10^{17}$ M^{-1}), however no additional MRI properties have been reported.[74]

Combining the well established DOTA ligand sphere with a calix[4]arene skeleton provides access to an efficient contrast agent. Tetra-gadolinium-calixarene **10a** has a relatively high relaxivity (18.3 s^{-1} mM^{-1}, 20 MHz, 37°C) thanks to micelle formation (critical micelle concentration [cmc] = 0.21 mM, 37 °C).[75,76] The relaxivity even increases (24.6 s^{-1} mM^{-1}, 20 MHz, 37 °C) upon HSA binding ($K_{ass} = 1.2 \times 10^3$ M^{-1}) due to a more efficient increase in τ_R. Replacing the monoamido chelates by four pyridine-*N*-oxide chelating units with favorably fast water-exchange kinetics leads to calix[4]arene **11**. It is more prone to micelle formation owing to the increased hydrophobicity and has an enhanced relaxivity (cmc = 35 μM, 31.2 s^{-1} mM^{-1}, 20 MHz, 25 °C), which can be further enhanced by binding to HSA. For supramolecular chemists, it is interesting that micelle formation as well as binding to targets can be studied using simple T_1 measurements since the formation of aggregates leads to decreased molecular tumbling, thus higher relaxivity and shorter T_1 relaxation times.

Co-assembly of the bis-octadecyl calixarene derivative **10b** with DSPC, cholesterol and DSPE-PEG2000 leads to stable liposomes. The relaxivity per Gd is unusually high in this system. Detailed mechanistic investigations have shown that **10b** shows an atypical distribution within the lipid bilayer: at least 80% of this compound is oriented at the outer surface of the bilayer. This overcomes an almost intrinsic problem of liposome-based T_1 agents: due to the limited amount of water within liposomes, the relaxation effect of paramagnetic compounds cannot be transduced to a sufficient amount of water molecules. As a result, the relaxivity of Gd-agents entrapped inside liposomes is generally very low. However, preferred orientation at the outer surface leads to an unprecedently high relaxivity of stable liposomes.[77]

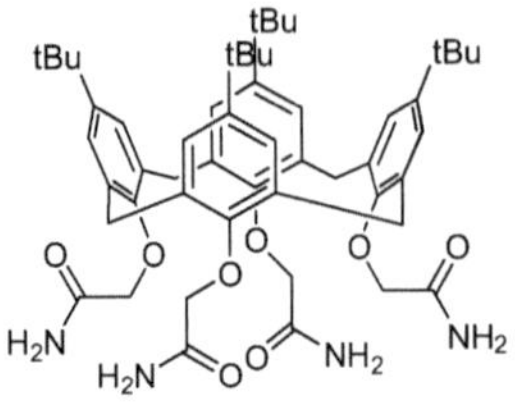

Figure 10.6 shows calixarene-based ligands **6**, 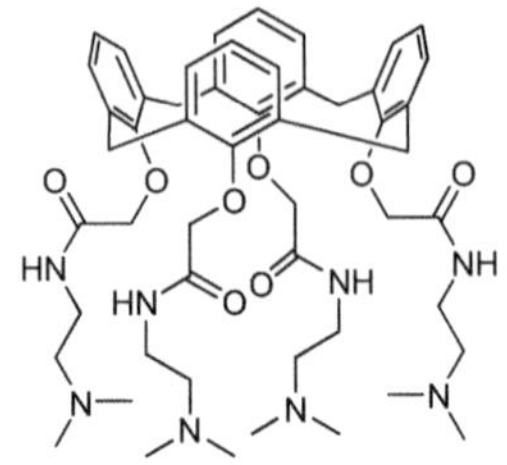 **7**, 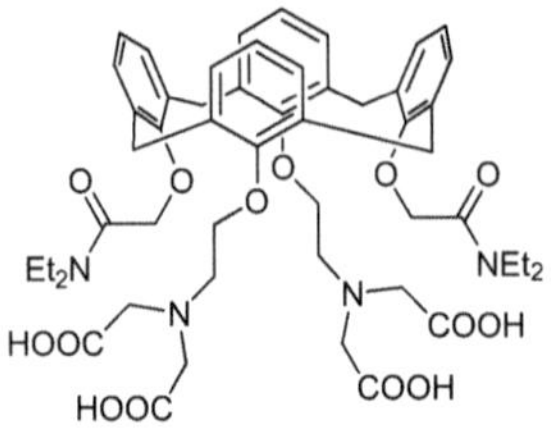 **8**, 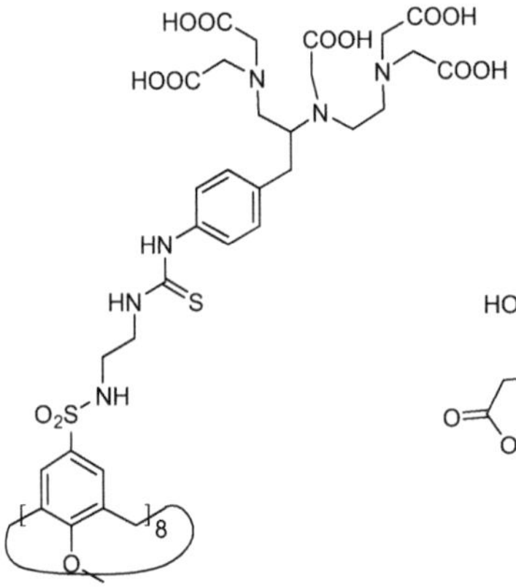 **9**, 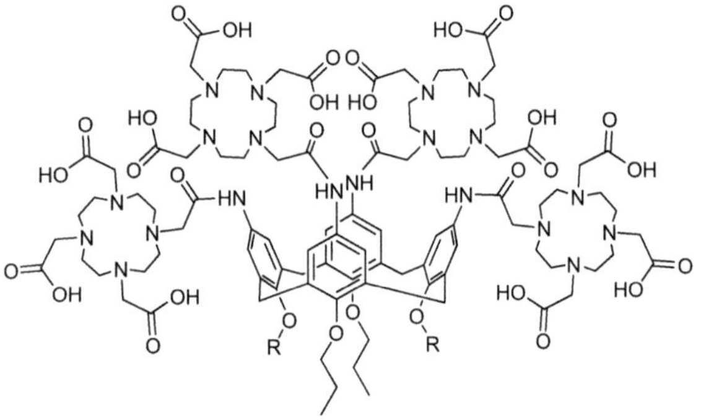 **10a** (R = C$_3$H$_7$), **10b** (R = C$_{18}$H$_{37}$), and **11**.

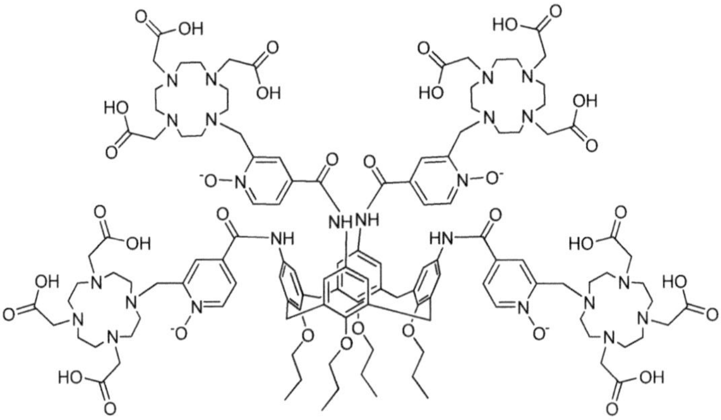

Figure 10.6 Calixarene-based ligands for Gd(III) complexation.

10.2.1.3 Dendritic, Oligomeric, and Polymeric MRI Contrast Agents

The number of gadolinium cores per molecule is a key parameter that influences the relaxivity of an MRI contrast agent. Therefore, one approach to enhance the contrast is just to increase the number of signaling units, for example, using oligomers or polymers as synthetic platforms.

Dendrimers are polymeric, highly branched structures and contain functionally tunable peripheral groups. Because of their low polydispersity, high degree of molecular uniformity, and precisely controlled structure, dendrimers are excellent platforms for therapeutic, nanomedical, and imaging applications.[78–82]

Often polypropylene imine (PPI) or PAMAM dendrimers (Figure 10.7) serve as macromolecules for improved contrast agents.[83] The terminal amino functions (*cf.* Figure 10.7, R $=NH_2$) can be used to attach gadolinium chelates via established amide or (thio)urea linkages. For example, for Gd(III)DOTA-based PAMAM dendrimer **12**, the relaxivity was studies as a function of the number of generations.[84] It was possible to create structures that ranged from an average of 127 chelates and 96 Gd ions per G = 5 dendrimer to an average of

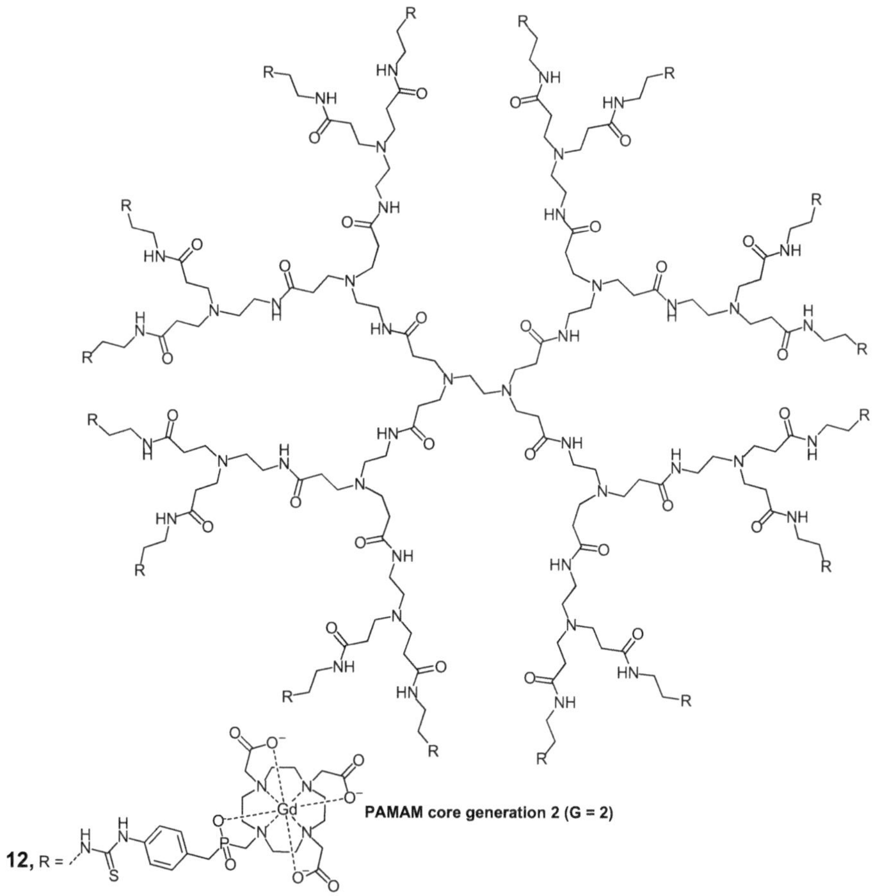

Figure 10.7 Dendritic MRI contrast agents.

3727 chelates and1860 Gd ions per G = 10 dendrimer. The relaxivity per Gd (III) plateaued at generation G = 7 ($r = 36\,\text{mM}^{-1}\,\text{s}^{-1}$, 20 MHz, 23 °C).

However, the application of PAMAM based contrast agents seems to be somewhat limited because of reports about cytotoxic effects of lower generations of PPI-based dendrimers.[85] However, there are indications that endcapping of unreacted amino groups at the surface with acetic anhydride decreases cytotoxicity significantly.

Gadomer-17, which contains 24 gadolinium centers per molecule, was tested in humans by Bayer-Schering and demonstrated the feasibility of dendrimer-based contrast agents for the first time (Figure 10.8).[86] Interestingly, its further development was stopped when it became apparent that MS-325—which relies on supramolecular interactions with HSA rather than on molecular size to restrict the agent to the cardiovascular system—advanced in clinical studies, too.

Generally, to achieve target-specific MRI contrast agents,[87] targeting units such as polysaccharides,[88] peptides,[89,89] antibodies,[80] oligonucleotides,[90] steroids,[91,92] and boronic acids[93] can be attached at the periphery of

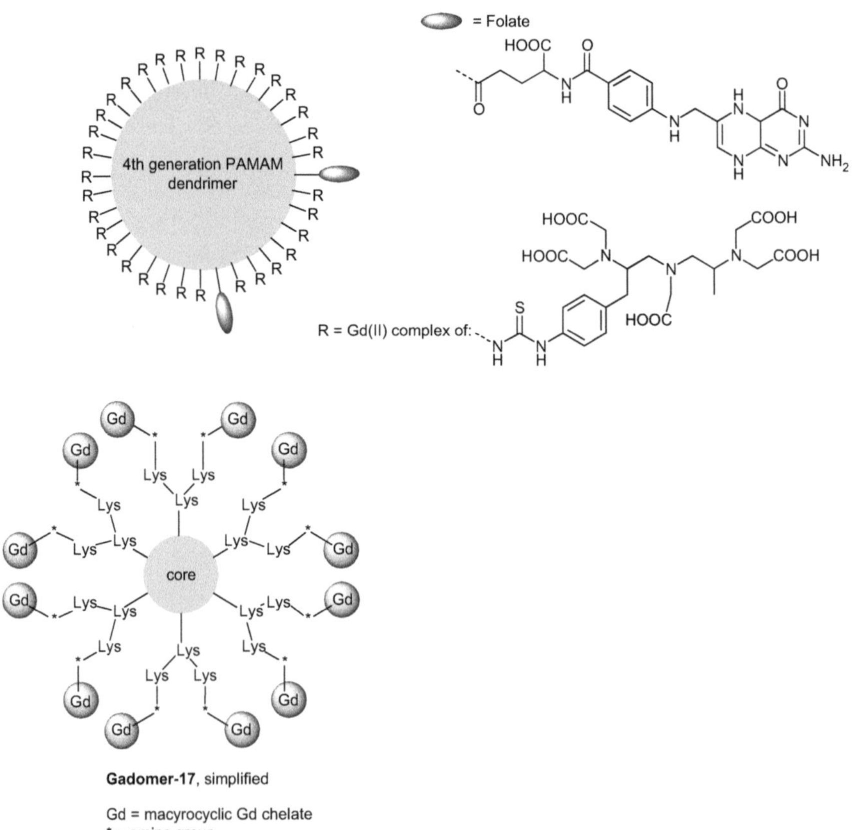

Figure 10.8 Dendritic MRI contrast agent with high gadolinium loadings.

dendrimer-based gadolinium complexes. One example (see Figure 10.8) uses a fourth generation PAMAM dendritic core combined with folate targeting groups, which allows a significant signal enhancement in case of folate-expressing tumors due to a local increase of concentration (targeting) and increased relaxivity of the bound agent compared with the free compound.[94,95]

A problem with covalent, relatively large compounds *in vivo* is their long-term fate. Gadolinium(III) is a potentially toxic, non-endogenous ion and there is no mechanism to eliminate it from the body. Therefore, full elimination must be insured by probe design. One approach is to include chemical bonds in such large systems that are broken over time. Such biocompatible dendrimers can be based on esteramide cores, which are functionalized with long ethylene glycol chains $(CH_2CH_2O)_{115}Me$ for solubility reasons (Figure 10.9).[96,97] Binding of the TACN-N1 complex to the ester amide dendrimer, triples the relaxivity due to the reduced tumbling of the complex in the dendritic environment (Gd-TACN-N1 + dendrimer: $r_1 = 31$, Gd-TACN-N1: $9.9\,s^{-1}\,mM^{-1}$ per Gd ion, 60 MHz, 37 °C). The ester bonds—or alternatively disulfide bridges[98–102]—were included to adjust biodegradability.

The dendritic approach also allows the study of macromolecular, nanosized contrast agents which are still biodegradable. Again, click chemistry can be a versatile tool for the surface modification of core structures.[103,104] Based on a biodegradable polycaprolactone-poly(ethylene oxide) block copolymer, relaxivities up to $26.1\,mM^{-1}\,s^{-1}$ can be achieved. Even virus capsids,[105,106] nanodiamonds,[107] or silica microparticles[108,109] can be used as nanoscale assemblies for high relaxivity contrast agents.

Generally, signal in MR images can be enhanced in two ways. Most commonly, multiple gadolinium moieties are located at the surface of a globular structure, for example a dendrimer. In a case where the assembly contains targeting sides for biologically relevant targets, a large signal of the now-labeled target can be expected, thanks to the presence of a large number of Gd(III) centers. In other words, signal is enhanced by increasing the local concentration of Gd(III) at the side of interest.

For low molecular weight probes another approach can also be useful.[110] For example, in compound **13** (Figure 10.10), a Gd(III) core is decorated with

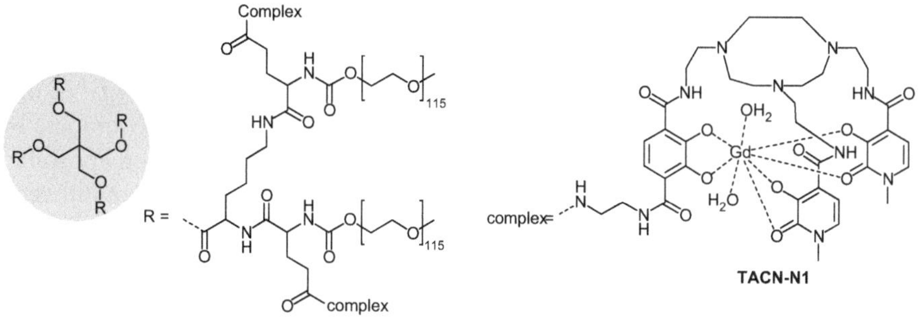

Figure 10.9 Dendritic esteramide MRI contrast agents.

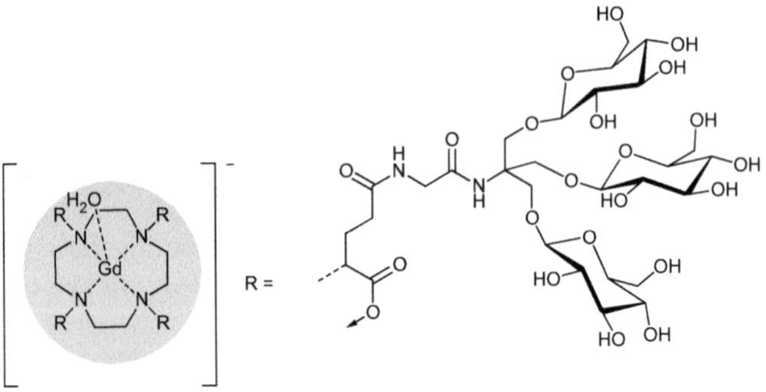

Figure 10.10 Dendritic MRI contrast agent **13** with "reversed arrangement".[111]

carbohydrate groups intended as targeting vectors.[111] Here, the focus lies on the improvement of the interaction of the sugar binding sites towards the biological target; improved binding now results in signal enhancement. In other words, signal is enhanced not only by increasing the local concentration, but also by increasing the relaxivity of the bound probe by increasing its molecular tumbling time (τ_R). For larger agents like dendrimer conjugates, targeting usually does not result in a significant increase in relaxivity anymore, since τ_R is already very long for the unbound conjugates and the relaxivity plateaus for long τ_R values.

10.2.1.4 Other Approaches

In the aforementioned approaches, covalent linkages between the gadolinium complex and the supramolecular backbone are used. Internal mobility and number of signaling units per molecule can be effectively controlled by this methodology. However, using covalent bonds to assemble high molecular weight systems means an enormous synthetic effort. Contrary to this time-consuming approach, non-covalent, self-assembly to structures such as micelles, lipid bilayers, or liposomes can be a valuable, simple alternative to control key physical properties such as τ_R and keep the synthetic effort minimal.[61,75,76,91,92,112–114]

The highest relaxivity for a micellar Gd(III) complex so far was realized by attaching a lipid chain to a Gd-AAZTA complex (Figure 10.11).[115] As mentioned above, liposome-based MRI agents have the intrinsic problem that the slow water diffusion through the lipid bilayer quenches relaxivity of any Gd(III) that is entrapped inside the vesicles. However, this can be overcome by developing systems that are oriented mostly towards the outside of liposomes (see under calixarenes) or by using vesicles that allow faster water diffusion through their bilayers. One such example was published by Botta and colleagues.[116] Transformation of Gd-DOTAGA to an amphiphile that can be incorporated into liposomal bilayers increases the relaxivity due to an increase in τ_R. When compound **14** bearing one hydrophobic chain attached to the Gd-core is co-assembled to liposomes, the relaxivity is increased by a factor of *ca.* 3

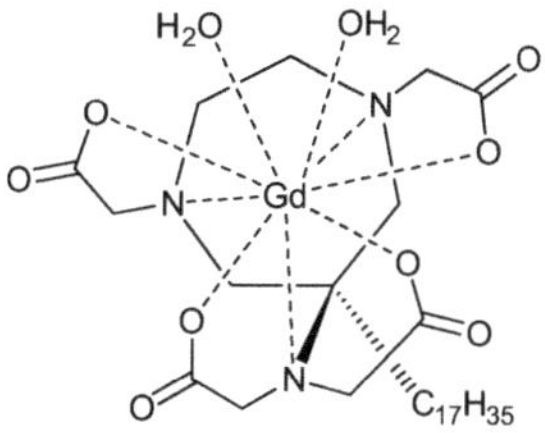

$Gd(AAZTA\text{-}C_{17})(H_2O)_2]^-$

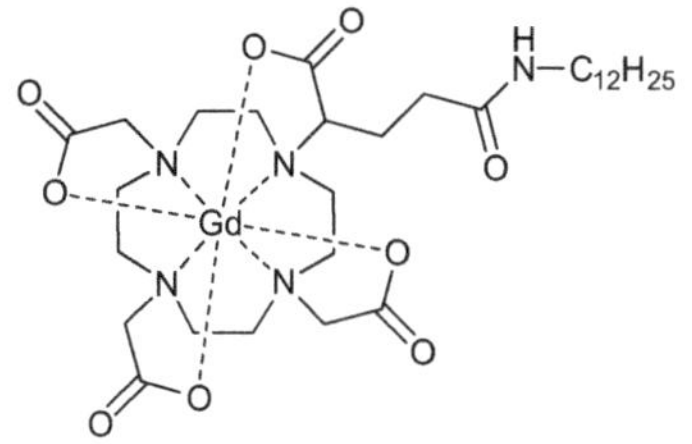

14

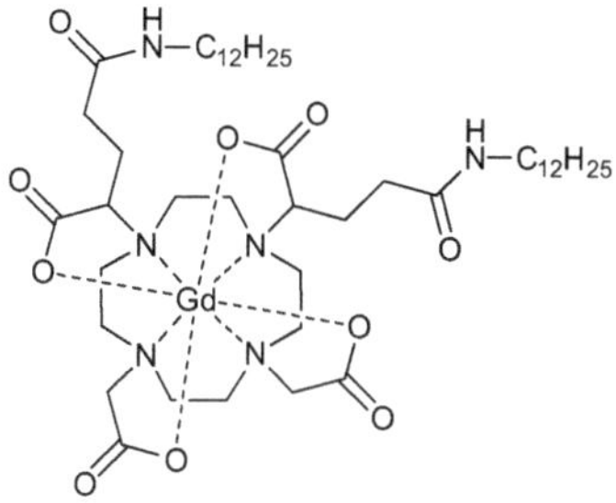

15

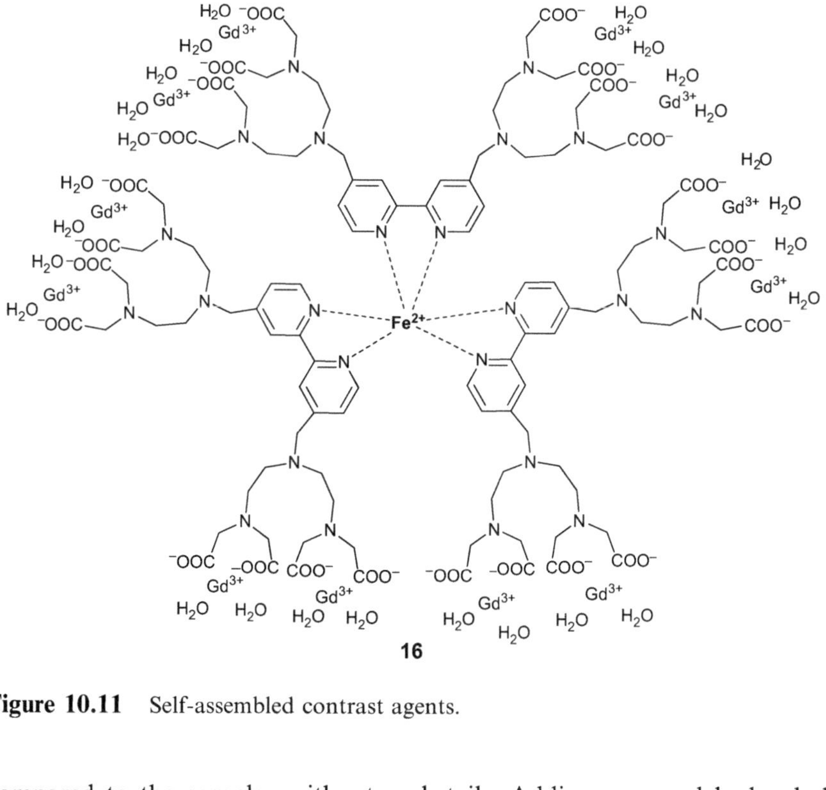

16

Figure 10.11 Self-assembled contrast agents.

compared to the complex without such tails. Adding a second hydrophobic alkyl chain (**15**) optimizes the intercalation in the bilayer and the relaxivity is doubled compared to the monoalkylated derivative (15.4 and 34.8 *vs.*

5.9 mM^{-1} s^{-1}).[116] Restriction of the mobility of the complex in the lipid bilayer and the resulting decreased local rotational dynamics is the reason for this large effect. It needs to be emphasized that liposomes are usually not stable under physiological conditions. Therefore, there is still an ongoing quest for liposomes that can deliver high Gd(III) payloads, have high per Gd(III) relaxivity, and are stable *in vivo* at the same time.

Solvophobic and lipophilic (*e.g.* dispersive) interactions are the driving force for the formation of higher aggregates such as micelles or liposomes. If stronger interactions are necessary for the formation of stable assemblies, transition metal–ligand interactions can be exploited.[117] A starburst-shaped bimetallic complex **16** may serve as an example (see Figure 10.11).[118–121] When Fe(III) is used for the self-assembly, the relaxivity of the monomeric bipyridine chelates is increased four-fold (16.4 *vs.* 4.0 mM^{-1} s^{-1}, 200 mHz, 25 °C) upon formation of the trimer.[119]

10.2.1.5   Responsive MRI Contrast Agents

All clinically approved contrast agents are blood-pool agents that typically show non-specific distribution in the body. Whereas targeted agents, contradistinction, bind to regions containing biomarkers and lead to better signals in MR images of those areas.

Another way of providing physiologically relevant information for the radiologist is to design agents that change their relaxivities in the presence of stimuli. Low molecular weight agents that bind to targets are one example of such responsiveness: the change in τ_R upon binding accompanied by the increase in local concentration can lead to astonishingly accurate differentiation of pathologies (see, *e.g.*, Makowski *et al.*,[122] where atherosclerotic blood vessels could be imaged in mice). Another way of using supramolecular interaction to visualize stimuli is to use enzyme activity to convert prodrugs to compounds with increased τ_R and thus increased relaxivity. This principle was explored in a study that was aimed at imaging the activity of the enzyme known as thrombin-activatable fibrinolysis inhibitor (TAFI) (Figure 10.12). This enzyme cleaves the lysine residues of the prodrug leading to a species with a significantly higher affinity for HSA. This leads to a significant increase in τ_R and thus to an about three-fold increase in relaxivity.[123]

Another way of employing supramolecular interactions to enhance MR response is to have endogenous metal ions induce self-assembly. An example is a bisphosphonate-based MRI contrast agent that has high affinity for Ca(II).[124]

An interesting targeting concept is the LIPOCEST approach where liposomes that contain a shift reagent in the aqueous core and a Gd-complex on the external surface are conjugated to the vesicles through a biodegradable linker. The whole assembly can generate T_1 contrast exclusively, but after the cleavage and removal of the Gd-coating, the CEST (Chemical Exchange Saturation Transfer) contrast is switched on. Multicontrast agents are accessible following this strategy.[125]

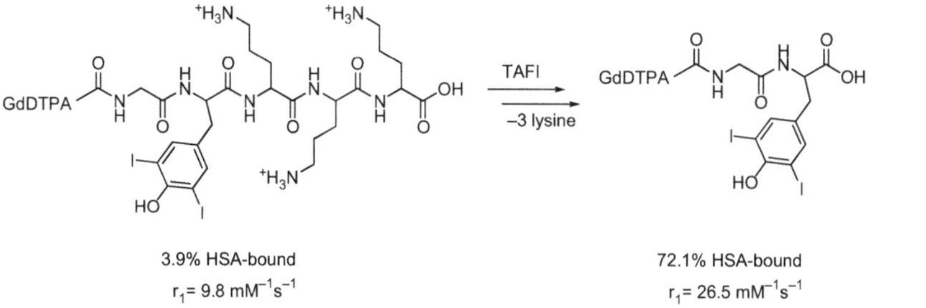

Figure 10.12 MRI agent responsive to the enzyme thrombin-activatable fibrinolysis inhibitor (TAFI). The prodrug (*left*) contains three lysine residues which inhibit HSA binding. Upon cleavage of these residues by TAFI, binding to HSA and relaxivity increases markedly.[123]

10.2.2 Optical Imaging

The basis of optical imaging in biology or medicine is the detection of light transmitted through biological tissue or cells. Both absorption and scattering can occur simultaneously in biological systems altering the transmission properties. In recent years several *in vivo* and *in vitro* molecular imaging methods have emerged as powerful tools in biochemistry. Diffuse optical tomography (DOT) relies on the detection of transmitted near-infrared (NIR) light using an array of sources combined with detectors. Optical projection tomography (OPT) uses visible light which is shined on a 360° rotated ex-vivo target surrounded by an organic solvent. Near-infrared fluorescence imaging exploits the excited NIR fluorescence of a dye within the target biological sample. General aspects can be found in recent reviews on optical imaging.[55,126–128] An interesting rather new approach is Cherenkov luminescence imaging, where the light source is "inside" the patient. This technique is based on the emission of photons when particles (emitted by radioactive sources) are decelerated from the velocity of light *in vacuo* to the velocity of light in the corresponding biological matrix where they are emitted.[129] Some of the emitted photons have wavelengths in the NIR, so this technique enables deep-tissue penetration, since no excitation that generally occurs with light of rather short wavelengths is needed. Many agents that can be used for nuclear imaging are therefore also suitable for Cherenkov luminescence imaging.

Although, many potential applications of a plethora of molecular structures has been proposed, "real" biochemical or medicinal applications of metal-based supramolecular optical imaging system are still scarce.[130]

Unlike MR imaging, where endogenous contrast exists, an artificial fluorescent contrast agent is the prerequisite for NIR imaging.[131] Additional functionalization of such dyes[131] with a supramolecular entity can help to increase both sensitivity and selectivity.

Again, cyclodextrins are ideal, biocompatible supramolecular platforms for this purpose.[132] For example, the release of zinc ions in cells can be detected by a fluorescent functionalized β-cyclodextrin **17** (Figure 10.13) in aqueous

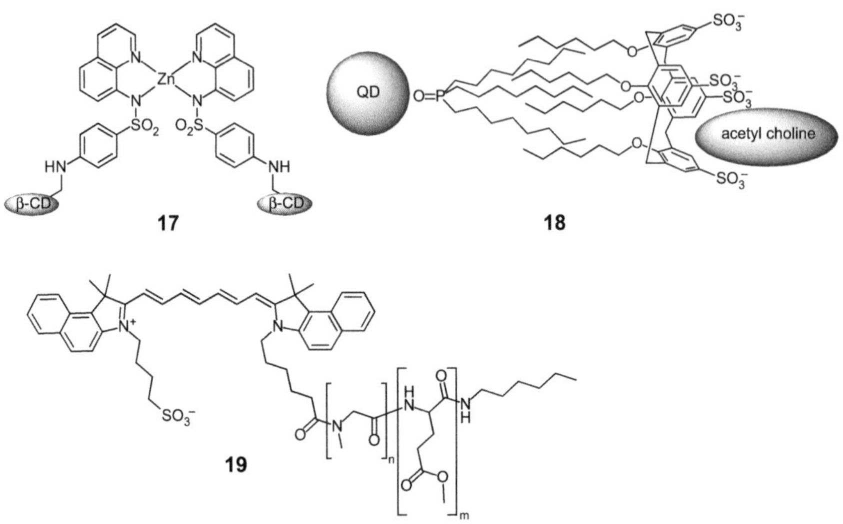

Figure 10.13 Contrast agents for near-infrared imaging.

solution. Under the same conditions, several metal ions commonly present in a physiological environment show little interference with the fluorescence response to Zn(II).[133]

Quantum dots (QDs) can be an appealing alternative to organic dye molecules.[134] To create selectivity in a "receptor–linker–reporter" system, the surface of QDs can be decorated with crown ethers,[135–138] cyclodextrins,[139] calixarenes,[140–144] or porphyrins.[145]

Water-soluble, sulfonated calix[4]arene was used to form a non-covalently linked detector–reporter array, **18,** for the optical detection of the neurotransmitter acetylcholine.[146] A CdSe/ZnS quantum dot (QD) was surface-coated with trioctyl phosphine oxide (TOPO) which forms the linker to lower-rim alkylated *p*-sulfonatocalix[4]arene. Adding acetylcholine to the non-covalent receptor system led to significant quenching of the fluorescence.

Polymers or oligomers can also be used as core structures of NIR imaging materials.[147–149] Various amphiphilic polypeptides which consist of hydrophilic polysarcosine and hydrophobic amino acid blocks were labeled with a suitable NIR dye (**19**). In buffered solution, the NIR dye was infiltrated in vesicles forming labeled peptosomes. Using this array NIR imaging of a small cancer on mouse was possible proofing the applicability of this general approach.

Lanthanides such as Eu^{3+} and Tb^{3+} are commonly used as luminescent probes. Similarly to gadolinium(III), such ions are highly toxic and strong complexation is necessary to overcome this problem. Bipyridyls are therefore often used to provide a reasonably stable coordination sphere around the lanthanides. Calixarenes (**20–24**, Figure 10.14) show nice luminescence properties in organic solution, but the water solubility of these complexes is too low for biological application.[150,151] (Solubility in aqueous media is a

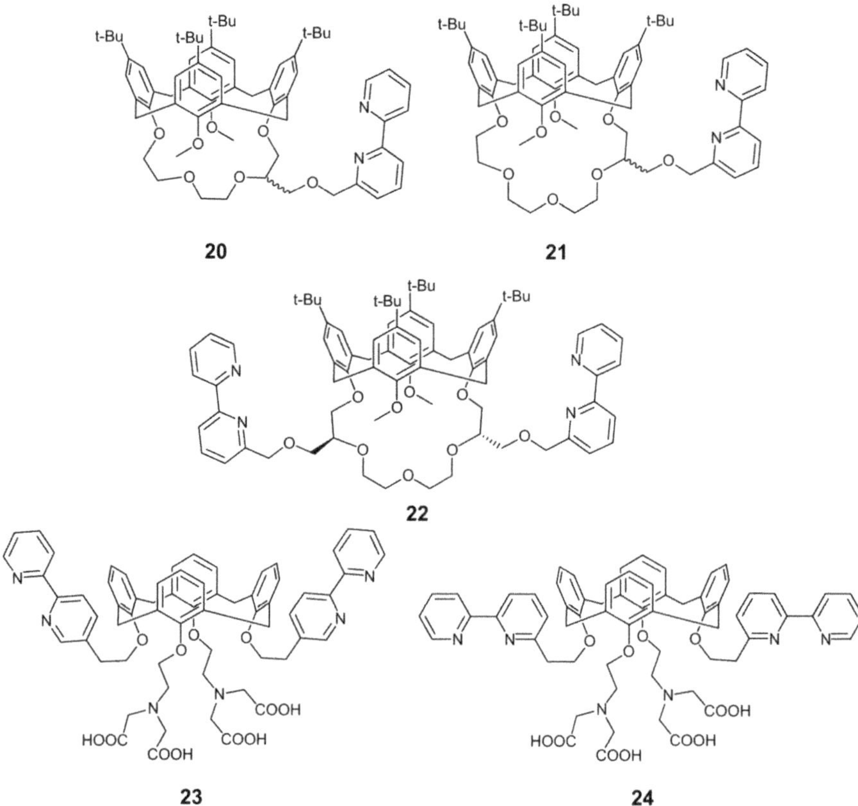

Figure 10.14 Calixarene-based chelators for luminescent lanthanide complexes.

prerequisite for biological application.) In addition, the luminescence should not be quenched in this environment and the stability and lability of the complexes needs to be very high.

10.3 Supramolecular Chemistry in Radioimaging and Radiotherapy

In radiotherapy ionizing radiation is used to treat cancer effectively.[152,153] Especially, Technetium (^{99m}Tc from ^{98}Mo) and Gallium (^{67}Ga from ^{68}Zn) are often used for such purposes.[154] It is very important in this field that the harmful radiation does not damage healthy tissue—or in the worst case, leads to cancer of originally healthy tissue due to the radiation. Therefore, the control of the biodistribution of radioactive metal complexes is essential in order to deliver the radiation only to the diseased side and not to harm healthy tissue. This can be achieved by binding the radioactive metal ion by a chelator that contains a targeting vector. This modular approach is very general and both covalent and non-covalent linkages can be used for attaching the targeting vector. Examples for supramolecular systems are co-assembly of imaging reporters with, for example, drug delivery systems such as liposomes or micelles making image-guided therapy possible. Taking the next step, drugs that are tailored to individuals might become desirable in the future (personalized medicine). Self-assembly might be the only way to provide doctors with affordable drugs that can be prepared simply by mixing without the necessity of (covalent) synthesis.

10.3.1 Technetium

The radionuclide ^{99m}Tc is available from commercial generators from its mother nucleus, ^{99}Mo. Its easy accessibility makes ^{99m}Tc an ideal metal for radiopharmaceuticals. Several examples of low-molecular-weight ^{99m}Tc-coordination complexes are commercially available, for example Cardiolite, Miraluma, Cardiotec, Myoview, Neurolite, Glucoscan, and Technescan (Figure 10.15). Applications range from cardiac, cancer, renal to bone single-photon emission computed tomography (SPECT).[155]

Technetium-based imaging complexes are commonly divided in two categories: technetium-essential and technetium-tagged systems. For the first, incorporation of ^{99m}Tc is essential for the accumulation, *etc.* For the second, the distribution of the pharmaceutical is based on the receptor attached to the ^{99m}Tc-complex.[156]

Recently, diethylenetriamine pentaacetic acid–polyethylene glycol–folate (DTPA-PEG-folate) was labeled with ^{99m}Tc and tested as a radiopharmaceutical

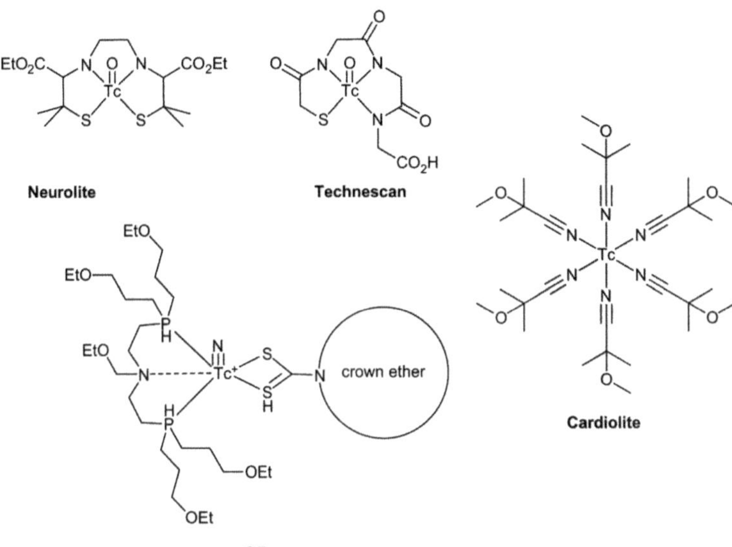

Figure 10.15 Examples of commercially available ^{99m}Tc radiopharmaceuticals.

agent, which targeted the lymphatic system with metastatic tumor. Lymphatic vessels could be readily visualized using SPECT.[157]

Crown ethers, such as **25** (Figure 10.15), exhibit a high affinity towards metal ions such as K^+ or Na^+ combined with low toxicity. In principle, this affinity can be exploited to influence the biodistribution of ^{99m}Tc-radiotracers in biological samples. However, total concentrations of cations are usually low, thus making it difficult to transfer this concept into real biological targets.[158]

Rhenium[159] or indium-111 can be used as alternative metals in SPECT instead of technetium. A PAMAM dendrimer conjugated to a pyridine-*N*-oxide DOTA ligand was labeled with ^{111}In (for core structure, see Figure 10.7). *In vivo*, the generation 1 core leads to fast renal elimination whereas generation 4 PAMAM resulted in a significant and long-term uptake in the kidney.[160] Specificity for ovarian carcinomas can again be created by attaching peptide chains to an indium-doped DOTA-type ligand.[161]

10.3.2 Gallium

From the 30 different gallium isotopes, only ^{66}Ga, ^{67}Ga, and ^{68}Ga are radionuclides that possess suitable decay properties. Gallium-67 is produced using a cyclotron from ^{68}Zn. The half-life of ^{67}Ga is sufficiently long (78.5 h). Gallium-68 is a positron emitter (half-life 67.7 min) with a positron energy higher than ^{18}F, making it an interesting isotope for positron emission tomography (PET). The advantage of the ^{68}Ga system is that it is accessible *via* a non-cyclotron-based process, using ^{68}Ge as a source.[154]

As already shown for gadolinium, azamacrocycles are ideal chelators for metals and are therefore also ligands of choice for gallium radiopharmacauticals.[162–164]

As a structural analog of the DOTA ligand (see p. 303), the triazacyclononane skeleton is effective for gallium ions. Attaching carboxylic acids (Figure 10.16, R =CH$_2$COOH) gives rise to the NOTA-ligand. Other acid functionalities such as R = CH$_2$PO$_2$H$_2$ (TRAP) are sometimes beneficial.[155] *N*-Substituents can easily be used to introduce further functionalities. In case of **26** (Figure 10.16) click chemistry[63] was used to attach cyclic peptides with a

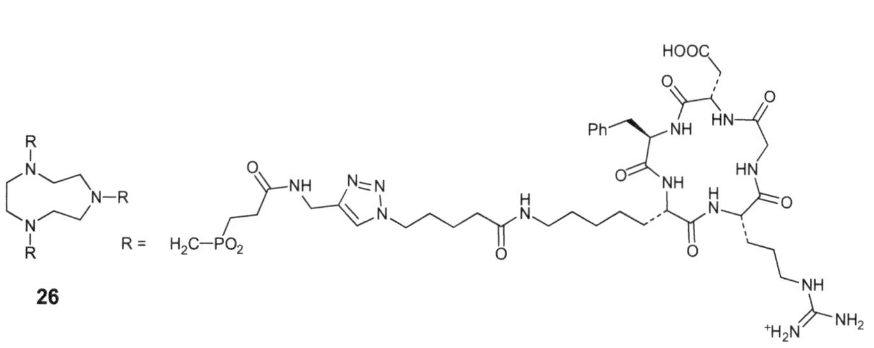

26

Figure 10.16   TRAP-based ligand for targeted radiotherapy.

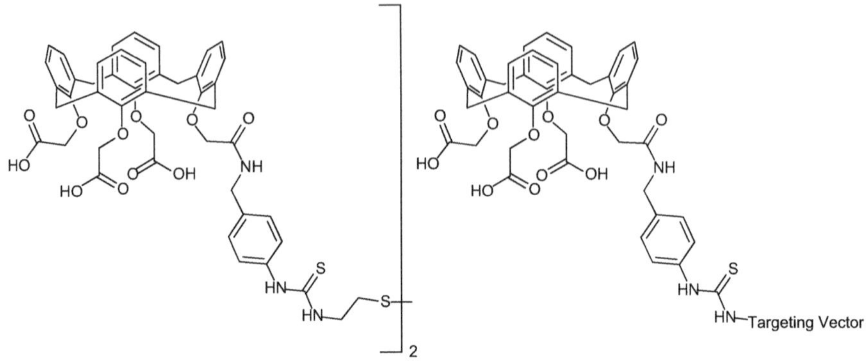

Figure 10.17   Calixarene-based ligands for Ac(III) complexation with potential use as radiopharmaceuticals.[167]

high integrin affinity with the intention to address the integrin status in angiogenic, oncological, and cardiological questions. The modular system could be used in microPET of nude mice bearing tumors. Biotin-labeling exploiting the avidin–biotin targeting system is similarly possible.[163]

10.3.3   Other Metal Ions

Copper isotopes can be used for PET imaging. Half-lives for the positron-decay are in the range of minutes to 12.7 hours for ^{64}Cu. The longer half-life compared to ^{11}C or ^{18}F makes this particular copper isotope an interesting alternative for these more established nuclei. The need for good targeting of the diseased side with ^{64}Cu(II) can be met by attaching, for example, oligopeptides to bis(semicarbazone), cyclam,[165] or other standard ligands for copper(II) ions.[159,166]

An alternative isotope for radiotherapy is the α-emitter ^{225}Ac^{3+} (half-life 10 days). Calixarenes of type **27** (Figure 10.17) were prepared and their Ac(III) coordination was studied.[167] *In vivo* tests of **27a/b** in mice showed that the binding to the target strongly depends on the nature of the targeting vector (antibodies or peptides), as well as on the dosage and the injection method.

References

1. K.-J. Chen, M. A. Garcia, H. Wang and H.-R. Tseng, Supramolecular nanoparticles for molecular diagnostics and therapeutics, in *Supramolecular Chemistry*, ed. P. A. Gale and J. W. Steed, 2012, 1st edn, pp. 3809–3824.
2. C. J. Jones and J. R. Thornback, *Medical Applications of Coordination Chemistry*, RSC, Cambridge, 2007.

3. C. S. Bonnet and É. Tóth, Magnetic resonance imaging contrast agents, in *Supramolecular Chemistry*, ed. P. A. Gale and J. W. Steed, Wiley-VCH, Weinheim, 2012, pp. 2693–2723.

4. A. E. Merbach and É. Toth, Eds., *The Chemistry of Contrast Agents in Medical Magnetic Resonance Imaging*, Wiley-VCH, Weinheim, 1st edn., 2001.

5. D. T. Schühle and P. Caravan, *Metal-Based MRI Contrast Agents*, in *Comp. Inorg. Chem. II*, 2013, chapter 3.36, 2013.

6. M. Rohrer, H. Bauer, J. Mintorovitch, M. Requardt and H.-J. Weinmann, *Invest. Radiol.*, 2005, **40**, 715–724.

7. M. Port, J.-M. Idée, C. Medina, C. Robic, M. Sabatou and C. Corot, *Biometals*, 2008, **21**, 469–490.

8. J. Rudovský, P. Cígler, J. Kotek, P. Hermann, P. Vojtísek, I. Lukes, J. A. Peters, L. Vander Elst and R. N. Muller, *Chem. Eur. J.*, 2005, **11**, 2373–2384.

9. R. A. Wallace, J. Haar, J. P., D. B. Miller, S. R. Woulfe, J. A. Polta, K. P. Galen, M. R. Hynes and K. Adzamli, *Magn. Reson. Med.*, 1998, **40**, 733–739.

10. F. M. Cavagna, F. Maggioni, P. M. Castelli, M. Dapra, L. G. Imperatori, V. Lorusso and B. G. Jenkins, *Invest. Radiol.*, 1997, **32**, 780–796.

11. V. V. Martin, W. H. Ralston, M. R. Hynes and J. F. W. Keana, *Bioconjug. Chem.*, 1995, **6**, 616–623.

12. R. B. Shukla, K. Kumar, R. Weber, X. Zhang and M. Tweedle, *Acta Radiologica*, 1997, **38**, 121–123.

13. S. Aime, M. Botta, G. Ermondi, E. Terreno, P. L. Anelli, F. Fedeliand and F. Uggeri, *Inorg. Chem.*, 1996, **35**, 2726–2736.

14. S. Aime, M. Botta, M. Fasano, S. G. Crich and E. Terreno, *J. Biol. Inorg. Chem.*, 1996, **1**, 312–319.

15. S. Aime, M. Botta, M. Panero, M. Grandi and F. Uggeri, *Magn. Reson. Chem.*, 1991, **29**, 923–927.

16. C. Geze, C. Mouro, F. Hindre, M. Le Plouzennec, C. Moinet, R. Rolland, L. Alderighi, A. Vacca and G. Simonneaux, *Bull. Soc. Chim. Fr.*, 1996, **133**, 267–272.

17. D. H. Powell, O. M. N. Dhubhghaill, D. Pubanz, L. Helm, Y. S. Lebedev and W. Schlaepfer, *J. Am. Chem. Soc.*, 1996, **7863**, 9333–9346.

18. S. W. A. Bligh, A. H. M. S. Chowdhury, M. McPartlin, J. I. Scowen and R. A. Bulman, *Polyhedron 1995, 14, 567-9*, 1995, **14**, 567–569.

19. M. Periasamy, D. White, L. DeLearie, D. Moore, R. Wallace, W. Lin, J. Dunn, W. Hirth and W. Cacheris, *Invest. Radiol.*, 1991, **26**, S217–S220.

20. Y.-M. Wang, T.-H. Cheng, R.-S. Sheu, I.-T. Chen and M. Y. Chiang, *J. Chin. Chem. Soc.*, 1997, **44**, 123–128.

21. J. A. Varadarajan, S. P. Crofts, J. F. Carvalho, J. D. Fellmann, S.-H. Kim, C. A. Chang and A. D. Watson, *Invest. Radiol.*, 1994, **29**, S18–S20.

22. M. B. Inoue, P. Oram, M. Inoue, Q. Fernando, A. L. Alexander and E. C. Unger, *Magn. Reson. Imaging*, 1994, **12**, 429–432.

23. M. B. Inoue, R. E. Navarro, M. Inoue and Q. Fernando, *Inorg. Chem.*, 1995, **34**, 6074–6079.

24. S. I. Kang, R. S. Ranganathan, J. E. Emswiler, K. Kumar, J. Z. Gougoutas, M. F. Malley and M. F. Tweedle, *Inorg. Chem.*, 1993, **32**, 2912–2918.

25. S. Aime, P. L. Anelli, M. Botta, F. Fedeli, M. Grandi, P. Paoli and F. Uggeri, *Inorg. Chem.*, 1992, **31**, 2422–2428.

26. C. F. G. C. Geraldes, A. D. Sherry, I. Lázar, A. Miseta, P. Bogner, E. Berenyi, B. Sumegi, G. E. Kiefer, K. McMillan, F. Maton and R. N. Muller, *Magn. Reson. Med.*, 1993, **30**, 696–703.

27. W. D. Kim, G. E. Kiefer, F. Maton, K. McMillan, R. N. Muller and A. D. Sherry, *Inorg. Chem.*, 1995, **34**, 2233–2243.

28. D. Parker, K. Pulukkody, T. J. Norman, A. Harrison, L. Royle and C. Walker, *J. Chem. Soc., Chem. Commun.*, 1992, 1441–1443.

29. S. Aime, M. Botta, D. Parker and J. A. G. Williams, *J. Chem. Soc. Dalton Trans.*, 1995, 2259–2266.

30. W. D. Kim, G. E. Kiefer, J. Huskens and A. D. Sherry, *Inorg. Chem.*, 1997, **36**, 4128–4134.

31. I. Lazar, A. D. Sherry, R. Ramasamy, E. Brucher and R. Kiraly, *Inorg. Chem.*, 1991, **30**, 5016–5019.

32. P. Caravan, J. J. Ellison, T. J. McMurry and R. B. Lauffer, *Chem. Rev.*, 1999, **99**, 2293–2352.

33. P. Caravan, *Chem. Soc. Rev.*, 2006, **35**, 512–523.

34. S. Dumas, V. Jacques, W.-C. Sun, J. S. Troughton, J. T. Welch, J. Chasse, H. Schmitt-Willich and P. Caravan, *Invest. Radiol.*, 2010, **45**, 600–612.

35. S. Aime, M. Botta and E. Terreno, *Adv. Inorg. Chem.*, 2005, **57**, 173–237.

36. A. Accardo, D. Tesauro, L. Aloj, C. Pedone and G. Morelli, *Coord. Chem. Rev.*, 2009, **253**, 2193–2213.

37. M. Botta, *Eur. J. Inorg. Chem.*, 2000, **2000**, 399–407.

38. S. Aime, M. Botta, M. Fasano and E. Terreno, *Chem. Soc. Rev.*, 1998, **27**, 19–29.

39. R. B. Lauffer, *Chem. Rev.*, 1987, **87**, 901–927.

40. P. Hermann, J. Kotek, V. Kubícek and I. Lukes, *J. Chem. Soc. Dalton Trans.*, 2008, **9226**, 3027–3047.

41. S. Zhang, M. Merritt, D. E. Woessner, R. E. Lenkinski and A. D. Sherry, *Acc. Chem. Res.*, 2003, **36**, 783–790.

42. S. J. Dorazio and J. R. Morrow, *Eur. J. Inorg. Chem.*, 2012, **2012**, 2006–2014.

43. P. Caravan and Z. Zhang, *Eur. J. Inorg. Chem.*, 2012, **2012**, 1916–1923.

44. I. Lukes, J. Kotek, P. Vojtı and P. Hermann, *Coord. Chem. Rev.*, 2001, **216-217**, 287–312.

45. A. Accardo, D. Tesauro and G. Morelli, in *Applications of Supramolecular Chemistry*, H.-J. Schneider, CRC, Boca Raton, FL, 2012.

46. S. J. Archibald, *Annual Reports Section "A" (Inorganic Chemistry)*, 2011, **107**, 274.

47. G. Giannarini, G. Petralia and H. C. Thoeny, *Eur. Urol.*, 2012, **61**, 326–340.

48. J. C. Gore, H. C. Manning, C. C. Quarles, K. W. Waddell and T. E. Yankeelov, *Magn. Reson. Imaging*, 2011, **29**, 587–600.

49. A. A. Neves and K. M. Brindle, *Biochim. Biophys. Acta*, 2006, **1766**, 242–261.

50. L. Helm, *Future Med. Chem.*, 2010, **2**, 385–396.

51. G. H. Lee, Y. Chang and T.-J. Kim, *Eur. J. Org. Chem.*, 2012, 1924–1933.

52. S. Laurent, C. Henoumont, L. Vander Elst and R. N. Muller, *Eur. J. Inorg. Chem.*, 2012, **2012**, 1889–1915.

53. J. L. Major and T. J. Meade, *Acc. Chem. Res.*, 2009, **42**, 893–903.

54. T. E. McCann, N. Kosaka, B. Turkbey, M. Mitsunaga, P. L. Choyke and H. Kobayashi, *NMR Biomed.*, 2011, **24**, 561–568.

55. M. Herranz and A. Ruibal, *J. Oncol.*, 2012, 1–10.

56. R. E. Mewis and S. J. Archibald, *Coord. Chem. Rev.*, 2010, **254**, 1686–1712.

57. Z. Zhou and Z.-R. Lu, *WIREs Nanomed. Nanobiotechnol.*, 2013, **5**, 1–18.

58. S. Geninatti Crich, D. Alberti, I. Szabo, S. Aime and K. Djanashvili, *Angew. Chem. Int. Ed.*, 2013, **52**, 1161–1164.

59. A. Méndez-Ardoy, N. Guilloteau, C. Di Giorgio, P. Vierling, F. Santoyo-González, C. Ortiz Mellet and J. M. García Fernández, *J. Org. Chem.*, 2011, **76**, 5882–5894.

60. Y. Song, E. K. Kohlmeir and T. J. Meade, *J. Am. Chem. Soc.*, 2008, **130**, 6662–6663.

61. C. Reaction, H. Micelles, C. Vanasschen, N. Bouslimani, D. Thonon and J. F. Desreux, *Inorg. Chem.*, 2011, **50**, 8946–8958.

62. Y. Song, H. Zong, E. R. Trivedi, B. J. Vesper, E. Waters, A. G. M. Barrett, J. Radosevich, B. M. Hoffman and T. J. Meade, *Bioconjug. Chem.*, 2010, **21**, 2267–2275.

63. T. L. Mindt, C. Müller, F. Stuker, J.-F. Salazar, A. Hohn, T. Mueggler, M. Rudin and R. Schibli, *Bioconjug. Chem.*, 2009, **20**, 1940–1949.

64. H. Struthers, B. Spingler, T. L. Mindt and R. Schibli, *Chem. Eur. J.*, 2008, **14**, 6173–6183.

65. T. L. Mindt, C. Mueller, M. Melis, J. de Marion and R. Schibli, *Bioconjug. Chem.*, 2008, **19**, 1689–1695.

66. D. J. Mastarone, V. S. R. Harrison, A. L. Eckermann, G. Parigi, C. Luchinat and T. J. Meade, *J. Am. Chem. Soc.*, 2011, **133**, 5329–5337.

67. Z. Kotková, J. Kotek, D. Jirák, P. Jendelová, V. Herynek, Z. Berková, P. Hermann and I. Lukes, *Chem. Eur. J.*, 2010, **16**, 10 094–10 102.

68. Z. Kotková, L. Helm, J. Kotek, P. Hermann and I. Lukeš, *J. Chem. Soc. Dalton Trans.*, 2012, **41**, 13 509–13 519.

69. J. M. Bryson, W.-J. Chu, J.-H. Lee and T. M. Reineke, *Bioconjug. Chem.*, 2008, **19**, 6–10.

70. B. S. Creaven, D. F. Donlon and J. McGinley, *Coord. Chem. Rev.*, 2009, **253**, 893–962.

71. S. B. Nimse and T. Kim, *Chem. Soc. Rev.*, 2012, **42**, 366–386.

72. E. M. Georgiev and D. M. Roundhill, *Inorg. Chim. Acta*, 1997, **258**, 93–96.

73. S. Aime, A. Barge, M. Botta, A. Casnati, M. Fragai, C. Luchinat and R. Ungaro, *Angew. Chem. Int. Ed.*, 2001, **40**, 4737–4739.

74. A. M. Krishnan and R. Lohrmann, *Calixarene conjugates useful as MRI and CT diagnostic imaging agents, and their preparation*, 1996, WO9614878, Molecular Biosystems, Inc., USA.

75. D. T. Schühle, M. Polasek, I. Lukes, T. Chauvin, E. Toth, J. Schatz, U. Hanefeld, M. C. A. Stuart and J. A. Peters, *Dalton Trans.*, 2010, **39**, 185–191.

76. D. T. Schühle, J. Schatz, S. Laurent, L. Vander Elst, R. N. Muller, M. C. A. Stuart and J. A. Peters, *Chem. Eur. J.*, 2009, **15**, 3290–3296.

77. D. T. Schühle, P. van Rijn, S. Laurent, L. Vander Elst, R. N. Muller, M. C. A. Stuart, J. Schatz and J. A. Peters, *Chem. Commun.*, 2010, **46**, 4399–4401.

78. A. R. Menjoge, R. M. Kannan and D. A. Tomalia, *Drug Discov. Today*, 2010, **15**, 171–185.

79. S. Langereis, A. Dirksen, T. M. Hackeng, M. H. P. van Genderen and E. W. Meijer, *New J. Chem.*, 2007, **31**, 1152.

80. S. El Kazzouli, S. Mignani, M. Bousmina and J.-P. Majoral, *New J. Chem.*, 2012, **36**, 227.

81. M. Shen and X. Shi, *Nanoscale*, 2010, **2**, 1596–1610.

82. A. Bump, M. W. Brechbiel and P. Choyke, *Acta Radiol.*, 2010, **51**, 751–767.

83. V. J. Venditto, C. A. S. Regino and M. W. Brechbiel, *Mol. Pharm.*, 2005, **2**, 302–311.

84. L. H. Bryant, M. W. Brechbiel, C. Wu, J. W. Bulte, V. Herynek and J. A. Frank, *J. Magn. Res. Imag.*, 1999, **9**, 348–352.

85. B. H. Zinselmeyer, S. P. Mackay, A. G. Schatzlein and I. F. Uchegbu, *Pharm. Res.*, 2002, **19**, 960–967.

86. B. Misselwitz, H. Schmitt-Willich, W. Ebert, T. Frenzel and H. J. Weinmann, *Magma*, 2001, **12**, 128–134.

87. M. Woods, Z. Kovacs and A. D. Sherry, *J. Supramol. Chem.*, 2002, **2**, 1–15.

88. L. Cipolla, M. Gregori and P.-W. So, *Curr. Med. Chem.*, 2011, **18**, 1002–1018.

89. R. C. Strauch, D. J. Mastarone, P. A. Sukerkar, Y. Song, J. J. Ipsaro and T. J. Meade, *J. Am. Chem. Soc.*, 2011, **133**, 16 346–16 349.

90. R. Mishra, W. Su, R. Pohmann, J. Pfeuffer, M. G. Sauer, K. Ugurbil and J. Engelmann, *Bioconjug. Chem.*, 2009, **20**, 1860–1868.

91. P. Sukerkar, K. W. MacRenaris, T. J. Meade and J. E. Burdette, *Mol. Pharm.*, 2011, **8**, 1390–1400.

92. P. A. Sukerkar, K. W. MacRenaris, T. R. Townsend, R. A. Ahmed, J. E. Burdette and T. J. Meade, *Bioconjug. Chem.*, 2011, **22**, 2304–2316.

93. L. Frullano, J. Rohovec, S. Aime, T. Maschmeyer, M. I. Prata, J. J. P. de Lima, C. F. G. C. Geraldes and J. A. Peters, *Chem. Eur. J.*, 2004, **10**, 5205–5217.

94. S. D. Konda, M. Aref, S. Wang, M. Brechbiel and E. C. Wiener, *MAGMA*, 2001, **12**, 104.

95. S. D. Konda, S. Wang, M. Brechbiel and E. C. Wiener, *Invest. Radiol.*, 2002, **37**, 199.

96. P. J. Klemm, W. C. Floyd III, C. M. Andolina, J. M. J. Fréchet and K. N. Raymond, *Eur. J. Inorg. Chem.*, 2012, **2012**, 2108–2114.

97. P. J. Klemm, W. C. Floyd, D. E. Smiles, J. M. J. Fréchet and K. N. Raymond, *Contrast Media Mol. Imag.*, 2012, **7**, 95–9.

98. A. M. Mohs and Z.-R. Lu, *Expert Opin. Drug Deliv.*, 2007, **4**, 149–164.

99. Y. Zong, X. Wang, E.-K. Jeong, D. L. Parker and Z.-R. Lu, *Magn. Reson. Imaging*, 2009, **27**, 503–511.

100. R. Xu, Y. Wang, X. Wang, E. Jeong, D. L. Parker and Z. Lu, *Exp. Biol. Med.*, 2007, **232**, 1081–1089.

101. Z. Ye, X. Wu, M. Tan, J. Jesberger, M. Grisworld and Z.-R. Lu, *Contrast Media Mol. Imag.*, 2013, **8**, 220–228.

102. Z.-R. Lu and X. Wu, *Israel J. Chem.*, 2010, **50**, 220–232.

103. A. Nazemi, F. Martínez, T. J. Scholl and E. R. Gillies, *RSC Adv.*, 2012, **2**, 7971.

104. F. Fernández-Trillo, J. Pacheco-Torres, J. Correa, P. Ballesteros, P. Lopez-Larrubia, S. Cerdán, R. Riguera and E. Fernandez-Megia, *Biomacromolecules*, 2011, **12**, 2902–2907.

105. M. Botta and L. Tei, *Eur. J. Inorg. Chem.*, 2012, **2012**, 1945–1960.

106. P. D. Garimella, A. Datta, D. W. Romanini, K. N. Raymond and M. B. Francis, *J. Am. Chem. Soc.*, 2011, **133**, 14704–14709.

107. L. M. Manus, D. J. Mastarone, E. A. Waters, X.-Q. Zhang, E. a Schultz-Sikma, K. W. Macrenaris, D. Ho and T. J. Meade, *Nano lett.*, 2010, **10**, 484–489.

108. A. K. Duncan, P. J. Klemm, K. N. Raymond and C. C. Landry, *J. Am. Chem. Soc.*, 2012, **134**, 8046–8049.

109. J. A. Peters and K. Djanashvili, *Eur. J. Inorg. Chem.*, 2012, **2012**, 1961–1974.

110. M. Port, C. Corot, O. Rousseaux, I. Raynal, L. Devoldere, J. M. Idée, A. Dencausse, S. Le Greneur, C. Simonot and D. Meyer, *MAGMA*, 2001, **12**, 121–127.

111. D. A. Fulton, E. M. Elemento, S. Aime, L. Chaabane, M. Botta and D. Parker, *Chem. Commun.*, 2006, 1064–1066.

112. M. Vaccaro, A. Accardo, D. Tesauro, G. Mangiapia, D. Löf, K. Schillén, O. Söderman, G. Morelli and L. Paduano, *Langmuir*, 2006, **22**, 6635–6643.

113. G. S. van Bochove, H. M. H. F. Sanders, M. de Smet, H. M. Keizer, W. J. M. Mulder, R. Krams, G. J. Strijkers and K. Nicolay, *Eur. J. Inorg. Chem.*, 2012, **2012**, 2115–2125.

114. O. M. Evbuomwan, G. Kiefer and A. D. Sherry, *Eur. J. Inorg. Chem.*, 2012, **2012**, 2126–2134.

115. E. Gianolio, G. B. Giovenzana, D. Longo, I. Longo, I. Menegotto and S. Aime, *Chem. Eur. J.*, 2007, **13**, 5785–5797.

116. F. Kielar, L. Tei, E. Terreno and M. Botta, *J. Am. Chem. Soc.*, 2010, **132**, 7836–7837.

117. V. C. Pierre, M. Botta, S. Aime and K. N. Raymond, *J. Am. Chem. Soc.*, 2006, **128**, 9272–9273.
118. J. B. Livramento, A. Sour, A. Borel, A. E. Merbach and E. Tóth, *Chemistry*, 2006, **12**, 989–1003.
119. J. B. Livramento, C. Weidensteiner, M. I. M. Prata, P. R. Allegrini, C. F. G. C. Geraldes, L. Helm, R. Kneuer, A. E. Merbach, A. C. Santos, P. Schmidt and E. Tóth, *Contrast Media Mol. Imag.*, 2006, **1**, 30–39.
120. J. B. Livramento, L. Helm, A. Sour, C. O'Neil, A. E. Merbach and E. Tóth, *Dalton Trans.*, 2008, 1195–1202.
121. L. Moriggi, A. Aebischer, C. Cannizzo, A. Sour, A. Borel, J.-C. G. Bünzli and L. Helm, *Dalton Trans.*, 2009, 2088–2095.
122. M. R. Makowski, A. J. Wiethoff, U. Blume, F. Cuello, A. Warley, C. H. P. Jansen, E. Nagel, R. Razavi, D. C. Onthank, R. R. Cesati, M. S. Marber, T. Schaeffter, A. Smith, S. P. Robinson and R. M. Botnar, *Nat. Med.*, 2011, **17**, 383–388.
123. V. Heeswijk, C. Res, A. L. Nivorozhkin, A. F. Kolodziej, P. Caravan, M. T. Greenfield, R. B. Lauffer and T. J. Mcmurry, *Angew. Chem. Int. Ed.*, 2001, **40**, 2903–2906.
124. V. Kubíček, T. Vitha, J. Kotek, P. Hermann, L. Vander Elst, R. N. Muller, I. Lukeš and J. A. Peters, *Contrast Media Mol. Imag.*, 2010, **5**, 294–296.
125. E. Terreno, C. Boffa, V. Menchise, F. Fedeli, C. Carrera, D. D. Castelli, G. Digilio and S. Aime, *Chem. Commun.*, 2011, **47**, 4667–4669.
126. D. Parker, *Comp. Supramol. Chem.*, 1996, **10**, 487–536.
127. P. J. Cassidy and G. K. Radda, *Interface*, 2005, **2**, 133–144.
128. V. Pansare, S. Hejazi, W. Faenza and R. K. Prud'homme, *Chem. Mater.*, 2012, **24**, 812–827.
129. R. Robertson, M. S. Germanos, C. Li, G. S. Mitchell, S. R. Cherry and M. D. Silva, *Physics Med. Biol.*, 2009, **54**, N355–N365.
130. E. A. Schultz-Sikma and T. J. Meade, in *Supramolecular Chemistry: from Molecules to Nanomaterials*, ed. P. A. Gale and J. W. Steed, John Wiley & Sons, Chichester, UK, 2012, pp. 1851–1876.
131. R. N. Dsouza, U. Pischel and W. M. Nau, *Chem. Rev.*, 2011, **111**, 7941–7980.
132. Y. Chen and Y. Liu, *Chem. Soc. Rev.*, 2010, **39**, 495–505.
133. Y. Liu, N. Zhang, Y. Chen and L.-H. Wang, *Org. Lett.*, 2007, **9**, 315–318.
134. C. Han and H. Li, *Anal. Bioanal. Chem.*, 2010, **397**, 1437–1444.
135. M.-L. Ho, J.-M. Hsieh, C.-W. Lai, H.-C. Peng, C.-C. Kang, I.-C. Wu, C.-H. Lai, Y.-C. Chen and P.-T. Chou, *J. Phys. Chem. C*, 2009, **113**, 1686–1693.
136. C.-Y. Chen, C.-T. Cheng, C.-W. Lai, P.-W. Wu, K.-C. Wu, P.-T. Chou, Y.-H. Chou and H.-T. Chiu, *Chem. Commun.*, 2006, 263–265.
137. S. Banerjee, S. Kar and S. Santra, *Chem. Commun.*, 2008, **1**, 3037–3039.
138. M. J. Ruedas-Rama and E. A. H. Hall, *Anal. Chem.*, 2008, **80**, 8260–8268.

139. R. Freeman, T. Finder, L. Bahshi and I. Willner, *Nano Lett.*, 2009, **9**, 2073–2076.

140. H. Li, Y. Zhang, X. Wang, D. Xiong and Y. Bai, *Mater. Lett.*, 2007, **61**, 1474–1477.

141. T. Jin, F. Fujii, H. Sakata, M. Tamura and M. Kinjo, *Chem. Commun.*, 2005, 2829–2831.

142. T. Jin, F. Fujii, E. Yamada, Y. Nodasaka and M. Kinjo, *J. Am. Chem. Soc.*, 2006, **128**, 9288–9289.

143. X. Wang, J. Wu, F. Li and H. Li, *Nanotechnology*, 2008, **19**, 205501.

144. H. Li and X. Wang, *Photochem. Photobiol. Sci.*, 2008, **7**, 694–699.

145. D. Vlascici, E. F. Cosma, E. M. Pica, V. Cosma, O. Bizerea, G. Mihailescu and L. Olenic, *Sensors*, 2008, **8**, 4995–5004.

146. T. Jin, F. Fujii, H. Sakata, M. Tamura and M. Kinjo, *Chem. Commun.*, 2005, 4300–4302.

147. H. Tanisaka, S. Kizaka-Kondoh, A. Makino, S. Tanaka, M. Hiraoka and S. Kimura, *Bioconjug. Chem.*, 2008, **19**, 109–17.

148. P. P. Ghoroghchian, P. R. Frail, K. Susumu, D. Blessington, A. K. Brannan, F. S. Bates, B. Chance, D. A. Hammer and M. J. Therien, *Prod. Natl Acad. Sci. USA*, 2005, **102**, 2922–2927.

149. C.-K. Lim, S. Kim, I. C. Kwon, C.-H. Ahn and S. Y. Park, *Chem. Mater.*, 2009, **21**, 5819–5825.

150. C. Fischer, G. Sarti, A. Casnati, B. Carrettoni, I. Manet, R. Schuurman, M. Guardigli, N. Sabbatini and R. Ungaro, *Chem. Eur. J.*, 2000, **6**, 1026–1034.

151. A. Casnati, L. Baldini, F. Sansone, R. Ungaro, N. Armaroli, D. Pompei and F. Barigelletti, *Supramol. Chem.*, 2002, **14**, 281–289.

152. V. Carroll, D. W. Demoin, T. J. Hoffmann and S. S. Jurisson, *Radiochim. Acta.*, 2012, **100**, 653–667.

153. A. M. Rey, *Curr. Med. Chem.*, 2010, **17**, 3673–3683.

154. M. D. Bartholomä, A. S. Louie, J. F. Valliant and J. Zubieta, *Chem. Rev.*, 2010, **110**, 2903–2920.

155. S. R. Banerjee, K. P. Maresca, L. Francesconi, J. Valliant, J. W. Babich and J. Zubieta, *Nuclear Med. Biol.*, 2005, **32**, 1–20.

156. A. Boschi, C. Bolzati, E. Benini, E. Malago, L. Uccelli, A. Duatti, A. Piffanelli, F. Refosco and F. Tisato, *Bioconjug. Chem.*, 2001, **12**, 1035–1042.

157. M. Liu, W. Xu, L.-J. Xu, G.-R. Zhong, S.-L. Chen and W.-Y. Lu, *Bioconjug. Chem.*, 2005, **16**, 1126–1132.

158. S. Liu, *Dalton Trans.*, 2007, 1183–1193.

159. P. S. Donnelly, *Dalton Trans.*, 2011, **40**, 999–1010.

160. V. Biricová, A. Láznicková, M. Láznícek, M. Polásek and P. Hermann, *J. Pharm. Biomed. Anal.*, 2011, **56**, 505–12.

161. S. L. Deutscher, S. D. Figueroa and S. R. Kumar, *J. Labelled Compd. Radiopharm.*, 2009, **52**, 583–590.

162. J. Simecek, O. Zemek, P. Hermann, H.-J. Wester and J. Notni, *Chem. Med. Chem*, 2012, **7**, 1375–1378.

163. J. Notni, J. Simecek, P. Hermann and H.-J. Wester, *Chem. Eur. J.*, 2011, **17**, 14 718–14 722.
164. K. Pohle, J. Notni, J. Bussemer, H. Kessler, M. Schwaiger and A. J. Beer, *Nucl. Med. Biol.*, 2012, **39**, 777–784.
165. X. Sun, M. Wuest, G. R. Weisman, E. H. Wong, D. P. Reed, C. A. Boswell, R. Motekaitis, A. E. Martell, M. J. Welch and C. J. Anderson, *J. Med. Chem.*, 2002, **45**, 469–477.
166. G. Liu and D. J. Hnatowich, *Anti-Cancer Agents in Medicinal Chemistry*, 2007, **7**, 367–377.
167. M. H. B. Grote Gansey, A. S. De Haan, E. S. Bos, W. Verboom and D. N. Reinhoudt, *Bioconjug. Chem.*, 1999, **10**, 613–623.

Supramolecular Gels for Pharmaceutical and Biomedical Applications

JUAN F. MIRAVET* AND BEATRIU ESCUDER*

Departament de Química Inorgànica i Orgànica, Universitat Jaume I, 12071 Castelló, Spain
*Email: miravet@uji.es; escuder@uji.es

11.1 Supramolecular Gel: Definition and Properties

Although the true definition of a gel can be debatable,[1] for the purposes of this chapter a gel can be considered as a solid-like viscoelastic material formed by a major component, the solvent, and a minor component, the gelator. The gelator forms a solid-like microfibrillar network that percolates the solvent and avoids its fluxion due to capillary forces (Figure 11.1). Commonly the concentration of gelator is *ca.* 0.1–1 wt% and the entrapped solvent forms large pools within the gel which behave in many instances as the pure solvent.

The defining attribute of the term "supramolecular gel" is that the gel is not formed by macromolecules or polymers (see Chapter 16), but by small molecules that provoke gelation upon supramolecular aggregation.[2–9] Having this idea in mind, the term "molecular gel", as opposed to macromolecular gel, can be considered a more accurate expression. However, the names supramolecular gel, molecular gel and self-assembled gel are used quite vaguely in the literature. In any case, the focus of this chapter *is not* on physical gels formed by non-covalent (supramolecular) crosslinking among covalent polymers or

Monographs in Supramolecular Chemistry No. 13
Supramolecular Systems in Biomedical Fields
Edited by Hans-Jörg Schneider
© The Royal Society of Chemistry 2013
Published by the Royal Society of Chemistry, www.rsc.org

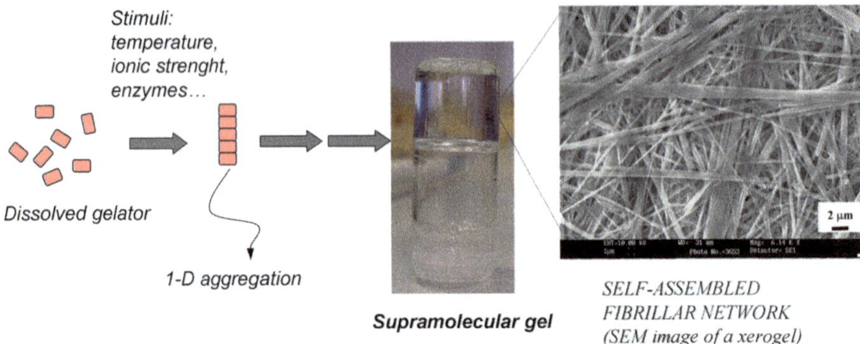

Figure 11.1 Schematic representation of the self-assembly process that produces supramolecular gels.

macromolecules, which could be defined as "supramolecular polymer gels".[10] The molecules that form supramolecular gels are known as "low molecular weight gelators". Obviously the borderline between low molecular weight and high molecular weight species is indistinct but the frontier could lie somewhere between 1000 and 3000 Da.

There is a wide structural variety of supramolecular gelators. Common structural units include, on the one hand, hydrogen (H)-bonding groups such as amides, ureas and carbamates and, on the other hand, apolar moieties such as long alkyl tails and extended aromatic surfaces. In many cases gelators derive from amino acids or sugars. The different moieties regulate both intermolecular interactions (H-bond, aromatic stacking, solvophobic) and solubility in such a way that self-assembled fibrillar networks can be formed. The relevance of the various intermolecular interactions, as expected, depends strongly on the solvent. For organic solvents of moderate to low polarity, H-bonding or ionic interactions are crucial but, for example, in water the hydrophobic interactions can direct the self-assembly.[11]

A rigorous way to assess the formation of gels is to perform rheological measurements. A gel is present when the viscoelastic modulus, G', is bigger than the loss modulus, G''. The difficulties associated with the measurement of these rheological parameters by non specialists are circumvented in many cases by using what has been called "tabletop rheology".[12] Basically, a gel is present if, when placed inside a vial, the solvent does not flow upon turning the vial upside down. Interestingly, this simple approach works very well for the determination of properties of gels, such as the minimum concentration required for gelation or thermal stability.

Within the realm of supramolecular chemistry, supramolecular gels differ from classic supramolecular systems: instead of discrete, well defined super-molecules, the gelators form extended, infinite, one-dimensional (1-D) supramolecular polymers. In addition, a two-phase system is present in supramolecular gels when the solid-like fibrillar network is formed. As a matter of fact, in some cases the formation of the fibrillar network can be seen as a crystallization process and the fibers are shown to be microcrystalline when analyzed by X-ray diffraction.[13] In this sense, the supramolecular chemistry of gels is related to crystal engineering.[14] On the basis of this comparison, it can be understood that a major challenge in the study of supramolecular gels is to rationally design gelators.[15] As in the case of crystal engineering, the difficulties associated with the conformational mobility of the molecules and the different docking modes for their supramolecular aggregation are major issues. For example, a hot topic in crystal engineering, such as polymorphism, seems to be also commonly present in supramolecular gel chemistry.[16,17] Overall, design parameters generally accepted for the preparation of supramolecular gelators are the capability of forming 1-D aggregates and an appropriate solubility balance, which should avoid compounds that are too soluble or too insoluble. However, some important parameters are not fully understood, such as the relationship between gelator structure and fiber nucleation and entanglement, or the relevance of the kinetics. Using Monte Carlo simulations, the stiffness of the 1-D aggregates formed upon gelator assembly and the strength of the intermolecular bonds between the resulting threads have been shown to be decisive for the formation of fibrillar networks. Should the threads be too stiff or their mutual attraction forces be too strong, disconnected bundles of threads would be formed instead of the connected network of bundles required for gel formation.[18]

The number of papers dealing with supramolecular gels has grown exponentially in the last 20 years (Figure 11.2). One may wonder what the origin of this growing interest is, and a way to answer this question is to compare supramolecular gels with conventional macromolecular gels. Some acknowledged advantages of supramolecular gels over polymeric gels include their highly reversible nature, the easy control of functionalization and composition, their dynamic behavior, and superior biodegradability and biocompatibility. Supramolecular gels are intrinsically reversible due to the non-covalent nature

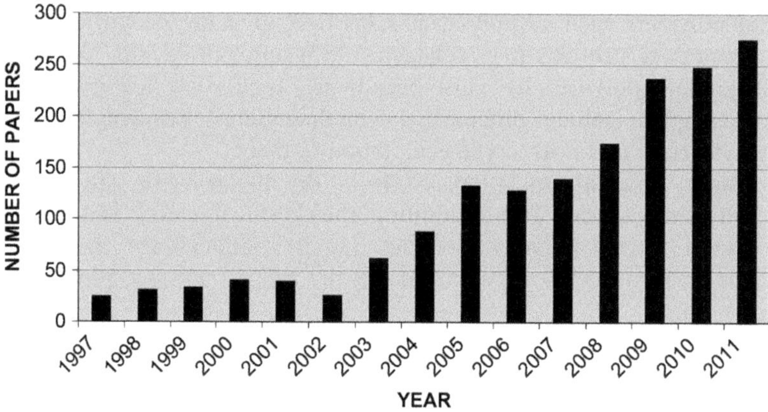

Figure 11.2   Evolution with time of the number of papers dealing with supramolecular gels (based on SciVerse Scopus database searches).

of the self-assembled networks. Therefore in many cases they can be considered as stimuli-responsive, smart, soft materials.[19] The formation/dissociation of supramolecular gels has been regulated by stimuli such as temperature, concentration, light or different chemicals. The control of supramolecular gelation by changes in concentration and/or temperature is a general phenomenon resulting from the non-covalent nature of the gels. The introduction of photoresponsive units such as, for example, azobenzene, affords light sensitive gels.[20] In addition, different chemical stimuli have been used which include pH changes,[21] red/ox reagents,[22] the presence of different anions or cations,[23] enzymes[24] and the use of species that bind covalently to the gelator.[25] Regarding biodegradability, the fact that supramolecular gels are composed of small molecules offers a clear advantage for *in vivo* disposal of the gel when compared with polymeric counterparts. Furthermore, many examples of low molecular weight gelators are biocompatible because they are built from natural products such as amino acids or sugars. Up to now, quite interesting applications of supramolecular gels have been described, outside the biomedical field, in diverse areas such as electronic[26] and photonic materials,[27] catalysis,[28] and templates for inorganic nanoshaped materials.[29]

11.2 Supramolecular Gels for Pharmaceutical and Biomedical Applications

In the context of supramolecular gels in medical applications it seems reasonable to comment briefly on their polymeric counterparts. As mentioned above, the interest in supramolecular gels is relatively recent but the use of polymeric hydrogels for biomedical applications can be dated back to the early 1950s with the development of contact lenses or implants.[30] Polymeric hydrogels have been used in a variety of pharmaceutical and medical applications which include tissue engineering,[31] controlled drug release,[32,33] and vascular prostheses,[34] among others. The gelators used for the preparation of these hydrogels can be natural polymers such as polysaccharides (*e.g.* alginate, pectin, carrageenan and dextran) or proteins (*e.g.* collagen, fibrin and polylysine). On the other hand, synthetic polymers have also been widely used in medical hydrogels and some examples include different polymethacrylates, polyvinylpyrrolidone, polyvinylacetate and polyacrylic acid, among others.[35]

Polymeric physical gels prepared from organic solvents are rather less common than hydrogels and, in addition, their medical interest is not so broad. For example, it has been pointed out that it is difficult for common vinyl polymers to form a 3-D network due to the lack of suitable physical cross-linking points.[36]

11.2.1 Emerging Therapeutic Properties of Supramolecular Gels

One of the most interesting features of supramolecular complexes is their new properties when compared to the isolated components of the assembly.[37] In the case of supramolecular gels, the functional groups introduced in the gelators are distributed in well organized arrays with interesting potential properties. Effects such as cooperation or intermolecular interactions among functional groups, high local concentration, conformational changes associated to aggregation or solvation changes could afford new functions.[38,39] For instance, upon gel formation a gelator with L-proline moieties became a basic catalyst for the nitro-aldol Henry reaction between 4-nitrobenzaldehyde and nitromethane. In this system the spatial proximity of the L-proline units afforded a boost in the basicity.[40]

A very significant case of new therapeutic properties gained upon gel formation was reported for a gelator derived from vancomycin. Vancomycin is a broad-spectrum antibiotic capable of inhibiting methicillin-resistant *Staphylococus aureus* (MRSA). The derivatization of vancomycin with a pyrene group afforded, serendipitously, a hydrogelator (**1**, Scheme 11.1) capable of forming gels at 0.36 wt%.[41,42] Structural analysis suggested that $\pi - \pi$ interactions among pyrene units and H-bonding among vancomycin moieties are responsible for the aggregation. Remarkably, the hydrogel exhibited much higher antibiotic activity (nearly three orders of magnitude) against vancomycin-resistant enterococci than the originally isolated, unmodified vancomycin. The enhanced activity might be ascribed to the self-assembly of the vancomycin–pyrene conjugate on the surface of bacteria cells.

A milestone in the development of therapeutic supramolecular gels has been the selective differentiation of neural progenitor cells within hydrogels formed by peptide amphiphiles (progenitor cells are said to be in a further stage of cell differentiation than stem cells).[43] This type of molecule contains a polar moiety

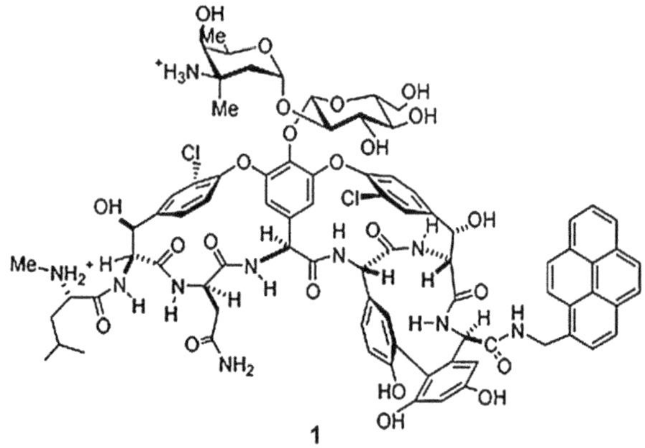

Scheme 11.1   Chemical structure of compounds **1** and **2**.

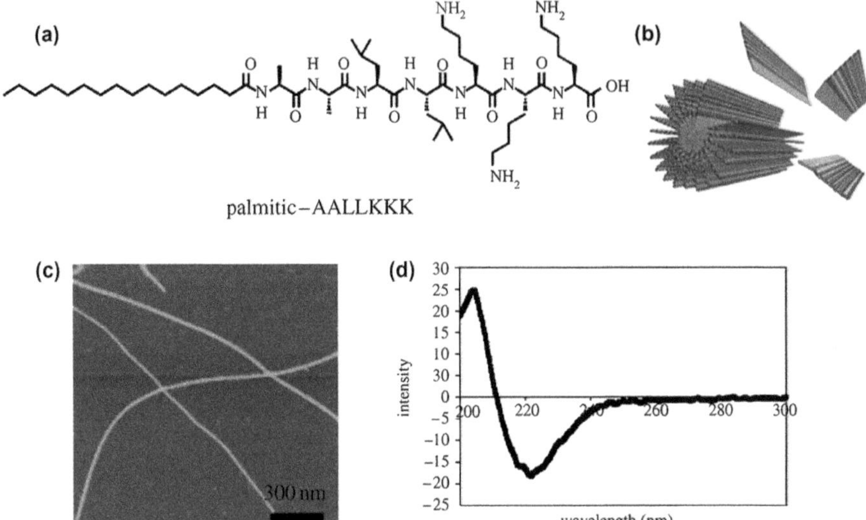

Figure 11.3   (a) Structure of a typical peptide–amphiphile monomer. (b) Schematic illustration of cylinder formed from β-sheets. (c) AFM image showing nanofibers from a 1 : 1 mixture of palmitic-IEIE and palmitic-IKIK. (d) Circular dichroic spectrum of palmitic-LLLGGKK indicating β-sheet formation.[45]

constituted by an ionizable peptide that provides water solubility linked to a hydrophobic linear alkyl chain. Under the appropriate conditions the dissolved peptide amphiphile aggregates into fibrillar micelles that evolve to a network capable of provoking hydrogelation (Figure 11.3). The driving forces for gelation are the hydrophobic effect associated with the alkyl chains and intermolecular H-bonding among the peptide units. It has to be noted that the fibrillar objects present a hydrophilic surface with a high local concentration of peptide moieties.

Neural progenitor cells were encapsulated *in vitro* within a hydrogel formed by an amphiphile (**2**, Scheme 11.1) which contained the laminin epitope Val-Ala-Val-Lys-Ile in the peptidic moiety. (Laminins are proteins present in the basal lamina, a layer of the extracellular matrix, which influences cell differentiation, migration and adhesion.) The self-assembly into a hydrogel took place, triggered by the pH and salinity changes, by mixing cell suspensions with dilute aqueous solutions of the gelator, and it was observed that the cells survived the growth of the fibrillar network. Relative to the isolated laminin epitope or the non-aggregated peptide amphiphile, the fibrillar network induced very rapid differentiation of cells into neurons, while discouraging the development of astrocytes (star-shaped, non-neuronal cells found in the brain and spinal cord). This rapid and selective differentiation was ascribed to the amplification of the bioactive epitope associated to its high local concentration in the surface of the nanofibers. As a continuation of this work, the same peptide amphiphile was used to treat a mouse model of spinal cord injury.[44]

A hydrogelator solution was injected into the injured spinal cord and the gel was formed as a result of ionic strength changes. Very remarkably, significant behavioral improvement, hindlimb movement, was observed in mice treated with the peptide amphiphile as compared to control experiments. These examples serve to demonstrate the amplification of properties associated with fiber formation, which constitutes cases of tissue engineering, a topic that is discussed in more detail below.

11.2.2 Supramolecular Gels in Cell Culture and Tissue Engineering

Tissue engineering has been defined as "an interdisciplinary field that applies the principles of engineering and life sciences toward the development of biological substitutes that restore, maintain, or improve tissue function".[46] Hydrogels are used as scaffolds in tissue engineering to mimic the extracellular matrix and provide support for cell adhesion, migration and proliferation. Some requirements for the hydrogel to be of use in tissue engineering include appropriate rheological properties (similar to real tissue), stability under physiological conditions for the period of time required for their function, and biodegradability allied to the production of "inert" degradation products. Most of the work reported in tissue engineering is based on polymeric hydrogels formed by polymers or macromolecules such as hyaluronic acid, chitosan, alginate, collagen, gelatin or poly(2-hydroxyethyl methacrylate), among others.[47] However, quite recently, the use of supramolecular gels, based on some advantageous properties over polymeric gels cited above, has begun to be explored with some noticeable results that justify the growing interest in the use of this approach for tissue engineering.

Fmoc-dipeptides constitute a very relevant family of supramolecular gelators used for cell culture. Although discovered recently,[48] Fmoc-protected dipeptides and amino acids probably constitute the most widely studied class of supramolecular hydrogelators. These molecules can form hydrogels by dissolution of the carboxylic acid-terminated gelator at basic pH values and posterior controlled acidification.[49] The involvement of the fluorenyl moeity in the aggregation through $\pi - \pi$ stacking interactions is decisive for the self-assembly process. A $\pi-\pi$ interlocked antipararell β-sheet assembly has been proposed to take place in these types of hydrogelators.[50] In one work, a variety of Fmoc-protected dipeptides were shown to form hydrogels (Fmoc-Gly-Gly, Fmoc-Ala-Gly, Fmoc-Ala-Ala, Fmoc-Leu-Gly, Fmoc-Phe-Phe and Fmoc-Phe-Gly). The Fmoc-dipeptides were dissolved in water by addition of 0.5 M NaOH and hydrogels were obtained by posterior acidification with dropwise addition of 0.1 M HCl. Quite interestingly, the pH required for gelation to take place was below 4 in all the cases except for Fmoc-Phe-Phe, which could form gels at pH values below 8, being of interest for the formation of gels at physiological pH. This remarkable difference of pH values required for aggregation should be ascribed to a considerable aggregation-induced pK_a shift of the terminal carboxylic acid.[51,52] It can be mentioned here that the use of δ-gluconolactone—a water-soluble compound whose hydrolysis in water produces a gradual acidification of the medium—represents a very convenient mild and controllable alternative to the use of HCl solutions for the preparation of this type of hydrogel.[53] Hydrogels formed from Fmoc-dipeptides at physiological pH presented a network of fibers with similar dimensions to that of the extracellular matrix components and were stable under cell-culture conditions. In this case chondrocytes (cartilage cells) were incorporated in the gel in both 2-D and 3-D cell culture experiments.[49] For the latter a fluorescent dye for visualization of the cell nuclei allowed confirmation of the presence of cells throughout the gel matrix.

In a closely related work published somewhat earlier by a different group, Fmoc-Phe-Phe hydrogels were prepared by diluting hexafluoroisopropanol stock solutions in aqueous media.[54] Chinese hamster ovary cells were suspended above the gel and, when compared to the control experiment, the elongated shape of the adhered cells was nearly identical to the cells that were grown on the hydrogel. A third group also reported the use of Fmoc-Phe-Phe for cell culture, replacing the use of hexafluoroisopropanol by dimethyl sulfoxide (DMSO), a more biocompatible solvent, and prepared *in situ* three different cell cultures (astrocytes, kidney cells and a fibroblast-like cell line).[55]

One step further in the use of Fmoc-Phe-Phe as hydrogelator was its combination with Fmoc-Arg-Gly-Asp (Fmoc-RGD using the one letter code for amino acids; Figure 11.4).[56] The tri-peptide sequence RGD is known to bind transmembrane integrins forming linkages between the cytoskeleton and the extracellular matrix. As a prerequisite of the regulation of cell behavior within a scaffold, cell adhesion is often promoted by introducing the RGD ligand into polymeric or supramolecular substrates. The two-component hydrogel formed by Fmoc-Phe-Phe and Fmoc-Arg-Gly-Asp generates cylindrical fibers whose concentration of the RGD ligand can be regulated. The hydrogels were observed to promote adhesion of encapsulated dermal fibroblasts through specific RGD–integrin binding, with subsequent cell spreading and proliferation.

A very interesting piece of work has been reported on the 3-D encapsulation and injectable delivery of cells.[57] For this purpose a hydrogel was prepared using a peptide formed by 20 amino acids (this molecule lies at the borderline between molecular and macromolecular gelators). The peptide contained the sequence Val-(D)Pro-Pro-Thr which is known to promote a type β-II′ turn. The presence of several positively charged lysine units in the peptide precludes folding in low ionic strength aqueous media. However when the ionic strength was increased by addition of DMEM (Dulbecco/Vogt modified Eagle's minimal essential medium), a cell culture medium with a high concentration of salts, the charges in the peptide were sufficiently screened to allow folding. Remarkably, the folded structures self-assembled into fibrillar structures resulting in a hydrogel. A smart utilization of these systems permitted, for example, gelation in the presence of cells, which became homogeneously impregnated within the fibrillar network.

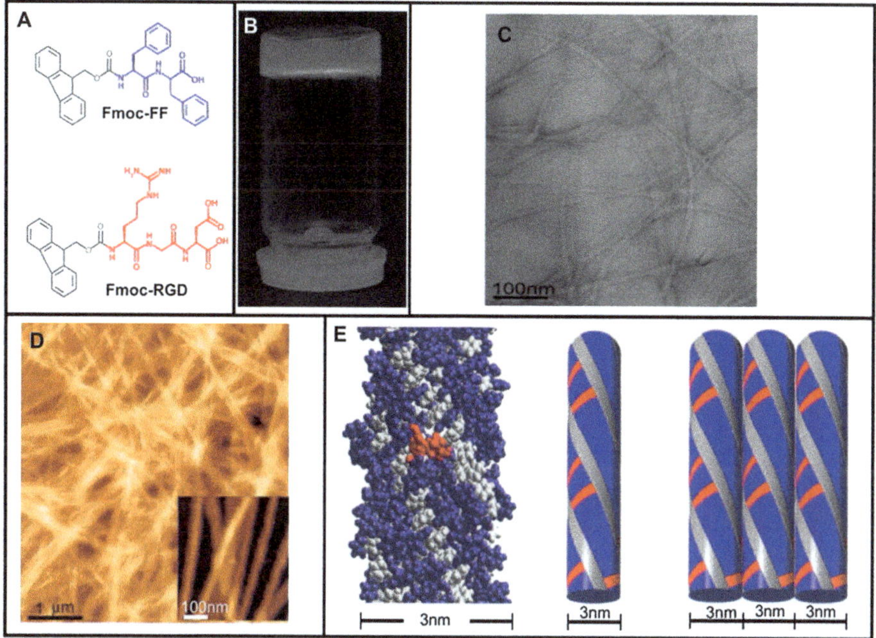

Figure 11.4 The self-assembled hydrogel of Fmoc-FF/RGD, its nanofibrous structure and the supramolecular model. (A) The chemical structures of the hydrogel building blocks: Fmoc-FF and Fmoc-RGD. (B) The mixture of Fmoc-FF and Fmoc-RGD self-assembles into a translucent hydrogel at 37 °C. (C) The AFM height image of the hydrogel shows an overlapping mesh of nanofibers, with bundles and entanglements. (D) The TEM image shows that the nanofibers are "flat ribbons". (E) The proposed supramolecular model demonstrates the formation of the 3 nm fibrils and their further lateral assembly into larger ribbons. RGD sequences are presented on the fiber surface, enhancing their accessibility and bioavailability.
(Adapted from Zhou *et al.*[56] with permission from Elsevier.)

A key property of the hydrogel is thixotropy: under stress it shears thin, resulting in a low-viscosity sample which can be injected. Once the shearing is stopped, the gel rapidly recovers to its original state. In this way, gel/cell constructs could be injected with syringe precision to the targeted site. The studies revealed that homogeneous cellular distribution and cell viability were not affected by the injection procedure. This is an outstanding advantage when compared to other gels that require surgical intervention and can be used, for example, for *in situ* tissue regeneration.

In a different approach, a hydrogelator containing a zwiterionic amino acid and a photoisomerizable fumaric amide (**3**, Scheme 11.2) unit was used smartly as a mold for 2D/3D microarchitectures of live cells.[58] Once the hydrogel had formed, UV light irradiation isomerized the fumaric amide unit into maleic amide, yielding a molecule that was no longer a hydrogelator. The stiffness of the gel permitted its manipulation and patterns could be traced with the use of a laser UV-light. For example, a cubic hydrogel of *ca.* 3 mm^3 was irradiated in such a way that a cross-shaped channel was built inside the gel. The mold containing the cross-shaped channel was filled with a cell culture medium and a collagen sol, which then afforded a hydrogel which was included inside the former hydrogel. The imbedded collagen hydrogel could be used, for example, for the differentiation of PC12 cells (a cell line derived from a tumor of the rat adrenal medulla). Remarkably, in this work it was also shown that the hydrogel used as a mold could be disassembled in basic media liberating, for example, a cross-shaped agarose gel.

11.2.3 Supramolecular Gels for Drug Delivery and Controlled Release

The use of synthetic polymeric hydrogels has opened up a convenient and effective way to administer drugs. The approach is based on the encapsulation of the desired drug within the 3D network of the gel, which after being applied starts releasing the drug steadily by diffusion. This diffusion out of the gel mesh affords a steady liberation, which in addition can be regulated in by an appropriate design of the hydrogel pore size, or, more interestingly, by changes in the porosity associated to external stimuli such as pH, temperature, light, magnetic fields, electric current or ultrasound. Hydrogel-based drug delivery can be used for oral, rectal, ocular, epidermal and subcutaneous application.[32]

Regarding supramolecular gels, their use in drug delivery and controlled release is gaining popularity due the interesting properties ascribed to the supramolecular nature of these materials. Some selected examples are shown below which reflect different approaches for the use of supramolecular gels in drug delivery: (i) the drug itself is a gelator; (ii) the drug is encapsulated within the gel network; and (iii) the drug is conjugated and converted into a gelator. In addition to these examples, which in most cases deal with hydrogels, the use of supramolelcular organogels (gels formed in organic solvents) for controlled release will be discussed.

Lanreotide (**4**, Scheme 11.2) is a synthetic octapeptide which is aimed to mimic a natural hormone, somatostatin. Somatostatin is known to lower the levels of growth hormones and therefore the synthetic analogue lanreotide can be used for the treatment of acromegaly, a syndrome that results when the body produces an excess of growth hormone. Interestingly, lanreotide self-assembles in aqueous media and affords gels. A detailed study revealed that self-assembly through H-bonding yields bilayers which then evolve to form nanotubes.[59] Lanreotide constitutes a paradigmatic case of a drug which self-assembles to yield hydrogels. This property can be used for the controlled sustained release of the drug associated to the progressive disassembly of the gel in the biological environment. As a matter of fact, this drug received approval as a subcutaneous, long-acting implant for the treatment of acromegaly. For example, a recent report revealed that home administration of lanreotide Autogel®

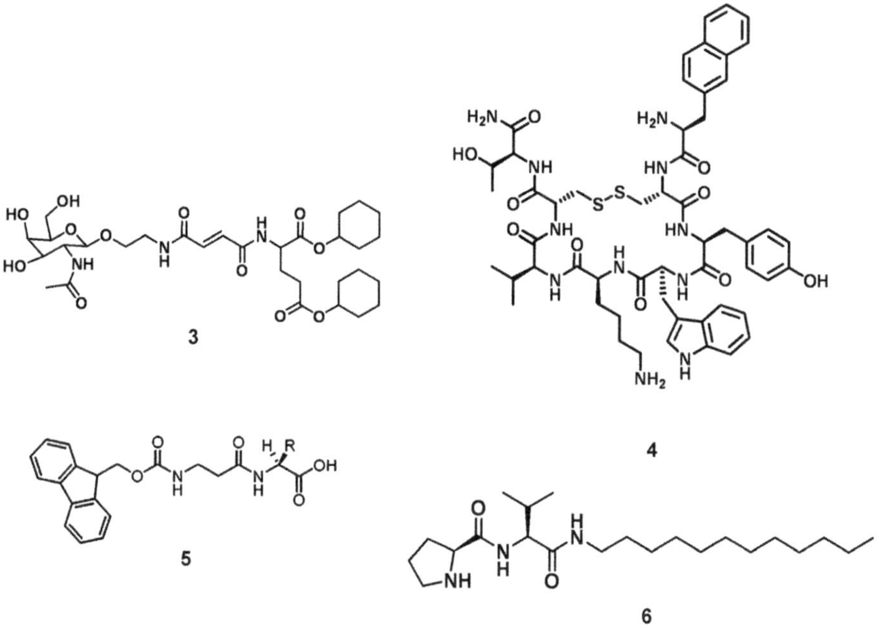

3

4

5

6

Scheme 11.2	Chemical structure of compounds **3**, **4**, **5** and **6**.

by patients with acromegaly is safe and effective. Lanreotide Autogel® is prepared as a viscous aqueous formulation which is supplied in prefilled syringes for deep subcutaneous injection. In this way a consistent drug release is achieved for 28 days which results in optimal control of biochemical markers and symptoms of acromegaly.[60]

A very relevant example of a self-assembling compound for therapeutic applications is the peptide (Arg-Ala-Asp-Ala)$_4$-CONH$_2$, known commercially as PuraMatrix™. This peptide was reported to self-assemble into interwoven nanofibers with diameters between 10 nm and 20 nm and to afford hydrogels with pore size diameters between 5 nm and 200 nm (Figure 11.5).[61] The peptide presents alternating hydrophobic (Ala) and hydrophilic, ionizable, amino acids of opposite charge (Arg and Asp). This repeating motif results in a very efficient self-assembly under physiological conditions associated to the pH and salinity changes, which is a key property for the practical use. Commonly, a solution of this peptide is placed in biological media, for example in phosphate-buffered saline at pH 7.4, and a hydrogel is formed immediately. PuraMatrix™ has been used in tissue engineering for cartilage and bone reconstruction and heart tissue regeneration.[62,63] In the context of controlled release, the self-assembled network provides an appropriate medium for the sustained controlled release of active species by progressive diffusion out of the gel network. For example, PuraMatrix™ is an efficient carrier for slow delivery of proteins such as lysozyme or bovine serum albumin, among others, and it was shown that the gel matrix did not impair the function of the proteins.

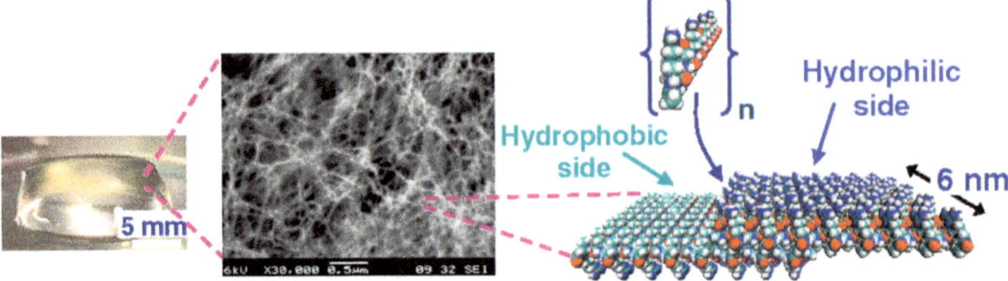

Figure 11.5 Ac-(Arg-Ala-Asp-Ala)$_4$-CONH$_2$ peptide monomer and the corresponding nanofibers. Color scheme for proteins and peptides: blue, positively charged; red, negatively charged; light blue, hydrophobic. (Reprinted with permission from Koutsopoulos *et al.*[61])

In recent example, an injectable formula containing a solution of PuraMatrix™ and insulin was delivered subcutaneously.[64] It was observed that the solution became a gel *in vitro* and *in vivo* as a result of the increased concentration of phosphate and pH change in the medium as compared to the original solution. Sustained release pharmacokinetics of insulin was observed *in vivo*, resulting in a markedly decrease of glucose level. The release rate could be regulated by changing the concentration of PuraMatrix™ in the formula. Higher concentrations resulted a in a more dense fibrillar network and lower release rates. When a formula containing a 2 wt% of peptide was used in rats, the lowered glucose levels could maintained for 24 h with a single injection.

In another example a Fmoc-dipeptide containing a β-alanine unit was prepared (**5**, Scheme 11.2). With this approach, proteolytically stable dipeptide-based hydrogels could be prepared. These hydrogels were obtained from concentrated solutions of the dipeptide in DMSO which were diluted in phosphate-buffer saline. The hydrogels were found to be resistant to proteinase K and provided *in vitro* a sustained release of vitamins B$_2$ and B$_{12}$.[65]

A step further in the development of drug delivery systems based on supramolecular gels aims to take advantage of the stimulus-sensitive nature of these materials: the presence of the appropriate stimuli would induce gel disassembly and trigger the release of entrapped species. This approach is superior to the mere use of the hydrogel network as a mesh to regulate the diffusion of active species outside of the gel. The use of stimuli responsive, smart, supramolecular gels would permit, for example, the liberation of drugs when the concentration of the trigger (ion, metabolite, enzyme, *etc.*) had reached a threshold.

In one example, a peptide amphiphile formed by the dipeptide Pro-Val and a C$_{12}$ alkyl chain (**6**) formed gels that released entrapped methylene blue when aldehydes were present in the medium.[25] This system is based on the reactivity of the terminal L-proline unit present in the gelator which reacts with aldehydes to afford hemiaminals and related species. This chemical transformation, taking place on the gelator fibers, provokes the disassembly of the gel and the corresponding release of entrapped species. Remarkably, the system is quite sensitive to the nature of the aldheyde. In particular, the hydrophobic character of the aldehyde is key, being the most reactive. This behavior is thought to arise from the favored incorporation of the more hydrophobic molecules into the hydrophobic core of the fibers. For example, when acetaldehyde was present in the medium, 11% of the entrapped dye was released after 30 min but for the same concentration of 3-phenylpropanal, a quantitative release was detected after the same time period. The same system could be used for the *in vitro* controlled release of ketoprofen, a non-steroidal antiinflamatory drug (NSAID) with poor solubility in water (0.01 wt%). Translucid hydrogels were prepared from a mixture of ketoprofen and the hydrogelator, revealing the increased drug solubility in this medium. After addition of 3-phenylpropanal disassembly of the hydrogel and quantitative release of the drug were achieved after 30 min.

Neutral molecules can also be used as a stimulus for the release of gel-entrapped active species. In a model study, a tetraphenylporphyrin was released from a gel formed in toluene by a C2-symmetrical L-valine derivative containing terminal isonicotinyl moieties (**7**, Scheme 11.3).[66] This gelator forms self-assembled fibrilar networks in a variety of solvents by means of multiple intermolecular H-bonding interactions among the four amide units present in the structure. Remarkably, the release took place selectively in the presence of catechol but not when related compounds such as resorcinol or hydroquinone, among others, were present. Catechol derivatives are interesting compounds in

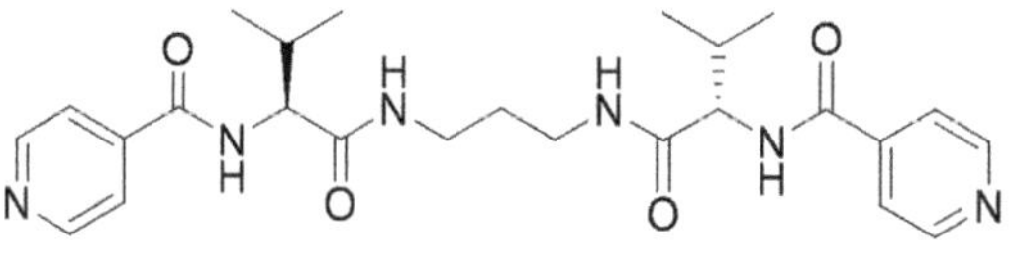

7

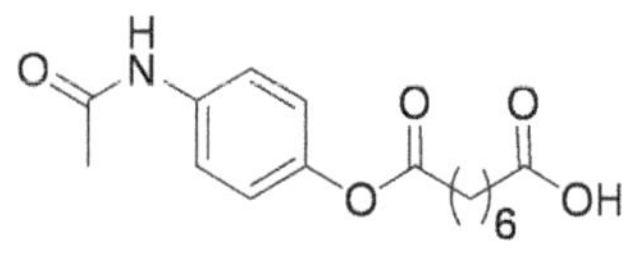

8

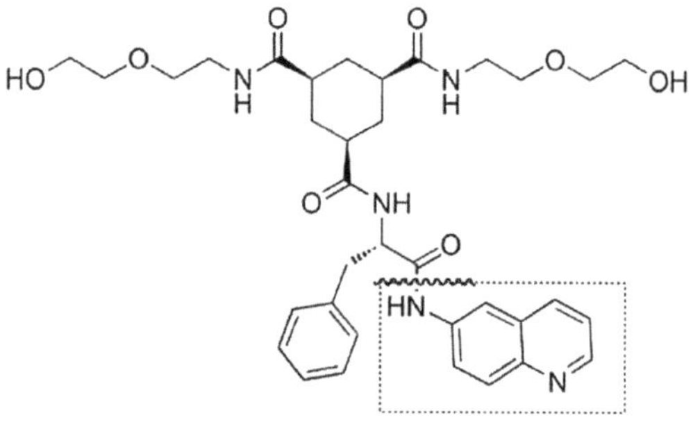

9

Scheme 11.3   Chemical structure of compounds **7**, **8** and **9**.

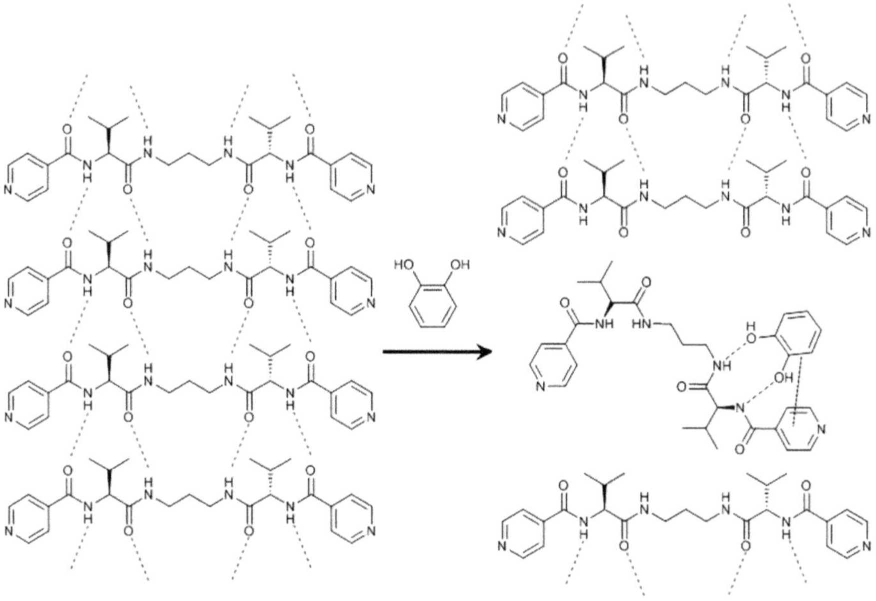

Figure 11.6 Schematic representation of the proposed interaction of the gel fibers formed by **7** with catechol.[66]

the biomedical context owing to the relevance of substances such as catecholamines (neurotransmitters) or urushiol, an organic allergen found in plants. In the described system catechol binds in a complementary way to the gelator through H-bonding and π–π stacking affording gel disassembly. The presence of an electron-poor isonicotinic unit favors the π–π stacking interactions with the electron-rich catechol moiety. When an analogue gelator built by the replacement of the pyridine unit by phenyl was prepared, no response to the presence of catechol was observed (Figure 11.6).

A rather interesting approach for drug delivery involves the use of enzymes as a trigger. For example, the antipyretic and analgesic drug acetaminophen was reacted with octanedioic acid to yield a monoester which is a hydrogelator (**8**, Scheme 11.3). The gelator was dissolved in water by gentle heating up to 80 °C and then upon cooling down to room temperature a gel was obtained. The ester group of the gelator could be hydrolyzed in the presence of a lipase and therefore acetaminophen was delivered to the medium. A step further involved the use of a second drug that was entrapped in the gel network during the gel preparation. For this purpose, curcumin, an anti-inflammatory drug, was chosen. Upon enzyme-triggered degradation, the hydrogel released both curcumin and acetaminophen. An interesting point is that the degradation products of the hydrogelator were acetaminophen and a biocompatible fatty acid.[67]

In a study published earlier, a model conjugated prodrug that behaves as hydrogelator was studied for enzyme-stimulated release. For this purpose a

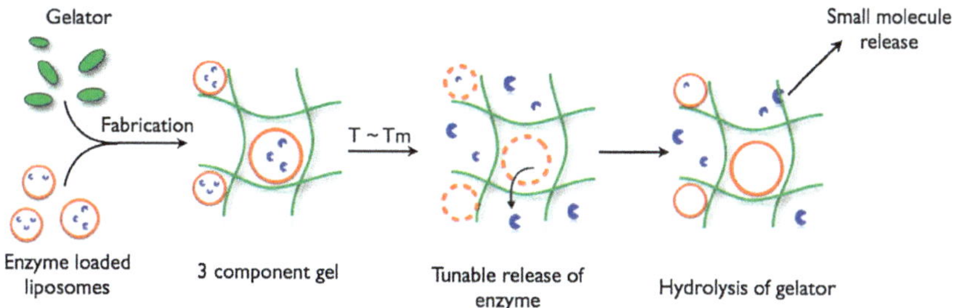

Figure 11.7 Schematic representation of the temperature-triggered drug release system formed by compound **9**, enzymes and liposomes.
(Reprinted with permission from Boekhoven *et al.*[69] Copyright 2012 American Chemical Society.)

cyclohexane trisamide-gelating scaffold was used to which an L-phenylalanyl–amidoquinoline moiety as well as two ethylene glycol chains were connected (**9**). This molecule formed gels at a concentration as low as 0.03 wt%, being therefore a so-called supergelator. The enzyme chymotrypsin hydrolyzed the amido units liberating the fluorescent model drug 6-aminoquinoline. Noticeably, a two-step release system was described because the enzyme was not able to hydrolyze the molecules in the fibrillar network. Therefore, the enzyme could only transform the free gelator molecules that coexist in equilibrium with the fibrillar network. In this way as the enzymatic reaction proceeded; new gelator molecules were progressively released to solution affording a steady gel disassembly and release of the model compound.[68]

A smart elaboration of the previous concept consisted in the preparation of a delivery platform based on the same hydrogelator (**9**) and enzyme, but in the presence of liposomes.[69] Chymotrypsin was entrapped in the liposomes and afterwards the gel prepared. No model drug release was observed at this stage indicating that the enzyme and the gelator were isolated from each other. Temperature was used as a stimulus to trigger release of the model drug. When the system was heated for 5 min at 42 °C, the phase transition temperature for the liposomes, the enzyme was liberated and the hydrolysis of the gelator in solution afforded a steady release of the fluorescent model molecule (Figure 11.7).

The conjugation of a drug with a gelating scaffold represents an interesting strategy for controlled release. In one study taxol® (paclitaxel), a drug that inhibits mitosis and is used in cancer therapy, was conjugated to a hydrogelator scaffold consisting in a tetrapeptide acylated at the N-terminal position with a 2-naphthaleneacetyl group (Nap-Phe-Phe-Lys-Tyr) (**10**, Scheme 11.4).[70] A taxol® derivative was connected to the gelating scaffold by amide formation with the pendant amino group of the lysine amino acid. Although Nap-Phe-Phe-Lys-Tyr and its taxol® derivative are good hydrogelators, their low solubility precluded direct gel formation. To circumvent this problem, the tyrosine moiety was phosphorylated, affording a water-soluble derivative (taxol®

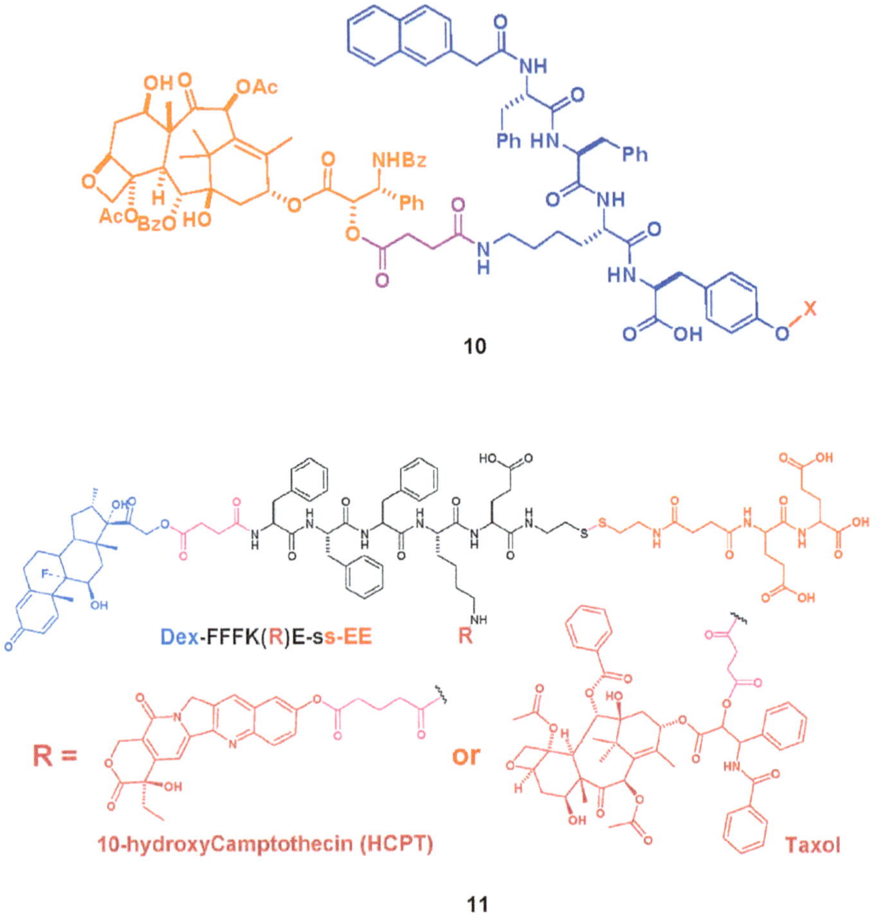

10

Dex-FFFK(R)E-ss-EE

R =

10-hydroxyCamptothecin (HCPT) or **Taxol**

11

Scheme 11.4   Chemical structure of compounds **10** (X = H) and **11**. (Adapted with permissions from Gao *et al.*,[70] copyright 2009 American Chemical Society, and Mao *et al.*[71])

conjugated to Nap-Phe-Phe-Lys-Tyr-PO$_3^{2-}$). This compound, using a previously reported approach, could be enzymatically dephosphorylated to yield the desired hydrogel. It was found that the gel released the taxol$^{®}$ conjugate at a rate of 0.13% of the total load per hour and, noticeably, the cytotoxicity of this conjugate was very similar to that of taxol$^{®}$ itself.

In a related approach two drugs were conjugated to a hydrogelator scaffold constituted with a derivative of the peptide Phe-Phe-Phe-Lys-Glu.[71] Dexamethasone, an immunosuppressant and anti-inflamatory drug, was conjugated forming an amide bond at the N-terminal position of the peptide. In addition, taxol$^{®}$ or 10-hydroxy-camptothecin (a cytotoxic alkaloid which inhibits DNA topoisomerase I and is used in cancer chemotherapy) was conjugated to the gelator by amide formation with the pendant amino group of the lysine residue (**11**). Importantly, the spacer used for the conjugation of the drugs to the peptide was anchored to both drugs by a hydrolysable ester bond. Further elaboration of the gelator included C-terminal modification with a diglutamate moiety linked to the peptide by a disulfide containing spacer. The diglutamate unit provided the conjugate with water solubility and could be removed upon disulfide bridge rupture to afford the water-insoluble peptidic hydrogelator. Therefore, gels were prepared by dissolving the conjugate in phosphate buffer and then, upon disulfide bridge breakage with dithiothreitol, gelation took place for systems with concentrations of gelator above 0.25 wt%. The gels successfully released taxol or camptothecin at a rate of *ca.* 5 µg mL^{-1} per hour for several hours. It has to be noted that the free drugs without the linkers and the gelator moiety were released as a result of ester hydrolysis taking place under physiological conditions (37 °C, 100 mM phosphate buffer, pH 7.4).

Most of the examples reported above deal with supramolecular hydrogels. However, their organic counterparts, supramolecular organogels, also deserve to be mentioned due to their importance in drug delivery. Commonly, organogels are used for topical or transdermal delivery. This approach for drug delivery is gaining importance because it represents a non invasive method with good tolerance. Supramolecular gels are advantageous for topical delivery when compared to other systems that are more fluid and therefore provide less contact time with the skin, such as microemulsions.[72] Pharmaceutical supramolecular organogels are usually formed in biocompatible oils or alcohols such as *n*-butanol, *n*-octanol, ethyl oleate, glycerine, isopropyl miristate, or eucaliptus oil, among many others. These substances in general are permeation enhancers which facilitate the migration of the drugs through the epidermis.[73] Different examples of supramolecular organogels loaded with drugs for topical or subcutaneous delivery have been reported and a few examples are described below.

Piroxicam (**12**, Scheme 11.5), an NSAID, was loaded into an organogel formed in Mygliol® 812 (a mixture of cocunut oil and glyceryl caprylate) using glycerylmonostearate (**13**) as gelling reagent. The drug-loaded organogel was applied topically to paw oedemas in rats and the results revealed that this formulation was successful in oedema inhibition.[74]

In another example, delivery of haloperidol (**14**), an antipsychotic drug, was studied. This hydrophobic drug is a suitable candidate for transdermal delivery. The only lasting formulation available for haloperidol is its decanoate ester which is administered by injection, but has some drawbacks such as injection pain and a complex administration regime. Haloperidol-loaded organogels were thought to be a possible alternative which could be potentially used as a maintenance therapy to prevent psychosis. Organogels were prepared in propylene glycol or isostearyl alcohol using a 5 wt% solution of *N*-laurory-L-glutamic acid di-n-butylamide (**15**) as organogelator. The gelator was dissolved in the hot alcohol and to this dissolution haloperidol was added to reach a final concentration of 3 mg mL^{-1}. Upon cooling down, organogels were formed which were studied for *in vitro* permeation through human

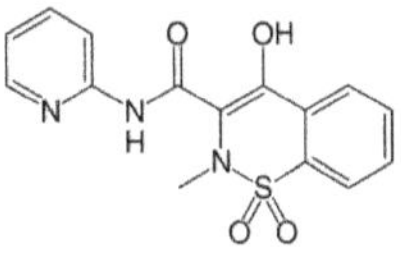

Scheme 11.5   Chemical structure of compounds **12**, **13**, **14**, **15**, **16** and **17**.

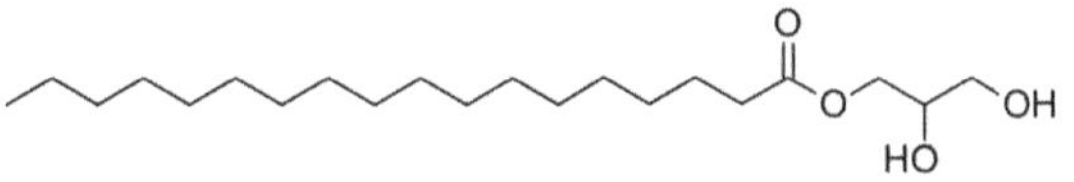

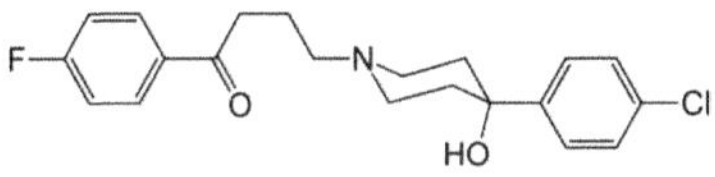

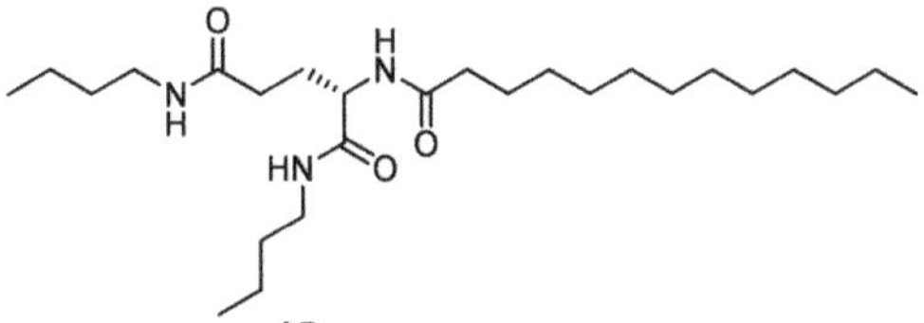

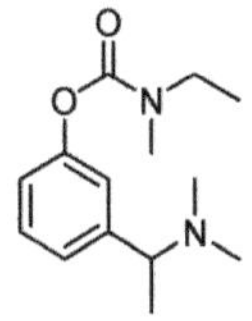

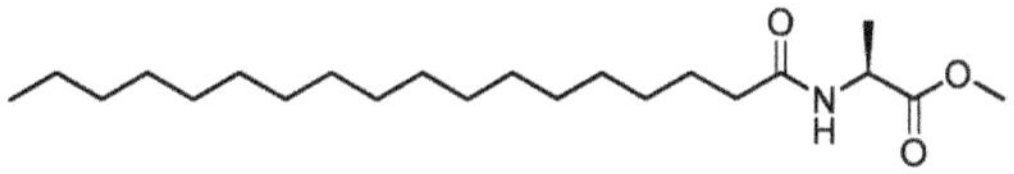

epidermis, revealing that these gels would be suitable for topical or transdermal delivery.[75]

An organogel implant forming *in situ* was used for the controlled release of rivastigmine (**16**), a cholinergic agent used for the treatment of Alzheimer's type dementia. The gelator, *N*-steaoryl-L-alanine methyl ester (**17**) (10 wt%), was dissolved in hot safflower oil and rivastigmine was added (up to 5 wt%). Pharmacokinetic studies demonstrated that this approach was feasible, producing a sustained release of rivagstigmine at therapeutic levels for up to 11 days. This dosing regime could be used to alleviate caregiver responsibilities and adverse effects caused by high variations in plasma drug concentration.[76]

11.2.4   Other Biomedical Applications

The fascinating properties of supramolecular gels can be of use for other biomedical applications aside from drug delivery and tissue engineering. Two examples are discussed in the next paragraph to highlight the potential that these materials may have in the future.

Peptide or protein arrays represent a very appealing strategy for the high-throughput analysis of protein activity. These arrays can be useful for a rapid assessment of the type and relative amount of different enzymes in a biological sample, which is a direct indicator of many diseases. However, the preparation of protein arrays can be difficult due to the fragility of proteins under the drying

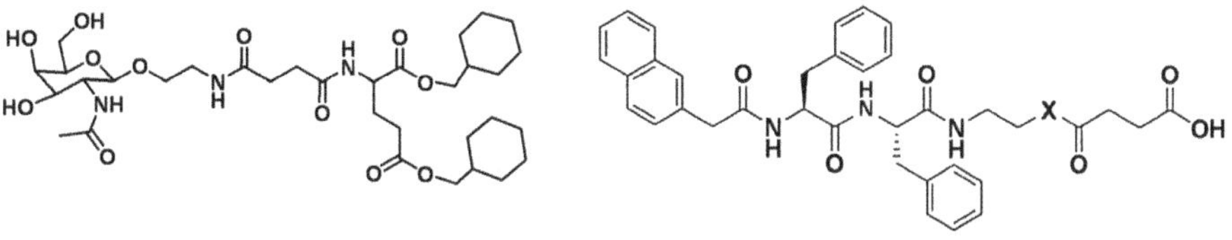

Scheme 11.6 Chemical structure of compounds **18** and **19** (X = O).

conditions. The immobilization of proteins in supramolecular hydrogels has been shown to be a smart solution for this problem. In one example hydrogels were prepared with a gelator that contained a succinic acid diamide core with glutamate diester and aminosaccharide terminal groups (**18**). This gelator formed a supramolecular hydrogel after preparing a homogeneous solution by gentle heating and posterior resting to room temperature for several minutes. For example, a supramolecular hydrogel-based peptide chip was prepared in this way. The objective of the chip was to assess rapidly which enzymes are capable of hydrolysing a given peptide. To detect peptide hydrolysis a fluorescent tag was attached to the peptide which only after peptide hydrolysis interacts with an added fluorescence resonance energy transfer (FRET) acceptor. In essence, the system was designed to emit bright green fluorescence only if and when peptide hydrolysis is taking place. A chip was prepared for the analysis of five proteins: leptin, bovine serum albumin, chymotrypsin, V8 protease and concanavalin A, which were placed in separated wells which in all the cases contained the target peptide. Remarkably, this analysis revealed that fluorescence was observed only in the wells containing leptin, showing that the studied peptide is sensitive only to this enzyme. It has to be noted that in this chip the hydrogelator is not a mere scaffold but plays an active role in capturing the fluorescent molecules into the hydrophobic part of the fibers. The spatial proximity of the fiber-adsorbed fluorophores gives rise to the observed FRET phenomena.[77]

In an outstanding study, a hydrogel was formed intracellularly regulating in this way cell death.[78] The hydrogelator used was based on the dipeptide Phe-Phe containing a 2-naphthaleneacetyl group at the N-terminus and a succinic acid monoester moiety at the C-terminus (**19**, Scheme 11.6). In this system hydrogelation is triggered in the presence of an esterase which hydrolyzes the ester bond formed by the succinic acid unit. The water soluble precursor of the hydrogelator was incubated for 3 days in a culture with HeLa cells. It was observed that the HeLa cells incorporated the prohydrogelator, which contained the fluorescent naphthyl unit. Remarkably, the researchers were able to demonstrate that hydrogelation took place inside the cells as a result of endogenous esterases, causing cell death. This result indicates that self-assembly may be used to amplify differences in the expression of enzymes such as esterases in different cell types (Figure 11.8).

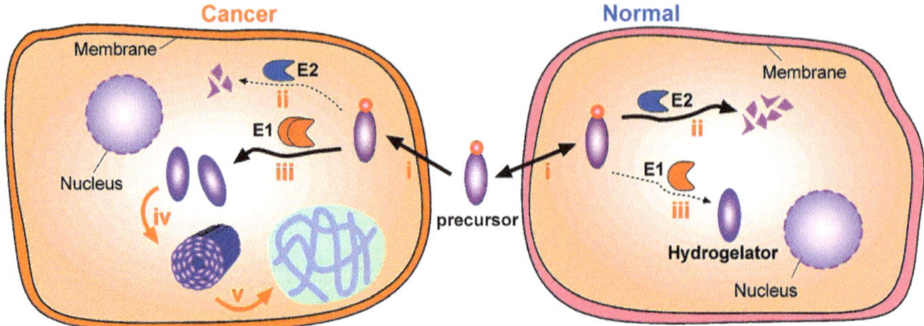

Figure 11.8   A possible mechanism for the formation of intracellular nanofibers leading to hydrogelation and cell death (E1 = phosphatase; E2 = other enzymes): (i) the precursor (**19**, see Scheme 11.6) is taken up by the cells; (ii) some molecules of **19** degrade, catalyzed by other enzymes; (iii) some molecules of **19** are converted into the hydrogelator; (iv) the hydrogelator self-assembles into nanofibers; (v) the nanofibers form 3-D networks and result in a gel, which induces inhibition of cell growth. (Reprinted with permission from Yand *et al.*[79] Copyright 2008 American Chemical Society.)

11.3   Final Remarks

The use of supramolecular gels in a biomedical context is quite recent, but the results obtained up to now, such as those described in this chapter, indicate much potential for this type of material. One key property of these gels is probably the intrinsic reversibility which is associated to the supramolecular nature of their interactions. One may assume that in the future the development of smart, stimulus-responsive gels could have a big impact. It seems feasible to design supramolecular gelators that respond *in vivo* to specific stimuli, which could include biologically relevant species such as ionic or neutral metabolites. In addition, the possibility of inducing gel formation inside cells, as shown in one of the works cited above, seems to open a whole world of possibilities for the modification of cellular processes. In the area of tissue engineering, the complete control of gelator composition as compared to the polymeric counterparts can be achieved by the tailoring of gelators with the desired properties related, for example, to cell adhesion and growth stimulation. Importantly, it could be said that a challenge for the development of this field is to gain a deeper understanding of the structure–property relation in supramolecular gelation, namely to predict, at least to a certain degree, properties such as thermal stability, stimulus responsiveness or the rheological profile, which are very relevant for the applications of these soft materials in biomedicine.

Finally, the authors want to point out that this chapter is not intended to represent a comprehensive account of the work done so far with supramolecular gels in biomedicine, which is not possible due to space limitations.

A subjective selection of examples has been made to acquaint the reader with this area of research. Necessarily, some nice and relevant reports are not mentioned, but the reader is urged to consult the various available reviews for further examples.[33,80–89]

References

1. K. Nishinari, *Prog. Colloid Polym. Sci.*, 2009, **136**, 87.
2. P. Terech and R. G. Weiss, *Chem. Rev.*, 1997, **97**, 3133.
3. J. H. van Esch and B. L. Feringa, *Angew. Chem.*, 2000, **39**, 2263.
4. D. J. Abdallah and R. G. Weiss, *Adv. Mater.*, 2000, **12**, 1237.
5. O. Gronwald, E. Snip and S. Shinkai, *Curr. Opin. Colloid In.*, 2002, **7**, 148.
6. L. A. Estroff and A. D. Hamilton, *Chem. Rev.*, 2004, **104**, 1201.
7. M. De Loos, B. L. Feringa and J. H. van Esch, *Eur. J. Org. Chem.*, 2005, 3615.
8. S. Banerjee, R. K. Das and U. Maitra, *J. Mater. Chem.*, 2009, **19**, 6649.
9. J. W. Steed, *Chem. Commun.*, 2011, **47**, 1379.
10. A. Noro, M. Hayashi and Y. Matsushita, *Soft Matter*, 2012, **8**, 6416.
11. V. J. Nebot, J. Armengol, J. Smets, S. F. Prieto, B. Escuder and J. F. Miravet, *Chem. Eur. J.*, 2012, **18**, 4063.
12. A. Ballabh, T. K. Adalder and P. Dastidar, *Cryst. Growth Des.*, 2008, **8**, 4144.
13. B. Escuder, S. Martí and J. F. Miravet, *Langmuir*, 2005, **21**, 6776.
14. P. Dastidar, *Chem. Soc. Rev.*, 2008, **37**, 2699.
15. J. H. van Esch, *Langmuir*, 2009, **25**, 8392.
16. D. S. Tsekova, J. A. Sáez, B. Escuder and J. F. Miravet, *Soft Matter*, 2009, **5**, 3727.
17. S. van der Laan, B. L. Feringa, R. M. Kellogg and J. van Esch, *Langmuir*, 2002, **18**, 7136.
18. M. Sayar and S. I. Stupp, *Phys. Rev. E*, 2005, **72**, article number: 011803.
19. X. Yang, G. Zhang and D. Zhang, *J. Mat. Chem.*, 2012, **22**, 38.
20. K. Murata, M. Aoki, T. Suzuki, T. Harada, H. Kawabata, T. Komori, F. Ohseto, K. Ueda and S. Shinkai, *J. Am. Chem. Soc.*, 1994, **116**, 6664.
21. J. F. Miravet and B. Escuder, *Chem. Commun.*, 2005, **41**, 5796.
22. S.-I. Kawano, N. Fujita and S. Shinkai, *J. Am. Chem. Soc.*, 2004, **126**, 8592.
23. M. O. M. Piepenbrock, G. O. Lloyd, N. Clarke and J. W. Steed, *Chem. Rev.*, 2010, **110**, 1960.
24. Y. Gao, Z. Yang, Y. Kuang, M. L. Ma, J. Li, F. Zhao and B. Xu, *Biopolymers*, 2010, **94**, 19.
25. M. D. Segarra-Maset, V. J. Nebot, J. F. Miravet and B. Escuder, *Chem. Soc. Rev.*, 2013, DOI: 10.1039/C2CS35436E.
26. A. Ajayaghosh and V. K. Praveen, *Acc. Chem. Res.*, 2007, **40**, 644.
27. A. Ajayaghosh, V. K. Praveen and C. Vijayakumar, *Chem. Soc. Rev.*, 2008, **37**, 109.

28. B. Escuder, F. Rodríguez-Llansola and J. F. Miravet, *New J. Chem.*, 2011, **34**, 1044.
29. J. H. Jung, K. Yoshida and T. Shimizu, *Langmuir*, 2002, **18**, 8724.
30. O. Wichterle and D. Lim, *Nature*, 1960, **185**, 117.
31. B. Balakrishnan and R. Banerjee, *Chem. Rev.*, 2011, **111**, 4453.
32. N. A. Peppas, P. Bures, W. Leobandung and H. Ichikawa, *Eur. J. Pharm. Biopharm.*, 2000, **50**, 27.
33. T. Vermonden, R. Censi and W. E. Hennink, *Chem. Rev.*, 2012, **112**, 2853.
34. J. L. West and J. A. Hubbell, *Proc. Natl Acad. Sci. USA*, 1996, **93**, 13188.
35. A. S. Hoffman, *Adv. Drug Delivery. Rev.*, 2012, **64**, 18.
36. M. Suzuki and K. Hanabusa, *Chem. Soc. Rev.*, 2010, **39**, 455.
37. J. M. Lehn, *Science*, 2002, **295**, 2400.
38. J. Puigmartí-Luis, V. Laukhin, Á. Pérez del Pino, J. Vidal-Gancedo, C. Rovira, E. Laukhina and D. B. Amabilino, *Angew. Chem. Int. Ed.*, 2007, **46**, 238.
39. K. Sugiyasu, N. Fujita and S. Shinkai, *Angew. Chem. Int. Ed.*, 2004, **43**, 1229.
40. F. Rodríguez-Llansola, B. Escuder and J. F. Miravet, *J. Am. Chem. Soc.*, 2009, **131**, 11478.
41. B. Xing, C. W. Yu, K. H. Chow, P. L. Ho, D. Fu and B. Xu, *J. Am. Chem. Soc.*, 2002, **124**, 14846.
42. B. Xing, P. L. Ho, C. W. Yu, K. H. Chow, H. Gu and B. Xu, *Chem. Commun.*, 2003, 2224.
43. G. A. Silva, C. Czeisler, K. L. Niece, E. Beniash, D. A. Harrington, J. A. Kessler and S. I. Stupp, *Science*, 2004, **303**, 1352.
44. V. M. Tysseling, V. Sahni, E. T. Pashuck, D. Birch, A. Hebert, C. Czeisler, S. I. Stupp and J. A. Kessler, *J. Neurosci. Res.*, 2010, **88**, 3161.
45. L. C. Palmer, Y. S. Velichko, M. Olvera de la Cruz and S. I. Stupp, *Phil. Trans. R. Soc. A*, 2007, **365**, 1417.
46. R. Langer and J. P. Vacanti, *Science*, 1993, **260**, 920.
47. Y. Li, J. Rodrigues and H. Tomas, *Chem. Soc. Rev.*, 2012, **41**, 2193.
48. Y. Zhang, H. Gu, Z. Yang and B. Xu, *J. Am. Chem. Soc.*, 2003, **125**, 13680.
49. V. Jayawarna, M. Ali, T. A. Jowitt, A. F. Miller, A. Saiani, J. E. Gough and R. V. Ulijn, *Adv. Mater.*, 2006, **18**, 611.
50. A. M. Smith, R. J. Williams, C. Tang, P. Coppo, R. F. Collins, M. L. Turner, A. Saiani and R. V. Ulijn, *Adv. Mater.*, 2008, **20**, 37.
51. C. Tang, A. M. Smith, R. F. Collins, R. V. Ulijn and A. Saiani, *Langmuir*, 2009, **25**, 9447.
52. L. Chen, S. Revel, K. Morris, L. C. Serpell and D. J. Adams, *Langmuir*, 2012, **26**, 13466.
53. L. Chen, K. Morris, A. Laybourn, D. Elias, M. R. Hicks, A. Rodger, L. Serpell and D. J. Adams, *Langmuir*, 2010, **26**, 5232.
54. A. Mahler, M. Reches, M. Rechter, S. Cohen and E. Gazit, *Adv. Mater.*, 2006, **18**, 1365.
55. T. Liebmann, S. Rydholm, V. Akpe and H. Brismar, *BMC Biotechnol.*, 2007, **7**, article number: 88.

56. M. Zhou, A. M. Smith, A. K. Das, N. W. Hodson, R. F. Collins, R. V. Ulijn and J. E. Gough, *Biomaterials*, 2009, **30**, 2523.
57. L. Haines-Butterick, K. Rajagopal, M. Branco, D. Salick, R. Rughani, M. Pilarz, M. S. Lamm, D. J. Pochan and J. P. Schneider, *Proc. Natl Acad. Sci. USA*, 2007, **104**, 7791.
58. H. Komatsu, S. Tsukiji, M. Ikeda and I. Hamachi, *Chem. Asian J.*, 2011, **6**, 2368.
59. C. Valery, M. Paternostre, B. Robert, T. Gulik-Krzywicki, T. Narayanan, J.-C. Dedieu, G. R. Keller, M.-L. Torres, R. Cherif-Cheikh, P. Calvo and F. Artzner, *Proc. Natl Acad. Sci. USA*, 2003, **100**, 10258.
60. J. S. Bevan, J. Newell-Price, J. A. H. Wass, S. L. Atkin, P. M. Bouloux, J. Chapman, J. R. E. Davis, T. A. Howlett, H. S. Randeva, P. M. Stewart and A. Viswanath, *Clin. Endocrinol.*, 2008, **68**, 343.
61. S. Koutsopoulos, L. D. Unsworth, Y. Nagai and S. Zhang, *Proc. Natl Acad. Sci. USA*, 2009, **106**, 4623.
62. J. L. Drury and D. J. Mooney, *Biomaterials*, 2003, **24**, 4337.
63. Q. Hou, P. A. De Bank and K. M. Shakesheff, *J. Mater. Chem.*, 2004, **14**, 1915.
64. A. Nishimura, T. Hayakawa, Y. Yamamoto, M. Hamori, K. Tabata, K. Seto and N. Shibata, *Eur. J. Pharm. Sci.*, 2012, **45**, 1.
65. J. Nanda and A. Banerjee, *Soft Matter*, 2012, **8**, 3380.
66. J. A. Sáez, B. Escuder and J. F. Miravet, *Chem. Commun.*, 2010, **46**, 7996.
67. P. K. Vemula, G. A. Cruikshank, J. M. Karp and G. John, *Biomaterials*, 2009, **30**, 383.
68. K. J. C. Van Bommel, M. C. A. Stuart, B. L. Feringa and J. van Esch, *Org. Biomol. Chem.*, 2005, **3**, 2917.
69. J. Boekhoven, M. Koot, T. A. Wezendonk, R. Eelkema and J. H. van Esch, *J. Am. Chem. Soc.*, 2012, **134**, 12908.
70. Y. Gao, Y. Kuang, Z. F. Guo, Z. Guo, I. J. Krauss and B. Xu, *J. Am. Chem. Soc.*, 2009, **131**, 13576.
71. L. Mao, H. Wang, M. Tan, L. Ou, D. Kong and Z. Yang, *Chem. Commun.*, 2012, **48**, 395.
72. H. Chen, D. Mou, D. Du, X. Chang, D. Zhu, J. Liu, H. Xu and X. Yang, *Int. J. Pharm.*, 2007, **341**, 78.
73. T. Sreedevi, D. Ramya Devi and B. N. Vedha Hari, *Int. J. Drug Dev. Res.*, 2012, **4**, 35.
74. T. Penzes, G. Blazso, Z. Aigner, G. Falkay and I. Eros, *Int. J. Pharm.*, 2005, **298**, 47.
75. L. Kang, X. Y. Liu, P. D. Sawant, P. C. Ho, Y. W. Chan and S. Y. Chan, *J. Control. Release*, 2005, **106**, 88.
76. A. Vintiloiu, M. Lafleur, G. Bastiat and J. C. Leroux, *Pharm. Res.*, 2008, **25**, 845.
77. M. Ikeda, R. Ochi and I. Hamachi, *Lab Chip*, 2010, **10**, 3325.
78. Z. Yang, K. Xu, Z. Guo, Z. Guo and B. Xu, *Adv. Mater.*, 2007, **19**, 3152.
79. Z. Yang, G. Liang and B. Xu, *Acc. Chem. Res.*, 2008, **41**, 315.
80. A. Vintiloiu and J. C. Leroux, *J. Control. Release*, 2008, **125**, 179.

81. F. Zhao, M. L. Ma and B. Xu, *Chem. Soc. Rev.*, 2009, **38**, 883.
82. M. C. Branco and J. P. Schneider, *Acta Biomater.*, 2009, **5**, 817.
83. A. M. Jonker, D. W. P. M. Lowik and J. C. M. Van Hest, *Chem. Mater.*, 2012, **24**, 759.
84. W. T. Truong, Y. Su, J. T. Meijer, P. Thordarson and F. Braet, *Chem. Asian J.*, 2011, **6**, 30.
85. J. B. Matson and S. I. Stupp, *Chem. Commun.*, 2012, **48**, 26.
86. D. M. Ryan and B. L. Nilsson, *Polym. Chem.*, 2012, **3**, 18.
87. A. S. Hoffman, *Adv. Drug Delivery. Rev.*, 2012, **64**, 18.
88. Y. Loo, S. Zhang and C. A. E. Hauser, *Biotechnol. Adv.*, 2012, **30**, 593.
89. M. Guvendiren, H. D. Lu and J. A. Burdick, *Soft Matter*, 2012, **8**, 260.

Supramolecular Enzyme Assays

ANDREAS HENNIG

BAM Federal Institute for Materials Research and Testing,
Richard-Willstätter-Str. 11, D-12489 Berlin, Germany
Email: andreas.hennig@bam.de; andreas-hennig@web.de

12.1 Introduction

Enzymes mediate a wide array of different functions in physiological processes: in digestion, blood coagulation, and fertilization; and in cell growth, death, and differentiation, to give just a few examples. Defects in the enzymes of metabolic pathways are involved in numerous major diseases including parasitic infections, arthritis, Alzheimer's disease, inflammation, osteoporosis, arteriosclerosis, and cancer.[1–7,8–14] Furthermore, enzymes serve as sustainable reagents for "green" organic synthesis in water and are thus an integral part of the development of "white" biotechnology.[15] The detection of enzyme activity by enzyme assays is therefore of utmost importance in academia and in industrial research,[16] for example in kinetic and mechanistic investigations of various substrates, co-factors, inhibitors and activators,[17] as well as for screening libraries of genetically engineered enzymes,[18–21] and potential drug lead structures in pharmaceutical–industrial high-throughput screening (HTS).[9]

A large variety of enzyme assays has therefore been developed, including indirect methods based on the measurement of physicochemical parameters such as infrared thermography,[22,23] infrared and Raman imaging,[24] calorimetry,[25,26] and direct detection methods such as high performance liquid chromatography (HPLC) and mass spectrometry (MS). Among the available methods, fluorescence-based assays stand out due to their high sensitivity, convenient access, short detection times, and their suitability for continuous

Monographs in Supramolecular Chemistry No. 13
Supramolecular Systems in Biomedical Fields
Edited by Hans-Jörg Schneider

Published by the Royal Society of Chemistry, www.rsc.org

monitoring.[9,16] In particular, for screening applications, the widespread availability of efficient and affordable enzyme assays with a rapid and automatable readout is required to evaluate the potency of libraries of new enzyme inhibitors or catalysts. Consequently, fluorescence-based assays represent the largest category of enzyme assays, and fluorescence microplate readers are an integral part of automated HTS systems in the pharmaceutical industry.

In this context, the most desirable features for enzyme assays have been summarized, which call for them to be (1) non-radioactive, (2) affordable, (3) label-free, (4) antibody-free, and (5) generic (*i.e.* applicable to many enzymes).[27] With conventional fluorescence-based assays, concerns related to radioactivity are irrelevant, but the remaining issues are challenging to address. For example, most fluorescent enzyme assays are based on synthetically modified fluorogenic substrates or rely on specific antibody–antigen interactions.

The aim of this contribution is to introduce and comprehensively review supramolecular approaches to enzyme assays based on chemosensors (Section 12.2), membrane transport systems (Section 12.3), and macrocyclic host–guest systems (Section 12.4). So far, relatively little effort has been put into the development of supramolecular approaches to enzyme assays, and this overview will demonstrate that supramolecular approaches are particularly well-suited to address the demands of screening applications on enzyme assays. In fact, the largest proportion of supramolecular enzyme assays are based on fluorescence readouts, rely on commercially available substances, are label- and antibody-free, and can be applied to many different enzymes.

Issues that may have hampered the development of supramolecular approaches to enzyme assays in the past are becoming less and less relevant with current advances. For example, supramolecular chemistry in water has already enabled numerous applications of supramolecular chemistry to biology,[28–34] and the continuous monitoring of enzymatic reactions in absence of organic solvents, which would interfere with the enzyme reaction, has been demonstrated (see *e.g.* Section 12.4). Furthermore, a high sensitivity, which is a common goal in the design of supramolecular receptors, is not necessarily a particularly relevant issue in the development of *in vitro* enzyme assays. Actually, it is sometimes even desirable that the assay can be performed at high substrate concentrations: for example, if the goal is to identify inhibitors of low-affinity enzymes; or is to identify enzymes for white biotechnology, in which a high turnover rate at high substrate concentrations is required for rapid conversion into the product.[35,36] Also, a particularly high selectivity is not necessary for all *in vitro* enzyme assays. For example, buffer ingredients may bind to supramolecular receptors to a certain extent (*e.g.* Na^+ to *p*-sulfonatocalixarenes), but the buffer composition is generally well-defined and does not change during the reaction, such that the differentiation between substrate and product of an enzymatic reaction can still be achieved (albeit the binding constants are reduced). Actually, the relatively low selectivity of supramolecular receptors compared to antibodies and proteins enables the application of a particular supramolecular enzyme assay to many different enzymes, which affords sensor arrays for the detection of a variety of related analytes (see Section 12.5.3).

12.2 Chemosensors

Early on, sporadic examples of supramolecular enzyme assays were mainly based on chemosensors, specifically designed for recognition of the substrate or product (Figure 12.1). Therein, the focus was receptor design, and detection of enzyme activity was mainly demonstrated as an interesting application. This means that the enzymatic reaction was not comprehensively characterized and the full potential of the supramolecular enzyme assay was thus not further explored. For example, Czarnik developed the pyrophosphate sensor **1** and demonstrated that it could be used for detection of pyrophosphatase activity.[37] Kikuchi introduced the Cd^{2+}-cyclen-based sensor **2**, which binds nucleoside mono-, di-, and triphosphates, but not the cyclic analogues of nucleoside

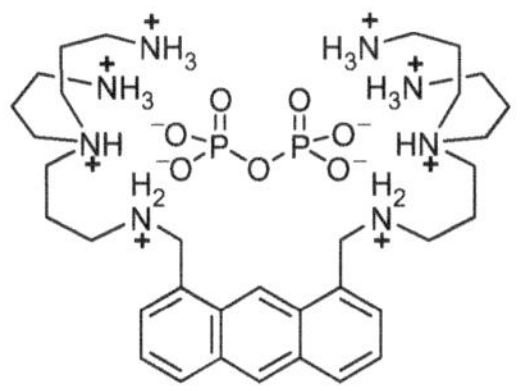

1
Czarnik
(1994)

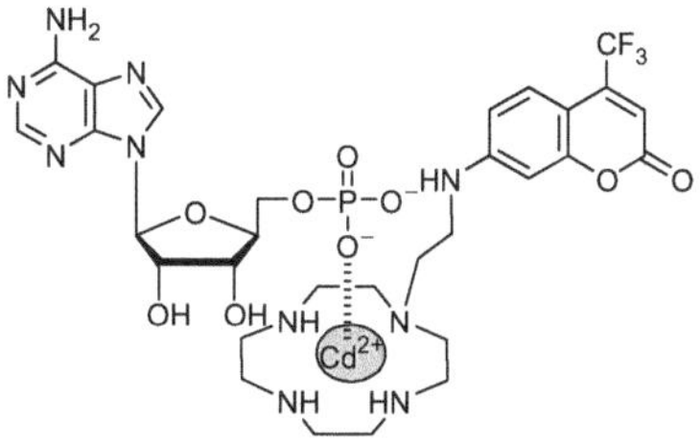

2
Kikuchi
(2002)

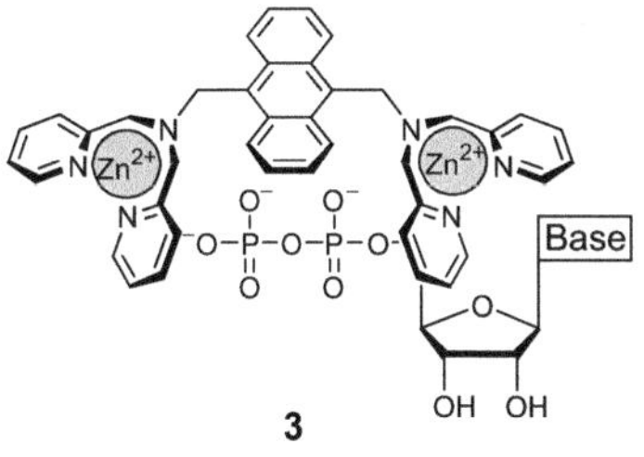

3
Hamachi
(2006)

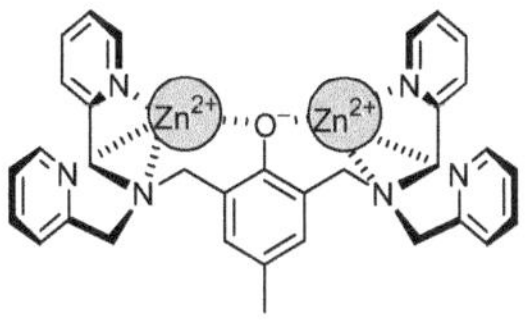

4
Kim
(2003)

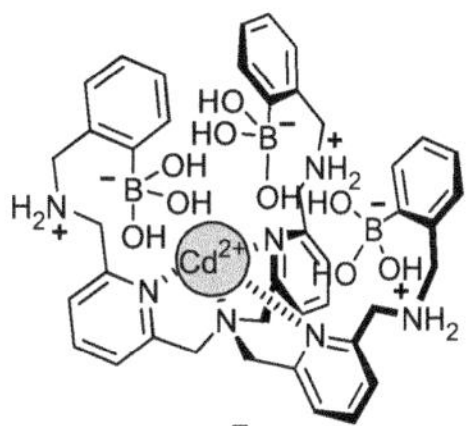

5
Anslyn
(2007)

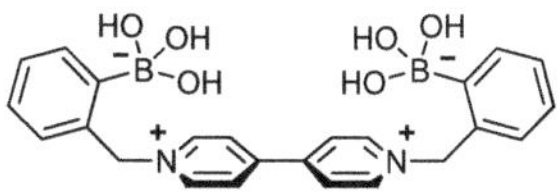

6
Schiller & Singaram
(2009)

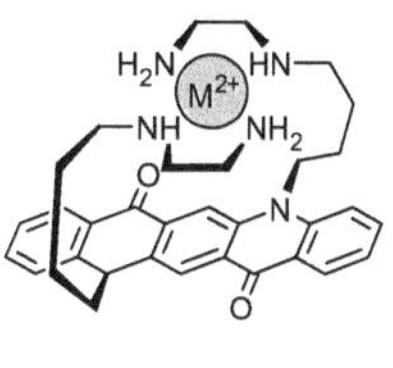

7
Reymond
(2001)

Figure 12.1 Chemosensors used in supramolecular enzyme assays.

monophosphates, such that **2** could signal the conversion of cyclic AMP (cAMP) to AMP by phosphodiesterase.[38]

The first examples of supramolecular enzyme assays, in which the enzymatic reaction was more comprehensively investigated, were reported by Kim[39] and by Hamachi.[40] Hamachi developed an assay for glycosyltransferases, which transfer a glycosyl residue from a glycosylated nucleotide (the glycosyl donor) to a glycosyl acceptor—for example, other carbohydrates, lipids, or proteins. The result of the reaction is the glycosylated product and a nucleotide, which binds to receptor **3,** thereby leading to an increase in fluorescence intensity. The wide applicability of **3** for assaying glycosyltransferases was demonstrated by β-1,4-galactosyltransferase (β-1,4-GalT), α-1,3-galactosyltransferase (α-1,3-GalT), and α-2,3-sialyltransferase. A substrate screening including chitobiose, *N*-acetylglucosamine, glucose, *N*-acetyllactosamine, lactose, and galactose was performed for β-1,4-GalT and α-1,3-GalT, which reproduced literature-known trends in substrate affinity and demonstrated the assay's potential for elucidation of enzyme kinetic parameters. Therein, uridine 5′-diphospho-galactose (UDP-Gal) served as the glycosyl donor, which was expanded to cytidine 5′-monophospho-*N*-acetylneuraminic acid (CMP-NeuAc) as a glycosyl donor for α-2,3-sialyltransferase. In addition, a small library of twelve inhibitors was screened for their activity to inhibit β-1,4-GalT.

This early example nicely demonstrates that supramolecular enzyme assays may in principle address all desires of scientists involved in high-throughput screening of enzymes; that is, assays based on **3** are extremely low-priced compared to antibody-based assays, work with unlabeled substrates, afford a fluorescent read-out and are applicable to many related, but different, enzymes.

Kim applied receptor **4** to the detection of the conversion of cAMP to AMP by phosphodiesterase and determined enzyme kinetics and inhibition constants.[39] The sensing mechanism in **4** is worth mentioning, because it involved the indicator displacement strategy, initially used to detect anion binding to α-cyclodextrin[41] and more recently popularized by Anslyn,[42,43] which will be referred to in more detail in Section 12.4. Herein, pyrocatechol violet (PV, **13,** see Figure 12.2 for the structure) is used as the indicator dye, which is non-fluorescent and therefore not compatible with conventional pharmaceutical–industrial HTS. However, the combination of PV and **4** still affords a colorimetric enzyme assay, which would be in principle compatible with HTS when absorption-based microplate readers are used.

Alternatively, PV could be replaced by a suitable fluorescent dye as demonstrated in supramolecular tandem assays (Section 12.4). Nonetheless, binding of PV to **4** leads to the appearance of a low energy absorption band around 600–650 nm, which is typical for metal ion complexes of this established colorimetric metal ion indicator. AMP, either added directly or generated *in situ* by phosphodiesterase, displaces PV from **4** and thereby regenerates the absorption spectrum of free PV in solution, such that the activity of phosphodiesterase could be conveniently monitored by this sensing ensemble.

PV was also applied as the indicator dye in the glucuronic acid receptor **5**.[44] Glucuronic acid is the product of the enzymatic reaction of glucose with glucose oxidase, which has allowed determination of the amount of glucose in blood serum. The main goal was thus, again, the colorimetric sensing of glucose. The combination of a highly selective signal generation by an enzyme (the oxidation of glucose) with a more general signal transduction (the displacement of PV from **5** and the concomitant change in its absorption spectrum) gave the desired highly selective suprabiomolecular sensing ensemble with a simple read-out. More examples of combining highly selective enzymes and a readout scheme by a supramolecular enzyme assay will be presented in Section 12.5.3.

A different combination of receptor and indicator for assaying enzymes involved in carbohydrate and oligosaccharide metabolism by a supramolecular indicator displacement-based method has recently been introduced by Schiller, Singaram and colleagues.[45] They used relatively simple bis(benzylboronic acid)-bipyridines such as **6** in combination with the fluorescent dye 8-hydroxypyrene-1,3,6-trisulfonate (HPTS, **9**, see Figure 12.2 for the structure). The strongly negatively charged HPTS binds to **6** *via* electrostatic interactions and thereby reduces the fluorescence. Addition of fructose or generation of fructose by sucrose phosphorylase leads to the displacement of HPTS from **6** and can thus be signaled as an increase in fluorescence. Moreover, a related 3,3′-bipyridinium-based receptor was selective for glucose-6-phosphate over glucose-1-phosphate, which allowed monitoring of their isomerization by phosphoglucomutase.[45] Both assays were comprehensively characterized and their potential for inhibitor and activator screening was shown.

Another interesting example of an enzyme assay based on metal ion complexation, which can also be considered supramolecular, has been introduced by Reymond.[46] The quinacridone-derived indicator **7** is strongly quenched in the presence of Cu^{2+} or Ni^{2+}, most likely by formation of a macrocyclic chelate complex as shown in Figure 12.1. Enzymatic conversion of a non-chelating substrate gives a chelating product, which competes with **7** for metal ion binding, thereby leading to dissociation of the quenched metal ion complex and a regeneration of the quinacridone fluorescence. It has been proposed that the chelate effect in **7** is due to the 17-membered ring being sufficiently weak to release the metal ion in presence of other weak chelators. This principle was used to monitor the hydrolysis of *N*-acetyl-L-methionine and L-leucinamide into L-methionine and L-leucine by acylase I and leucine aminopeptidase, respectively. Solubilization of **7** required the addition 40% dimethyl sulfoxide (DMSO) or dimethyl formamide (DMF), which resulted in significantly lower enzyme reaction rates when the reaction was directly monitored compared with an assay in which aliquots from the enzymatic reaction in neat buffered water were taken and subsequently mixed with the sensor system.

12.3 Membrane Transport

About the same time when the first supramolecular approaches to enzyme assays appeared in the literature, Matile and co-workers were developing artificial membrane pores (see Section 12.3.2),[47–49] and subsequently applied them to the detection of enzyme activity.[50] This concept proved to be so versatile that it afforded sensor arrays for biosensing,[51,52] and was expanded to membrane-active polymers as a synthetically more accessible substitute for the sophisticated membrane pores.[53] The following section summarizes these developments by first familiarizing the reader with assays that are used to characterize membrane transport systems and the general principles of sensing with membrane transporters (Section 12.3.1). Then, enzyme assays with artificial membrane pores (see Section 12.3.2) and membrane-active polymers (see Section 12.3.3) are introduced, whereas more advanced concepts and applications of artificial membrane pores and membrane-active polymers will be addressed separately in Section 12.5.

12.3.1   Transport Assays and Sensing Concepts

12.3.1.1   *Characterization of Transport Activity*

To visualize the activity of membrane transporters such as membrane pores, ion channels, carriers or membrane-active polymers, fluorescence-based assays have proven most versatile, because they are highly sensitive and straightforward to implement, which allows rapid screening of transport activities under varying conditions. Examples of alternative methods include colorimetric assays,[54] assays based on circular dichroism,[55] or electrical conductance measurements in black lipid membranes.[56] Noteworthy is that the latter has been extensively investigated in the context of biosensing (*e.g.* for oligonucleotide sequencing; see Chapter 5, Section 12.3) because of its sensitivity down to the single-molecule level, but still has to be exploited for enzyme assays. Detailed guidelines for comprehensive characterization of a membrane transporter's activity can be found in the literature,[56,57] such that herein the focus will be on those membrane transport assays that have so far been used in the context of enzyme assays and sensing.

Most relevant in the context of enzyme assays, fluorescence-based membrane transport assays are set up by preparing phospholipid vesicles, in which the interior of the vesicle is labeled with a fluorescent dye (Figure 12.2a). The entrapped dye can be either responsive to a molecule transported into the vesicle, or the dye itself can be transported out of the vesicle and thereby respond with a change in its fluorescence spectroscopic properties. This change is recorded in a time-dependent manner during addition of varying concentrations of the membrane transporter (Figure 12.2b). In a typical series of experiments, a constant initial fluorescence, I_0, is recorded before addition of the transporter. The latter causes a change in fluorescence, most commonly an increase, which levels off with time. Addition of a detergent like Triton X-100 at the end of the reaction destroys the vesicle and gives the final fluorescence intensity, I_∞, which is used for endpoint calibration of the fluorescence traces. Normalization by subtraction of I_0 and division by $(I_\infty - I_0)$ gives the parameters I_{MAX}, I_{MIN}, and Y, that is, the maximal and minimal normalized fluorescence intensities in absence and with excess transporter and the fractional activity. A plot of Y *versus* the membrane transporter concentration c_M (Figure 12.2c) is then fitted by the Hill equation (1), which gives the minimal and maximal detectable activities Y_0 and Y_∞, the effective concentration EC_{50}, and the Hill coefficient, n, which are characteristic parameters of a specific transport system.

$$Y = Y_\infty + (Y_0 - Y_\infty) / [1 + (c_M / EC_{50})^n] \tag{1}$$

When the fluorescence kinetic traces have been normalized with I_0 and I_∞ as described above, the fractional activity varies from $Y_0 = I_{MIN}$ to $Y_\infty = I_{MAX}$. Alternatively, normalization of the fluorescence kinetic traces has also been performed with I_{MAX} and I_{MIN}, which leads to Hill plots in which the fractional activity varies from $Y_0 = 0$ to $Y_\infty = 1$.

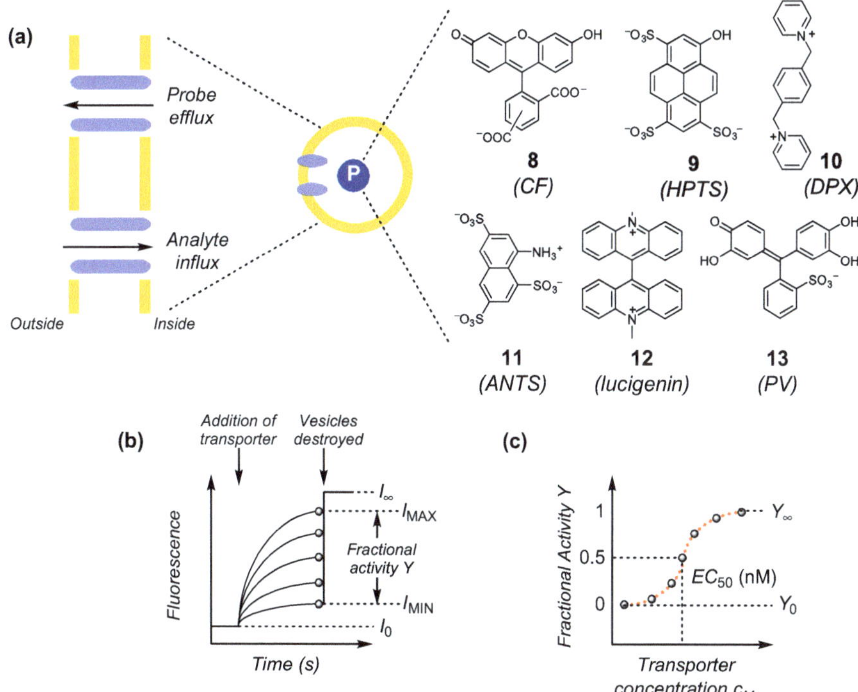

Figure 12.2 Characterization of membrane transport activity as used in supramolecular enzyme assays based on membrane transport systems. (a) General scheme for a membrane transport assay and the respective fluorescent probes and quenchers. (b) Fluorescence kinetic traces at varying concentrations of membrane transporter and parameters used for normalization of the fluorescence traces. (c) Schematic representation of the resulting dose–response curves and a Hill analysis according to eqn (1). (Adapted from Matile *et al.*[56] with permission. Copyright 2012 Wiley-VCH.)

The most often used transport assay is based on the encapsulation of carboxyfluorescein (CF, **8**) within vesicles at self-quenching concentrations. Addition of a membrane transporter or membrane pore allows the hydrophilic CF to move across the membrane into the external medium. Because the volume of the surrounding medium is much larger than the interior of the vesicle, this leads to a dilution effect, such that self-quenching no longer applies and membrane transport activity can be followed by a regeneration of CF fluorescence.

A limitation of the CF assay lies in the negative charge of CF—that is, the activity of a transporter selective for cationic molecules cannot be detected with the CF assay. In this case, a mixture of a negatively charged fluorescent dye (*e.g.* HPTS **9**, or 8-aminonaphthalene-1,3,6-trisulfonate [ANTS, **11**]) and a positively charged quencher (*p*-xylene-bis(*N*-pyridinium bromide) [DPX, **10**]) is encapsulated in the vesicle interior at sufficiently high concentrations to afford fluorescence quenching. Addition of a membrane transporter to these vesicles can either allow the dye, the quencher, or both to cross the membrane depending on the selectivity of the transporter. Regardless of which molecule has been transported in an HPTS/DPX or ANTS/DPX assay, the dilution effect leads to an increase in fluorescence, which signals the transport activity.

A third line of transport assays uses fluorescent dyes, which respond to a molecule transported into the vesicle by a change in its fluorescence spectroscopic properties. For example: HPTS is a fluorescent pH indicator with a pK_a value of *ca.* 7.3 and thus responds sensitively to proton influx around neutral pH; lucigenin **12** is quenched by chloride anions and thus responds to chloride influx with a decrease in fluorescence; and PV **13** is a colorimetric metal ion indicator and responds to influx of transition metal cations with change in its absorption spectrum. Details of these assays and on other methods for the characterization of membrane transport systems have been summarized elsewhere.[56]

12.3.1.2 *Blockage, Activation, and Reaction Monitoring with Membrane Transporters*

The activity of membrane transporters can be regulated by addition of an activator (ligand gating, opening) or a blocker (blockage, closing), both of which can represent an analyte of interest. The interaction of the analyte with the membrane transporter is characterized analogously to the determination of the transporter's activity itself (see Section 12.3.1.1), that is, by recording the time-dependent fluorescence change during addition of the membrane transporter in presence of varying concentrations of the blocker c_{BLOCKER} or activator $c_{\text{ACTIVATOR}}$ (Figure 12.3a).

Normalization with I_0 and I_∞ as described in Section 12.3.1.1 and analysis of the dependence of the fractional activity, Y, on the analyte concentration by the Hill equation gives the minimal and maximal detectable activities, Y_0 and Y_∞, and the Hill coefficient, n. Depending on whether the analyte activates or

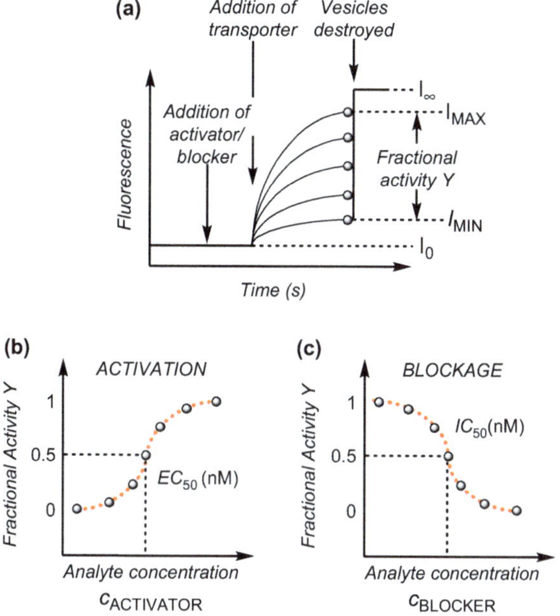

Figure 12.3 Characterization of activation and blockage (inhibition) of membrane transport activity. (a) Fluorescence kinetic traces at varying activator or blocker concentrations and parameters used for normalization of the fluorescence traces. (b,c) Schematic representation of the resulting dose–response curves and a Hill analysis for activation (b) according to eqn (3) and blockage (c) according to eqn (2)
(Adapted from Matile *et al.*[56] with permission. Copyright 2012 Wiley-VCH.)

inhibits the transport activity, Hill analysis gives the effective concentration EC_{50} (Figure 12.3b, eqn 2) or inhibitory concentration IC_{50} (Figure 12.3c, eqn 3), which refer in most cases to the dissociation constant of the pore/analyte complex (see the discrimination of IP_6 and IP_7 at the end of Section 12.3.2 for an exceptional example).

$$Y = Y_\infty + (Y_0 - Y_\infty)/[1 + (c_{BLOCKER}/IC_{50})^n] \qquad (2)$$

$$Y = Y_\infty + (Y_0 - Y_\infty)/[1 + (c_{ACTIVATOR}/EC_{50})^n] \qquad (3)$$

This activation and blockage of membrane transport forms the basis for monitoring enzyme-catalyzed reactions with membrane transporter systems, and a schematic example is shown in Figure 12.4a. Therein, the substrate of the enzymatic reaction is an efficient blocker of transport activity, and the product does not inactivate membrane transport. In practice, the enzyme assay is carried out such that the enzyme reaction mixture is incubated under appropriate conditions and aliquots are taken at varying time intervals. These aliquots are added to vesicles containing a fluorescent probe and incubated. Then, the membrane transporter is added and the change in fluorescence intensity during

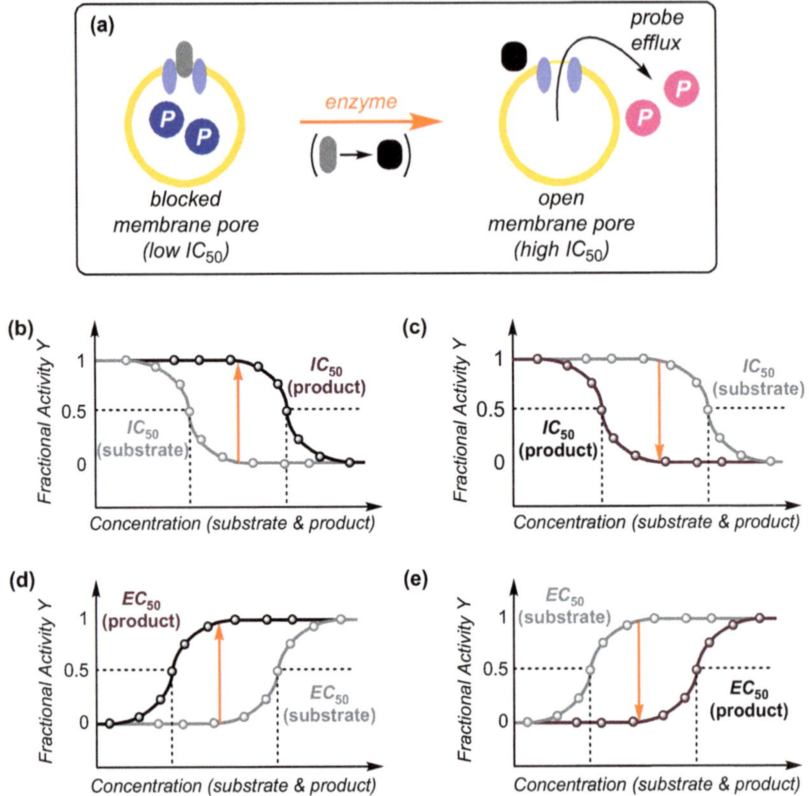

Figure 12.4   Principle of enzyme reaction monitoring by membrane transporters. (a) Membrane transporters are added to a solution containing fluorescently labeled vesicles and a pore blocker. Addition of an enzyme converts the good blocker into a poor blocker such that enzyme activity is signaled by an increasing transport activity. (b–e) Dependence of the resulting response on whether transport blockers (b, c) or activators (d, e) are used and on whether the substrate (b, e) or product (c, d) interacts more efficiently with the membrane transporter.
(Adapted from Matile *et al.*[56] with permission. Copyright 2012 Wiley-VCH.)

addition of the membrane transport system is followed with the fluorometric transport assays described in Section 12.3.1.1. As such, enzyme assays with membrane transporters report on the changing ability of the reaction mixture to activate or inactivate the membrane transport system.

The prerequisite for setting up an enzyme assay based on membrane transporters is that substrate and product of the enzymatic reaction activate or inactivate membrane transport to different extent, or in other words the EC_{50} and IC_{50} values of substrate and product must be sufficiently different. Depending on whether membrane transport is blocked or activated, and on whether substrate or product is more efficient, four different situations result

(Figures 12.4b–e). In detail, the substrate can be a more efficient blocker (Figure 12.4b) or a less efficient activator (Figure 12.4d) than the product; in both cases enzyme activity would lead to an increase in transport activity. Alternatively, the product can be a more efficient blocker (Figure 12.4c) or a less efficient activator (Figure 12.4e) than the substrate, such that enzyme activity leads to a decrease in transport activity. In all cases, determination of the respective EC_{50} or IC_{50} values allows assessment of the performance of the membrane transport-based enzyme assay. It has been concluded from an exploratory study that a factor of approximately three between the EC_{50} or IC_{50} values of substrate and product suffices to qualitatively monitor a reaction with appreciable accuracy at a fixed concentration.[58] However, a larger difference leads to a higher accuracy and allows a more flexible choice of substrate or product concentrations.

12.3.2 Synthetic Multifunctional Pores

The first demonstration of a membrane transport-based enzyme assay utilized artificial membrane pores designed by Matile and co-workers (Figure 12.5).[50] These pores are based on a *p*-octiphenyl scaffold, which matches the length of a lipid bilayer membrane. Eight side arms can be attached to the *p*-octiphenyl scaffold, and in the context of enzyme assays and molecular recognition, short β-sheet-forming peptides with $(2i+1)$ amino acids with alternating hydrophobic and hydrophilic side chains have proven most versatile.

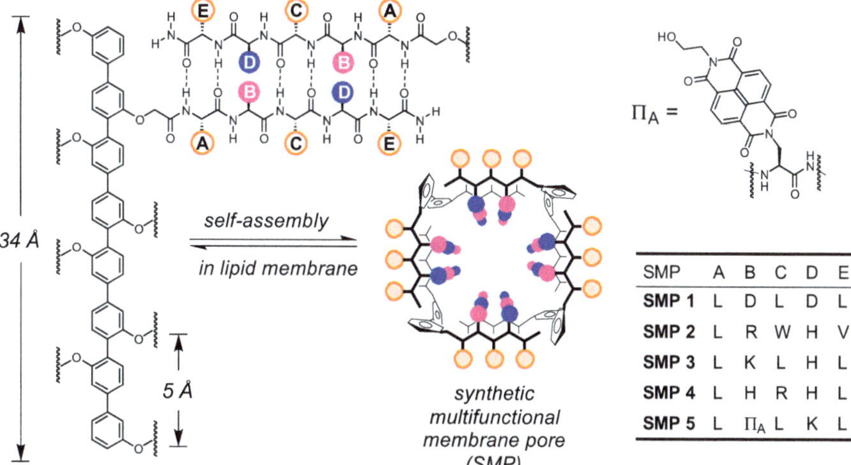

SMP	A	B	C	D	E
SMP 1	L	D	L	D	L
SMP 2	L	R	W	H	V
SMP 3	L	K	L	H	L
SMP 4	L	H	R	H	L
SMP 5	L	Π_A	L	K	L

Figure 12.5 Structural model for synthetic multifunctional pores (SMPs) based on *p*-octiphenyl staves and β-sheet-forming peptide side arms. Incorporation into a lipid bilayer membrane leads to their self-assembly. The peptide sequences used in the context of enzyme assays are shown in the table. D = L-aspartic acid, H = L-histidine, L = L-leucine, R = L-arginine, V = L-valine, W = L-tryptophan, and the structure of the unnatural amino acid π_A is shown.

Incorporation of these molecules into lipid bilayer membranes leads to the self-assembly of membrane pores with vastly different functions depending on the peptide sequence, such that the name "synthetic multifunctional pore" (SMP) was coined. The most conclusive structural model of SMPs is a tetrameric, cylindrical barrel-stave supramolecule,[47] in which the *p*-octiphenyl staves are oriented perpendicular to the lipid membrane surface, and in which the peptide side arms interdigitate with chains from another stave to form short antiparallel β-sheets. Thereby, the side chains of the hydrophobic amino acid, $i, i+2, \ldots$, are placed at the outer barrel surface in contact with the lipid membrane, whereas hydrophilic residues, $i+1, i+3, \ldots$, form the pore interior. These pores are in the open, active form when they are incorporated into the membrane, but can be closed or blocked by addition of a molecule that binds to the pore interior.

Formation of this suprastructure is enabled by the torsion mode of the phenyl groups, which enforce a non-planar arrangement of the side arms, and by the distance between adjacent side arms, which matches the distance between peptide strands in β-sheets in proteins. Overall, the resulting membrane pores are reminiscent of natural β-barrels, the most common structural motif of naturally occurring membrane proteins, such that SMPs, have also been termed artificial β-barrels.[48,49]

The first membrane transport-based enzyme assay focused on enzymatic reactions involving phosphate-containing substrates such as the hydrolysis of ATP into AMP and thiamine pyrophosphate into thiamine by potato apyrase, the hydrolysis of UDP to uridine by alkaline phosphatase, and disaccharide formation by galactosyl transfer from UDP-Gal to *N*-acetyl glucosamine by galactosyl transferase (Figure 12.6).[50] These negatively charged substrates and products were detected with **SMP 1** to which Mg^{2+} had been added. Mg^{2+} binds to the negatively charged aspartic acid residues on the inner surface of **SMP 1** and converts it into the cationic metallopore **SMP 1** $\supset Mg^{2+}$. Overall, the sensitivity of metallopore **SMP 1** $\supset Mg^{2+}$ was rather limited, such that also **SMP 2** has been explored. The latter has turned out to be much more sensitive, revealing an excellent IC_{50} of $2\,\mu M$ for blockage by ATP, for example. In addition to the above enzymes belonging to the enzyme class of hydrolases (EC 3), the retroaldol reaction converting D-fructose 1,6-diphosphate into the triose phosphates dihydroxyacetone phosphate and glyceraldehyde 3-phosphate catalyzed by aldolase could also be followed. To shift the equilibrium to the product side, triosephosphate isomerase, converting glyceraldehyde 3-phosphate into another molecule of dihydroxyacetone phosphate, was also present in the enzyme reaction mixture, such that the activity of triosephosphate isomerase could principally also be detected with **SMP 2** (see Section 12.5.1).

A striking example of using SMPs for the possibility to monitor certain enzymatic reactions, which are otherwise not easily accessible with conventional fluorometric detection schemes, involves polymeric substrates.[59] Therein, the highly efficient blockage (low nanomolar range) of **SMP 2** with cationic residues in the pore interior by negatively charged DNA, RNA,

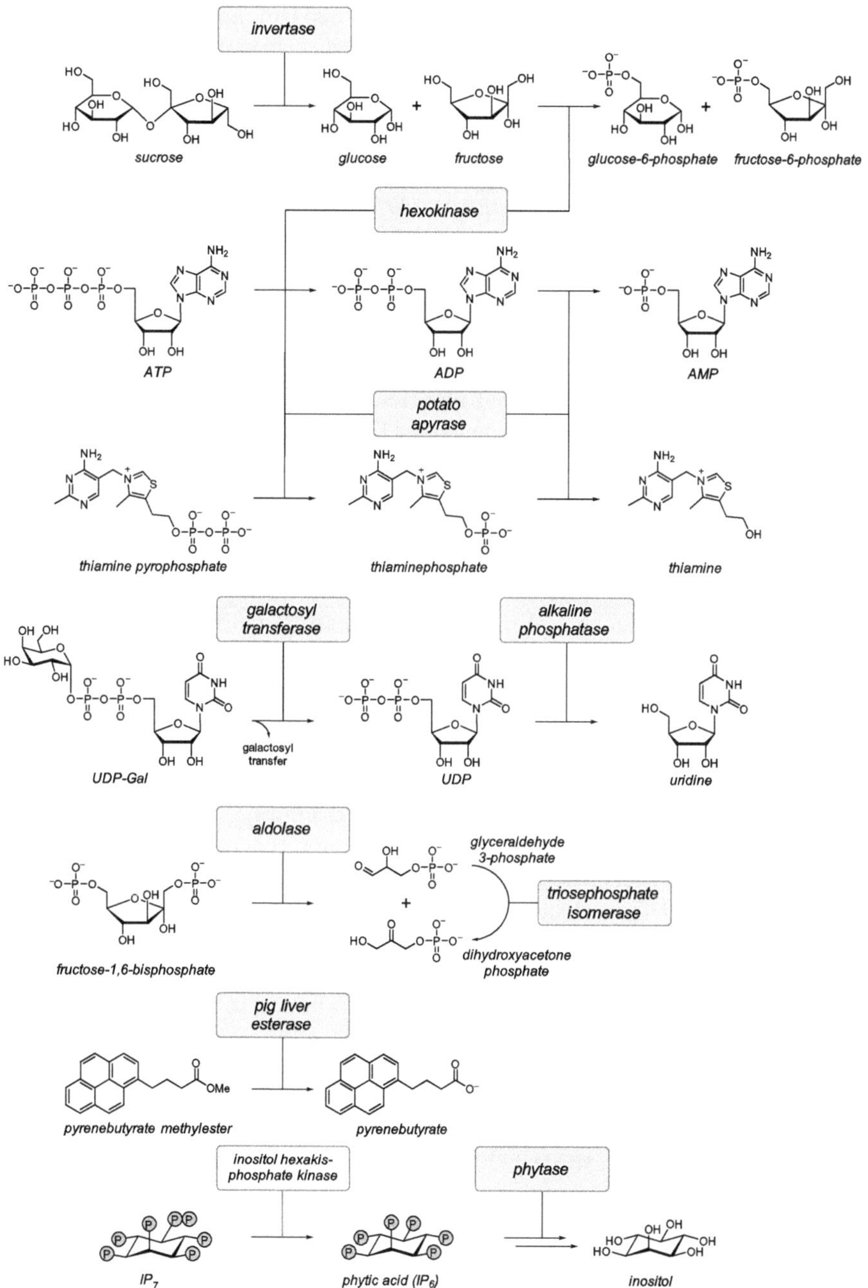

Figure 12.6 Realized examples of enzyme reaction monitoring with **SMP 1** to **SMP 4**. For the conversion of inositol phosphates IP$_7$ into IP$_6$, only the sufficient selectivity of **SMP 3** was demonstrated.

polyglutamate and polysaccharides such as heparin and hyaluronan, was utilized. Incubation of these substrates with the respective hydrolytic or degradative enzymes (DNA exonuclease III, RNase A, pronase, papain, ficin, elastase, subtilisin, heparinase I, hyaluronidase) converts the efficient polymeric blockers into much less efficient monomers, such that enzyme activity can be signaled by pore activation during the reaction. Conversely, the formation of DNA strands by the Klenow fragment of DNA polymerase I could be followed with **SMP 2**, and in analogy of detecting the degradation of negatively charged polypeptides by cationic **SMP 2**, hydrolysis of cationic polylysine and polyarginine by papain could be monitored with anionic **SMP 1**. Notably, the pore-forming bee toxin melittin, which is known to be rich in lysine and arginine residues, could not be blocked by anionic polyglutamate, albeit that binding of polyglutamate to melittin has been reported.[58]

Despite the success in developing assays for the hydrolysis of ATP to AMP, the selectivity of **SMP 2** was not sufficient to reliably discriminate between ATP and ADP. The key to success was substitution of arginine by lysine in **SMP 3**, which afforded the required selectivity and gave IC_{50}'s of 22 µM and 224 µM for ATP and ADP, respectively.[60] This enabled detection of the activity of hexokinase, which transfers a phosphate residue from ATP to the primary hydroxyl group of glucose. Moreover, sucrose could be detected and reliably quantified in a variety of soft drinks by incubating soft drink samples with invertase, which converts sucrose into glucose and fructose, and subsequent phosphorylation by hexokinase and ATP. Noteworthy is that this sequential reaction of invertase and hexokinase does also enable assaying the activity of invertase by using high concentrations of hexokinase, such that the conversion of sucrose into glucose and fructose becomes rate-limiting. Such an exploitation of two sequential enzymatic reactions involving the detection of reaction products of the latter reaction is known as enzyme-coupled enzyme assay (see Section 12.5.1 for a detailed description and more examples).

Despite the wide variety of enzymatic reactions that could be followed with synthetic multiple pores **SMP 1** to **SMP 3**, continuous reaction monitoring was not possible with these pores, because substrates bound to the pore interior did not appear to be accessible to the enzyme and pore closing upon enzymatic reaction is not compatible with the fluorescence assays described in Section 12.3.1.1.

An elegant example of how this limitation might be overcome, however, has been demonstrated with **SMP 4**.[61] In this pore one of the external hydrophobic amino acid residues has been exchanged with an arginine residue, rendering the pore more water-soluble such that it less efficiently partitions into the lipid membrane of the vesicle. However, **SMP 4** becomes activated in presence of amphiphilic anions, which bind to the arginine side chain and neutralize the excess charge such that the amphiphilic anion –**SMP 4** complex more efficiently partitions into the vesicle membrane, thereby forming an active pore. Among others, pyrene butyrate turned out to be an excellent activator of membrane transport of **SMP 4**. This has been exploited by generating pyrenebutyrate from pyrene butyrate methylester by pig liver esterase, which activated **SMP 4** and led to a continuous time-dependent increase in fluorescence intensity. Increasing concentrations of pig liver esterase in the reaction mixture resulted in steeper fluorescence curves, and a plot of the initial change in fluorescence intensity *versus* enzyme concentration was linear. Strictly speaking, this pyrene butyrate-based approach is not a label-free assay, because pyrene is not part of the natural substrates of pig liver esterase, but it nevertheless demonstrates that continuous monitoring is, in principle, possible with SMP-based enzyme assays.

Another elegant example of how inherent limitations of SMP-based enzyme assays could be overcome has been demonstrated for the detection of phytase activity.[62] This enzyme hydrolyses the phosphate ester bonds in inositol phosphates (IP_n's). Of particular interest was the discrimination of D-*myo*-5PP-IP_5 (IP_7, see bottom line of Figure 12.6) containing seven phosphate groups against inositol hexaphosphate (IP_6). The IP_7/IP_6 couple has been proposed as an equivalent to ATP/ADP in bioenergetics and protein phosphorylation, such that an enzyme assay would be high in demand. The problem of discriminating IP_7 against IP_6 originates from the very strong binding of both inositol phosphates to **SMP 3**, that is, the true dissociation constants of the inclusion complexes of IP_6 and IP_7 with **SMP 3** are much lower than the dissociation constant of the active tetrameric suprastructure of **SMP 3** itself. This leads to the situation where the working concentrations necessary to form active **SMP 3** are too high and all added analyte binds to the membrane pore. In other words, lowering the working concentrations below the dissociation constants of the $SMP-IP_n$ complex was not possible, because at such low concentrations **SMP 3** was not active.

This problem has been referred to in the field of membrane transporters as "stoichiometric binding" and could be bypassed by addition of Zn^{2+} to the detection mixture. Zn^{2+} ions compete with the internal lysine residue in **SMP 3** for phosphate binding, such that the apparent dissociation constants of the SMP–IP complexes increased from *ca.* 72 nM for IP_7 and 45 nM for IP_6 to 2.6 μM for IP_7 and 15 μM for IP_6 at 120 μM Zn^{2+}. This difference by a factor of about six would be sufficient to discriminate IP_7 against IP_6 in an enzyme assay.[58]

12.3.3 Membrane-Active Polymers

Another membrane transport system that has been successfully applied in enzyme assays and as a supramolecular sensor system is based on counterion activation and deactivation of membrane-active polymers. The development of this system originated from the peculiar role of guanidinium groups in cell-penetrating peptides (CPPs) such as those derived from the human immuno-deficiency virus transactivator of transcription (HIV Tat), in other arginine-rich protein transduction domains, in DNA transfection, as well as in the voltage gating of biological K^+ ion channels.[63] A potential solution to this puzzling phenomenon of how a charged group mediates transport across a hydrophobic membrane has been proposed by considering the role of counterions.

Guanidinium-rich oligomers and polymers tend to form dynamic complexes with anions to neutralize their overall positive charge. This phenomenon is reminiscent of the formation of an electrical double layer on the surface of colloids, which also bind counterions from the bulk solution to counterbalance the high charge density on their surface. Clear support for this hypothesis was provided by the notion that in the presence of suitable counteranions, poly-arginine partitions into chloroform and does even transport hydrophilic, positively charged molecules across the bulk chloroform phase in U-tube experiments.[63] Numerous counteranions have been screened for their capability to activate transmembrane transport by polyarginine, and it has been concluded that amphiphilic counteranions are most active, for example phosphatidylglycerol lipids, as well as polyaromatic hydrocarbons, fullerenes, and calixarenes with negatively charged groups.[63–66] In contrast, hydrophilic polymeric counteranions, for example heparin, inhibited otherwise active polyarginine/amphiphilic counteranion complexes.[66]

This interplay of activation and inactivation of membrane transport has later been extended to negatively charged polymers such as DNA, RNA, and polyphosphate. The most conclusive mechanism including the transport of fluorescent probes for sensing is shown in Figure 12.7. Therein, amphiphilic activator molecules partition into the hydrophobic lipid membrane with their lipophilic tails, and expose their charged head groups on the vesicle surface, thereby creating a high charge density on the vesicle surface to which polymeric

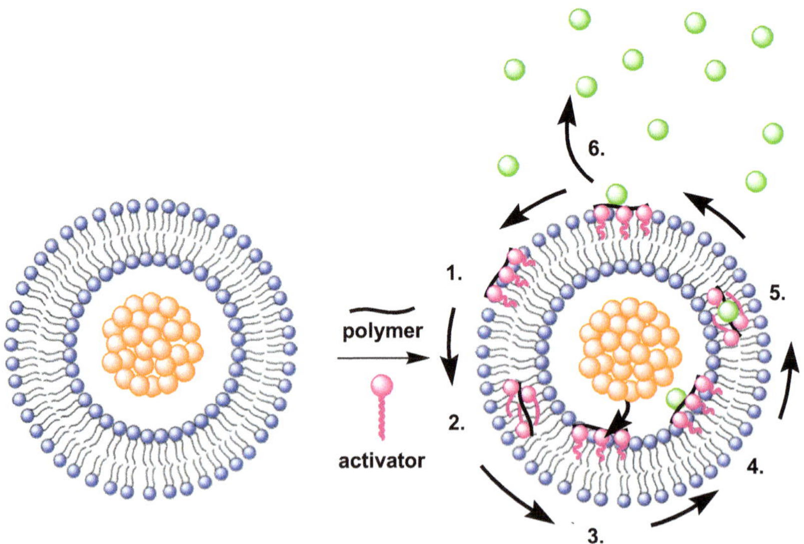

Figure 12.7   Mechanism of membrane transport by membrane-active polymers including counterion activation and efflux of an encapsulated fluorescent dye (see main text for details).
(Adapted with permission from Miyatake *et al.*,[53] Copyright 2006 American Chemical Society.)

anions can bind (step 1 in Figure 12.7). This leads to mutual charge neutralization and the resulting polymer–activator complex is sufficiently hydrophobic to partition into the hydrophobic alkyl region of the membrane (step 2). On the inner membrane surface, some of the amphiphilic counterions will exchange with fluorescent probes of suitable charge from the vesicle interior (steps 3 and 4). The resulting ternary complex of polymer, activator, and fluorescent probe can again move across the hydrophobic alkyl region of the membrane (step 5) and release the fluorescent dye on the outer membrane surface (Step 6). This mechanism of membrane transport and the inhibition thereof affords a cost-effective enzyme assay method purely based on commercially available substances or on synthetically easily accessible polymers.

The first demonstration of this enzyme assay principle was developed for detecting the activity of hyaluronidase.[53] The substrate hyaluronan **15** is a high-molecular-weight polysaccharide and has become prominent as an ingredient in skin-care cosmetics. Hyaluronidase cleaves the glycosidic bond in hyaluronan and degrades it, mainly into tetrasaccharides (Figure 12.8a). In the hyaluronidase assay, polyarginine served as the membrane-active polymer, dodecylphosphate **14** as the amphiphilic counteranion activator, and the CF assay was used for detection of membrane transport activity. The hydrophilic, negatively charged hyaluronan efficiently competes with dodecylphosphate for binding to polyarginine and inhibits membrane transport with an IC_{50} of 280 pM. The degradation products are shorter and compete less efficiently with dodecylphosphate/polyarginine binding, such that the activity of this enzyme can be followed as a turn-on of transmembrane transport activity.

Since the detection of enzymatic reactions with polymeric substrates by fluorescence methods is generally challenging with conventional enzyme assay strategies, a fluorescent hyaluronidase assay was also highly desirable for inhibitor screening. This has been demonstrated with disodium cromoglycerate as an established hyaluronidase inhibitor.[53] More challenging was detecting the inhibitory activity of heparin on hyaluronidase. Heparin is also a hydrophilic, negatively charged high-molecular-weight polysaccharide, and thus interferes with the detection scheme of the membrane transport-based enzyme assay by polyarginine binding.

Remedy was found by immobilizing the substrate hyaluronan on a polymer resin. This resin was used to extract polyarginine from solution, such that the remaining supernatant did not activate membrane transport. Hyaluronidase cleaves hyaluronan on the solid support and thereby deactivates the ability of the resin to extract polyarginine, such that sufficient amounts of polyarginine remain in solution to turn on membrane transport by hyaluronidase. To ultimately determine the inhibition of hyaluronidase by heparin, a mixture of enzyme, inhibitor and resin-bound hyaluronan was incubated, the resin was collected by filtration and washed, and remaining activity of the resin to extract polyarginine was tested in a transport assay. Increasing amounts of the inhibitor in the reaction mixture thus led to more intact hyaluronan on the solid support, thus to more efficient extraction of polyarginine and consequently to a lower membrane transport activity.[53]

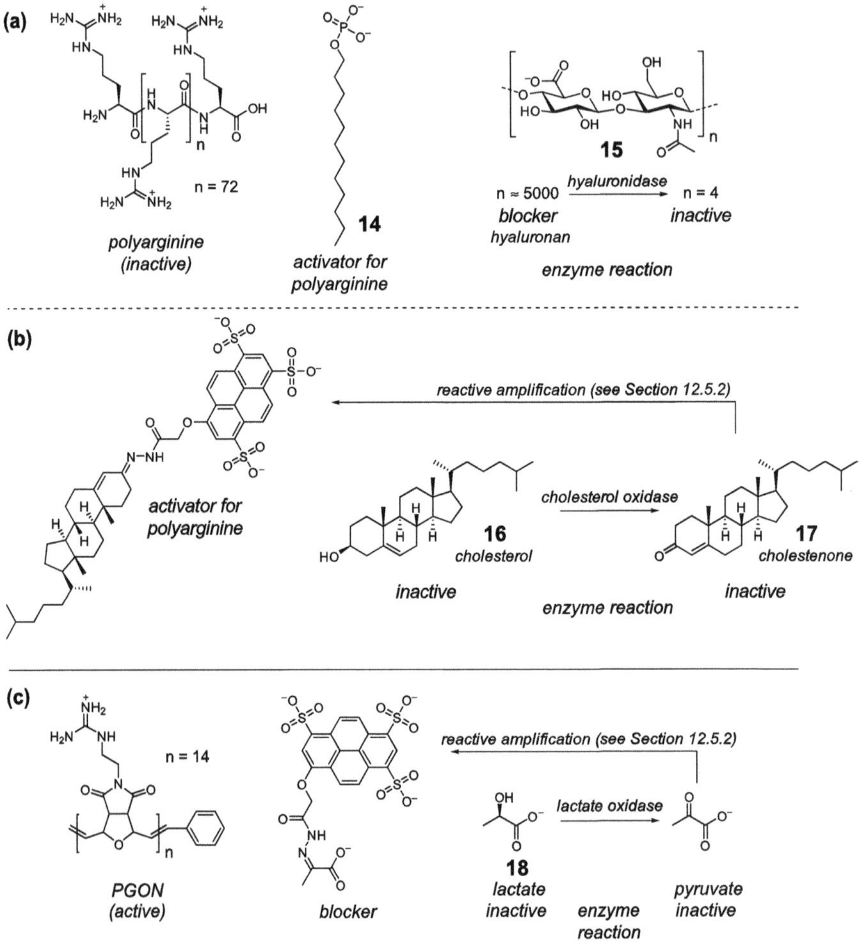

Figure 12.8 Membrane-active polymer systems used in supramolecular enzyme assays and the respective enzymatic reactions. (a) Polyarginine is activated by dodecylphosphate **14** and blocked by hyaluronan. Degradation by hyaluronidase restores transport activity. (b) Cholesterol oxidase affords cholestenone, which is converted into a polyarginine activator by a reactive amplification step (see Section 12.5.2). (c) Polyguanidino-oxanorbornenes (PGONs) are active membrane transporters and can be blocked by the amplified reaction product of lactate oxidase.

Noteworthy is that the polyarginine/dodecylphosphate transport system could also be used for discrimination of ATP and ADP during the enzymatic reaction of hexokinase. However, the sensitivity and selectivity was lower than that with **SMP 3** (see Section 12.3.2), such that this enzyme assay was not further investigated. As an alternative to dodecylphosphate, a Cascade Blue (CB)-hydrazide–cholestenone conjugate could also be used as amphiphilic activator (Figure 12.8b). The latter is the result of the reaction of cholesterol **16** with cholesterol oxidase and a subsequent reactive amplification step (see Section 12.5.2 for details).[67]

As an example for synthetic polymers, polyguanidino-oxanorbornenes (PGONs) have been used for membrane transport-based enzyme assays (Figure 12.8c).[68] In contrast to polyarginine, PGONs are already sufficiently hydrophobic to mediate CF efflux from vesicles and thus do not require further activation by addition of amphiphilic counterions. Moreover, addition of a CB-hydrazide–pyruvate conjugate (designated as *blocker* in Figure 12.8c) leads to the formation of PGON–amplifier complexes, which are too hydrophilic to cross the lipid bilayer membrane. Analogous to cholesterol, the CB-hydrazide–pyruvate conjugates result from the reaction of lactate **18** with lactate oxidase and a subsequent reactive amplification step (see Section 12.5.2), such that overall, the activity of lactate oxidase can be monitored as deactivation of PGON transport activity.[68]

12.4 Supramolecular Tandem Assays

The most recent development in supramolecular enzyme assays is based on macrocyclic receptors, which are commonly less selective than specifically designed chemoreceptors and thus bind a large variety of different but related guests. By coupling the differential binding of substrate and product of an enzymatic reaction to a macrocyclic receptor with the competitive displacement of a fluorescent dye, "supramolecular tandem assays" result. They combine the advantages of chemoreceptors (Section 12.2), in particular the possibility for continuous monitoring, and the relatively unselective molecular recognition of membrane transport systems (Section 12.3). The latter affords the possibility to apply one sensing system to a wide variety of different but related enzymatic reactions (see Section 12.5.3 on biosensing).[69–79]

The following section introduces the basic conceptual approach of supramolecular tandem assays (Section 12.4.1) and its more elaborate extension to macrocyclic receptors, which bind the substrate more efficiently than the product of the enzymatic reaction (Section 12.4.2). Moreover, it will be briefly discussed how the complex interrelated binding equilibria of supramolecular tandem assays can result in different shapes of the kinetic fluorescence progress curves, and how this may influence enzyme kinetic measurements (Section 12.4.3).

12.4.1 Conceptual Approach

The concept of supramolecular tandem assays is inspired by the indicator displacement assay strategy (see also Section 12.2, chemoreceptors **4** to **6**).[42,43] In general, indicator displacement assays rely on indicator dyes, which change their spectroscopic characteristics upon binding to a supramolecular receptor. Addition of an analyte competing with the dye for binding to the receptor leads to a recovery of the spectroscopic properties of the unbound dye, which forms the basis for detecting the presence of an analyte.

Macrocyclic receptors such as cucurbiturils, calixarenes, cyclodextrins, and cyclophanes are prominent examples of supramolecular receptors, which affect the fluorescence spectroscopic properties of a large variety of fluorescent dyes, and an excellent, comprehensive review has recently been published (see also Chapters 3, 4 and 10).[80] Certain combinations of macrocycles and dyes have proven particularly suitable for supramolecular sensing, because the resulting spectroscopic changes are large, and these combinations have been referred to in the context of supramolecular tandem assays as a library of reporter pairs (Figure 12.9). Most common is a change in fluorescence intensity, whereby the fluorescence may be either quenched (Figure 12.10a) or enhanced (Figure 12.10b) by complex formation with the macrocycle. When an analyte, which more strongly binds to the receptor, is added to a reporter pair solution, a displacement of the dye from the receptor results, and the spectroscopic properties of the unbound dye in solution are regenerated. This signal change affords the analyte concentration *via* the respective binding isotherms (Figure 12.10c, d) and forms the basis of indicator displacement assays.

The step from indicator displacement assays to supramolecular tandem assays is then that the competing analyte is not added but generated from a weakly bound substrate during the course of an enzymatic reaction. Supramolecular tandem assays thus present a time-resolved version of indicator displacement assays. This key idea is indicated by the red arrows in the binding isotherms shown in Figure 12.10(c, d) and the result of the enzymatic reaction is a continuous change in fluorescence intensity, which may be either an increase (switch-ON assay, Figure 12.10e) or a decrease (switch-OFF assay, Figure 12.10f), depending on the photophysical properties of the reporter pair. Noteworthy is that ratiometric tandem assays have also been developed, in which two absorption or fluorescence bands simultaneously show both a switch-ON and a switch-OFF response in distinct regions of their spectrum (*e.g.* the carbazole **22**).[72]

Supramolecular tandem assays following the principle depicted in Figure 12.10 have been initially developed for various amino acid decarboxylases (lysine, arginine, histidine, ornithine, tyrosine, and tryptophan decarboxylases).[70–72] Therein, the CB7/dapoxyl **19**, the CB6/carbazole **22**, and the CX4/DBO **23** reporter pairs were used (Figure 12.11). In all cases, the more positively charged decarboxylation product bound more strongly to the macrocyclic receptors and the resulting assays led to a fluorescence switch-OFF for the CB7/dapoxyl reporter pairs, to a switch-ON response for CX4/DBO and to a ratiometric response for the combination of CB6 with carbazole **22**.

In the decarboxylase assay, the products of the enzymatic reactions were relatively small molecules, which were nearly completely immersed upon binding into the macrocyclic cavity. As an extension, it has also been demonstrated that the selectivity of "rather unselective" macrocyclic receptors suffices to monitor enzymatic transformations at a single functional group of much larger substrates. For example, the protease thermolysin cleaves the peptide substrate H-Thr-Gly-Ala-Phe-Met-NH$_2$ between the alanine and phenylalanine residues, which liberates an N-terminal phenylalanine group.[75]

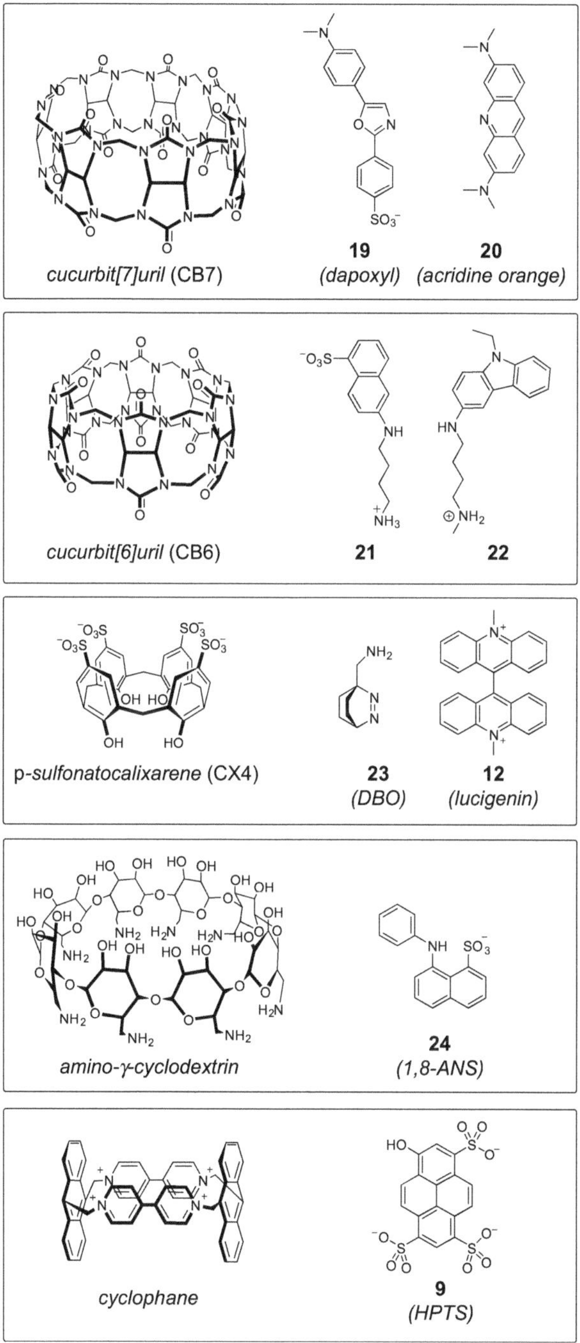

Figure 12.9 Reporter pair library consisting of macrocyclic host molecules and fluorescent dyes particularly well-suited for supramolecular enzyme assays.

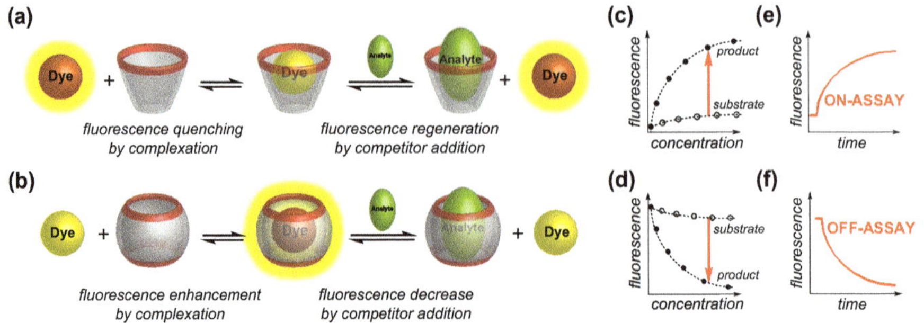

Figure 12.10   Principle of supramolecular tandem enzyme assays. Binding of fluorescent dyes to macrocyclic receptors commonly leads to (a) fluorescence quenching or (b) fluorescence enhancement. Addition of a competitor displaces the dye and restores the fluorescence properties of the free dye. This indicator displacement enables conventional competitive binding titrations, in which the fluorescence either increases (c) or decreases (d). Alternatively, the transformation of a weak competitor (open circles) into a strong competitor (filled circles), *e.g.* by an enzymatic reaction, can be followed as indicated by the red arrow, such that the progress of the reaction is signaled by (e) an increase (ON-assay) or (f) a decrease (OFF-assay) in fluorescence intensity.
(Reproduced from Dsouza *et al.*[79] with permission, Copyright 2012 Wiley-VCH.)

This residue is known to be efficiently bound by CB7 through combined hydrophobic and electrostatic interactions, whereas the other peptide residues only weakly bind to CB7.[81] Therefore, it is only this one residue of the peptide chain, which complexes with differential affinity to the macrocycle and affords the required selectivity to signal the activity of thermolysin by displacement of acridine orange. The same principle was used to monitor the proteolytic activity of trypsin, which cleaves the peptide bond at the carboxyl group of the arginine residue in the peptide H-Leu-Ser-Arg-Phe-Ser-Trp-Gly-Ala-OH and thereby liberates an N-terminal phenylalanine group.[77]

Another example is a supramolecular tandem assay for histone lysine methyl transferases, which play a key role in gene transcription, but for which so far efficient fluorescent enzyme assays have been lacking.[78] These enzymes sequentially methylate the ε-amino group of lysine residues in peptides up to the quarternary alkylammonium ion product. Because CX4 is well known to bind to quarternary alkylammonium ions, it was chosen as macrocyclic receptor. Binding titrations with Lys(Me)$_n$ ($n = 0$, 1, 2, 3) indicated an increasing binding constant with each methylation step ($K_a < 10^3$ M^{-1}, 2×10^4 M^{-1}, 6×10^4 M^{-1}, and 1.3×10^5 M^{-1}), and in fact, the activity of histone lysine methyl transferases could be reliably signaled by displacement of lucigenin and the concomitant fluorescence increase. Owing to the presence of numerous other lysine and arginine residues in the peptide substrate, which can also bind to CX4, this represents a striking example of how ''rather unselective'' macrocycles can detect complex transformations.

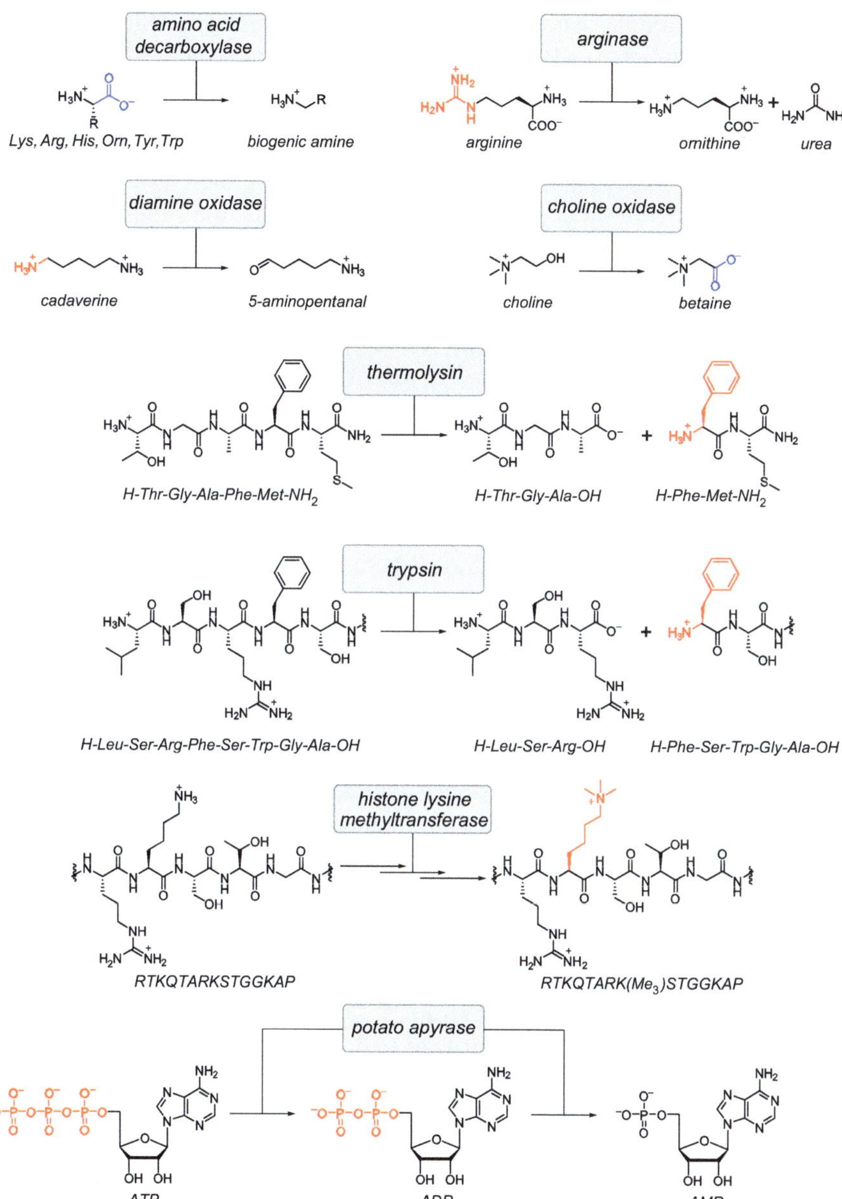

Figure 12.11 Realized examples of supramolecular tandem enzyme assays. Note that truncated structures are partially shown for the peptide substrates; the complete peptide sequence is given below the structure. The primary supramolecular recognition element modulating the selectivity is highlighted in red, if attractive forces are altered, or in blue, if repulsive forces are altered during the enzymatic reaction.

In a third example, leucine aminopeptidase was assayed with the peptide substrate H-Leu-Ser-Arg-Phe-Ser-Trp-Gly-Ala-OH and the CB7/acridine orange reporter pair.[77] Leucine aminopeptides hydrolyzes stepwise the complete peptide, starting from the amino terminus. This first leads to a strongly binding peptide fragment with an N-terminal phenylalanine group, but subsequent cleavage of the phenylalanine residue leads to weakly binding products. This stepwise degradation was confirmed by electrospray ionization mass spectrometry and showed a fluorescence time trace, in which the intensity first decreased owing to the formation of the strongly binding intermediate and a subsequent increase owing to its degradation. An extended version of the substrate containing two phenylalanine residues, namely H-Leu-Ser-Arg-Phe-Ser-Trp-Gly-Ala-Leu-Ser-Arg-Phe-Ser-Trp-Gly-Ala-OH, even gave an alternating OFF/ON/OFF/ON fluorescence response. Such a stepwise degradation is reminiscent of the Edman degradation protocol used for peptide sequencing, and such an assay might, although perhaps a little far-fetched, be useful for peptide sequencing.[77] This could principally work with an exopeptidase for peptide degradation (*e.g.* leucine aminopeptidase) and different reporter pairs, which are selective for a specific terminal amino acid. Successive cleavage would then result in time-dependent alternating fluorescence response traces for each reporter pair reminiscent of a chromatogram from an automated DNA sequencer.

12.4.2 Substrate-Selective Tandem Assays

In the initial demonstration of supramolecular tandem assays with amino acid decarboxylases, the overall charge was increased during the conversion of the substrate into the product. Because electrostatic interactions are commonly one of the main driving forces for supramolecular host–guest interactions in water, an increased net charge most frequently leads to a stronger binding to a complementary supramolecular receptor. For example, the doubly positively charged decarboxylation product cadaverine binds more strongly to the cation receptors CX4 and CB7 than the substrate lysine carrying an additional negatively charged carboxylate group.[68] The resulting "product-selective" tandem assays are most straightforward to develop and they are reminiscent of antibody-based immunoassays, in which the product of the enzymatic reaction binds to an antibody and, for example, displaces a fluorescently labeled antigen from the antibody-binding pocket.[82,83]

Owing to the fact that numerous enzymatic reactions also decrease the charge status of the analyte, it was naturally desirable to develop "substrate-selective" tandem assays, in which the more highly charged substrate binds more strongly to the macrocyclic receptor than the product. This required, however, a more elaborated design strategy, because complexation of enzyme substrates frequently leads to an efficient inhibition of the enzymatic reaction, which is advantageous for drug delivery applications, but must clearly be avoided in enzyme assays.[84–86]

The key to "substrate-selective" tandem assays thus lies in finding conditions in which the macrocyclic receptor does not influence the enzymatic conversion.

This can, for example, be achieved by selecting macrocycles that bind the substrate relatively weakly, such that an excess of substrate is required in the reaction mixture to significantly displace the fluorescent dye from the macrocycle. The reason is that the overall capability of a substrate (or any other guest) to bind to a host is not only determined by the association constant K_{HS} but by the product of K_{HS} and the substrate concentration [S].[79] In the case of weakly binding substrates, only a minor fraction of the substrate is actually complexed by the receptor, while the major fraction is free in solution and can thus be recognized and converted by the enzyme (Figure 12.12).

The enzymatic conversion results in a non-equilibrium situation leading to dissociation of the remaining macrocycle–substrate complex, which will replenish unbound substrate in solution. Subsequently, the free macrocycle can bind the fluorescent dye in the reaction mixture, which affords the fluorescence change required to signal the enzymatic activity (lower part of Figure 12.12). The prerequisite—that the time-dependent change in fluorescence reflects the kinetics of the enzymatic reaction—is met if the enzymatic conversion is the rate-limiting step, that is, the enzymatic turnover rate k_{enz} must be significantly larger than the dissociation rate of the substrate–macrocycle complex $k_{diss,HS}$ and the association rate $k_{ass,HD}$ of dye and macrocyclic host. More detailed guidelines for the development of supramolecular tandem assays, including a comprehensive thermodynamic and kinetic analysis as well as an accompanying computer program for simulation and optimization' have recently been published.[79] It is worth noting that such a substrate-selective assay design is commonly not feasible for antibody-based assays, because of the slow dissociation kinetics of antibody–antigen complexes, whereas the exchange kinetics of macrocyclic host–guest complexes are commonly of the microsecond to millisecond timescale and thus much faster than the enzymatic reaction.

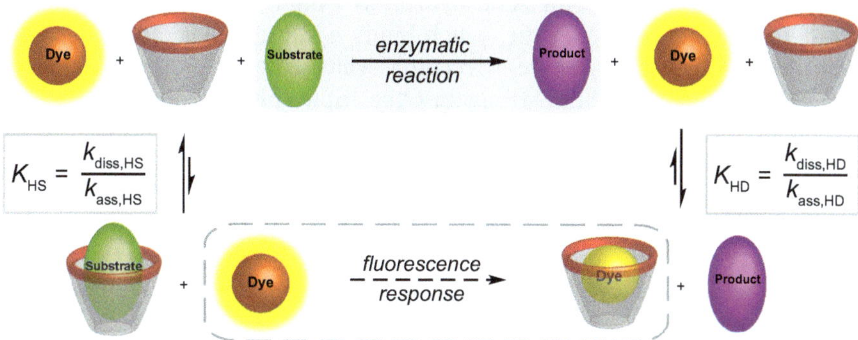

Figure 12.12 Principle of substrate-selective supramolecular tandem enzyme assays. Dye and substrate are in a rapid dynamic competitive equilibrium for encapsulation within the macrocycle, whereas only a small fraction of the total substrate displaces the fluorescent dye. Enzymatic conversion of the substrate gives product, which binds less efficiently to the macrocycle. The dye can thus bind to the cavity, which is accompanied by a modulation of the fluorescence.

Substrate-selective supramolecular tandem enzyme assays have been developed for arginase, which is a hydrolase converting arginine into ornithine,[73] for diamine oxidase, which converts diamines into the corresponding imines,[73] for the hydrolysis of ATP by potato apyrase,[74] and for the detection of choline oxidase activity (Figure 12.11).[76] The arginase assay is based on CX4, which binds the substrate arginine and the product ornithine relatively weakly and displays a relatively low selectivity ($K_a = 6400$ M^{-1} and 550 M^{-1}, respectively). This affords the working conditions specified above, in which only a minor fraction of the substrate is bound to the macrocycle. The enzymatic conversion leading to the hydrolysis of arginine into ornithine is signaled by binding of the fluorescent dye DBO into the vacant cavity of CX4 and the concomitant fluorescence quenching, such that overall a substrate-selective switch-OFF assay results. The qualitative change in fluorescence at a fixed substrate concentration was used for the determination of the inhibition constants of two known arginase inhibitors, namely *S*-(2-boronoethyl)-L-cysteine and 2-(*S*)-amino-6-boronohexanoic acid, and the obtained inhibition constants were in good agreement with literature values. Similar results were obtained with a supramolecular tandem assay for diamine oxidase based on the CB7/acridine orange reporter pair. Therein, lower substrate concentrations were chosen owing to the generally higher affinity of cucurbiturils compared with calixarenes. Nevertheless, this assay could reliably signal the activity of diamine oxidase and enabled the determination of the inhibition constant of a known diamine oxidase inhibitor.

The development of supramolecular tandem assays for ATP-dependent enzymes presents a striking example for the utility of the concept of substrate-selective assays.[74] ATP is a ubiquitous co-substrate for numerous enzymatic reactions, such that a label-free and continuously monitored fluorescent enzyme assay is highly desirable. However, the most challenging part in designing a product-selective assay for ATP hydrolysis would be to find a simple and synthetically accessible receptor, which binds AMP in water more effectively than ATP. This would require a receptor, which efficiently discriminates the monophosphate group of AMP against the triphosphate group of ATP. Although this selectivity has been achieved in mixed aqueous/organic solvent,[87] a general strategy for selective AMP or ADP recognition in water is currently not available.[88] In contrast, a stronger binding of ATP than AMP is simply achieved by exploiting the increased electrostatic interactions between a cationic anion receptor and the more negatively charged ATP. In fact, there are so many potential macrocycles which bind ATP more strongly than ADP or AMP that a library of macrocycle–dye reporter pairs was screened to identify the most suitable combination.[74] This approach afforded two complementary reporter pairs, namely octaamino-γ-cyclodextrin/1,8-ANS and a 4,4'-bipyridine-based cyclophane with HPTS as dye (see Figure 12.9), which could be applied in different substrate concentration ranges. Octaamino-γ-cyclodextrin had a high affinity ($K_a = 1.0 \times 10^8$ M^{-1}) and the cyclophane a relatively low affinity ($K_a = 4.5 \times 10^3$ M^{-1}) for ATP, such that substrate concentrations in the micromolar and millimolar range could be used, respectively. These two substrate concentration ranges are below and above the K_M of the model enzyme, potato apyrase, which is useful for determining enzyme kinetic parameters in fundamental biochemical research and for determining the maximum rate of conversion as an indicator of enzyme purity in manufacturer's quality control, respectively.[74]

Another example of a substrate-selective tandem assay is the detection of choline oxidase activity.[76] Therein, the positively charged substrate choline, which is well-known to bind to CX4 (K_a *ca.* $1.0 \times 10^5 \, M^{-1}$), is oxidized to the zwitterionic betaine, which binds much more weakly ($K_a < 500 \, M^{-1}$). This assay has also been used in an enzyme-coupled assay for monitoring the activity of acetylcholine esterase and will be described in more detail in the respective Section 12.5.1.

12.4.3 Kinetics, Progress Curves and Inhibition Constants

The supramolecular tandem assays from sections 12.4.1 and 12.4.2 could all be continuously monitored, regardless of the spectroscopic response or whether the assay was product- or substrate-selective. However, as a result of the complexity of the multiple and interrelated reaction equilibria and kinetics, it has been noted that the fluorescence intensity progress curves can have markedly different shapes (Figure 12.13). In particular, sigmoidal progress curves can result when: (i) the binding constants are moderate, (ii) the difference between substrate and product binding constants is small, or (iii) concentrations are inappropriately adjusted.[68,79] Furthermore, progress curves have been obtained, in which the initial slopes were steepest at intermediary

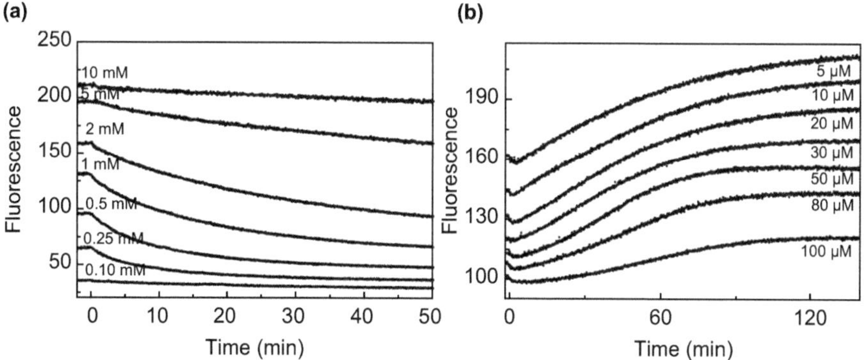

Figure 12.13 (a) Evolution of fluorescence intensity of the CX4/DBO (200 µM/ 100 µM) system during enzymatic hydrolysis of arginine by arginase (140 nM) at different substrate concentrations. (b) Evolution of fluorescence intensity of the CB7/AO (8 µM/0.5 µM) system during enzymatic oxidation of cadaverine to 5-aminopentanal by diamine oxidase (1 U mL^{-1}) at different substrate concentrations. (Experimental data from Nau *et al.*,[73] reproduced with permission. Copyright 2009 American Chemical Society.)

concentrations far below the K_M value. As a consequence, it is thus not always guaranteed that the observed change in fluorescence intensity is directly proportional to a change in substrate or product concentration, but as a potential remedy for progress curve analysis and as assistance for assay optimization, a simulation program is available.[79]

Nonetheless, it has been shown that the relative change in fluorescence at a given substrate concentration readily allows a qualitative comparison of enzyme activities as well as a quantitative determination of effective inhibition and activation concentrations, regardless of the proportionality between fluorescence change and concentration.[73–75,79] This is because at a specific and fixed substrate concentration the same change in fluorescence refers to the same amount of substrate consumed, and thus provides an indirect measure for the reaction progress in terms of a relative reaction rate.

By this approach the inhibition constants of numerous literature-known inhibitors of arginase, diamine oxidase, thermolysin, and acetylcholine esterase, also the order of activation potential for several divalent metal ions of potato apyrase, as well as the substrate selectivity of thermolysin[75] and lysine methyltransferase[78] could be reliably reproduced. Moreover, when the binding is strong and when the difference between substrate and product binding constants is large, tandem assays can also be used quantitatively to determine enzyme kinetic parameters. This clearly demonstrates that tandem assays are useful tools for rapid screening of relative enzymatic activity, which can be affected, for example, by the presence of pharmaceutically relevant inhibitors or by the biotechnological production of enzyme mutants.

Supramolecular tandem assays thus afford a very simple and accessible way to screen for enzyme activities with a high potential for up-scaling to HTS formats owing to the fluorescence readout. In an HTS environment, compounds to be screened for may present a source of some uncertainty, because they could potentially also interact with the macrocycle and thereby displace the fluorescent dye, or their activation or inhibition potential could be artificially reduced because of their interaction with the macrocycle. However, the concentrations of common pharmaceutically relevant compounds in HTS are generally too low to significantly interfere with the detection scheme of supramolecular tandem assays, in particular, because of the low affinity of most macrocyclic receptors. Moreover, control experiments can alleviate the probability of false negative hits in HTS; for example, the interaction of a potential inhibitor with the macrocycle would be indicated by different fluorescence intensities of a mixture of macrocycle and dye with or without the inhibitor.

12.5 Extended Applications and Concepts

This last section summarizes concepts in supramolecular enzyme assays, which have been introduced with the aim of expanding the application scope of supramolecular enzyme assays. For example, enzyme-coupled assays (Section 12.5.1) have been used to visualize the activity of enzymes, for which it would have been very challenging to develop a direct supramolecular enzyme assay, and reactive amplifiers (Section 12.5.2) have been introduced to decrease the detection limit (i.e. the EC_{50} or IC_{50} values) in supramolecular enzyme assays based on synthetic multifunctional pores and polymeric membrane transport systems as described in Section 12.3.

Furthermore, the applications of supramolecular enzyme assays are not limited to the detection of enzyme activity and the inhibition or activation thereof, but supramolecular enzyme assays can be utilized as a means of detecting the substrate itself, which leads to potential applications of supramolecular enzyme assays such as biosensing and chiral discrimination (Section 12.5.3).

12.5.1 Coupled Enzyme Assays

Coupled enzyme assays describe a classical strategy to assay enzymatic reactions, which are otherwise difficult to monitor directly. Therein, the spectroscopic signal change is caused by a secondary enzymatic reaction, which is specific for the product of the primary enzymatic reaction. To give a well-established example, peptidylprolyl isomerases (PPIases, EC 5.2.1.8), which have recently been suggested as molecular timers of cellular processes and as a new target for therapeutic interventions, catalyze the *cis–trans* isomerization around the proline amide bond.[89] This reaction is very challenging to follow by a direct spectroscopic method with a large signal change. However, it has been noted that the protease chymotrypsin specifically hydrolyzes substrates with a *trans* proline amide bond,[90] and this hydrolysis is very straightforward to detect with a chromogenic or fluorogenic substrate. The overall cleavage rate is initially determined by the concentration of peptides with a *trans* proline bond and, after their complete hydrolysis, by the *cis–trans* isomerization rate, which replenishes substrates with a *trans* proline bond. This reaction can consequently be utilized to monitor the isomerization rate around the proline amide bond as well as its increase by the catalytic activity of PPIases.

In general, the kinetic analysis of coupled enzymatic reactions has often been simplified by adding a large excess of the secondary enzyme such that the primary enzymatic reaction became the rate-limiting step, but more complex models for analyzing complete progress curves involving biphasic kinetics have also been developed.[91]

This coupled enzyme assay strategy has also been adapted for supramolecular enzyme assays. A specific example for membrane transport-based assays involves invertase, which hydrolyzes the glycosidic bond in sucrose leading to glucose and fructose. The latter can subsequently be phosphorylated by hexokinase (Figure 12.6).[58] Thereby, ATP, which is a good blocker of **SMP 3** (see Section 12.3.2), is converted into less active ADP, such that the activity of invertase can be followed by activation of membrane transport in a coupled enzyme assay.[60] This strategy serves also as a basis for sucrose sensing in soft drinks (see Section 12.5.3). Although not explicitly demonstrated in practice, a membrane transport-based coupled enzyme assay is also conceivable for triosephosphate isomerase, which converts glyceraldehyde 3-phosphate into dihydroxyacetone phosphate. Substrate and product block transport activity with similar efficiency, but its activity could be followed by monitoring the consumption of the substrate by aldolase with **SMP 2** (see Section 12.3.2 and Figure 12.6).

An example of an enzyme-coupled supramolecular tandem assay has been realized with the CX4/lucigenin reporter pair. Acetylcholine and choline, the reaction product of its hydrolysis by acetylcholine esterase, bind to many macrocyclic cation receptors with similar efficiency, including calixarenes, resorcinarenes, and cavitands.[76] This is primarily due to their similar charge status, such that a receptor would be needed, which efficiently discriminates a free against an acetylated hydroxyl group (see also previously published guidelines for receptor selection).[79]

Such a kind of selectivity is difficult to achieve, but choline can be oxidized to the zwitterionic betaine by choline oxidase. This reduces the binding constant with CX4 from $K_a \approx 1.0 \times 10^5$ M^{-1} for the positively charged choline to $K_a < 500$ M^{-1} for the zwitterionic betaine, because of the repulsive electrostatic interaction between the betaine carboxylate and the sulfonato groups of CX4. As a consequence, acetylcholine esterase activity could be detected in an enzyme-coupled tandem assay by utilizing a secondary enzymatic reaction, which oxidizes the reaction product choline.[76] Furthermore, this assay provided a means to quantify mixtures of choline and acetylcholine at biologically relevant, low micromolar concentrations (see Section 12.5.3).

Additional potential applications of a series of enzymatic reactions in combination with supramolecular tandem assays have been proposed as "domino tandem assays".[73] For example, in the biochemical degradation pathway, the decarboxylation of lysine to cadaverine by lysine decarboxylase is followed by its subsequent oxidation to 5-aminopentanal by diamine oxidase. This reaction cascade can be monitored with the same reporter pair in the same solution. The product-selective tandem assay of lysine decarboxylase leads to a decrease in fluorescence (see Section 12.4.1) and addition of diamine oxidase to a subsequent recovery of the fluorescence, because the strong competitor is converted into a weak competitor during this substrate-selective assay (see Section 12.4.2). Such a fluorescence switching in response to different enzymes might be useful, for example, in the context of logic gate operations for biocomputing.[73]

12.5.2 Reactive Amplifiers

A related approach for expanding the utility of supramolecular enzyme assays by coupling a secondary reaction to the primary enzymatic reaction has been introduced by the concept of reactive amplification to membrane transport-based enzyme assays. Reactive amplifiers have been defined as molecules that can react spontaneously with the functional group altered during an enzymatic reaction. Thereby, the reactive amplification step provides *in situ* analytes, which are much more active and selective as membrane transport inhibitors or activators than the enzyme reaction substrates or products themselves. This strategy is inspired from pre- or postcolumn derivatization procedures known from chromatography and electrophoresis, such as in amino acid analysis after peptide hydrolysis.

The concept was initially demonstrated with Cascade Blue (CB) hydrazide (Figure 12.14) as amplification reagent for aldehydes and ketones as functional groups.[52] The trisulfonated pyrene core of CB hydrazide was known to

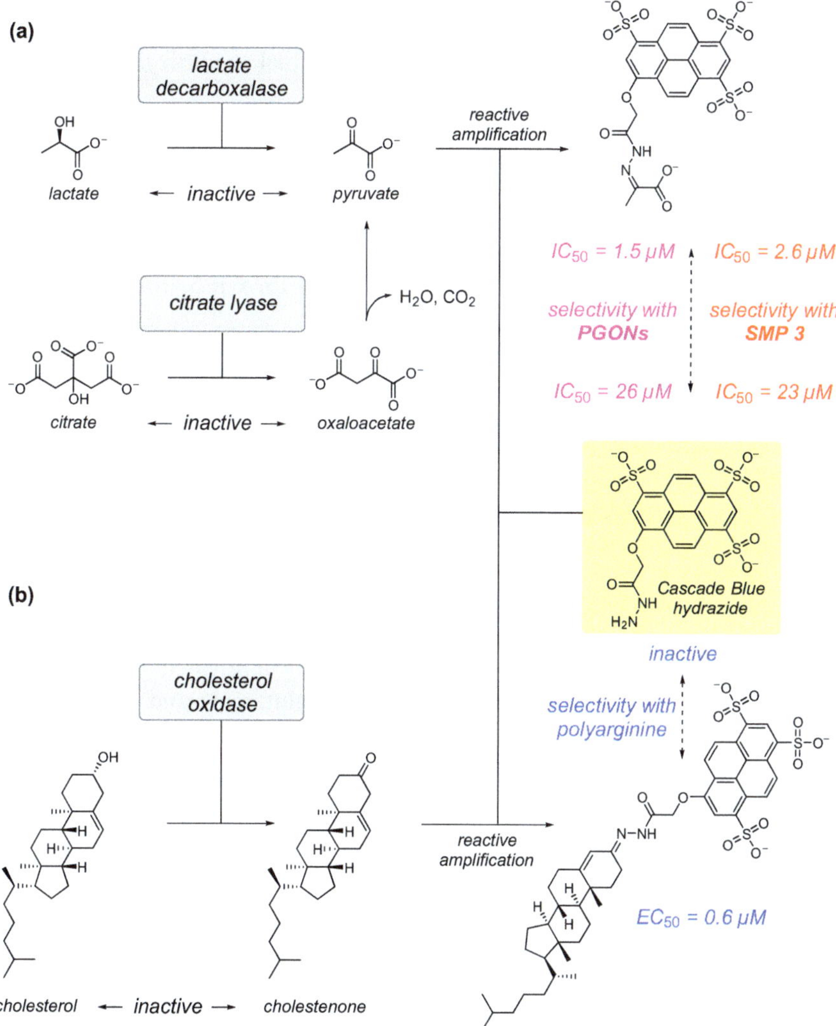

Figure 12.14 Reactive amplification with Cascade Blue (CB) hydrazide. (a) Amplification of the inactive reaction product, pyruvate, into synthetic multifunctional pore (SMP) blockers by hydrazone formation with CB hydrazide. (b) Amplification of hydrophobic, inactive cholestenone into an amphiphilic polyarginine activator by CB hydrazide.

efficiently block **SMP 3**, and hydrazone formation is known as a selective, bioorthogonal reaction. Following this strategy, the activity of lactate oxidase (converting lactate into pyruvate) and citrate lyase (converting citrate into acetate and oxaloacetate) could be monitored. Because oxaloacetate spontaneously decarboxylates into pyruvate, the CB hydrazide derivate of pyruvate was the common analyte to be detected in both enzymatic reactions. The sensitivity increased dramatically from an IC_{50} around 40 mM for pyruvate to $IC_{50} = 2.6$ μM for the amplified product, while the substrates lactate and citrate had no detectable activity. The selectivity, however, needs to be judged on basis of the CB hydrazide amplifier, which had an $IC_{50} = 23$ μM, yet was sufficiently different from the amplified enzyme reaction product to allow the detection of enzyme activity without much effort.

Also the activity of synthetic polymers such as PGONs (Section 12.3.3) could be modulated by reactive amplification of pyruvate by CB hydrazide. PGONs are membrane-active without an additional amphiphilic activator, and the PGON–amplifier complexes are too hydrophilic to cross the lipid bilayer membrane. In this case, the selectivity of PGONs was even superior to **SMP 3** with similar sensitivity ($IC_{50} = 26$ μM and 1.5 μM for CB hydrazide and amplified pyruvate, respectively).[68]

In another elegant example, amphiphilic activators of membrane transport by polyarginine were generated *in situ* by reaction of CB hydrazide with cholestenone (Figure 12.14b). The latter is the product of the enzymatic reaction of cholesterol with cholesterol oxidase, such that cholesterol oxidase activity ultimately led to activation of membrane transport.[67]

As alternative reactive amplifiers, electron-rich dialkoxynaphthalene (DAN) and dialkoxyanthracene (DAA) hydrazides have been introduced. DANs and DAAs efficiently block **SMP 5** (Figure 12.5), which contains the unnatural amino acid π_A derived from naphthalene diimide (NDI). The latter is electron-poor and efficiently binds DANs and DAAs through formation of charge transfer complexes.[52,92–94] This modification was motivated by the goal to develop an enzyme assay for the conversion of glutamate and pyruvate to α-ketoglutarate and alanine by transaminase. The challenge there was to discriminate the amplified pyruvate substrate from the amplified product α-ketoglutarate, which could not be achieved with the CB hydrazide amplifier and **SMP 3**. The combination of **SMP 5** with DAN amplifiers theoretically provided the necessary selectivity and sensitivity (see Figure 12.15a), but the enzyme assay was not carried out in practice.[52]

A modular possibility for extending hydrazide amplifiers to analytes other than aldehydes and ketones was achieved by *in situ* conversion of reactive DAN and DAA hydrazides into boronic acids (Figure 12.15b). The latter served to capture catechols from green tea leaves, but did not react with catechol-free polyphenols such as resveratrol—the famous polyphenol from red wine associated with its antioxidant activity.[93,94] Addition of tyrosinase oxidizes phenols to catechols, which can be subsequently detected by boronic acid amplifiers. As such, this combination affords an enzyme assay for tyrosinase.[93,95]

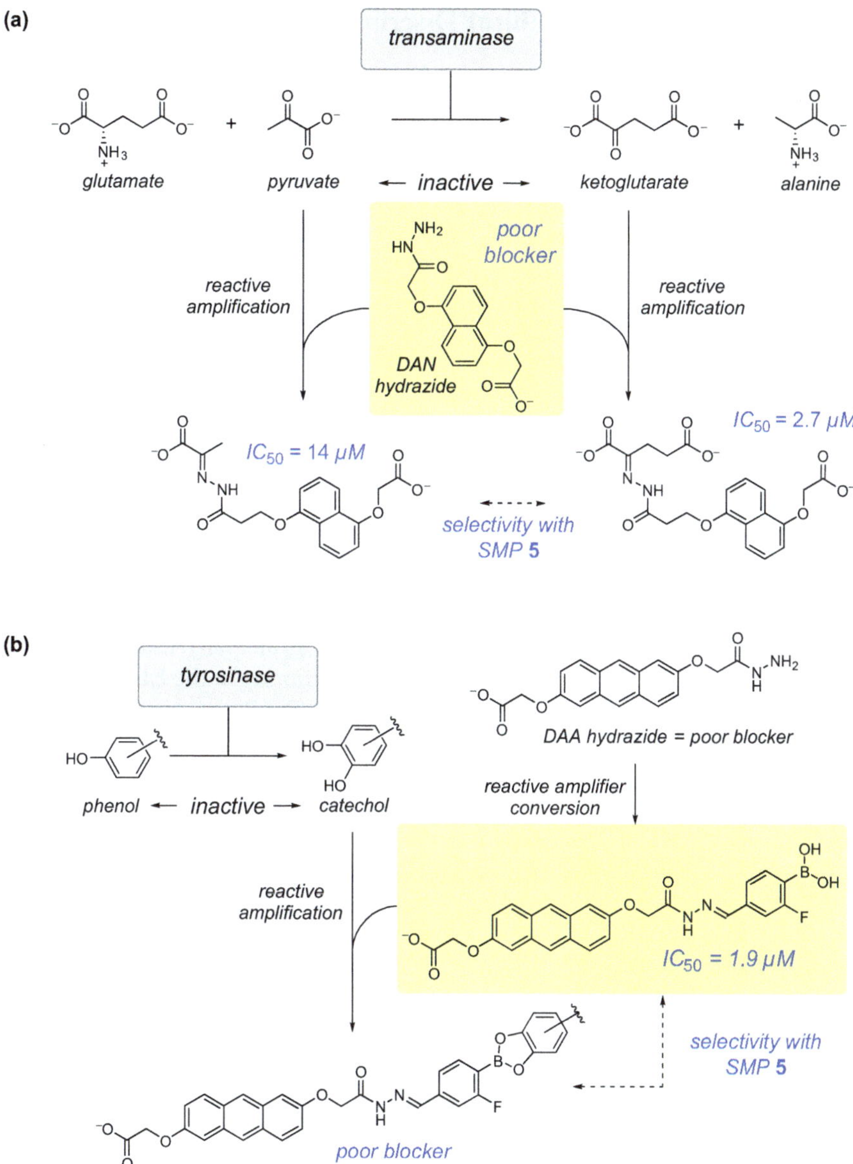

Figure 12.15 Alternative reactive amplifiers used in supramolecular enzyme assays. (a) Conversion of inactive pyruvate and ketoglutarate into their dialkoxynaphthalene (DAN) hydrazones affords active pore blockers with sufficient selectivity to discriminate between both reaction products. (b) Conversion of dialkoxyanthracene (DAA) hydrazide into a boronic acid derivative for amplification of catechols.

12.5.3 Biosensing and Chiral Discrimination

The notion that the selectivity in most supramolecular enzyme assays is based on relatively basic supramolecular interactions, mainly on alteration of the charge status during enzymatic conversion, led to the idea that supramolecular enzyme assays could be used for the development of multianalyte sensor arrays.[51,71,79] The outcome of this is the conception that all enzyme assays demonstrated so far may also be considered as a suprabiomolecular sensing ensemble.

The overall concept of considering supramolecular enzyme assays as suprabiomolecular sensing ensembles is gleaned from the basic principle of a biosensor as shown in Figure 12.16. Therein, a highly specific biomolecular recognition is used for signal generation and coupled to a general signal transduction mechanism. The most prominent example is the glucose biosensor, in which the current generated by glucose oxidase is transduced into an amperometric signal. Another example, although not a biosensor according to the IUPAC definition, is the enzyme-linked immunsorbant assay (ELISA). Therein, the signal is specifically generated by antigen binding of the Fab fragment of the primary antibody and transduced by binding of the secondary antibody equipped with an enzyme. In both cases, the transducer is universal, for eample, the amperometric transducer can be principally used in combination with any other oxidase and the secondary antibody in the ELISA binds to the constant Fc region of any primary antibody. Similarly, in suprabiomolecular sensing ensembles, the same supramolecular detection system can be coupled to a variety of different enzymatic reactions.

Remarkable examples realized with supramolecular enzyme assays include a sensor array for amino acids based on the amino acid decarboxylases and their detection by a supramolecular tandem assay,[71] an artificial tongue based on SMPs for the sensation of sweet, sour and the umami flavor,[52] as well as analyte detection in matrices as complex as, for example, various beverages,[52,60] sour milk,[68] or blood serum.[67]

Another elegant example is the quantification of acetylcholine and choline by varying the order of reagent addition and carefully adjusting enzyme concentrations in an enzyme-coupled tandem assay involving acetylcholine esterase and choline oxidase. Choline could be quantified by first adding choline oxidase to convert all choline into betaine and acetylcholine could be quantified by subsequent addition of acetylcholine esterase. Generally, in supramolecular tandem assays, it appeared faster and more reproducible to determine analyte

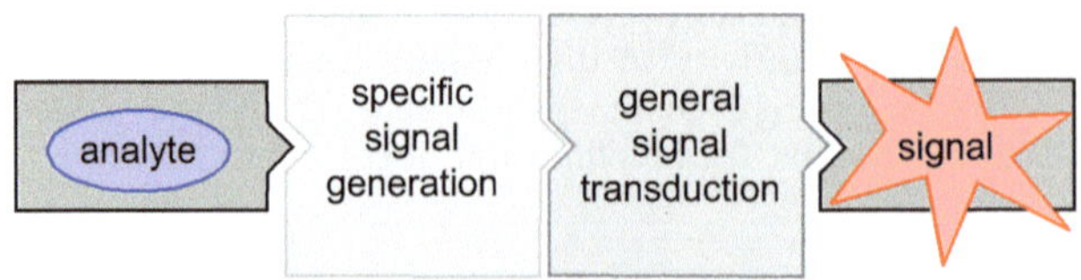

Figure 12.16　General working principle of biosensors.

concentrations from a linear correlation between the initial rates of fluorescence response and the analyte concentration than from endpoint fluorescence values.[71,76]

Another application of the sensing principle depicted in Figure 12.16 is the determination of the enantiomeric excess (ee) of a reaction mixture or, more general, chiral discrimination. This area is of interest to bioengineers and synthetic organic chemists for discovering enzymes, chiral catalysts, and auxiliaries for asymmetric functional group transformations. To advance this scientific area, true HTS approaches are urgently needed, in particular a rapid and reliable determination of the enantiomeric excess of several thousand samples per day.[96] A key requirement is that the assay works with unlabeled substrates, because screening libraries of modified, labeled substrates is unlikely to provide full access to all catalysts for the desired transformation of the unlabeled substrates. The combination of a highly enantioselective, biomolecular signal generation and a supramolecular enzyme assay as transducer could thus become highly useful in this research area. Demonstrated examples of *ee* determination with supramolecular enzyme assays include enantiomerically enriched mixtures of polyglutamate,[97] lactate,[98] and various amino acids,[71] and the assays with the respective enzymes subtilisin A, lactate oxidase, and amino acid decarboxylases have been described in previous sections. Strikingly, all assays have been extremely sensitive and accurate and provided *ee* values in a range not accessible with conventional methods, for example, up to 99.98% for amino acid mixtures and >99.99% for polyglutamate.[71,98]

12.6 Summary and Conclusions

In view of ongoing advancements in applying supramolecular systems to biology, current approaches to supramolecular enzyme assays have just taken the first step beyond proof-of-principle studies. So far, the principal viability of supramolecular enzyme assays has been demonstrated for nearly all enzyme classes except EC6 (see Table 1 for an overview of all currently established supramolecular enzyme assays). Thereby, supramolecular enzyme assays, nearly ideally, address all needs for timely biomolecular screening applications, that is, they are fluorescence-based, affordable, label-free, antibody-free and can be applied to a relatively large variety of enzymatic reactions. This affords assays for enzymes, which are otherwise difficult to address with more conventional methods, for example, enzyme assays involving polymeric substrates as well as enzyme assays for highly selective enzymes involving small substrates have been demonstrated.

Regarding future developments, it is highly likely that lessons learned in the design of supramolecular enzyme assays will also be applicable to the monitoring of other catalytic activities, such as for screening applications in organic synthesis.[99–102] It would also be highly useful to develop and utilize receptors for ubiquitous co-substrates, like FAD or NADP. This would pave the way for assaying a wide variety of different enzymes as demonstrated with supramolecular enzyme assays involving ATP as a co-substrate. If desired, macrocycles

Table 12.1 Literature overview of demonstrated supramolecular enzyme assays.

EC number	Name	Subclass	Examples	Method	References
1	Oxidoreductases	1.1. (CH–OH)	choline oxidase	SuMoTa	76
			cholesterol oxidase	polymer with amplifier	67
			glucose oxidase	chemosensor	44
			lactate oxidase[a]	SMP/polymer with amplifier	52, 68
		1.4. (CH–NH)	diamine oxidase	SuMoTa	73
		1.14 (CH–OH, paired)	tyrosinase	SMP with amplifier	95
2	Transferases	2.1. (C-transfer)	histone-lysine N-methyltransferase	SuMoTa	78
		2.4. (glycosyl)	galactosyl transferase	SMP	50, 59
			α-1,3-galactosyltransferase	chemosensor	40
			β-1,4-galactosyltransferase	chemosensor	40
			α-2,3-sialyltransferase	chemosensor	40
			sucrose phosphorylase	chemosensor	45
		2.6. (N-transfer)	transaminase	SMP with amplifier	52
		2.7. (P-transfer)	acetate kinase	SMP	52
			DNA polymerase	SMP	59
			hexokinase	SMP, polymer	52, 53, 58, 60
			phosphofructokinase	SMP	58
3	Hydrolases	3.1. (ester)	acetylcholine esterase	SuMoTa	76
			alkaline phosphatase	SMP	50
			DNA exonuclease	SMP	59
			phosphodiesterase	chemosensor	38, 39
			phytase	SMP	62
			pig liver esterase	SMP	61
			RNase A	SMP	59

	3.2. (glycosylase)	β-galactosidase	SMP coupled assay	52
		hyaluronidase	SMP, polymer	53, 59
		invertase	SMP coupled assay	52, 60
	3.4. (peptide)	leucine amino peptidase	SuMoTa, pM sensor	46, 77
		trypsin	SuMoTa	77
		thermolysine	SuMoTa	75
		pronase, papain, ficin, elastase, subtilisin	SMP	59
	3.5. (C–N, not peptide)	acylase I	pM sensor	46
		arginase	SuMoTa	73
	3.6. (anhydride)	apyrase	SuMoTa, SMP	50, 74
		pyrophosphatase	chemosensor	37
4	Lyases	4.1. (C–C)	amino acid decarboxylase	SuMoTa	69, 71, 72
			aldolase A	SMP	50
			citrate lyase	SMP with amplifier	52
		4.2. (C–O)	heparinase	SMP	59
5	Isomerases	5.3. (intra OxRed)[b]	triose-phosphate isomerase	SMP coupled assay	50
		5.4. (mutases)	phosphoglucomutase	chemosensor	45
6	Ligases		n.a.	n.a.	n.a.

[a]Lactate oxidase has meanwhile been transferred to EC 1.13.12.4.
[b]intra OxRed = intramolecular oxidoreductase.
n.a. = not available. A supramolecular tandem assay for monitoring butyrylcholinesterase has also recently been published (see ref. 103).

may also be tailored to afford highly selective and sensitive host molecules for a specific product of an enzymatic reaction. One successful example of an engineered macrocyclic host, albeit for therapeutic applications, is the cyclodextrin-based drug sugammadex, which has been described in Chapter 5, Section 11.6. Moreover, related concepts can be transferred from one field to another field; for instance, reactive amplification established for membrane transport-based enzyme assays may also be useful for supramolecular tandem assays to increase their sensitivity or selectivity, or the molecular recognition capabilities of cell-penetrating macrocycles (see *e.g.* cyclodextrin-based non-viral gene vectors in Chapter 5, Section 9) may be utilized in membrane transport-based enzyme assays. Ultimately, other supramolecular receptors and physicochemical readouts other than fluorescence or absorption spectroscopic measurements may be used.

References

1. M. J. Blackman, J. E. T. Corrie, J. C. Croney, G. Kelly, J. F. Eccleston and D. M. Jameson, *Biochemistry*, 2002, **41**, 12 244–12 252.
2. K. Licha and C. Olbrich, *Adv. Drug Delivery Rev.*, 2005, **57**, 1087–1108.
3. J.-O. Deguchi, M. Aikawa, C.-H. Tung, E. Aikawa, D.-E. Kim, V. Ntziachristos, R. Weissleder and P. Libby, *Circulation*, 2006, **114**, 55–62.
4. Q. Wang, S. M. Cahill, M. Blumenstein and D. S. Lawrence, *J. Am. Chem. Soc.*, 2006, **128**, 1808–1809.
5. M. Kamiya, H. Kobayashi, Y. Hama, Y. Koyama, M. Bernardo, T. Nagano, P. L. Choyke and Y. Urano, *J. Am. Chem. Soc.*, 2007, **129**, 3918–3929.
6. J.-H. Kim, S. Lee, K. Park, Hae Y. Nam, Soon Y. Jang, I. Youn, K. Kim, H. Jeon, R.-W. Park, I.-S. Kim, K. Choi and I. C. Kwon, *Angew. Chem. Int. Ed.*, 2007, **46**, 5779–5782.
7. D. R. Elias, D. L. J. Thorek, A. K. Chen, J. Czupryna and A. Tsourkas, *Cancer Biomark.*, 2008, **4**, 287–305.
8. P. Libby, *Nature*, 2002, **420**, 868–874.
9. J. Neefjes and N. P. Dantuma, *Nat. Rev. Drug Discov.*, 2004, **3**, 58–69.
10. D. A. Fidock, P. J. Rosenthal, S. L. Croft, R. Brun and S. Nwaka, *Nat. Rev. Drug Discov.*, 2004, **3**, 509–520.
11. J. Zhang, P. L. Yang and N. S. Gray, *Nat. Rev. Cancer*, 2009, **9**, 28–39.
12. B. De Strooper, R. Vassar and T. Golde, *Nat. Rev. Neurol.*, 2010, **6**, 99–107.
13. C. Lopez-Otin and T. Hunter, *Nat. Rev. Cancer*, 2010, **10**, 278–292.
14. K. Naylor and R. Eastell, *Nat. Rev. Rheumatol.*, 2012, **8**, 379–389.
15. F. Bordusa, *Chem. Rev.*, 2002, **102**, 4817–4868.
16. J.-L. Reymond, V. S. Fluxà and N. Maillard, *Chem. Commun.*, 2009, 34–46.
17. L. Hedstrom, *Chem. Rev.*, 2002, **102**, 4501–4524.

18. A. Aharoni, A. D. Griffiths and D. S. Tawfik, *Curr. Opin. Chem. Biol.*, 2005, **9**, 210–216.
19. K. L. Tee and U. Schwaneberg, *Comb. Chem. High Throughput Screening*, 2007, **10**, 197–217.
20. C. Jäckel and D. Hilvert, *Curr. Opin. Biotechnol.*, 2010, **21**, 753–759.
21. M. T. Reetz, *Angew. Chem. Int. Ed.*, 2011, **50**, 138–174.
22. S. J. Taylor and J. P. Morken, *Science*, 1998, **280**, 267–270.
23. M. T. Reetz, M. H. Becker, M. Liebl and A. Fürstner, *Angew. Chem. Int. Ed.*, 2000, **39**, 1236–1239.
24. E. Stavitski and B. M. Weckhuysen, *Chem. Soc. Rev.*, 2010, **39**, 4615–4625.
25. F. E. Torres, P. Kuhn, D. De Bruyker, A. G. Bell, M. V. Wolkin, E. Peeters, J. R. Williamson, G. B. Anderson, G. P. Schmitz, M. I. Recht, S. Schweizer, L. G. Scott, J. H. Ho, S. A. Elrod, P. G. Schultz, R. A. Lerner and R. H. Bruce, *Proc. Natl Acad. Sci. USA*, 2004, **101**, 9517–9522.
26. M. I. Recht, F. E. Torres, D. D. Bruyker, A. G. Bell, M. Klumpp and R. H. Bruce, *Anal. Biochem.*, 2009, **388**, 204–212.
27. J. Comley, *Drug Discov. World*, 2006, **Winter 2006/7**, 27–50.
28. G. V. Oshovsky, D. N. Reinhoudt and W. Verboom, *Angew. Chem. Int. Ed.*, 2007, **46**, 2366–2393.
29. T. Rehm and C. Schmuck, *Chem. Commun.*, 2008, 801–813.
30. D. A. Uhlenheuer, K. Petkau and L. Brunsveld, *Chem. Soc. Rev.*, 2010, **39**, 2817–2826.
31. S. Kubik, *Chem. Soc. Rev.*, 2010, **39**, 3648–3663.
32. Z. Laughrey and B. C. Gibb, *Chem. Soc. Rev.*, 2010, **40**, 363–386.
33. J. M. Zayed, N. Nouvel, U. Rauwald and O. A. Scherman, *Chem. Soc. Rev.*, 2010, **39**, 2806–2816.
34. H.-J. Schneider, *Angew. Chem. Int. Ed.*, 2009, **48**, 3924–3977.
35. I. Beis and E. A. Newsholme, *Biochem. J.*, 1975, **152**, 23–32.
36. E. A. Weitz, J. Y. Chang, A. H. Rosenfield and V. C. Pierre, *J. Am. Chem. Soc.*, 2012, **134**, 16099–16102.
37. D. H. Vance and A. W. Czarnik, *J. Am. Chem. Soc.*, 1994, **116**, 9397–9398.
38. S. Mizukami, T. Nagano, Y. Urano, A. Odani and K. Kikuchi, *J. Am. Chem. Soc.*, 2002, **124**, 3920–3925.
39. M. S. Han and D. H. Kim, *Bioorg. Med. Chem. Lett.*, 2003, **13**, 1079–1082.
40. J. Wongkongkatep, Y. Miyahara, A. Ojida and I. Hamachi, *Angew. Chem. Int. Ed.*, 2006, **45**, 665–668.
41. F. Cramer, W. Saenger and H. C. Spatz, *J. Am. Chem. Soc.*, 1967, **89**, 14–20.
42. S. L. Wiskur, H. Ait-Haddou, J. J. Lavigne and E. V. Anslyn, *Acc. Chem. Res.*, 2001, **34**, 963–972.
43. B. T. Nguyen and E. V. Anslyn, *Coord. Chem. Rev.*, 2006, **250**, 3118–3127.
44. T. Zhang and E. V. Anslyn, *Org. Lett.*, 2007, **9**, 1627–1629.

45. B. Vilozny, A. Schiller, R. A. Wessling and B. Singaram, *Anal. Chim. Acta*, 2009, **649**, 246–251.

46. G. Klein and J.-L. Reymond, *Angew. Chem. Int. Ed.*, 2001, **40**, 1771–1773.

47. S. Matile, *Chem. Soc. Rev.*, 2001, **30**, 158–167.

48. N. Sakai, J. Mareda and S. Matile, *Acc. Chem. Res.*, 2005, **38**, 79–87.

49. N. Sakai, J. Mareda and S. Matile, *Acc. Chem. Res.*, 2008, **41**, 1354–1365.

50. G. Das, P. Talukdar and S. Matile, *Science*, 2002, **298**, 1600–1602.

51. G. Das and S. Matile, *Chem. Eur. J*, 2006, **12**, 2936–2944.

52. S. Litvinchuk, H. Tanaka, T. Miyatake, D. Pasini, T. Tanaka, G. Bollot, J. Mareda and S. Matile, *Nat. Mater.*, 2007, **6**, 576–580.

53. T. Miyatake, M. Nishihara and S. Matile, *J. Am. Chem. Soc.*, 2006, **128**, 12420–12421.

54. S. M. Butterfield, A. Hennig and S. Matile, *Org. Biomol. Chem.*, 2009, **7**, 1784–1792.

55. A. Hennig and S. Matile, *Chirality*, 2008, **20**, 932–937.

56. S. Matile, N. Sakai and A. Hennig, in *Supramolecular Chemistry: From Molecules to Nanomaterials*, P. A. Gale and J. W. Steed, Wiley, Chichester, 2012, pp. 473–500.

57. S. Matile and N. Sakai, in *Analytical Methods in Supramolecular Chemistry*, ed. C. A. Schalley, Wiley, Weinheim, 2012, pp. 711–742.

58. N. Sordé and S. Matile, *Biopolymers*, 2004, **76**, 55–65.

59. N. Sordé, G. Das and S. Matile, *Proc. Natl Acad. Sci. USA*, 2003, **100**, 11964–11969.

60. S. Litvinchuk, N. Sordé and S. Matile, *J. Am. Chem. Soc.*, 2005, **127**, 9316–9317.

61. V. Gorteau, F. Perret, G. Bollot, J. Mareda, A. N. Lazar, A. W. Coleman, D.-H. Tran, N. Sakai and S. Matile, *J. Am. Chem. Soc.*, 2004, **126**, 13 592–13 593.

62. S. M. Butterfield, D.-H. Tran, H. Zhang, G. D. Prestwich and S. Matile, *J. Am. Chem. Soc.*, 2008, **130**, 3270–3271.

63. N. Sakai and S. Matile, *J. Am. Chem. Soc.*, 2003, **125**, 14 348–14 356.

64. F. Perret, M. Nishihara, T. Takeuchi, S. Futaki, A. N. Lazar, A. W. Coleman, N. Sakai and S. Matile, *J. Am. Chem. Soc.*, 2005, **127**, 1114–1115.

65. M. Nishihara, F. Perret, T. Takeuchi, S. Futaki, A. N. Lazar, A. W. Coleman, N. Sakai and S. Matile, *Org. Biomol. Chem.*, 2005, **3**, 1659–1669.

66. N. Sakai, S. Futaki and S. Matile, *Soft Matter*, 2006, **2**, 636–641.

67. S. M. Butterfield, T. Miyatake and S. Matile, *Angew. Chem. Int. Ed.*, 2009, **48**, 325–328.

68. A. Hennig, G. J. Gabriel, G. N. Tew and S. Matile, *J. Am. Chem. Soc.*, 2008, **130**, 10338–10344.

69. A. Hennig, H. Bakirci and W. M. Nau, *Nat. Meth.*, 2007, **4**, 629–632.

70. A. Hennig, H. Bakirci and W. M. Nau, *Nat. Protoc.*, 2007, DOI: 10.1038/nprot.2007.281.

71. D. M. Bailey, A. Hennig, V. D. Uzunova and W. M. Nau, *Chem. Eur. J.*, 2008, **14**, 6069–6077.

72. A. Praetorius, D. M. Bailey, T. Schwarzlose and W. M. Nau, *Org. Lett.*, 2008, **10**, 4089–4092.

73. W. M. Nau, G. Ghale, A. Hennig, H. Bakirci and D. M. Bailey, *J. Am. Chem. Soc.*, 2009, **131**, 11558–11570.

74. M. Florea and W. M. Nau, *Org. Biomol. Chem.*, 2010, **8**, 1033–1039.

75. G. Ghale, V. Ramalingam, A. R. Urbach and W. M. Nau, *J. Am. Chem. Soc.*, 2011, **133**, 7528–7535.

76. D.-S. Guo, V. D. Uzunova, X. Su, Y. Liu and W. M. Nau, *Chem. Sci.*, 2011, **2**, 1722–1734.

77. G. Ghale, N. Kuhnert and W. M. Nau, *Nat. Prod. Commun.*, 2012, **7**, 343–348.

78. M. Florea, S. Kudithipudi, A. Rei, M. J. González-Álvarez, A. Jeltsch and W. M. Nau, *Chem. Eur. J.*, 2012, **18**, 3521–3528.

79. R. N. Dsouza, A. Hennig and W. M. Nau, *Chem. Eur. J.*, 2012, **18**, 3444–3459.

80. R. N. Dsouza, U. Pischel and W. M. Nau, *Chem. Rev.*, 2011, **111**, 7941–7980.

81. J. M. Chinai, A. B. Taylor, L. M. Ryno, N. D. Hargreaves, C. A. Morris, P. J. Hart and A. R. Urbach, *J. Am. Chem. Soc.*, 2011, **133**, 8810–8813.

82. P. Geymayer, N. Bahr and J.-L. Reymond, *Chem. Eur. J.*, 1999, **5**, 1006–1012.

83. D. M. Jameson and J. A. Ross, *Chem. Rev.*, 2010, **110**, 2685–2708.

84. K. Uekama, F. Hirayama and T. Irie, *Chem. Rev.*, 1998, **98**, 2045–2076.

85. A. Hennig, G. Ghale and W. M. Nau, *Chem. Commun.*, 2007, 1614–1616.

86. I. Ghosh and W. M. Nau, *Adv. Drug Deliv. Rev.*, 2012, **64**, 764–783.

87. X.-F. Shang, H. Su, H. Lin and H.-K. Lin, *Inorg. Chem. Commun.*, 2010, **13**, 999–1003.

88. Y. Zhou, Z. Xu and J. Yoon, *Chem. Soc. Rev.*, 2011, **40**, 2222–2235.

89. K. P. Lu, G. Finn, T. H. Lee and L. K. Nicholson, *Nat. Chem. Biol.*, 2007, **3**, 619–629.

90. G. Fischer, H. Bang and C. Mech, *Biomed. Biochim. Acta*, 1984, **43**, 1101–1111.

91. A. C. Storer and A. Cornish-Bowden, *Biochem. J.*, 1974, **141**, 205–209.

92. P. Talukdar, G. Bollot, J. Mareda, N. Sakai and S. Matile, *J. Am. Chem. Soc.*, 2005, **127**, 6528–6529.

93. S. Hagihara, L. Gremaud, G. Bollot, J. Mareda and S. Matile, *J. Am. Chem. Soc.*, 2008, **130**, 4347–4351.

94. S. Hagihara, H. Tanaka and S. Matile, *J. Am. Chem. Soc.*, 2008, **130**, 5656–5657.

95. S. Hagihara, H. Tanaka and S. Matile, *Org. Biomol. Chem.*, 2008, **6**, 2259–2262.

96. D. Leung, S. O. Kang and E. V. Anslyn, *Chem. Soc. Rev.*, 2012, **41**, 448–479.

97. H. Tanaka and S. Matile, *Chirality*, 2008, **20**, 307–312.
98. A. Hennig, S. Hagihara and S. Matile, *Chirality*, 2009, **21**, 826–835.
99. C. Portal and M. Bradley, *Org. Biomol. Chem.*, 2007, **5**, 587–592.
100. J. Montgomery, *Science*, 2011, **333**, 1387–1388.
101. E. Jung, S. Kim, Y. Kim, S. H. Seo, S. S. Lee, M. S. Han and S. Lee, *Angew. Chem. Int. Ed.*, 2011, **50**, 4386–4389.
102. J. A. Friest, S. Broussy, W. J. Chung and D. B. Berkowitz, *Angew. Chem. Int. Ed.*, 2011, **50**, 8895–8899.
103. D.-S. Guo, J. Yang, Y. Liu, *Chem. Eur. J.*, 2013, doi:10.1002/chem.201300980.

Constitutional Dynamic Chemistry for Bioactive Compounds

YAN ZHANG, LEI HU AND OLOF RAMSTRÖM*

KTH Royal Institute of Technology, Department of Chemistry,
Teknikringen 30, S-10044 Stockholm, Sweden
*Email: ramstrom@kth.se

13.1 Introduction

Supramolecular chemistry has over the last decades been established as an efficient method to generate and study molecular complexation primarily based on non-covalent interactions of discrete molecules, generating systems that possess adaptive features. In a wider perspective, adopting the dynamic nature of supramolecular chemistry, constitutional dynamic chemistry (CDC) has more recently emerged as a concept based on reversible-covalent or non-covalent interactions, thus encompassing both the supramolecular and the molecular levels.[1–10] With all the components mutually undergoing reversible interactions in the systems, any internal pressure, such as parameter changes, crystallization, or external selectors (*e.g.* metal ions, enzymes), can cause rearrangement of the system's constitution and finally lead to amplifications of the optimal components.

The establishment of the dynamic systems generally includes three important elements:[1,9] first, initial building blocks able to reversibly interact with each other; second, reversible reactions/interactions to make the system dynamic;

Monographs in Supramolecular Chemistry No. 13
Supramolecular Systems in Biomedical Fields
Edited by Hans-Jörg Schneider

Published by the Royal Society of Chemistry, www.rsc.org

and third, selection pressures to amplify the fittest or expel the weakest constituents from the system.

When choosing the initial building blocks for dynamic systems, not only their structural similarities should be considered—thus having the same functional groups to undergo reversible reactions/interactions—but also their variations need to be addressed in order to generate larger diversity and thus maximize the discrimination under the chosen selection pressures. Furthermore, when a target is present in the system as the selector, the structure of the building blocks should be designed to interact with the active sites of the targets in a complementary way.

As the key point of the dynamic system, a range of reversible reactions/ interactions has been developed. Table 13.1 represents a summary of most of

Table 13.1 Reversible reaction/interaction types used in dynamic systems.

Reversible covalent reactions	
Imine-related reactions	
Imine formation	$R_1CHO + R_2-NH_2 \rightleftharpoons R_1CH=N-R_2 + H_2O$
Transimination	$R_1CH=N-R_2 + R_3-NH_2 \rightleftharpoons R_1CH=N-R_3 + R_2-NH_2$
Imine metathesis	$R_1CH=N-R_2 + R_3CH=N-R_4 \rightleftharpoons R_1CH=N-R_4 + R_3CH=N-R_2$
Disulfide-related reactions	
Disulfide formation	$R_1-SH + R_2-SH \rightleftharpoons R_1-S-S-R_2$
Disulfide exchange	$R_1-S-S-R_2 + R_3-SH \rightleftharpoons R_1-S-S-R_3 + R_2-SH$
Disulfide metathesis	$R_1-S-S-R_2 + R_3-S-S-R_4 \rightleftharpoons R_1-S-S-R_4 + R_3-S-S-R_2$
Aldol-type reactions	
Aldol addition	$R_1CHO + R_2CH_2COR_3 \rightleftharpoons R_1CH(OH)CH(R_2)COR_3$
Nitroaldol reaction	$R_1CHO + R_2R_3CH-NO_2 \rightleftharpoons R_1CH(OH)C(R_1R_2)NO_2$
Hemiacetal-type reactions	
Hemiacetal formation	$R_1CHO + R_2-OH \rightleftharpoons R_1CH(OH)O-R_2$

Table 13.1   (*Continued*)

Reversible covalent reactions

Hemithioacetal formation	R_1-CHO + R_2-SH $\rightleftharpoons$ R_1-CH(OH)-S-R_2
Acetal exchange	R_3O-C(R_1)(R_2)-OR_4 + R_5-OH $\rightleftharpoons$ R_3O-C(R_1)(R_2)-OR_5 + R_4-OH
Transthiolesterification	R_1-C(O)-S-R_2 + R_3-SH $\rightleftharpoons$ R_1-C(O)-S-R_3 + R_2-SH
Transesterification	R_1-C(O)-O-R_2 + R_3-OH $\rightleftharpoons$ R_1-C(O)-O-R_3 + R_2-OH

Cyanide addition reactions

Cyanohydrin formation	R_1-CHO + HCN $\rightleftharpoons$ R_1-CH(OH)-CN
Strecker reaction	R_1-CH=N-R_2 + HCN $\rightleftharpoons$ R_1-CH(NH-R_2)-CN
Michael-type reactions	CH_2=C(R_2)-C(O)-R_1 + R_3-SH $\rightleftharpoons$ R_3-S-CH_2-CH(R_2)-C(O)-R_1
Boronate ester formation	R_1-B(OH)$_2$ + HO-CH(R_2)-CH(R_3)-OH $\rightleftharpoons$ R_1-B(O-CH(R_2)-CH(R_3)-O) + 2 H_2O
Diels–Alder reactions	diene (R_1, R_2) + dienophile (R_3, R_4) $\rightleftharpoons$ cyclohexene (R_1, R_2, R_3, R_4)
Alkene metathesis reactions	R_1,R_1-C=CH$_2$ + R_2,R_2-C=CH$_2$ $\rightleftharpoons$ R_1-CH=CH-R_2 + R_1-CH=CH-R_2

Reversible non-covalent interactions

Metal coordination	M^{m+} + nL $\rightleftharpoons$ $[ML_n]^{m+}$
Electrostatic interactions	R_1-COO^- + R_2-$\overset{+}{N}H_3$ $\rightleftharpoons$ R_1-$COO^-\cdots H_3\overset{+}{N}$-$R_2$
Hydrogen bonding	R_1(R_2)C=O + H-N(R_4)-C(O)-R_3 $\rightleftharpoons$ R_1(R_2)C=O$\cdots$H-N(R_4)-C(O)-R_3

the reaction/interaction types used in CDC, among which imine and disulfide-related reversible reactions are the most widely used methods for dynamic chemistry, often characterized by milder reaction conditions, reasonably rapid exchange rates and, for disulfide chemistry, straightforward means to halting the equilibrations. These reaction types are also generally compatible with many biomolecular targets, making them attractive for use in bio-related dynamic systems. On the other hand, some of the reversible reaction/interaction types are rarely used for bioactive compounds, for example, Diels–Alder reactions, because of the comparatively harsh conditions. Moreover, some reversible reactions have recently been established and applied to dynamic systems, for example, hemithioacetal formation, nitroaldol reactions, cyanide addition reactions and boronic ester formation reactions, which have expanded the scope of reversible reactions used in dynamic chemistry, generating more diversity for this field. However, discovering new reversible reaction types is still one of the biggest challenges in CDC.

Once the dynamic systems have been generated, various selection pressures can be applied to the system depending on the different purposes. In this context, there are essentially three major mechanisms for the selection:[11] receptor-induced substrate assembly; substrate-induced receptor assembly; and induced self-assembly processes to form molecular entities and networks. When the fittest structure has been stabilized by the target or by itself, the other constituents will generate more of this optimal structure, and in most cases will reach an amplification effect.

13.2   Applications of CDC for Bioactive Compounds

Among the fields that CDC has been applied to, biological systems are very attractive. Targeting dynamic systems with biological entities, like proteins, nucleotides and even whole cells, can provide more information about the targets together with their substrates, inhibitors, *etc.*, leading to a better understanding of the biological system and enabling potential drug discovery.

Owing to the inherent properties of many biological entities targeted by dynamic systems, the reactions should be conducted in aqueous buffered solutions under mild conditions. In part for this reason, disulfides, imines and metal coordination are the most frequently used reversible reactions/interactions in CDC for biological applications, other reasons being for example the lack of experience with other reaction formats. Although other reaction types have been exemplified, they have been less preferred. However, other reactions do have the potential to be useful also with biological targets, and, for example, the use of biocatalysis in organic solvents has opened up several new reaction types for CDC, including the reversible nitroaldol and Strecker reactions. These examples extend the list of possible reaction types and the performing conditions of CDC for bioactive compounds.

In this section, a list of recent examples where dynamic systems have been applied to bioactive compounds, with their various biological targets, will be illustrated grouped according to their respective reaction/interaction types.

13.2.1 CDC with Reversible Disulfide Reactions

Disulfide formation, exchange, and metathesis reactions are among the most robust reversible reactions used in CDC, especially attractive for biological systems, owing to their mild reaction conditions, stability of the constructs and ease of operation.[12–14] Disulfide formation from thiols can be performed by mild oxidation, for example by air, and disulfide exchange can simultaneously occur under basic conditions. Alternatively, the exchange can be established directly from the disulfides, either initiated by mild reducing agents such as dithiothreitol (DTT),[15] or using metathesis mediation by phosphines.[16] The reactions are generally pH-dependent. Under neutral conditions (pH 7–9) the disulfide exchange is fast enough to generate dynamic systems, while acidic conditions allow the reactions to be easily halted for analyses, which is also an advantage when applied in dynamic chemistry.

A resin-bound dynamic chemistry strategy based on disulfide formation and exchange reactions has been exemplified. Starting from resin-attached components, the dynamic system with disulfide exchanges was allowed to reach equilibrium in the presence of a fluorescently labeled target. With physical removal of resin beads, the active components with high-affinity could be easily isolated and identified using subsequent spectral analysis.[17] This method provides a good solution to the problem of analyzing considerably larger dynamic systems, and has been applied to various biological targets. In a recent example, a dynamic system with a theoretical size of 11 325 members has been generated from 150 cysteine-containing components.[18] Disulfide exchange was carried out in this system with the presenting target (CUG) repeat RNA, which is a splicing regulator for an RNA-mediated disease called myotonic dystrophy type 1 (DM1). As a result, four selected compounds with K_i values in the micromolar range were detected from this system. Furthermore, using one of the hit compounds (**1**, Figure 13.1) as starting point, replacing the disulfide bridge with an olefin resulted in the synthesis and evaluation of several modified compounds.[19] Some of them possessed high-affinity to (CUG) repeat RNA both *in vitro* and *in vivo*. Better selectivity and lower toxicity effects were also observed by one of the compounds (**2**, Figure 13.1), which showed good potential as a lead drug for myotonic dystrophy therapy.

A concept of phase-transfer dynamic system using disulfide formation reaction has also been developed.[20] Starting with water-soluble building block **3**, chloroform-soluble building blocks **4** and **5**, disulfide formation and exchange reactions were performed in a two-phase system (Figure 13.2), using an unbuffered aqueous phase at pH 7.4 and a chloroform organic phase. High performance liquid chromatography (HPLC) analyses were chosen in this case to monitor the reactions. Following slow oxidization and simultaneous equilibration, water-soluble building block **3** was transferred to the organic phase in the form of trimers **4·3·4**, **4·3·5** and **5·4·5**. Different organic bases such as *N*-methylmorpholine (NMM), triethylamine (TEA), tetra-methylethylenediamine (TMEDA), Tris–HCl buffer (1 M, pH 7.4) and dime-thylpiperazine were also tested, resulting in variation of building block **3**

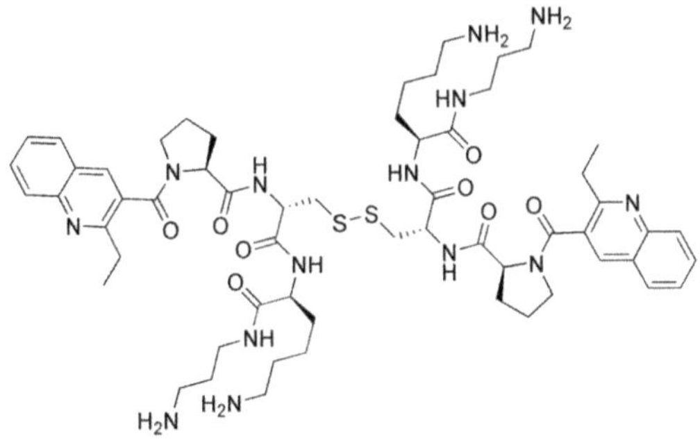

1

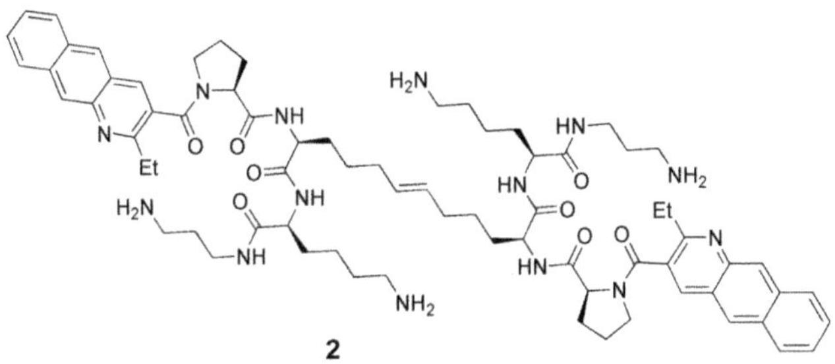

2

Figure 13.1 Hit compound identified from resin-bound dynamic system and the modified lead-drug targeting myotonic dystrophy.

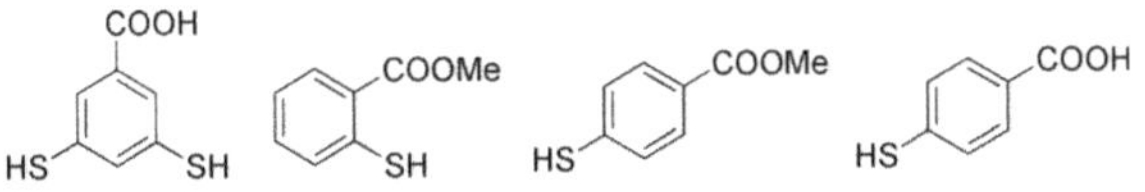

3   **4**   **5**   **6**

Figure 13.2 Thiols as initial building blocks used in the phase-transfer dynamic system.

transfer between 4% and 33%. However, quantitative transfer of building block **3** could be achieved by addition of tributylamine. This dynamic system was further applied to a bulk liquid membrane, with building blocks **3** and **6** (Figure 13.2) distributed in two aqueous phases, separated by chloroform with added tributylamine. The appearance of most building blocks in the organic phase in the form of **6 · 1 · 6** indicated the transfer of hydrophilic building blocks through the bulk organic solvent.

Thiol-disulfide exchange reactions were also used for probing adenosine recognition sites.[21] Based on building block **7**, which was envisaged to interact with the protein's active site, eight other thiols were selected in the dynamic system, both of hydrophobic character (**8a–d**) and with potential ability to participate in charge–charge and hydrogen-bond interactions (**8e–h**) (Figure 13.3). The equilibrium was reached in a glutathione redox buffer (at pH 8.5 under an inert atmosphere) within 24 h. HPLC-assisted monitoring of the

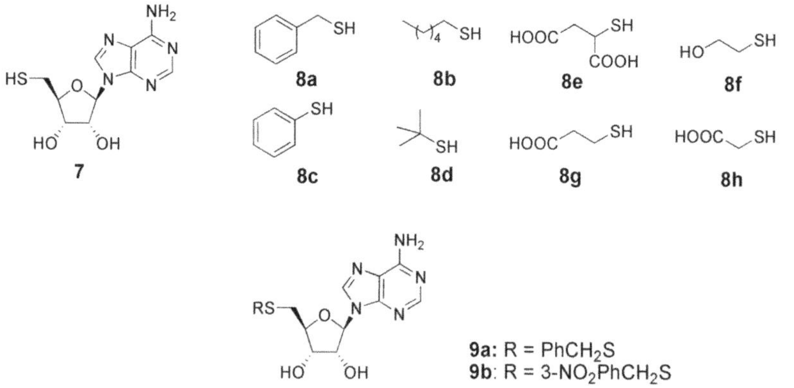

Figure 13.3 Thiols as building blocks used in the dynamic system for probing adenosine recognition sites, and the modified, improved ligand.

dynamic system showed that the disulfide **7**-glutathione was the major constituent when no protein target was present. However, when adding pantothenate synthetase, an adenosine-binding enzyme, into the system, the disulfide (**9a**) formed between **7** and **8a** was also amplified. Further biophysical studies of **9a**, and structural modifications, led to an even better binder (**9b**) of pantothenate synthetase from *Mycobacterium tuberculosis*, with a K_D value of 80 µM (Figure 13.3).

Besides these examples, several other dynamic systems using disulfide formation and exchange for biological applications were recently reported, such as exploration of differential recognition of DNA G-quadruplexes,[22] self-assembly of high-affinity receptors for spermine,[23] discovery of bifunctional ligands for calmodulin,[24] identification of ligands for the Tau exon 10 splicing regulatory element RNA, and synthetic receptor for trimethyl lysine.[25,26]

13.2.2 CDC with Reversible Transimination Reactions

Transimination represents a well studied, thermodynamically controlled reaction, based on the reversible nature of imine bond formation. In aqueous media, transimination is usually catalyzed by the addition of a Brønsted acid or base, such as acetic acid or aniline.[27,28] In organic solvent, the process can also be efficiently catalyzed by a Lewis acid, such as for example scandium triflate or zinc bromide.[29,30] The responsive behavior of imines to external stimuli makes transimination a suitable model for CDC studies.[31,32]

A series of dynamic glycopolymers has been generated by polycondensation *via* acylhydrazone formation and transimination between components bearing bioactive oligosaccharide chains.[33,34] Five possible glycodynamers **10a11a**, **10a11b**, **10b11a**, **10b11b** and **10a12** were formed by mixing a diluted solution of dialdehyde **11a**, **11b** or **12** and dihydrazide **10a**, **10b** in D$_2$O under weakly acidic conditions at pD 4–6 (Figure 13.4). The dynamic character of the glycopolymers was demonstrated by adding equimolar amounts of **10b** to

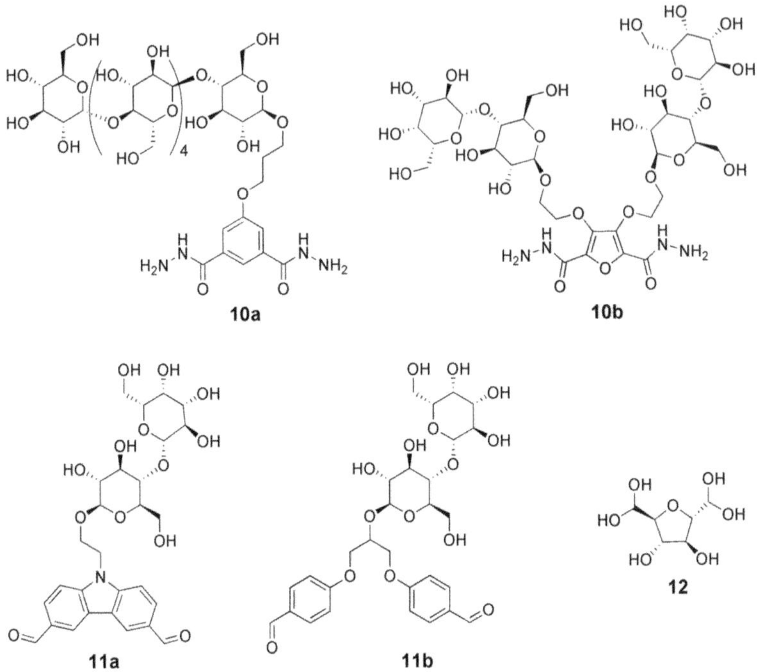

Figure 13.4 Dialdehydes and dihydrazides used in the dynamic glycodynamer system.

10a11b, and monitoring the corresponding changes using ^{1}H-NMR and fluorescence spectroscopy. In both experiments, a change in polymer constitution from **10a11b** to **10b11b** at the expense of **10a** was observed, indicating the incorporation of **10b** into **10a11b**. In later research, peanut agglutinin (PNA) was applied to the system as the target. In this case, surface plasmon resonance (SPR) was used to monitor the exchange and binding activity of the constituents. The results showed that polymers **10a11a** and **10a11b** underwent no significant increase with respect to the corresponding monomers, probably due to the steric hindrance caused by the long maltohexose side chain, while **10b11a** and **10b11b** displayed much higher affinities with PNA than the starting monomers.

A nucleophilically catalyzed acylhydrazone equilibration for protein-directed dynamic covalent chemistry has been addressed.[28] *S*-linked glutathione- (GS-) conjugated aldehyde **13** and 10 different hydrazides (**14a–j**) were chosen as starting materials to form the corresponding acylhydrazones, and equilibration was reached within four days (Figure 13.5). However, by adding excess of aniline, the equilibration was completed within six hours. Proteins were then introduced to the dynamic system, where two recombinant glutathione *S*-transferase (GST) isozymes, SjGST and hGST P1-1, were chosen as targets. Clear amplification could be observed for both GST targets: thiophene (**15g**) was selected by SjGST, while *t*-butylphenyl (**15c**) was selected by hGST P1-1.

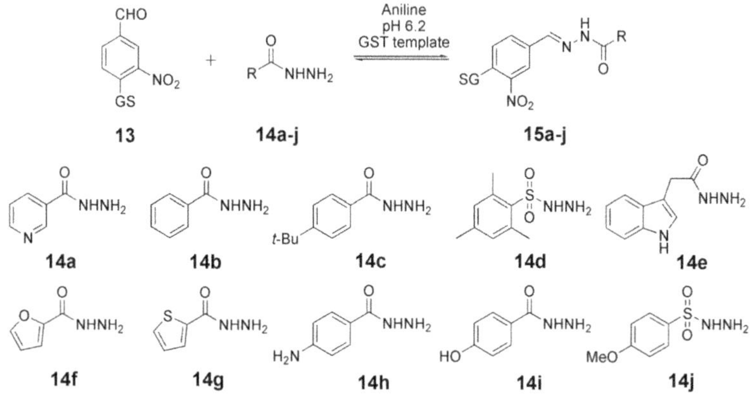

Figure 13.5 Dynamic system used in GST directed ligand discovery.

Both compounds were amplified to over 300% in relation to their concentrations in the corresponding blank system, at the expense of almost all the other competitors, especially for **15b**, **15f** and **15i** (SjGST) and **15f**, **15g** and **15i** (hGST P1-1). Further bioactivity studies showed that thiophene **15g** displayed the lowest IC_{50} against SjGST (22 μM), whereas *t*-butylphenyl **15c** exhibited the lowest value against hGST P1-1 (57 μM), demonstrating the efficiency of this approach in the context of protein–ligand discovery.

Transimination has also been applied to screen various amines and polyamines for their capacity to stabilize the triplex-forming oligonucleotide (TFO).[35] To build the dynamic systems, seven amines **A**–**G** were chosen, having different chain lengths and number of positive charges present at physiological pH, and exhibiting various potentials for electrostatic interaction and hydrogen bonding (Figure 13.6). A triplex between TFO **16** and target **17** was formed before the addition of amines and sodium cyanoborohydride in order to avoid a possible kinetic bias of the reduction upon imine formation. After generating the dynamic system, MALDI-T of mass spectrometry was used to analyze the mixtures obtained from both untemplated and templated dynamic systems. A clear amplification of conjugates **16E** (9%) and **16G** (29%) at the expense of conjugates **16A** and **16F** (−33% and −29%, respectively) could be observed. This amplification indicates that the main driving force of the selection is the electrostatic interactions between positively charged 2′-*O*-appending chains at pH 6 and the polyanionic DNA target. A terminal guanidinium group led to a higher affinity than a terminal ammonium group, and diamines with shorter aliphatic chains showed improved binding of conjugates compared with those of longer ones.

Transimination-based CDC has also been used in several other applications, for example, the discovery of isozyme inhibitors,[36] target-induced selection of a glycosyltransferase inhibitors,[37] thermodynamically controlled, modular synthesis of mechanically interlocked dendrimers,[38] and also the synthesis of nucleobase-appended dynamic polycationic polymers.[39]

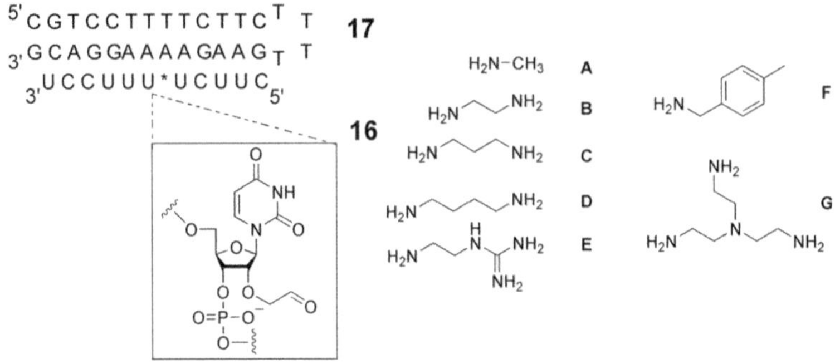

Figure 13.6 Building blocks used in dynamic systems to stablize the triplex-forming oligonucleotide.

13.2.3 CDC with Reversible C–C Bond-Forming Reactions

The reversible chemistry is a limiting factor for the progress of CDC, and most systems have relied upon reversible reactions based on imines and disulfides. For this reason, new reactions need to be explored in order to further advance the technology. Especially CDC systems based on reversible C–C bond formation provide efficient and easy access to the synthesis of organic carbon skeletons in dynamic systems.

A direct asymmetric lipase-mediated screening of a dynamic nitroaldol system was recently established in this context.[40] Equimolar amount of different aromatic aldehydes (**18–22**, Scheme 13.1), showing similar reactivities to achieve isoenergetic behavior of the system, were allowed to react with 2-nitropropane **23**, generating a dynamic set of nitroaldol racemic adducts in presence of TEA. Subsequent kinetic resolution was conducted by adding lipase PS-C I from *Burkholderia* (formerly *Pseudomonas*) *cepacia* to the system together with *p*-chlorophenyl acetate. Clear amplification could be observed for the main acetylated product **20–23**, while the corresponding intermediate of which was one of the lowest in the system in the absence of the enzyme. Ester **18–23** was found to be a minor product deriving from aldehyde **18** and 2-nitropropane **23**. Besides the amplification of optimal substrates, chiral discrimination was also obtained, with 99% and 98% *ee* of the *R*-enantiomers of ester **20–23** and ester **18–23**, respectively.

A lipase-mediated amidation of a double dynamic covalent system has been exemplified based on domino reversible transimination and cyanation reactions (Strecker).[30] A transimination system, generated by mixing equimolar amounts of three imines, **24-A1**, **25-A2** and **26-A3**, together with amine **A4** (Scheme 13.2), catalyzed by addition of ZnBr₂, was thus exposed to trimethylsilyl cyanide (TMSCN, **27**) in the presence of acetic acid as activator for the nucleophilic cyanide addition, resulting in a dynamic α-aminonitrile system. The whole double dynamic system could reach equilibrium within hours. The following kinetic resolution was achieved by applying PS-C I and phenyl acetate at 0 °C in toluene. Amide **25-A1-27** was found to be the major product

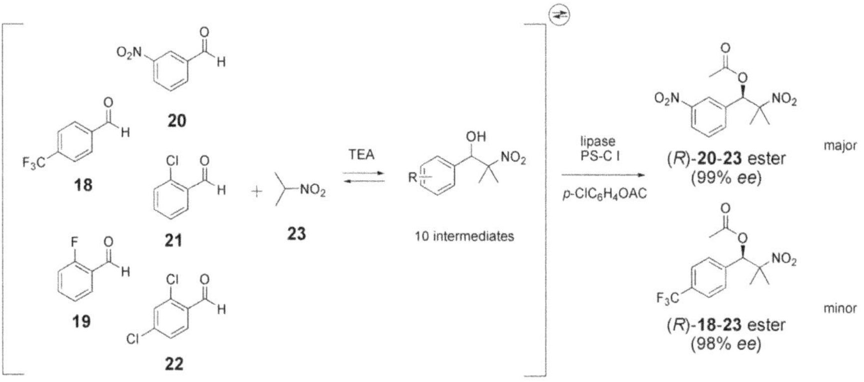

Scheme 13.1 Dynamic nitroaldol system coupled to enzymatic resolution.

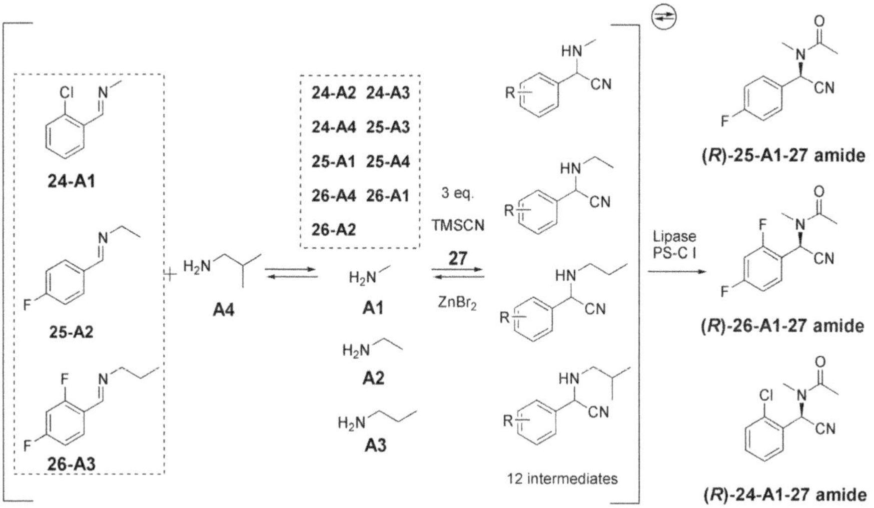

Scheme 13.2 Double dynamic Strecker system coupled to enzymatic resolution.

while the concentration of its corresponding substrate, **25-A1-27** α-amino-nitrile, was not among the highest in the double dynamic system, indicating a clear amplification during the kinetic resolution. An overall yield of 65% could be recorded with 37% of **25-A1-27** *amide*, 15% of **26-A1-27** *amide*, and 13% of **24-A1-27** *amide* respectively. According to HPLC analysis, higher enantios-electivities were observed by changing the solvent from toluene to *tert*-butyl methyl ether (TBME); 90% *ee* (**25-A1-27** *amide*), 92% *ee* (**24-A1-27** *amide*) and 73% *ee* (**26-A1-27** *amide*).

A dynamic resolution protocol based on multiple reversible cyanohydrin C–C bond formation coupled to lipase-mediated transesterification has also been demonstrated.[41] The dynamic system was generated from aldehydes

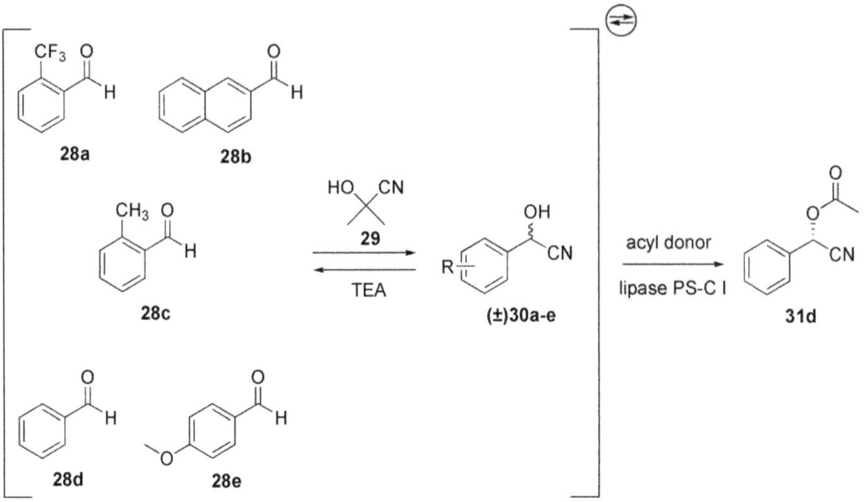

Scheme 13.3 Dynamic cyanohydrin system coupled to enzymatic resolution.

28a–e and acetone cyanohydrin **29** in presence of TEA (Scheme 13.3), where equilibration was reached within three hours. Lipase-mediated resolution was subsequently applied to the dynamic cyanohydrin system, resulting in rapid and efficient evaluation of different resolution parameters. It was shown, for example, that higher selectivities could be obtained by using lower loading of PS-C I at lower temperature (0 °C) at the cost of longer reaction time. Ester **31d** was found to be the preferred product while cyanohydrin **30a** was the dominant intermediate in the system. Enantiomeric enrichment could be obtained for all the selected cyanohydrins, and the solvent effect proved important for the enantioselectivity, although without changing the overall product distribution. When dynamic kinetic resolution was conducted in toluene, lower enantioselectivities were recorded compared to those in chloroform.

13.2.4 CDC with Reversible Transthiolesterification Reactions

Reversible transthiolesterification reactions hold several advantages for applications in CDC, such as mild reaction conditions, rapid reaction rates, and compatibility with aqueous solutions and biological targets. This reaction type has thus shown to be very useful in dynamic chemistry, especially for life sciences.[42–45]

Transthiolesterification reactions were first investigated in aqueous solutions with acetylcholinesterase as the target enzyme.[42] The dynamic system was established from a homologous series of thiolesters (**32a–e**) and thiocholine (**33**) (Scheme 13.4). At pD 7.0 and room temperature, as monitored by ^{1}H NMR, the equilibrium was reached with a $t_{1/2}$ value of approximately 50 min. After introducing acetylcholinesterase, the formed acetylthiocholine (**34b**) and propionylthiocholine (**34c**) were the main active substrates for the enzyme. The overall hydrolysis rate for formation of acetic acid ($t_{1/2} = 210$ min) was around

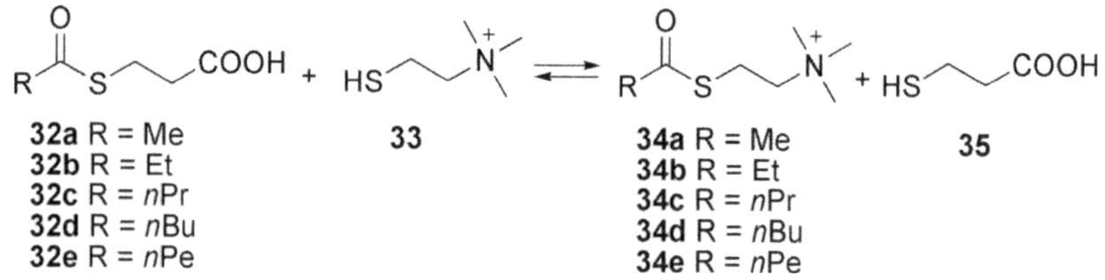

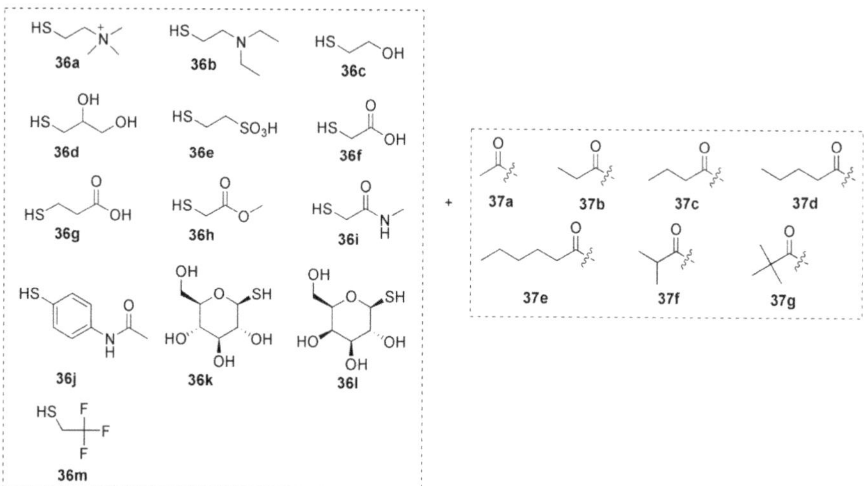

Scheme 13.4 Dynamic transthiolesterification system for acetylcholinesterase substrate screening.

Figure 13.7 Building blocks used in dynamic transthiolesterification system for further substrates screening of enzymes.

20% faster than that of propionic acid ($t_{1/2} = 270$ min). Butyrate ester **34d** was also hydrolyzed by acetylcholinesterase, following a lag phase, although at a much slower rate ($t_{1/2} = ca.$ 1100 min), while the other acyl groups remained untouched in the whole process.

A more detailed exploration of the characteristics of the self-screening dynamic system based on transthiolesterification reactions was later pursued.[43] Transthiolesterification between a range of thiols (**36a–m**) and acetylthiocholine was tested at pD 7.0 in buffer solution (Figure 13.7). The results showed a correlation between the exchange rate and the pK_a value of the individual thiols. The lower the pK_a value, the higher was the reaction rate. For the thiols with pK_a values lower than 8.5, the equilibria were reached rapidly ($t_{1/2} < 15$ min). Also, aromatic- (**36j**) and secondary thiols (**36k, l**) shifted the equilibrium toward the reactants. Evaluation of different acyl groups (**37a–g**) (Figure 13.7) indicated that branched acyl groups, as expected, reduced the exchange rate of the system. Besides acetylcholinesterase, six other enzymes belonging to the hydrolase family were also tested under the same conditions. While butyrylcholinesterase acted on all acyl groups with lower selectivity,

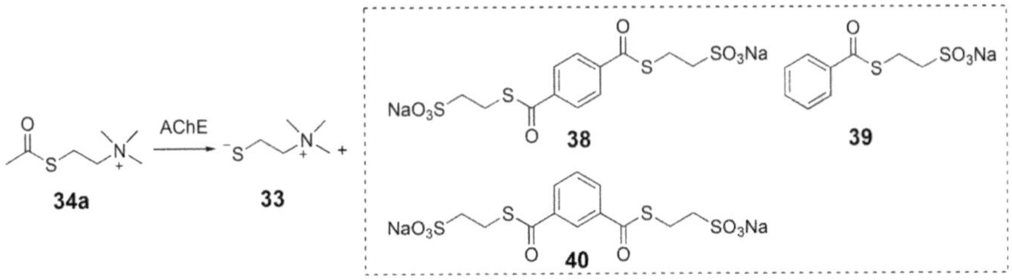

Scheme 13.5 Dynamic transthiolesterification system for self-inhibition of acetylcholinesterase.

horse liver esterase had a slight preference for longer acyl chains. The lipase from *Candida rugosa*, a hydrolase, also acting on carboxylic ester bonds, was able to promote only very modest hydrolysis. Of the two proteases tested, namely trypsin and subtilisin, only subtilisin showed some activity. Moreover, β-galactosidase (β-Gal), which belongs to the family of glycosidases, did not give any hydrolysis product.

A tandem-driven dynamic self-inhibition process has furthermore been devised, based on reversible transthiolesterification.[44] In this strategy, hydrolysis of acetylthiocholine by acetylcholinesterase led to the product thiocholine, which was *in situ* undergoing transthiolesterification with non-active, stealth inhibitors, forming real inhibitors of acetylcholinesterase (Scheme 13.5). The dynamic systems were generated by adding acetylthiocholine with three thioesters **38**, **39** and **40**, respectively, in the presence of acetylcholinesterase under neutral buffer conditions. Compared to the control experiments in the absence of thiolesters, products from all three stealth inhibitors clearly showed inhibitory effects toward acetylcholinesterase, in the order **38**>**40**>**39**. This result was in accordance with the previous study which showed that *bis*-quaternary ligands, as potent inhibitors, can bridge the two binding sites of acetylcholinesterase. The inhibitory potency of the final product from stealth inhibitor **38** was further studied, showing a competitive inhibition constant (K_i) of 47 nM, and a non-competitive constant (αK_i) of 103 nM.

13.2.5 CDC with Reversible Hemithioacetal Reactions

Reversible hemithioacetal reactions are the most recently applied reversible reactions used in dynamic systems, already utilized in both aqueous and organic solutions. Fast reversibility of this reaction can be detected in water; however, providing virtual dynamic systems, where the starting materials are thermodynamically favored and the products are not easily distinguishable in the absence of the target entity.[46] When the reaction is performed in organic solvents, base catalysis has been shown to accelerate the equilibration rates, resulting in real dynamic systems.[47]

Dynamic hemithioacetal (HTA) systems were first explored, in combination with saturation transfer difference NMR spectroscopy (STD-NMR), to

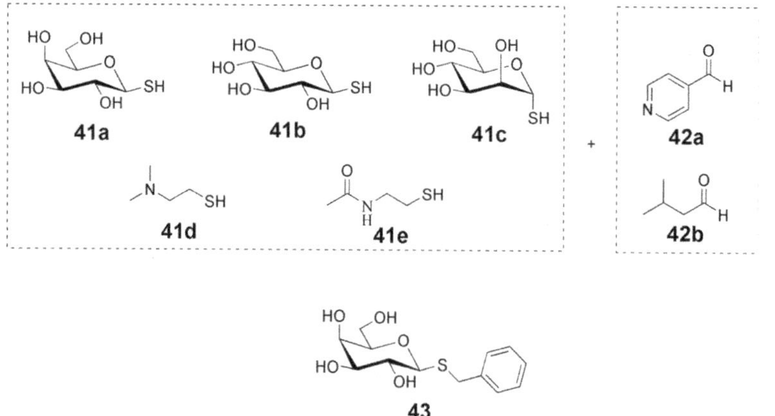

Figure 13.8 Components used for dynamic hemithioacetal (HTA) systems and the identified non-reversible inhibitor of β-galactosidase.

identify β-galactosidase inhibitors.[46] The system was formed from different aldehydes **42a–b** and thiol derivatives **41a–e** in buffer solution (pH 7.2) (Figure 13.8), resulting in transient, chiral HTA intermediates. Aldehydes of both aromatic and aliphatic types were chosen due to their similarities to known substrates and inhibitors of the enzyme. The reversible reaction proved very rapid, yielding a virtual dynamic system under neutral conditions. After subjecting the system to the target enzyme β-galactosidase, the bound ligands could be exclusively observed by STD-NMR spectroscopy: only constituents based on galactose analogue **41a** showed specific binding to β-galactosidase. Inhibition studies for each system were also performed, further corroborating that both of the thiogalactose-containing derivatives inhibited the β-galactosidase activity. Based on this dynamic system and the selected hemithioacetal compounds, two series of non-reversible glycosides were designed and produced, thioglycoside and C-glycoside derivatives, respectively.[48] The subsequent inhibition studies revealed that aromatic thiogalactoside **43** had the strongest inhibitory effect, with an IC_{50} value of 18.4 μM.

The hemithioacetal formation reaction has also been applied in organic solvents with dynamic systemic resolution.[47] Using lipases as external selectors, the hemithioacetal intermediates were irreversibly transformed through γ-lactonization, resulting in both chemo- and enantioselectivities. The dynamic systems were generated from different aldehydes **44a–c**, of both aliphatic and aromatic character, and thiols **45a, b** (Scheme 13.6). The thiols were chosen to enable the formation of γ- (five-member rings) and δ- (six-member rings) lactonized products, respectively. Triethylamine was chosen as catalyst owing to its accelerating of the reversible reaction while being compatible with the target enzyme. After adding *Candida antarctica* lipase B (CAL-B) into the system, only lactonizations from γ-hydroxyester intermediates were observed, with the enzyme preference sequence **48a** > **48b** > **48c**. Compound **48a** was the most amplified product from the system.

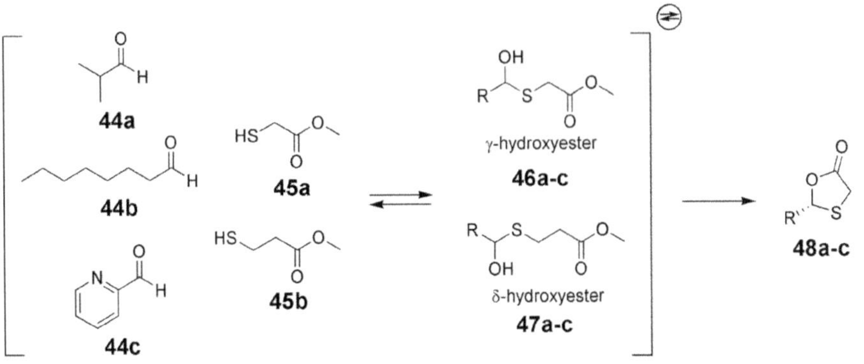

Scheme 13.6 structure region.

Scheme 13.6 Dynamic hemithioacetal system coupled to enzyme-catalyzed resolution.

13.2.6 CDC with Other Reversible Reactions

13.2.6.1 Boronate Ester Reactions

Boronate ester formation is a well known reversible reaction that has been used in dynamic chemistry,[49,50] and also its applications for biological targets have been illustrated recently. The reaction is reversible in aqueous solution under neutral or slightly basic conditions, making it compatible with protein or other targets. However, several factors need to be considered concerning the reversibility and the equilibrium compound distribution of this reaction, such as the acidities of the boronic acids and the alcohols, the nature of the buffer, and its concentration, *etc.*[51,52]

A recent approach to apply boronate ester formation reactions to a biological target was an inhibition study of a serine protease, α-chymotrypsin (αCT).[53] This was chosen as the model enzyme in this case due to its having a serine residue in the active site, which can reversibly form a ternary protein–ligand complex and thus be monitored by NMR. When investigating different carbohydrates, including D-fructose, D-glucose, L-glucose and L-fructose, with boronic acid **49a** (pK_a ∼7.4) (Figure 13.9) in a solution of pH 5.8, only D-fructose was found to improve the inhibitory effect of **49a**. The selectivity of αCT for boronic acids was also revealed after adding another boronic acid inhibitor **49b**, giving a preference of αCT-**49a**-D-fructose over the αCT-**49b**-D-fructose ternary complex. Despite the affinity of the boronate ester for the target protein, the propensity of the boronate ester to be readily formed in solution also enabled the observation of the ternary complex by NMR. For example, in the system of four boronic acids **49a, 50a–c**, together with D-fructose, only the formation of αCT-**49a**-D-fructose ternary complex was observed due to the comparatively high pK_a value of the other boronic acids. Besides ^{11}B NMR, ^{1}H water-LOGSY (water-Ligand Observed *via* Gradient SpectroscopY, a technique where water mediates magnetization transfer from a

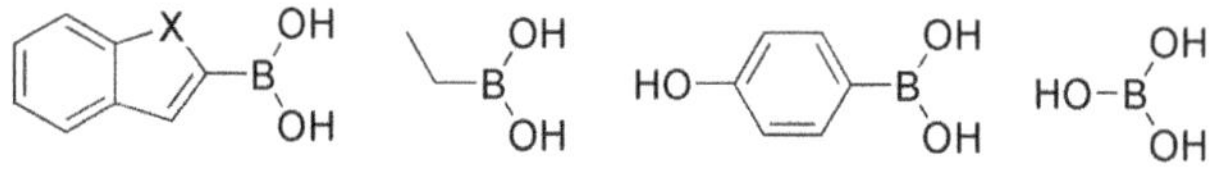

Figure 13.9 Building blocks used in dynamic boronate ester system for inhibition of α-chymotrypsin.

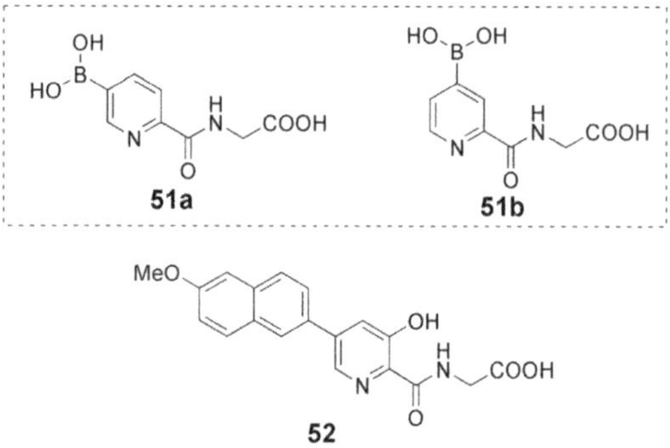

Figure 13.10 Building blocks used in dynamic boronate ester formation system and the modified potent oxygenase inhibitor.

macromolecular target to a ligand.) with superior sensitivity was further explored in this system, clearly confirming the results.

Another approach to using this dynamic system for identification of protein ligands by non-denaturing protein mass spectrometry has been demonstrated lately.[54] In this study, boronic acids **51a**, **b** (Figure 13.10) were tested, respectively, with different sets of diols in the presence of prolyl hydroxylase domain isoform 2 (PHD2). In accordance with the previous report—that is, that compound **51a** can fit into the active-site while **51b** clashes with the active-site wall—mass analysis of the systems showed better *in situ* interaction between **51a** and diols than with compound **51b**. The results were also validated by NMR-based water relaxation experiments, and further modifications of the selected boronate esters led to potent oxygenase inhibitors such as compound **52** ($IC_{50} = 13$ nM).

13.2.6.2 Michael-type Reversible Reactions

Michael-type reactions are often exothermic processes showing stable product formations, but can sometimes be sufficiently reversible for use in CDC. An example of applying a Michael-type reaction to dynamic chemistry was recently illustrated, used to discover glutathione *S*-transferase (GST) inhibitors.[55,56] A biased model system was first established, with amplification of compound **53** (Figure 13.11) from a system of ethacrynic acid and glutathione (GSH)

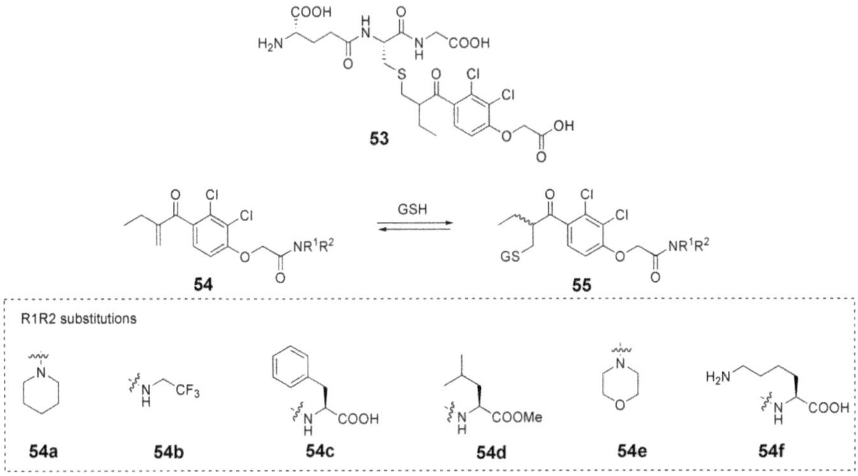

Figure 13.11 Selected compounds from a dynamic Michael-type system with glutathione *S*-transferases.

analogues, demonstrating that GST was able to successfully select the best binder from the model system. Expansion of the system subsequently led to the selection of a subset of adducts (**54a–d**) at the expense of the others. To ascertain the inhibitory effects of the selected compounds, IC_{50} values of adducts **54b** and **54d**, together with non-amplified adducts **54e** and **54f**, were measured. Modest differences in IC_{50} values were observed with compounds **54b, d** *versus* **54e, f**.

13.2.6.3 *Metathesis Reactions*

The alkene metathesis reaction is a well-known reversible reaction used in dynamic chemistry. The whole process includes a series of [2 + 2] cycloaddition/cycloreversion steps, with all individual steps being reversible.[57] Both first- and second-generations of Grubbs' catalysts have been used to initiate the reversible reactions, though the second generation catalysts proved to be compatible with more functional groups.[58,59] Furthermore, immobilized Grubbs' catalyst has been developed to stop the alkene metathesis easily by filtration.[60]

In an application of fragment-based drug discovery of carbonic anhydrase II (CA II) inhibitors, a dynamic system of alkene cross metathesis was generated.[61] In this case, a start–stop methodology was achieved by adding and filtering the immobilized first generation Grubbs' catalyst, forming a dynamic alkene metathesis system that could be quenched and re-initiated effectively. The initial building block, **56** (Figure 13.12), was selected for its dual functionality of an aromatic sulfonamide moiety and an allyl substituent, which were envisioned as bCA II recognition fragments. Together with 10 additional building blocks, **57a–j** (each processing a terminal alkene functional group), the

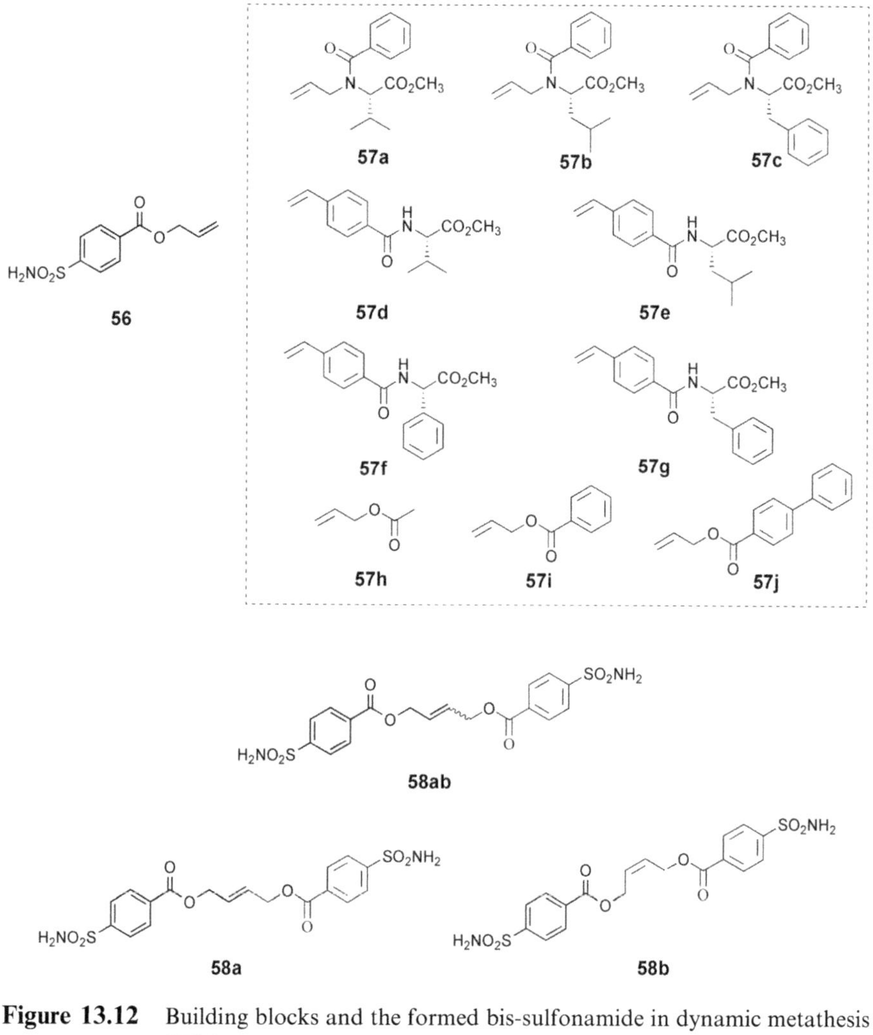

Figure 13.12 Building blocks and the formed bis-sulfonamide in dynamic metathesis system.

dynamic system was generated and assessed for bCA II binding by a fluorescence-based assay (Figure 13.12). Since the bis-sulfonamide **58ab** showed background bCA II binding affinity, 10 equivalents of **57a–j** were elected to increase the proportion of **56-57a–j**. From the systemic screening results, **56–57h** was revealed to be the most potent inhibitor, with **56-57i** and **56-57j** following on as less potent compounds. The affinity study of pure compounds further confirmed the relative order of K_i's as **58a, b ∼ 58a ∼ 56-57h >** **56-57i > 56-57j > 58b**, with K_i values all in the low to mid nanomolar range.

The alkene metathesis reactions with Grubbs' catalysts were generally processed in organic solvents, such as dichloromethane and dichloroethane. Probably due to this limitation, biological targets were rarely introduced to this dynamic system, and the applications were mainly focused on self-assembly of the building blocks, such as the templated synthesis of porphyrin boxes.[58] However, there are still some exceptions. For example, in a dynamic way of discovering antibiotics against vancomycin-resistant enterococci, Grubbs' catalyst together with phase-transfer agent were first dissolved in dichloromethane, and by subsequently removing the organic solvent, only water was left as solvent to conduct the dynamic process and target recognition.[62]

13.3   Conclusion

This chapter has illustrated that CDC efficiently enables the selection of optimal bioactive compounds for various external targets, providing information of the biological entities together with the substrates. The concept has been much explored and expanded during the last decade, as shown in all studies illustrated, where reversible reactions constitute the key step in establishing dynamic systems. In combination with different biological targets, optimal ligands and inhibitors have been selected and amplified and, in some cases, potential drug leads have been identified and further designed from the dynamic systems. It is thus clear that CDC-based strategies show strong potential for future pharmaceutical applications.

References

1. B. J. Miller, ed., *Dynamic Combinatorial Chemistry—in Drug Discovery, Bioorganic Chemistry, and Materials Science*, Wiley, New York, 2010.
2. J. Reek and S. Otto, eds., *Dynamic Combinatorial Chemistry*, Wiley-VCH, Weinheim, 2010.
3. M. Barboiu, ed., *Constitutional Dynamic Chemistry, Topics in Current Chemistry*, Springer-Verlag, Berlin, 2012, p. 322.
4. B. d. Bruin, P. Hauwert and J. N. H. Reek, *Angew. Chem. Int. Ed.*, 2006, **45**, 2660.
5. O. Ramström, T. Bunyapaiboonsri, S. Lohmann and J.-M. Lehn, *Biochim. Biophys. Acta.*, 2002, **1572**, 178.
6. R. A. R. Hunt and S. Otto, *Chem. Commun.*, 2011, **47**, 847.
7. P. T. Corbett, J. Leclaire, L. Vial, K. R West, J.-L. Wietor, J. K. M. Sanders and S. Otto, *Chem. Rev.*, 2006, **106**, 3652.
8. J.-M. Lehn, *Chem. Soc. Rev.*, 2007, **36**, 151.
9. O. Ramström and J.-M. Lehn, *Nat. Rev. Drug Discov.*, 2002, **1**, 26.
10. J.-M. Lehn, *Top. Curr. Chem.*, 2012, **322**, 1.
11. I. Huc and J.-M. Lehn, *Proc. Natl. Acad. Sci. USA*, 1997, **94**, 2106.
12. O. Ramström and J.-M. Lehn, *ChemBioChem*, 2000, **1**, 41.
13. S. Sando, A. Narita and Y. Aoyama, *Bioorg. Med. Chem. Lett.*, 2004, **14**, 2835.
14. A. M. Whitney, S. Ladame and S. Balasubramanian, *Angew. Chem., Int. Ed.*, 2004, **43**, 1143.
15. P. A. Fernandes and M. J. Ramos, *Chem. Eur. J*, 2004, **10**, 257.

16. R. Caraballo, M. Sakulsombat and O. Ramström, *Chem. Commun.*, 2010, **46**, 8469.
17. B. R. McNaughton and B. L. Miller, *Org. Lett.*, 2006, **8**, 1803.
18. P. C. Gareiss, K. Sobczak, B. R. McNaughton, P. B. Palde, C. A. Thornton and B. L. Miller, *J. Am. Chem. Soc.*, 2008, **130**, 16254.
19. L. O. Ofori, J. Hoskins, M. Nakamori, C. A. Thornton and B. L. Miller, *Nucleic Acids Res.*, 2012.
20. R. Perez-Fernandez, M. Pittelkow, A. M. Belenguer and J. K. M. Sanders, *Chem. Commun.*, 2008, 1738.
21. D. E. Scott, G. J. Dawes, M. Ando, C. Abell and A. Ciulli, *ChemBioChem*, 2009, **10**, 2772.
22. A. Bugaut, K. Jantos, J.-L. Wietor, R. Rodriguez, J. K. M. Sanders and S. Balasubramanian, *Angew. Chem., Int. Ed.*, 2008, **47**, 2677.
23. L. Vial, R. F. Ludlow, J. Leclaire, R. Pérez-Fernández and S. Otto, *J. Am. Chem. Soc.*, 2006, **128**, 10253.
24. L. Milanesi, C. A. Hunter, S. E. Sedelnikova and J. P. Waltho, *Chem. Eur. J*, 2006, **12**, 1081.
25. P. López-Senín, I. Gómez-Pinto, A. Grandas and V. Marchán, *Chem. Eur. J*, 2011, **17**, 1946.
26. L. A. Ingerman, M. E. Cuellar and M. L. Waters, *Chem. Commun.*, 2010, **46**, 1839.
27. J. L. Hogg, D. A. Jencks and W. P. Jencks, *J. Am. Chem. Soc.*, 1977, **99**, 4772.
28. V. T. Bhat, A. M. Caniard, T. Luksch, R. Brenk, D. J. Campopiano and M. F. Greaney, *Nat. Chem*, 2010, **2**, 490.
29. N. Giuseppone, J.-L. Schmitt, E. Schwartz and J.-M. Lehn, *J. Am. Chem. Soc.*, 2005, **127**, 5528.
30. P. Vongvilai and O. Ramström, *J. Am. Chem. Soc.*, 2009, **131**, 14419.
31. M. Hochgürtel, H. Kroth, D. Piecha, M. W. Hofmann, C. Nicolau, S. Krause, O. Schaaf, G. Sonnenmoser and A. V. Eliseev, *Proc. Natl. Acad. Sci. USA*, 2002, **99**, 3382.
32. N. Hafezi and J.-M. Lehn, *J. Am. Chem. Soc.*, 2012, **134**, 12861.
33. Y. Ruff and J.-M. Lehn, *Angew. Chem., Int. Ed.*, 2008, **47**, 3556.
34. Y. Ruff, E. Buhler, S.-J. Candau, E. Kesselman, Y. Talmon and J.-M. Lehn, *J. Am. Chem. Soc.*, 2010, **132**, 2573.
35. L. Azéma, K. Bathany and B. Rayner, *ChemBioChem*, 2010, **11**, 2513.
36. G. Nasr, E. Petit, D. Vullo, J.-Y. Winum, C. T. Supuran and M. Barboiu, *J. Med. Chem.*, 2009, **52**, 4853.
37. A. Valade, D. Urban and J.-M. Beau, *ChemBioChem*, 2006, **7**, 1023.
38. K. C. F. Leung, F. Aricó, S. J. Cantrill and J. F. Stoddart, *Macromolecules*, 2007, **40**, 3951.
39. N. Sreenivasachary, D. T. Hickman, D. Sarazin and J.-M. Lehn, *Chem. Eur. J*, 2006, **12**, 8581.
40. P. Vongvilai, M. Angelin, R. Larsson and O. Ramström, *Angew. Chem., Int. Ed.*, 2007, **46**, 948.
41. M. Sakulsombat, P. Vongvilai and O. Ramström, *Org. Biomol. Chem.*, 2012, **9**, 1112.

42. R. Larsson, Z. Pei and O. Ramström, *Angew. Chem., Int. Ed.*, 2004, **43**, 3716.
43. R. Larsson and O. Ramström, *Eur. J. Org. Chem*, 2006, **2006**, 285.
44. Y. Zhang, M. Angelin, R. Larsson, A. Albers, A. Simons and O. Ramström, *Chem. Commun.*, 2010, **46**, 8457.
45. J. Leclaire, L. Vial, S. Otto and J. K. M. Sanders, *Chem. Commun.*, 2005, 1959.
46. R. Caraballo, H. Dong, J. P. Ribeiro, J. Jiménez-Barbero and O. Ramström, *Angew. Chem., Int. Ed.*, 2010, **49**, 589.
47. M. Sakulsombat, Y. Zhang and O. Ramström, *Chem. Eur. J*, 2012, **18**, 6129.
48. R. Caraballo, M. Sakulsombat and O. Ramström, *ChemBioChem*, 2010, **11**, 1600.
49. J. Zhao, T. M. Fyles and T. D. James, *Angew. Chem., Int. Ed.*, 2004, **43**, 3461.
50. G. T. Morin, M.-F. Paugam, M. P. Hughes and B. D. Smith, *J. Org. Chem.*, 1994, **59**, 2724.
51. J. Yan, G. Springsteen, S. Deeter and B. Wang, *Tetrahedron*, 2004, **60**, 11205.
52. G. Springsteen and B. Wang, *Tetrahedron*, 2002, **58**, 5291.
53. I. K. H. Leung, T. Brown Jr, C. J. Schofield and T. D. W. Claridge, *MedChemComm*, 2011, **2**, 390.
54. M. Demetriades, I. K. H. Leung, R. Chowdhury, M. C. Chan, M. A. McDonough, K. K. Yeoh, Y.-M. Tian, T. D. W. Claridge, P. J. Ratcliffe, E. C. Y. Woon and C. J. Schofield, *Angew. Chem., Int. Ed.*, 2012, **51**, 6672.
55. B. Shi and M. F. Greaney, *Chem. Commun.*, 2005, 886.
56. B. Shi, R. Stevenson, D. J. Campopiano and M. F. Greaney, *J. Am. Chem. Soc.*, 2006, **128**, 8459.
57. A. Fürstner, *Angew. Chem., Int. Ed.*, 2000, **39**, 3012.
58. P. C. M. van Gerven, J. A. A. W. Elemans, J. W. Gerritsen, S. Speller, R. J. M. Nolte and A. E. Rowan, *Chem. Commun.*, 2005, 3535.
59. C. W. Lee and R. H. Grubbs, *J. Org. Chem.*, 2001, **66**, 7155.
60. S. T. Nguyen and R. H. Grubbs, *J. Organomet. Chem.*, 1995, **497**, 195.
61. S.-A. Poulsen and L. F. Bornaghi, *Bioorg. Med. Chem.*, 2006, **14**, 3275.
62. K. C. Nicolaou, R. Hughes, S. Y. Cho, N. Winssinger, C. Smethurst, H. Labischinski and R. Endermann, *Angew. Chem., Int. Ed.*, 2000, **39**, 3823.

Molecular Imprinted Polymers for Biomedical Applications

ADNAN MUJAHID[a,b] AND FRANZ L. DICKERT*[a]

[a] Department of Analytical Chemistry, University of Vienna, Waehringer Strasse 38 A-1090, Vienna, Austria; [b] Institute of Chemistry, University of the Punjab, Quaid-i-Azam Campus, Lahore 54590, Pakistan
*Email: Franz.Dickert@univie.ac.at

14.1 Introduction

Molecular imprinting[1,2] is an emerging technology that offers an alternative route to the crafting of synthetic receptor materials.[3,4] In molecular imprinting, monomers along with the cross-linkers are polymerized in the presence of template molecules. The polymer chains are self-organized around the template and in this way analyte-analogous cavities are generated that can recognize the target species. Molecular imprinted polymers (MIPs) have found numerous applications in various fields including catalysis,[5,6] liquid chromatography stationary phases,[7,8] enantiomeric separations,[9,10] solid phase micro-extraction (SPME) sorbents,[11,12] drug delivery systems,[13,14] immunoassays,[15,16] designing sensor coatings[17,18] and many others.

The foremost advantage of molecular imprinted materials is their versatile applications in many paradigms of the biomedical field. MIPs can be adopted in a number of ways according to the desired purpose; for instance, they have been reported in new drug development or designing, separation of active medicinal components from natural herbs, controlled drug release and in the extraction of drug metabolites from complex matrices including blood and urine. This list represents a complete cycle of MIP application in the medicinal

Monographs in Supramolecular Chemistry No. 13
Supramolecular Systems in Biomedical Fields
Edited by Hans-Jörg Schneider

Published by the Royal Society of Chemistry, www.rsc.org

context, from synthesis of bioactive compounds to their extraction from human body fluids.

During the last few years, MIPs have been recognized as highly valuable tools for selective screening as well as isolation of new compounds showing appreciable bioactivity. This has been reported in a number of studies in which, after identifying potential anticancer drugs in natural products, subsequently they can be used as lead compound for further derivatization to improve their efficiency. Considering sensor applications, MIPs can bind selectively to target analytes in complex matrices *via* a variety of chemical interactions.[19] Molecular imprinting technology allows the design of sensor layers for a wide range of analytes such as bacteria,[20] yeast,[21] red blood cells,[22] proteins,[23] hormones,[24] and extremely small viruses[25] having dimensions of just a few nanometers. MIP coatings are robust and rigid and can endure harsh environments; thus the geometrical features of designed cavities are not disturbed. These coatings can tolerate solvents, temperature and pH changes without losing performance. The binding with analyte is completely reversible, which allows their use in multiple measurements, thus ultimately reducing the analysis cost.

In order to realize a complete sensor setup, MIPs can be integrated with suitable transducers that can monitor layer characteristics upon analyte interaction. As a result the physical or chemical properties of the sensor layer are changed when the target analyte interacts. A wide range of transducers can be used for sensing applications; these include electrical, optical and mass-sensitive devices.[26,27] In fact, transducers also play a very important role in achieving high sensitivity, low detection limits and fast sensor response. In addition, the compatibility with coating materials and the working environment should also be considered. The small size of the measuring device is highly desirable in respect of portability, so that online monitoring can be performed without trained operators.

Apart from typical sensing applications, MIPs have been extensively studied for analyzing body fluids,[28,29] degradation products of engine oil,[30,31] food samples,[32,33] and drug extraction or separations[34,35] from real complex matrices. More recently, MIPs have been considered highly valuable in tissue engineering applications[36] and advanced drug delivery systems. It has been discovered that nanostructured MIPs can be used as advanced synthetic support materials for selective cell adhesion and proliferation. The main concern in implanting a synthetic material inside the body is its response to specific tissues, which has been apparent from selective linkages of functionalized MIPs towards target proteins or peptides. *In vitro* studies have demonstrated that the MIP interface facilitates target-cell adhesion and growth. Moreover, the development of biocompatible imprinted polymers has already shown minimal cytotoxic activity. These findings are somewhat new, though very encouraging in the field of biomaterials. In this chapter, modern trends in imprinting[37] strategies will be explained in detail and the contribution of MIPs will be highlighted explicitly in the biomedical field by describing typical applications.

14.2 Molecular Imprinting Strategies

Recent developments in molecular imprinting are largely attributed to the pioneering efforts of Mosbach and co-workers.[38] The imprinting phenomenon was first investigated by Polyakov[39,40] while he observed the recognition properties of a silica matrix towards the solvent that was used during synthesis. However, this work remained unnoticed for many years, until Pauling[41,42] had explained the molecular recognition in synthetic antibodies. Later, the experiments of Frank Dickey[43] on silica gel for rebinding a patterning dye led to the foundation of molecular imprinting.

Molecular imprinting principally can be classified into two major types. These are formally designated as covalent and non-covalent imprinting, since the classification depends on the interactions between host and guest molecules. The covalent approach was first introduced by Wulff and Sarhan in 1972,[44] by which an adaptable covalent bond is formed between template and monomer molecules. This technique is advantageous since the splitting-off of template molecules during polymerization can be avoided. However, in covalent imprinting the bond between polymer and template has to be cleaved in order to remove the template molecules. Covalent imprinting has found useful application in chromatographic separations[45,46] with respect to resolution issues and peak symmetry; however, this technique has been relatively less explored for sensing purposes. On the other hand, non-covalent imprinting[47] is a relatively flexible, versatile and straightforward approach, since it does not require any special synthetic procedures to introduce template molecules in the pre-polymer matrix. Moreover, the template removal is also very feasible without cleaving the bonds. Monomers and cross-linkers along with the template are mixed together and polymerized under appropriate conditions, whereby polymer chains undergo self-organization[48] around the template. After the polymerization, the template molecules can be removed by heating or washing with a suitable solvent.

The main condition applicable to molecular imprinting is that the template molecule should not interfere with the polymer system. In addition, high proportions of cross-linkers are used during the polymerization reaction. This imparts the desired stability in order to avoid any collapse of polymer chains when template is removed. After the template removal, the polymer matrix retains the shape of the adapted cavities which can reversibly incorporate the target analytes through a large variety of chemical interactions, ranging from van der Waals forces and hydrogen bonding to ionic contributions. Some hybrid or semi-covalent imprinting schemes have also been reported where the template is bonded covalently in polymer matrix and after its removal it can bind with imprinting sites in a reversible fashion through non-covalent forces.

In molecular imprinting, selecting the appropriate size of template is also very important. In *bulk imprinting*,[49] the template molecules are added in the bulk of polymer at the start of reaction and the interaction sites are distributed in the whole matrix. The approach is useful for smaller size analytes as the exclusion of template is unproblematic and thus the reversibility is not disturbed by sterical aspects. In addition, the sensor response would be high due to large number of interaction centers. However, if the template belongs to biomolecules of relatively larger size, the bulk imprinting approach is not complementary. The constraints concerning imprinting of bio-analytes and some recent advances in this field are discussed in the following section.

14.2.1   Imprinting of Biomolecules

For imprinting with large size bio-analytes, bulk processing of MIPs is not recommended. Owing to the geometrical constraints of the three-dimensional polymer network, the complete release of larger size bioanalytes would be not favorable. In addition, the longer diffusion pathways would lead to sluggish response times. Therefore, the recognition of macromolecules or bio-analytes can be accomplished by *surface imprinting*[50] where the binding event is exclusively monitored only at the MIP interface. A schematic representation of surface imprinting is shown in Figure 14.1.

A closely packed stamp of template is pressed over the pre-polymer matrix layer and then allowed to harden under controlled conditions. At this stage, polymer chains grow around the template and develop non-covalent interactions with the outer cell wall of the bio-analyte. After polymer curing, the templates are washed away by using an appropriate solvent. This leads to generation of patterned analyte-analogous cavities at the surface of the polymer. The strategy is very useful and has been exploited for imprinting of various bio-analytes[51] such as yeast, bacteria and remarkably small sized viruses. The highly cross-linked imprinted polymer cavities are capable of reversible reincorporation and do not crumple after analyte removal.

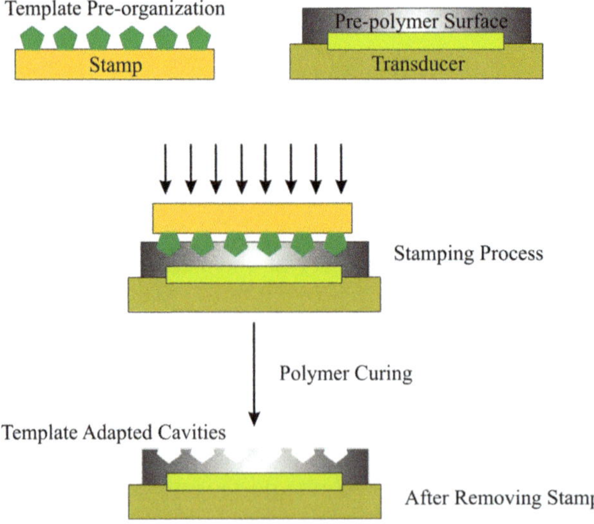

Figure 14.1   Schematic representation of surface imprinting by stamping technique.

Instead of using whole cells for templating, a small sequence of specific protein can be introduced that exclusively interacts with the target bio-analyte. This particular protein or peptide part is similar to the terminal part of the target. The approach is generally named as epitope imprinting[52,53] since it is similar to protein recognition by natural antibodies. The principle of epitope imprinting is displayed in Figure 14.2.

The minimum length of the peptide sequence has to be defined, as it is very important to accomplish the desired recognition. Nishino *et al.*[54] reported that a minimum of 9 amino acids is necessary for this type of imprinting. However, in some cases this length could be different depending upon the epitope of bio-analyte. The system can be considered for manifold measurements by following the washing protocols to remove attached proteins. This technique is very attractive for separation of target proteins with high selectivity, similar to that of

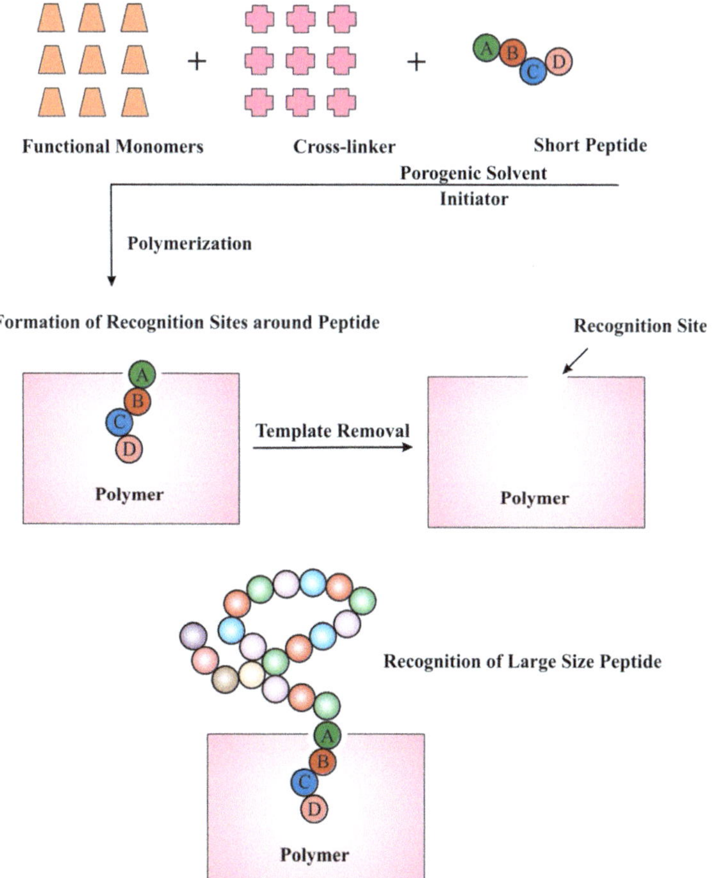

Figure 14.2 Principle of epitope imprinting. (After Rachkov and Minoura.[52])

natural antibodies. In addition, owing to epitope imprinting, it is not necessary to retain the conformation of the whole protein.

More recently, molecular imprinting technology has been extended to cast the image of natural antibodies. In this method, monomer solutions are pre-polymerized in the presence of the desired natural antibody, and after polymerization under control conditions the antibodies are washed away. In the next step, the nanoparticle stamp having the impression of the antibodies is pressed onto the pre-polymer surface. This indeed is a double imprinting approach that enables the crafting of MIPs as real plastic antibodies.[55] This technique has two advantages over the former surface imprinting. First, the interaction surface area is considerably increased, which results in improved sensitivity. Second, the recognition mechanism is also different, since in the previous scheme the whole template was recognized, however here, only the epitopes are sensed, which is beneficial for the detection of larger size biomolecules. Moreover, due to epitope interactions, the sensing can be performed in plasma or other complex mixtures. Previous detection systems were mostly concerned with aqueous media. The double imprinting technique as shown in Figure 14.3, which can be used for proteins separations as well as chemosensor coatings, is very promising.

In terms of sensitivity and selectivity, molecular imprinted coatings are highly competitive with other synthetic sensor materials and natural antibodies. During the last two decades, molecular imprinting technology has been explored progressively for designing advanced chemical sensors.[56] The integration of imprinted polymers with a suitable transducer is very important so that the binding event can be translated into a measurable electronic signal. MIPs can be integrated with various transducers which include optical, mass sensitive and some electrochemical devices, thus to develop chemical sensors.

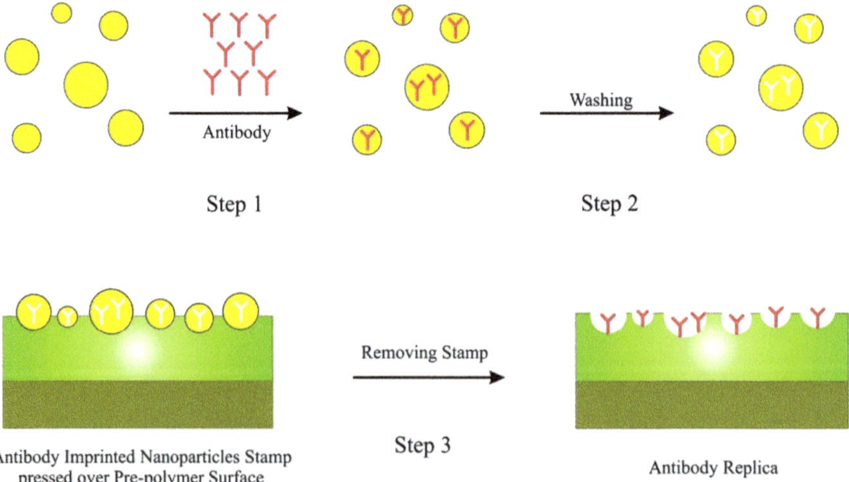

Figure 14.3 Double imprinting method for generating artificial antibody replica. (After Schirhagl *et al.*[55])

The selection of a transducer primarily relies on the nature of the MIP's properties, which changes upon analyte binding. For instance, MIPs integrated with optical devices[57] have shown considerable potential as a highly sensitive sensor system. The binding of analyte with imprinted polymer could also bring about some electrochemical changes, for example, shifts in current, surface charge or potential.[58] These changes can be monitored through an electronic device with suitable precision. Nevertheless, if an analyte does not alter any optical or electrical characteristics of the MIP, then it can preferentially be recognized by gravimetric or acoustic devices.[59] (As we know, mass is the most fundamental feature of any analyte, which can be determined by acoustic or gravimetric devices with an accuracy of a few picograms.) In general, the choice of transducer ultimately depends on the nature of binding at the MIP interface. Low detection limits, high signal-to-noise ratio, prompt response, low drift and ease of MIP processing are customary features of an ideal transduction system. In the following sections, MIP-based chemical sensors will be exclusively discussed, particularly in view of their clinical interest; in addition to their contribution and achievements in the biomedical field.

14.3 Typical Applications of Molecular Imprinted Polymers

14.3.1 Microorganism Recognition

One of the most significant achievements of molecular imprinting is its extension of scope from molecular species to bio-analytes. This allows the transformation of complementary cell details into synthetic polymers for recognition purposes. For successful imprinting of cells and various microorganisms,[60] the selection of suitable polymers is crucial in order to accomplish the desired interactions. Vulfson and co-workers[61] were the first to report on the surface imprinting of bacteria. They exploited the fact that bacteria prefer to assemble at water and oil surfaces; therefore, they adopted a non-polar monomer system along with aqueous suspension of bacteria. They followed the emulsion polymerization which generates polymer beads of micro-size having bacteria on their surface. The bacteria could be washed off after polymerization whereas the surface of polymer beads retained the memory of the template bacteria.

Highly cross-linked polyurethanes have proved to be very handy in designing patterned surfaces for the detection of various cell species.[62] For example, *Escherichia coli* is a Gram-negative bacterium that has a rod-like shape; it can be templated on a polyurethane surface following the glass stamping procedure. It has been found that the geometrical configuration of *E. coli* can be imprinted with suitable perfection on a polymer layer. The stamp should be densely packed, thus to capture the patterned surface. In addition, the polymer composition, curing time and the humidity content of cell assembly play a vital role in generating optimized cell interaction centers. The dimensions of cavities are very similar to templates, thus the target *E. coli* strain can selectively be reincorporated. For instance, the W strain of *E. coli* was surface-imprinted on a polyurethane surface coated on a 10 MHz quartz crystal microbalance (QCM). The mass-sensitive measurements[63] revealed that imprinted the polyurethane layer exhibited a response which was about five times higher than the non-imprinted channel, thus indicating selective interaction centers at the MIP interface. In addition, the selectivity of the imprinted polyurethane film was also investigated by exposing equal concentration of B strain of *E. coli* under similar conditions. It was observed that B strain showed a comparatively smaller effect, as the sensor response for W strain was much higher than with the B strain, which indicates the strong affinity of polyurethane towards the imprinted strain. The respective sensor effects are displayed in Figure 14.4.

Yeasts play an important role in various biological processes such as fermentation, and are suitable model microorganisms for designing MIP sensor setups. Molecular imprinted polyurethanes can be grafted to monitor the concentration, growth stage and viability of yeast cells in order to optimize the process conditions. Although the shape and size of the yeast cells change during their growth stage, the surface chemical properties remain the same. Molecular imprinting offers both the geometrical and chemical fit, which is helpful in the discriminating the cells of other strains and at different growth stages. Polyurethanes are suitable materials that can be crafted for surface imprinting of synchronized yeast cells. The chemical properties of cell surface are mimicked at the polymer surface *via* stamping technique. This leads to a certain chemical fit between microorganism cell and polyurethane surface, and thus a marked distinction between two different yeast strains could be observed.[64] It was shown[65] that the sensor response of *Saccharomyces cerevisiae* imprinted polyurethane was twice as large as for *S. bayanus*, which is similar in size and

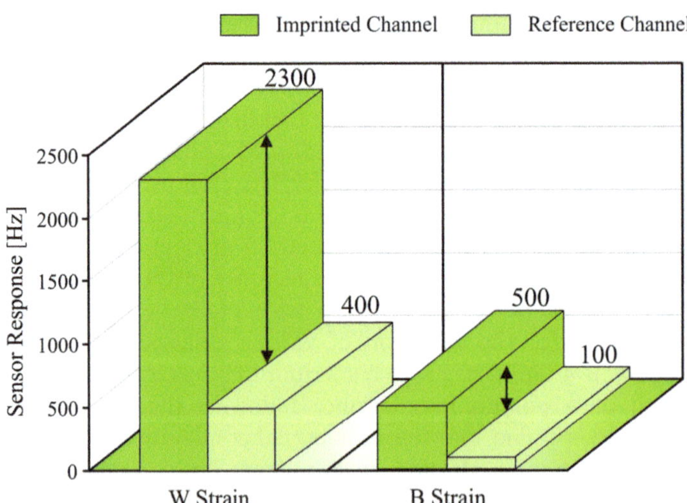

Figure 14.4　Selectivity evaluation of polyurethane surface imprinted with *Escherichia coli* W strain while exposed to W and B strains. (Data from Bajwa *et al.*[63])

geometry. The response of the non-imprinted polymer layer was observed to be similar for both strains, indicating unselective adsorption behavior.

This strategy can be extended to distinguish individual cell-growth stages,[66] that is, between single and duplex yeast cells, which have similar chemical surface properties but are different in size and shape. The imprinting of duplex cells can be accomplished first by treating the cell culture with *N*-hydroxyurea which inhibits the DNA replication, thus ultimately stopping the cell division. At this moment, the cell exists in duplex form having size of 9 μm, which is slightly less than the size of two individual singlet yeast cells. These duplex cells were imprinted on a polyurethane surface following spin-coating to ensure the homogenous distribution of cells over the thin film. Mass sensitive measurements revealed that respective singlet and duplex imprinted polyurethanes films showed preferential responses to the imprinted yeast cells as demonstrated in Figure 14.5.

The singlet and duplex imprinted polyurethane films can be integrated with a tetra-electrode QCM,[67] respectively, whereas the remaining the two electrodes work as a control covered by non-imprinted material. All these four channels work simultaneously which is promising for monitoring the fermentation process and for online process control applications.[68]

Moving on, detection of red blood cells (erythrocytes) through synthetic materials[69] is another interesting use of imprinted polymers. The ABO blood group system is classified into blood groups A, B, AB and O based on the differences in the surface antigens of the red blood cells among individuals. For blood transfusion it is obligatory to match the blood groups of donor and recipient otherwise blood agglutination may occur. The image of a particular blood group can be casted on polyurethane film by the stamping procedure as described in surface imprinting. Blood cells of a target group are deposited on

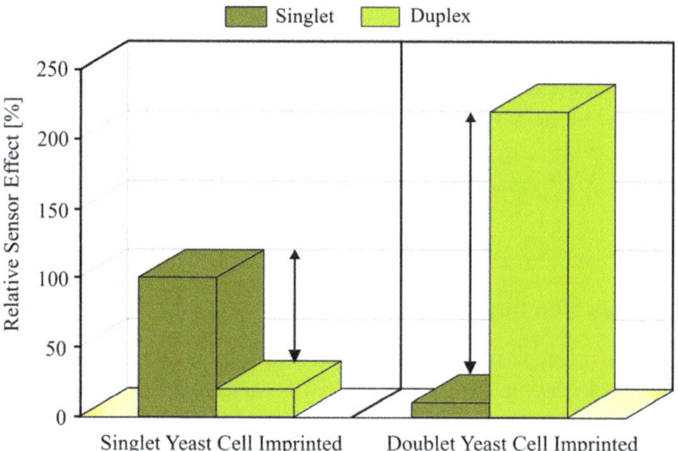

Figure 14.5 Relative sensor effects for singlet and duplex cells against their respective imprinted polyurethane layers coated on quartz microbalance. (Data from Seidler *et al.*[66])

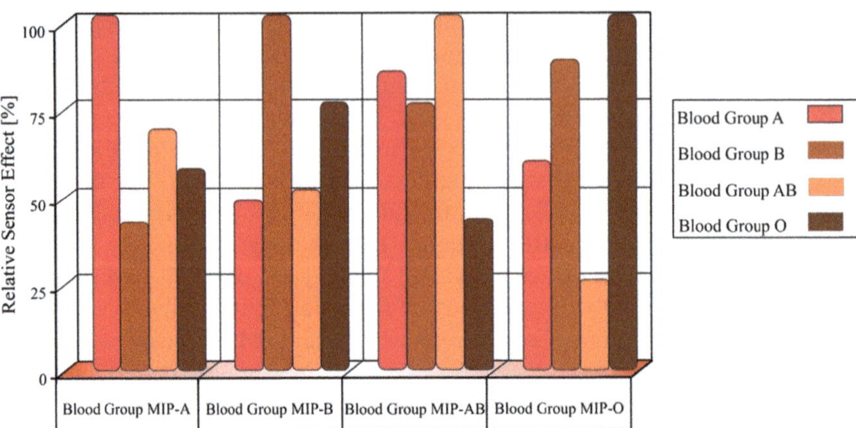

Figure 14.6   Selectivity pattern obtained for ABO blood group system. It is evident that each polymer layer shows highest sensor response for that blood group which was used as template during imprinting.
(Data from Hayden *et al.*[69])

a silicone stamp and then pressed over a pre-polymerized polyurethane surface, which is allowed to cure at controlled temperature. This leads to generation of highly tailored cavities for red blood cell recognition with perfect selectivity. A comparison of the imprinted layers of different blood groups along with their relative sensor effects is shown in Figure 14.6.

Each surface imprinted polyurethane layer shows highest response to their respective template blood group. Although the size of all four blood cells is in the same range of 6–8 µm, and they also possess high flexibility, the difference in their surface chemistry makes their affiliation different with the MIP surface. The surface antigens of blood groups A, B and O differ according to the terminal sugar molecules. The sensor effects can be seen in Figure 14.6; in each case the responses are optimized when the analyte is identical to the template. In further studies, sub-group blood types, such as A1 and A2, were differentiated using the same strategy.[70] These polymer layers possess the desired selectivity and stability after several rounds of regeneration, which means that in future low cost MIP sensors could be used for blood group matching in clinical laboratories.

14.3.2   Virus Sensing

Molecular imprinting can be extended from micro-size to nano-size bio-species, the best example being the recognition of various viruses through artificial polymeric receptor materials. The imprinting of such small analytes is a challenging task, and therefore, different imprinting techniques were explored including bulk, surface and, recently, double imprinting as well. Tobacco mosaic virus (TMV) was the first model virus to be considered for imprinting, as its structure and properties have already been established. It has a rod-like shape, length of around 300 nm and diameter of about 18 nm. Kofinas and co-workers[71,72] adopted the bulk imprinting methodology, as in this case a TMV suspension was mixed together with polyallylamine hydrochloride in aqueous solution and then subjected to polymerization. The imprinting was followed by a non-covalent approach, where the TMV was extracted from a cured hydrogel by treating with suitable solvent. The washed matrix could selectively incorporate the target virus; the binding capacity for TMV was twice as high in comparison to non-target tobacco necrosis virus (TNV). The non-imprinted hydrogel exhibited much less response for both TMV and TNV. In a following study, the authors further optimized the conditions for TMV recognition. Aside from bulk imprinting, viruses can also be imprinted exclusively at a polymer surface by applying a suitable template stamp. Initially,[73] the TMV stamp was pressed on a pre-polymerized mixture of methacrylic acid, styrene and divinyl benzene. The surface imprints of TMV were studied by atomic force microscopic (AFM) images followed by mass-sensitive measurements, which revealed that MIPs contain the complementary adapted cavities for TMV to fit in. In later studies, the polyacrylate layers were substituted by functional polyurethanes[74] for better detection. The cross sensitivity was also impressive as evaluated by exposing TMV imprinted polyurethane[75] to other viruses.

The successful surface imprints of TMV paved the way for the recognition of other viruses, for example, human rhino virus (HRV), thus shifting the focus from plant diseases to human diseases. HRV is an infectious virus and is the major cause of the common cold. It belongs to the family of picornaviruses, having an icosahedral shape with an outer protein capsid. There are various types of HRV, which differ from each other due to their surface proteins. Molecular imprinting allows establishment of intragroup and intergroup selectivity,[76] that is, viruses belonging to same group as well as to different groups can be differentiated. Interestingly, this was realized by the fact that the extent of variation in surface proteins of viruses belonging to same family is comparable to that of between two different virus families.

The ultimate goal is to mimic biological recognition phenomena, thus to interact with the entire capsid rather than with the receptor site only. The geometrical features of HRV were casted on a polyurethane surface as confirmed by AFM images. The relative sensor effects for different HRV serotypes, namely HRV 1A, HRV 2 and HRV 14, by their respective imprinted polymer coatings are summarized in Figure 14.7.

This result is an excellent demonstration of intragroup selectivity. In addition, the intergroup selectivity was also investigated, where an HRV-imprinted surface was treated with the foot and mouth disease virus (FMDV), which belongs to the same family of picornaviruses. It has been observed that FMDV has negligible cross-sensitivity, as the sensor effect for HRV was nine times higher. The reference channel exhibited a similar response for both HRV and FMDV, indicating their unselective adsorption.

Bovine leukemia virus (BLV) was the first animal virus to be imprinted in a polypyrrole matrix as doped by glycoprotein *gp* 51.[77] The marker-free detection of *gp* 51 was made by a pulse amperometric technique. The main concern

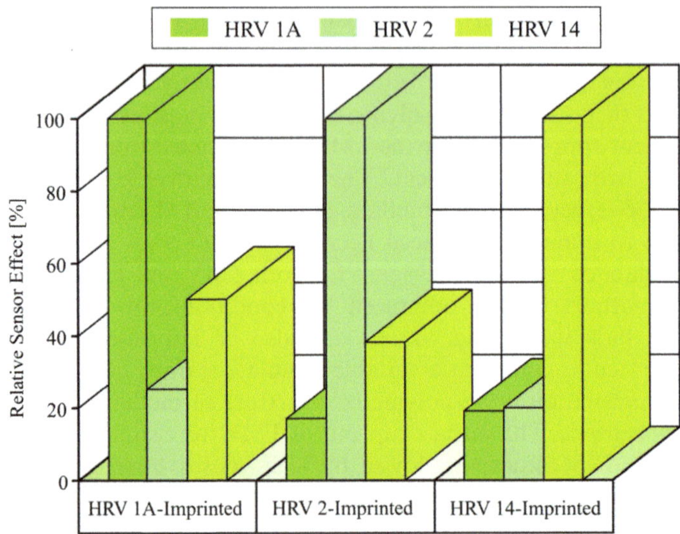

Figure 14.7 Comparison of different imprinted polymer layers of human rhino virus (HRV) serotypes along with their relative sensor effects. It is clear that the premium sensor effect is obtained only when the template and analyte was the same.
(Data from Jenik *et al.*[76])

regarding electrochemical sensing is the lack of selective interactions. Since in electrochemical sensing, non-specific adsorption cannot be distinguished from specific adsorption of unlabeled antigens, authors therefore tested a series of different solutions of KCl at different pH in glycine phosphate buffers, and ultimately concluded that by optimizing the incubation conditions, the non-selective electrochemical signals could be reduced. The re-usability of the polymer can be enhanced by optimized incubation conditions and additionally also improving the sensitivity.

Unlike TMV which has rigid structure, surface imprinting can be applied for the more delicate and fragile parapoxvirus, which is a major cause of various infections in cattle. Owing to the soft nature of virus membranes, the stamps can be made of flexible polydimethylsilicone instead of glass, thus avoiding any damage to membranes. Surface-imprinted polyurethane layers can distinguish remarkably well fresh and aged virus samples[78] stored under different conditions. As the aged virus samples exhibited smaller sensor response as compared to fresh ones, in this way, the extent of harmful effects by virus suspension could also be measured.

Dengue hemorrhagic fever[79] is an emerging disease in many tropical parts of the world and has become a serious threat to public health. This fever is caused by the dengue virus which belongs to the *Flavivirus* family. There is no effective vaccination or medication available in the market to counteract the disease. Therefore, a rapid detection protocol is highly desirable for early diagnosis with adequate accuracy and precision. Tai and colleagues[80] were the first to develop immunochips using two monoclonal antibodies for recognition of dengue virus. However, in later attempts, the authors adopted an epitope-mediated imprinting methodology[81] for dengue virus sensing. They introduced a pentadecapeptide as linear epitope of the dengue virus protein. The imprinted polymer film coated on QCM allows the *in vitro* label-free assay of virus protein. A major advantage of this setup is that the analysis can be performed without separation and purification procedures. The epitope imprinting approach is relatively novel and has a unique advantages over others, being of prime importance for viral recognition. Apart from dengue virus detection, Chen and co-workers[82] have reported a human immunodeficiency virus (HIV) type 1 related protein (glycoprotein 41, *gp* 41) sensor following epitope imprinting. A synthetic peptide similar to the residues 579–613 of *gp* 41 was selected as epitope along with the functional monomer dopamine, which altogether resulted in the formation of a hydrophilic MIP sensor. Mass sensitive measurements exhibited suitable affinity for *gp* 41 and the detection limit was also comparable to enzyme-linked immunosorbent assay (ELISA). The results of human urine sampling for HIV *gp* 41 were satisfactory, which indicates the suitability of this sensing scheme for real-time analysis. Recently, natural antibodies were used as template in the imprinting process, and then further using this MIP to generate a patterned plastic replica of the initially used antibody. This in fact is a double-imprinting strategy, which was used for HRV sensing[83] with acoustic devices. The recognition capability of this artificial surface was compared with natural antibodies for HRV detection, which revealed that the synthetic receptor is about six times more sensitive than the natural one. These findings clearly suggest that MIP receptors possess the desired potential to substitute for natural antibodies.

MIPs are mostly studied along with optical or mass sensitive devices for building up virus sensors; however, Wang *et al.*[84] used potentiometric devices for poliovirus detection and cancer biomarkers. Poliovirus enters human body through the digestive system and its fast rate of replication severely disturbs the functions of neurons. The authors combined the advantages of the self-assembled monolayer (SAM) technique with molecular imprinting for crafting sensor layer material. Hydroxyl terminated alkanethiol chains were used to produce the SAM and thus to imprint poliovirus. The potentiometric response increased linearly with the increasing concentrations of poliovirus, whereas the non-target adenovirus and reference layer exhibited negligible effect as shown in Figure 14.8. The potential stabilization time after each addition was about three minutes.

In principle, MIPs possess the desired potential for sensing of plant, animal and human viruses as evident from the above examples. MIP coatings are highly cost-effective while considering the production protocols of natural antibodies. However, online screening of viral contamination needs a better sensing platform for field analysis. The recently developed micro total analysis system (μTAS)[85] is very promising for continuous monitoring. The combination of contactless dielectric sensors with microfluidics[86] has proven highly suitable for understanding viral binding affinities and dissociation kinetics at

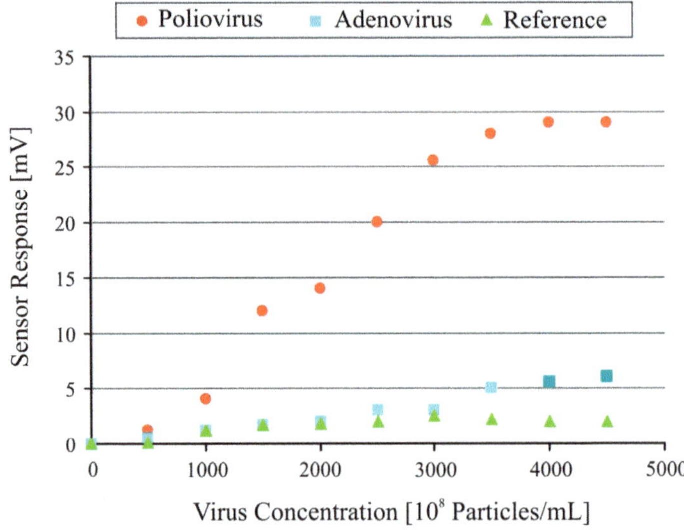

Figure 14.8   Potentiometric sensor response of poliovirus and non-targeted adenovirus against imprinted and non-imprinted sensor layers.
(Data from Wang *et al.*[84])

the MIP interface. This setup could be a real competitor to ELISA by properly
adjusting sensor design and geometry along with more sensitive and selective
MIP coatings.

14.3.3   Protein Recognition

Protein recognition by artificial receptors is potentially important in various
disciplines of the life sciences. A number of concise review articles[87,88] has been
published in recent years; however, in comparison to the imprinting of other,
smaller molecules, this field has a broader impact in biotechnology. There are
certain problems regarding the imprinting of proteins which bring complexities
and therefore, straightforward imprinting of proteins is not simple. For instance, proteins are highly sensitive to pH and temperature, not compatible
with organic solvents as they could be denatured, or some conformational
changes may occur. The ionic strength of the medium is also an important effect
in the case of protein imprinting. Thus, solvation by water molecules *via*
hydrogen bonding between template and functional monomers is altered.[89] In
addition, their large size is somewhat challenging for direct imprinting and their
release or rebinding by MIPs is slow. These issues make the imprinting procedure very complex and as a result non-selective binding becomes dominant,
which reduces both sensitivity and selectivity. Moreover, there are various
imprinting methods which can be applied, depending upon the nature and
structure of a protein.

In principle, the most simple and straightforward method is bulk imprinting,
where the whole protein structure is imprinted in a polymer matrix which can

be recognized as a whole target after washing or extraction of the MIPs. The size and number of pores at the MIP interface is crucial to allow proteins to access and fit into three-dimensional cavities. Polymer density is another question which is related to the selectivity. Acrylic polymers are best considered for bulk imprinting and have found useful applications; however, protein transfer reduces their density, ultimately disturbing polymer stability. In this regard, sol-gels[90] and hydrogels[91] are promising, as the highly cross-linked three-dimensional networks produce stable and more porous polymers with an improved number of interaction centers. These systems retain the size of imprinted cavities relatively better during regenerations. Nevertheless, it is difficult for the interfacial functional groups to incorporate large-size protein templates keeping adequate sensitivity and selectivity for next stages of regeneration.

In surface imprinting,[92] the recognition of proteins is exclusively made at the polymer interface, which ensures better access to interaction centers. In addition, the binding and releasing of proteins is relatively easier in comparison to bulk imprinted polymers. However, the sensitivity is less, as recognition sites are located only at the polymer surface whereas in bulk imprinting these are distributed all over the matrix. A wide variety of methods and materials is adopted for surface imprinting of proteins. These materials range from poly-siloxanes[93] to polyacrylamide,[94] nanowires[95] and beads.[96] The degree of complexity in the synthesis of these materials and their protein recognition ability varies from one to another. For example, the stamping technique is more useful for fast synthesis of MIPs taking small amounts of template with adequate solvent compatibility. The combination of sol-gel with self-assembly[97] has proven very effective in minimizing non-selective interaction, thus achieving high selectivity. In a similar way, imprinted nanomaterials are better in sensitivity owing to their enhanced surface area as compare to conventional polymeric materials. For instance, surface imprinted nanowires[98] are favorable for protein recognition as they possess high selectivity and binding capacity. Owing to their large surface area, there are more interaction sites in comparison to other surface imprinted materials. The extent of hydrogen bonding between template and target proteins is relatively higher which leads to superior sensitivity. Examples of different analytes tested by surface imprinted polymers are highlighted in Table 14.1.

In certain cases, the interaction between surface-imprinted polymers with proteins is not optimal due to unselective bindings, complexity and conformational flexibility in protein structure. Therefore, in order to address these issues an effective and improved technique was introduced, named epitope imprinting, which combines advantages of both bulk and surface imprinting. In epitope imprinting, it is not necessary to retain the conformation of the whole protein as it allows the implementing of harsh synthetic conditions including temperature and the use of organic solvents. The key factor is the selection of the right peptide sequence for the best recognition results, and to achieve this, the structure of target protein should be well understood.

In general, the terminal part of a target protein is used in epitope imprinting. The selectivity for protein recognition is enhanced by cooperative and

Table 14.1   Some selected examples of different protein imprinting approaches along with their sensing applications.

Protein imprinting approach	Polymer systems	Target protein analytes	Sensing platform	Refs
Bulk Imprinting	Xerogels	Human serum albumin (HSA), Oval Albumin	Fluorescence	101
	Acrylic acid and *N,N*-methylenebisacrylamide	Lysozyme	SPR	102
Surface Imprinting	Phenyltrimethoxysilane, Methyltrimethoxysilane, Tetraethylorthosilicate,	HSA	QCM / Impedance	97
	Styrene, divinyl benzene, methacrylic acid	Lysozyme	QCM	51
	Methacrylic acid, styrene, 1-vinyl-2-pyrrolidone	Trypsin, lysozyme	QCM	50
	Ternary lipid monolayer	Ferritin	QCM	103
	3-aminophenylboronic acid	Lysozyme, cytochrome *c*	QCM	104
Epitope Imprinting	Acrylic acid	Dengue virus NS-1 protein	QCM	81
	Dopamine	(HIV-1) glycoprotein 41	QCM	82

QCM, quartz crystal microbalance; SPR, surface plasmon resonance.

multivalent hydrogen bonding; consequently, the non-selective interactions of non-target proteins are reduced to a minimum, as reflects the unavailability of potential binding sites. Another interesting feature of this technique is the inexpensive synthesis of peptide sequence used as epitope since bulk and surface imprinting strategies often need proteins which are difficult to purify or synthesize which are costly.

Concerning the transducer for sensor applications, optical or mass sensitive systems are more widely studied. Electrochemical devices are mostly used for small electroactive compounds and not for proteins. However, there are few proteins which show redox behavior[99] in Nature and thus their detection can be possible by electrochemical sensors. Impedance and capacitance are other properties which can be exploited for protein sensing.[100] On the other hand, optical and acoustic devices offer better sensitivity and selectivity and most of the published articles describe the use of these two sensing platforms. Especially, acoustic sensors can provide label-free detection systems with minimal interferences; Table 14.1 show that most of the protein recognition systems adopt acoustic setups owing to their better detection limits and also the relative ease of polymer processing with the transducer.

14.3.4   Hormonal Sensing

Hormones are chemical messengers, as they deliver specific information from one part of the body to cells in another part in response to certain metabolic reactions. Hormonal disorder or imbalance often leads to serious problems, ultimately disturbing the normal functions of cells. For example, the hormone insulin is vital in regulating carbohydrate and fat metabolism, since it regulates the storage of glucose from the blood in liver as glycogen, thus preventing the use of body fat an as energy source. The imbalance of insulin in human body leads to diabetes, insulinoma and other related problems. Molecular imprinting enables the crafting of insulin-sensitive cavities, which are as selective as natural antibodies. This is of utmost importance since insulin detection is highly desirable not only in the human body, but also in its large-scale, industrial production where its level needs to be monitored. Synthetic insulin is produced by the fermentation reactions of yeast cells and bacteria. Sensor layers can be generated by the surface imprinting technique where stamps of assembled insulin crystals are pressed over pre-polymerized polyurethane.[105] Mass sensitive results indicated that these sensor layers are highly responsive to a wide concentration range of insulin, which is comparable to medical formulations. In addition, the temperature effect on insulin could also be measured by this setup, which is beneficial at the industrial level in order to evaluate the quality of product. Different insulin solutions of the same concentration were subjected to different temperatures, *viz.* 25 °C, 40 °C, 60 °C and 95 °C. It was observed that sensor response was highest for 25 °C and was lowest for 95 °C treated sample. In fact, with increasing temperature the insulin protein starts denaturing which is relatively little at 40 °C but becomes substantial at 95 °C. Denaturing has strong effects on size and shape of the protein as this make it unable to fit geometrically in the adapted cavities at the polyurethane surface.

In later studies, insulin sensing[106] was realized by the double imprinting strategy which gives an idea about the competitive recognition of this scheme in comparison to synthetic and natural antibodies. Whereas synthetic antibodies for insulin detection were synthesized by surface imprinting, in double imprinting the nanoparticles are precipitated in the presence of selected natural antibodies. These antibodies are removed by washing the nanoparticles with water. The washed nanoparticles, having the memory of imprint antibodies, were pressed over pre-polymer thin films. This would be lead to generate synthetic or plastic copies of relevant antibodies on the polymer surface with a high percentage of imprinting density. The three different layers for insulin recognition were investigated; these included synthetic antibodies, natural antibodies and antibody replica as crafted by double imprinting.

The relative sensor response for insulin and other non-target analytes, *viz.* glargine, lysozyme and pepsin, are displayed in Figure 14.9. All three receptor materials favor insulin recognition as indicated by their respective frequency shifts. The response of synthetic antibodies was somewhat comparable to natural antibodies; however, the antibody replica shows the highest sensor effect among all. This suggests that imprinting density is much higher in the antibody replica owing to the large surface area of nanoparticles, thus leading to better performance. The maximum difference between sensor effects of insulin and glargine is obtained when the receptor material is natural antibodies. However, it is least in case of synthetic antibodies and is intermediate for the

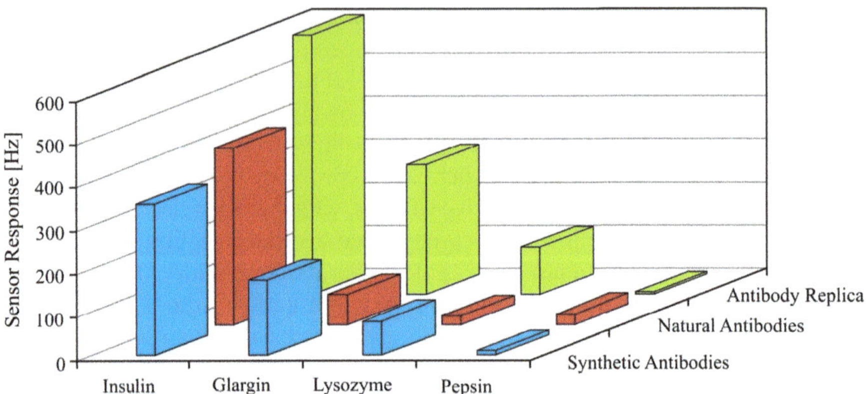

Figure 14.9   Comparison of sensor responses of synthetic antibody, natural antibody and antibody replica for insulin and the other non-target analytes, *viz.* glargine, lysozyme and pepsin.
(Data from Schirhagl *et al.*[106])

antibody replica. The frequency shifts for lysozyme were relatively less and pepsin makes negligible impression on any of the receptor materials. The superior selectivity of natural antibodies is due to the fact that the interaction sites are present on their surface, while the synthetic receptors interact with the entire protein surface. Nevertheless, the selectivity is still good enough, bearing in mind the minimal structural difference in insulin and glargine. Ultimately, the sensitivity of the antibody replica was larger by a factor of 1.5 in comparison to the natural receptor.

Following the same approach, a detection limit of $1\mu g$ mL^{-1} for insulin has been reported.[24] The hallmark of these antibody replicas is their resistive and robust character in tough conditions. Recently it has been shown that antibody replicas can effectively work in organic solvents (*e.g.* chloroform) for extracting the hormones[107] from aqueous media. The natural antibody cannot work in organic solvents as it is denatured. The other most exciting feature of sensing in organic solvents is that the detection limits can be improved astonishingly by factor of 10, since the target analytes can be extracted from aqueous solutions into organic solvents, thus achieving pre-concentration. Comparing the cost with natural antibodies, these materials are inexpensive and therefore always the first choice.

Liu and co-workers[108] reported surface capping of MIP layers to improve their selectivity in the detection of catecholamines, which are important neurotransmitters. Generally, dopamine, epinephrine and norepinephrine, which are the major catecholamines, are released in blood as a result of certain physiological changes. However, their detection is usually done in urine samples. The authors proposed silica–alumina gel as a sensitive recognition material with optimized conditions of pH and Si/Al ratio. The sensor films containing silanol groups were capped with hexamethyldisilazane, which resulted in the replacement of terminal –OH with trimethylsilyl groups.

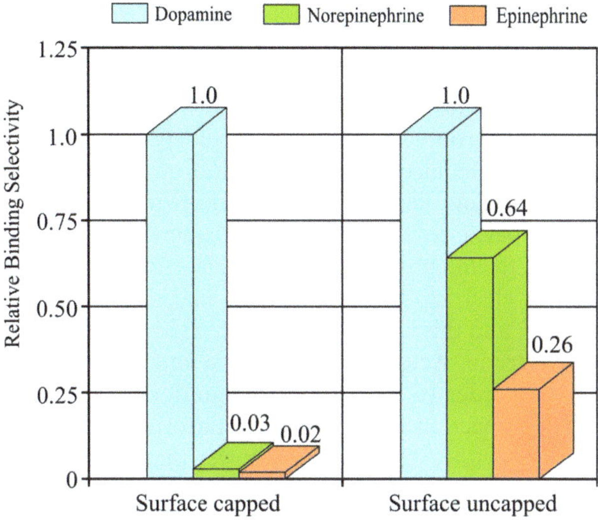

Figure 14.10 Comparison of relative binding capacities before and after surface capping. It is obvious that surface capping remarkably enhances selectivity.
(Data from Ling *et al.*[108])

The relative binding selectivity of dopamine-MIP with surface capped and uncapped layers is summarized in Figure 14.10. It can be seen from this graph that after surface capping the relative selectivity has been increased enormously as the non-specific interactions are very much reduced. In fact surface silanol groups are responsible for adsorption of different analytes, whereas the substitution of –OH with trimethylsilyl reduces this adsorption. In addition, the size of the trimethylsilyl group is large in comparison to the –OH group and subsequently this brings steric hindrance, which is beneficial in order to screen the adsorption of non-targeted analytes. In this way, the selectivity is improved; however, the sensitivity was lowered due to fewer interaction centers being available, as observed by *in situ* frequency shifts.

MIP-based fluorescence immunoassay of epinephrine has already been reported[109] following an oxidative polymerization technique using the functional monomer 3-aminophenylboronic acid. This functional monomer is capable of forming reversible covalent ester bonds between boronic acid and catechol, and in addition some electrostatic interactions with amino and hydroxyl groups of the template. It was observed that the dissociation constant of MIP-epinephrine was $9.2 \pm 2\,\mu M$ whereas for the non-imprinted species it was $150 \pm 15\,\mu M$. In a similar way the dissociation constants for closely related analytes showed much weaker binding affinity with MIPs. The authors suggest that the binding affinity of MIP-epinephrine is comparable to that of natural antibodies.

Piletsky and co-workers[110] designed a novel hybrid material for electrochemical sensing of catechols. The authors combined catalytic MIPs with a conducting polymer in order to provide a direct pathway for the conduction of electrons from the active sites of imprinted hybrid material to a sensing electrode surface. They selected the recently designed bi-functional monomer *N*-phenylethylene diamine methacrylamide which contains aniline and methacrylamide functional groups as both can be independently polymerized with different techniques. The electro-polymerization of the monomer leads to the generation of polyaniline films at the gold electrode surface, having double bonds which can participate in addition polymerization thus, developing a molecular network for integration of MIPs. This network may be considered as molecular wire for electrical communication, since the MIPs oxidize the template (*e.g.* catechol) and this information is delivered through polyaniline layers to the sensing electrode. Actually, the catalytic and conducting properties are merged together in order to accomplish recognition of analytes by oxidation and developing a connection with the sensor platform. This setup offers a lower detection limit of 228 nM, which is much lower than the previously described strategy. The non-imprinted reference electrode did not show any detectable signal below 100 µM. Furthermore, the selectivity of the hybrid material was significantly higher against potential interfering substances such as resorcinol, hydroquinone, serotonin and ascorbic acid. The hybrid sensor material sets the new trend for electrochemical recognition of various bio-analytes with un-matched specificity.

MIPs have also been applied for recognition of plant hormones,[111] such as indoleacetic acid, which plays an important role in plant growth. The authors followed the bulk imprinting strategy, taking methacrylic acid as the functional monomer. Initially, this scheme was designed for the extraction out of complex biological mixture and subsequently analyzed by solid phase extraction (SPE). Later these sensor layers were combined with QCM for developing an indole acetic acid sensor[112] for plant growth monitoring with high selectivity. In general, MIPs are equally suitable for hormonal detection like other bio-analytes and their applications are expected to expand from the biomedical field to industrial process control.

14.3.5 Body Fluid Analysis

The analysis of body fluids such as urine and blood serum or plasma is a well established diagnostic procedure. Generally, the goal is to measure the levels of distinctive metabolites present in human body fluids, which ultimately aids in disease diagnostics. The most obvious example is the elevated level of glucose in a blood sample, which is a straightforward sign that the patient is diabetic. The determination of certain metal ion concentrations in urine or blood is also gives direct indications about some particular diseases, for example, the presence of Ca^{2+} ions in high percentage is related to urinary stone formation.[113] Moreover, analysis of biological fluids by MIPs is gaining much interest in doping investigations, since some drugs are used to enhance the performance in competitive sports. Molecular imprinted materials are suitable for the recognition of various drug metabolites and metal ions in complex matrices such as urine or blood samples.

Several publications[114,115] are devoted to urine and blood serum analysis using MIPs as the sensing or recognition material. In order to extract the target analyte, MIPs are often used with SPE followed by spectroscopic or chromatographic analysis. It has been noticed that most SPE applications acquire a non-covalent imprinting approach, since it offers more flexibility in the selection of functional monomers. However, covalent imprinting has also been applied in the synthesis of selective adsorbents for different sugars.[116] The authors selected boronic acid derivatives as monomers due to their exclusive formation of cyclic esters with diols in aqueous media. Nevertheless, non-covalent imprinting[117] is the predominant approach over covalent schemes as indicated by number of published articles in this area. The extraction with MIP adsorbents[118] from body fluids is usually carried out with high performance liquid chromatography (HPLC); however, some other methods such as ion mobility spectrometry, gas chromatography–mass spectrometry (GCMS) and capillary electrophoresis were also investigated. The sensitivity of HPLC coupled with a fluorescence detector is quite significant, as indicated by extremely low detection limits; in respect of selectivity; however, capillary electrophoresis seems to be a better option. The sensitivity of capillary electrophoresis can be improved by adopting pre-concentration methods. Apart from optimized composition of imprinted polymer, the pH is also a very important factor which affects the recognition capacity of a particular MIP system, for example working with blood or urine samples. Table 14.2 gives an idea about the recognition of various drugs or their metabolites using molecular imprinted solid phase extraction (MISPE) systems along with respective analysis technique.

Apart from drugs, MISPE is also used for isolation and extraction of naturally occurring steroids[119] like progesterone, testosterone and 17-estradiol, which are important in the human reproduction system. Here MISPE was used for pre-concentration of target analytes and cleaning the matrix prior to HPLC analysis. These steps are extremely beneficial in order to improve sensitivity and specificity, thus achieving lower detection limits with less interference of matrix substances. In this way, MISPE applicability has expanded from body fluids to food samples,[120] for example, in analyzing the presence of chloramphenicol in milk and honey. Chloramphenicol is a restricted antibiotic for food production due its adverse effects on humans. The designed MISPE exhibited good performance for cleaning up the biological samples, namely urine, feed water, milk and honey, prior to their GCMS analysis.

14.3.6 Drug Delivery

One of the most significant contributions of molecular imprinted materials in the biomedical field is drug delivery or controlled drug release applications.[128,129] Specific molecular recognition at the cellular or subcellular level is the principal condition in various biological processes including neural transmittance, immune defense, cellular differentiation and many others.

Table 14.2 Applications of molecular imprinted solid phase extraction (MISPE) systems for recognition of various drugs in biological fluids.

Target drug / metabolite	Composition of MISPE system	Analysis technique	Biological fluids	Ref.
Tamoxifen	Methacrylic acid (MAA), ethylene glycol dimethacrylate (EGDMA)	HPLC	Urine	121
Salicylic Acid	4-vinylpyridine (4-VP), EGDMA	Electro spray ionization –ion mobility spectrometry (ESI-IMS)	Urine, blood plasma	122
Nicotine	MAA, EGDMA	Piezoelectric QCM sensor	Serum, urine	123
Ciprofloxacin	MAA, EGDMA	HPLC	Urine	124
Progesterone,Testosterone, 17-estradiol	4-VP, MAA, EGDMA, trimethylolpropane, trimethacrylate	HPLC/diode array detector	Urine	119
Dopamine, 3-methoxytyramine, 5-hydroxytryptamine	AFFINIMIP by POLYNTELL[a]	Capillary electrophoresis	Urine	125
Chloramphenicol	SupelMIP SPE	GCMS	Urine, feed water, milk, honey	120
Metoclopramide	MAA, EGDMA	HPLC	Serum, urine	114
Tramadol	MAA, EGDMA	HPLC	Plasma, urine	126
Carbamazepine	MAA, divinyl benzene	HPLC	Urine, waste water	127

[a]AFFINIMIP, SPE cartridge based on MIPs by POLYNTELL company.

The interaction of a low molecular weight compound with its macromolecular target host is also one of the important processes in living systems which relies on selective recognition phenomena. In drug release, molecular imprinting has two advantages over other synthetic systems: first is the desired flexibility in the synthesis and design; and second is the preferred selective recognition with the host cells. The main concern about drug releasing systems is provision of the right dosage of drug and its interaction at the target tissues or cells. In general drug carrier systems, the drug is dispersed within a matrix which is designed to release under the prescribed physiological conditions. Therefore, the delayed or controlled release of drugs ensures their prolonged residence in patients and aims to ensure the desired dosage to achieve better results. In case of certain drugs the therapeutic window is narrow and it is highly recommended that the concentration should be below optimal dosage in order to avoid harmful side effects. In addition, the drug concentration in the blood would be drastically altered if an abrupt breakdown of the carrier matrix occurs. Therefore, it is imperative to design such systems which have the desired characteristics of intelligent drug liberation.[130] In the literature, intelligent drug release is generally considered as release in a controlled and predictable fashion.

Principally, MIPs are ideally suited for advance drug delivery applications with the inherent characteristics of specificity. A typical illustration of molecular imprinted materials as drug carriers is demonstrated in Figure 14.11. The cross-linked polymer matrix is taken as the reservoir of low molecular weight compounds which can be delivered at projected sites. Generally, the cross-linking percentage[131] is quite high, that is, 25–90% in imprinted materials, which is considered to be essential for retaining the complementary spatial orientation. Consequently, this avoids the immediate collapse of inbuilt cavities and preserves molecular recognition. Interestingly, for drug delivery applications, relatively fewer cross-linker polymers are recommended, especially for developing water-compatible systems. The increasing percentage of cross-linking increases the rigidness and subsequently results in minimal changes in polymer geometry. The selective recognition should not depend on the medium or external variables, ensuring that the drug regulating properties are conserved even in complex biological matrices. In addition, there are certain compatibility issues between the drug administrative site and the highly cross-linked polymer. For instance, the rigid polymer chains will not adapt the shape of target drug administrative site and moreover, there could be some mechanical friction with nearby tissues.[129] Therefore, the imprinted polymer network should be flexible enough to facilitate drug dissociation and to achieve fast equilibrium. It is recommended that there should be a balance between rigidity and flexibility of polymer. The introduction of hydrophilic cross-linkers and monomers would be advantageous in designing water-compatible polymer systems.

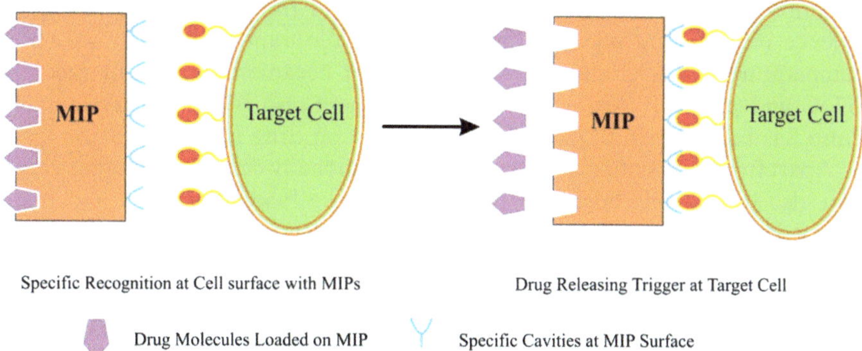

Figure 14.11 Schematic representation of molecular imprinted polymers (MIPs) as drug carriers and their interaction with cells in the body. (After Sellergren and Allender.[14])

The choice of molecular imprinting technique, whether covalent or non-covalent, has a considerable impact on the drug release mechanism. The sensitivity and binding kinetics are also associated with the choice of imprinting strategy. Covalent imprinting is reputed to achieve more selective recognition; however, the equilibrium kinetics concerning binding and dissociation of the template are relatively slow, and furthermore, the numbers of reversible covalent interaction sites are limited. Conversely, in non-covalent imprinting the equilibrium kinetics are comparatively much faster, with a greater number of interaction centers. Non-covalent imprinting is usually carried out in organic solvents. For drug delivery applications, the use of organic solvents is not suitable, as the residual solvents in the MIP drug carrier could lead to serious problems. This can be avoided by using inorganic solvents and precipitation polymerization.

However, molecular imprinting in aqueous media is still in the development phase. This is because of the fact that, in water, the solvent effects become dominant over the weak intermolecular interactions, which influence the pre-polymer monomer template complex. MIP hydrogels[132] are moderately cross-linked polymer networks that are highly suitable for drug delivery applications. This is an emerging field in drug delivery systems and a number of research papers are devoted to this area. Acrylic acid based MIP hydrogels[133] have been studied for the controlled release of 5-fluorouracil which is an important anticancer drug. It has been observed that by adjusting the percentage of cross-linker the desired flexibility can be reached, which ultimately affects the swelling properties of polymer. Moreover, with the increase in template concentration, the swelling also increases, which is better for controlled drug release. Recently, molecular imprinted silicone hydrogels[134] were used to deliver an antibiotic drug, namely ciprofloxacin, using acetic acid and acrylic acid as functional monomers. This study indicated that the lower monomer–template ratio tends to release the drugs in larger amounts within a shorter time, while a relatively high monomer–template ratio leads to a longer release time. Following another example of controlled release of citalopram,[135] anti-depressant drug, by MIP, it was noticed that the pH and the temperature of the medium had strong influence on the drug releasing mechanism. Citalopram release was slower at temperature of 25 °C and pH 4.3, whereas it became considerably faster at 37 °C and pH 10.1. Again like other cases, here also the monomer–template ratio has been shown to have a strong impact on drug release behavior.

Apart from conventional materials, computationally designed nanoparticles[136] can also be a good choice for an improved drug loading capacity and better selectivity. This is advantageous over the trial-and-error method for selecting the optimal polymer system. In this study acrylic acid and methacrylic acid nanoparticles were synthesized after computational modeling for naltrexone, which is a drug used in the treatment of narcotic addiction. An acrylic acid based MIP showed a high drug loading capacity of 75 mg of drug per gram of MIP comparing to methacrylic acid MIP. Furthermore, the selectivity was also evaluated by exposing methadone, which is analogous to naltrexone. The respective binding capacities of acrylic acid and methacrylic acid MIPs are shown in

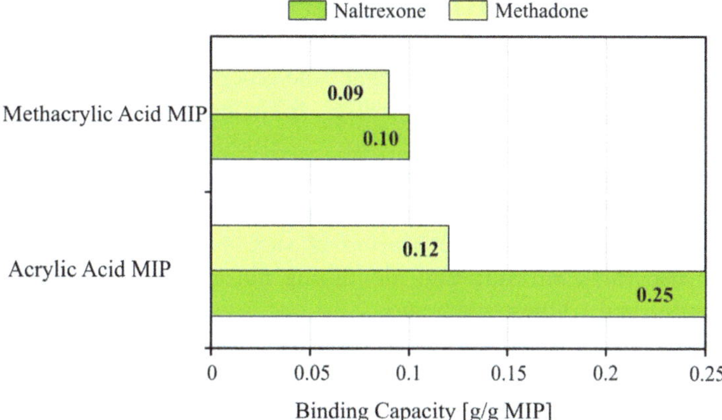

Figure 14.12 Comparison of acrylic acid and methacrylic acid polymers for their loading capacities for naltrexone and its analogue methadone. (Data from Rostamizadeh *et al.*[136])

Figure 14.12. From this figure it is clear that the acrylic acid-MIP is much more selective in comparison to the methacrylic acid-MIP. This clearly suggests that the selection of monomer itself is highly important for optimal selectivity.

Molecular imprinting is proving to be a promising and competitive technique in the drug delivery field; however, *in vivo* studies are somewhat limited, which is a major concern. There are a number of factors which need to be considered, such as the complex nature of the drug releasing medium, susceptibility due to attack by enzymes that could influence drug release before the target is reached, pH and temperature, degradation products of polymer, residual impurities or un-reacted monomers and their impact on sensitive cells and tissues. Therefore, the safety of drug carrier material should be assured and the probable toxicological effects fully investigated. In addition, before introducing an entirely new MIP for drug releasing applications, its biocompatibility should be studied in detail to understand its impact or behavior in the body.

The solution to existing problems of MIP design can be made by following a computational approach as well for establishing optimal conditions in polymer synthesis. This includes the development of hydrophilic functional monomers and cross-linkers, thus adjusting polymer polarity. In a typical polymer system, the selection of a more suitable polymerization technique is also important for generating polymer beads and nanoparticles of uniform size and morphology. An intense research of the above-mentioned factors would lead to the development of new generation MIP materials for advanced drug delivery applications.

14.4   Concluding Remarks and Future Prospects

As discussed above, molecular imprinted materials have found a huge number of applications, ranging from clinical diagnostics to drug delivery systems.

Among synthetic materials, MIPs are one of the best-suited materials to mimic various biological processes, as they offer unmatched selectivity and specificity. Following a typical imprinting approach, the recognition interface can be developed for microorganisms, cells, viruses, biomacromolecules and small size drug molecules. MISPE has been extensively studied for the analysis of body fluids, which includes blood plasma, serum and urine samples. Moreover, MIP membranes have also been used for blood purification;[137] they are equally suitable for the enantiomeric separation of various drugs from racemic mixtures and also for separating and identifying natural products such as those used in traditional Chinese medicine.[138]

However, concerning clinical analysis or biosensing,[139] there are two focal points: one is related to the transduction of measurable signals, and the second is the optimization of the MIP interface for more selective recognition. Transducer selection should consider the adaptability with sensor material, ability to measuring small changes on interfacial coatings and miniaturized design with the ease of portability for field analysis under extreme conditions. All these parameters are somewhat difficult to integrate at one place while considering the real-time measurements. Most of the first step clinical diagnostics or screening is performed by commercially available low-cost test strips. Nevertheless, some of the optical or acoustic devices can offer comparable sensitivity to existing analytical tools; until now, however, their commercial applications are limited.

Regarding MIPs, the first point is the selection of right technique for the target analyte. It can be seen from previous discussed examples that surface or epitope imprinting methods are most widely used for improved recognition. The next task is the overall composition of the imprinted polymer matrix, which includes monomer, cross-linker, solvent system, initiator and other reaction conditions. We have seen that the degree of cross-linking is decisive in controlling flexibility or rigidity which ultimately determines their end-use as sensing material or in drug delivery systems. The optimal molar ratio between monomer and template plays a significant role in sensitivity and selectivity, which is generally adjusted by trial and error; however, introducing combinatorial approaches would be beneficial for achieving better results. The polarity of polymer matrix strongly affects the weak intermolecular forces and generally organic solvents favor the formation of a pre-polymer complex between monomer and template. However, due to their toxicological effects, organic solvents are not suited to crafting of MIPs as drug carriers. On the other hand, solvents like water strongly compete in hydrogen bonding and, therefore, do not favor the complex formation. In this scenario, solvents with moderate polarity and suitable biocompatibility should be used in MIP synthesis. From the previous examples, it is well understood that methyl acrylic acid- and EGDMA-derived polymers (see Table 14.2) are the best suited system for their recognition characteristics and compatibility with aqueous solutions. In principle, soft polymer matrices can be best recommended for drug delivery systems, whereas sensing applications require more rigid polymers. To sum up, molecular imprinting has the potential to be the finest technique for selective recognition in the biomedical field and will likely become a mature technology in the coming years.

References

1. K. Mosbach, *Trends Biochem. Sci.*, 1994, **19**, 9–14.
2. K. Mosbach and O. Ramstrom, *Nat. Biotech.*, 1996, **14**, 163–170.
3. J. H. Hartley, T. D. James and C. J. Ward, *J. Chem. Soc., Perkin Trans.*, 2000, **1**, 3155–3184.
4. F. L. Dickert, P. Lieberzeit, S. G. Miarecka, K. J. Mann, O. Hayden and C. Palfinger, *Biosens. Bioelectron.*, 2004, **20**, 1040–1044.
5. G. Wulff, *Chem. Rev.*, 2002, **102**, 1–27.
6. C. Alexander, L. Davidson and W. Hayes, *Tetrahedron*, 2003, **59**, 2025–2057.
7. Z. J. Tan and V. T. Remcho, *Electrophoresis*, 1998, **19**, 2055–2060.
8. E. Turiel and A. Martin-Esteban, *Anal. Bioanal. Chem.*, 2004, **378**, 1876–1886.
9. D. J. O'Shannessy, B. Ekberg, L. I. Andersson and K. Mosbach, *J. Chromatogr., A*, 1989, **470**, 391–399.
10. L. I. Andersson and K. Mosbach, *J. Chromatogr., A*, 1990, **516**, 313–322.
11. S.-G. Hu, S.-W. Wang and X.-W. He, *Analyst*, 2003, **128**, 1485–1489.
12. C. He, Y. Long, J. Pan, K. Li and F. Liu, *J. Biochem. Biophys. Methods*, 2007, **70**, 133–150.
13. J. Z. Hilt and M. E. Byrne, *Adv. Drug Delivery Rev.*, 2004, **56**, 1599–1620.
14. B. Sellergren and C. J. Allender, *Adv. Drug Delivery Rev.*, 2005, **57**, 1733–1741.
15. N. Lavignac, C. J. Allender and K. R. Brain, *Anal. Chim. Acta*, 2004, **510**, 139–145.
16. I. Surugiu, L. Ye, E. Yilmaz, A. Dzgoev, B. Danielsson, K. Mosbach and K. Haupt, *Analyst*, 2000, **125**, 13–16.
17. K. Haupt and K. Mosbach, *Chem. Rev.*, 2000, **100**, 2495–2504.
18. F. L. Dickert and O. Hayden, *TrAC, Trends Anal. Chem.*, 1999, **18**, 192–199.
19. A. Ersöz, A. Denizli, İ. Şener, A. Atılır, S. Diltemiz and R. Say, *Sep. Purif. Technol.*, 2004, **38**, 173–179.
20. F. Dickert, O. Hayden, P. Lieberzeit, C. Palfinger, D. Pickert, U. Wolff and G. Scholl, *Sens. Actuators, B*, 2003, **95**, 20–24.
21. O. Hayden, R. Bindeus and F. L. Dickert, *Meas. Sci. Technol.*, 2003, **14**, 1876.
22. P. Lieberzeit, G. Glanznig, M. Jenik, S. Sylwia Gazda-Miarecka, F. Dickert and A. Leidl, *Sensors*, 2005, **5**, 509–518.
23. A. Bossi, F. Bonini, A. P. F. Turner and S. A. Piletsky, *Biosens. Bioelectron.*, 2007, **22**, 1131–1137.
24. R. Schirhagl, U. Latif, D. Podlipna, H. Blumenstock and F. L. Dickert, *Anal. Chem.*, 2012, **84**, 3908–3913.

25. A. Mujahid and F. Dickert, in *Handbook of Molecular Imprinting - Advanced Sensor Applications*, ed. S.-W. Lee and T. Kunitake, Pan Stanford Publishing, Singapore, 2012, pp. 527-570.
26. A. Mujahid and F. Dickert, in *Nano-Bio-Sensing*, ed. S. Carrara, Springer, New York, Dordrecht, Heidelberg, London, 2010, pp. 45-82.
27. A. Mujahid and F. Dickert, in *Molecularly Imprinted Sensors Overview and Applications*, ed. S. Li, Y. Ge, S. Piletsky and J. Lunec, Elsevier, Oxford, UK, 2012, pp. 125-160.
28. M. M. Moein, M. Javanbakht and B. Akbari-adergani, *J. Chromatogr., B*, 2011, **879**, 777–782.
29. M. Javanbakht, S. E. Fard, A. Mohammadi, M. Abdouss, M. R. Ganjali, P. Norouzi and L. Safaraliee, *Anal. Chim. Acta*, 2008, **612**, 65–74.
30. A. Mujahid, A. Afzal, G. Glanzing, A. Leidl, P. A. Lieberzeit and F. L. Dickert, *Anal. Chim. Acta*, 2010, **675**, 53–57.
31. A. Mujahid and F. Dickert, *Anal. Bioanal. Chem.*, 2012, **404**, 1197–1209.
32. M. Ávila, M. Zougagh, A. Escarpa and Á. Ríos, *Talanta*, 2007, **72**, 1362–1369.
33. M. B. Gholivand, M. Torkashvand and G. Malekzadeh, *Anal. Chim. Acta*, 2012, **713**, 36–44.
34. X.-h. Gu, R. Xu, G.-l. Yuan, H. Lu, B.-r. Gu and H.-p. Xie, *Anal. Chim. Acta*, 2010, **675**, 64–70.
35. Y. Li, M.-j. Ding, S. Wang, R.-y. Wang, X.-l. Wu, T.-t. Wen, L.-h. Yuan, P. Dai, Y.-h. Lin and X.-m. Zhou, *Appl. Mater. Interfaces*, 2011, **3**, 3308–3315.
36. E. Rosellini, N. Barbani, P. Giusti, G. Ciardelli and C. Cristallini, *Biomed. Materials*, 2010, **5**, 065007.
37. S. A. Piletsky, S. Alcock and A. P. F. Turner, *Trends Biotechnol.*, 2001, **19**, 9–12.
38. P. A. G. Cormack and K. Mosbach, *React. Funct. Polym.*, 1999, **41**, 115–124.
39. M. Polyakov, *Zhur. Fiz. Khim.*, 1931, **2**, 799–805.
40. M. V. Polyakov, P. M. Stadnik, M. W. Paryckij and I. M. Malkin, *Zhur. Fiz. Khim.*, 1933, **4**, 454–456.
41. L. Pauling, *J. Am. Chem. Soc.*, 1940, **62**, 2643–2657.
42. L. Pauling and D. H. Campbell, *Science*, 1942, **95**, 440–441.
43. F. H. Dickey, *Proc. Natl. Acad. Sci., USA*, 1949, **35**, 227–229.
44. G. Wulff and A. Sarhan, *Angew. Chem.*, 1972, **84**, 364–364.
45. C.-C. Hwang and W.-C. Lee, *J. Chromatogr., A*, 2002, **962**, 69–78.
46. T. Ikegami, W.-S. Lee, H. Nariai and T. Takeuchi, *J. Chromatogr., B*, 2004, **804**, 197–201.
47. B. Sellergren, in *Techniques and Instrumentation in Analytical Chemistry*, ed. S. Börje, Elsevier, Amsterdam, 2000, vol. 23, pp. 113–184.
48. A. Mujahid, M. Keppler, P. A. Lieberzeit and F. L. Dickert, *Mater. Sci. Eng., C*, 2011, **31**, 553–557.
49. J. L. Defreese and A. Katz, *Chem. Mater.*, 2005, **17**, 6503–6506.

50. O. Hayden, C. Haderspock, S. Krassnig, X. Chen and F. L. Dickert, *Analyst*, 2006, **131**, 1044–1050.

51. O. Hayden, R. Bindeus, C. Haderspöck, K.-J. Mann, B. Wirl and F. L. Dickert, *Sens. Actuators, B*, 2003, **91**, 316–319.

52. A. Rachkov and N. Minoura, *Biochim. Biophys. Acta, Protein Struct. Mol. Enzymol.*, 2001, **1544**, 255–266.

53. A. Rachkov, M. Hu, E. Bulgarevich, T. Matsumoto and N. Minoura, *Anal. Chim. Acta*, 2004, **504**, 191–197.

54. H. Nishino, C.-S. Huang and K. J. Shea, *Angew. Chem., Int. Ed.*, 2006, **45**, 2392–2396.

55. R. Schirhagl, A. Seifner, F. T. Husain, M. Cichna-Markl, P. A. Lieberzeit and F. L. Dickert, *Sensor Lett.*, 2010, **8**, 399–404.

56. A. Mujahid, P. A. Lieberzeit and F. L. Dickert, *Materials*, 2010, **3**, 2196–2217.

57. O. Y. F. Henry, D. C. Cullen and S. A. Piletsky, *Anal. Bioanal. Chem.*, 2005, **382**, 947–956.

58. M. C. Blanco-López, M. J. Lobo-Castañón, A. J. Miranda-Ordieres and P. Tuñón-Blanco, *TrAC Trends Anal. Chem.*, 2004, **23**, 36–48.

59. P. A. Lieberzeit and F. Dickert, in *Piezoelectric Sensors*, ed. C. Steinem and A. Janshoff, Springer-Verlag Berlin, Heidelberg, New York, 2007, pp. 173-210.

60. F. Dickert, P. Lieberzeit and O. Hayden, *Anal. Bioanal. Chem.*, 2003, **377**, 540–549.

61. A. Aherne, C. Alexander, M. J. Payne, N. Perez and E. N. Vulfson, *J. Am. Chem. Soc.*, 1996, **118**, 8771–8772.

62. A. Afzal, X. Chen, M. Jenik, S. Krassnig and F. L. Dickert, in *Commercial and Pre-Commercial Cell Detection Technologies for Defence Against Bioterror*, ed. L. M. Lechuga, F. P. Milanovich, P. Skládal, O. Ignatov and T. R. Austin, IOS Press, Amsterdam, 2008, pp. 60-76.

63. S. Z. Bajwa, G. Mustafa, R. Samardzic, T. Wangchareansak and P. A. Lieberzeit, *Nanoscale Res. Lett.*, 2012, **7**, 328.

64. O. Hayden and F. L. Dickert, *Adv. Mater.*, 2001, **13**, 1480–1483.

65. M. Jenik, A. Seifner, S. Krassnig, K. Seidler, P. A. Lieberzeit, F. L. Dickert and C. Jungbauer, *Biosens. Bioelectron.*, 2009, **25**, 9–14.

66. K. Seidler, M. Polreichová, P. Lieberzeit and F. Dickert, *Sensors*, 2009, **9**, 8146–8157.

67. U. Latif, A. Mujahid, A. Afzal, R. Sikorski, P. Lieberzeit and F. Dickert, *Anal. Bioanal. Chem.*, 2011, **400**, 2507–2515.

68. K. Seidler, P. A. Lieberzeit and F. L. Dickert, *Analyst*, 2009, **134**, 361–366.

69. O. Hayden, K.-J. Mann, S. Krassnig and F. L. Dickert, *Angew. Chem., Int. Ed.*, 2006, **45**, 2626–2629.

70. A. Seifner, P. Lieberzeit, C. Jungbauer and F. L. Dickert, *Anal. Chim. Acta*, 2009, **651**, 215–219.

71. L. D. Bolisay, J. N. Culver and P. Kofinas, *Biomaterials*, 2006, **27**, 4165–4168.

72. L. D. Bolisay, J. N. Culver and P. Kofinas, *Biomacromolecules*, 2007, **8**, 3893–3899.
73. F. L. Dickert, O. Hayden, P. Lieberzeit, C. Haderspoeck, R. Bindeus, C. Palfinger and B. Wirl, *Synth. Met.*, 2003, **138**, 65–69.
74. F. L. Dickert, P. Lieberzeit, S. Gazda-Miarecka, K. Halikias and K.-J. Mann, *Sens. Actuators, B*, 2004, **100**, 112–116.
75. O. Hayden, P. A. Lieberzeit, D. Blaas and F. L. Dickert, *Adv. Funct. Mater.*, 2006, **16**, 1269–1278.
76. M. Jenik, R. Schirhagl, C. Schirk, O. Hayden, P. Lieberzeit, D. Blaas, G. Paul and F. L. Dickert, *Anal. Chem.*, 2009, **81**, 5320–5326.
77. A. Ramanaviciene and A. Ramanavicius, *Biosens. Bioelectron.*, 2004, **20**, 1076–1082.
78. P. A. Lieberzeit, S. Gazda-Miarecka, K. Halikias, C. Schirk, J. Kauling and F. L. Dickert, *Sens. Actuators, B*, 2005, **111–112**, 259–263.
79. I. Kurane, *Comp. Immunol. Microbiol. Infect. Dis.*, 2007, **30**, 329–340.
80. C.-C. Su, T.-Z. Wu, L.-K. Chen, H.-H. Yang and D.-F. Tai, *Anal. Chim. Acta*, 2003, **479**, 117–123.
81. D.-F. Tai, C.-Y. Lin, T.-Z. Wu and L.-K. Chen, *Anal. Chem.*, 2005, **77**, 5140–5143.
82. C.-H. Lu, Y. Zhang, S.-F. Tang, Z.-B. Fang, H.-H. Yang, X. Chen and G.-N. Chen, *Biosens. Bioelectron.*, 2012, **31**, 439–444.
83. R. Schirhagl, P. A. Lieberzeit and F. L. Dickert, *Adv. Mater.*, 2010, **22**, 2078–2081.
84. Y. Wang, Z. Zhang, V. Jain, J. Yi, S. Mueller, J. Sokolov, Z. Liu, K. Levon, B. Rigas and M. H. Rafailovich, *Sens. Actuators, B*, 2010, **146**, 381–387.
85. G. M. Birnbaumer, P. A. Lieberzeit, L. Richter, R. Schirhagl, M. Milnera, F. L. Dickert, A. Bailey and P. Ertl, *Lab Chip*, 2009, **9**, 3549–3556.
86. M. Bäcker, M. Raue, S. Schusser, C. Jeitner, L. Breuer, P. Wagner, A. Poghossian, A. Förster, T. Mang and M. J. Schöning, *Phys. Status Solidi A*, 2012, **209**, 839–845.
87. Y. Ge and A. P. F. Turner, *Trends Biotechnol.*, 2008, **26**, 218–224.
88. M. J. Whitcombe, I. Chianella, L. Larcombe, S. A. Piletsky, J. Noble, R. Porter and A. Horgan, *Chem. Soc. Rev.*, 2011, **40**, 1547–1571.
89. O. Ramström and R. J. Ansell, *Chirality*, 1998, **10**, 195–209.
90. D. L. Venton and E. Gudipati, *Biochim. Biophys. Acta, Protein Struct. Mol. Enzymol.*, 1995, **1250**, 126–136.
91. M. E. Byrne, K. Park and N. A. Peppas, *Adv. Drug Delivery Rev.*, 2002, **54**, 149–161.
92. A. Bossi, S. A. Piletsky, E. V. Piletska, P. G. Righetti and A. P. F. Turner, *Anal. Chem.*, 2001, **73**, 5281–5286.
93. K. Lee, R. R. Itharaju and D. A. Puleo, *Acta Biomater.*, 2007, **3**, 515–522.
94. X. Pang, G. Cheng, R. Li, S. Lu and Y. Zhang, *Anal. Chim. Acta*, 2005, **550**, 13–17.
95. Y. Li, H.-H. Yang, Q.-H. You, Z.-X. Zhuang and X.-R. Wang, *Anal. Chem.*, 2005, **78**, 317–320.

96. F. Bonini, S. Piletsky, A. P. F. Turner, A. Speghini and A. Bossi, *Biosens. Bioelectron.*, 2007, **22**, 2322–2328.

97. Z. Zhang, Y. Long, L. Nie and S. Yao, *Biosens. Bioelectron.*, 2006, **21**, 1244–1251.

98. H.-H. Yang, S.-Q. Zhang, F. Tan, Z.-X. Zhuang and X.-R. Wang, *J. Am. Chem. Soc.*, 2005, **127**, 1378–1379.

99. I. Willner and E. Katz, *Angew. Chem., Int. Ed.*, 2000, **39**, 1180–1218.

100. A. Poghossian, M. H. Abouzar, M. Sakkari, T. Kassab, Y. Han, S. Ingebrandt, A. Offenhäusser and M. J. Schöning, *Sens. Actuators, B*, 2006, **118**, 163–170.

101. Z. Tao, E. C. Tehan, R. M. Bukowski, Y. Tang, E. L. Shughart, W. G. Holthoff, A. N. Cartwright, A. H. Titus and F. V. Bright, *Anal. Chim. Acta*, 2006, **564**, 59–65.

102. M. Odabaşi, R. Say and A. Denizli, *Mater. Sci. Eng., C*, 2007, **27**, 90–99.

103. N. W. Turner, B. E. Wright, V. Hlady and D. W. Britt, *J. Colloid Interface Sci.*, 2007, **308**, 71–80.

104. J. Rick and T.-C. Chou, *Anal. Chim. Acta*, 2005, **542**, 26–31.

105. P. A. Lieberzeit, A. Afzal, D. Podlipna, S. Krassnig, H. Blumenstock and F. L. Dickert, *Sens. Actuators, B*, 2007, **126**, 153–158.

106. R. Schirhagl, D. Podlipna, P. A. Lieberzeit and F. L. Dickert, *Chem. Commun.*, 2010, **46**, 3128–3130.

107. R. Schirhagl, J. Qian and F. L. Dickert, *Sens. Actuators, B*, 2012.

108. T.-R. Ling, Y. Z. Syu, Y.-C. Tasi, T.-C. Chou and C.-C. Liu, *Biosens. Bioelectron.*, 2005, **21**, 901–907.

109. S. A. Piletsky, E. V. Piletska, A. Bossi, K. Karim, P. Lowe and A. P. F. Turner, *Biosens. Bioelectron.*, 2001, **16**, 701–707.

110. D. Lakshmi, A. Bossi, M. J. Whitcombe, I. Chianella, S. A. Fowler, S. Subrahmanyam, E. V. Piletska and S. A. Piletsky, *Anal. Chem.*, 2009, **81**, 3576–3584.

111. A. Kugimiya and T. Takeuchi, *Anal. Chim. Acta*, 1999, **395**, 251–255.

112. A. Kugimiya and T. Takeuchi, *Electroanalysis*, 1999, **11**, 1158–1160.

113. S. Beging, D. Mlynek, S. Hataihimakul, A. Poghossian, G. Baldsiefen, H. Busch, N. Laube, L. Kleinen and M. J. Schöning, *Sens. Actuators, B*, 2010, **144**, 374–379.

114. M. Javanbakht, N. Shaabani and B. Akbari-adergani, *J. Chromatogr., B*, 2009, **877**, 2537–2544.

115. M. Javanbakht, M. H. Namjumanesh and B. Akbari-adergani, *Talanta*, 2009, **80**, 133–138.

116. B. Okutucu and S. Önal, *Talanta*, 2011, **87**, 74–79.

117. F. G. Tamayo, E. Turiel and A. Martín-Esteban, *J. Chromatogr., A*, 2007, **1152**, 32–40.

118. N. Masqué, R. M. Marcé and F. Borrull, *TrAC Trends Anal. Chem.*, 2001, **20**, 477–486.

119. R. Gadzała-Kopciuch, J. Ričanyová and B. Buszewski, *J. Chromatogr., B*, 2009, **877**, 1177–1184.

120. M. Rejtharová and L. Rejthar, *J. Chromatogr., A*, 2009, **1216**, 8246–8253.

121. B. Claude, P. Morin, S. Bayoudh and J. de Ceaurriz, *J. Chromatogr., A*, 2008, **1196–1197**, 81–88.

122. M. T. Jafari, Z. Badihi and E. Jazan, *Talanta*, 2012, **99**, 520–526.

123. Y. Tan, J. Yin, C. Liang, H. Peng, L. Nie and S. Yao, *Bioelectrochemistry*, 2001, **53**, 141–148.

124. H. Yan, K. H. Row and G. Yang, *Talanta*, 2008, **75**, 227–232.

125. B. Claude, R. Nehmé and P. Morin, *Anal. Chim. Acta*, 2011, **699**, 242–248.

126. M. Javanbakht, A. M. Attaran, M. H. Namjumanesh, M. Esfandyari-Manesh and B. Akbari-adergani, *J. Chromatogr., B*, 2010, **878**, 1700–1706.

127. A. Beltran, E. Caro, R. M. Marcé, P. A. G. Cormack, D. C. Sherrington and F. Borrull, *Anal. Chim. Acta*, 2007, **597**, 6–11.

128. D. Cunliffe, A. Kirby and C. Alexander, *Adv. Drug Delivery Rev.*, 2005, **57**, 1836–1853.

129. C. Alvarez-Lorenzo and A. Concheiro, *J. Chromatogr., B*, 2004, **804**, 231–245.

130. A. L. Hillberg, K. R. Brain and C. J. Allender, *Adv. Drug Delivery Rev.*, 2005, **57**, 1875–1889.

131. G. Wulff, *Angew. Chem., Int. Ed.*, 1995, **34**, 1812–1832.

132. S. Venkatesh, J. Saha, S. Pass and M. E. Byrne, *Eur. J. Pharm. Biopharm.*, 2008, **69**, 852–860.

133. B. Singh and N. Chauhan, *Acta Biomater.*, 2008, **4**, 1244–1254.

134. A. Hui, H. Sheardown and L. Jones, *Materials*, 2012, **5**, 85–107.

135. M. Abdouss, E. Asadi, S. Azodi-Deilami, N. Beik-mohammadi and S. Aslanzadeh, *J. Mater. Sci. Mater. Med.*, 2011, **22**, 2273–2281.

136. K. Rostamizadeh, M. Vahedpour and S. Bozorgi, *Int. J. Pharm.*, 2012, **424**, 67–75.

137. M. Ulbricht, *J. Chromatogr., B*, 2004, **804**, 113–125.

138. L. Peng, Y. Wang, H. Zeng and Y. Yuan, *Analyst*, 2011, **136**, 756–763.

139. M. H. Abouzar, A. Poghossian, J. R. Siqueira, O. N. Oliveira, W. Moritz and M. J. Schöning, *Phys. Status Solidi A*, 2010, **207**, 884–890.

Supramolecular Approach for Tumor Imaging and Photodynamic Therapy

ANURAG GUPTA AND RAVINDRA K. PANDEY*

Photodynamic Therapy Center, Department of Cell Stress Biology, Roswell
Park Cancer Institute, Buffalo, NY 14263, USA
*Email: ravindra.pandey@roswellpark.org

15.1 Molecular Imaging

Molecular imaging is an emerging technique for the "noninvasive, real-time
visualization of biochemical events at the cellular and molecular level
within living cells, tissues and/or intact subjects".[1–4] Traditionally, molecular
imaging has been used for diagnostics, for example disease detection, therapy
monitoring, drug discovery and development, and for probing biochemical
events to better understand mechanisms of disease formation and progression.
The new area of research involving molecular imaging is in the design and
implementation of multimodal tumor imaging agents with therapeutic
capabilities (theranostics), which is the combination of disease diagnosis,
treatment, and disease monitoring (Figure 15.1).

The field of molecular imaging is deeply rooted in nuclear medicine. In
nuclear medicine, radionuclides have been developed along with sophisticated
imaging systems such as single photon emission computed tomography
(SPECT) and positron emission tomography (PET) for the non-invasive
detection of disease, staging of tumors, and for monitoring the progression

Monographs in Supramolecular Chemistry No. 13
Supramolecular Systems in Biomedical Fields
Edited by Hans-Jörg Schneider

Published by the Royal Society of Chemistry, www.rsc.org

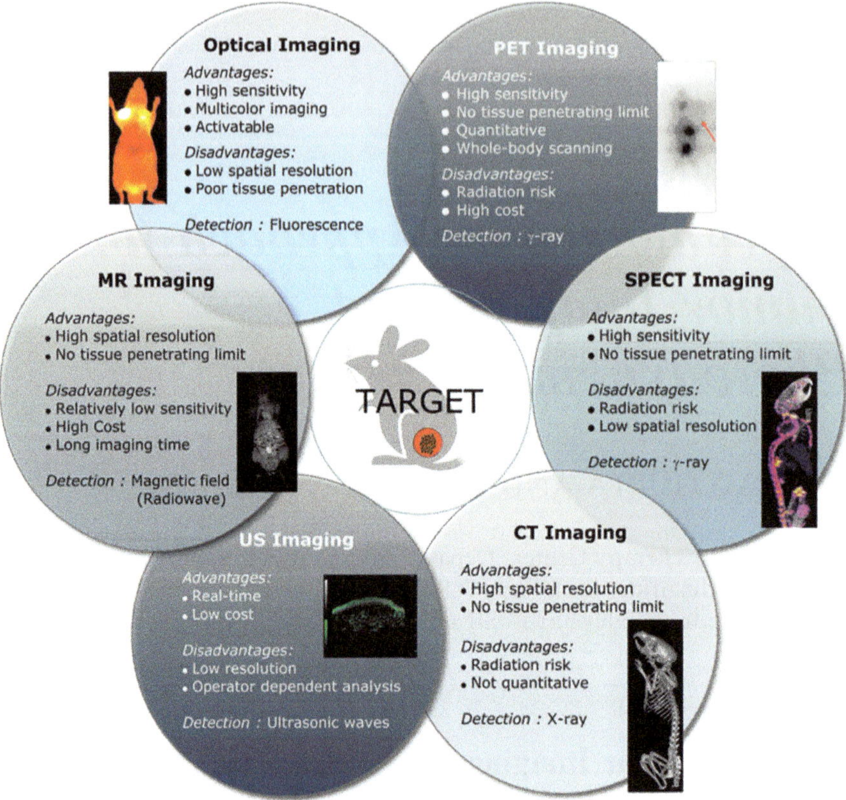

Figure 15.1   Characterization of the major preclinical imaging modalities used with their advantages and disadvantages.[5] For explanation to modality abbreviations, see text.

of the disease with and without treatment. Over the last two decades, two-dimensional (2D) and 3D fluorescence imaging have seen a tremendous increase in use for molecular imaging. Both PET and fluorescence imaging have certain advantages and limitations as follows.

15.1.1 Positron Emission Tomography (PET)

PET imaging, a nuclear imaging modality, utilizes radioisotopes such as ^{18}F ($t_{1/2} = 109.8$ min), ^{64}Cu ($t_{1/2} = 12.7$ h), or ^{124}I ($t_{1/2} = 4.2$ days) that decay *via* positron emission. The positrons annihilate electrons that are present in the surrounding tissue to produce two 511 keV gamma photons that travel in equal and opposite directions and are detected by a ring of detectors. Because the 511 keV gamma photons have 10-fold more energy than X-rays, they are likely to leave the body for external detection (limitless depth of penetration).[1]

Clinically, the most common radiotracer used for the detection, staging, and monitoring of cancer is ^{18}F-fluoro-2-deoxy-glucose (^{18}F-FDG). Because

^{18}F-FDG is an analogue of glucose, its uptake within cancer cells is due to the higher metabolic rate present in these cells compared to normal cells. Preclinically, researchers utilize PET imaging to assess the imaging potential of new PET probes for detecting tumors, studying the organ biodistribution of these new probes, for treatment planning, and to assess the organ biodistribution of cancer therapeutics, such as taxol.[6,7,8]

Several key strengths and limitations exist with PET imaging. The strengths are (i) biochemical changes in tumors occur before anatomical changes; therefore, PET imaging will depict these changes before CT and MRI can, (ii) its high sensitivity (radioactivity can be detected at concentrations as low as 10^{-11}–10^{-12} M), and (iii) the ability to scan the whole-body, and (v) its quantification capabilities. The limitations include (i) no anatomical reference, (ii) potential harm due to ionizing radiation, (iii) high cost, and (iv) timing resolution is poor (image acquisition can last from 20 to 30 minutes depending on the radioisotope used). Because nanograms to micrograms of radioisotopes are used, it is not expected that the radioisotope would exert any physiological change of the corresponding non-labeled analogue.[1,9]

15.1.2 Whole-body Fluorescence Imaging

Unlike PET, whole-body fluorescence imaging is a non-ionizing imaging modality. In fluorescence imaging, a CCD camera is used to detect low energy photons (fluorescence) that are emitted by endogenous or exogenous fluorophores. When a fluorophore is excited by an external light source, it jumps to a higher energy state and then loses a portion of its energy through a process called internal conversion prior to relaxing to the ground state by emitting photons. The loss of energy due to internal conversion causes the fluorescence wavelength to be red-shifted compared to the absorbance wavelength. Because low energy (non-ionizing) photons are being detected, fluorophores are considered relatively safe; however, the low energy photons limit the depth of penetration to a few millimeters due to absorption and scattering of light by the deoxy- and oxyhemoglobin and water. To minimize the absorption and scattering of light and autofluorescence from tissue, and to maximize the depth of penetration, near-infrared light in the 650–900 nm bandwidth is used. This is often called the "optical or biological window." Figure 15.2 shows that as the wavelength increases, the depth of penetration increases due to less absorption and scattering.[1]

In cancer, fluorescence imaging is used for image-guided tumor resection, tumor detection, treatment planning, and for monitoring of disease progression by fluorophore-labeled cancer therapeutic or antibodies. The advantages of fluorescence imaging are (i) it is relatively inexpensive, (ii) it can be multiplexed (fluorophores with different absorbance and emission wavelength can be used), and (iii) with the advent of fluorescence molecular tomography, the depth of penetration can increase to beyond 1 cm. The limitations include (i) being limited to 1 cm depth of penetration (reflectance mode), (ii) not quantitative, (iii) being surface weighted, (iv) requiring more mass than for PET imaging

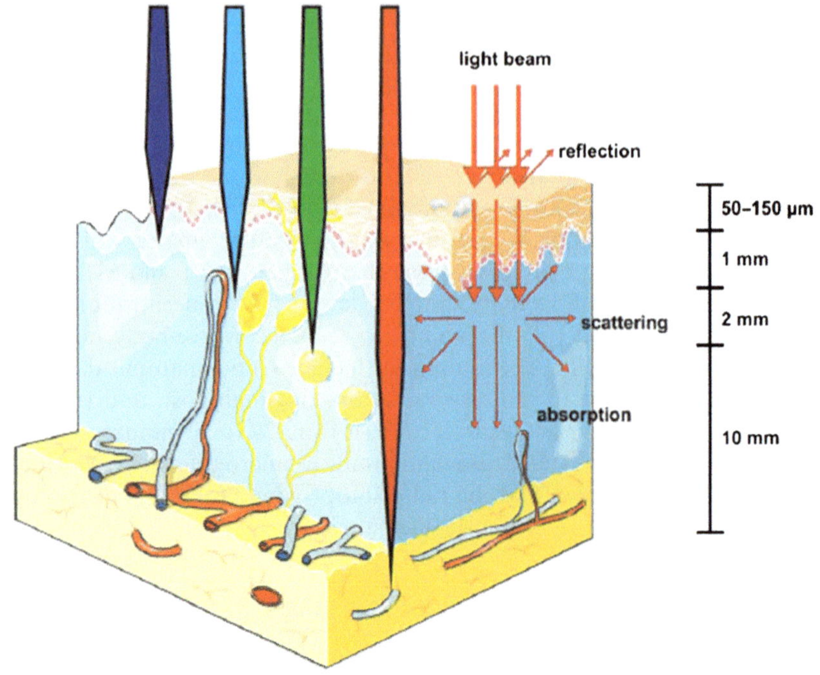

Figure 15.2   Propagation of light through tissue.[9]

$(10^{-9}–10^{-12}$ M), and (v) decrease in sensitivity, compared to PET imaging due to autofluorescence.[1,5]

15.2   Multimodal Tumor Imaging Probes

Early cancer detection is imperative for improving cancer therapy and survival outcomes. Multimodal tumor imaging probes combine two or more imaging modalities in a single agent. These probes are advantageous for a number of reasons including (i) being able to confirm the presence and extent of a tumor by multiple imaging modalities, (ii) more accurate treatment planning can occur, (iii) they can determine if the probe is homogeneously or non-homogeneously distributed throughout the tumor, especially when MRI is used, and (iv) real-time fluorescence based image-guided surgery can be combined with PET for tumor staging and disease monitoring. In recent years, mainly synthetic and nanoparticle-based approaches have been investigated for the development of multifunctional (combined) imaging probes (Figure 15.3).

15.2.1   Synthetic Approach to Design of Multifunctional Tumor Imaging Probes

The synthetic approach relies on the conjugation of two or more contrast probes to form a single probe capable of providing contrast in multiple imaging

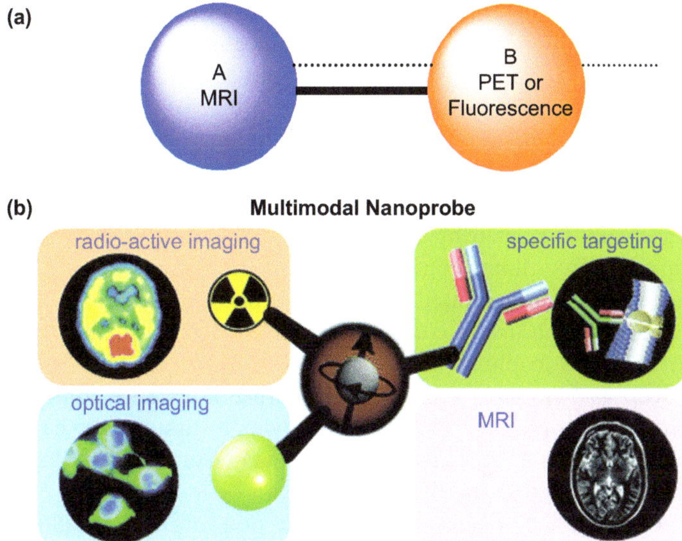

Figure 15.3 Multimodal tumor imaging probes. (1) Synthetic based: A and B are tumor imaging probes, MRI and PET or near-infrared fluorophore, conjugated together. (2) Multimodal nanoprobes encompassing MRI, PET, and/or near-infrared fluorophores.[10]

platforms. With this approach, it is often difficult to conjugate the two molecular imaging probes at a ratio other than 1 : 1. This approach has yielded contrast agents for PET and fluorescence imaging, MRI and fluorescence imaging, and PET/fluorescence and MRI. The advantages for synthesizing a multimodal tumor imaging probe over a single modality imaging probe include (i) deriving more information in less time, for example, simultaneous multimodal imaging (MR/PET, MR/fluorescence, PET/fluorescence), (ii) if information from multiple imaging modalities is required, then the multimodal tumor imaging probe requires only one injection, (iii) complementary imaging techniques can be combined to overcome limitations present in each imaging modality, separately (*e.g.*, the highly sensitive PET images can be combined with high resolution MR imaging), and finally (iv) the multimodal imaging probe in which the two different imaging moieties are combined in one will share the same biodistribution and clearance.[1]

15.2.2 Multimodal PET, Fluorescence Imaging, and MRI

PET imaging is routinely used in the clinic for the detection of tumors because it has high sensitivity, can image deeply seated tumors, can stage tumors, and it can provide quantitative organ biodistribution of the radiotracer. Because *in vivo* biochemical changes occur prior to anatomical changes during disease progression, PET imaging can detect these changes earlier than conventional diagnostic imaging, for example MR imaging. However, PET is an imaging modality with a low temporal resolution (seconds to minutes) that requires a gamma-emitting radioisotope such as ^{18}F, ^{124}I, or ^{64}Cu for contrast. Even though gamma-emitting radioisotopes produce ionizing radiation, a fixed amount is injected, thereby limiting the amount of ionizing radiation that the patient receives.

Fluorescence imaging is a non-ionizing imaging modality that can complement PET by providing real-time detection of the tumor with high sensitivity. Real-time detection is important for image-guided resection of tumors, especially for ensuring complete resection of the tumor including the tumor margins. Two-dimensional fluorescence imaging is a surface-weighted imaging technique that is limited to imaging a few millimeters below the tissue surface with fluorophores excited with visible light. Utilizing near-infrared fluorophores, or fluorophores with peak excitation in the 750–950 nm range, tissue as deep as 1 cm can be visualized. With fluorescence tomography and near-infrared fluorophores, the depth at which tissue can be imaged is dependent on the tissue type. Regardless, it has been reported that depths of several centimeters below the tissue surface can be imaged.[1,8]

Magnetic resonance imaging (MRI) is a non-ionizing imaging modality that can provide anatomical information along with highly detailed soft tissue imaging. In recent years, researchers have combined contrast for MR imaging with fluorescence imaging. Combining fluorescence imaging with MR imaging enables the correlation of molecular information with high resolution, high contrast structural/functional information Because MR imaging is less sensitive to exogenous contrast than other molecular imaging modalities, such as fluorescence and PET, the concentration of exogenous contrast needed is in the millimolar range.

15.2.2.1 PET and Fluorescence Imaging

It has been shown that the radioisotope ^{64}Cu or the fluorophore CY5.5 can be conjugated to cyclic RGD (cRGD or cyclic arginylglycylaspartic acid) for PET or fluorescence imaging and can target the $\alpha_v\beta_3$, $\alpha_v\beta_5$, and $\alpha_{ii}\beta_3$ integrins that are overexpressed on tumor vasculature. Kimura *et al.*[11] found that the IC_{50} value for cyclic RGD binding was 150 ± 10 nM for CY5.5-cRGD and 380 ± 190 nM for the ^{64}Cu-DOTA-cRGD in U87MG cells, a human glioblastoma cell line. Therefore, they wanted to create a formulation that could further improve integrin targeting by lowering the IC_{50} for the RGD motif. Using the cystine knot peptide, which comes from the squash family of protease inhibitors, the authors were able to incorporate RGD for integrin binding at low IC_{50} values ranging from 10–30 nM. With two different variants of the modified knot peptide RGD complex (2.5D: PQGRGDWAPTS and 2.5F: PRPRGDNPPLT), it was found by PET and fluorescence imaging that the modified knot peptide was more effective for targeting than the corresponding unmodified knot peptide or the conventional cRGD complexes. Based on the gamma-based (^{64}Cu) biodistribution, the 2.5F showed the highest accumulation within the tumor, while the 2.5D showed the highest tumor to normal tissue ratio by whole-body fluorescence imaging. Both 2.5F and 2.5D showed lower liver accumulation compared to the cRGD complexes. These formulations show that the modified knot peptide can be used to enhance the imaging potential of ^{64}Cu and CY5.5 for the detection of U87MG human glioblastoma tumors. With this formulation several limitations exists and they are (i) both PET and fluorescence capabilities were not integrated into one construct, (ii) by utilizing a longer wavelength fluorophore such as CY7, a near-infrared cyanine dye with an absorbance and fluorescence of 747 nm and 774 nm, respectively, the depth of penetration or depth below the tissue surface that can be imaged will increase, and (iii) complex peptide chemistry was required for the synthesis of the targeted imaging probe.[11]

Sampath *et al.*[12] have shown by using a monoclonal antibody, trastuzamab, that ^{64}Cu and IRDye 800 can be conjugated for PET and near-infrared fluorescence imaging of Her-2 positive primary and metastatic tumors.[12] To show the enhanced capabilities of their multimodal imaging agent, they compared the PET and near-infrared fluorescence capabilities against ^{18}F-FDG. In mice bearing the 4T1.2neu/R tumors, the dual-labeled probe and ^{18}F-FDG were able to detect the primary tumor by PET and near-infrared fluorescence imaging. In addition, PET and fluorescence imaging using the dual-labeled agent could detect lung metastasis, whereas ^{18}F-FDG could not. Through *ex vivo* fluorescence imaging, the authors were able to detect metastasis not only in the lung, but also in the skin, skeletal muscle, and lymph nodes. The presence of cancer cells in these tissues was confirmed by hematoxylin and eosin (H&E) staining. For imaging other tumor types, different monoclonal antibodies will have to be used. The implication is that this imaging probe is only applicable for tumors that overexpress the Her-2 receptor.[12]

Because antibodies are cleared through the liver, it is difficult to image liver metastasis due to the high background to tumor ratio. In addition, the antibody probe, specifically HER2/neu cannot be used for other types of tumors. Affibody molecules—a novel class of antibody moieties (size, 6 kDa) with superior characteristics surpassing mononucleal antibodies (mAbs) and antibody fragments that have been engineered to target HER2/neu positive tumors—are also restricted in their use. For other tumor types, such as colon cancer, different affibody molecules are currently under investigation.[13]

15.2.2.2 Fluorescence Imaging and MRI

Lin *et al.*[14] have synthesized a fluorescence and MRI imaging agent that can target the gastrin-releasing peptide (GRP) receptor *via* the Bombesin peptide. The Gd-TTDA-NP-BN-Cy5.5 complex was imaged in mice double-tumored with the PC-3 (human prostate cancer cell, GRP receptor positive) and the KB (human nasopharyngeal epidermal carcinoma cell line, GRP receptor negative) tumors. Fluorescence imaging showed that the Bombesin peptide was able to target the GRP receptor present on PC-3 tumors with the maximal uptake seen at 0.5 hours post-injection. MR imaging also showed that the peptide conjugate is selective for PC-3 tumors over KB tumors. However, the maximal enhancement in signal was seen 24 h post-injection.[14]

The two interesting aspects of the conjugate are first, that the Bombesin targeted conjugate shows less uptake in the liver as compared to other targeting peptides such as cRGD, and second, that the time post-injection for maximum contrast for fluorescence and MR imaging was significantly different. With active targeting, it is often seen that the accumulation of the targeted payload arrives at the tumor site significantly earlier than the passively targeted formulation.[14–16] MR imaging poses some unique advantages in that it provides an anatomical reference, and secondly, when combined with contrast imaging, it can delineate if the distribution of the probe is homogeneous within the tumor.

15.2.3   Nanoparticle Approach for Design of Multifunctional Tumor Imaging Probes

The National Nanotechnology Initiative (NNI) and the National Cancer Institute (NCI) have defined nanoparticles as particles that are 1–100 nm in size (diameter).[17] They can be used to deliver antibodies, drugs, imaging agents, or other substances to desired sites within the body. Nanoparticles are poised to bridge the gap between solubilizing hydrophobic agents in an aqueous environment and improving the selective delivery of the payload to tumors.

Nanoparticles are advantageous in that they can formulate a hydrophobic agent into an aqueous environment, can target malignant tissues through the "enhanced permeability and retention (EPR) effect" (Figure 15.4), or by active targeting *via* peptides or antibodies, and can be synthesized to evade the

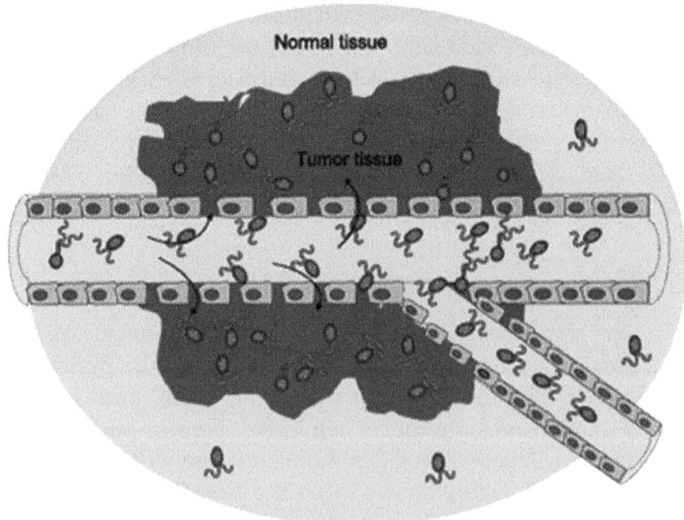

Figure 15.4   Passive targeting of nanoparticles *via* the enhanced permeability and retention effect.
(Adapted from Cho *et al.*[18])

reticuloendothelial system (RES) by modification of its size, surface charge, or by cloaking the nanoparticles with polymers such as polyethylene glycol (PEG).[18]

Because a difference in sensitivity exists between different imaging modalities such as MR and fluorescence or PET imaging, it becomes difficult to chemically conjugate an MR imaging agent with a fluorophore or radiotracer at the ideal dosing for each individual imaging modality. For this reason among others, the nanoparticle approach for the design of multimodal imaging probes is exciting and flourishing. The formation and loading of nanoparticles with multiple imaging agents is often easier than the synthetic approach and provides several advantages (i) a multi-step synthesis is not required, (ii) the nanoparticles can be purified in rather simple methods, such as centrifuge filtration, dialysis, pelleting, syringe filtration or by desalting columns, and (iii) the loading of a single or multiple imaging agents can be precisely tailored to the dosing required for each individual imaging modality. Depending on the material composition of the nanoparticle, they can be used themselves as imaging agents. For example, iron oxide nanoparticles, quantum dots and semi-conducting crystals can be used as contrast agents for MR and fluorescence imaging, respectively, and can be loaded with other contrast agents to create a multimodal nanoprobe. In such a probe, the dosing for each imaging modality can be tailored to the ideal dose requirements.[19]

15.2.4 Multimodal Nanoprobes for Tumor Imaging

Nanoparticles can be engineered to be either non-degradable or biodegradable. The most common non-degradable nanoparticles used are silica, quantum dots, and gold. Through the use of biodegradable polymers such as poly(lactic-co-glycolic acid) (PLGA), poly(lactic acid) (PLA), and poly(3-acryloyloxy-2-hydroxypropylmethacrylate) (AHM) cross-linked poly-acrylamide, biodegradable nanoparticles can be synthesized. Biodegradable nanoparticles pose several unique advantages over non-degradable nano-particles, which include (i) improved availability of the payload, (ii) controlled release of the payload, and (iii) can be degraded to non-toxic monomers that can be cleared from the body with minimal side effects.[20]

Non-degradable Multimodal Nanoprobes for PET and Fluorescence Imaging. Benezra *et al.*[21] have highlighted the importance of being able to combine real-time fluorescence and PET imaging for the detection of melanoma. This was accomplished by using *ca.* 7 nm targeted core-shell silica nanoparticles conjugated to the radioisotope, ^{124}I, and encapsulating the fluorophore, CY5. The multimodal nanoparticle probe was designed to target primary and metastatic melanoma tumors *via* cyclic RGD. The PET imaging results indicate that the targeted nanoparticles showed tumor specificity for M21 (human melanoma, $\alpha_v\beta_3$ positive) but not for M21L (human melanoma, α_v negative). The M21 human melanoma tumors can metastasize to the lymph nodes. Because the lymph nodes in mice are 1–2 mm in diameter, which is the spatial resolution limit of positron emission tomography, PET imaging may not adequately resolve the signal in the lymph nodes. Therefore, image-guided fluorescence was used to detect the presence of metastasis in the lymph nodes, such as the draining and sentinel lymph nodes. The combination of PET and fluorescence imaging was instrumental in the detection of the primary tumor and metastasis in the lymph nodes.[21]

However, the silica nanoparticle formulation is not without limitations, which include (i) the nanoparticles may not be able to traverse the tight endothelial gap junctions that are present within blood vessels, which have a mean diameter between 5 nm and 10 nm, due to the mean hydrodynamic diameter of the core-shell silica nanoparticles being *ca.* 7 nm, (ii) because melanin present in melanoma tumors, silica nanoparticles loaded with CY5 can absorb light up to *ca.* 700 nm, which could lower the sensitivity for CY5 detection, and (iii) the depth of penetration for CY5 is lower than NIR fluorophores such as CY7 and IRDye 800, a near-infrared cyanine dye that has an absorbance of 745 nm and 800 nm, respectively.[22,23]

Quantum dots (QDs) or semiconducting nanocrystals are an alternative class of fluorescent nanoparticles that are of interest due to their high quantum yield of fluorescence and photostability. Because *in vivo* fluorescence imaging is rather limited in its ability to provide quantitative information, Chen *et al.*[24] synthesized a vascular endothelial growth factor receptor (VEGFR) targeted with QDs that was also conjugated to the PET radioisotope ^{64}Cu-DOTA for quantitation. The PET imaging probe will allow for quantitative organ bio-distribution. The QDs were targeted to the VEGFR with the protein VEGF conjugated to the nanoparticle. VEGFRs are overexpressed on U87MG cancer cells. The authors found that VEGF significantly increased the uptake of the QDs in U87MG tumors, $4.2 \pm 0.5\%$ ID g^{-1} *versus* <1%, for the targeted and non-targeted nanoparticles, respectively.[24]

The first limitation with the formulation is the % ID per gram in the liver for both the targeted QDs and non-targeted QDs was approximately 50%, indicating high uptake in the RES system. The whole-body and *ex vivo* organ fluorescence and PET imaging showed similar imaging results. The second limitation is that since QDs are primarily synthesized by using the heavy metals cadmium and selenium, long-term organ toxicity needs to be tested. The third and final limitation is that the QDs used for fluorescence imaging were excited between 575 nm and 605 nm, which not only limits the depth of penetration, but limits the sensitivity of detection due to the absorption peaks being outside the "optical window".

15.2.4.1   Biodegradable Multimodal Nanoprobes for PET and Fluorescence Imaging

One class of biodegradable nanoparticles that has been extensively investigated is liposomes. They can be designed using non-toxic phospholipids and cholesterol that are found naturally in humans. Such formulations can be rendered temperature and/or pH sensitive and can be engineered for controlled

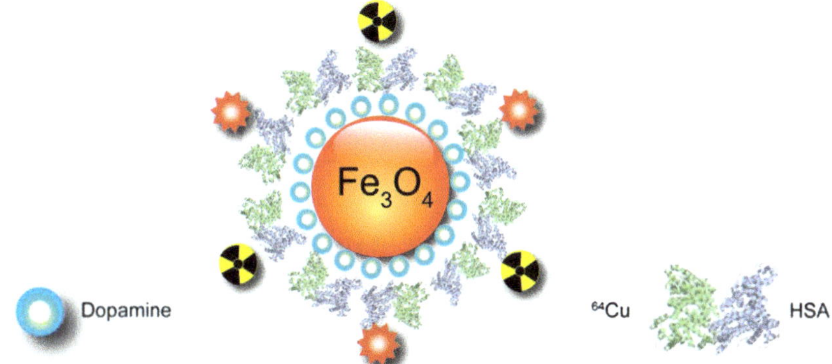

Figure 15.5 Structure of BAT. For full chemical name, see text.

Figure 15.6 A trifunctional human serum albumin (HSA) linked iron oxide nanoparticles for PET/near-infrared fluorescence (NIRF)/MRI (HSA-IONP).[26]

release. Paoli *et al.*[25] have designed liposomes that are sensitive to mild hyperthermic temperatures (*e.g.*, 42 °C). These nanoparticles can be conjugated to either ^{18}F or ^{64}Cu-6-[*p*-(bromoacetamido)benzyl]-1,4,8,11-tetraazacyclotetradecane-*N*,*N*′,*N*″,*N*‴-tetraacetic acid (BAT) (Figure 15.5) and can encapsulate the fluorophore, AF-750 (structure not reported), for multimodal PET and near-infrared fluorescence imaging of Met-1, mouse mammary tumor cell line, tumors. The authors found that by using long-circulating liposomes and long-circulating temperature-sensitive liposomes, 24 hours post-injection, a 177- and 101-fold increase in the fluorescence of AF-750 was seen in the tumor, respectively. The PET imaging and organ biodistribution corroborated with the fluorescence imaging data that the accumulation of the long-circulating liposomes in the tumor increased over time.[25]

Xie *et al.*[26] have extended the nanoparticle-based probe design for PET and fluorescence imaging to include contrast for MR imaging by using biodegradable iron oxide nanoparticles. The nanoparticles were coated with dopamine to render them moderately hydrophilic and then they were layered with human serum albumin (HSA) (see Figure 15.6). The fluorophore, CY5.5, and the PET radioisotope, ^{64}Cu, were conjugated to the human serum albumin iron oxide nanoparticles (HSA-IONPs). These nanoparticles were tested in mice bearing subcutaneous U87MG tumors. The nanoparticles showed high accumulation in the tumor by PET/fluorescence/MRI. Because fluorescence imaging, even in the near-infrared portion of the electromagnetic spectrum, is depth limited and can be affected by the absorption and scattering of photons in differently pigmented tissues/organs, such as tumor and liver, it did not show the same high signal in the liver as seen by PET and MR imaging. However, PET and near-infrared fluorescence imaging showed a similar pharmacokinetic profile with contrast enhancement increasing from 1 to 4 h and peaking at 18 h post injection. MRI also showed significant contrast enhancement 18 h post-injection. The authors show that by PET and fluorescence imaging the distribution of the HSA-IONPs seemed homogeneous when in fact it was not. The inhomogeneity was clearly shown by MR imaging. In addition, the images clearly show that the nanoparticles can accumulate in U87MG tumors through the EPR effect. Because iron oxide nanoparticles tend to accumulate in the liver, a high background was seen by all three imaging modalities, especially by PET imaging, thus limiting the tumor to background contrast.[26]

15.3 Theranostic Agents for Multimodal Tumor Imaging and Therapy

The Holy Grail in cancer treatment is to have a single formulation that can detect tumors, provide the clinician with the option to perform real-time image-guided surgery, be used for the treatment of the tumor, and also be used for monitoring disease progression. A formulation, such as a single agent or a nanoparticle that can integrate these functions is known as a "theranostic" agent/nanoparticle. The above sections show that multimodal probes/nanoprobes can be designed for the diagnosis of tumors. As theranostics involve the detection and treatment of tumors, tumor detection or imaging *via* whole-body fluorescence and positron emission tomography will be discussed in the coming sections along with photodynamic therapy (PDT) for the treatment of tumors. In recent years, two different methodologies have been employed for the development of theranostic agents: (i) classical synthetic approach and (ii) nanoformulations.[27,28]

15.3.1 Photodynamic Therapy

Classical cancer therapies include surgical intervention, radiation therapy, and chemotherapy. Unfortunately, these treatment modalities cause damage to normal cells and tissues, which leads to toxic side effects. Treatments are sought that can improve upon these shortcomings. Photodynamic Therapy (PDT), one such alternative, is an attractive option because localized and superficial tumors can be treated by activating a light-sensitive cancer therapeutic known as a photosensitizer by non-ionizing radiation, for example, laser light of a specific wavelength. Upon exciting the photosensitizer, it generates singlet oxygen, the key cytotoxic agent for PDT. The advantages of PDT are (i) it is precisely targeted by selective illumination (non-ionizing radiation), (ii) treatment can be repeated, as necessary, (iii) it is associated with low morbidity, and (iv) it is less invasive than surgery.[29]

15.3.2 Synthetic Approach for Development of a Theranostic Agent

The synthetic approach utilizes photosensitizers as a substrate for the development of theranostic agents. Multiple conjugation approaches exist for the design of theranostic agents. They include (i) enzyme cleavable and (ii) conjugation of a photosensitizer to an imaging agent through a non-flexible or flexible linker.

Photosensitizers, such as pyropheophorbide-*a*, 2-[1-hexyloxyethyl]-2-devinylpyropheo phorbide-*a* (HPPH), chlorin e6, and purpurinide and other tetrapyrrolic systems (Figure 15.7) have been used as substrates for the synthesis of theranostic agents for fluorescence imaging and/or PET and PDT as they are tumor-avid.[31–33]

15.3.2.1 Enzyme-Cleavable Photosensitizers

Fluorescence Imaging and PDT: Enzyme-cleavable linkers have been investigated for many years for cancer imaging and therapeutics. They have shown great potential for tumor-specific delivery of flurophores for tumor imaging, because upon conjugating the fluorophore with an enzyme-cleavable linker, its fluorescence is quenched. This in turn will cause the signal in background tissue to be minimal or low in comparison to sites that have a high expression of a particular enzyme responsible for cleaving the linker used. In the last two decades, enzyme-cleavable photosensitizers have gained interest because site-specific delivery and activation of the photosensitizer will minimize sunlight-based skin phototoxicity. Proteases such as cathepsin B and matrix metalloproteinase-7 (MMP7) are overexpressed in a variety of cancers and have

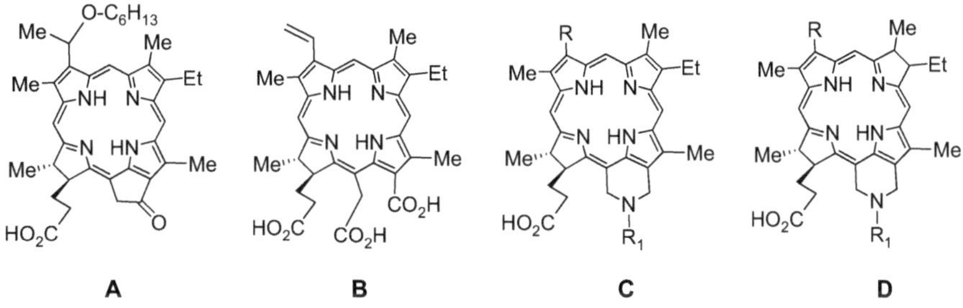

Figure 15.7 Structure of various photosensitizers: HPPH (**A**, 660 nm), Chlorin e$_6$ (**B**, 660 nm), purpurinide (**C**, 700 nm), and bacteriopurpurinimide (**D**, 787 nm).

been investigated for their ability to cleave enzymes. Examples highlighting this approach for tumor imaging and therapy will be discussed herein.

Two major issues with PDT are (i) selectively in destroying tumors while minimizing collateral damage, and (ii) minimizing sunlight-induced skin phototoxicity. An approach that Zheng *et al.*[33] used to address these issues was to create a Photodynamic Molecular Beacon (PMB) that consists of a pyropheophorbide-*a* based photosensitizer that is conjugated to the black-hole quencher 3 (BHQ3) (Figure 15.8) *via* an enzyme cleavable linker, GPLGLARK. The molecular beacon can silence the fluorescence and singlet oxygen production of the photosensitizer through the energy transfer mechanism known as Förster Resonance Energy Transfer or FRET (illustrated in Figure 15.7). To restore the fluorescence and singlet oxygen capabilities, the linker is cleaved by the tumor-associated protease, matrix metalloproteinase-7. By measuring the singlet oxygen emission at 1270 nm they found that the singlet oxygen production for the molecular beacon was 18-fold lower than for the photosensitizer when conjugated to the cleavable linker. By adding MMP7 to the molecular beacon containing the quencher, a 19-fold increase in singlet oxygen was found. Because the molecular beacon can be cleaved by MMP7, its PDT efficacy was tested against KB cells that are positive for MMP7 and the human breast carcinoma, BT20, which is negative for MMP7. The molecular beacon showed significant photodynamic activity for KB cells but not for the BT20 cells. In mice bearing two KB tumors (right and left flank), it was found that 3 hours post-injection provided the highest fluorescence in the tumors. Three hours post-injection, the tumor was treated *via* PDT. The PDT results were compared against a control mouse bearing a KB tumor that was treated with the same light parameter as the photosensitizer-injected mouse, but not injected with the PMB. Thirty days post-PDT it was found that the mouse treated with the PMB showed no signs of tumor regrowth. The data shows that a PMB accumulates and is photodynamically activated in MMP7$^+$ tumors.[34]

Since PDT was only performed on one mouse in this study, a determination cannot be made as to how effective the treatment would be in a larger cohort of mice. The major limitation with this approach to theranostics is that

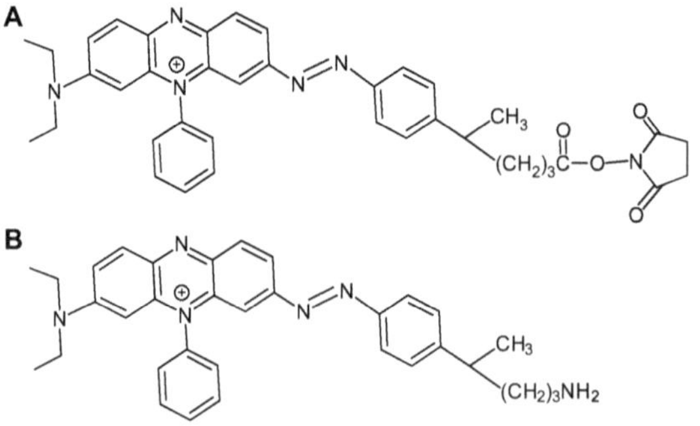

Figure 15.8 Structure of BHQ-3 carboxylic acid, succinimidyl ester (A) and BHQ3 amine (B).

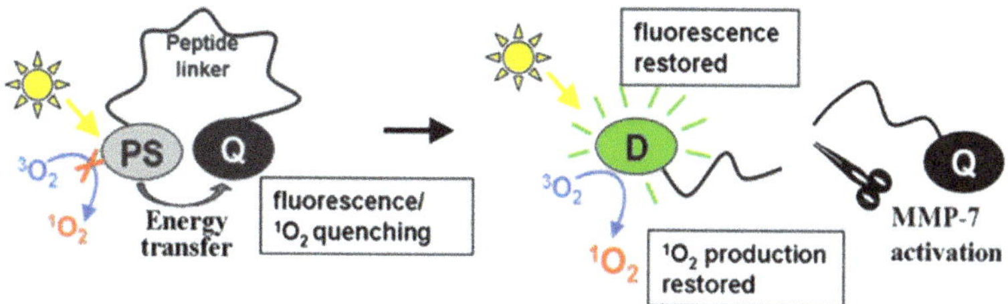

Figure 15.9 Schematic of fluorescence and singlet oxygen quenching and restoration using Photodynamic Molecular Beacons.[34]

since MMP7 plays an important role in normal tissue as well as diseased tissue, therefore the molecular beacon may be cleaved in normal tissues as well as diseased tissue (Figure 15.9). From the *in vivo* fluorescence imaging, a higher fluorescence signal could be seen in various regions along the back of the mouse than in the tumor that was treated *via* PDT, indicating that the molecular beacon is also being cleaved in normal tissue. For increased depth of penetration for treatment and imaging, longer wavelength photosensitizers such as bacteriochlorophyll-*a* and bacteriopurpurinimides should be utilized.

Choi *et al.*[33] were among the earliest groups to study enhancing PDT through the use of enzyme-cleavable linkers. The photosensitizer chlorin e6 was conjugated onto biodegradable poly-L-lysine grafted with monomethoxy-polyethylene glycol. In this state the photosensitizer aggregated and the fluorescence and singlet oxygen product was quenched. After incubation of the photosensitizer with cathepsin B, L, and S, it was found that the fluorescence intensity and singlet oxygen generation capabilities increased 4-fold over the control. *In vivo* fluorescence tomography imaging of HT 1080 (human fibrosarcoma) tumors showed that the cathepsins present in the tumor, specifically cathepsin B, enhanced the fluorescence and detection capabilities compared to chlorin e6. Since the photosensitizer is cleaved within HT 1080 tumors, their PDT efficacy was also evaluated. *In vivo* PDT efficacy of the enzyme-cleavable photosensitizer showed superiority over the free photosensitizer; however, it did not produce long-term cures.[33]

With enzyme-cleavable photosensitizers comes the promise of tumor-specific activation of the photosensitizer. Several limitations exists with such an approach: (i) proteases such as MMP7 are found in normal tissue as well as in cancer cells, which will lead to the recovery of the photosensitizer-quenched state in normal tissue, (ii) tumor to background fluorescence *in vivo* will decrease, and (iii) for effective *in vivo* PDT, a high dose of the photosensitizer (80 nmol) is required for long-term cures.

15.3.2.2 Conjugation of a Photosensitizer to an Imaging Agent

Fluorescence Imaging and PDT: In 2005, Chen *et al.*[31] published the first example of a photosensitizer conjugated to a near-infrared cyanine dye

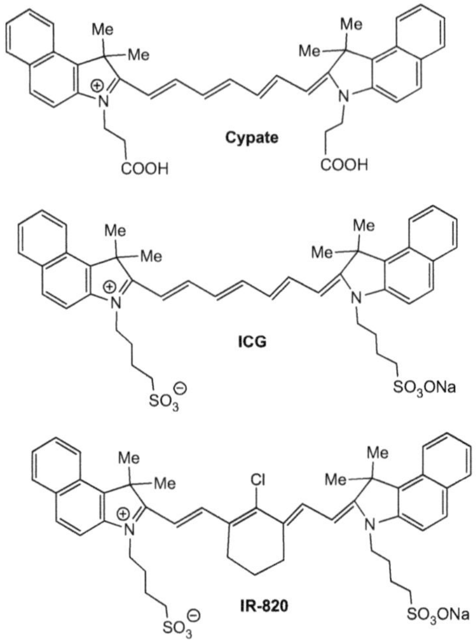

Figure 15.10   Structures of some of commercially available cyanine dyes (CDs).

for whole-body fluorescence imaging and PDT. The photosensitizer, HPPH—which is in phase I/II clinical trials for a variety of cancers including but not limited to head and neck, esophageal, advanced obstructing endo-bronchial non-small cell lung cancer, and invasive cancer of the larynx—was conjugated to a near-infrared cyanine dye, modified from the commercially available IR-820 or new indocyanine green (Figure 15.10).

The fluorophore portion of the conjugate is for deep tumor imaging, treatment planning, and the photosensitizer portion is for tumor therapy *via* PDT. The conjugate can successfully detect and treat U87, RIF (radiation induced fibrosarcoma) and Colon26 (murine colon tumors) tumors. However, the dose required for near-infrared fluorescence imaging and therapy can differ by as much as 5 to 8. The reason for this is that when the photo-sensitizer is activated, a portion of the excited energy is non-radiatively transferred to the cyanine dye. This in turn lowers the photosensitizer's quantum yield of singlet oxygen. The authors showed that photosensitizers can be used as a platform for creating theranostic agents for tumor imaging and therapy.[30,31]

To improve upon the dose differential between imaging and therapy, Williams *et al.*[31] synthesized purpurinimide–cyanine dye conjugates (Figure 15.11) by varying the length of the linker between the photosensitizer and cyanine dye. The cyanine dye conjugated to HPPH is tumor-avid, whereas the cyanine dye used for conjugating with the *N*-substituted purpurinimide is not. Therefore, by utilizing the non-specific cyanine dye in conjunction

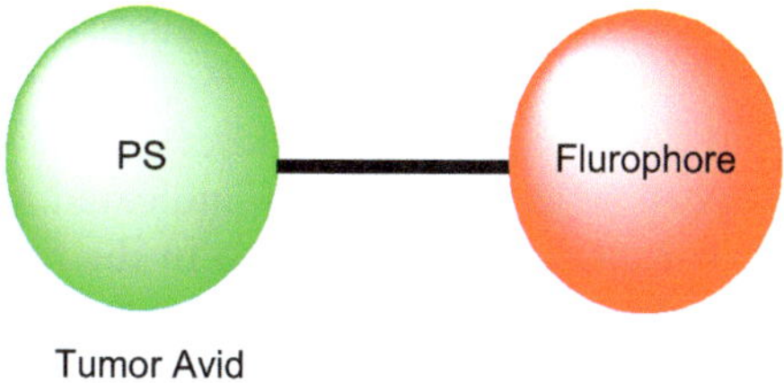

Figure 15.11　A tumor-avid photosensitizer used as a substrate to deliver a fluorophore to the tumor.

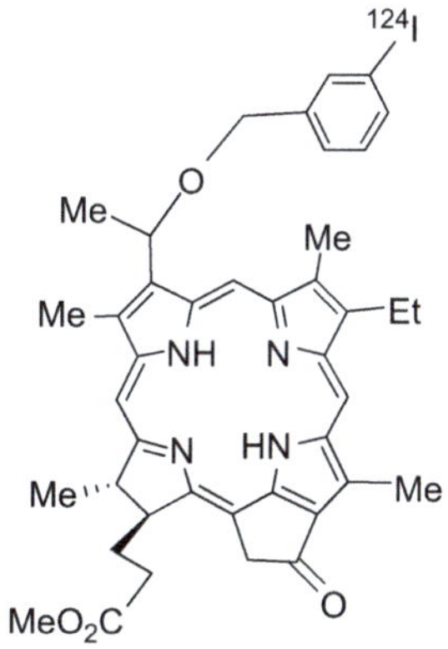

Figure 15.12　Structure of ^{124}I-531.

with purpurinimides, the photosensitizer can be used as a vehicle to deliver the near-infrared fluorophore to the tumor. The length of the linker used for conjugation was (i) a direct amide bond, (ii) a two-carbon linker, or (iii) a four-carbon PEG linker. In theory, non-radiative energy transfer decreases as the distance between the two chromophores increases. Based on the laser flash photolysis experiments, the four-carbon PEG linker showed the least energy transfer. Interestingly the HPPH–cyanine dye conjugate can detect Colon26 tumors using the direct amide formulation, but for the purpurinimide–cyanine dye conjugates, Colon26 tumors could only be detected when the length of the linker was two or four. The *ex vivo* fluorescence intensity in the tumor for the two-carbon and four-carbon linkers was similar. The *in vivo* PDT showed that the two-carbon linker cured 100% of the tumors at a dose lower than required for the HPPH-cyanine dye directly conjugated, yet the difference in dose required for tumor imaging and therapy did not significantly change.[31]

Multimodality Imaging and PDT: Pandey *et al.*[35] showed that the tumor-avid and fluorescent photosensitizer, methyl 3-(1′-*m*-iodobenzyloxyethyl)pyropheophorbide-*a*, can be converted to a PET radiotracer by replacing the cold iodine with the radioactive ^{124}I (^{124}I-531) (see Figure 15.12). By having a trifunctional agent, tumor imaging, image-guided surgery, and cancer therapy can be accomplished. Fluorescence imaging was used for the detection of tumors and for image-guided surgery of Colon26 tumors. The fluorescence images post-surgical resection indicated a non-complete resection of the tumor. The tumor that remained in the tumor bed may not be visible to the surgeon unless tumor resection is being monitored in real-time *via* fluorescence. The fluorescence imaging capability of the compound is extremely useful but limited to depths of a few millimeters below the tissue surface; therefore, the PET imaging functionality was incorporated. Upon labeling the photosensitizer with ^{124}I, it could image RIF and Colon26 tumors. Through quantitative *ex vivo* organ biodistribution, it was found that over 4% (%ID g^{-1}) was present in Colon26 tumors 24 hours post-injection and over 15% and 9% in the spleen and liver, respectively. This makes it difficult to discern the location of the tumor 24 hours post-injection. Seventy-two hours post-injection of the radiotracer, the tumor was clearly visible because the radiotracer rapidly cleared from the body. The fluorescence component can aid in overcoming the high background to tumor signal seen in PET imaging, 24 hours post-injection. To show the theranostic application of the trifunctional agent, PDT was performed on mice bearing either RIF or Colon26 tumors. The results were extremely encouraging and provided a platform for a possible "see and treat" approach to cancer detection and therapy.[34]

Stefflova *et al.*[35] have extended the work of Pandey *et al.*[34] to create a pyropheorbide-*a* based photosensitizer that targets the folate receptors that are overexpressed in a variety of cancers. The photosensitizer was tested *in vivo* on mice bearing both the KB (folate receptor positive) and the HT 1080 (human fibrosarcoma, folate receptor negative) tumors by whole-body fluorescence imaging. The fluorescence in the KB tumor was found to be on average 2.5 times higher than in the HT 1080 tumor. To extend pyropheophorbide-*a* based photosensitizer to PET imaging, Zheng and his colleagues have incorporated ^{64}Cu to afford a targeted PET imaging radiotracer. The results showed that the targeted photosensitizer was significantly more effective in killing KB cells than HT 1080 cells. In this study, a targeted theranostic agent was synthesized that showed applications in PET/fluorescence imaging and PDT.[35,36]

The uptake (%ID g^{-1}) of the targeted photosensitizer was found to be $3.02 \pm 0.55\%$ in the tumor, 4 h post injection and the % ID per gram in the kidney was close to 40%. Folic acid that was linked regioselectively through its α- and γ-carboxyl groups to 4-fluorobenzylamine (FBA) and radiolabeled with ^{18}F showed higher uptake within the tumor, $6.56 \pm 1.80\%$ID g^{-1}, compared with the non-targeted pyropheophorbide-*a*.[37] However, the uptake within the kidney was found to be $40.65 \pm 12.81\%$ID g^{-1}. In both cases, the uptake within the kidney was the highest among all organs and tissues. The high uptake within the kidney would hinder the clinician's ability to detect/distinguish small lesions as being cancerous because the spatial resolution limits of clinical PET systems is 5–7 mm.[1]

15.3.3   Theranostic Nanoparticles Approach

For nanoparticles to be effective *in vivo* for tumor imaging and therapy, they should possess the following characteristics:

A. They should be biodegradable:
   a. The nanoparticle should degrade to nontoxic monomers or higher order oligomers that can be cleared out of the body.
   b. This can impart a controlled release of its payload.
B. They should be less than 200 nm in size because the diffusion of singlet oxygen is rather short, 10–55 nm *in vivo* and because the initial photodynamic effect is limited to the site of the photosensitizer.
C. The nanoparticle matrix should allow for the diffusion of singlet oxygen.
D. The diameter of the nanoparticles should be larger than 10 nm, as the tight endothelial gap junctions found in normal blood vessels are in the range of 5–10 nm.
E. They should not be immunogenic or not capable of eliciting an immune response.
F. The nanoparticles should be optically transparent in the 650–950 nm range, where the maximum q-band absorbance for the photosensitizer and near-infrared fluorophore occur.
G. The nanoparticle platform should allow for the loading of multiple agents such as near-infrared fluorophore or radiotracer and photosensitizer for tumor detection and therapy.[38]

With nanotechnology comes the possibility of creating theranostic nanoparticles that are capable of detecting cancerous tissue *via* fluorescence, PET, and MRI,[39,40,41] and treating tumors *via* PDT.[40,41] The co-loading of a nanoparticle with a photosensitizer and a fluorophore or radiotracer provides several advantages over the synthetic approach for the creation of a theranostic agent. The advantages are (i) the laborious multi-step synthesis and purification used for the synthesis of a theranostic agent is not required, (ii) the contrast and therapeutic agent can be formulated in an aqueous environment at the ideal dose for imaging and therapy, (iii) the payload can be released from the nanoparticle in a controlled manner, and (iv) the energy transfer between the photosensitizer and imaging agent can be minimized for enhanced PDT. The two major classes of theranostic nanoparticles that will be discussed are (1) ORMOSIL (ORganically MOdified SILica) and (2) Polyacrylamide (PAA).[42]

15.3.4   ORMOSIL Nanoparticles for Tumor Imaging and Photodynamic Therapy

ORMOSIL (organically modified silica) nanoparticles have gained interest because of their relative ease of synthesis and purification and biocompatibility. Their mean diameter is in the 20–30 nm range, allows singlet oxygen to diffuse freely through the pores on the nanoparticle, and are stable against pH and temperature changes. Because they are optically inert in the UV, visible, and near-infrared portion of the electromagnetic spectrum,[39] ORMSOIL nanoparticles have been utilized for *in vitro* cellular imaging and PDT. In our lab's first report utilizing ORMOSIL nanoparticles, the photosensitizer HPPH was encapsulated within the nanoparticle. *In vitro* PDT was performed on UCI-107 (human epithelial ovarian cancer cell line from the University of California, Irvine) and HeLa (human cervix epitheloid carcinoma) cells and compared against HPPH formulated in 1% Tween-80 / 5% dextrose. The *in vitro* PDT results as measured by the MTT assay were found to be similar. Confocal laser scanning microscopy showed high uptake of the ORMOSIL nanoparticles in both cell lines. This shows that for *in vitro* theranostics, ORMOSIL nanoparticles show promise, but the goal is to translate these results *in vivo*.[43]

To enhance the depth at which HPPH can treat tumors *in vivo*, Kim *et al.*,[44] co-encapsulated HPPH with the two-photon absorbing dye, 9,10-bis[4'-(4''-aminostyryl)styryl]anthracene (BDSA) dye that can be excited at 850 nm. For the BDSA dye to efficiently transfer energy non-radiatively to HPPH upon two-photon excitation, its fluorescence spectra needs to be up-converted such that it significantly overlaps the absorption spectra for HPPH. One photon excitation of BDSA at 425 nm showed that the fluorescence of BDSA significantly overlaps the absorption of HPPH; therefore, two-photon excitation of BDSA can be used to excite HPPH. *In vitro* cellular uptake and PDT studies were performed using HeLa cells. Confocal laser scanning microscopy showed that 3 hours after incubation of the nanoparticles in HeLa cells and using two-photon excitation, an intense HPPH signal was present in the cytoplasm. This indicates that these nanoparticles were highly taken up and that *in vitro* two-photon excitation can be used to excite HPPH. The *in vitro* PDT potential for the co-encapsulated ORMOSIL nanoparticles was evaluated by transmission imaging of live HeLa cells under two-photon excitation (850 nm). The cells were treated overnight with 20 wt% BDSA or 1.1 wt% HPPH and 20 wt% BDSA co-encapsulated within ORMOSIL nanoparticles, and then exposed to an 850 nm pulsed laser for a total of 90 s. Fifteen minutes post-treatment, necrosis could be seen in the cells treated with the nanoparticles co-encapsulating BDSA and HPPH but not with the cells treated with BDSA containing nanoparticles. This showed that by using BDSA, two-photon PDT can be achieved *in vitro*.[44]

The above shows that HPPH can be encapsulated within ORMOSIL nanoparticles and are capable of producing similar photodynamic efficacy as the free photosensitizer. These formulations have not been tested *in vivo* for their photodynamic efficacy. *In vivo*, ORMOSIL nanoparticles have been tested for their organ distribution in nude mice utilizing ORMOSIL nanoparticles conjugated with a ^{124}I-labeled bolton–hunter reagent by Kumar *et al.*[45] Twenty-four hours after intravenous injection, over 40 and 50%ID g^{-1} were found in the liver and spleen, respectively. This poses a problem, in that this may hamper our ability to use ORMOSIL nanoparticles for the detection of tumors, or the ratio of the signal in the tumor to liver may be low. Qian *et al.*[46] encapsulated the fluorophore IR-820 in ORMOSIL nanoparticles, which were conjugated with polyethylene glycol (PEG) to the surface of the nanoparticle, and tested these nanoparticles for their ability to target HeLa tumors in nude mice. Twenty-four hours post-intravenous injection, a higher amount of nanoparticles were found in the liver than in the tumor even though they were conjugated with PEG.[46] These results show that ORMOSIL nanoparticles may pose problems in their use as theranostic nanoparticles.

PET/Fluorescence Imaging and PDT. When a photosensitizer is encapsulated within ORMOSIL nanoparticles, it may be released upon incubation with cell culture media or upon systemic circulation. Compagnin *et al.*[47] showed that when the photosensitizer meta-tetra(hydroxyphenyl)chlorin (mTHPC) is encapsulated in silica nanoparticles conjugated with a near-infrared cyanine dye, the photosensitizer is released from the nanoparticle. The authors found that serum proteins mediate this release. To prevent the release of a photosensitizer from the ORMOSIL nanoparticles, Ohulchanskyy *et al.*[48] covalently linked a photosensitizer, methyl 3-(1'-*m*-iodobenzyloxyethyl)pyropheophorbide-*a* carboxylic acid, to ORMOSIL nanoparticles. Their study showed that 80–90% of the photosensitizer was retained within the nanoparticle (NY-363). Among the nanoparticle formulations tested for cellular uptake in Colon26 cells, the nanoparticle formulation in which the molar ratio of the photosensitizer and the ORMOSIL nanoparticle precursor, vinyltriethoxysilane, was 1 to 480 showed the highest uptake. They also showed the highest *in vitro* PDT efficacy against RIF-1 cancer cells. When compared to the free photosensitizer formulated in 1% Tween-80 / 5% dextrose, the nanoparticle formulation required a higher photosensitizer dose and longer treatments with light in order to achieve similar cell-kill. This indicates that the nanoparticle formulation does not provide an advantage over free photosensitizer for *in vitro* PDT.[48]

15.3.5 Polyacrylamide Nanoparticles for Tumor Imaging and Photodynamic Therapy

Compared to ORMOSIL nanoparticles, polyacrylamide (PAA) nanoparticles can be rendered biodegradable by cross-linking poly(3-acryloyloxy 2-hydroxypropyl methacrylate) with polyacrylamide (Figure 15.13). The biodegradability is an important feature, because it can impart controlled release of the payload through degradation of the nanoparticle matrix within the tumor microenvironment. PAA nanoparticles can be degraded by esterases at the degradable points shown in the PAA structure. This was demonstrated by Wang *et al.*[51] when they performed an *in vitro* experiment using porcine liver esterases to study their biodegradation products.

Magnetic Resonance Imaging and PDT: Polyacrylamide nanoparticles have shown great potential as theranostic nanoparticles.[40,51] Reddy *et al.*[52] have shown that F3 conjugated polyacrylamide nanoparticles co-encapsulated with iron oxide nanoparticles and Photofrin (photosensitizer) can be used for T2

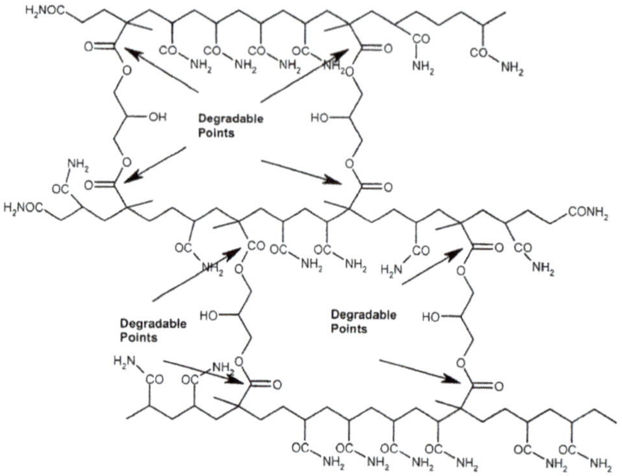

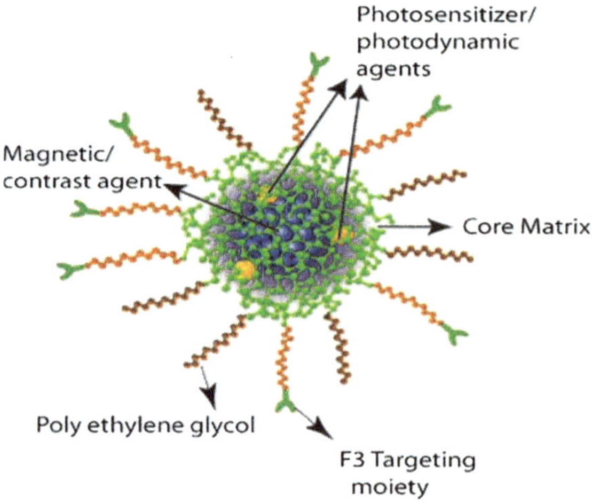

Figure 15.13 Structure of amine functionalized polyacrylamide nanoparticles.[50]

Figure 15.14 Multifunctional targeted nanoparticles for MR imaging and photodynamic therapy.

weighted MR imaging and PDT for targeting intracerebral rat 9L glioma tumors. F3 peptide used for targeting is a 31-amino acid synthetic linear peptide derived from a fragment of the high mobility group protein-2 (Figure 15.14) that targets the nucleolin receptor that is overexpressed on glioma tumors. The *in vivo* MR imaging data showed that upon systemic circulation the targeted nanoparticles can localize within i.c. rat 9L glioma tumors. Because the MR signal enhancement for the targeted group was two-fold higher (significant enhancement), as compared with the non-targeted group, it was expected that the therapeutic efficacy would be higher as well. Rats bearing i.c. 9L glioma tumors were treated with the targeted and non-targeted nanoparticle platform, Photofrin, and with the laser light alone. The rats treated with the targeted nanoparticle showed 40% long-term survival at 60 days post-PDT. These results clearly show that polyacrylamide nanoparticles can be used successfully for *in vivo* theranostic applications.[52] Various synthetic approaches were used for developing multifunctional PAA nanoparticles for imaging and therapy and are illustrated in Figure 15.15.

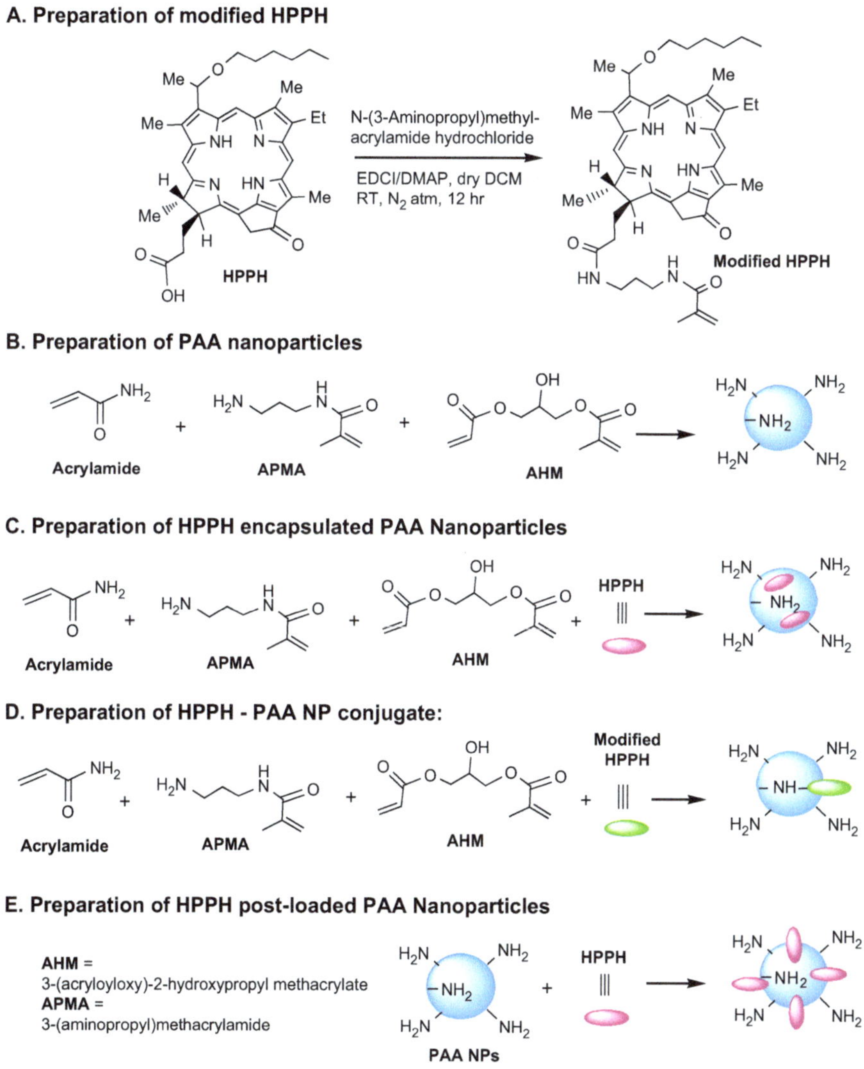

A. Preparation of modified HPPH

B. Preparation of PAA nanoparticles

C. Preparation of HPPH encapsulated PAA Nanoparticles

D. Preparation of HPPH - PAA NP conjugate:

E. Preparation of HPPH post-loaded PAA Nanoparticles

AHM =
3-(acryloyloxy)-2-hydroxypropyl methacrylate
APMA =
3-(aminopropyl)methacrylamide

Figure 15.15 The synthesis of **A**: Modified HPPH, **B**: Blank polyacrylamide nanoparticles (PAA NPs), **C**: HPPH encapsulated PAA NPs, **D**: HPPH internally conjugated to PAA NPs, and **E**: HPPH post-loaded to PAA NPs.

Fluorescence Imaging and PDT. Due to the fact that PAA nanoparticles can be rendered theranostic and biodegradable, Wang *et al.*[49] have loaded the second generation photosensitizer, HPPH, within PAA nanoparticles by internally conjugating the photosensitizer to *N*-(3-aminopropyl)methacrylamide hydrochloride then forming the nanoparticle, encapsulating the photosensitizer within the nanoparticle, or by post-loading the photosensitizer to preformed nanoparticles for fluorescence imaging and PDT. The three different loading methodologies employed are shown in Figure 15.15. Among the different loading methods studied, they found the post-loading method to be a facile method that provided a loading efficiency of 95%. This method did not inhibit the fluorescence emission and singlet oxygen production capabilities of the photosensitizer. Compared to the other loading strategies, the loading efficiency was 16% and 4% for the internally conjugated and encapsulated approaches, respectively. It was expected that the formulation that provided the highest loading efficiency would provide the best phototoxicity, *in vitro* and *in vivo*. To measure the *in vitro* phototoxicity, the MTT assay was used. The *in vitro* and *in vivo* assessments were conducted in Colon26 cells or in mice implanted with Colon26 tumors. As expected the post-loading method showed the highest phototoxicity among the nanoparticle formulations and was similar to the photosensitizer alone formulated in 1% Tween-80/D5W (Figure 15.16). From these results, the post-loaded nanoparticle and the free photosensitizer were tested *in vivo*. The free photosensitizer and the post-loaded nanoparticle formulation were compared for their ability to target to Colon26 tumor subcutaneously implanted in BALB/c mice by fluorescence imaging (Ex 665 nm/Em ≥ 695 nm).

The mean fluorescence intensity of the photosensitizer in the tumor was higher for the nanoparticle formulation but was not statistically significant. Owing to the similar levels of fluorescence in the tumor for the nanoparticle formulation and the free photosensitizer, the PDT efficacy was not expected to differ. The fluorescence-based tumor imaging and Kaplan–Meier plot in Figure 15.17 shows similar levels of fluorescence for both the nanoparticle and non-nanoparticle formulations and that 40% of the mice treated with either formulation were tumor-free at 60 days after PDT. These results clearly indicate that the post-loading approach provides an alternative to internally conjugating a photosensitizer or for the encapsulation method, while retaining the photophysical characteristics and *in vitro* and *in vivo* efficacy of the photosensitizer. For imaging deeply seated tumors using fluorescence, further red-shifted fluorophores, such as cyanine dyes that emit in the 800–900 nm bandwidth should be used. This, as described above, increases the depth at which a tumor can be imaged (from millimeters to >1 cm). Because of this, Gupta and co-workers extended the work of Wang's group by simultaneously post-loading a near-infrared cyanine dye that maximally absorbs at approximately 830 nm and emits at 870 nm. IR-820, which has limited tumor-imaging capabilities but contains the desired photophysical properties, was modified to the tumor-avid cyanine dye by replacing the chloro group of IR-820 with a *p*-aminothiol functionality. The photosensitizer, HPPH, and the modified

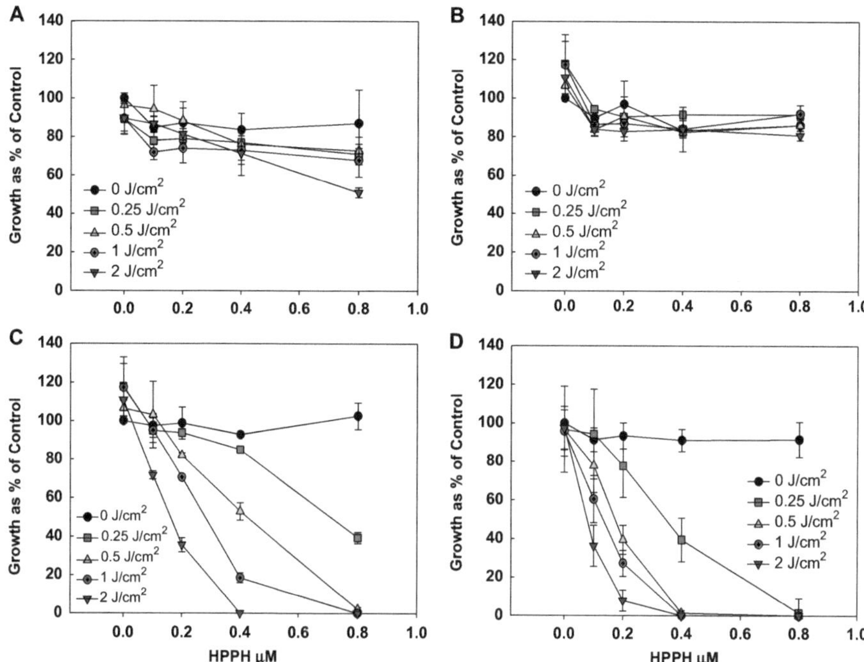

Figure 15.16 Phototoxicity of Colon26 cells after incubation with the different preparations of polyacrylamide nanoparticles (PAA NPs) for 4 h followed by light at a series of doses at a fluence rate of 3.2 mW cm^{-2}. Growth was assayed by the MTT method. A: HPPH Encapsulated PAA NPs, B: HPPH conjugated PAA NPs C: HPPH post-loaded PAA NPs, and D: free HPPH. Values are the mean ± standard deviation.

cyanine dye were co-post-loaded to biodegradable PAA nanoparticles and were separately post-loaded to individual nanoparticles and combined together such that the molar ratios of the photosensitizer to cyanine dye were 1 : 1, 2 : 1, 3 : 1, and 4 : 1. Compared to the synthetic approach mentioned above, in which HPPH was conjugated to a near-infrared cyanine dye, this approach provides flexibility in controlling the ratio of the photosensitizer to cyanine dye and does not require a laborious synthetic process.

The nanoformulations obtained by following the methodology shown in Figure 15.18 (Approach 1 and Approach 2) were tested in mice bearing Colon26 tumors for their ability to detect and destroy tumors. It was found that the nanoparticle formulation in which the photosensitizer and cyanine dye were co–post-loaded at a molar ratio of 2 : 1, respectively, was the best for tumor imaging and therapy, with 60% of the mice tumor-free at 60 days after PDT. In addition, PDT was performed at the standard dose of the free photosensitizer. When a photosensitizer is in close proximity to a near-infrared cyanine dye and the fluorescence emission spectra of the two overlap, energy can be transferred from the photosensitizer to the cyanine dye upon excitation. It causes a decrease in singlet oxygen that is available for tumor destruction. Among

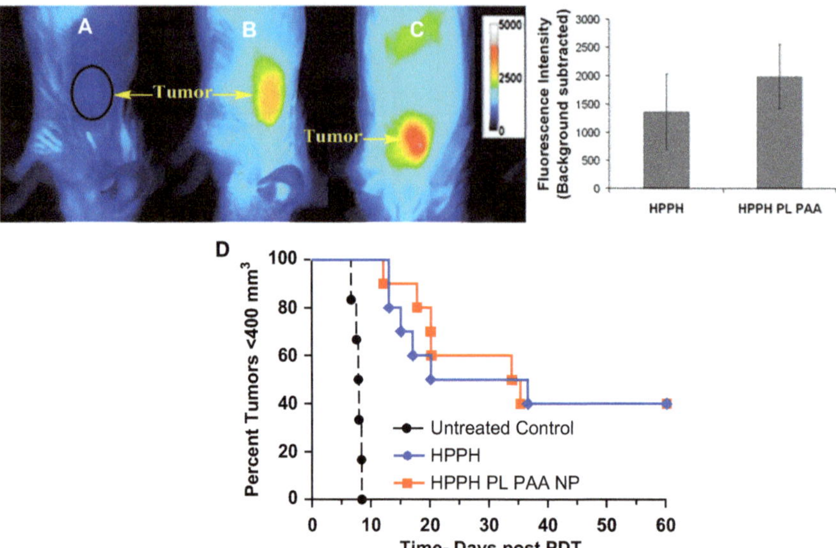

Figure 15.17   **A–C:** Whole-body fluorescence imaging of HPPH and HPPH post-loaded polyacrylamide nanoparticles (PAA NPs) using a 12-bit Nuance (CRi, Woburn, MA) scientific CCD camera. The fluorescence was excited at 665 nm with a pumped argon-dye laser and the emission was collected beyond 695 nm. **D:** The Kaplan–Meier plot for 10 BALB/c mice per group bearing subcutaneous Colon26 tumors treated with either HPPH or HPPH post-loaded (PL) PAA NPs. The fluence and fluence rate were 135 J cm^{-2} and 75 mW cm^{-2}. The dose for HPPH was 0.47 µmoles kg^{-1} for both the fluorescence imaging and photodynamic therapy.

these nano-formulations, with the co–post-loaded nanoparticles—that is the formulation in which 2 molar equivalents of HPPH were co–post-loaded with 1 molar equivalent of cyanine dye (Figure 15.19)—the amount of FRET generated was the least. This explains why the formulation provided the best tumor response. Gupta *et al.*[50] clearly show that biodegradable PAA nanoparticles can be used for near-infrared fluorescence imaging and therapy at the therapeutic dose required for the free photosensitizer.

Wang *et al.*[51] have shown that PAA nanoparticles internally conjugated to HPPH can be rendered multifunctional through external conjugation with a fluorophore such as fluorescein isothiocyanate (FITC), rhodamine, or a near-infrared cyanine dye modified from IR-820. To impart "stealth" or the ability to evade the reticuloendothelial system, polyethylene glycol (PEG) was grafted onto the nanoparticles conjugated with both HPPH and a fluorophore. For selective delivery, the nanoparticles were targeted to the nucleolin receptor, in the same fashion as Reddy *et al.*[52], by conjugating the nanoparticles with the F3 peptide. *In vitro* targeting experiments showed that these nanoparticles can actively target the human breast tumor cells, MDA-MB-435, and rat 9L glioma cells.[51,52]

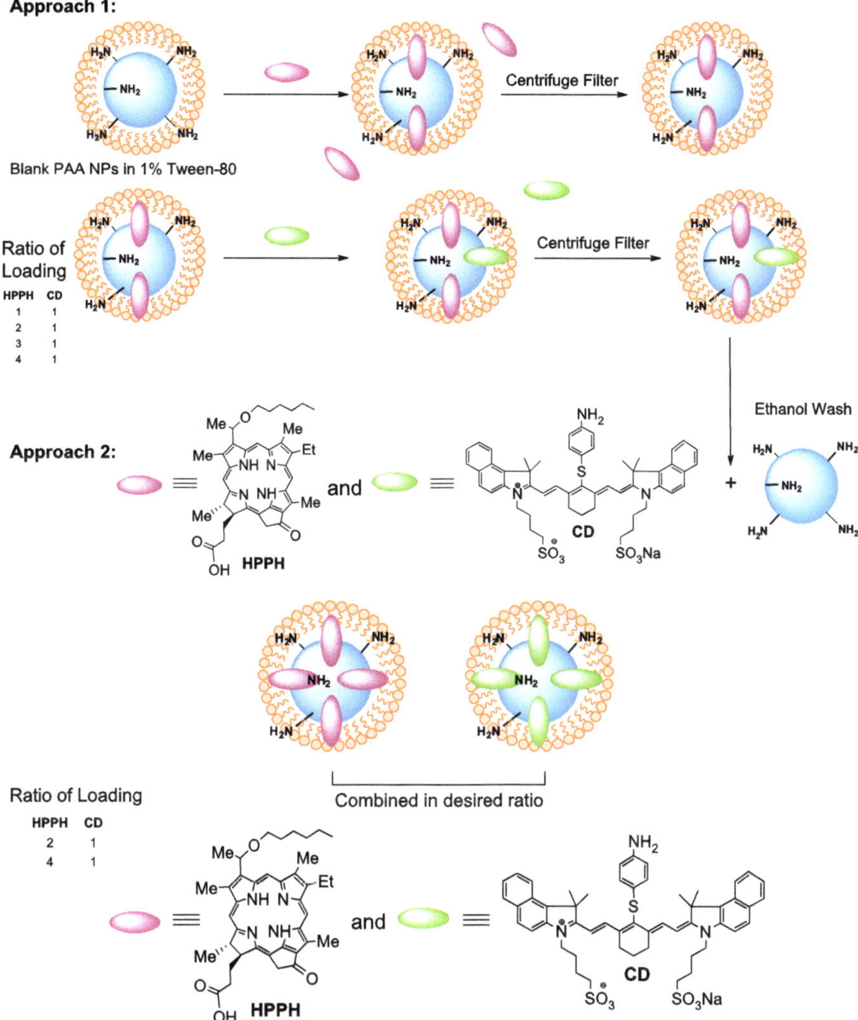

Figure 15.18 Two different approaches used for the synthesis of multifunctional polyacrylamide nanoparticle for whole-body fluorescence imaging and photodynamic therapy. The ratio of photosensitizer (HPPH) to cyanine dye (CD) used was 1 : 1, 2 : 1, 3 : 1, and 4 : 1 for Approach 1 and 2 : 1 and 4 : 1 for Approach 2.

Because the photosensitizer HPPH is internally conjugated to the nanoparticle matrix, it is expected that a higher payload will reach the tumor compared with the co-encapsulated nanoparticle, because the photosensitizer is free to release through the pores of the nanoparticle. Through the live/dead cellular assay, it was found that the targeted nanoparticles are capable of a high level of phototoxicity.

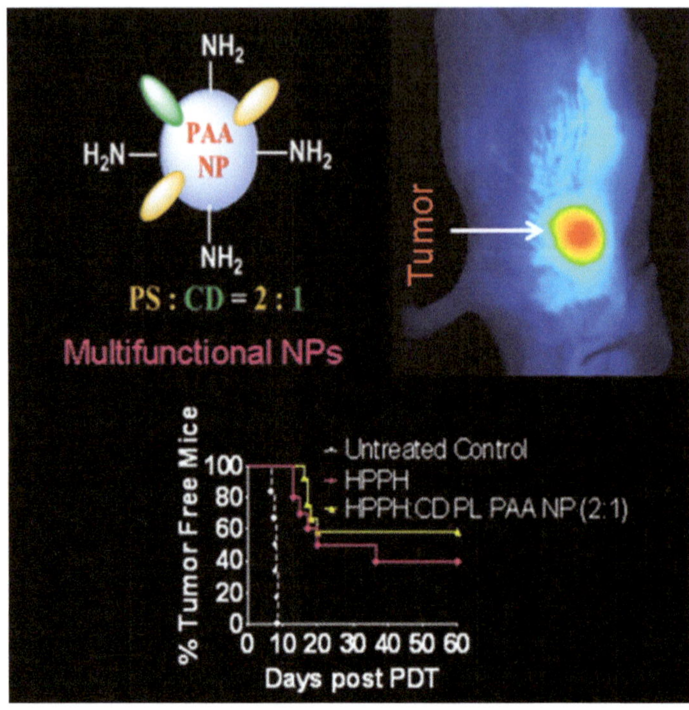

Figure 15.19 Multifunctional polyacrylamide nanoparticles (PAA NPs) co–post-loaded with photosensitizer (**HPPH**) and cyanine dye (**CD**) with the molar ratio of **HPPH** to **CD** = 2 : 1. The tumor imaging and photo-dynamic therapy (PDT) were performed 24 h post-injection in BALB/c mice bearing Colon26 tumors. The concentration of **HPPH** injected intravenously was 0.47 µmoles kg^{-1}. For fluorescence imaging, the NPs were excited at 782 nm and the fluorescence beyond 830 nm was collected and for PDT; the excitation wavelength was 665 nm. The fluence and fluence rate were 135 J cm^{-2} and 75 mW cm^{-2}.

15.3.6 Other Classes of Nanoparticles for Photodynamic Therapy and Tumor Imaging

McCarthy *et al.*[53] used the biodegradable poly(lactic-*co*-glycolic acid), (PLGA, 50 : 50), nanoparticles to encapsulate and deliver the photosensitizer, *meso*-tetraphenylporpholactol, to 9L tumors *in vivo*. They found that upon encapsulation, the photosensitizer aggregate causes efficient quenching of the excited states (fluorescence and singlet oxygen). The advantages of this formulation are similar to those outlined by Zheng *et al.*[33] as above, when they conjugated a pyropheophorbide-*a* to the black-hole quencher 3 *via* a MMP7-sensitive linker. *In vitro* cellular uptake studies showed that when the nanoparticle was internalized in 9L glioblastoma cells, the fluorescence recovered, indicating that the photosensitizer is being released from the nanoparticle.[33]

The limitations with the formulation is that for effective *in vitro* and *in vivo* PDT, a high concentration of the photosensitizer, $10\,\mu M$ and $780\,\mu M$ respectively, is required and a high laser fluence rate of $42\,mW\,cm^{-2}$ is needed for *in vitro* phototoxicity. The nanoparticle formulation was not compared against the free photosensitizer for *in vitro* or *in vivo* PDT; therefore, it cannot be determined whether the nanoparticle formulation is an improvement over the free photosensitizer.

Lovell *et al.*[54] showed, through a liposomal-type nanoparticle targeted to KB cells *via* folic acid, named porphysomes, that pyropheophorbide-*a* or bacteriochlorophyll can be loaded within the bilayers of the nanovesicle for tumor imaging, photoacoustic tomography, and fluorescence imaging and therapy. As shown by Weissleder's group, quenching of the photosensitizer's excited fluorescence and singlet state can be favorable for preventing collateral damage and sunlight-induced skin phototoxicity. Exciting the photosensitizer, in its quenched state, showed that the excitation energy is dissipated thermally. This was interesting because the authors showed that mice treated 24 h post-injection of the porphysomes ($42\,mg\,kg^{-1}$) with a 658 nm laser outputting 750 mW (power density $1.9\,mW\,cm^{-2}$) for 60 s, the temperature within the tumor increased to 60 °C. This is indicative of the fact that photothermal therapy rather than PDT is the primary means for tumor destruction. Photothermal therapy showed 100% survival or tumor cure at 28 days post-treatment. To highlight the theranostic application of such nanoparticles, the authors showed that due to the aggregation of the photosensitizer, photoacoustic tomography could be used for lymph node mapping, and as the photosensitizer released from the nanoparticle, whole-body near-infrared fluorescence imaging could be used to detect KB tumors. The limitation with the porphysome formulation is that a very high dose of the photosensitizer, namely $42\,mg\,kg^{-1}$, is required for tumor imaging and therapy.[54]

15.4 Summary

The concept of nanotechnology is of great scientific interest, because the nanostructures engage with cells and organisms in a completely new way and with this approach one can cross the biological barriers that could be difficult to cross by synthetic approaches. Properly designed supramolecular nanotechnology approaches could also provide a high payload of more than one moiety in desired quantity for imaging and therapy to the target site. Nanotechnology research has come a long way from early days. It is now widely accepted in the research community. However, long-term toxicity is the main issue with most of the nanoconstructs and should be investigated carefully.

Two distinct methodologies, synthetic and nanoplatform-based supramolecular approaches, exist for the design of multimodal tumor imaging probes and theranostic agents for tumor imaging and photodynamic therapy. The biodegradable nanoparticle approach, specifically polyacrylamide, is a promising alternative to the synthetic approach for the development of theranostics. These nanoparticles fulfill a majority of the requirements for an "ideal" theranostic nanoparticle formulation. In our laboratory, efforts are underway to move this technology to Phase I human clinical trials.

Acknowledgements

We are thankful to NIH (CA119358, CA127369 and CA55791) and the shared resources of the RPCI support grant (P30CA16056).

References

1. M. L. James and S. S. Gambhir, *Physiol. Rev.*, 2012, **92**, 897–965.
2. T. F. Massoud and S. S. Gambhir, *Genes Dev.*, 2003, **17**, 545–580.
3. E. A. Osborn and F. A. Jaffer, *Curr. Opin. Cardiol.*, 2008, **23**, 620–628.
4. R. Weissleder, *Science*, 2006, **312**, 1168–1171.
5. D.-E. Lee, H. Koo, I.-C. Sun, J. H. Ryu, K. Kim and I. C. Kwon, *Chem. Soc. Rev.*, 2012, **41**, 2656–2672.
6. M. Sajjad, U. Riaz, R. Yao, R. J. Bernacki, M. Abouzied, D. A. Erb, N. D. Chaudhary, J. M. Veith, G. I. Georg and H. A. Nabi, *Appl. Radiat. Isotopes*, 2012, **70**, 1624–1631.
7. E. J. Roh, Y. H. Park, C. E. Song, S.-J. Oh, Y. S. Choe, B. -T. Kim, D. Y. Chi and D. Kim, *Bioorg. Med. Chem.*, 2000, **8**, 65–68.
8. R. Weissleder and M. J. Pittet, *Nature*, 2008, **452**, 580–589.
9. P. Agostinis, K. Berg, K. A. Cengel, T. H. Foster, A. W. Girotti, S. O. Gollnick, S. M. Hahn, M. R. Hamblin, A. Juzeniene, D. Kessel, M. Korbelik, J. Moan, P. Mroz, D. Nowis, J. Piette, B. C. Wilson and J. Golab, *CA: Cancer J. Clinic.*, 2011, **61**, 250–281.
10. J. Cheon and J.-H. Lee, *Acc. Chem. Res.*, 2008, **41**, 1630–1640.
11. R. H. Kimura, Z. Cheng, S. S. Gambhir and J. R. Cochran, *Cancer Res.*, 2009, **69**, 2435–2442.
12. L. Sampath, S. Kwon, M. A. Hall, R. E. Price and E. M. Sevick-Muraca, *Transl. Oncol.*, 2010, **3**, 307–217.
13. S. Ahlgren, H. Wållberg, T. A. Tran, C. Widström, M. Hjertman, L. Abrahmsén, D. Berndorff, L. M. Dinkelborg, J. E. Cyr, J. Feldwisch, A. Orlova and V. Tolmachev, *J. Nucl. Med.*, 2009, **50**, 274–283.
14. Y.-H. Lin, K. Dayananda, C.-Y. Chen, G.-C. Liu, T.-Y. Luo, H.-S. Hsu and Y.-M. Wang, *Bioorg. Med. Chem.*, 2011, **19**, 1085–1096.
15. A. Srivatsan, M. Ethirajan, S. K. Pandey, S. Dubey, X. Zheng, T.-H. Liu, M. Shibata, J. Missert, J. Morgan and R. K. Pandey, *Mol. Pharm.*, 2011, **8**, 1186–1197.
16. A. O. Abu-Yousif, A. C. E. Moor, X. Zheng, M. D. Savellano, W. Yu, P. K. Selbo and T. Hasan, *Cancer Lett.*, 2012, **321**, 120–127.
17. *What is Nanotechnology?*, Available at: http://www.nano.gov/nanotech-101/what/definition. Accessed 25 June 2012.
18. K. Cho, X. Wang, S. Nie, Z. Chen and D. M. Shin, *Clin. Cancer Res.*, 2008, **14**, 1310–1316.

19. E. Pöselt, C. Schmidtke, S. Fischer, K. Peldschus, J. Salamon, H. Kloust, H. Tran, A. Pietsch, M. Heine, G. Adam, U. Schumacher, C. Wagener, S. Förster and H. Weller, *ACS Nano*, 2012, **6**, 3346–3355.
20. P. Jason and F. Tarek, in *Handbook of Nanophysics*, CRC Press, Boca Raton, FL, 2010, pp. 1–14.
21. M. Benezra, O. Penate-Medina, P. B. Zanzonico, D. Schaer, H. Ow, A. Burns, E. DeStanchina, V. Longo, E. Herz, S. Iyer, J. Wolchok, S. M. Larson, U. Wiesner and M. S. Bradbury, *J. Clin. Invest.*, 2011, **121**, 2768–2780.
22. *Absorption spectrum of melanin*, Available at: http://www.cl.cam.ac.uk/~jgd1000/melanin.html. Accessed 8 May 2012.
23. Y. Wu, W. Cai and X. Chen, *Mol. Imaging Biol.*, 2006, **8**, 226–236.
24. K. Chen, Z.-B. Li, H. Wang, W. Cai and X. Chen, *Eur. J. Nucl. Med. Mol. Imaging*, 2008, **35**, 2235–2244.
25. E. E. Paoli, D. E. Kruse, J. W. Seo, H. Zhang, A. Kheirolomoom, K. D. Watson, P. Chiu, H. Stahlberg and K. W. Ferrara, *J. Control. Release*, 2010, **143**, 13–22.
26. J. Xie, K. Chen, J. Huang, S. Lee, J. Wang, J. Gao, X. Li and X. Chen, *Biomaterials*, 2010, **31**, 3016–3022.
27. M. Marites, W. Xiaoxia, Z. Guodong and L. Chun, in *Nanoimaging, Pan Stanford Series on Biomedical Nanotechnology*, Pan Stanford Publishing, 2011, pp. 171–195.
28. J. A. Moss, A. L. Vāvere and A. Azhdarinia, *Curr. Med. Chem.*, 2012, **19**, 3255–3265.
29. M. Olivo, R. Bhuvaneswari, S. S. Lucky, N. Dendukuri and P. Soo-Ping Thong, *Pharmaceuticals*, 2010, **3**, 1507–1529.
30. Y. Chen, K. Ohkubo, M. Zhang, E. Wenbo, W. Liu, S. K. Pandey, M. Ciesielski, H. Baumann, T. Erin, S. Fukuzumi, K. M. Kadish, R. Fenstermaker, A. Oseroff and R. K. Pandey, *Photochem. Photobiol. Sci.*, 2007, **6**, 1257–1267.
31. M. P. Williams, M. Ethirajan, K. Ohkubo, P. Chen, P. Pera, J. Morgan, W. H. White 3rd, M. Shibata, S. Fukuzumi, K. M. Kadish and R. K. Pandey, *Bioconjug. Chem.*, 2011, **22**, 2283–2295.
32. Y. Choi, R. Weissleder and C. H. Tung, Protease-mediated phototoxicity of a polylysine-chlorin(E6) conjugate, *ChemMedChem*, 2006, **1**, 698–701.
33. G. Zheng, J. Chen, K. Stefflova, M. Jarvi, H. Li and B. C. Wilson, *Proc. Natl Acad. Sci. USA*, 2007, **104**, 8989–8994.
34. S. K. Pandey, M. Sajjad, Y. Chen, A. Pandey, J. R. Missert, C. Batt, R. Yao, H. A. Nabi, A. R. Oseroff and R. K. Pandey, *Bioconjug. Chem.*, 2009, **20**, 274–282.
35. K. Stefflova, H. Li, J. Chen and G. Zheng, *Bioconjug. Chem.*, 2007, **18**, 379–388.
36. J. Shi, T. W. B. Liu, J. Chen, D. Green, D. Jaffray, B. C. Wilson, F. Wang and G. Zheng, *Theranostics*, 2011, **1**, 363–370.
37. D. Bechet, P. Couleaud, C. Frochot, M.-L. Viriot, F. Guillemin and M. Barberi-Heyob, *Trends Biotechnol.*, 2008, **26**, 612–621.

38. F. Yan and R. Kopelman, *Photochem. Photobiol.*, 2003, **78**, 587–591.
39. B. A. Moffat, G. R. Reddy, P. McConville, D. E. Hall, T. L. Chenevert, R. R. Kopelman, M. Philbert, R. Weissleder, A. Rehemtulla and B. D. Ross, *Mol. Imaging*, 2003, **2**, 324–332.
40. Y.-E. K. Lee and R. Kopelman, in *Multifunctional Nanoparticles for Drug Delivery Applications*, ed. Svenson, S. and Prud'homme, R. K., Springer, New York, pp. 225–255.
41. H. Xu, S. M. Buck, R. Kopelman, M. A. Philbert, M. Brasuel, B. D. Ross and A. Rehemtulla, *Isr. J. Chem.*, 2004, **44**, 317–337.
42. Y. E. Koo, G. R. Reddy, M. Bhojani, R. Schneider, M. A. Philbert, A. Rehemtulla, B. D. Ross and R. Kopelman, *Adv. Drug Deliv. Rev.*, 2006, **58**, 1556–1577.
43. I. Roy, T. Y. Ohulchanskyy, H. E. Pudavar, E. J. Bergey, A. R. Oseroff, J. Morgan, T. J. Dougherty and P. N. Prasad, *J. Am. Chem. Soc.*, 2003, **125**, 7860–7865.
44. S. Kim, T. Y. Ohulchanskyy, H. E. Pudavar, R. K. Pandey and P. N. Prasad, *J. Am. Chem. Soc.*, 2007, **129**, 2669–2675.
45. R. Kumar, I. Roy, T. Y. Ohulchanskky, L. A. Vathy, E. J. Bergey, M. Sajjad and P. N. Prasad, *ACS Nano*, 2010, **4**, 699–708.
46. J. Qian, D. Wang, F. Cai, Q. Zhan, Y. Wang and S. He, *Biomaterials*, 2012, **33**, 4851–4860.
47. C. Compagnin, L. Baù, M. Mognato, L. Celotti, G. Miotto, M. Arduini, F. Moret, C. Fede, F. Selvestrel, I. M. R. Echevarria, F. Mancin and E. Reddi, *Nanotechnology*, 2009, **20**, 345101.
48. T. Y. Ohulchanskyy, I. Roy, L. N. Goswami, Y. Chen, E. J. Bergey, R. K. Pandey, A. R. Oseroff and P. N. Prasad, *Nano Lett.*, 2007, **7**, 2835–2842.
49. S. Wang, W. Fan, G. Kim., H. J. Hah, Y. E. Lee, R. Kopelman, M. Ethirajan, A. Gupta, L. N. Goswami, P. Pera, J. Morgan and R. K. Pandey, *Lasers Surg. Med.*, 2011, **43**, 686–695.
50. A. Gupta, S. Wang, P. Pera, K. V. Rao, N. Patel, T. Y. Ohulchanskyy, J. Missert, J. Morgan, Y. E. Koo-Lee, R. Kopelman and R. K. Pandey, *Nanomedicine*, 2012, **8**, 941–950.
51. S. Wang, G. Kim, Y-E. K. Lee, H. J. Hah, M. Ethirajan, R. K. Pandey and R. Kopelman, *ACS Nano*, 2012, **6**, 6843–6851.
52. G. R. Reddy, M. S. Bhojani, P. McConville, J. Moody, B. A. Moffat, D. E. Hall, G. Kim, Y. E. Koo, M. J. Woolliscroft, J. V. Sugai, T. D. Johnson, M. A. Philbert, R. Kopelman, A. Rehemtulla and B. D. Ross, *Clin. Cancer Res.*, 2006, **12**, 6677–6686.
53. J. R. McCarthy, J. M. Perez, C. Bruckner and R. Weissleder, Polymeric nanoparticle preparation that eradicates tumors, *Nano. Lett.*, 2005, **5**, 2552–2556.
54. J. F. Lovell, C. S. Jin, E. Huynh, H. Jin, C. Kim, J. L. Rubinstein, W. C. Chan, W. Cao, L. V. Wang and G. Zheng, *Nat. Mater.*, 2011, **10**, 324–332.

Designing Polymeric Binders for Pharmaceutical Applications

NICOLAS BERTRAND,[a] PATRICK COLIN,[b]
MAXIME RANGER[b] AND JEANNE LEBLOND*[b]

[a] Current address: David H. Koch Institute for Integrative Cancer Research at Massachusetts Institute of Technology, 500 Main Street, Cambridge, MA 02139, USA; [b] Faculty of pharmacy, University of Montreal, PO box 6128, Downtown station, Montréal, QC, H3C 3J7, Canada
*Email: jeanne.leblond-chain@umontreal.ca

16.1 Introduction

Polymers have been used for a long time in biomedical fields such as drug delivery or tissue regeneration. However, their use as therapeutic agents by themselves is relatively recent.[1] The Food and Drug Administration's (FDA) approval of glatiramer acetate (Copaxone®) in 1996,[2] or the FDA's fast track designation Investigational New Drug (IND) application of SPL7013 in 2006,[3] the first dendrimer-based therapeutic, has attracted the interest on the distinctive properties of these molecules. Due to their high molecular weight, they can be designed to mimic biomacromolecules. In addition, they represent an attractive alternative to biological therapeutics, since they are easier to manufacture and their synthetic structures can be easily tuned. Significant progress has been made on the precise control of polymer synthesis, functionalization and characterization[4] and opened new therapeutic applications of polymers as drugs.

Monographs in Supramolecular Chemistry No. 13
Supramolecular Systems in Biomedical Fields
Edited by Hans-Jörg Schneider
© The Royal Society of Chemistry 2013
Published by the Royal Society of Chemistry, www.rsc.org

Nature's strong interactions between macromolecules rely on multiple weak interactions.[5] The singular three-dimensional (3D) combination of electrostatic, hydrogen bonding, hydrophobic, van der Waals interactions and metal coordination leads to unrivaled affinity and selectivity of biological systems. Scientists have imitated these complex interactions using multivalent interactions, that is, additive interactions of multiple binding sites. Using polymeric multivalent ligands instead of their monovalent counterparts generally results in dramatically increased interaction forces. Multiplying relatively weak interactions can drive affinity to rise above those of some antibody–antigen couples (180 μmol g^{-1} [polymer] *vs.* 13 μmol.g^{-1} [IgG] for a toxic peptide binding[6]). This is explained by the required force to dissociate multipoint interactions, especially when both players are polyvalent. For instance, the affinity constant of sialic acid for isolated hemagglutinin ($K_a \sim 4 \times 10^2$ M^{-1}) was enhanced by polymerization of sialic acid moieties, and further magnified (up to 10^7 M^{-1}) using poly(sialic acid) with a viral surface presenting many hemagglutinin proteins.[7]

Polymeric binders are macromolecules designed to exert a pharmacological effect by selectively interacting with exogenous or endogenous substrates. The targeted substrates range from small molecules (ions, peptides, proteins) to larger ones (virus, bacteria, cells).[8] Upon binding to the target, the polymer can inhibit or trigger a biological response, for instance preventing bacterial invasion or stimulating the immune system. The pharmacological efficacy is directly related to the affinity and the selectivity of the polymer for the target. The design of such complementary binder/substrate pairs is an exciting interdisciplinary challenge, combining elements of supramolecular chemistry and pharmacology. Moreover, because the end goal is to create new therapies for human use, scientists must also keep track of the toxicological and clinical hurdles that must be overcome.

This chapter presents the step-by-step development of polymeric binders, from the structure to the clinical trials. It will focus on the therapeutic applications of polymers as new drug entities; thereby it does not cover drug delivery or imaging applications. These subjects have been partially dealt with in chapters 5, 6, 7 and 15 of this book and reviewed recently.[1] Through diversified examples, this work reviews the potential applications, the chemical structures developed and the methods employed to assess efficacy and safety of the polymeric binders, both *in vitro* and *in vivo*. Finally, the clinical challenges related to this particular class of drugs are also discussed.

16.2 Finding Targets

Polymeric binders present multivalent properties which enable them to bind selectively to a substrate. They can be designed to sequester irritants or bind to the surface of cells or pathogens. In all cases, their binding prevents or promotes biological interactions and leads to a therapeutic response.

16.2.1 Sequestrants

16.2.1.1 Overloads

Many common substances turn out to be toxic as soon as their burden overloads the natural capacity of the host. For example, hypercholesterolemia increases cardiovascular risks;[9] hyperphosphatemia is associated with increased calcification and subsequent mortality in patients with chronic kidney disease;[10] and acute hyperkalemia can cause cardiac arrhythmias and fatal outcomes.[11] Orally administered polymeric binders are able to sequester the toxin in the gastrointestinal (GI) tract, thereby reducing intestinal absorption and restoring normal serum levels. Kayexalate® has been used since 1959 for the treatment of hyperkalemia. Composed of polystyrene sulfate, this resin binds dietary potassium in the GI tract through ionic interactions, and the complex is then excreted in the feces.[11] More recently, sevelamer hydrochloride (Renagel®), a cationic cross-linked polymer, has been commercialized to sequester alimentary phosphate in patients with chronic kidney disease, whose phosphate blood levels cannot be fully normalized by dietary reduction and dialysis. This polyaliphatic amine resin reduces phosphate serum levels and decreases associated vascular risks, while exhibiting less toxicity than other phosphate binders, for example aluminium or magnesium salts.[10] Anecdotally, sevelamer hydrochloride has also been shown to bind up to 8 mmol of bile acids per gram of polymer, probably by co-binding of fatty acids and bile acids in tight complexes, resulting in decreased cholesterol levels in patients.[12] Other multivalent cross-linked structures, such as cholestyramine (Questran®), Colestilan (BindRen®)[13] or colesevelam (Welchol®)[14] offer efficient sequestration and retention of bile acids until the excretion of the complex. Their high molecular weights prevent absorption in the GI and the safety concerns associated with systemic exposure (Figure 16.1A). Binding is mainly driven by ion pairing and electrostatic interactions, but hydrophobic pockets can be introduced to improve the binding of the hydrophobic part of bile acids, such as in the case of colesevelam.[14] This sequestration strategy has been extended to diverse toxic ions, such as heavy metals[8] or radioactive materials.[15] For instance, an orally-administered polymeric binder has been proposed as a first-aid treatment for radiation emergency.[15] The polymeric beads bearing phosphonic acid groups were able to sequester up to 0.2 mmol of strontium-90 per gram of polymer *in vitro* and reduce the bone accumulation of strontium-90 in rats. New generations of oral polymeric binders will present decreased GI effects and increased specificity (see Section 16.5). As an example, polymeric binders for iron have been designed using specific chelators of ferric ions grafted onto a polymeric or dendritic backbone.[16] The resulting binder demonstrated a rapid and efficient complexation of ferric ions *in vitro*, maintaining the high affinity ($K' = 3.5 \times 10^{26}$ M^{-1}) and selectivity of the ligands.

16.2.1.2 Exogenous Irritants

Our body is continuously exposed to irritants. Most of the time, it successfully controls these aggressions and maintains homeostasis. However, it sometimes

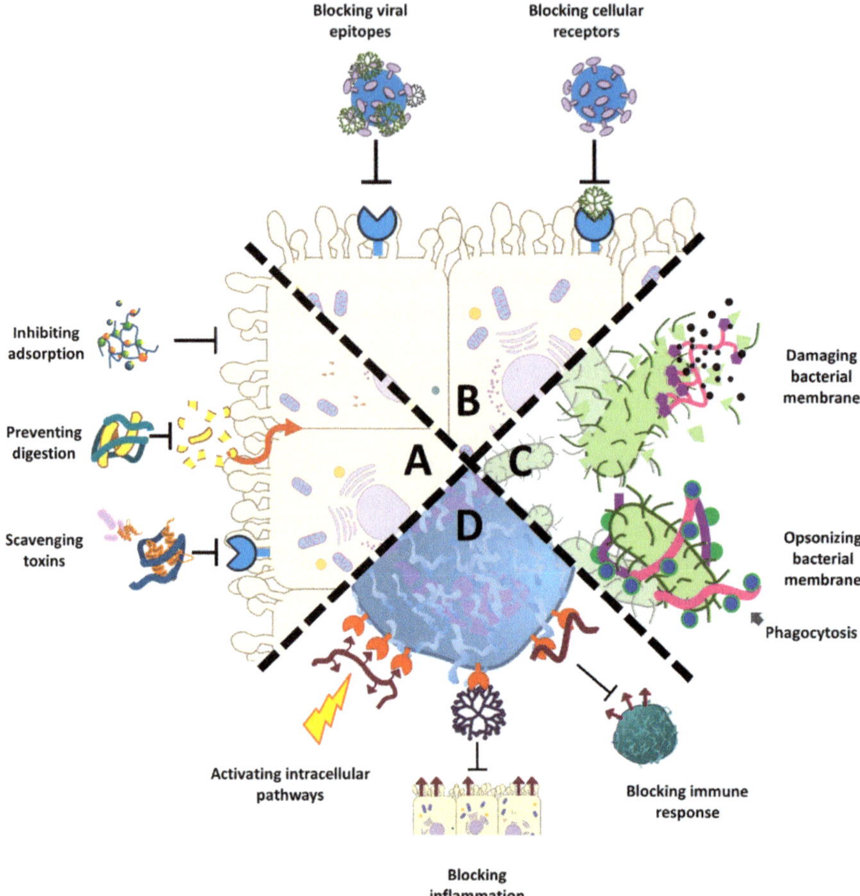

Figure 16.1 Therapeutic applications of polymeric binders include: (A) scavenging overloads, irritants or toxins in the gastrointestinal tract; (B) blocking viral invasion through viral epitope or cell receptor binding; (C) binding to the bacterial membrane to induce microbicidal activity or improve phagocytosis; (D) binding to the host-cell receptor to trigger or alter a biological response.

requires additional support to cope with sustained or more intense assaults. Polymeric binders can prove versatile tools to sequester part of these irritants and alleviate the body's burden.

In some cases, specific disorders or genetic predispositions can activate uncontrolled or amplified reactions to somehow inoffensive stimuli. For instance, the protein α-gliadin has been identified as the main trigger of celiac disease (gluten intolerance). In some people, the digestion of this protein from gluten produces non-degradable peptides that are responsible for intense inflammatory and immune responses. Pinier *et al.*[17] engineered a polymer able to sequester α-gliadin in the GI tract in order to reduce the production of these peptides (Figure 16.1.A). This binder mitigates the inflammatory response to gluten in a mouse model *in vivo*, and *ex vivo* on human biopsy specimens taken from patients with celiac disease. Following the same concept, Hoshino *et al.*[6] have designed a polymeric binder able to bind melittin, a toxic peptide present in the bee venom, with high affinity ($K_a = 6.6 \times 10^7$ M^{-1}). They demonstrated that administration of the polymer 20 seconds after injection of lethal doses of melittin was able to significantly prolong the survival of mice. The lethal dose of the peptide in the presence of the binder was increased two-fold.[6] For both of these polymeric binders, the polymer balance of hydrophobic and hydrogen-bonding or negatively-charged monomers was adjusted to maximize inter-actions and biocompatibility.

Polymeric scavengers have also been developed to target bacterial toxins, the main virulence factors responsible for the symptoms associated with bacterial infections.[18] For example, synsorb 90 and tolevamer are macromolecules which are able to sequester the *Clostridium difficile* toxin in the GI tract to prevent the diarrhea associated with this opportunistic infection (Figure 16.1A). These binders reached Phase III clinical trials, however both have failed because of insufficient activity.[19] Nevertheless, they paved the way for the rational design of polymeric sequestrants for other bacterial toxins.

Bundle and co-workers discovered the potency of Starfish, a perfectly matching pentavalent inhibitor of pentameric B5 unit of *Shiga* toxins, reaching subnanomolar inhibitory concentrations.[20] Based on this concept, they engin-eered a polymeric heterofunctional ligand able to bind *Shiga* toxin on one side and an endogenous human serum protein on the other side. The formation of the ternary complex protected mice from *Shiga* toxin intoxication, while the endogenous protein redirected it to the spleen and liver for degradation.[21] Kane and co-workers have proposed inhibitory peptides of anthrax toxin that showed amplified inhibition (IC$_{50}$ varied from 250 µM to 20 nM) when grafted to a polymeric backbone.[22,23] Through detailed structure–activity studies, they investigated various supports (liposomes,[24] polymers,[23,25,26] cyclodextrins[27]) and optimized the structure to match the heptameric geometry of the toxin's cell binding unit.

Finally, polymeric binders have been designed to neutralize the toxicity of lipopolysaccharides (LPS) of Gram-negative bacteria responsible for septic shock. Interestingly, the conjugation of inhibitor on the poly(ethylene glycol) support led to the colloidal stabilization and neutralization of the circulating LPS micelles. The polymer-bound inhibitors therefore showed better ability to decrease the toxicity of the lipids than free inhibitors.[28]

16.2.2 Surface Binders

16.2.2.1 Pathogens

Polymeric binders have also been designed to help the body fight against in-fections by targeting the pathogen themselves, either to eradicate them or simply to diminish their virulence. For obvious reasons of size, the polymer in this case is not able to sequester the whole organism, but rather binds to its surface to exert some beneficial effect.

Viruses present a range of surface epitopes which enable them to dock to cell walls through multiple weak interactions. Masking the virus epitopes with shielding polymers has been used to impair the colonization capacity of viruses (Figure 16.1B). The most successful story is illustrated by SPL7013 (VivaGel®), a poly(lysine) dendrimer with broad-spectrum activity which has been designed for prevention of sexually transmitted infections.[3] After the establishment of its safety and tolerability in humans,[29] recent clinical studies proved a total inhibitory activity against HIV-1 and HSV-2 maintained for 3 hours after vaginal administration in women, even in the presence of seminal fluid.[30] This activity is attributed to the docking of the dendrimer to the gp20 epitope of viruses[31] (Figure 16.1B), in addition to intrinsic virucidal activity of SPL7013.[32] Polymeric biguanides, used for a long time as antiseptic agents, is another type of cationic polymer, which demonstrated interesting antiviral activity against HIV-1.[33] In this latter case, therapeutic efficacy also involves direct interaction with host-cell receptors[34] (Figure 16.1B).

The bacterial membrane has also been an interesting target for surface binders. Charge-switching polymeric nanoparticles were used to bind bacteria at low pH and increase the local concentration of the delivered antibiotics to improve their efficacy (Figure 16.1C). This approach is particularly attractive, because most bacterial colonies thrive in slightly acidic environments where the effectiveness of antibiotics is decreased.[35] Sugar or protein epitopes can also bind to the surface of bacteria[7] and be incorporated into heterofunctional polymers to trigger a therapeutic response (Figure 16.1C). Krishnamurthy *et al.*[36] designed a bifunctional poly(acrylamide) polymer to fictively introduce haptens on the surface of bacteria, thus promoting opsonization and phagocytosis of multiple species of Gram-positive bacteria (Figure 16.1.C). Templating on their ternary complexes developed for *Shiga* toxins, Cui *et al.*[37] developed a similar "homing device", a polymer presenting two preorganized ligands: a trisaccharide moiety responsible for CD22 membrane receptor recognition, and nitrophenol, able to recruit antihapten IgM. This polymer was able to induce supramolecular formation between B-cells and CD-22 conjugated magnetic beads *in vitro*.

Numerous polymers have been developed for microbicidal applications (Figure 16.1C). The diversity of these materials has been extensively reviewed recently.[38] Among them, some have been designed to mimic natural host-defense peptides, which adopt α-helix folding patterns.[39–43] The activity of these peptides is attributed to the spatial segregation of cationic charges on one side of the helix and of hydrophobic moieties on the other side. Tuning the balance between cationic and hydrophobic motives, synthetic polymers have been designed to assemble in a secondary structure that promotes the binding and disruption of the bacterial membrane.[41] For example, Nederberg *et al.*[40] have prepared polymers that exhibited comparable activity to conventional antimicrobial agents over five bacteria and fungi strains, including multiresistant *Staphlococcus aureus*, with a good safety profile *in vivo*. Dendrimers have also revealed their potential as antimicrobial agents. Among them, cationic structures generally exhibit the best activity.[44] The globular architecture and the multiple cationic surface charges allow a strong interaction with the peptidoglycan of the bacterial membrane and can lead to membrane disruption.[45] Overall, these synthetic polymers are easier to prepare in large quantities at lower cost than the natural peptides.[41]

16.2.2.2 Host Receptors

Surface binders can neutralize pathogen colonization by binding epitopes present on the host's own cells to protect them against bacterial or viral aggressions. For instance, poly(hexamethylene biguanide) exhibited specific interactions with the HIV-1 co-receptor CXCR4[34] (Figure 16.1B). This persistent perturbation of the receptor extended the antiviral protection up to 24 h after the removal of the polymer.[34] Neutralizing the host receptors for microorganisms (viruses or bacteria) protects the host from various mutant forms, offering a promising strategy to overcome the resistance issue in bacterial infections.[46]

Various cell receptors have been targeted and exploited in alternative strategies against cancer. In an elegant study, Wu *et al.*[47] have targeted B-cell lymphoma using a two-step biorecognition mechanism. In the first step, CD20 B-cell receptor is recognized by anti-CD20 Fab' antibody conjugated to a coiled-coil peptide. The subsequent administration of the complementary pentaheptad peptide grafted onto a polymeric backbone induced biorecognition of both peptides, the formation of a coiled-coil heterodimer, and triggered the cell apoptosis secondary to the crosslinking of the CD20 receptors. *In vivo*, the intravenous administration of both agents, combined or successively, has produced long-term survivors in SCID mice bearing B-cell lymphoma xenografts.[48] Selective affinity for the cancer cell can also be achieved using aptamer engineering. Polymerization of the aptamer motifs demonstrated that the polymer scaffold increased the binding affinity for cells by 7.5-fold and altered the internalization process, resulting in an improved and selective cytotoxicity against lymphoblastic cell lines (Figure 16.1D).[49]

Mediators of inflammation have also been targeted by polymeric binders. Dernedde *et al.*[50] designed dendritic polyglycerol sulfates (dPGS), which present high affinity for selectins, and inhibit the leukocyte recruitment by endothelial cells (Figure 16.1D). Intraperitoneal administration of dPGS in a contact dermitis mouse model reduced the inflammatory response as efficiently as glucocorticoids. In this system, dPGS bind both L-selectin on endothelial cells and P-selectin on leukocytes. These interactions result in reduced contacts between endothelial cells and leukocytes because of steric repulsions.[51]

Targeting host-cell receptors with engineered polymers enables modification of the cell surface and induces a desired biological response, for example, immunomodulation. Polymeric antigens of the B-cell antigen receptor have been able to trigger the B-cell signaling pathway, which was not activated in the presence of a high concentration of single ligands.[52] Conversely, the immune response may be suppressed by polymeric ligands (Figure 16.1D). Glatiramer acetate (Copaxone®), commercialized since 1996 for reducing the frequency of relapses of multiple sclerosis, has been the first therapeutic agent to have a copolymer as its active ingredient. This random copolymer of four amino acids (Y, E, A, K) has a broad immunomodulatory effect. The polypeptide sequence is complex and shows important variability; therefore, the true structure and the mechanisms responsible for its activity are still unclear. The main hypotheses involve interactions with cells which compete with certain major histocompatibility complex (MHC) class II molecules or antagonism at specific T-cell receptors (Figure 16.1D). The polymer has been shown to down-regulate the T-cell response, stimulate neuronal repair and lead to a neuroprotective effect in the central nervous system (CNS).[2]

16.3　Designing Polymeric Ligands

Once the target has been selected, the challenge consists in designing a multivalent ligand that maintains high affinity and selectivity for the target in biological conditions. Two methods, radically opposed in principle, allow these criteria to be met: rationale design and serendipity. In the former, a screening method (*e.g.* phage display,[22] aptamer generation,[49] computational design[27] or constitutional dynamic chemistry (see Chapter 12 for details) is used to select the ligand with the best binding properties; this ligand is then grafted to a polymeric support. In the second approach, the identification of a good candidate is attributable to coincidence, when a molecule is found to exhibit notable affinity for certain substrates. In those cases, which still represent the vast majority of known polymeric binders,[2,3,33] detailed structure–activity relationships are conducted subsequently to optimize the multivalent structure.[31,53,54] Notwithstanding the method leading to the identification of the candidate, key aspects pertaining to the nature of the ligand, the structure of the polymeric support and spatial display can be put forward.

16.3.1　Polyvalent Ligands

Given that many biological processes involve multivalent binding on the cell surface, polymers offer a simple and accessible scaffold to present such multiple binding sites. Indeed, an increasing the number of ligands results in improved affinity along with selectivity provided that one interaction is dominating.[5] This improvement in binding constant by increasing valency does not depend on the type of ligand and has been observed for aptamers,[49] polysaccharides,[55] cationic[33] or hydrophobic[56] repeating units as well as peptide[26] or peptide-like moieties.[2] For example, when comparing monovalent aptamers to their multivalent counterparts, Yang *et al.*[49] observed a 7.5-fold increase in the binding affinity. This also translates into improved efficacy, for example, a 4 orders of magnitude increase in anthrax toxin inhibition.[23] This phenomenon represents positive cooperative interactions of multivalent binding, and can be explained by a favorable enthalpy gain or a decreased entropy cost. In the former case, the binding of one ligand increases the enthalpy of binding of the neighboring ligand. This case, exemplified in monovalent systems by the successive binding of four molecules of oxygen on hemoglobin, is seldom identified in polyvalent systems.[7] In the latter case, the first binding event decreases the entropy sacrifice of subsequent bindings, because the loss of conformational movement is already accounted for in the early interactions. This cost reduction is all the more amplified when the polymeric architecture well matches the structure of the target.[1] This strategy has been largely exploited in inhibitors of bacterial toxins, such as Shiga[57] or anthrax.[27]

Besides using multiple copies of a single ligand, the affinity of a polymeric binder can be improved by adding a second recognition moiety to diversify the type of interactions with the substrate. The overall binding improvement through multivalent attachments between biomolecules has been reported as improved avidity.[7] For example, Tran *et al.*[55] designed a two-headed ligand possessing a galactose moiety and a second variable headgroup, to target the cholera toxin B subunit. The screening of their polymeric ligand library generated a nanomolar inhibitor of cholera toxin (IC_{50} up to 10 nM), while the single galactose-headed ligand did not show any activity. In other cases, combining hydrophobic interactions with electrostatic interactions improved the affinity as well as the biocompatibility of the α-gliadin sequestrant.[58] Finally, bifunctional ligands can be used to trigger the formation of ternary complexes.[21,59] In this case, the polyvalent ligand serves as a binding agent to adequately position two biomolecules, and improves the selectivity of the whole biorecognition process. This is exemplified by Starship, a polyvalent inhibitor for Shiga toxin. Its binding affinity rises from 10^{-3} to 10^{-7} M in the presence of albumin, which acts as a template protein to preorganize ligand display on Shiga toxin.[57]

16.3.2 Polymeric Scaffolds

In single "key–lock" recognition systems such as enzyme-substrate recognition, the 3D conformation of the ligand determines the quality of the binding, for example, the activity of the enzyme. In polyvalent recognition, the spatial display of multiple ligands is crucial to the global avidity for the biological target. Therefore, the challenge resides in selecting the scaffold adapted to the target. Table 16.1 presents some of the diverse structures that have been developed for polymeric binders. Thanks to advances in polymeric chemistry, the synthesis of supramolecular materials with controlled structures has opened the possibility of targeting more and more complex targets.[4]

The simplest way of generating polyvalent ligands is to polymerize the binding unit, which might be derivatized with polymerizable functions (Table 16.1). Acrylamide derivatives of amino acid binding groups[60] or aptamers[49] have been prepared and subsequently polymerized by free-radical polymerization. Ring opening polymerization of β-lactams has also yielded nylon-3 copolymers mimicking natural antibacterial peptides.[39] When the binding ability is due to diverse monomers, random copolymerization has been

Table 16.1 Examples of polymeric binder's scaffolds.

Structure	Example	Variability	Ref.
Poly(ligand)	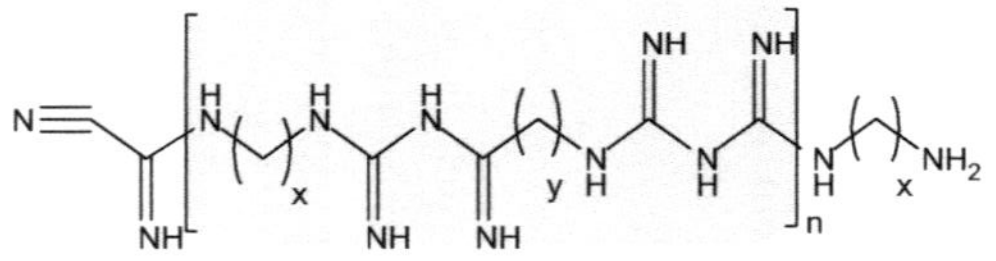	• Molecular weight (n) • Hydrophobic content (x,y) • Spatial display (x,y)	33
Linear copolymer	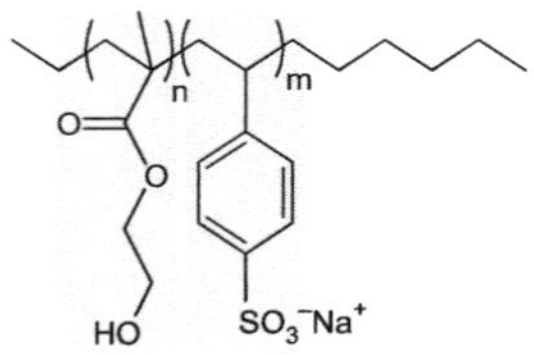	• Molecular weight (n, m) • Ratios (n, m)	59
	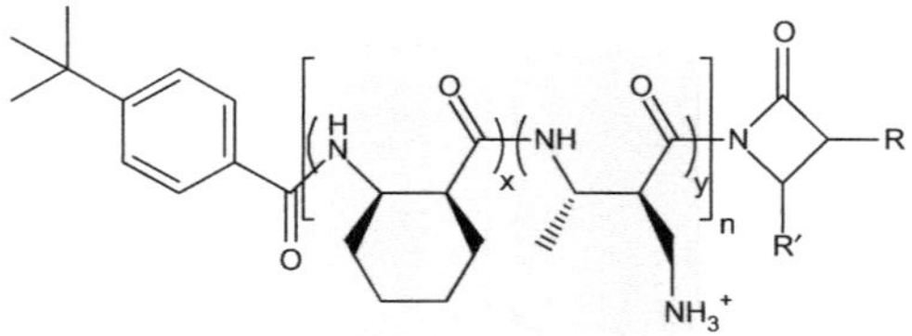	• Molecular weight (n) • Ratios (x, y) • Spatial display (x,y)	40

Cross-linked polymer

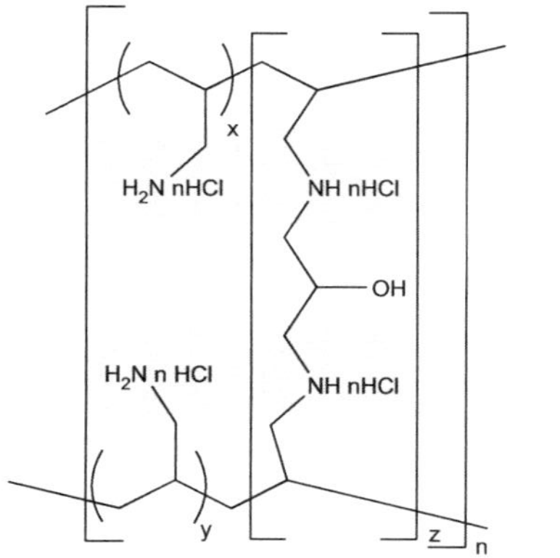

- Molecular weight (n) 97
- Ligands density (x,y)
- Cross-linking (z)

Grafted polymer

- Molecular weight (n,m) 24
- Ligand density (n)

Table 16.1 (*Continued*)

Structure	Example	Variability	Ref.
Heterofunctional polymer	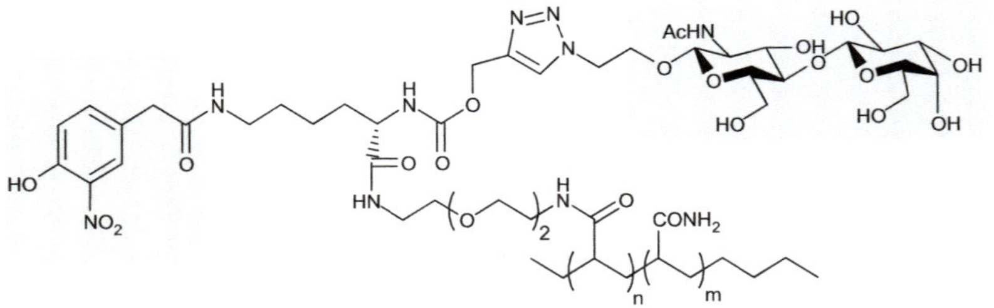	• Molecular weight (n, m) • Ligands density (n)	38
Dendrimer[a]	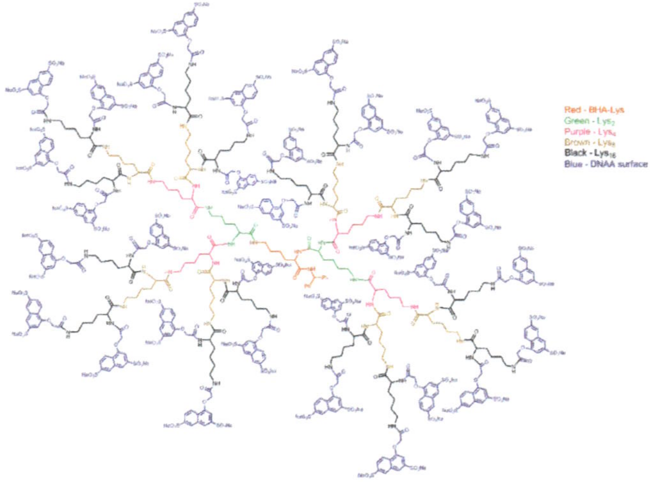	• Generation • Ligand density	31

[a] Reproduced with permission from Tyssen *et al.*[31]

frequently used to yield linear copolymers[2,6,42,56] (Table 16.1). In these cases, the ratio of various monomers is optimized according to the binding affinity and the biocompatibility of the resulting polymers. Through rational design, Hoshino *et al.*[6] demonstrated that an equimolar ratio of hydrophobic (*N-t*-tributylacrylamide) and negatively charged monomer (acrylic acid) resulted in the best activity and biocompatibility to scavenge a toxic peptide. The balance of hydrophilic and hydrophobic moieties is particularly significant for anti-microbial polymers, since it affects the antimicrobial activity as well as the selectivity towards bacteria.[61] Mowery *et al.*[62] have demonstrated that in-creasing the overall hydrophobicity of the polymers has a stronger effect on the hemolytic activity than on the inhibition of bacterial growth. These findings are in agreement with several studies where hydrophobic monomers have to be balanced with more hydrophilic monomers to ensure colloidal stability and biocompatibility.[40,42,56] Moreover, the positioning of the hydropohobic and charged groups within the polymer chain impacts the selectivity towards bac-teria.[61] Passic *et al.*[54] demonstrated that an optimal distance of 6 carbon atoms between the charged groups (biguanides) provided the best charge density and decreased the cytotoxicity of the antimicrobial agent. Finally, crosslinking the linear polymers creates a porous matrix able to trap substrates according to their 3D structure in addition to their affinity. This enables oral scavengers to sequester high quantities of ions or small molecules—up to 8 mmol of bile acids per gram of polymer, in the case of sevelamer (Table 16.1).

Grafting the ligand on polymeric backbones is an alternative strategy to avoid direct polymerization of the ligands (Table 16.1). Activated poly-acrylamides, such as poly(*N*-acryloyloxysuccinimide)[22] are particularly ap-pealing for this approach. If the nature of the backbone has minimal influence on the binding activity, the acrylamide scaffold can be substituted to improve biocompatibility (*e.g.* to *N*-(2-hydroxypropyl)methacryamide[21]) or to ensure biodegradability (*e.g.* to poly(glutamic acid [PGA][23] or polycarbonates[40]). In some cases, the backbone plays a critical role and cannot be substituted. For example, Vicent *et al.*[28] decorated two different hydrophilic polymers, namely poly(ethylene glycol) (PEG) and PGA, with the same hydrophobic, cytotoxic ligand and only the PEG derivative was able to adequately mitigate the ligand's toxicity.

In most cases, the molecular weight of the polymer barely impacts the ac-tivity of the binder.[21,25] Mowery *et al.*[62] pointed out that increasing molecular weight over 30 subunits of their nylon-3 copolymers did not affect the anti-bacterial activity *per se* but significantly increased the hemolytic activity and toxicity of their polymers. Oppositely, the ligand density can be of crucial importance for certain targets. Kane and co-workers demonstrated that in-creasing peptide density on a polymeric anthrax toxin inhibitor improved the IC_{50} until it reaches a plateau.[25] After that threshold, further increase in density did not translate into improved efficacy and even decreased the relative activity of each pendant ligand. The reported transition value was moderate (between 6 and 10 units per polymeric chain) and corresponded to the number of binding sites on the toxin, suggesting a stoichiometric interaction between the binder and the biological target.[23,25] Other groups did not find any impact of the ligand density on the affinity of the binder.[21,22] These last observations have been done at high ligand densities where the saturation might have already been reached, or some binding mechanism impaired such as ion pairing. In summary, the ligand density impacts the number and the spatial organization of the ligands on the 3D structure of the polymer, which determines the avidity for the target. In particular, preorganization of heterofunctional ligands on the polymer side chain has been proven a decisive factor for the formation of ternary complexes.[37,59]

In a similar manner, dendrimers are attractive scaffolds for binders because they present hyperbranched monodisperse architectures, which allow functionalization with a high density of surface groups with relatively precise spatial arrangements. They have been extensively used in drug delivery and biomedical applications[63] and have raised particular enthusiasm when SPL7013 (VivaGel®), became the first dendrimer-based drug submitted to the FDA.[3]

The properties of dendrimers are mainly related to their generation, that is, the number of repeating branching cycles, as well as the nature of their surface ending groups.[45] Expanding the generation increases the molecular weight and the number of terminal groups that can be reacted with ligands. As with grafted polymers, increasing the ligand density generally results in increased activity until a plateau is reached.[44] For example, second or third generation of L-lysine dendrimers was required to inhibit HSV-1 strains; however, dendrimers of the fourth generation exhibited decreased antiviral activity.[31] It has to be noted that the value of the plateau depends on the biological target.[31] Weinhart *et al.*[51] demonstrated that large polyglycerol cores (250 kDa) exhibited better L-selectin inhibition than their smaller counterparts (6 kDa), for an even number of ligands. The increase in molecular weight was associated with improved flexibility and allowed a better interaction with the target. Finally, capping groups determine the surface exposed to the biological target, and therefore their mutual interactions. Tyssen *et al.*[31] demonstrated that surface groups with the greatest anionic charge and hydrophobicity exhibited the best antiviral activity. Increasing the number of surface groups on the same dendritic scaffold also improved L-selectin inhibition.[51]

16.4   Assessing Efficacy

The type of experiments required to assess the efficacy of newly-designed polymeric binders are as diverse as the number of possible targets. Like drug candidates with small molecular weights, the effectiveness of the approach must be evaluated both *in vitro* and *in vivo* during preclinical development.

Polymeric binders are bound to be employed in biological environments containing ions, small molecular weight solutes, proteins and cells. These entities can also bind the polymer or hinder accessibility of the substrate to the docking sites. To successfully design a macromolecular binder for pharmaceutical applications, it is essential to ascertain that non-specific interactions are minimal and do not interfere with the binding capacity of the molecule.

Specificity is therefore intimately related to efficacy. A lack of specificity can also translate into safety concerns as the binding of off-target substrates can trigger toxicities.

Assessing the efficacy and specificity of a polymeric binder is challenging, because very few models and standard procedures are available. In the following section we propose a summary of diverse approaches that have been used for the multiple possible therapeutic classes of polymeric binders.

16.4.1 *In Vitro* Experiments

16.4.1.1 *Macromolecular Scavengers*

The *in vitro* proof-of-concept for polymeric scavengers can be established by measuring the binding properties of the polymer with various concentrations of the substrate. Further information about the specificity can be obtained by co-incubating the polymer and its target with potentially interfering substances. The binding can be measured by qualitative (*e.g.* turbidity,[56] dynamic light scattering [DLS][28]) or quantitative approaches (*e.g.* UV/visible spectrophotometry,[8] isothermal calorimetry,[64] surface plasmon resonance [SPR],[64] Fourier-transform infrared spectrometry [FTIR], quartz crystal microbalance[6]). For example, the peptide binding properties of the α-gliadin binder poly(HEMA-*co*-SS) can be visualized by observing the precipitate obtained by mixing simple solutions of polymer and α-gliadin[56] (Figure 16.2A, *top panel*). The colloidal complexes formed can be further characterized by circular dichroism and DLS measurements.[58] For binders intended for oral administration, because both the polymer and the substrate might have different ionization/solubility properties, the interactions must be observed at different pHs to account for the range of environments encountered during transit in the GI tract (*i.e.* from 1.2 to 6.8). However, for scavengers intended for parenteral administration, a single pH around 7.4 is usually suitable.

Polymers do not always form an insoluble, turbid complex upon binding to their target. Sometimes, the association constant between the binder and the substrate can be monitored directly by infrared-spectrometry, isothermal titration calorimetry or SPR.[64] However, these techniques are perfected to study one single substrate with one specific binder, and might be harder to implement in biological environments, which can contain interfering substances. In those cases, it might be easier to follow the free substrate instead of the formed complex. The amount of bound molecules can be calculated from the remaining quantity of free substrate. Among others, gel electrophoresis,[56] centrifugation,[35,56] (ultra)filtration,[65] dialysis[64] or size-exclusion chromatography[66] are ways to separate the binder–substrate complexes from the free molecules. The concentrations in individual fractions can then be measured by conventional analytical methods such as UV/visible spectrophotometry, high performance liquid chromatography or mass spectrometry. Finally, *in silico* analysis has also been utilized to optimize polymer design and determine the potential affinity of a binder for its substrate.[27] While this method might be

A. Visualizing the binding

Aggregation with the substrate

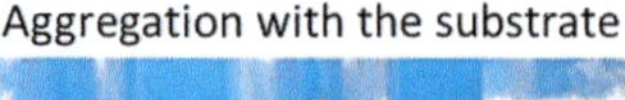

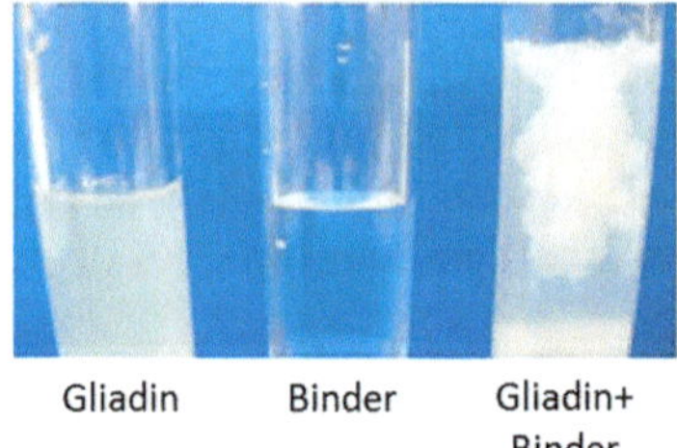

Solubilizing the cell wall

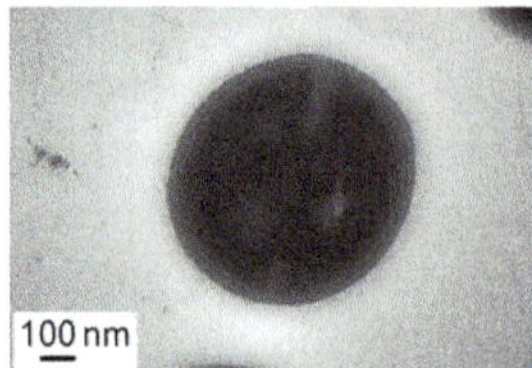

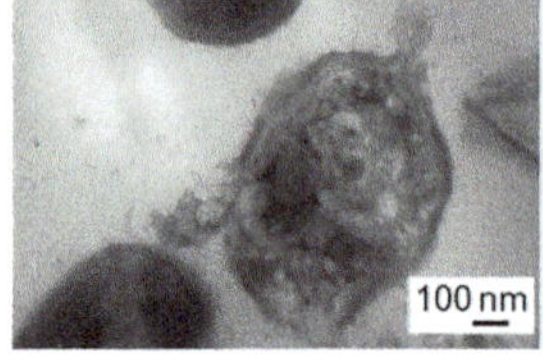

E. Faecalis before...   ... and after incubation with the binder

B. Measuring the selectivity

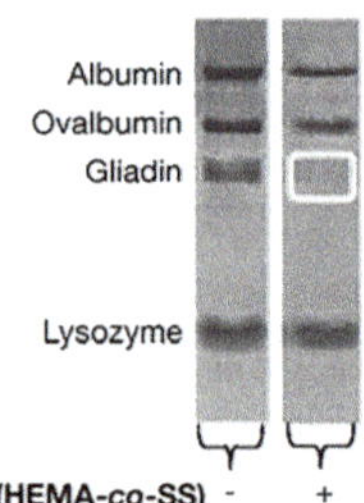

Protein	Binding (%) (±SD)
Gliadin	67±3
Ovalbumin	4±6
Lysozyme	< 5
BSA	18±3
IgG (heavy chains)	19±6
Myoglobin	6±6
Casein	< 5
Thrombin	19±5
Collagenase	27±8
Pepsin	12±9
Trypsin	< 5

Figure 16.2 (**A**) *In vitro,* the interactions between a polymeric binder and its substrate can be visualized in different ways. *Top panel*: the α-gliadin binder forms an insoluble precipitate when bound to the protein (reproduced with authorization from Pinier *et al.*[56]). *Bottom panel*: a polymer lyses the cell wall of bacteria upon co-incubation (reproduced with authorization from Nederberg *et al.*[40]). (**B**) The selectivity of a binder will determine how it will maintain its binding capacity when in the presence of *off-target* molecules (reproduced with authorization from Pinier *et al.*[56]).

interesting to increase the screening throughput and identify attractive candidates early in the development process, it strongly depends on the artificial parameters chosen for calculations and can hardly account for all the non-specific interactions that can compete for the binding sites in biological samples.

Whatever their type, polymeric binders need some degree of specificity to be efficacious: locally applied binders must avoid being saturated by the physiological bacterial flora; orally administered polymeric scavengers should not interfere with the digestive and metabolic functions of GI enzymes; and systemic treatments must not trigger adverse effects by interacting with the wrong substrates. The binding properties must therefore be evaluated in presence of these potentially interfering substances. Pinier *et al.*[56] monitored the binding properties of their polymer in presence of different proteins to account for any potential interference. Although other proteins were bound by the polymer, the observed binding efficacy of α-gliadin was 4 times that of albumin (Figure 16.2B).

One way of assessing the efficacy of polymeric scavengers in complex media is to design an *in vitro* experiment where the functional properties of the substrate are monitored directly. For example, the hemolytic properties of toxic peptides can be monitored in presence and absence of the detoxifying binder.[6,67] Although working with hemolysis as an endpoint has its limitations (see Section 16.4.3 for more details), this approach has the advantage of observing a direct toxic manifestation of the substrate and provide a frank evaluation of the protecting effect of the binder. A similar approach was applied when working with scavengers of cytotoxic toxins, and the half-maximal inhibitory concentration (IC_{50}) of the binder after pre-incubation with the toxic substrate can be determined.[20,24,59] In more relevant experimental settings, the IC_{50} is obtained by putting the substrate in contact with the cells and the polymer at the same time. In those cases, the concentrations of binder required to mitigate the substrate's toxicity are usually higher, as the polymer needs to compete with the cells for the binding of the toxin.

16.4.1.2 Surface Binders

Polymeric surface binders differ from polymeric scavengers because they interact with substrates that are at least two to three orders of magnitude larger. In opposition to smaller molecules (ions, peptides, proteins), the targets (viruses, bacteria, cells) possess heterogeneous surfaces with multifaceted and evolving patterns of epitope presentation. Due to their complex pathological pathways, designing *in vitro* experiments relevant to their *in vivo* behaviors remains a particular challenge.

Microbicide polymers are among the more common examples of macromolecular surface binders.[44] SPL7013 is one example of a microbicide dendrimer, initially designed to prevent viral infection of healthy cells by human immunodeficiency virus (HIV-1) and herpes simplex virus (HSV). The *in vitro* assessment of the polymer efficacy was first established against HSV types 1 (HSV-1) and 2 (HSV-2) in cell culture.[68] In these experiments, Vero cells (a strain of primate kidney cells commonly used to study viral infection) were inoculated with both types of HSV and it was established that solutions of dendrimer were able to abolish viral adsorption and cell entry, at concentrations above 3 μg mL^{-1}. This study also showed that the macromolecule could treat already infected cells, even when the viral strains were resistant to common antiviral therapies. The preventive effect of different concentrations of the dendrimer formulated in carbopol gel was also demonstrated for the infection by HIV-1 of peripheral monocytes, macrophages, urogenital and colorectal cells.[69]

Other polymeric microbicides are designed to bind and permeabilize the cellular membrane of bacteria.[39–41,43] This approach is attractive because it is believed that it is harder for the pathogens to develop resistance to these membrane-disruptive properties. The *in vitro* evaluation of this kind of polymer consists in determining the minimal inhibitory concentration that prevents the bacterial growth, and comparing it to that of common antibiotics.[39–41,43] To further ascertain the mechanism by which the polymers exert their effects, the bacteria are usually imaged in the presence or absence of the binding polymer[40,43] (Figure 16.2A, *bottom panel*). In most cases, distinct irregularities and defects are detectable in the cell membrane, confirming the disruptive mechanism. The membrane integrity can also be evaluated by exposing the bacteria to macromolecular fluorescent dyes that do not permeate through intact barriers.[43]

When polymeric binders target cells, flow cytometry, where the fluorescence signal on single cells is monitored, has proven a useful tool to quantify the amount of polymer bound to cells. This approach helped Yang and co-workers to identify polymeric aptamers exerting cell-specific cytotoxicity, by screening which candidates showed the most specific interactions.[49] In applications where the binder is expected to kill its cell substrate, conventional assays measuring cellular viability are uncomplicated ways of measuring the effect of the binder.[49] To further evaluate which pathway is responsible for cell death (apoptosis or necrosis), more specific tests evaluating the loss of membrane asymmetry,[49] the presence of apoptotic enzymes or the fragmentation of DNA can be used.[47]

16.4.2 *In Vivo* Experiments

The experience from development of phosphate and cholesterol binders offers a solid basis for the design of *in vivo* experiments to evaluate the performance of orally administered polymeric scavengers. The decrease in absorption can unmistakably be evaluated by monitoring the serum concentrations.[70] Nevertheless, the most important thing is to limit the pathological manifestations secondary to the initial overload. Hence, sevelamer hydrochloride can decrease the oral absorption of phosphate, but more importantly, it restores parathyroid hormone balance and reduces organ calcification.[70] Similarly, the α-gliadin binder can certainly prevent digestion of its substrate into the inflammatory peptides, but most notably it protects against immunogenic reactions and cell damage.[56] In α-gliadin-sensitized mice challenged with gluten, the polymeric binder prevented alteration of the GI barrier (Figure 16.3, *top panel*), hindered the recruitment of inflammatory cells and the replication of splenocytes.[17,56]

The evaluation of the performance of systemically injected scavengers *in vivo* can be challenging. In theory, colloidal scavengers should be able to modify the blood circulation kinetics of the toxins and change their biodistribution.[66,71] Given that most toxins have very short plasmatic half-lives when injected

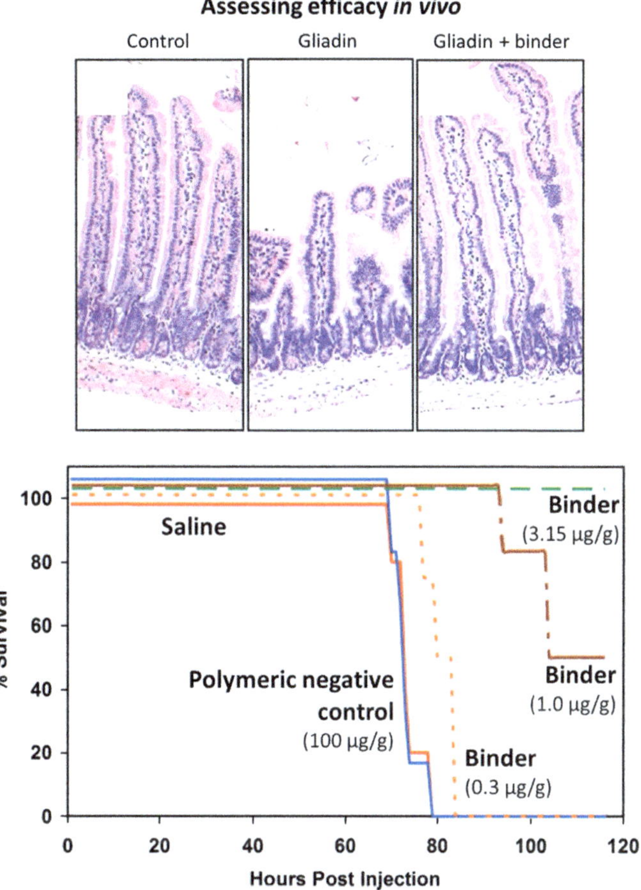

Figure 16.3 The ideal way of assessing the efficacy of a polymeric binder *in vivo* is directly monitoring the physiological manifestations of effective binding. *Top panel*: a gliadin binder prevents the histological abnormalities in intestinal epithelial cells induced by long-term administration of gluten in gliadin-sensitized mice (reproduced with authorization from Pinier *et al.*[17]). *Bottom panel*: a macromolecular binder prolongs survival in mice challenged with *Shiga* toxin in a dose-dependent manner (reproduced with authorization from Kitov *et al.*[59]).

intravenously, the time lapse for interactions between the binder and its substrate is limited. The usual experimental design, where the toxins are administered after pre-incubation with their binder,[24,59] overestimates the protective effects of the binder by enhancing the interactions between the polymer and its target in a protein-free, non-diluted environment. In a more realistic manner, the toxin can be administered intravenously, then quickly followed by the scavenging binder (within minutes).[6] In other instances, the toxin can be injected subcutaneously and followed by an intravenous injection of the polymeric binder (Figure 16.3, *bottom panel*).[59] In that case, the delayed distribution of the toxin from the subcutaneous space to the systemic circulation more closely mimics the development of infection and the slow release of the toxin in the body. It also provides increased opportunities for the long-circulating binder to interact with the substrate in the bloodstream.

For ethical reasons and to minimize distress and discomfort to the animals, the endpoints of toxicity experiments are based on which animals are moribund and must be euthanized.[24,59] Nevertheless, definitive toxicity criteria should be precisely defined *a priori* (*e.g.* loss of >20% of initial weight, lesions larger than 10 mm^2) and ideally be evaluated by a blinded researcher, to diminish the subjectivity and minimize potential biases.

16.5   Establishing Innocuousness

During preclinical development, the toxicological profile of new biomaterials must be established. In complement to the efficacy data, the toxicology studies will help ascertain the potential of the material for a specific clinical application. Factors like the severity of the target disease and the availability of other treatments will help researchers and authorities to establish the benefit-to-risk ratio of using the material in humans. Untreatable, deadly conditions might better tolerate treatments with more adverse effects than chronic diseases for which many alternatives exist. The administration routes, frequency of dosing and doses used for a specific therapeutic application will also dictate the degree of innocuousness required. For example, antidotes which are used punctually usually necessitate higher doses, but present less chronic toxicity concerns. The requirements for the duration of preclinical toxicological studies stipulate that toxicological data will have to be obtained from both rodent and non-rodent animal models before proceeding to exploratory clinical phases in humans.[72,73]

16.5.1   Mucosal Application

Macromolecular microbicides used to protect mucosae from bacterial or viral aggressions[3] are intended to interact with epithelia for extended periods of time. During the development of such polymeric binders, the impact of lengthy tissue exposure to the new material must be thoroughly ascertained.

Assessing the local toxicity upon mucosal application of the binder is reasonably straightforward. *In vitro* cytotoxicity studies can help determine the concentrations at which co-incubation of the polymer with cells will start to have a deleterious effect on their survival.[68] *In vivo*, the effect of the mucosal application on the histology of the epithelia can be surveyed after one or more applications. Toxicological studies on the antiviral dendrimer SPL7013 in non-human primates examined the impact of four daily vaginal or rectal applications of increasing amounts of the dendrimer on mucosal pH and microbiologic flora.[74] Comparison with placebo hydrogels helped establish the safest dendrimer concentration (*i.e.* 3% of SPL7013) to use in subsequent clinical investigations. The safety and tolerability of the gels were confirmed in ensuing clinical trials.[29]

16.5.2 Oral Administration

The GI barrier allows the absorption of food and nutrients, while blocking the penetration of harmful compounds and pathogens. Orally administered binders can exert toxicity by interfering with these barrier functions.

When a binder is used for the sequestration of its substrate in the GI lumen, it is essential to confirm that the polymer does not cross the epithelium. In the context where a polymer would be administered punctually, a permeation study on a high dose of polymer can be conducted.[75,76] The lack of GI absorption can be evidenced by tracking the amounts of radiolabeled polymer in the excreta (feces and urine) after one episode of gavage. For example, in the case of the poly(SS-*co*-HEMA) α-gliadin binder, more than 95% of the administered dose was retrieved in the feces over 72 hours and only minimal amounts ($< 1\%$) were isolated from the urine, the blood, or the liver and kidneys in both mice and rats.[17]

When more than one administration of the polymer is expected, the impact of chronic exposure on the barrier properties of the GI epithelium must also be ascertained. The non-absorption of the macromolecular binder must also be confirmed after chronic dosing for sustained periods (*e.g.* up to 28 successive days).[13,16] Finally, in certain cases (*e.g.* celiac disease), the disease itself can trigger inflammation and enhanced permeation of the GI mucosa. For example, the absorption of the α-gliadin binder was investigated in gluten-sensitized HCD4/DC8 mice with defective GI epithelia and showed similar results as those obtained in healthy animals.[17]

Non-specific binders can also impair absorption of essential nutrients by preventing their diffusion through the GI barrier. This is particularly un-desirable in the case of treatments designed to be administered chronically. Nevertheless, directly measuring the influence of a compound on the absorption of nutrients in general is not an easy task. In some cases, specific interactions can be predicted *a priori* because of the chemical structure of the binder. For example, the anionic nature of sevelamer hydrochloride raised concerns that the resin might interfere with oral adsorption of positively-charged minerals (*e.g.* iron, cobalt, zinc, copper) or vitamins (*e.g.* vitamins B6, B12, C, K and folic acid).[65] When simple aqueous solutions of these nutrients were incubated with the polymer, some *off-target* binding was detected *in vitro*. Because the resin was already approved and widely used in patients, the impact of these potential interactions could be directly evaluated in a clinical setting.[77] These findings now act as warning signs to better define the future usage of the binder in patient cohorts.

In contrast, in the context of preclinical development, when non-specific binding with some nutrients is detected *in vitro*, studies must be designed to monitor the systemic exposure to those nutrients and ascertain the physio-logical relevance of the interactions. For example, the influence of the acute and chronic GI administration of polystyrene nanoparticles on the absorption of iron was investigated in chicken.[78] This study correlated the results obtained *in vitro* on the transport of iron through a cell monolayer (*i.e.* a model mimicking the intestinal epithelial layer) with *in vivo* observations showing that the oral administration of negatively-charged particles affected the kinetic of absorption of a ^{58}Fe-enriched solution. Although the long-term consequences of such interactions are still not clear, this study highlights the complexity involved in designing *in vivo* experiments to evaluate the influence of a material on the GI adsorption of a specific compound.

It is much more straightforward to indirectly measure the effect of the binder administration on the general homeostasis of the animal. For example, Pinier *et al.*[56] monitored the effect of their α-gliadin binder on the adsorption of nutrients by monitoring the weight gain and the blood biochemistry of healthy mice after a chronic, daily administration of the polymer (15 mg kg^{-1} per day) over a period of 22 days. The similarities of the parameters observed in the treated and control groups confirmed that the polymer did not cause flagrant toxicities or interfere with nutrient absorption in a critical way. Nevertheless, the sensitivity of this type of study directly depends on the length of time that the animals are observed; lesser toxicities might not develop within the observed time frame. For example, metabolic acidosis and GI tract necrosis upon release of the protons of sevelamer hydrochloride were observed in some patients only after commercialization of the resin. This side-effect led to the development of carbonate-buffered sevelamer carbonate[79] and other phosphate binders.[80]

16.5.3 Systemic Toxicities

Polymeric binders injected systemically can interfere with physiological function, just like oral scavengers. However, this administration route being less accessible and practical, polymeric binders designed for systemic injections are usually intended to treat acute and/or life-threatening conditions (*e.g.* exposure to a toxin, bacterial infection or cancer). Therefore, risk–benefit evaluations might be more tolerant to potential toxicities.

Here again, a general way of evaluating acute toxicity is to monitor the body weight and blood biochemistry of animals after one injection or more.[67] For instance, Nederberg *et al.*[40] confirmed that the administration of their antimicrobial polymer did not trigger severe systemic toxicities by monitoring the blood biochemistry: aspartate and alanine aminotransferases (AST, ALT), creatinine and urea nitrogen levels, as well as potassium and sodium ion balance. Major discrepancies in these indicators would denote important liver (AST, ALT) or kidney toxicities. In that case, no significant variations from baseline were detected at 2 and 14 days after injection.

Besides, histological evaluations of target organs can also be carried out.[6] These toxicological experiments largely depend on the number of animals followed, the duration of the monitoring and the choice of organs for evaluation; they are mostly reliable to detect major and obvious toxicities. Using more than one dose of polymer might help broaden the range of detectable adverse effects and studying the biodistribution of the polymeric binders might identify which tissues are more at risk of toxicities.

Polymers that act as surface antagonists bind substrates of larger sizes; possible interactions with off-target cells must also be considered. For example, polymeric antimicrobial agents synthesized to disrupt bacterial membranes must demonstrate that their lytic effects are restricted to pathogens and are not conveyed to healthy cells. Measuring the degree of hemolysis caused by these polymers is a frequent and straightforward way of showing the specificity for the polymers' interactions.[39,40,42,43] Nevertheless, the release of hemoglobin—a 64 kDa, colored tetramer protein present in the red blood cells (RBCs)—is sometimes an oversimplistic model of cell damage. First, the RBCs of different animal species are more or less susceptible to hemolysis,[81] complexifying the correlation with human data. Second, hemoglobin can be released by osmotic shock in a hypotonic environment, even in the absence of hemolytic molecules. Third, hemoglobin release is not necessarily a primary indicator of cell membrane damage. For example, in mammal cells, the alpha-toxin of *Staphylococcus aureus* creates pores with a diameter of 1–2 nm and a cut-off between 1 kDa and 4 kDa.[81] These pores being too small to let hemoglobin leak out of the cells, hemolysis is a consequence of the osmotic shock that follows the unrestricted passage of small molecular weight solutes through the membrane. Hence, to supplement the hemolysis assay, researchers routinely use alternative methods, like visually assessing the association of the polymer with the cell membrane by fluorescence[42] or studying the release from model liposomal vesicles.[39]

Finally, when designing systemically administered polymeric binders, the metabolism and elimination pathways of the macromolecules must also be considered. In the simplest systems, the polymers are excreted unchanged in the urine after being filtrated by the kidneys. This avoids their accumulation in the body after repeated dosing. The maximum size at which macromolecules can be filtered depends on their hydration, flexibility and architecture.[82,83] Although some reports have determined the glomerular *cut-off* of specific macromolecules and can be used as caveats,[84–88] the elimination of each new polymeric binder must be studied independently. In most cases, it is sufficient to show that the most promising candidate is excreted in the urine in adequate amounts after systemic injection.[8] The glomerular filtration threshold is not as critical for polymers, which can be biodegraded into small molecular weight oligo- or monomers (*e.g.* polypeptides or polyesters).[40,43] However, in that case, the toxicity and the fate of all the degradation products must be accounted for.

16.6 Challenges to Clinical Development of Polymeric Binders

16.6.1 Clinical Examples

Polymeric binders present particular challenges in terms of clinical development. These drugs are defined as New Molecular Entities (NMEs) and must comply with the same regulations as small molecules both from efficacy and safety assessment points of view. These NMEs are usually classified as systemic or non-systemic agents: systemic being defined clinically as being absorbed, reaching and circulating in the bloodstream, and distributing in peripheral tissues. Since most polymeric binders currently in clinical trials belong to the second category, this section will mainly focus on the clinical development of these various non-systemic compounds.

Non-systemic polymeric binders include the unique subset of orally administered agents that exert their therapeutic effects locally in the GI tract but affect overall homeostasis systemically. They are designed to exert intraluminal effects in the GI tract, either to prevent the absorption or enterohepatic recycling of endogenous elements, or to interact directly with GI mucosal constituents. Early examples of non-systemic drugs are cation and anion exchange resins that remove bile acids or potassium ions from the lumen of the gut for the treatment of hypercholestemia[89] and hyperkalemia.[11] The development of non-absorbable drugs has expanded greatly and now encompasses a variety of agents, polymers, small molecules, or peptides used to treat systemic metabolic and mineral disorders, as well as diseases of the GI tract.[90] Furthermore, non-systemic binders include topically active drugs acting on the skin or mucosa, preventing the deleterious effects of pathogens. Compounds designed to exert only local effects need to show they are not (or minimally) absorbed through the skin or mucosal epithelial barrier. Non-absorbable drugs including polymeric binder offer the advantage of minimizing off-target systemic effects and thereby greatly reduce the risks of drug–drug interaction, toxicity, or side-effects.[91] As a result, several innovative products at various stages of clinical development exploit increasingly sophisticated approaches, such as SBR159,[92] AST-120,[93–95] PA21,[80] and CLP[96] as presented in Table 16.2.

The binding capacity and selectivity of non-absorbable drugs are keys to their therapeutic success. Cholestyramine or colestipol could exemplify this situation, since therapeutic doses of this bile acid sequestrant are in the multi-grams range (*i.e.* 12–16 g per day).[9] More recent compounds have greater potency; colesevelam is an anion-exchange resin with a seven-fold higher bile acid-binding capacity and fewer side-effects than cholestyramine.[12]

16.6.2 Phase I

A challenge of the clinical development of this type of drug is the interaction between the binder and the physiological intestinal pathways. For example, the suboptimal clinical efficacy of phosphate binders is caused at least in part by the competition from active intestinal phosphate transport.[11] Alternative strategies target the intestinal sodium/phosphate co-transporter NaP2b (SLC34A3), expressed throughout the upper GI tract and mediating about half of dietary phosphate uptake.[97] Because NaP2b is expressed in other tissues, it is imperative that the inhibitor be non-absorbed and maintained within the intestinal lumen. With a lower clinical dose than current phosphate binders, this new class of phosphate lowering drugs allow its administration to a larger

Table 16.2 Polymeric binders in clinical development, as of December 2012.

Drug name or code	Indication	Stage	Company	Comments	Ref.
AST-120	Indoxyl sulfate intoxication in patients with chronic kidney disease (CKD)	Phase IV	Ewha Womans University (South Korea)	Orally administered, non-absorbable, spherical carbon adsorbent, which removes indole, the precursor of indoxyl sulfate, a uremic toxin	93
AST-120	Mild hepatic encephalopathy in patients with liver cirrhosis	Phase II	Ocera Therapeutics (USA)	Orally administered, non-absorbable, spherical carbon adsorbent, which removes ammonia and bile acids	94
AST-120	Pouchitis and irritable bowel syndrome	Phase II	Ocera Therapeutics (USA)	Orally administered, non-absorbable, spherical carbon adsorbent, which adsorbs various substances, such as serotonin, histamine, bile acids	95
SPL7013 (Vivagel®)	Bacterial vaginosis	Phase III	Starpharma (Australia)	Vaginally-administered, non-absorbable poly(lysine)-based dendrimer, which retains HIV-1 and HSV-2 antiviral activity	30
SBR759	Hyperphosphatemia in patients with CKD on hemodialysis	Phase III	Novartis (Switzerland)	Orally administered, calcium-free, iron(III)-based polymeric binder, which removes phosphate	92
PA21	Hyperphosphatemia in patients with CKD on hemodialysis	Phase III	Vifor Pharma (Switzerland)	Orally administered, chewable, iron-based polymeric binder containing a mixture of polynuclear iron(III)-oxyhydroxide, sucrose and starches, which removes phosphate	80
AMG223 ASP1585	Hyperphosphatemia in patients with CKD on hemodialysis	Phase III	Amgen (USA) Astella Pharma (Japan)	Orally administered, non-absorbable polymeric binder, which removes potassium	US NIH NCT01742585
Colestilan Colestimide Cholebine (BindRen®)	Hyperphosphatemia in patients with CKD on hemodialysis or peritoneal dialysis.	Phase III	Mitsubishi Tanabe Pharma Corp. (Japan)	Orally administered, non-calcium, non-metallic polymeric binder, which removes phosphate. It is also recognized to bind bile acids and potassium.	11

Table 16.2 (*Continued*)

Drug name or code	Indication	Stage	Company	Comments	Ref.
Colestipol (Collestid®)	Erythropoietic protoporphyria (mild form of porphyria)	Phase II/III	Brigham and Women's Hospital (USA)	Orally administered, non-calcium, non-metallic polymeric binder, which binds porphyrin.	US NIH NCT01422915
RLY5016	Hyperkalemia in patients with diabetic neuropathy and CKD	Phase IIb	Relypsa (USA)	Orally administered, high-capacity cation polymeric binder, which removes potassium	107
RLY5016	Hyperkalemia in heart failure patients	Phase IIb	Relypsa (USA)	Orally administered, high-capacity cation polymeric binder, which removes potassium	107
CLP	Hyperkalemia & hypernatremia in patients with congestive heart failure (CHF) and with CKD	Phase II	Sorbent Therapeutics (USA)	Orally administered, non-absorbable polymeric binder, which removes potassium, sodium and fluid	96
CLP	Hyperkalemia in patients with CKD, end stage renal disease	Phase II	Sorbent Therapeutics (USA)	Orally administered, non-absorbable polymeric binder, which removes potassium	US NIH NCT01265524
CLP	Hypernatremia in patients with CHF (Fluid overload)	Phase II	Sorbent Therapeutics (USA)	Orally administered, non-absorbable polymeric binder, which removes sodium and fluid	US NIH NCT01736735

pre-dialysis population of patients with chronic kidney disease.[98] Such considerations are important in the design of phase II proof-of-concept studies, since they are usually dose-ranging.

Non-specific polymeric binders may interfere with essential nutrients, induce electrolytes disturbance and impair their normal absorption (see Section 16.5.2). Binding to fatty acids, vitamins, and hormones can lead to malnutrition and pathological disturbances. Consequently, it is important to monitor blood levels of these nutrients, whenever feasible, as well as the overall nutritional status of patients in case of long-term clinical trials. Binders have also the potential to interact with other drugs, forming insoluble and non-absorbable complexes, for example cholestyramine and digoxin, which may modify the bioavailability of the concomitant drug.[99] Human proof-of-efficacy phase IIa studies should pay attention to potential safety risks, using strict exclusion criteria and allowing no concomitant drugs. Prior *in vitro* binding studies should also be carried out with physiological, endogenous intraluminal constituents, to find out if interactions can take place (see Section 16.5.1).

Another particular aspect of developing polymeric drugs that act in the GI tract is the potential mucosal toxicity of these compounds. Prolonged treatment duration may lead to cumulative toxicity, translating into chronic inflammation, impaired absorption of nutrients, *etc.* Direct physical damage of the GI mucosa has been reported with Kayexalate®, for example. Kayexalate® beads present as irregularly shaped particles, mixed with sorbitol, an osmotic laxative, to favor transit and mitigate constipation. Cases of potentially fatal intestinal necrosis and other serious GI adverse events (bleeding, ischemic colitis, perforation) have been reported in association with the use of Kayexalate®.[11] These observations suggest that during early during clinical development, direct assessment of potential mucosal damage should be assessed using upper and lower endoscopy.

16.6.3 Pharmacokinetic–Pharmacodynamic Relationship

The Pharmacokinetic–Pharmacodynamic (PK-PD) relationship is another important dimension in the clinical development of polymeric binding drugs.[100] For non-systemic drugs, biological fluid refers to the intestinal lumen, a body compartment usually not explored in drug development. However, understanding the PK-PD relationship is critical for optimizing drug performance, dosage regimen, and formulation. For example, agents blocking absorption or digestion of dietary components must be bioavailable in the upper part of the gut and provide optimal exposure to the gut target, whether apical or luminal. As opposed to systemic targets continuously perfused by biological fluids and drug at relatively constant composition, the lumen of gut is highly heterogeneous. Pockets of fluids revealed by magnetic resonance imaging (MRI) techniques showed rapid changes in luminal free water as a function of diet composition or bowel disorders.[101] For non-systemic small molecule drugs, interaction with chyme components can reduce drug availability and potency,[102] a situation analogous to that of systemic drugs binding to plasma proteins. Thus, an early, good understanding of the PK-PD profile of a new polymeric binder, explored during Phase I clinical development is essential in the design of Phase II and III clinical trials.

16.6.4   Phase II and III

Another challenge during clinical development of polymeric binding drugs is the design of human efficacy proof-of-concept Phase II trials. Typically, such a study should provide early evidence that the NME is truly effective when compared to a matching placebo.

Several approaches have been reported for directly neutralizing *Clostridium difficile* toxin activity in the intestinal tract using multi-gram quantities of anion exchange resins (cholestyramine, colestipol). Cholestyramine, a cationic resin that binds to *C. difficile* toxins, has been used as a treatment for *C. difficile colitis* in some patients. This resin has shown only modest activity and is not recommended for use in patients with severe cases of the colitis. Cholestyramine also binds to vancomycin and must be dosed separately if it is used in combination with this antibiotic. In a rat ileum *C. difficile* toxin A challenge assay, cholestyramine showed weak activity compared to linear high molecular weight polystyrene sulfonate, the toxin-binder,[103] which binds toxin A with relatively high affinity, *via* a combination of electrostatic and hydrophobic forces.[104]

Despite encouraging results in early clinical trials, neither Synsorb pk nor Tolevamer toxin-binders had confirmed clinical efficacy in later stage development to prevent hemolytic uremic syndrome (HUS) in patients infected with *Escherichia coli*. In spite of the promising advancements of polymeric binders, concomitant antibiotics like poorly-absorbed fidaxomicin and vancomycin remain today the best option to treat *C. difficile* infection.[105] Another illustration of the difficulty to extrapolate Phase II data to the general patient population is RLY5016, a polymeric scavenger designed to bind potassium ions in the GI tract and thereby reduce serum potassium levels. In clinical practice, where patient selection and serial monitoring of serum potassium may be less rigorous, the incidence of hyperkalemia is increased, at times resulting in kidney failure and death.[106] To anticipate the potential risk of inducing hyperkalemia in patients with concomitant chronic kidney disease (CKD), a clinical study has explored a dosage lower than the therapeutic target or maximally-tolerated doses. The aim of this study is to determine the efficacy, safety and tolerability of RLY5016 in a prospective randomized double-blind pilot study in patients receiving standard therapy for heart failure, who were initiating aldosterone antagonists therapy with spironolactone.[106]

As exemplified by these clinical failures, the choice of primary efficacy endpoint is crucial and should be done according to the therapeutic goal of the drug candidate. For example, the assessment of a bile sequestrant requires a complete lipid profile as an endpoint, to ensure that binder doesn't interact with intraluminal chylomicrons or pancreatic enzymes, which may lead in turn to unexpected variations of plasma triglycerides.[107] Moreover, sometimes a surrogate marker is the ideal primary efficacy endpoint, especially when long-term therapy is needed to show significant results, or too many patients would be needed to obtain an adequate sample. A good example is celiac disease where there is a serious lack of surrogate markers to be used as primary or secondary endpoint.[108] Although BL-7010 (licensed to BiolineRx)[56] is only at the preclinical stage, it appears to be a promising non-absorbable polymeric binder to neutralize α-gliadin. For its clinical development, the evaluation of the intestinal mucosa integrity by measuring the villus height–crypt depth ratio remains the most appropriate clinical outcome measure.[109]

Recognizing that the main pathway for sodium re-import present throughout the GI tract is the Na/H antiporter 3 (NHE3) has led to the development of highly potent and non-systemic NHE3 inhibitors as an alternative approach to managing sodium overload. Representative of this class is RDX5791, a molecule that modulates the uptake of intestinal sodium from dietary or endogenous sources. A Phase II clinical study carried out on this compound used Complete Spontaneous Bowel Movement as the primary efficacy endpoint, since sodium concentrations affect intestinal motility (NIH Registration NCT01340053).

The selected endpoint is similar to the one that should be used in Phase III. In the latter case, there is no surrogate marker that could be used instead of the classical irritable bowel syndrome (IBS) clinical endpoint selected, imposing a sample size of almost 200 patients. Following this example, target indications of polymeric binders could be extended to chronic GI diseases. AST-120 (spherical carbon adsorbent) can adsorb small-molecular-weight toxins, inflammatory mediators, and harmful bile acid products, that are liable to induce IBS and pouchitis.[95] Classical Pouchitis Disease Activity Index (PDAI) is used as the outcome measure, and AST-120 seemed to be efficacious in the induction of remission or response in patients with active pouchitis.[95]

16.7 Conclusion

Since their discovery, non-absorbable polymeric binders have been privileged to systemically-absorbed binders in drug development. While non-absorbable molecules were historically designed to treat local GI disorders, non-absorbable binders show great potential to treat systemic diseases or conditions. Several examples clearly demonstrate that a non-absorbed binder can induce systemic electrolytes regulation or prevent the absorption of putative substances causing a whole body disease. At the same time, rational design of the polymeric structure helps to optimize the selectivity and to limit off-target effects. The recent development of such polymeric binders for various applications has required the refinement of the traditional assessment methods. Innovative methods for screening selectivity of polymeric binders, in preclinical as well as clinical development, will be required to ensure the safety and efficacy of these new macromolecular drug entities.

Acknowledgements

This work was partly founded by Natural Sciences and Engineering Research Council of Canada and the Faculty of pharmacy of University of Montreal. N.B. acknowledges a postdoctoral fellowship from the Canadian Institutes of Health Research (CIHR).

References

1. S. Liu, R. Maheshwari and K. L. Kiick, *Macromolecules*, 2009, **42**, 3.
2. Aharoni, R. *Autoimmun. Rev.*, 2013, **12**, 543.
3. R. Rupp, S. L. Rosenthal and L. R. Stanberry, *Int. J. Nanomed.*, 2007, **2**, 561.
4. B. Rybtchinski, *ACS Nano*, 2011, **5**, 6791.
5. H.-J. Schneider, *Angew. Chem., Int. Ed.*, 2009, **48**, 3924.
6. Y. Hoshino, H. Koide, K. Furuya, W. W. I. Haberaecker, S.-H. Lee, T. Kodama, H. Kanazawa, N. Oku and K. J. Shea, *Proc. Natl. Acad. Sci. USA*, 2012, **109**, 33.
7. M. Mammen, S. K. Choi and G. M. Whitesides, *Angew. Chem., Int. Ed.*, 1998, **37**, 2755.
8. N. Bertrand, M. A. Gauthier, C. Bouvet, P. Moreau, A. Petitjean, J. C. Leroux and J. Leblond, *J. Control. Release*, 2011, **155**, 200.
9. M. Yamaoka-Tojo, T. Tojo and T. Izumi, *Curr. Vasc. Pharmacol.*, 2008, **6**, 271.
10. A. J. Hutchison, C. P. Smith and P. E. C. Brenchley, *Nat. Rev. Nephrol.*, 2011, **7**, 579.
11. M. Watson, K. C. Abbott and C. M. Yuan, *Clin. J. Am. Soc. Nephrol.*, 2010, **5**, 1723.
12. W. Braunlin, E. Zhorov, A. Guo, W. Apruzzese, Q. Xu, P. Hook, D. L. Smisek, W. H. Mandeville and S. R. Holmes-Farley, *Kidney Int.*, 2002, **62**, 611.
13. S. Kurihara, Y. Tsuruta and T. Akizawa, *Nephrol., Dial., Transplant.*, 2005, **20**.
14. M. Florentin, E. N. Liberopoulos, D. P. Mikhailidis and M. S. Elisaf, *Curr. Med. Res. Opin.*, 2008, **24**, 995.
15. M. Haratake, E. Hatanaka, T. Fuchigami, M. Akashi and M. Nakayama, *Chem. Pharm. Bull.*, 2012, **60**, 1258.
16. T. Zhou, X. L. Kong, Z. D. Liu, D. Y. Liu and R. C. Hider, *Biomacromolecules*, 2008, **9**, 1372.
17. M. Pinier, G. Fuhrmann, H. J. Galipeau, N. Rivard, J. A. Murray, C. S. David, H. Drasarova, L. Tuckova, J. C. Leroux and E. F. Verdu, *Gastroenterology*, 2012, **142**, 316.
18. M. E. Ivarsson, J.-C. Leroux and B. Castagner, *Angew. Chem., Int. Ed.*, 2012, **51**, 4024.
19. K. Weiss, *Int. J. Antimicrob. Agents*, 2009, **33**, 4.

20. P. I. Kitov, J. M. Sadowska, G. L. Mulvey, G. D. Armstrong, H. Ling, N. S. Pannu, R. J. Read and D. R. Bundle, *Nature*, 2000, **403**, 669.

21. P. I. Kitov, E. Paszkiewicz, J. M. Sadowska, Z. C. Deng, M. Ahmed, R. Narain, T. P. Griener, G. L. Mulvey, G. D. Armstrong and D. R. Bundle, *Toxins*, 2011, **3**, 1065.

22. M. Mourez, R. S. Kane, J. Mogridge, S. Metallo, P. Deschatelets, B. R. Sellman, G. M. Whitesides and R. J. Collier, *Nat. Biotechnol.*, 2001, **19**, 958.

23. A. Joshi, A. Saraph, V. Poon, J. Mogridge and R. S. Kane, *Bioconjug. Chem.*, 2006, **17**, 1265.

24. P. Rai, C. Padala, V. Poon, A. Saraph, S. Basha, S. Kate, K. Tao, J. Mogridge and R. S. Kane, *Nat. Biotechnol.*, 2006, **24**, 582.

25. K. V. Gujraty, A. Joshi, A. Saraph, V. Poon, J. Mogridge and R. S. Kane, *Biomacromolecules*, 2006, **7**, 2082.

26. K. V. Gujraty, M. J. Yanjarappa, A. Saraph, A. Joshi, J. Mogridge and R. S. Kane, *J. Polym. Sci., Part A: Polym. Chem.*, 2008, **46**, 7249.

27. A. Joshi, S. Kate, V. Poon, D. Mondal, M. B. Boggara, A. Saraph, J. T. Martin, R. McAlpine, R. Day, A. E. Garcia, J. Mogridge and R. S. Kane, *Biomacromolecules*, 2011, **12**, 791.

28. M. J. Vicent, L. Cascales, R. J. Carbajo, N. Cortés, A. Messeguer and E. Pérez Payá, *J. Control. Release*, 2010, **142**, 277.

29. J. O'Loughlin, I. Y. Millwood, H. M. McDonald, C. F. Price, J. M. Kaldor and J. R. A. Paull, *Sex. Transm. Dis.*, 2010, **37**, 100.

30. C. F. Price, D. Tyssen, S. Sonza, A. Davie, S. Evans, G. R. Lewis, S. Xia, T. Spelman, P. Hodsman, T. R. Moench, A. Humberstone, J. R. A. Paull and G. Tachedjian, *PLoS One*, 2011, **6**, e24095.

31. D. Tyssen, S. A. Henderson, A. Johnson, J. Sterjovski, K. Moore, J. La, M. Zanin, S. Sonza, P. Karellas, M. P. Giannis, G. Krippner, S. Wesselingh, T. McCarthy, P. R. Gorry, P. A. Ramsland, R. Cone, J. R. A. Paull, G. R. Lewis and G. Tachedjian, *PLoS One*, 2010, **5**, e12309.

32. S. Telwatte, K. Moore, A. Johnson, D. Tyssen, J. Sterjovski, M. Aldunate, P. R. Gorry, P. A. Ramsland, G. R. Lewis, J. R. A. Paull, S. Sonza and G. Tachedjian, *Antiviral Res.*, 2011, **90**, 195.

33. F. C. Krebs, S. R. Miller, M. L. Ferguson, M. Labib, R. F. Rando and B. Wigdahl, *Biomed. Pharmacother.*, 2005, **59**, 438.

34. N. Thakkar, V. Pirrone, S. Passic, W. Zhu, V. Kholodovych, W. Welsh, R. F. Rando, M. E. Labib, B. Wigdahl and F. C. Krebs, *Antimicrob. Agents Chemother.*, 2009, **53**, 631.

35. A. F. Radovic-Moreno, T. K. Lu, V. A. Puscasu, C. J. Yoon, R. Langer and O. C. Farokhzad, *ACS Nano*, 2012, **6**, 4279.

36. V. M. Krishnamurthy, L. J. Quinton, L. A. Estroff, S. J. Metallo, J. M. Isaacs, J. P. Mizgerd and G. M. Whitesides, *Biomaterials*, 2006, **27**, 3663.

37. L. Cui, P. I. Kitov, G. C. Completo, J. C. Paulson and D. R. Bundle, *Bioconjug. Chem.*, 2011, **22**, 546.

38. A. Muñoz-Bonilla and M. Fernández-García, *Prog. Polym. Sci.*, 2012, **37**, 281.

39. B. P. Mowery, S. E. Lee, D. A. Kissounko, R. F. Epand, R. M. Epand, B. Weisblum, S. S. Stahl and S. H. Gellman, *J. Am. Chem. Soc.*, 2007, **129**, 15 474.

40. F. Nederberg, Y. Zhang, J. P. Tan, K. Xu, H. Wang, C. Yang, S. Gao, X. D. Guo, K. Fukushima, L. Li, J. L. Hedrick and Y.-Y. Yang, *Nat. Chem.*, 2011, **3**, 409.

41. G. N. Tew, D. Liu, B. Chen, R. J. Doerksen, J. Kaplan, P. J. Carroll, M. L. Klein and W. F. DeGrado, *Proc. Natl Acad. Sci. USA*, 2002, **99**, 5110.

42. V. Sambhy, B. R. Peterson and A. Sen, *Angew. Chem., Int. Ed.*, 2008, **47**, 1250.

43. K. Fukushima, J. P. K. Tan, P. A. Korevaar, Y. Y. Yang, J. Pitera, A. Nelson, H. Maune, D. J. Coady, J. E. Frommer, A. C. Engler, Y. Huang, K. Xu, Z. Ji, Y. Qiao, W. Fan, L. Li, N. Wiradharma, E. W. Meijer and J. L. Hedrick, *ACS Nano*, 2012, **6**, 9191.

44. A. Castonguay, E. Ladd, van de Ven and T. G. M. Kakkar, *A.,* New J. Chem., 2012, **36**, 199.

45. M. A. Mintzer, E. L. Dane, G. A. O'Toole and M. W. Grinstaff, *Mol. Pharm.*, 2011, **9**, 342.

46. S. Basha, P. Rai, V. Poon, A. Saraph, K. Gujraty, M. Y. Go, S. Sadacharan, M. Frost, J. Mogridge and R. S. Kane, *Proc. Natl Acad. Sci. USA*, 2006, **103**, 13 509.

47. K. Wu, J. Liu, R. N. Johnson, J. Yang and J. Kopeček, *Angew. Chem., Int. Ed.*, 2010, **49**, 1451.

48. K. Wu, J. Yang, J. Liu and J. Kopeček, *J. Control. Release*, 2012, **157**, 126.

49. L. Yang, L. Meng, X. Zhang, Y. Chen, G. Zhu, H. Liu, X. Xiong, K. Sefah and W. Tan, *J. Am. Chem. Soc.*, 2011, **133**, 13 380.

50. J. Dernedde, A. Rausch, M. Weinhart, S. Enders, R. Tauber, K. Licha, M. Schirner, U. Zügel, A. von Bonin and R. Haag, *Proc. Natl Acad. Sci. USA*, 2010, **107**, 19 679.

51. M. Weinhart, D. Gröger, S. Enders, S. B. Riese, J. Dernedde, R. K. Kainthan, D. E. Brooks and R. Haag, *Macromol. Biosci.*, 2011, **11**, 1088.

52. E. B. Puffer, J. K. Pontrello, J. J. Hollenbeck, J. A. Kink and L. L. Kiessling, *ACS Chem. Biol.*, 2007, **2**, 252.

53. Z. Illés, J. N. H. Stern, J. Reddy, H. Waldner, M. P. Mycko, C. F. Brosnan, S. Ellmerich, D. M. Altmann, L. Santambrogio, J. L. Strominger and V. K. Kuchroo, *Proc. Natl Acad. Sci. USA*, 2004, **101**, 11 749.

54. S. R. Passic, M. L. Ferguson, B. J. Catalone, T. Kish-Catalone, V. Kholodovych, W. Zhu, W. Welsh, R. Rando, M. K. Howett, B. Wigdahl, M. Labib and F. C. Krebs, *Biomed. Pharmacother.*, 2010, **64**, 723.

55. H.-A. Tran, P. I. Kitov, E. Paszkiewicz, J. M. Sadowska and D. R. Bundle, *Org. Biomol. Chem.*, 2011, **9**, 3658.
56. M. Pinier, E. F. Verdu, M. Nasser-Eddine, C. S. David, A. Vézina, N. Rivard and J. C. Leroux, *Gastroenterology*, 2009, **136**, 288.
57. P. I. Kitov, T. Lipinski, E. Paszkiewicz, D. Solomon, J. M. Sadowska, G. A. Grant, G. L. Mulvey, E. N. Kitova, J. S. Klassen, G. D. Armstrong and D. R. Bundle, *Angew. Chem., Int. Ed.*, 2008, **47**, 672.
58. L. Liang, M. Pinier, J.-C. Leroux and M. Subirade, *Biopolymers*, 2010, **93**, 418.
59. P. I. Kitov, G. L. Mulvey, T. P. Griener, T. Lipinski, D. Solomon, E. Paszkiewicz, J. M. Jacobson, J. M. Sadowska, M. Suzuki, K.-i. Yamamura, G. D. Armstrong and D. R. Bundle, *Proc. Natl Acad. Sci. USA*, 2008, **105**, 16 837.
60. S. J. Koch, C. Renner, X. Xie and T. Schrader, *Angew. Chem., Int. Ed.*, 2006, **45**, 6352.
61. A. C. Engler, N. Wiradharma, Z. Y. Ong, D. J. Coady, J. L. Hedrick and Y.-Y. Yang, *Nano Today*, 2012, **7**, 201.
62. B. P. Mowery, A. H. Lindner, B. Weisblum, S. S. Stahl and S. H. Gellman, *J. Am. Chem. Soc.*, 2009, **131**, 9735.
63. C. C. Lee, J. A. MacKay, J. M. J. Frechet and F. C. Szoka, *Nat. Biotechnol.*, 2005, **23**, 1517.
64. Z. Zeng, J. Patel, S.-H. Lee, M. McCallum, A. Tyagi, M. Yan and K. J. Shea, *J. Am. Chem. Soc.*, 2012, **134**, 2681.
65. K. Takagi, K. Masuda, M. Yamazaki, C. Kiyohara, S. Itoh, M. Wasaki and H. Inoue, *Clin. Nephrol.*, 2010, **73**, 30.
66. N. Bertrand, C. Bouvet, P. Moreau and J. C. Leroux, *ACS Nano*, 2010, **4**, 7552.
67. Y. Hoshino, H. Koide, T. Urakami, H. Kanazawa, T. Kodama, N. Oku and K. J. Shea, *J. Am. Chem. Soc.*, 2010, **132**, 6644.
68. E. Gong, B. Matthews, T. McCarthy, J. Chu, G. Holan, J. Raff and S. Sacks, *Antiviral Res.*, 2005, **68**, 139.
69. C. S. Dezzutti, V. N. James, A. Ramos, S. T. Sullivan, A. Siddig, T. J. Bush, L. A. Grohskopf, L. Paxton, S. Subbarao and C. E. Hart, *Antimicrob. Agents Chemother.*, 2004, **48**, 3834.
70. N. Nagano, S. Miyata, M. Abe, S. Wakita, N. Kobayashi and M. Wada, *Nephrol. Dial. Transplant.*, 2006, **21**, 634.
71. J. C. Leroux, *Nat. Nanotechnol.*, 2007, **2**, 679.
72. ICH, H. T. G., Guidance on nonclinical safety studies for the conduct of human clinical trials and marketing authorization for pharmaceuticals M3(R2.), 2009.
73. ICH, H. T. G., Guideline on the need for carcinogenicity studies of pharmaceuticals S1A, 1995.
74. D. Patton, Y. Cosgrove Sweeney, T. McCarthy and S. Hillier, *Antimicrob. Agents Chemother.*, 2006, **50**, 1696.
75. R. C. Thomas, R. S. Hsi, H. Harpootlian, T. D. Johnson and R. W. Judy, *Atherosclerosis*, 1978, **29**, 9.

76. M. A. Plone, J. S. Petersen, D. P. Rosenbaum and S. K. Burke, *Clin. Pharmacokinet.*, 2002, **41**, 517.
77. S. Goto, H. Fujii, J.-I. Kim and M. Fukagawa, *Clin. Nephrol.*, 2010, **73**, 420.
78. G. J. Mahler, M. B. Esch, E. Tako, T. L. Southard, S. D. Archer, R. P. Glahn and M. L. Shuler, *Nat. Nanotechnol.*, 2012, **7**, 264.
79. M. M. Barna, T. Kapoian and N. B. O'Mara, *Ann. Pharmacother.*, 2010, **44**, 127.
80. R. P. Wuthrich, M. Chonchol, A. Covic, S. Gaillard, E. Chong and J. A. Tumlin, *Clin. J. Am. Soc. Nephrol.*, 2012, doi: 10.2215/CJN.08230811.
81. S. Bhakdi and J. Tranum-Jensen, *Microbiol. Rev.*, 1991, **55**, 733.
82. M. E. Fox, F. C. Szoka and J. M. J. Fréchet, *Acc. Chem. Res.*, 2009, **42**, 1141.
83. N. Bertrand and J. C. Leroux, *J. Control. Release*, 2012, **161**, 152.
84. N. Bertrand, J. G. Fleischer, K. M. Wasan and J. C. Leroux, *Biomaterials*, 2009, **30**, 2598.
85. T. Yamaoka, Y. Tabata and Y. Ikada, *J. Pharm. Sci.*, 1994, **83**, 601.
86. T. Yamaoka, Y. Tabata and Y. Ikada, *J. Pharm. Pharmacol.*, 1995, **47**, 479.
87. T. Yamaoka, Y. Tabata and Y. Ika da, *Drug Deliv.*, 1993, **1**, 75.
88. L. W. Seymour, R. Duncan, J. Strohalm and J. Kopecek, *J., Biomed. Mater. Res.*, 1987, **21**, 1341.
89. K. Takebayashi, Y. Aso and T. Inukai, *World J. Diabetes*, 2010, **1**, 146.
90. M. L. Lahdeaho, K. Lindfors, L. Airaksinen, K. Kaukinen and M. Maki, *Expert Opin. Biol. Ther.*, 2012, **12**, 1589.
91. D. Charmot, *Curr. Pharm. Des.*, 2012, **18**, 1434.
92. J. B. Chen, S. S. Chiang, H. C. Chen, S. Obayashi, M. Nagasawa, J. M. Hexham, A. Balfour, G. Junge, T. Akiba and M. Fukagawa, *Nephrology*, 2011, **16**, 743.
93. H. Shimizu, S. Okada, O. I. Shinsuke and M. Mori, *Diabetes Care*, 2005, **28**, 2590.
94. C. R. Bosoi, C. Parent-Robitaille, K. Anderson, M. Tremblay and C. F. Rose, *Hepatology*, 2011, **53**, 1995.
95. B. Shen, D. S. Pardi, A. E. Bennett, E. Queener, P. Kammer, J. P. Hammel, C. LaPlaca and M. S. Harris, *Am. J. Gastroenterol.*, 2009, **104**, 1468.
96. M. R. Costanzo, J. T. Heywood, B. M. Massie, J. Iwashita, L. Henderson, M. Mamatsashvili, H. Sisakian, H. Hayrapetyan, P. Sager, D. J. van Veldhuisen and D. Albrecht, *Eur. J. Heart Failure*, 2012, **14**, 922.
97. Y. Sabbagh, S. P. O'Brien, W. Song, J. H. Boulanger, A. Stockmann, C. Arbeeny and S. C. Schiavi, *J. Am. Soc. Nephrol.*, 2009, **20**, 2348.
98. M. Tonelli, N. Pannu and B. Manns, *N. Engl J. Med.*, 2010, **362**, 1312.
99. D. Brown, R. Juhl and S. Warner, *Circulation*, 1978, **58**, 164.
100. G. Levy, *Ann. Pharmacother.*, 2006, **40**, 520.

101. C. L. Hoad, L. Marciani, S. Foley, J. J. Totman, J. Wright, D. Bush, E. F. Cox, E. Campbell, R. C. Spiller and P. A. Gowland, *Phys. Med. Biol.*, 2007, **52**, 6909.
102. F. Carriere, C. Renou, S. Ransac, V. Lopez, J. De Caro, F. Ferrato, A. De Caro, A. Fleury, P. Sanwald-Ducray, H. Lengsfeld, C. Beglinger, P. Hadvary, R. Verger and R. Laugier, *Am. J. Physiol. Gastrointest. Liver Physiol.*, 2001, **281**, G16.
103. C. B. Kurtz, E. P. Cannon, A. Brezzani, M. Pitruzzello, C. Dinardo, E. Rinard, D. W. Acheson, R. Fitzpatrick, P. Kelly, K. Shackett, A. T. Papoulis, P. J. Goddard, R. H. Barker, G. P. Palace, Jr. and J. D. Klinger, *Antimicrob. Agents Chemother.*, 2001, **45**, 2340.
104. W. Braunlin, Q. Xu, P. Hook, R. Fitzpatrick, J. D. Klinger, R. Burrier and C. B. Kurtz, *Biophysical J.*, 2004, **87**, 534.
105. K. M. Mullane, M. A. Miller, K. Weiss, A. Lentnek, Y. Golan, P. S. Sears, Y. K. Shue, T. J. Louie and S. L. Gorbach, *Clin. Infect. Dis.*, 2011, **53**, 440.
106. B. Pitt, S. D. Anker, D. A. Bushinsky, D. W. Kitzman, F. Zannad, I. Z. Huang and P.-H. Investigators, *Eur. Heart J.*, 2011, **32**, 820.
107. E. Ros, *Atherosclerosis*, 2000, **151**, 357.
108. L. M. Sollid and C. Khosla, *J. Intern. Med.*, 2011, **269**, 604.
109. M. M. Walker and J. A. Murray, *Histopathology*, 2011, **59**, 166.

Subject Index